Applied Mathematical Sciences
Volume 170

For further volumes:
http://www.springer.com/series/34

Zeev Schuss

Theory and Applications of Stochastic Processes

An Analytical Approach

Springer

Zeev Schuss
Department of Applied Mathematics
School of Mathematical Science
Tel Aviv University
69 978 Tel Aviv
Israel

Editors:

S.S. Antman
Department of Mathematics
and
Institute for Physical Science
and Technology
University of Maryland
College Park
MD 20742-4015
USA
ssa@math.umd.edu

J.E. Marsden
Control and Dynamical
Systems, 107-81
California Institute
of Technology
Pasadena, CA 91125
USA
marsden@cds.caltech.edu

L. Sirovich
Laboratory of Applied
Mathematics
Department of
Biomathematical Sciences
Mount Sinai School of Medicine
New York, NY 10029-6574
USA
Lawrence.Sirovich@mssm.edu

ISSN 0066-5452
ISBN 978-1-4614-2542-7 e-ISBN 978-1-4419-1605-1
DOI 10.1007/978-1-4419-1605-1
Springer New York Dordrecht Heidelberg London

Mathematics Subject Classification (2000): 60-01, 60J60, 60J65, 60J70, 60J75, 60J50, 60J35, 60J05, 60J25, 60J27, 58J37, 37H10, 60H10, 65C30, 60G40, 46N55, 47N55, 60F05, 60F10

Printed on acid-free paper

Springer is part of Springer Science+Business Media (www.springer.com)

Preface

Stochastic processes and diffusion theory are the mathematical underpinnings of many scientific disciplines, including statistical physics, physical chemistry, molecular biophysics, communications theory, and many more. Many books, reviews, and research articles have been published on this topic, from the purely mathematical to the most practical. Some are listed below in alphabetical order, but a Web search reveals many more.

- Mathematical theory of stochastic processes [14], [25], [46], [47], [53], [58], [57], [72], [76], [74], [82], [206], [42], [101], [106], [115], [116], [117], [150], [153], [161], [208], [234], [241]
- Stochastic dynamics [5], [80], [84], [171], [193], [204], [213]
- Numerical analysis of stochastic differential equations [207], [59], [132], [131], [103], [174], [206], [4]
- Large deviations theory [32], [52], [54], [68], [77], [110], [55]
- Statistical physics [38], [82], [100], [98], [99], [185], [206], [196], [242]
- Electrical engineering [108], [148], [199]

This book offers an analytical approach to stochastic processes that are most common in the physical and life sciences. Its aim is to make probability theory in function space readily accessible to scientists trained in the traditional methods of applied mathematics, such as integral, ordinary, and partial differential equations and asymptotic methods, rather than in probability and measure theory.

The book presents the basic notions of stochastic processes, mostly diffusions, including fundamental concepts from measure theory in the space of continuous functions (the space of random trajectories). The rudiments of the Wiener measure in function space are introduced not for the sake of mathematical rigor, but rather as a tool for counting trajectories, which are the elementary events in the probabilistic description of experiments. The Wiener measure turns out to be appropriate for the statistical description of nonequilibrium systems and replaces the equilibrium Boltzmann distribution in configuration space. The relevant stochastic processes are mainly stochastic differential equations driven by Einstein's white or the Ornstein–Uhlenbeck (OU) colored noise, but include also continuous-time Markovian jump processes and discrete-time semi-Markovian processes, such as renewal processes. Continuous- or discrete-time jump processes often contain more information about the origin of randomness in a given system than diffusion (white noise) models. Approximations of the more microscopic discrete models by their continuum limits (diffusions) are the cornerstone of modeling random systems.

The analytical approach relies heavily on initial and boundary value problems for partial differential equations that describe important probabilistic quantities, such as probability density functions, mean first passage times, density of the mean time spent at a point, survival probability, probability flux density, and so on. Green's function and its derivatives play a central role in expressing these quantities analytically and in determining their interrelationships. The most prominent equations are the Fokker–Planck and Kolmogorov equations for the transition probability density, which are the well-known transport-diffusion partial differential equations of continuum theory, the Kolmogorov and Feynman–Kac representation formulas for solutions of initial boundary value problems in terms of conditional probabilities, the Andronov–Vitt–Pontryagin equation for the conditional mean first passage time, and more.

Computer simulations are the numerical realizations of stochastic dynamics and are ubiquitous in computational physics, chemistry, molecular biophysics, and communications theory. The behavior of random trajectories near boundaries of the simulation impose a variety of boundary conditions on the probability density and its functionals. The quite intricate connection between the random trajectories and the boundary conditions for the partial differential equations is treated here with special care. The Wiener path integral and Wiener measure in function space are used extensively for determining these connections. Some of the simulation-oriented topics are discussed in one dimension only, because the book is not about simulations. Special topics that arise in Brownian dynamics simulations and their analysis, such as simulating trajectories between constant concentrations, connecting a simulation to the continuum, the albedo problem, behavior of trajectories at higher-dimensional partially reflecting boundaries, the estimate of rare events due to escape of random trajectories through small openings, the analysis of simulations of interacting particles and more are discussed in [214]. The important topic of nonlinear optimal filtering theory and performance evaluation of optimal phase trackers is discussed in [215].

The extraction of useful information about stochastic processes requires the solution of the basic equations of the theory. Numerical solutions of integral and partial differential equations or computer simulations of stochastic dynamics are often inadequate for the exploration of the parameter space or for studying rare events. The methods and tricks of the trade of applied mathematics are used extensively to obtain explicit analytical approximations to the solutions of complicated equations. The methods include singular perturbation techniques such as the WKB expansion, borrowed from quantum mechanics, boundary layer theory, borrowed from continuum theory, the Wiener–Hopf method, and more. These methods are the main analytical tools for the analysis of systems driven by small noise (small relative to the drift). In particular, they provide analytical expressions for the mean exit time of a random trajectory from a given domain, a problem often encountered in physics, physical chemistry, molecular and cellular biology, and in other applications. The last chapter, on stochastic stability, illustrates the power of the analytical methods for establishing the stochastic stability, or even for stabilizing unstable structures by noise.

The book contains exercises and worked-out examples. The hands-on training in stochastic processes, as my long teaching experience shows, consists of solving the exercises, without which understanding is only illusory. The book is based on lecture notes from a one- or two-semester course on stochastic processes and their applications that I taught many times to graduate students of mathematics, applied mathematics, physics, chemistry, computer science, electrical engineering, and other disciplines. My experience shows that mathematical rigor and applications cannot coexist in the same course; excessive rigor leaves no room for in-depth development of modeling methodology and turns off students interested in scientific and engineering applications. Therefore the book contains only the minimal mathematical rigor required for understanding the necessary measure-theoretical concepts and for enabling the students to use their own judgment of what is correct and what requires further theoretical study. The student should be familiar with basic measure and integration theory as well as with rudiments of functional analysis, as taught in most graduate analysis courses (see, e.g., [79]). More intricate measure-theoretical problems are discussed in theoretical texts, as mentioned in the first list of references above. The first course can cover the fundamental concepts presented in the first six chapters and either Chapter 8 on Markov processes and their diffusion approximations, or Chapters 7 or 11, which develop some of the mentioned methods of applied mathematics and apply them to the study of systems driven by small noise. Chapters 7 and 9–12, which develop and apply the analytical tools to the study of the equations developed in Chapters 1–6 and 8, can be chosen for the second semester. A well-paced course can cover most of the book in a single semester. The two books [214], on Brownian dynamics and simulations, and [215], on optimal filtering and phase tracking, can be used for one-semester special topics followup courses.

It is recommended that students of Chapters 7 and 9–12 acquire some independent training in singular perturbation methods, for example, from classical texts such as [194], [121], [20], [195], [251]. These chapters require the basic knowledge of Chapters 1–6 and a solid training in partial differential equations of mathematical physics and in the asymptotic methods of applied mathematics.

Acknowledgment. Much of the material presented in this book is based on my collaboration with many scientists and students, whose names are listed next to mine in the author index. Much of the work done in the nearly 30 years since my first book appeared [213], was initiated in the Department of Applied Mathematics and Engineering Science at Northwestern University and in the Department of Molecular Biophysics and Physiology at Rush Medical Center in Chicago, for whose hospitality I am grateful. The scientific environment provided by Tel-Aviv University, my home institution, was conducive to interdisciplinary cooperation. Its double-major undergraduate program in mathematics and physics, as well as its M.Sc. program in E.E. produced my most brilliant graduate students and collaborators in applied mathematics.

Zeev Schuss

List of Figures

2.1 Four Brownian trajectories sampled at discrete times. Three cylinders are marked by vertical bars. The trajectories were sampled according to the scheme (2.21). 29
2.2 Three Brownian trajectories sampled at discrete times according to the Wiener measure $\Pr_0\{\cdot\}$ by the scheme (2.21). 41
2.3 The graphs of $X_1(t)$ (dots), its first refinement $X_2(t)$ (dot-dash), and second refinement $X_3(t)$ (dash). 50
2.4 A Brownian trajectory sampled at 1024 points. 50

6.1 The potential $U(x) = x^2/2 - x^4/4$ (solid line) and the force $F(x) = -U'(x)$ (dashed line). The domain of attraction of the stable equilibrium $x = 0$ is the interval $D = (-1, 1)$. 186
6.2 The solution (6.31) of the noiseless dynamics (6.30) with $x(0) = 0.99$ inside (bottom line) and $x(0) = 1.01$ outside (top line) the domain of attraction D. 187

8.1 The potential $U(x) = 3(x-1)^2 - 2(x-1)^3$. The well is $x < x_C = 2$, the bottom is at $x = x_A = 1$, the top of the barrier is $C = (x_C, U(x_C))$, and the height of the barrier is $E_C = U(x_C) - U(x_A) = 1$. 258
8.2 The energy contours $y^2/2 + U(x) = E$ for the potential $U(x) = x^2/2 - x^3/3$. 277
8.3 The washboard potential $U(x) = \cos x - Ix$ for $I = 0.3$. 286
8.4 The stable running solution (top) and separatrices, 2π apart in the phase space (x, y), for $\gamma = 0.1$ and $I = 0.3$. The thick curve is the critical trajectory $S = S_C$ that touches the x-axis. The energy of the system on the running solution slides to $-\infty$ down the washboard potential in Figure 8.3. 287
8.5 In the limit $\gamma \to \infty$ the stable running solution of Figure 8.4 disappears to infinity and the critical energy contour $S = S_C$ and the separatrices coalesce. 288
8.6 The effective global potential $\Phi(A)$ (8.131) in the steady-state FPE (8.130). 297

10.1 The three-term outer solution (dashed line), the uniform expansion (dotted line), and the exact solution (solid line). 342
10.2 A potential $U(x)$ with two wells separated by a sharp barrier at x_C. 348
10.3 At the barrier of the left well of the potential in Figure 10.2 the (negative) drift $a(x_C-) = -U'(x_C-)$ points into the well (toward the attractor x_A). 349
10.4 Potential well with a single saddle point on the boundary. 380
10.5 Equipotential contours of the potential in Figure 10.4. 381
10.6 The saddle point region in Figure 10.5. The boundary is the trajectory that converges to the saddle point (on the x-axis). The unstable direction is the x-axis, which is normal to the boundary. 381
10.7 The domain of attraction D of the stable equilibrium point $A = (x_A, 0)$ in the phase plane (x, y) is bounded by the separatrix Γ. The saddle point is at $C = (x_C, 0)$. The potential is $U(x) = (x+1)^2/2 - (x+1)^3/3$. 383
10.8 Comparison of formulas for reaction rate κ as a function of γ, for the potential $U(x) = 3x^2 - 2x^3$ with $\varepsilon = 0.2$. All rates are normalized by κ_{TST}. The underdamped (dashed) rate is (10.163), Kramers' rate (solid) is (10.153), and the uniform approximation (dot-dashed) is (10.166). 390

11.1 The angle of the load $P(t)$ is $\eta\varphi$ relative to vertical; that of the tangent line at the end of the deflected column is φ. 426
11.2 The straight and deflected pendulum. 430

List of Symbols

We use interchangeably $\langle \cdot \rangle$ and $\mathbb{E}(\cdot)$ to denote expectation (average) of a random variable, but $\mathbb{E}(\cdot \mid \cdot)$ and $\Pr\{\cdot \mid \cdot\}$ to denote conditional expectation and conditional probability, respectively.

$\boldsymbol{A}, \boldsymbol{B} \ldots$	matrices — bold upper case letters
$\boldsymbol{A}^T$	the transpose of $\boldsymbol{A}$
$\boldsymbol{A}^{-1}$	the inverse of $\boldsymbol{A}$
$\delta(\boldsymbol{x})$	Dirac's delta function (functional)
$\det(\boldsymbol{A})$	the determinant of the matrix $\boldsymbol{A}$
$\mathbb{E}(x)$, $\langle x \rangle$	the expected value (expectation) of x
Δ, $\Delta_{\boldsymbol{x}}$	Greek *Delta*, the Laplace operator (with respect to $\boldsymbol{x}$): $\frac{\partial^2}{\partial x_1^2} + \frac{\partial^2}{\partial x_2^2} + \cdots + \frac{\partial^2}{\partial x_d^2}$
$\boldsymbol{e}(t)$	the estimation error process: $\hat{\boldsymbol{x}}(t) - \boldsymbol{x}(t)$
$\mathcal{F}$	the sample space of Brownian events
$J[\boldsymbol{x}(\cdot)]$	functional of the trajectory $\boldsymbol{x}(\cdot)$
$L^2[a, b]$	Square integrable functions on the interval $[a, b]$
$\mathbb{M}_{n,m}$	space of $n \times m$ real matrices
$m_1 \wedge m_2$	the minimum $\min\{m_1, m_2\}$

$n \sim \mathcal{N}(\mu, \sigma^2)$	the random variable n is normally distributed with mean μ and variance σ^2
$\boldsymbol{N} \sim \mathcal{N}(\boldsymbol{\mu}, \boldsymbol{\Sigma})$	the random vector $\boldsymbol{N}$ is normally distributed with mean $\boldsymbol{\mu}$ and covariance matrix $\boldsymbol{\Sigma}$
∇, $\nabla \boldsymbol{x}$	Greek *nabla*, the gradient operator (with respect to $\boldsymbol{x}$): $\left(\frac{\partial}{\partial x_1}, \frac{\partial}{\partial x_2}, \dots, \frac{\partial}{\partial x_d}\right)^T$
$\nabla \cdot \boldsymbol{J}$	the divergence operator $\frac{\partial J_1(\boldsymbol{x})}{\partial x_1} + \frac{\partial J_2(\boldsymbol{x})}{\partial x_2} + \cdots + \frac{\partial J_d(\boldsymbol{x})}{\partial x_d}$
$\Pr\{event\}$	the probability of $event$
$p_{\boldsymbol{X}}(\boldsymbol{x})$	the probability density function of the vector $\boldsymbol{X}$
$\mathbb{Q}$	the rational numbers
$\mathbb{R}, \mathbb{R}^d$	the real line, the d-dimensional Euclidean space
V_x	the partial derivative of V with respect to x : $\partial V \partial x$
$\mathrm{tr}(\boldsymbol{A})$	trace of the matrix $\boldsymbol{A}$
$\mathrm{Var}(x)$	the variance of x
$\boldsymbol{w}(t)$, $\boldsymbol{v}(t)$	vectors of independent Brownian motions
x, $f(x)$	scalars — lower case letters
$\boldsymbol{x}$, $\boldsymbol{f}(\boldsymbol{x})$	vectors — bold lower case letters
x_i	the ith element of the vector $\boldsymbol{x}$
$\boldsymbol{x}(\cdot)$	trajectory or function in function space
$\lvert\boldsymbol{x}\rvert^2 = \boldsymbol{x}^T\boldsymbol{x}$	L_2 norm of $\boldsymbol{x}$
$\boldsymbol{x} \cdot \boldsymbol{y}$	dot (scalar) product of the vectors $\boldsymbol{x}$ and $\boldsymbol{y}$: $\boldsymbol{x} \cdot \boldsymbol{y} = x_1 y_1 + x_2 + y_2 + \cdots + x_d y_d$
$\dot{\boldsymbol{x}}(t)$	time derivative: $d\boldsymbol{x}(t)/dt$

List of Acronyms

BKE	backward Kolmogorov equation
CKE	Chapman–Kolmogorov equation
FPE	Fokker–Planck equation
FPT	first passage time
i.i.d.	independent identically distributed
i.o.	infinitely often
ODE	ordinary differential equation
OU	Ornstein–Uhlenbeck (process)
pdf	probability density function
PDE	partial differential equation
PDF	probability distribution function
RMS	root mean square
SDE	stochastic differential equation

Contents

1 The Physical Brownian Motion: Diffusion And Noise **1**

1.1 Einstein's theory of diffusion 1

1.2 The velocity process and Langevin's approach 5

1.3 The displacement process 10

1.4 Classical theory of noise 13

1.5 An application: Johnson noise 16

1.6 Linear systems 21

2 The Probability Space of Brownian Motion **25**

2.1 Introduction 25

2.2 The space of Brownian trajectories 27

2.2.1 The Wiener measure of Brownian trajectories 37

2.2.2 The MBM in $\mathbb{R}^d$ 44

2.3 Constructions of the MBM 46

2.3.1 The Paley–Wiener construction of the Brownian motion . . 46

2.3.2 P. Lévy's method and refinements 49

2.4 Analytical and statistical properties of Brownian paths 52

2.4.1 The Markov property of the MBM 55

2.4.2 Reflecting and absorbing walls 56

2.4.3 MBM and martingales 60

3 Itô Integration and Calculus **63**

3.1 Integration of white noise 63

3.2 The Itô, Stratonovich, and other integrals 66

3.2.1 The Itô integral 66

3.2.2 The Stratonovich integral 68

3.2.3 The backward integral 73

3.3 The construction of the Itô integral 74

3.4 The Itô calculus 81

4 Stochastic Differential Equations **92**

4.1 Itô and Stratonovich SDEs 93

4.2 Transformations of Itô equations 97

4.3 Solutions of SDEs are Markovian 101
4.4 Stochastic and partial differential equations 104
4.4.1 The Andronov–Vitt–Pontryagin equation 109
4.4.2 The exit distribution 111
4.4.3 The PDF of the FPT 114
4.5 The Fokker–Planck equation 119
4.6 The backward Kolmogorov equation 124
4.7 Appendix: Proof of Theorem 4.1.1 125
4.7.1 Continuous dependence on parameters 131

5 The Discrete Approach and Boundary Behavior 133
5.1 The Euler simulation scheme and its convergence 133
5.2 The pdf of Euler's scheme in $\mathbb{R}$ and the FPE 137
5.2.1 Unidirectional and net probability flux density 145
5.3 Boundary behavior of diffusions 150
5.4 Absorbing boundaries 151
5.4.1 Unidirectional flux and the survival probability 155
5.5 Reflecting and partially reflecting boundaries 157
5.5.1 Total and partial reflection in one dimension 158
5.5.2 Partially reflected diffusion in higher dimensions 165
5.5.3 Discontinuous coefficients 168
5.5.4 Diffusion on a sphere 168
5.6 The Wiener measure induced by SDEs 169
5.7 Annotations 173

6 The First Passage Time of Diffusions 176
6.1 The FPT and escape from a domain 176
6.2 The PDF of the FPT 180
6.3 The exit density and probability flux density 184
6.4 The exit problem in one dimension 185
6.4.1 The exit time 191
6.4.2 Application of the Laplace method 194
6.5 Conditioning 197
6.5.1 Conditioning on trajectories that reach A before B 198
6.6 Killing measure and the survival probability 202

7 Markov Processes and their Diffusion Approximations 207
7.1 Markov processes 207
7.1.1 The general form of the master equation 211
7.1.2 Jump-diffusion processes 218
7.2 A semi-Markovian example: Renewal processes 222
7.3 Diffusion approximations of Markovian jump processes 230
7.3.1 A refresher on solvability of linear equations 230
7.3.2 Dynamics with large and fast jumps 231

7.3.3 Small jumps and the Kramers–Moyal expansion 236
7.3.4 An application to Brownian motion in a field of force 241
7.3.5 Dynamics driven by wideband noise 244
7.3.6 Boundary behavior of diffusion approximations 247
7.4 Diffusion approximation of the MFPT 249

8 Diffusion Approximations to Langevin's Equation 257
8.1 The overdamped Langevin equation 257
8.1.1 The overdamped limit of the GLE 259
8.2 Smoluchowski expansion in the entire space 265
8.3 Boundary conditions in the Smoluchowski limit 268
8.3.1 Appendix 275
8.4 Low-friction asymptotics of the FPE 276
8.5 The noisy underdamped forced pendulum 285
8.5.1 The noiseless underdamped forced pendulum 286
8.5.2 Local fluctuations about a nonequilibrium steady state . . . 290
8.5.3 The FPE and the MFPT far from equilibrium 295
8.5.4 Application to the shunted Josephson junction 299
8.6 Annotations 301

9 Large Deviations of Markovian Jump Processes 302
9.1 The WKB structure of the stationary pdf 302
9.2 The mean time to a large deviation 308
9.3 Asymptotic theory of large deviations 322
9.3.1 More general sums 328
9.3.2 A central limit theorem for dependent variables 333
9.4 Annotations 337

10 Noise-Induced Escape From an Attractor 339
10.1 Asymptotic analysis of the exit problem 339
10.1.1 The exit problem for small diffusion with the flow 343
10.1.2 Small diffusion against the flow 348
10.1.3 The MFPT of small diffusion against the flow 352
10.1.4 Escape over a sharp barrier 353
10.1.5 The MFPT to a smooth boundary and the escape rate 356
10.1.6 The MFPT eigenvalues of the Fokker–Planck operator . . . 359
10.2 The exit problem in higher dimensions 359
10.2.1 The WKB structure of the pdf 361
10.2.2 The eikonal equation 362
10.2.3 The transport equation 364
10.2.4 The characteristic equations 365
10.2.5 Boundary layers at noncharacteristic boundaries 366
10.2.6 Boundary layers at characteristic boundaries in the plane . . 369
10.2.7 Exit through noncharacteristic boundaries 371
10.2.8 Exit through characteristic boundaries in the plane 376

10.2.9 Kramers' exit problem 378
10.3 Activated escape in Langevin's equation 382
10.3.1 The separatrix in phase space 382
10.3.2 Kramers' exit problem at high and low friction 384
10.3.3 The MFPT to the separatrix Γ 386
10.3.4 Uniform approximation to Kramers' rate 387
10.3.5 The exit distribution on the separatrix 389
10.4 Annotations 397

11 Stochastic Stability **399**
11.1 Stochastic stability of nonlinear oscillators 403
11.1.1 Underdamped pendulum with parametric noise 404
11.1.2 The steady-state distribution of the noisy oscillator 407
11.1.3 First passage times and stability 410
11.2 Stabilization with oscillations and noise 417
11.2.1 Stabilization by high-frequency noise 417
11.2.2 The generating equation 418
11.2.3 The correlation-free equation 419
11.2.4 The stability of (11.72) 421
11.3 Stability of columns with noisy loads 425
11.3.1 A thin column with a noisy load 426
11.3.2 The double pendulum 429
11.3.3 The damped vertically loaded double pendulum 434
11.3.4 A tangentially loaded double pendulum (follower load) . . . 436
11.3.5 The N-fold pendulum and the continuous column 438

Bibliography **442**

Index **459**

Chapter 1

The Physical Brownian Motion: Diffusion And Noise

1.1 Einstein's theory of diffusion

This chapter reviews the elementary phenomenology of diffusion and Fick's derivation of the diffusion equation. It recounts Einstein's theory that connects diffusion with the Brownian motion and Langevin's extension of that theory. It presents the early mathematical theories of the Brownian motion.

In his *Encyclopedia Britannica* article Maxwell describes the phenomenon of diffusion as follows:

> When two fluids are capable of being mixed, they cannot remain in equilibrium with each other; if they are placed in contact with each other the process of mixture begins of itself, and goes on till the state of equilibrium is attained, which, in the case of fluids which mix in all proportions, is a state of uniform mixture.
>
> This process of mixture is called diffusion. It may be easily observed by taking a glass jar half full of water and pouring a strong solution of a coloured salt, such as sulphate of copper, through a long-stemmed funnel, so as to occupy the lower part of the jar. If the jar is not disturbed we may trace the process of diffusion for weeks, months, or years, by the gradual rise of the colour into the upper part of the jar, and the weakening of the colour in the lower part.
>
> This, however, is not a method capable of giving accurate measurements of the composition of the liquid at different depths in the vessel. . . .
>
> M. Voit has observed the diffusion of cane-sugar in water by passing a ray of plane-polarized light horizontally through the vessel and determining the angle through which the plane of polarization is turned by

Z. Schuss, *Theory and Applications of Stochastic Processes: An Analytical Approach*, Applied Mathematical Sciences 170, DOI 10.1007/978-1-4419-1605-1_1,

> the solution of sugar. . . .
>
> The laws of diffusion were first investigated by Graham. The diffusion of gases has recently been observed with great accuracy by Loschmidt, and that of liquids by Fick and Voit. . . .
>
> If we observe the process of diffusion with our most powerful microscopes, we cannot follow the motion of any individual portions of the fluids. We cannot point out one place in which the lower fluid is ascending, and another in which the upper fluid is descending. There are no currents visible to us, and the motion of the material substances goes on as imperceptibly as the conduction of heat or electricity. Hence the motion which constitutes diffusion must be distinguished from those motions of fluids which we can trace by means of floating motes. It may be described as a motion of the fluids, not in mass but by molecules. . . .

The laws of diffusion were first formulated by Fick. His first law of diffusion, formulated in 1856 by analogy with Fourier's first law of heat conduction, asserts that *the diffusion flux between two points of different concentrations in the fluid is proportional to the concentration gradient between these points.* The constant of proportionality is called *the diffusion coefficient* and it is measured in units of area per unit time. Fick's first law was verified experimentally in many different diffusion experiments. A straightforward consequence of Fick's first law and the principle of mass conservation is Fick's second law, which asserts that *the rate of change of the concentration of a solute diffusing in a solvent equals minus the divergence of the diffusion flux.* Fick's two laws of diffusion can be written in the form

$$\boldsymbol{J} = -D\nabla\rho, \quad \frac{\partial\rho}{\partial t} = -\nabla\cdot\boldsymbol{J}, \tag{1.1}$$

where $\boldsymbol{J}$ denotes the diffusion flux density vector, ρ is the concentration of the diffusing substance (e.g., sugar in water), and D is the diffusion coefficient. Fick's two equations combine to give the *diffusion equation*

$$\frac{\partial\rho}{\partial t} = \nabla\cdot D\nabla\rho. \tag{1.2}$$

The observation in 1827 of the irregular motion of small particles immersed in fluid, the so-called *Brownian motion* [30],

> . . . played almost no role in physics until 1905, and was generally ignored even by the physicists who developed the kinetic theory of gases, though it is now frequently remarked that Brownian movement is the best illustration of the existence of random molecular motion.

In 1905 Einstein [63] and, independently, in 1906 Smoluchowski [228] offered an explanation of the Brownian motion based on kinetic theory and demonstrated, theoretically, that the phenomenon of diffusion is the result of Brownian motion. Einstein's theory was later verified experimentally by Perrin [200] and Svedberg [236].

That of Smoluchowski was verified by Smoluchowski [226], Svedberg [237], and Westgren [246], [247].

Einstein approached the problem of diffusion from two directions. On the one hand, assuming the Brownian particles to be in thermodynamical equilibrium, he used the ideal gas equation for the osmotic pressure

$$p = \frac{RT}{V^*}\frac{n}{N} = \frac{RT}{N}\rho, \tag{1.3}$$

where T = absolute (Kelvin) temperature, n = number of suspended particles in a volume V^* (partitioned from a larger volume V), N = Avogadro's number, and the concentration is $\rho = n/V^*$.

In equilibrium, the suspended particles are diffusing in the liquid in such a way that the osmotic force originating in a concentration gradient is balanced by the viscous force, which retards the motion of the particle according to hydrodynamics. Thus,

$$K\rho \propto \frac{\partial p}{\partial x}, \tag{1.4}$$

where $K\rho$ is the viscous force. Using eq.(1.3), this gives

$$K\rho = \frac{RT}{N}\frac{\partial \rho}{\partial x}. \tag{1.5}$$

Assuming that the Stokes formula for the velocity of a particle moving through a viscous medium can be applied in the case at hand,

$$v = \frac{K}{6\pi a\eta}, \tag{1.6}$$

where a = radius of the particle and η = coefficient of dynamical viscosity, the coefficient K can be eliminated to give

$$6\pi a\eta v\rho = \frac{RT}{N}\frac{\partial \rho}{\partial x}. \tag{1.7}$$

Actually, as Einstein noted, eq. (1.6) indicates that $\rho K/6\pi a\eta$ is the number of particles crossing a unit area per unit time. This can be equated to $-D\,(\partial\rho/\partial x)$, where D is the diffusion coefficient, according to Fick's law. Thus, in addition to eq. (1.5), Einstein's second equation is

$$\frac{K\rho}{6\pi a\eta} = D\frac{\partial \rho}{\partial x}, \tag{1.8}$$

hence Einstein's expression for the diffusion coefficient is

$$D = \frac{RT}{N}\frac{1}{6\pi a\eta}. \tag{1.9}$$

Equation (1.9) was obtained from similar considerations by Sutherland in 1904 and published in 1905 [235].

To connect this theory with the "irregular movement which arises from thermal molecular movement," Einstein made the following assumptions: (1) the motion of each particle is independent of the others and (2) "the movements of one and the same particle after different intervals of time must be considered as mutually independent processes, so long as we think of these intervals of time as being chosen not too small." Assuming that the particle is observed at consecutive time intervals τ, $2\tau, \ldots,$ consider the number of particles whose displacement in a time interval lies between Δ and $\Delta + d\Delta$, in the form

$$dn = n\phi(\Delta)\, d\Delta, \tag{1.10}$$

where

$$\int_{-\infty}^{\infty} \phi(\Delta)\, d\Delta = 1, \quad \phi(\Delta) = \phi(-\Delta), \tag{1.11}$$

n is the total number of particles, and ϕ is only different from zero for small values of Δ. Equation (1.10) in effect defines the function $\phi(\Delta)$. The value of the concentration $\rho(x,t)$ after time τ has elapsed can be computed from the values of $\rho(x+\Delta,t)$ for all possible values of Δ, weighted by $\phi(\Delta)$,

$$\rho(x,t+\tau) = \int_{-\infty}^{\infty} \rho(x+\Delta,t)\phi(\Delta)\, d\Delta. \tag{1.12}$$

Expanding ρ in Taylor's series for small τ and Δ, the equation

$$\begin{aligned}\rho(x,t) + \frac{\partial \rho(x,t)}{\partial t}\tau + \cdots = \rho(x,t)\int_{-\infty}^{\infty} \phi(\Delta)\, d\Delta + \frac{\partial \rho(x,t)}{\partial x}\int_{-\infty}^{\infty} \Delta\phi(\Delta)\, d\Delta \\ + \frac{\partial^2 \rho(x,t)}{\partial x^2}\int_{-\infty}^{\infty} \frac{\Delta^2}{2}\phi(\Delta)\, d\Delta + \cdots \end{aligned} \tag{1.13}$$

is obtained. Using the conditions eq. (1.11) and neglecting terms of higher order, eq. (1.13) reduces to

$$\frac{\partial \rho(x,t)}{\partial t} = D\frac{\partial^2 \rho(x,t)}{\partial x^2}, \tag{1.14}$$

where

$$D = \frac{1}{\tau}\int_{-\infty}^{\infty} \frac{\Delta^2}{2}\phi(\Delta)\, d\Delta. \tag{1.15}$$

The solution of eq. (1.14),

$$\rho(x,t) = \frac{n}{\sqrt{4\pi Dt}} \exp\left\{-\frac{x^2}{4Dt}\right\}, \tag{1.16}$$

is described as the solution to the problem of diffusion from a single point (neglecting interactions between the particles). That is, the function

$$p(x,t) = \frac{1}{\sqrt{4\pi Dt}} \exp\left\{-\frac{x^2}{4Dt}\right\} \tag{1.17}$$

can be interpreted as the transition probability density of a particle from the point $x = 0$ at time 0 to the point x at time t.

If we denote by $x(t)$ the displacement of the particle at time t, then for any interval A,

$$\Pr\{x(t) \in A\} = \int_A p(x,t)\,dx. \tag{1.18}$$

It follows that the mean value

$$\mathbb{E}x(t) = \int x p(x,t)\,dx$$

and the variance of the displacement are

$$\mathbb{E}x(t) = 0, \quad \mathbb{E}x^2(t) = 2Dt. \tag{1.19}$$

Obviously, if the particle starts at $x(0) = x_0$, then

$$\begin{aligned} \mathbb{E}[x(t) \mid x(0) = x_0] &= x_0 \\ \mathrm{Var}[x(t) \mid x(0) = x_0] &= \mathbb{E}[(x(t) - x_0)^2 \mid x(0) = x_0] = 2Dt. \end{aligned} \tag{1.20}$$

Now, using eq. (1.9) in eq. (1.20), the mean square displacement of a Brownian particle along the x-axis is found as

$$\sigma = \sqrt{t}\sqrt{\frac{kT}{3\pi a\eta}}, \tag{1.21}$$

where $k = R/N$ is Boltzmann's constant. This formula was verified experimentally [236]. It indicates that the mean square displacement of a Brownian particle at times t not too short is proportional to the square root of time.

1.2 The velocity process and Langevin's approach

According to the Waterston-Maxwell equipartition theorem, the root mean square (RMS) velocity $\bar{v} = \sqrt{\langle v^2 \rangle}$ of a suspended particle should be determined by the equation

$$\frac{m}{2}\bar{v}^2 = \frac{3kT}{2}. \tag{1.22}$$

Each component of the velocity vector has the same variance, so that

$$\frac{m}{2}\bar{v}^2_{x,y,z} = \frac{kT}{2}, \tag{1.23}$$

which is the one-dimensional version of eq. (1.22). The RMS velocity comes out to be about 8.6 cm/sec for the particles used in Svedberg's experiment [236]. Einstein argued in 1907 and 1908 [63] that there is no possibility of observing this velocity, because of the very rapid viscous damping, which can be calculated from the Stokes formula. The velocity of such a particle would drop to 1/10 of its initial value in about 3.3×10^{-7} sec. Therefore, Einstein argued, in the period τ between observations the particle must get new impulses to movement by some process that is the inverse of viscosity, so that it retains a velocity whose RMS average is $\bar{v}$. Between consecutive observations these impulses alter the magnitude and direction of the velocity in an irregular manner, even in the extraordinarily short time of 3.3×10^{-7} sec. According to this theory, the RMS velocity in the interval τ has to be inversely proportional to $\sqrt{\tau}$; that is, it increases without limit as the time interval between observations becomes smaller.

In 1908 Langevin [147] offered an alternative approach to the problem of the Brownian motion. He assumed that the dynamics of a free Brownian particle is governed by the frictional force $-6\pi a\eta v$ and by a fluctuational force X, that results from the random collisions of the Brownian particle with the molecules of the surrounding fluid, after the frictional force is subtracted. This force is random and assumes positive and negative values with equal probabilities. It follows that Newton's second law of motion for the Brownian particle is given by

$$m\ddot{x} = -6\pi a\eta\dot{x} + \Xi. \tag{1.24}$$

Denoting $v = \dot{x}$ and multiplying eq. (1.24) by x, we obtain

$$\frac{m}{2}\frac{d^2}{dt^2}x^2 - mv^2 = -3\pi a\eta\frac{d}{dt}x^2 + \Xi x. \tag{1.25}$$

Averaging under the assumption that the fluctuational force Ξ and the displacement of the particle x are independent, we obtain

$$\frac{m}{2}\frac{d^2}{dt^2}\langle x^2\rangle + 3\pi a\eta\frac{d}{dt}\langle x^2\rangle = kT, \tag{1.26}$$

where eq. (1.23) has been used. The solution is given by $d\langle x^2\rangle/dt = kT/3\pi a\eta + Ce^{-6\pi a\eta t/m}$, where C is a constant. The time constant in the exponent is 10^{-8} sec, so the mean square speed decays on a time scale much shorter than that of observations. It follows that $\langle x^2\rangle - \langle x_0^2\rangle = (kT/3\pi a\eta)t$. This, in turn (see eq. (1.19)), implies that the diffusion coefficient is given by $D = kT/6\pi a\eta$, as in Einstein's equation (1.9).

Langevin's equation (1.24) is a *stochastic differential equation*, because it is driven by a random force Ξ. If additional fields of force act on the diffusing particles (e.g., electrostatic, magnetic, gravitational, etc.), Langevin's equation is modified to include the external force, $F(x,t)$, say, [140], [38],

$$m\ddot{x} + \Gamma\dot{x} - F(x,t) = \Xi, \tag{1.27}$$

where $\Gamma = 6\pi a\eta$ is the friction coefficient of a diffusing particle. We denote the dynamical friction coefficient (per unit mass) $\gamma = \Gamma/m$. If the force can be derived from a potential, $F = -\nabla U(x)$, Langevin's equation takes the form $m\ddot{x} + \Gamma\dot{x} + \nabla U(x) = \Xi$.

The main mathematical difference between the two approaches is that Einstein assumes that the displacements Δ are independent, whereas Langevin assumes that the random force Ξ and the displacement x are independent. The two theories are reconciled in Section 1.3 below.

To investigate the statistical properties of the fluctuating force Ξ, we make the following assumptions.

(i) *The fluctuating force Ξ is independent of the velocity v.*

(ii) *Ξ changes much faster than v.*

(iii) $\langle\Xi\rangle = 0$

(iv) *The accelerations imparted in disjoint time intervals Δt_1 and Δt_2 are independent.*

The conditional probability distribution function of the velocity process of a Brownian particle (PDF), given that it started with velocity v_0 at time $t = 0$, is defined as $P(v,t\,|\,v_0) = \Pr\{v(t) < v\,|\,v_0\}$ and the conditional probability density function is defined by

$$p\,(v,t\,|\,v_0) = \frac{\partial P(v,t\,|\,v_0)}{\partial v}.$$

In higher dimensions, we denote the displacement vector $\boldsymbol{x} = (x_1, x_2, \ldots, x_d)^T$, the velocity vector $\dot{\boldsymbol{x}} = \boldsymbol{v} = (v_1, v_2, \ldots, v_d)^T$, the random force vector $\boldsymbol{\Xi} = (\Xi_1, \Xi_2, \ldots, \Xi_d)^T$, the PDF

$$P(\boldsymbol{v},t\,|\,\boldsymbol{v}_0) = \Pr\{v_1(t) < v_1, v_2(t) < v_2, \ldots, v_d(t) < v_d\,|\,\boldsymbol{v}(0) = \boldsymbol{v}_0\},$$

and the pdf

$$p\,(\boldsymbol{v},t\,|\,\boldsymbol{v}_0) = \frac{\partial^n P(\boldsymbol{v},t\,|\,\boldsymbol{v}_0)}{\partial v_1 \partial v_2, \ldots \partial v_d}.$$

The conditioning implies that the initial condition for the pdf is $p\,(\boldsymbol{v},t\,|\,\boldsymbol{v}_0) \to \delta(\boldsymbol{v}-\boldsymbol{v}_0)$ as $t \to 0$. According to the Waterston-Maxwell theory, when the system is in thermal equilibrium, the velocities of free Brownian particles have the Maxwell–Boltzmann pdf; that is,

$$\lim_{t\to\infty} p\,(\boldsymbol{v},t\,|\,\boldsymbol{v}_0) = \left(\frac{m}{2\pi kT}\right)^{3/2} \exp\left\{-\frac{m|\boldsymbol{v}|^2}{2kT}\right\}. \tag{1.28}$$

The solution of the Langevin equation (1.24) for a free Brownian particle is given by

$$v(t) = v_0 e^{-\gamma t} + \frac{1}{m} \int_0^t e^{-\gamma(t-s)} \Xi(s)\, ds. \tag{1.29}$$

To interpret the stochastic integral in eq. (1.29), we make a short mathematical digression on the definition of integrals of the type $\int_0^t g(s)\Xi(s)\, ds$, where $g(s)$ is a deterministic integrable function. Such an integral is defined as the limit of finite Riemann sums of the form

$$\int_0^t g(s)\Xi(s)\, ds = \lim_{\Delta s_i \to 0} \sum_i g(s_i)\Xi(s_i)\, \Delta s_i, \tag{1.30}$$

where $0 = s_0 < s_1 < \cdots < s_N = t$ is a partition of the interval $[0, t]$. According to the assumptions about Ξ, if we choose $\Delta s_i = \Delta t = t/N$ for all i, the increments $\Delta b_i = \Xi(s_i)\, \Delta s_i$ are independent identically distributed (i.i.d.) random variables. Einstein's observation (see the beginning of Section 1.2) that the RMS velocity on time intervals of length Δt are inversely proportional to $\sqrt{\Delta t}$, implies that if the increments $\Xi(s_i)\, \Delta s_i$ are chosen to be normally distributed, their mean must be zero and their covariance matrix must be $\langle \Delta b_i \Delta b_i \rangle = q\Delta t$ with q a parameter to be determined. We write $\Delta b_i \sim \mathcal{N}(0, q\Delta t)$. Then $g(s_i)\Xi(s_i)\, \Delta s_i \sim \mathcal{N}\left(0, |g(s_i)|^2 q\Delta t\right)$, so that $\sum_i g(s_i)\Xi(s_i)\, \Delta s_i \sim \mathcal{N}(0, \sigma_N^2)$, where $\sigma_N^2 = \sum_i |g(s_i)|^2 q\Delta s_i$. As $\Delta t \to 0$, we obtain $\lim_{\Delta t \to 0} \sigma_N^2 = q \int_0^t g^2(s)\, ds$ and $\int_0^t g(s)\Xi(s)\, ds \sim \mathcal{N}(0, \sigma^2)$, where

$$\sigma^2 = q \int_0^t g^2(s)\, ds. \tag{1.31}$$

By considering Riemann sums of the form eq. (1.30), we find that the cross-correlation between the integrals of two deterministic functions is the expectation (average) of the Gaussian variables

$$\mathbb{E} \int_0^{t_1} f(s_1)\Xi(s_1)\, ds_1 \int_0^{t_2} g(s_2)\Xi(s_2)\, ds_2 = q \int_0^{t_1 \wedge t_2} f(s)g(s)\, ds, \tag{1.32}$$

where $t_1 \wedge t_2 = \min\{t_1, t_2\}$. We note that for the Heaviside function $H(t)$

$$\frac{\partial t_1 \wedge t_1}{\partial t_1} = H(t_2 - t_1), \quad \frac{\partial^2 t_1 \wedge t_2}{\partial t_2 \partial t_1} = \delta(t_2 - t_1)$$

$$\frac{\partial t_1 \wedge t_2}{\partial t_1} \frac{\partial t_1 \wedge t_2}{\partial t_2} = H(t_2 - t_1)H(t_1 - t_2) = 0,$$

so (1.32) means that

$$\langle \Xi(s_1)\, ds_1 \Xi(s_2)\, ds_2 \rangle = q\delta(s_1 - s_2)\, ds_1\, ds_2. \tag{1.33}$$

To interpret eq. (1.29), we use eq. (1.31) with $g(s) = e^{-\gamma(t-s)}$ and obtain

$$\sigma^2 = \frac{q}{2\gamma}\left(1 - e^{-2\gamma t}\right). \tag{1.34}$$

Returning to the velocity vector $\boldsymbol{v}(t)$, we obtain from the above considerations

$$\boldsymbol{v}(t) - \boldsymbol{v}_0 e^{-\gamma t} \sim \mathcal{N}\left(\boldsymbol{0}, \sigma^2 \boldsymbol{I}\right) \tag{1.35}$$

with σ^2 given by eq. (1.34). Finally, the condition (1.28) implies that $q = 2\gamma kT/m$, so that in 3-D the mean energy is as given in eq. (1.22).

In the limit $\gamma \to \infty$ the acceleration $\gamma \boldsymbol{v}(t)$ inherits the properties of the random acceleration $\boldsymbol{\Xi}(t)$ in the sense that for different times $t_2 > t_1 > 0$ the accelerations $\gamma \boldsymbol{v}(t_1)$ and $\gamma \boldsymbol{v}(t_2)$ become independent. In fact, from eqs.(1.29) and (1.33) we find that

$$\begin{aligned}
&\lim_{\gamma\to\infty} \langle \gamma \boldsymbol{v}(t_1) \cdot \gamma \boldsymbol{v}(t_2) \rangle \qquad\qquad (1.36)\\
&= \lim_{\gamma\to\infty} \frac{\gamma^2}{m^2} \int_0^{t_1}\int_0^{t_2} e^{-\gamma(t_1-s_1)} e^{-\gamma(t_2-s_2)} \langle \boldsymbol{\Xi}(s_1) \cdot \boldsymbol{\Xi}(s_2) \rangle\, ds_1\, ds_2\\
&= \lim_{\gamma\to\infty} \int_0^{t_1} e^{-\gamma(t_1-s_1)} e^{-\gamma(t_2-s_1)}\, ds_1 = \lim_{\gamma\to\infty} \frac{\gamma}{m^2}\left[e^{-\gamma(t_2-t_1)} - e^{-\gamma(t_2+t_1)}\right] = 0
\end{aligned}$$

and a similar result for $0 < t_2 < t_1$. It follows that

$$\langle \gamma \boldsymbol{v}(t_1) \cdot \gamma \boldsymbol{v}(t_2) \rangle = \frac{\gamma}{m^2} e^{-\gamma|t_2-t_1|}(1 + o(1)) \text{ for } \gamma(t_1 \wedge t_2) \gg 1. \tag{1.37}$$

We conclude that

$$\langle \gamma \boldsymbol{v}(t_1) \cdot \gamma \boldsymbol{v}(t_2) \rangle = \frac{2q}{m^2} \delta(t_2 - t_1)(1 + o(1)) \text{ for } \gamma(t_1 \wedge t_2) \gg 1, \tag{1.38}$$

because for $t_1 > 0$

$$\lim_{\gamma\to\infty} \int_0^{\infty} f(t_2) \frac{\gamma}{m^2} e^{-\gamma|t_2-t_1|}(1 + o(1))\Delta t_2 = \frac{2}{m^2} f(t_1) \tag{1.39}$$

for all test functions $f(t)$ in $\mathbb{R}^+$.

1.3 The displacement process

The displacement of a free Brownian particle is obtained from integration of the velocity process,

$$\boldsymbol{x}(t) = \boldsymbol{x}_0 + \int_0^t \boldsymbol{v}(s)\, ds. \tag{1.40}$$

Using the expression (1.29) in eq. (1.40) and changing the order of integration in the resulting iterated integral, we obtain

$$\boldsymbol{x}(t) - \boldsymbol{x}_0 - \boldsymbol{v}_0 \frac{1-e^{-\gamma t}}{\gamma} = \int_0^t g(s)\boldsymbol{\Xi}(s)\, ds, \tag{1.41}$$

where $g(s) = (1 - e^{-\gamma(t-s)})/m\gamma$.

Reasoning as above, we find that the stochastic integral in eq. (1.41) is a normal variable with zero mean and covariance matrix $\boldsymbol{\Sigma} = \sigma^2 \boldsymbol{I}$, where

$$\sigma^2 = q \int_0^t g^2(s)\, ds = \frac{q}{2\gamma^3}\left(2\gamma t - 3 + 4e^{-\gamma t} - e^{-2\gamma t}\right). \tag{1.42}$$

It follows that $\boldsymbol{x}(t) - \boldsymbol{x}_0 - \boldsymbol{v}_0(1 - e^{-\gamma t})/\gamma \sim \mathcal{N}\left(\boldsymbol{0}, \sigma^2 \boldsymbol{I}\right)$; that is,

$$p\left(\boldsymbol{x}, t \,|\, \boldsymbol{x}_0, \boldsymbol{v}_0\right) = \left\{\frac{m\gamma^2}{2\pi kT\left(2\gamma t - 3 + 4e^{-\gamma t} - e^{-2\gamma t}\right)}\right\}^{3/2}$$

$$\times \exp\left\{-\frac{m\gamma^2 \left|\boldsymbol{x} - \boldsymbol{x}_0 - \boldsymbol{v}_0 \dfrac{1-e^{-\gamma t}}{\gamma}\right|^2}{2\gamma t - 3 + 4e^{-\gamma t} - e^{-2\gamma t}}\right\}. \tag{1.43}$$

Next, we calculate the moments of the displacement. Obviously,

$$\mathbb{E}\left[\boldsymbol{x}(t) - \boldsymbol{x}_0 - \boldsymbol{v}_0 \frac{1-e^{-\gamma t}}{\gamma}\right] = 0 \tag{1.44}$$

and the conditional second moment of the displacement is

$$\begin{aligned} &\mathbb{E}\left(\left|\boldsymbol{x}(t) - \boldsymbol{x}_0\right|^2 \,|\, \boldsymbol{x}_0, \boldsymbol{v}_0\right) \\ &= \int |\boldsymbol{x} - \boldsymbol{x}_0|^2 p\left(\boldsymbol{x}, t \,|\, \boldsymbol{x}_0, \boldsymbol{v}_0\right) d\boldsymbol{x} \\ &= \frac{|\boldsymbol{v}_0|^2}{\gamma^2}\left(1 - e^{-\gamma t}\right)^2 + \frac{3kT}{m\gamma^2}\left(2\gamma t - 3 + 4e^{-\gamma t} - e^{-2\gamma t}\right), \end{aligned} \tag{1.45}$$

which is independent of $\boldsymbol{x}_0$. Using the Maxwell distribution of velocities (1.28), we find that the unconditional second moment is

$$\begin{aligned}\mathbb{E}\left|\boldsymbol{x}(t)-\boldsymbol{x}_0\right|^2 &= \mathbb{E}_{\boldsymbol{x}_0}\mathbb{E}_{\boldsymbol{v}_0}\left(\left|\boldsymbol{x}(t)-\boldsymbol{x}_0\right|^2 \mid \boldsymbol{x}_0,\boldsymbol{v}_0\right)\\ &= \frac{3kT}{m\gamma^2}\left(1-e^{-\gamma t}\right)^2+\frac{3kT}{m\gamma^2}\left(2\gamma t-3+4e^{-\gamma t}-e^{-2\gamma t}\right)\\ &= \frac{6kT}{m\gamma^2}\left(\gamma t-1+e^{-\gamma t}\right). \end{aligned} \tag{1.46}$$

The long time asymptotics of $\mathbb{E}\left|\boldsymbol{x}(t)-\boldsymbol{x}_0\right|^2$ is found from eq. (1.46) to be

$$\mathbb{E}\left|\boldsymbol{x}(t)-\boldsymbol{x}_0\right|^2 \sim \frac{6kT}{m\gamma}t=\frac{kT}{ma\eta}t \text{ for } t\gamma \gg 1; \tag{1.47}$$

that is, the displacement variance of each component is asymptotically $kT/3ma\eta$. It was this fact that was verified experimentally by Perrin [200]. The one-dimensional diffusion coefficient, as defined in eq. (1.19), is therefore given by $D=kT/6ma\eta$.

Equation (1.46) implies that the short time asymptotics of $\mathbb{E}\left|\boldsymbol{x}(t)-\boldsymbol{x}_0\right|^2$ is given by

$$\mathbb{E}\left|\boldsymbol{x}(t)-\boldsymbol{x}_0\right|^2 \sim \frac{3kT}{m}t^2=\langle|\boldsymbol{v}_0|^2\rangle t^2. \tag{1.48}$$

This result was first obtained by Smoluchowski.

To reconcile the Einstein and the Langevin approaches, we have to show that for two disjoint time intervals, (t_1,t_2) and (t_3,t_4), in the limit $\gamma\to\infty$, the increments $\Delta_1\boldsymbol{x}=\boldsymbol{x}(t_2)-\boldsymbol{x}(t_1)$ and $\Delta_3\boldsymbol{x}=\boldsymbol{x}(t_4)-\boldsymbol{x}(t_3)$ are independent zero mean Gaussian variables with variances proportional to the time increments. Equation (1.41) implies that in the limit $\gamma\to\infty$ the increments $\Delta_1\boldsymbol{x}$ and $\Delta_3\boldsymbol{x}$ are zero mean Gaussian variables and eq. (1.47) shows that the variance of an increment is proportional to the time increment.

To show that the increments are independent, we use eq. (1.38) in eq. (1.41) to obtain

$$\begin{aligned}\lim_{\gamma\to\infty}\gamma^2\langle\Delta_1\boldsymbol{x}\cdot\Delta_3\boldsymbol{x}\rangle &= \lim_{\gamma\to\infty}\int_{t_1}^{t_2}\int_{t_3}^{t_4}\langle\gamma\boldsymbol{v}(s_1)\cdot\gamma\boldsymbol{v}(s_2)\rangle\,ds_1\,ds_2\\ &= \frac{2q}{m^2}\int_{t_1}^{t_2}\int_{t_3}^{t_4}\delta(s_2-s_1)\,ds_1\,ds_2=0. \end{aligned} \tag{1.49}$$

As is well-known [72], uncorrelated Gaussian variables are independent. This reconciles the Einstein and Langevin theories of Brownian motion in liquid.

Introducing the dimensionless variables $s=\gamma t$ and $\boldsymbol{\xi}(s)=\sqrt{m/6kT}\gamma\boldsymbol{x}(t)$, we find from eq. (1.46) that

$$\lim_{\gamma\to\infty}\mathbb{E}\left|\boldsymbol{\xi}(s)-\boldsymbol{\xi}(0)\right|^2=s-1+e^{-s}\sim s \text{ for } s\gg 1 \tag{1.50}$$

and from eq. (1.49) that

$$\lim_{\gamma\to\infty} \mathbb{E}\Delta_1\boldsymbol{\xi}\cdot\Delta_3\boldsymbol{\xi} = 0. \tag{1.51}$$

Equations (1.50) and (1.51) explain (in the context of Langevin's description) Einstein's quoted assumption that "... the movement of one and the same particle after different intervals of time [are] mutually independent processes, so long as we think of these intervals of time as being chosen not too small."

Exercise 1.1 (The Maxwell distribution). Denote by $\boldsymbol{v} = (v_1, v_2, v_3)^T$ the velocity vector of a gas particle. Following Maxwell, assume that $\boldsymbol{v}$ is a three-dimensional random variable whose probability density function, $h(\boldsymbol{v})$, satisfies the following assumptions.

(i) v_j $(j = 1, 2, 3)$ are identically distributed independent random variables.

(ii) The probability density function of $\boldsymbol{v}$ is a function of the kinetic energy of a particle, that is,

$$h(\boldsymbol{v}) = g\left(\frac{1}{2}m|\boldsymbol{v}|^2\right) = p\left(|\boldsymbol{v}|\right),$$

where m is the mass of the particle. Show that $\boldsymbol{v}$ must be $N\left(\boldsymbol{0}, k\boldsymbol{I}\right)$, where k is some positive constant and $(\boldsymbol{I})_{ij} = \delta_{ij}$ (Kronecker's δ), that is, $\boldsymbol{v}$ is a Gaussian variable. (HINT: Set $h(\boldsymbol{v}) = f(v_1)f(v_2)f(v_3)$ and derive the differential equation $p'(s)/sp(s) = \text{constant}$.) □

Exercise 1.2 (The joint pdf of displacement and velocity). Prove that the joint pdf of displacement and velocity for a one-dimensional free Brownian motion in a constant external field $V'(x) = g$ is given by [38]

$$\begin{aligned}
&p_c(x, v, t \,|\, x_0, v_0) \\
&= \frac{1}{2\pi\sqrt{FG - H^2}} \exp\left\{-\left[GR^2 - 2HRS + FS^2\right]/2(FG - H^2)\right\},
\end{aligned} \tag{1.52}$$

where

$$\begin{aligned}
R &= x - x_0 - \gamma^{-1}v_0(1 - e^{-\gamma t}) + g\gamma^{-2}\left(\gamma t - 1 + e^{-\gamma t}\right),\\
S &= v - v_0 e^{-\gamma t} + g\gamma^{-1}\left(1 - e^{-\gamma t}\right),
\end{aligned} \tag{1.53}$$

and

$$\begin{aligned}
F &= \varepsilon\gamma^{-2}\left[2\gamma t - 3 + 4e^{-\gamma t} - e^{-2\gamma t}\right], \quad G = \varepsilon(1 - e^{-2\gamma t}),\\
H &= \varepsilon\gamma^{-1}(1 - e^{-\gamma t})^2, \quad \varepsilon = \frac{kT}{m}.
\end{aligned} \tag{1.54}$$

The marginal pdf of the velocity is

$$p\left(v, t \,|\, v_0\right) = \frac{1}{\sqrt{2\pi\varepsilon(1 - e^{-2\gamma t})}} \exp\left\{-\frac{[v - v_0 e^{-\gamma t} + g\gamma^{-1}(1 - e^{-\gamma t})]^2}{2\varepsilon(1 - e^{-2\gamma t})}\right\},$$

and the marginal pdf of the displacement is

$$p(x,t\,|\,x_0,v_0) = \sqrt{\frac{\gamma^2}{2\pi\varepsilon\left[2\gamma t - 3 + 4e^{-\gamma t} - e^{-2\gamma t}\right]}} \tag{1.55}$$

$$\times \exp\left\{-\frac{\gamma^2\left[x - x_0 - v_0\gamma^{-1}(1-e^{-\gamma t}) - g\gamma^{-2}(1-e^{-\gamma t}-\gamma t)\right]^2}{2\varepsilon\left[2\gamma t - 3 + 4e^{-\gamma t} - e^{-2\gamma t}\right]}\right\}.$$

□

Exercise 1.3 (The Brownian harmonic oscillator). Solve the Langevin equation for a Brownian harmonic oscillator (in one dimension) for both the displacement and the velocity. Find the joint pdf of the two processes. Calculate the long time limits of the first moments and the covariance matrix of the two processes (see [38]). □

Exercise 1.4 (The forced Brownian oscillator). Solve the same problem for a periodically forced Brownian harmonic oscillator. Consider in particular the limits of small friction and resonance [67]. □

Exercise 1.5 (Charged Brownian particles). Solve the same problem for a charged Brownian particle in a uniform electrostatic and constant magnetic field. □

Exercise 1.6 (Correlation). Calculate the limit (1.49) when the intervals (t_1, t_2) and (t_3, t_4) overlap. □

1.4 Classical theory of noise

The concept of Gaussian noise is widely used in physics, signal processing, communications theory, in modeling stock prices, and in many other applications. Gaussian noise is usually meant as a zero mean process whose all n-dimensional probability distributions are Gaussian ($n = 1, 2, \dots$). More specifically, a process $x(t)$ is a *Gaussian process* if for all $t_1 < t_2, \cdots < t_n$ and all $\boldsymbol{x} = (x_1, x_2, \dots, x_n)^T$ the joint PDF of $x(t_1), x(t_2), \dots, x(t_n)$ has the Gaussian density

$$\begin{aligned} p(x_1,t_1;x_2,t_2;\dots,x_n,t_n) &= \frac{\partial^n \Pr\{x(t_1)\le x_1, x(t_2)\le x_2,\dots,x(t_n)\le x_n\}}{\partial x_1\partial x_2\cdots\partial x_n} \\ &= (2\pi\det\boldsymbol{\sigma})^{-n/2}\exp\left\{-\frac{1}{2}\boldsymbol{x}^T\boldsymbol{\sigma}^{-1}\boldsymbol{x}\right\}, \end{aligned}$$

where $\boldsymbol{\sigma} = \left\{\sigma^{ij}(t_1,t_2,\dots,t_n)\right\}_{i,j=1}^n$ is the autocovariance matrix, defined by

$$\sigma^{ij}(t_1,t_2,\dots,t_n) = \mathbb{E}x(t_i)x(t_j). \tag{1.56}$$

The process is *stationary* if

$$p(x_1,t_1;x_2,t_2;\dots,x_n,t_n) = p(x_1,0;x_2,t_2-t_1;\dots,x_n,t_n-t_1). \tag{1.57}$$

This definition applies to vector-valued processes as well. Classical theory describes stationary Gaussian processes $\xi(t)$ in terms of their autocorrelation function

$$R_{\xi}(\tau) = \mathbb{E}\xi(t+\tau)\xi(t) \tag{1.58}$$

and its Fourier transform, called the *power spectral density function*,

$$S_{\xi}(\nu) = \int_{-\infty}^{\infty} e^{-i\nu\tau} R_{\xi}(\tau)\, d\tau. \tag{1.59}$$

When $\xi(t)$ is measured in volts, $S_{\xi}(\nu)$ is scaled with resistance and measured in watt/Hz. For example, if $S_{\xi}(\nu) = 1$, the process $\xi(t)$ is called "white noise", because its power output per Hz is the same, 1 watt/Hz at all frequencies ν.

We consider here two examples of stationary Gaussian noise. The velocity process of a free Brownian particle, as described in Section 1.2, is often used as a model of noise. It can be formally defined as the output of the "low pass filter"

$$\dot{x}(t) + ax(t) = b\Xi(t), \tag{1.60}$$

where $\Xi(t)\Delta t \sim \mathcal{N}(0, \Delta t)$, and $a, b > 0$ are constants. We set

$$w(t) = \int_0^t \Xi(s)\, ds, \tag{1.61}$$

which is a continuous function, and write the solution of eq. (1.60) as

$$\begin{aligned} x(t_2) =& x(t_1)e^{-a(t_2-t_1)} \\ &- ab\int_{t_1}^{t_2} e^{-a(t_2-s)} w(s)\, ds + b\left[w(t_2) - e^{-a(t_2-t_1)}w(t_1)\right]. \end{aligned} \tag{1.62}$$

Definition 1.4.1 (The Ornstein–Uhlenbeck process). *The process $x(t)$ defined in eq.* (1.62) *is called the* Ornstein–Uhlenbeck process *or colored noise.*

We can show now that $x(t)$ is a stationary Gaussian process whose n-dimensional autocovariance matrix (1.56) is given by (1.36) as

$$\sigma^{ij} = \frac{b^2}{2a}\left[e^{-a(t_j-t_i)} - e^{-a(t_j+t_i)}\right].$$

It follows that in the limit $t_i \to \infty$, such that $t_i - t_j \to \tilde{t}_i$, the n-dimensional pdf of $x(t)$ satisfies the stationarity condition (1.57).

This means that the limit (in distribution) $\xi(t) = \lim_{T\to\infty} x(t+\tau) - x(\tau)$ exists and is a stationary Gaussian process. The limit is defined as *colored noise* or *wideband noise*. Its autocorrelation function is given by

$$R_{\xi}(\tau) = \mathbb{E}\xi(t+\tau)\xi(t) = \frac{b^2}{2a}e^{-a|\tau|}. \tag{1.63}$$

Note that due to stationarity, $R_\xi(\tau)$ in eq. (1.63) is independent of t and can be written as $R_\xi(\tau) = E\xi(\tau)\xi(0)$. The *correlation time* of the process is the time $R_\xi(\tau)$ decays by a factor of e. For colored noise the decay time is $\tau_{\text{decay}} = 1/a$. In the n-dimensional case the power spectral density matrix is defined as the Fourier transform of the autocorrelation matrix.

The Fourier inversion formula gives

$$R_\xi(\tau) = \frac{1}{2\pi}\int_{-\infty}^{\infty} e^{i\nu\tau} S_\xi(\nu)\,d\nu.$$

The *total power output* of the process is defined as

$$P_\xi = \int_{-\infty}^{\infty} S_\xi(\nu)\,d\nu = 2\pi R_\xi(0); \tag{1.64}$$

that is, the total power output is the area under the graph of the power spectral density function. If $\xi(t,\omega)$ is measured in volts, P_ξ is scaled with resistance and measured in watts. The *spectral height* is defined as the maximum of $S_\xi(\nu)$. When it is achieved at $\nu = 0$, eq. (1.59) gives $S_\xi(0) = \int_{-\infty}^{\infty} R_\xi(\tau)\,d\tau$; that is, the spectral height is the area under the graph of the autocorrelation function.

For colored noise, eqs.(1.63) and (1.59) give

$$S_\xi(\nu) = \frac{b^2}{2a}\int_{-\infty}^{\infty} e^{-i\nu\tau}e^{-a|\tau|}\,d\tau = \frac{b^2}{a^2+\nu^2},$$

which is called the *Lorentzian power spectrum*. The total power output is, according to eq. (1.64), $P_\xi = \pi b^2/2a$. The spectral height of colored noise is $S_\xi(0) = b^2/a^2$. The *width* of the power spectral density is defined as the (two-sided) distance between the frequencies at which $S_\xi(\nu)$ decays to one half of its maximum. For colored noise the width is $2a$.

If the spectral height is scaled with the spectral width, $b = ab_0$, the power spectral density function becomes

$$S_\xi(\nu) = \frac{a^2b_0^2}{a^2+\nu^2} \to b_0^2 \text{ as } a \to \infty, \tag{1.65}$$

so that

$$\lim_{a\to\infty} R_\xi(\tau) = b_0^2\delta(\tau) \tag{1.66}$$

(see Section 2.2 below).

If the Brownian motion is the primary object that can be constructed mathematically, as described below, then eq. (1.61) identifies white Gaussian noise as the derivative (in some sense) of the Brownian motion; that is, $\Xi(t) = \dot{w}(t)$. In view of the Paley–Wiener–Zygmund theorem [198] (see Section 3.3.1 below) about the nondifferentiability of the Brownian paths, the concept of Gaussian white noise cannot be interpreted in the naïve differential calculus sense. The issue here is similar to that of the nonexistence of Dirac's $\delta(t)$ as a function, although it has many formal properties that are similar to those of differentiable functions. Both can, however, be interpreted in the sense of distributions (or generalized functions [152]). White noise can be defined as a limit in the sense of distributions [152] of the wideband (colored) noise $\xi(ta)$ defined above. We describe below formal properties of white noise, which can also be derived from the properties of colored noise.

Due to the stationarity of the Brownian increments, the formal definition of white noise implies that it is a stationary Gaussian process. Equation (2.36) below implies that the autocorrelation function of white noise is given by

$$E\dot{w}(t)\dot{w}(s) = \frac{\partial^2}{\partial t \partial s}(t \wedge s) = \delta(t-s), \tag{1.67}$$

or alternatively, for all t,

$$R_{\dot{w}}(\tau) = E\dot{w}(t+\tau)\dot{w}(t) = \delta(\tau). \tag{1.68}$$

For n-dimensional white noise $\dot{\boldsymbol{w}}(t)$ the autocorrelation function is found from (2.47) as

$$\mathbf{Cov}_{\dot{\boldsymbol{w}}}(t,s) = \boldsymbol{I}\delta(t-s) = \boldsymbol{I}\delta(\tau). \tag{1.69}$$

Now, eqs.(1.68) and (1.59) give

$$S_{\dot{w}}(\nu) = 1, \tag{1.70}$$

so that the power spectrum of $\dot{w}(t)$ is flat; that is, the total power output of $\dot{w}(t)$ is the same at all frequencies. Processes with a flat spectrum are called *white noises*. According to the definition (1.64) and eq. (1.70), the total power output of white noise is infinite.

Similarly, $\boldsymbol{S}_{\dot{\boldsymbol{w}}}(\nu) = \boldsymbol{I}$ is the power-spectral matrix of n-dimensional white noise. The concept of white noise is an idealization of a wise-band noise with infinite bandwidth, although it does not represent a mathematically well-defined object.

If we define white Gaussian noise as the primitive object, colored noise can be defined as low-pass filtered white noise and vice versa, if colored noise is the primitive object, white noise can be defined as the limit of colored noise as bandwidth becomes infinite.

1.5 An application: Johnson noise

In 1928 Johnson [111] measured the random fluctuating voltage across a resistor and found that the power-spectral density function of the random electromotive force produced by the resistor was white with spectral height proportional to resistance and temperature. A theoretical derivation of this result was presented by Nyquist [192] in the same issue of *Physical Reviews*. Here, we derive Nyquist's result for an ionic solution, where the ions are assumed identical independent Brownian particles in a uniform electrostatic field.

The Ramo–Shockley theorem [220], [205] relates the microscopic motion of mobile charges in a domain D to the electric current measured at any given electrode. For a single moving charge q at location $\boldsymbol{x}$ with velocity $\boldsymbol{v}$, the instantaneous current at the jth electrode is given by

$$I_j = q\boldsymbol{v} \cdot \nabla u_j(\boldsymbol{x}), \tag{1.71}$$

where u_j is the solution of the Laplace equation

$$\nabla \cdot [\varepsilon(\boldsymbol{x})\nabla u_j] = 0 \text{ for } \boldsymbol{x} \in D \tag{1.72}$$

with the boundary conditions

$$u_j\Big|_{\partial D_j} = 1, \quad u_j\Big|_{\partial D_i} = 0, \quad (i \neq j), \tag{1.73}$$

where ∂D_j is the boundary of the jth electrode. In addition, the normal component of the field is continuous at dielectric interfaces [107],

$$\varepsilon_1 \frac{\partial u_j}{\partial n} - \varepsilon_2 \frac{\partial u_j}{\partial n} = 0,$$

where derivatives are taken in the normal direction to the interface, and ε_1 and ε_2 are the dielectric coefficients on the two sides. In the case of many particles, due to superposition, the total current recorded at the jth electrode is given by $I_j = \sum_i q_i \boldsymbol{v}_i \cdot u_j(\boldsymbol{x}_i)$. Consider, for example, an infinite conducting parallel plates capacitor, shorted through an Ampère meter. The separation between the plates is L and a point charge q is moving with an instantaneous velocity $v(t)$ in a direction perpendicular to the electrodes. The solution of (1.72), (1.73) is $u(x) = x/L$ for $0 \leq x \leq L$, therefore, according to (1.71), the current on the Ampère meter is

$$I = \frac{qv}{L}. \tag{1.74}$$

Exercise 1.7 (Elementary derivation of (1.74)). The charges induced on the plates of the capacitor move through the shorting Ampère meter to equalize the electrostatic potential between the plates (at zero, say), giving rise to a current through the Ampère meter. To calculate this current (i) calculate first the induced charge density on the right plate, $\sigma(r, \theta)$. Use Maxwell's equation [107]

$$\text{Div}\, \boldsymbol{E} = -4\pi\sigma, \tag{1.75}$$

where $\boldsymbol{E}$ is the electrostatic field on the right plate. The surface divergence operator is defined as $\text{Div}\, \boldsymbol{E} = \boldsymbol{n} \cdot [\boldsymbol{E}]$, where $[\boldsymbol{E}]$ is the jump in the field across the plate and $\boldsymbol{n}$ is the unit normal to the plate. The electric field, given by

$$\boldsymbol{E} = -\nabla\phi, \tag{1.76}$$

where ϕ is the potential, is found by placing image charges q on the x-axis at the points $2nL + d$, and image charges $-q$ at the points $2(n+1)L - d$ to maintain zero potential on the plates. Introduce in the yz-plane of the electrode a system of polar coordinates (r, θ). Find from Coulomb's law that

$$\phi(d, r, \theta) = q \sum_{n=0}^{\infty} \left\{ \frac{1}{\left[(2nL + d)^2 + r^2\right]^{1/2}} - \frac{1}{\left[(2(n+1)L - d)^2 + r^2\right]^{1/2}} \right\},$$

hence eq. (1.76) gives the field on the plate as

$$\boldsymbol{E}(r, \theta) = q \sum_{n=0}^{\infty} \left\{ \frac{\boldsymbol{r}_n}{\left[(2nL + d)^2 + r^2\right]^{3/2}} - \frac{\boldsymbol{r}'_n}{\left[(2(n+1)L - d)^2 + r^2\right]^{3/2}} \right\},$$

where $\boldsymbol{r}_n$ and $\boldsymbol{r}'_n$ are unit vectors pointing from the positive and negative image charges to the point (r, θ) in the yz-plane, respectively. The component of the field parallel to the yz-plane vanishes, due to symmetry, and the normal component has magnitude

$$\text{Div}\, \boldsymbol{E} = 2q \sum_{n=0}^{\infty} \left\{ \frac{2nL + d}{\left[(2nL + d)^2 + r^2\right]^{3/2}} - \frac{2(n+1)L - d}{\left[(2(n+1)L - d)^2 + r^2\right]^{3/2}} \right\}.$$

Finally, the surface charge density is found from Maxwell's equation (1.75) as the θ independent expression

$$\sigma(r) = \frac{-q}{2\pi} \sum_{n=0}^{\infty} \left\{ \frac{2nL + d}{\left[(2nL + d)^2 + r^2\right]^{3/2}} - \frac{2(n+1)L - d}{\left[(2(n+1)L - d)^2 + r^2\right]^{3/2}} \right\}.$$

(ii) Calculate the total charge on the right plate $Q = 2\pi \int_0^\infty \sigma(r)\, r\, dr$. To evaluate this integral, use the definition of the improper integral

$$Q = \lim_{R \to \infty} 2\pi \int_0^R \sigma(r)\, r\, dr.$$

The latter integral can be evaluated by using termwise integration, because for each finite R the sum converges uniformly. Get

$$\int_0^R \frac{\alpha\, r\, dr}{[\alpha^2 + r^2]^{3/2}} = \frac{\alpha}{[\alpha^2 + r^2]^{1/2}} - 1, \tag{1.77}$$

with $\alpha = 2nL + d$ or $\alpha = 2(n+1)L - d$. Returning to the infinite sum, isolate the $n = 0$ term and rearrange the sum; then, integrating term by terms, get from (1.77)

$$\begin{aligned}
Q_R &= 2\pi \int_0^R \sigma(r)\, r\, dr \\
&= q \sum_{n=1}^{\infty} \left\{ \frac{2nL+d}{\left[(2nL+d)^2 + R^2\right]^{1/2}} - \frac{2(n+1)L-d}{\left[(2(n+1)L-d)^2 + R^2\right]^{1/2}} \right\} - q \\
&= q \sum_{n=1}^{\infty} \left\{ \frac{2nL+d}{\left[(2nL+d)^2 + R^2\right]^{1/2}} - \frac{2nL-d}{\left[(2nL-d)^2 + R^2\right]^{1/2}} \right\} - q \\
&= qR \sum_{n=1}^{\infty} \frac{1}{R} \left\{ \frac{\dfrac{2nL+d}{R}}{\left[\left(\dfrac{2nL+d}{R}\right)^2 + 1\right]^{1/2}} - \frac{\dfrac{2nL-d}{R}}{\left[\left(\dfrac{2nL-d}{R}\right)^2 + 1\right]^{1/2}} \right\} - q.
\end{aligned}$$

Write the infinite sum as the limit $\sum_{n=1}^{\infty} = \lim_{N\to\infty} \sum_{n=1}^{N}$ and note that the resulting sums are the Riemann sums that define the integral $\int_0^M \frac{x\,dx}{\sqrt{1+x^2}}$. Set $2N/R = M$. Now, for any smooth function $f(x)$, the difference between two Riemann sums is $\sum_j \left[f(x_j) - f(y_j)\right] \Delta x_j = \sum_j f'(\xi_j)(x_j - y_j)\Delta x_j$. In the case at hand, $x_j - y_j = 2d/R$, $\Delta x_j = 2L/R$, so that for $M >> 1$ we obtain

$$\begin{aligned}
\sum_j \left[f(x_j) - f(y_j\right]\Delta x_j &= \frac{2d}{R} \sum_j f'(\xi_j)\Delta x_j \\
&\approx \frac{2d}{R} \int_0^M f'(x)\, dx = \frac{2d}{R} \left.\frac{x}{\sqrt{1+x^2}}\right|_0^M \approx \frac{2d}{R}.
\end{aligned}$$

It follows that for large R and M,

$$\begin{aligned}
Q_R &= q \frac{R}{2L} \sum_{n=1}^{\infty} \frac{2L}{R} \left\{ \frac{\dfrac{2nL+d}{R}}{\left[\left(\dfrac{2nL+d}{R}\right)^2 + 1\right]^{1/2}} - \frac{\dfrac{2nL-d}{R}}{\left[\left(\dfrac{2nL-d}{R}\right)^2 + 1\right]^{1/2}} \right\} + q \\
&\approx q\left(1 - \frac{d}{L}\right),
\end{aligned}$$

so that $Q = q(1 - d/L)$. The distance of the charge from the right plate is a function of time, $d = d(t)$, because it is moving, and so is the total charge on the plate, $Q = Q(t)$. Thus, the current carried by the charge is obtained as

$$I_p = \frac{d}{dt}Q(t) = -\frac{q}{L}\frac{d}{dt}d(t) = \frac{qv}{L}, \tag{1.78}$$

where the velocity of the moving charge is $v = -\dot{d}(t)$. □

Now, an electrostatically neutral electrolytic solution of concentration ρ is placed between the plates of the capacitor and a voltage V is maintained across the plates. The concentrations on the plates are maintained equal and constant. We assume, for simplicity, that the positive ions have charge q and the negative ions have charge $-q$ and they have the same constant diffusion coefficient. Obviously, this is generally not the case (see Exercise 1.8 below). Thus, we do not distinguish between positive and negative ions, because they make the same contribution to the current and to the noise. Under these conditions the electrostatic field E in the solution is uniform.

The average motion of an ion in solution is described by

$$\frac{d^2\mathbb{E}x(t)}{dt^2} + \gamma\frac{d\mathbb{E}x(t)}{dt} = \frac{qE}{m}, \tag{1.79}$$

where γ is the friction coefficient. In the steady-state the velocity is given by $\lim_{t\to\infty} d\mathbb{E}x(t)/dt = qE/\gamma m$, so the steady-state average current per particle is given by $\bar{I}_p = q\mathbb{E}\dot{x}/L = q^2E/\gamma mL$. The voltage across the capacitor is $V = EL$, so that $\bar{I}_p$ can be written as $\bar{I}_p = q^2V/\gamma mL^2$. If N identical charges are uniformly distributed between the plates of the capacitor with density (per unit length) $\rho = N/L$, the average current is given by $\bar{I} = Nq^2V/\gamma mL^2 = q^2\rho V/\gamma mL$, so that Ohm's law gives the resistance

$$R = \frac{V}{\bar{I}} = \frac{\gamma mL}{q^2\rho}. \tag{1.80}$$

Thus the resistance of the one-dimensional ionic solution is proportional to the friction coefficient, to the mass of the moving charge, and to the length of the "resistor", and inversely proportional to the density of the ions and to the square of the ionic charge.

The motion of an ion of mass m in the solution is described by the overdamped Langevin equation

$$\ddot{x}(t) + \gamma\dot{x}(t) - \frac{qE}{m} = \sqrt{\frac{2\gamma kT}{m}}\,\dot{w}(t), \tag{1.81}$$

giving the average motion of eq. (1.79). Setting $\Delta x(t) = x(t) - \bar{x}(t)$, (1.81) takes the form $\Delta\ddot{x}(t) + \gamma\Delta\dot{x}(t) = \sqrt{2\gamma kT/m}\,\dot{w}(t)$, so that $\Delta\dot{x}(t)$ is the Ornstein–Uhlenbeck process. Writing the noisy current per particle as $I_p(t) = \bar{I}_p + \Delta I_p(t)$, we find from eq. (1.78) that $\Delta I_p(t) = q\Delta\dot{x}(t)/L$.

Thus the autocorrelation function of current fluctuations per particle is given by

$$\lim_{t\to\infty}\langle\Delta I_p(t)\Delta I_p(t+s)\rangle = \frac{q^2}{L^2}\lim_{t\to\infty}\langle\Delta\dot{x}(t)\Delta\dot{x}(t+s)\rangle. \tag{1.82}$$

According to (1.63), $\lim_{t\to\infty}\langle\Delta\dot{x}(t)\Delta\dot{x}(t+s)\rangle = (kT/m)e^{-\gamma|s|}$, so that (1.82) gives

$$\lim_{t\to\infty}\langle\Delta I_p(t)\Delta I_p(t+s)\rangle = \frac{q^2}{L^2}\frac{kT}{m}e^{-\gamma|s|}. \tag{1.83}$$

According to (1.38), eq. (1.83) can be approximated for large γ by

$$\lim_{t\to\infty}\langle\Delta I_p(t)\Delta I_p(t+s)\rangle = \frac{2q^2}{L^2}\frac{kT}{\gamma m}\delta(s). \tag{1.84}$$

For N identical noninteracting particles eqs.(1.84) and (1.80) give

$$\lim_{t\to\infty}\langle\Delta I(t)\Delta I(t+s)\rangle = N\lim_{t\to\infty}\langle\Delta I_p(t)\Delta I_p(t+s)\rangle = \frac{2kT}{R}\delta(s). \tag{1.85}$$

Thus, according to (1.65), the power spectrum of the current fluctuations is given by $S_I(\omega) = 2kT/R$. The power spectrum of the voltage fluctuations is given by $S_V(\omega) = R^2 S_I(\omega) = 2kTR$, which is Nyquist's formula for the random electromotive force of a resistor [192].

Exercise 1.8 (Resistance of solution). Assume the negative and positive ions have charges q^- and q^+ and diffusion coefficients D^- and D^+, respectively. Calculate the resistance of the solution and find the corresponding Nyquist formula assuming (a) $q^- = -q^+$, $D^- \neq D^+$; (b) $q^- \neq -q^+$ and $D^- \neq D^+$. □

1.6 Linear systems

In a linear system

$$\dot{\boldsymbol{x}}(t) = \boldsymbol{A}(t)\boldsymbol{x}(t) + \boldsymbol{f}(t), \quad \boldsymbol{x}(0) = \boldsymbol{x}_0, \tag{1.86}$$

where $\boldsymbol{A}(t)$ is a deterministic $n\times n$ matrix, the inhomogeneous term $\boldsymbol{f}(t)$ is called *the input signal* and the solution $\boldsymbol{x}(t)$ is *the output.* The output can be represented in terms of a *fundamental solution*, denoted $\boldsymbol{\Phi}(t,s)$, which is an $n\times n$ matrix that satisfies the matrix differential equation

$$\frac{\partial}{\partial t}\boldsymbol{\Phi}(t,s) = \boldsymbol{A}(t)\boldsymbol{\Phi}(t,s), \quad \boldsymbol{\Phi}(s,s) = \boldsymbol{I}, \tag{1.87}$$

where $\boldsymbol{I}$ is the identity matrix. The solution of (1.86) can be represented as

$$\boldsymbol{x}(t) = \boldsymbol{\Phi}(t,0)\boldsymbol{x}_0 + \int_0^t \boldsymbol{\Phi}(t,s)\boldsymbol{f}(s)\,ds. \tag{1.88}$$

When the input is white noise, the system

$$\dot{\boldsymbol{x}}(t) = \boldsymbol{A}(t)\boldsymbol{x}(t) + \boldsymbol{B}(t)\,\dot{\boldsymbol{w}}(t), \quad \boldsymbol{x}(0) = \boldsymbol{x}_0, \tag{1.89}$$

where $\boldsymbol{B}(t)$ is a deterministic differentiable matrix and $\boldsymbol{w}(t)$ is d-dimensional Brownian motion, can be converted into the well-defined integral equation

$$\begin{aligned}\boldsymbol{x}(t_2) = \boldsymbol{x}(t_1) + \int_{t_1}^{t_2} \boldsymbol{A}(s)\boldsymbol{x}(s)\,ds + \boldsymbol{B}(t_2)\boldsymbol{w}(t_2) - \boldsymbol{B}(t_1)\boldsymbol{w}(t_1)\\ - \int_{t_1}^{t_2} \dot{\boldsymbol{B}}(s)\boldsymbol{w}(s)\,ds\end{aligned}$$

whose solution is

$$\begin{aligned}\boldsymbol{x}(t_2) = \boldsymbol{\Phi}(t_2,t_1)\boldsymbol{x}(t_1) + \boldsymbol{B}(t_2)\boldsymbol{w}(t_2) - \boldsymbol{\Phi}(t_2,t_1)\boldsymbol{B}(t_1)\boldsymbol{w}(t_1)\\ - \int_{t_1}^{t_2} \frac{\partial}{\partial s}\left[\boldsymbol{\Phi}(t,s)\boldsymbol{B}(s)\right]\boldsymbol{w}(s)\,ds.\end{aligned} \tag{1.90}$$

Obviously, setting in eq. (1.90) $t_2 = t$ and $t_1 = 0$ and averaging gives $\mathbb{E}\boldsymbol{x}(t) = \boldsymbol{\Phi}(t,0)\mathbb{E}\boldsymbol{x}_0$, which is equivalent to $d\mathbb{E}\boldsymbol{x}(t)/dt = \boldsymbol{A}(t)\mathbb{E}\boldsymbol{x}(t)$ and $\mathbb{E}\boldsymbol{x}(0) = \mathbb{E}\boldsymbol{x}_0$, which can be obtained from (1.89) by formal averaging. To calculate higher-order moments, the formal representation $\boldsymbol{x}(t) = \boldsymbol{\Phi}(t,0)\boldsymbol{x}_0 + \int_0^t \boldsymbol{\Phi}(t,s)\boldsymbol{B}(s)\dot{\boldsymbol{w}}(s)\,ds$ and (1.69) can be used for the calculation of the autocovariance matrix of the output.

The constant coefficients case, where $\boldsymbol{A}(t)$ and $\boldsymbol{B}(t)$ are independent of t is widely used in filtering theory [243]. In this case the functionals of the output can be evaluated explicitly. The fundamental solution $\boldsymbol{\Phi}(t,s)$ is simply $\boldsymbol{\Phi}(t,s) = \exp\{\boldsymbol{A}(t-s)\}$. If the eigenvalues of $\boldsymbol{A}$ are in the left half of the complex plane, then $\lim_{s\to\infty}\exp\{\boldsymbol{A}(t-s)\} = \boldsymbol{0}$ and the convergence is exponential. The integral $\int_{-\infty}^t \exp\{\boldsymbol{A}(t-s)\}\boldsymbol{B}\boldsymbol{w}(s)\,ds$ exists, because the growth of the Brownian motion is not stronger than a fractional power of t (see Section 3.3.2 below) and the process

$$\begin{aligned}\boldsymbol{x}_{st}(t) &= \int_{-\infty}^{t} \exp\{\boldsymbol{A}(t-s)\}\boldsymbol{B}\dot{\boldsymbol{w}}(s)\,ds\\ &= \boldsymbol{B}\boldsymbol{w}(t) - \boldsymbol{A}\int_{-\infty}^{t} \exp\{\boldsymbol{A}(t-s)\}\boldsymbol{B}\boldsymbol{w}(s)\,ds\end{aligned} \tag{1.91}$$

is a solution of the system (1.90). The solution $\boldsymbol{x}_{st}(t)$ is a stationary process, because it is Gaussian, $\boldsymbol{x}_{st}(-\infty) = 0$, and its autocovariance matrix,

$$\boldsymbol{\sigma}(t,s) = \mathbb{E}\boldsymbol{x}_{st}(t)\boldsymbol{x}_{st}^T(s), \tag{1.92}$$

is a function of $t-s$. Indeed, the covariance matrix $\boldsymbol{\sigma}(t,t) = \mathbb{E}\boldsymbol{x}_{st}(t)\boldsymbol{x}_{st}^T(t)$ can be found, using (1.69), as

$$\boldsymbol{\sigma}(t,t) = \int_{-\infty}^{t} \exp\{\boldsymbol{A}(t-s)\}\boldsymbol{B}\boldsymbol{B}^T\exp\{\boldsymbol{A}^T(t-s)\}\,ds.$$

It follows that

$$\begin{aligned}\boldsymbol{A}\boldsymbol{\sigma}(t,t)+\boldsymbol{\sigma}(t,t)\boldsymbol{A}^T &= \int_{-\infty}^{t}\Big[\boldsymbol{A}\exp\{\boldsymbol{A}(t-s)\}\boldsymbol{B}\boldsymbol{B}^T\exp\{\boldsymbol{A}^T(t-s)\}\\ &\quad+\exp\{\boldsymbol{A}(t-s)\}\boldsymbol{B}\boldsymbol{B}^T\exp\{\boldsymbol{A}^T(t-s)\}\boldsymbol{A}^T\Big]\,ds\\ &= -\int_{-\infty}^{t}\frac{d}{ds}\exp\{\boldsymbol{A}(t-s)\}\boldsymbol{B}\boldsymbol{B}^T\exp\{\boldsymbol{A}^T(t-s)\}\,ds\\ &= -\boldsymbol{B}\boldsymbol{B}^T.\end{aligned}$$

Thus $\boldsymbol{\sigma}(t,t)$ is determined from a system of linear equations independent of t and it is therefore constant, denoted $\boldsymbol{\sigma}$. The equation

$$\boldsymbol{A}\boldsymbol{\sigma}+\boldsymbol{\sigma}\boldsymbol{A}^T+\boldsymbol{B}\boldsymbol{B}^T=\boldsymbol{0} \tag{1.93}$$

is called *Lyapunov's equation.*

Using (1.69), we find from (1.92) that

$$\boldsymbol{\sigma}(t,s)=\int_{-\infty}^{t\wedge s}\exp\{\boldsymbol{A}(t-\tau)\}\boldsymbol{B}\boldsymbol{B}^T\exp\{\boldsymbol{A}^T(s-\tau)\}\,d\tau. \tag{1.94}$$

For $t>s>\tau$, we write $\exp\{\boldsymbol{A}(t-\tau)\}=\exp\{\boldsymbol{A}(t-s)\}\exp\{\boldsymbol{A}(s-\tau)\}$ so that (1.94) takes the form

$$\begin{aligned}\boldsymbol{\sigma}(t,s) &= \exp\{\boldsymbol{A}(t-s)\}\int_{-\infty}^{t}\exp\{\boldsymbol{A}(s-\tau)\}\boldsymbol{B}\boldsymbol{B}^T\exp\{\boldsymbol{A}^T(s-\tau)\}\,d\tau\\ &= \exp\{\boldsymbol{A}(t-s)\}\boldsymbol{\sigma}.\end{aligned}$$

For $s>t$, we obtain $\boldsymbol{\sigma}(t,s)=\boldsymbol{\sigma}\exp\{\boldsymbol{A}^T(s-t)\}$. Thus, the autocovariance matrix of the Gaussian process $\boldsymbol{x}_{st}(t)$ is a function of $t-s$, which proves that all its multidimensional pdfs are functions of $t-s$; that is, the process $\boldsymbol{x}_{st}(t)$ is a stationary solution of the linear system in the constant coefficients case where the eigenvalues of $\boldsymbol{A}$ are in the left half of the complex plane (the *stable* case). We write in this case $\boldsymbol{\sigma}(t,s)=\boldsymbol{\sigma}(\tau)$, where $\tau=t-s$. Stable linear systems can be viewed as higher-dimensional generalizations of the one-dimensional Ornstein–Uhlenbeck process (see Section 1.4).

Exercise 1.9 (White noise as a limit of colored noise). Derive the properties of white noise from those of colored noise in the limit $a\to\infty$, as described at the end of Section 1.4. □

Exercise 1.10 (Integrals of white noise). Use the formal properties of white noise, described in this section, for the following calculations. Assume that $g(t)$ and $h(t)$

are integrable deterministic functions. Define

$$x(t) = \int_0^t g(s)\dot{w}(s)\,ds, \quad y(t) = \int_0^t h(s)\dot{w}(s)\,ds$$

and the two-dimensional process $\boldsymbol{z}(t) = (x(t), y(t))$. Show that $\boldsymbol{z}(t)$ is a Gaussian process, $E\boldsymbol{z}(t) = \boldsymbol{0}$, and that the covariance matrix is

$$E\boldsymbol{z}^T(t)\boldsymbol{z}(t) = \begin{pmatrix} A & B \\ B & C \end{pmatrix},$$

where

$$A = \int_0^t g^2(s)\,ds, \quad B = \int_0^t g(s)h(s)\,ds, \quad C = \int_0^t h^2(s)\,ds.$$

□

Exercise 1.11 (The pdf of a multidimensional BM). Find the joint pdf of the vector $\boldsymbol{w} = (w(t_1), w(t_2), \ldots, w(t_n))^T$. □

Exercise 1.12 (Lyapunov's equation in 2-D). Show that for 2×2 matrices Lyapunov's equation (1.93) gives

$$\boldsymbol{\sigma} = \frac{1}{2\,(tr\,\boldsymbol{A})\,(det\,\boldsymbol{A})}\left\{(det\,\boldsymbol{A})\,\boldsymbol{B}\boldsymbol{B}^T + [\boldsymbol{A} - (tr\,\boldsymbol{A})\,\boldsymbol{I}]\,\boldsymbol{B}\boldsymbol{B}^T\,[\boldsymbol{A} - (tr\,\boldsymbol{A})\,\boldsymbol{I}]^T\right\}.$$

□

Exercise 1.13 (Power spectrum). Show that the power spectral density matrix of the stationary solution,

$$\boldsymbol{S}(\nu) = \int_{-\infty}^{\infty} e^{-i\nu\tau}\boldsymbol{\sigma}(\tau)\,d\tau,$$

is given by $\boldsymbol{S}(\nu) = [\boldsymbol{A} + i\nu]^{-1}\,\boldsymbol{B}\boldsymbol{B}^T\left[\boldsymbol{A}^T + i\nu\right]^{-1}$. □

Exercise 1.14 (The stationary case). Show that in the stationary case $\dot{\boldsymbol{\sigma}}(\tau) = \boldsymbol{A}\boldsymbol{\sigma}(\tau)$. □

Exercise 1.15 (The autocorrelation of a linear system). Express the autocorrelation matrix of the solution of a linear system in terms of the fundamental solution. □

Chapter 2

The Probability Space of Brownian Motion

2.1 Introduction

According to Einstein's description, the Brownian motion can be defined by the following two properties: first, it has continuous trajectories (sample paths) and second, the increments of the paths in disjoint time intervals are independent zero mean Gaussian random variables with variance proportional to the duration of the time interval (it is assumed, for definiteness, that the possible trajectories of a Brownian particle start at the origin). These properties have far-reaching implications about the analytic properties of the Brownian trajectories. It can be shown, for example (see Theorem 2.4.1), that these trajectories are not differentiable at any point with probability 1 [198]. That is, the velocity process of the Brownian motion cannot be defined as a real-valued function, although it can be defined as a distribution (generalized function) [152]. Langevin's construction does not resolve this difficulty, because it gives rise to a velocity process that is not differentiable so that the acceleration process, $\boldsymbol{\Xi}(t)$ in eq. (1.24), cannot be defined.

One might guess that in order to overcome this difficulty in Langevin's equation all differential equations could be converted into integral equations so that the equations contain only well defined velocities. This approach, however, fails even in the simplest differential equations that contain the process $\boldsymbol{\Xi}(t)$ (which in one dimension is denoted $\Xi(t)$). For example, if we assume that $\Delta w(t) \equiv \int_t^{t+\Delta t} \Xi(s)\, ds \sim \mathcal{N}(0, \Delta t)$ and construct the solution of the initial value problem

$$\dot{x} = x\Xi(t), \quad x(0) = x_0 > 0 \tag{2.1}$$

by the Euler method

$$x_{\Delta t}(t + \Delta t) - x_{\Delta t}(t) = x_{\Delta t}(t)\Delta w(t), \quad x_{\Delta t}(0) = x_0 > 0, \tag{2.2}$$

Z. Schuss, *Theory and Applications of Stochastic Processes: An Analytical Approach*, Applied Mathematical Sciences 170, DOI 10.1007/978-1-4419-1605-1_2,

the limit $x(t) = \lim_{\Delta t \to 0} x_{\Delta t}(t)$ is not the function

$$x(t) = x_0 \exp\left\{ \int_0^t \Xi(s)\, ds \right\}.$$

It is shown below that the solution is

$$x(t) = x_0 \exp\left\{ \int_0^t \Xi(s)\, ds - \frac{1}{2}t \right\}.$$

It is evident from this example that differential equations that involve the Brownian motion do not obey the rules of the differential and integral calculus.

A similar phenomenon manifests itself in other numerical schemes. Consider, for example, three different numerical schemes for integrating eq. (2.1) (or rather (2.2)), an explicit Euler, semi implicit, and implicit schemes. More specifically, consider the one-dimensional version of eq. (2.2) with $\Delta w(t) \sim \mathcal{N}(0, \Delta t)$. Discretizing time by setting $t_j = j\Delta t$ for $j = 0, 1, 2, \ldots$, the random increments $\Delta w(t_j) = w(t_{j+1}) - w(t_j)$ are simulated by $\Delta w(t_j) = n_j\sqrt{\Delta t}$, where $n_j \sim \mathcal{N}(0,1)$ are independent (zero mean standard Gaussian random numbers taken from the random number generator). The explicit Euler scheme (2.2) is written as $x_{ex}(t_{j+1}) = x_{ex}(t_j) + x_{ex}(t_j)\, n_j\sqrt{\Delta t}$, with $x_{ex}(0) = x_0 > 0$, the semi implicit scheme is $x_{si}(t_{j+1}) = x_{si}(t_j) + \frac{1}{2}[x_{si}(t_j) + x_{si}(t_{j+1})]\, n_j\sqrt{\Delta t}$, with $x_{si}(0) = x_0 > 0$, and the implicit scheme is $x_{im}(t_{j+1}) = x_{im}(t_j) + x_{im}(t_{j+1})\, n_j\sqrt{\Delta t}$, with $x_{im}(0) = x_0 > 0$. In the limit $\Delta t \to 0,\ t_j \to t$ the numerical solutions converge (in probability) to the three different limits

$$\lim_{\Delta t \to 0,\ t_j \to t} x_{ex}(t_j) = x_0 \exp\left\{ \int_0^t \Xi(s)\, ds - \frac{1}{2}t \right\}$$

$$\lim_{\Delta t \to 0,\ t_j \to t} x_{si}(t_j) = x_0 \exp\left\{ \int_0^t \Xi(s)\, ds \right\}$$

$$\lim_{\Delta t \to 0,\ t_j \to t} x_{im}(t_j) = x_0 \exp\left\{ \int_0^t \Xi(s)\, ds + \frac{1}{2}t \right\}.$$

These examples indicate that naïve applications of elementary analysis and probability theory to the simulation of Brownian motion may lead to conflicting results. The study of the trajectories of the Brownian motion requires a minimal degree of mathematical rigor in the definitions and constructions of the probability space and the probability measure for the Brownian trajectories in order to gain some insight into stochastic dynamics. Thus Section 2.2 contains a smattering of basic measure theory that is necessary for the required mathematical insight.

In this chapter the mathematical Brownian motion is defined axiomatically by the properties of the physical Brownian motion as described in Chapter 1. Two constructions of the mathematical Brownian motion are presented, the Paley–Wiener Fourier series expansion and Lévy's method of refinements of piecewise linear approximations [150]. Some analytical properties of the Brownian trajectories are derived from the definition.

2.2 The space of Brownian trajectories

A continuous-time *random process* (or *stochastic process*) $x(t, \omega) : \mathbb{R}^+ \times \Omega \to \mathbb{R}$ is a function of two variables, a real variable t, usually interpreted as time, and ω in a *probability space* (or *sample space*) Ω, in which events are defined. More generally, the random process $x(t, \omega)$ can take values in a set X, called *the state space*, such as the real line $\mathbb{R}$, or the Euclidean space $\mathbb{R}^d$, or any other set. When t is interpreted as time, we write $x(t, \omega) : \mathbb{R}_+ \times \Omega \to X$. For each $\omega \in \Omega$ the stochastic process is a function of t, called a *trajectory*.

We assume henceforth that the state space of a stochastic process $x(t, \omega)$ is $X = \mathbb{R}^d$ and its trajectories are continuous functions; that is, for fixed $\omega \in \Omega$ the trajectories are continuous curves in $\mathbb{R}^d$. To assign probability to events connected to trajectories, it is necessary to describe the probability space Ω. We begin with the description of the probability space and the Einstein-Langevin requirement that the trajectories of the Brownian motion be continuous. Thus, we define events in a probability space for the Brownian motion in terms of continuous functions of time. We identify all possible paths of the Brownian motion as all continuous functions. Each continuous function is an elementary event in this space. Physically, this event can represent the path of a microscopic particle in solution. The path, and thus the event in the probability space, is the outcome of the experiment of continuous recording of the path of a particle diffusing according to the Einstein-Langevin description, namely, without jumps. If jumps were found experimentally, a different theory might be needed, depending on the properties of the paths, for example, as is the case for the paths of the Poisson jump process [199, p. 290], [116, p. 22, Example 2]. In many cases, we consider sets of elementary events, which are often called "events", for short.

We hardly ever consider elementary events, because their probability is zero. This, for example is the case of turning a roulette wheel with a needle pointing to a single point. The outcome of each spin is a single point or number (an elementary event). Each point must have probability zero, because in an honest roulette wheel all points are equally likely and there are an infinite number of them on the wheel. Of course, for every spin there is an outcome, so that events of probability zero do occur. The roulette wheel is partitioned into a finite number of intervals of finite lengths, each containing an infinite number of points (elementary events), because we want a finite nonzero estimate of the probability. Every time the roulette wheel is spun the needle comes to rest in only one interval, which is the outcome of the game or experiment. This outcome is called "an event", which is a composite event consisting of uncountably many elementary events, whose individual probabilities are zero; however, the probability of the outcome, the composite event, is a finite nonzero number.

In the same vein, a Brownian elementary event will have to be assigned probability zero. A typical Brownian event that corresponds to an experiment consists of (uncountably many) elementary events. It may be, for example, the set of all continuous functions that satisfy some given criteria. Thus, in an experiment one might record the ensemble of all Brownian paths that are found in a given region (under a microscope) at a given time. We formally define Brownian elementary events and events as follows. Denote by $\mathbb{R}$ and $\mathbb{R}_+$ the real numbers and the nonnegative real numbers, respectively, then

Definition 2.2.1 (The space of elementary events). *The* space of elementary events *for the Brownian motion is the set of all continuous real functions,*

$$\Omega = \{\omega(t) : \mathbb{R}_+ \mapsto \mathbb{R}\}.$$

Thus each continuous function is an elementary event. To define Brownian events that are more complicated than elementary events; that is, events that consist of uncountably many Brownian trajectories (each of which is an elementary event), we define first events called "cylinders".

Definition 2.2.2 (Cylinder sets). *A* cylinder set *of Brownian trajectories is defined by times* $0 \le t_1 < t_2 < \cdots < t_n$ *and real intervals* $I_k = (a_k, b_k)$, $(k = 1, 2, \dots, n)$ *as*

$$C(t_1, \dots, t_n;\ I_1, \dots, I_n) = \{\omega(t) \in \Omega \,|\, \omega(t_k) \in I_k, \text{ for all } 1 \le k \le n\}. \tag{2.3}$$

Obviously, for any $0 \le t_1 < t$ and any interval I_1,

$$C(t; \mathbb{R}) = \Omega, \quad C(t_1, t; I_1, \mathbb{R}) = C(t_1; I_1). \tag{2.4}$$

Thus, for the the cylinder $C(t_1, t_2, \dots, t_n;\ I_1, I_2, \dots, I_n)$ not to contain a trajectory $\omega(t)$ it suffices that for at least one of the times t_k the value of $\omega(t_k)$ is not in the interval I_k, for example, the dotted trajectory in Figure 2.1 belongs to the cylinder $C(126, [-0.1, 0.5])$, but neither to $C(132, [-0.4, 0.1])$ nor to $C(136, [-0.2, -0.10])$. Thus it does not belong to the cylinder $C(126, 132, [-0.1, 0.5], [-0.4, 0.1])$, which is their intersection.

For each real x, we set $I^x = (-\infty, x]$. Then the cylinder $C(t; I^x)$ is the set of all continuous functions $\omega(\cdot)$ such that $\omega(t) \le x$. It is the set of all Brownian trajectories that would be observed at time t to be below the level x. It is important to note that $C(t; I^x)$ consists of entire trajectories, not merely of their segments observed below the level x at time t.

The cylinder $C(t_1, t_2; I_1, I_2)$ consists of all continuous functions $\omega(\cdot)$ such that $a_1 < \omega(t_1) < b_1$ and $a_2 < \omega(t_2) < b_2$. That is, $C(t_1, t_2; I_1, I_2)$ consists of all

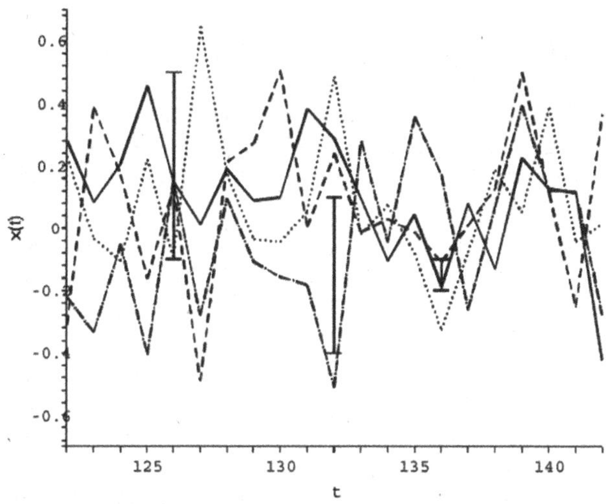

Figure 2.1. Four Brownian trajectories sampled at discrete times. Three cylinders are marked by vertical bars. The trajectories were sampled according to the scheme (2.21).

Brownian paths that are observed at time t_1 to be between the levels a_1 and b_1 and at time t_2 to be between the levels a_2 and b_2 (see Figure 2.1).

Definition 2.2.3 (Brownian events). Brownian events *are all sets of Brownian trajectories that can be obtained from cylinders by the operations of countable unions, intersections, and the operation of complement.*

These sets form the space of *Brownian events*, denoted $\mathcal{F}$. The space $\mathcal{F}$ is characterized by the property that if A_i are subsets of $\mathcal{F}$ $(i = 1, 2, \dots)$; that is, if A_i are Brownian events, then their (countable) union, $\bigcup_{i=1}^{\infty} A_i$, is also an event and so are the complements $A_i^c = \Omega - A_i$. This space of Brownian events is an example of a σ-*algebra.*

Definition 2.2.4 (σ-algebra). *A* σ-algebra *in Ω is a nonempty collection $\mathcal{F}$ of subsets of Ω such that*

1. $\Omega \in \mathcal{F}$.
2. *If* $A \in \mathcal{F}$ *then* $A^c = \Omega - A \in \mathcal{F}$.
3. *If* $A_i \in \mathcal{F}$, $(i = 1, 2, \dots)$, *then* $\bigcup_{i=1}^{\infty} A_i \in \mathcal{F}$.

In this notation the designation σ stands for the word "countable". If only a finite number of unions, intersections, and complement of events are considered, we refer to the resulting set of events as an "algebra". The elements (events) of $\mathcal{F}$ are called *measurable sets*. Examples exist of nonmeasurable sets [183], [58]. Sigma-algebras are introduced so we can easily keep track of past, present, and future in our study of Brownian events. Causality must be included in our description of the evolution of Brownian trajectories; that is, events occurring after time t do not affect events up to time t. There are noncausal problems in probability theory, such as the problem of estimating a given random signal in a given time interval, given its noisy measurements in the past and in the future (e.g., the entire signal is recorded on a CD). This is the smoothing problem (see [215] for some discussion).

The pair $(\Omega, \mathcal{F})$ is called *probability space*. The probability measure is defined on events. In the process of constructing a probability model of the Brownian motion (or any other process) both the space of elementary events and the relevant σ-algebra of events have to be specified. There is more than one way to specify structures of events (algebras or σ-algebras) in the same space of elementary events and different pairs of spaces and σ-algebras of events are considered different probability spaces, as described below. For example, if the roulette wheel is partitioned into arcs in two different ways so that one partition cannot be obtained from the other by the operations of union, intersection, and complement, then the different partitions form different algebras in the space of elementary events for the experiment of rotating the roulette wheel. Thus, partitioning the wheel into two equal arcs or into three equal arcs results in two different algebras of events. In general, each σ-algebra in Ω specifies a different way of selecting the elementary events to form composite events that correspond to different ways of handling the same raw experimental data, when the data are the elementary events.

Definition 2.2.5 (Brownian filtration). *The σ-algebra $\mathcal{F}_t$ is defined by cylinder sets confined to times $0 \leq t_i < t$, for some fixed t. Obviously, $\mathcal{F}_s \subset \mathcal{F}_t \subset \mathcal{F}$ if $0 \leq s < t < \infty$. The family of $\sigma-$algebras $\mathcal{F}_t$ for $t \geq 0$ is called* the Brownian filtration *and is said to be generated by the Brownian events up to time t.*

Note that the elementary events of the Brownian filtration $\mathcal{F}_t$ are continuous functions in the entire time range, not just the initial segments in the time interval $[0, t]$. However, only the initial segments of the Brownian paths in $\mathcal{F}_t$ that occur by time t are observed and so can be used to define the filtration. The pairs $(\Omega, \mathcal{F}_t)$ are different probability spaces for different values of t.

Up to now, we have considered only elementary events and sets of elementary events that were referred to as "events". The events we have defined mathematically have to be assigned probabilities, to represent some measure of our uncertainty about the outcome of a given experiment [53]. The assigned probabilities form a mathematical model for the statistical processing of collected data. This was the case for the recordings of paths of Brownian particles before mathematical models were constructed by Einstein, Smoluchowski, and Langevin.

Definition 2.2.6 (Random variables in $(\Omega, \mathcal{F})$). *A* random variable $X(\omega)$ *in $(\Omega, \mathcal{F})$ is a real function $X(\cdot) : \Omega \to \mathbb{R}$ such that $\{\omega \in \Omega \,|\, X(\omega) \leq x\} \in \mathcal{F}$ for all $x \in \mathbb{R}$.*

That is, $\{\omega \in \Omega \,|\, X(\omega) \leq x\}$ is a Brownian event that can be expressed by countable operations of union, intersection, and complement of cylinders. In mathematical terminology a random variable in Ω is a real $\mathcal{F}$-measurable function.

Example 2.1 (Random functions). For each $t \geq 0$ consider the random variable $X_t(\omega) = \omega(t)$ in Ω. This random variable is the outcome of the experiment of sampling the position of a Brownian particle (trajectory) at a fixed time t. Thus $X_t(\omega)$ takes different values on different trajectories. Obviously, $\{\omega \in \Omega \,|\, X_t(\omega) \leq x\} = \{\omega \in \Omega \,|\, \omega(t) \leq x\} = C(t; I^x) \in \mathcal{F}$, so that $X_t(\omega)$ is a random variable in $(\Omega, \mathcal{F})$. □

Example 2.2 (Average velocity). Although, as mentioned in Section 2.1, the derivative of the Brownian path does not exist as a real-valued function, the *average velocity process* of a Brownian trajectory ω in the time interval $[t, t + \Delta t]$ can be defined as $\bar{V}_t(\omega) = [\omega(t + \Delta t) - \omega(t)]/\Delta t$. The time averaging here is not expectation, because it is defined separately on each trajectory, therefore $\bar{V}_t(\omega)$ is a random variable, which takes different values on different trajectories. To see that $\bar{V}_t(\omega)$ is a random variable in $(\Omega, \mathcal{F})$, we have to show that for every real number v the event $\{\omega \in \Omega \,|\, \bar{V}_t(\omega) \leq v\}$ can be expressed by countable operations of union, intersection, and complement of cylinders. To do so, we assume that $\Delta t > 0$ and write

$$\{\omega \in \Omega \,|\, \bar{V}_t(\omega) \leq v\} = \{\omega \in \Omega \,|\, \omega(t + \Delta t) - \omega(t) \leq v\Delta t\} \equiv A.$$

We denote the set of rational numbers by $\mathbb{Q}$ and the set of positive rational numbers by $\mathbb{Q}^+$ and define in $\mathcal{F}$ the set of paths

$$B \equiv \bigcap_{\varepsilon \in \mathbb{Q}^+} \bigcup_{y \in \mathbb{Q}} C(t, t + \Delta t; [y - \varepsilon, y + \varepsilon], (-\infty, v\Delta t + y + \varepsilon]).$$

The set B is simply the set of paths in $\mathcal{F}$ such that for every rational $\varepsilon > 0$, there exists a rational y such that $|w(t) - y| \leq \varepsilon$ and $w(t + \Delta t) \leq y + v\Delta t + \varepsilon$. Showing that $A = B$ proves that $\bar{V}_t(\omega)$ is a random variable in Ω (i.e., $\bar{V}_t(\omega)$ is $\mathcal{F}$-measurable).

To show that $A = B$, we show that $A \subset B$ and $B \subset A$. If $\omega \in A$ then

$$\omega(t + \Delta t) - \omega(t) \leq v\Delta t \tag{2.5}$$

and as is well-known from the differential calculus, for every number $\omega(t)$ and every $\varepsilon \in \mathbb{Q}^+$ there exists $y \in \mathbb{Q}$ such that

$$y - \varepsilon \leq \omega(t) \leq y + \varepsilon. \tag{2.6}$$

It follows from eqs. (2.5) and (2.6) that both inequalities

$$y - \varepsilon \leq \omega(t) \leq y + \varepsilon, \quad \omega(t + \Delta t) \leq v\Delta t + y + \varepsilon \tag{2.7}$$

hold. They mean that for every $\varepsilon \in \mathbb{Q}^+$ there exists $y \in \mathbb{Q}$ such that

$$\omega \in C(t, t + \Delta t; [y - \varepsilon, y + \varepsilon], (-\infty, v\Delta t + y + \varepsilon]).$$

This in turn means that $\omega \in B$. Hence $A \subset B$.

Conversely, if $\omega \in B$, then for every $\varepsilon \in \mathbb{Q}^+$ there exists $y \in \mathbb{Q}$ such that $\omega \in C(t, t + \Delta t; [y - \varepsilon, y + \varepsilon], (-\infty, v\Delta t + y + \varepsilon])$, which implies that the inequalities (2.7) hold and consequently $\omega(t + \Delta t) - \omega(t) \le v\Delta t + 2\varepsilon$ for every $\varepsilon \in \mathbb{Q}^+$. It follows that $\omega(t + \Delta t) - \omega(t) \le v\Delta t$, so that $\omega \in A$, which implies that $B \subset A$, as claimed above. □

Example 2.3 (Integrals of random functions). A similar argument can be used to show, for example, that $X(\omega) = \int_0^T \omega(t)\, dt$ is a random variable in Ω, measurable with respect to $\mathcal{F}_T$. □

Definition 2.2.7 (Markov times). *A nonnegative random variable $\tau(\omega)$, defined on Ω, is called a* stopping time *or a* Markov time *relative to the filtration $\mathcal{F}_t$ for $t \ge 0$ if*

$$\{\omega \in \Omega \,|\, \tau(\omega) \le t\} \in \mathcal{F}_t \quad \text{for all } t \ge 0.$$

Example 2.4 (First passage times). The *first passage time* (FPT) of a Brownian trajectory through a given point is a random variable in Ω and a stopping time. Indeed, assume that $\omega(0) < y$ and set $\tau_y(\omega) = \inf\{t \ge 0 \,|\, \omega(t) > y\}$; that is, $\tau_y(\omega)$ is the first passage time of $\omega(t)$ through the value y. To show that τ_y is $\mathcal{F}_t$-measurable for every $t > 0$, we proceed in an analogous manner to that above. We denote by $\mathbb{Q}_t$ the set of all positive rational numbers that do not exceed t. Obviously, $\mathbb{Q}_t$ is a countable set. The event $\{\omega \in \Omega \,|\, \tau_y(\omega) \le t\}$ consists of all Brownian trajectories ω that go above the level y at some time prior to t. Thus, due to the continuity of the Brownian paths,

$$\{\omega \in \Omega \,|\, \tau_y(\omega) \le t\} = \bigcup_{r \in \mathbb{Q}_t} \{\omega \in \Omega \,|\, \omega(r) \ge y\},$$

which is a countable union of the cylinders $C(r, [y, \infty))$ for $r \le t$ and is thus in $\mathcal{F}_t$. □

Example 2.5 (Last passage time). On the other hand, the *last passage time* (LPT) of $x(t, \omega)$ to a given point y before time T, denoted $LPT(y, T, \omega)$, is not a stopping time, because at time t it is not a Brownian event that depends on the Brownian trajectories up to time t. Rather, it depends on events after time t, because the last passage may occur after that; that is,

$$\{\omega \in \Omega \,|\, LPT(y, T, \omega) \le t\} \notin \mathcal{F}_t \quad \text{for all } 0 \le t < T.$$

Although $LPT(y, T, \omega)$ is a random variable in $(\Omega, \mathcal{F})$, it is not a random variable in $(\Omega, \mathcal{F}_t)$ for $t < T$. Last passage times occur in practical problems. For example, if a Brownian particle is trapped in a finite potential well and escapes at a random time, the time between the last visit to the bottom of the well and the first passage time through the top of the well is a LPT. □

Example 2.6 (Indicators). For any set $A \in \Omega$ the *indicator* function of A is defined by

$$\mathbb{1}_A(\omega) = \begin{cases} 1 & \text{if } \omega \in A \\ 0 & \text{otherwise.} \end{cases} \tag{2.8}$$

For all $A \in \mathcal{F}$ the function $\mathbf{1}_A(\omega)$ is a random variable in $(\Omega, \mathcal{F})$. Indeed, if $x < 1$, then $\{\omega \in \Omega \,|\, \mathbf{1}_A(\omega) \leq x\} = \Omega - A = A^c$ and if $x \geq 1$, then $\{\omega \in \Omega \,|\, \mathbf{1}_A(\omega) \leq x\} = \Omega$ so that in either case the set $\{\omega \in \Omega \,|\, \mathbf{1}_A(\omega) \leq x\}$ is in $\mathcal{F}$. Thus, if A is not a measurable set, its indicator function $\mathbf{1}_A(\omega)$ is a nonmeasurable function so that $\mathbf{1}_A(\omega)$ is not a random variable. □

Exercise 2.1 (Positive random variables). For a random variable $X(\omega)$ define the functions $X^+(\omega) = \max\{X(\omega), 0\}$ and $X^-(\omega) = \min\{X(\omega), 0\}$. Show that $X^+(\omega)$ and $X^-(\omega)$ are random variables (i.e., they are measurable functions). □

Measurements recorded sequentially in time are often represented graphically as points in the d-dimensional real Euclidean space $\mathbb{R}^d$ $(d = 1, 2, \ldots,)$. When the points are sampled from a curve in $\mathbb{R}^d$, they form a path. For example, recordings of trajectories of Brownian particles in $\mathbb{R}^3$ reveal that they have continuous paths, however, repeated recordings yield different paths that look completely erratic and random. When tracking charged Brownian particles (e.g., ions in solution) in the presence of an external electrostatic field, the paths remain continuous, erratic, and random; however, they tend to look different from those of uncharged Brownian particles.

Definition 2.2.8 (Stochastic processes in $(\Omega, \mathcal{F})$). *A function $x(t, \omega) : \mathbb{R}_+ \times \Omega \mapsto \mathbb{R}$ is called a* stochastic process *in $(\Omega, \mathcal{F})$ with continuous trajectories if*

(i) $x(t, \omega)$ is a continuous function of t for every $\omega \in \Omega$,

(ii) for every fixed $t \geq 0$ the function $x(t, \omega) : \Omega \mapsto \mathbb{R}$ is a random variable in Ω.

The variable ω in the notation for a stochastic process $x(t, \omega)$ denotes the dependence of the value the process takes at any given time t on the elementary event ω; that is, on the particular realization of the Brownian path ω. Point (ii) of the definition means that the sets $\{\omega \in \Omega \,|\, x(t, \omega) \leq y\}$ are Brownian events for each $t \geq 0$ and $y \in \mathbb{R}$; that is, they belong to $\mathcal{F}$. When they do, we say that the process $x(t, \omega)$ is *measurable* with respect to $\mathcal{F}$ or simply $\mathcal{F}$-measurable.

Definition 2.2.9 (Adapted processes). *The process $x(t, \omega)$ is said to be* adapted *to the Brownian filtration $\mathcal{F}_t$ if $\{\omega \in \Omega \,|\, x(t, \omega) \leq y\} \in \mathcal{F}_t$ for every $t \geq 0$ and $y \in \mathbb{R}$. In that case we also say that $x(t, \omega)$ is $\mathcal{F}_t$-measurable.*

This means that the events $\{\omega \in \Omega \,|\, x(t, \omega) \leq y\}$ can be expressed in terms of Brownian events up to time t. Thus an adapted process at time t does not depend on the future behavior of the Brownian trajectories from time t on: an adapted process is nonanticipatory. For example, for any deterministic integrable function $f(t)$ the process $x(t, \omega)$, whose trajectories are $x(t, \omega) = \int_0^t f(s)\omega(s)\, ds$, is adapted to the Brownian filtration. An adapted process, such as $x(t, \omega)$ above, can be viewed as the output of a causal filter operating on Brownian trajectories (which may represent a random signal).

Example 2.7 (First passage times). The first passage time (FPT) $\tau_y(\omega)$ of an adapted continuous process $x(t, \omega)$ to a given point y is a Markov time relative to the Brownian filtration because it depends on the Brownian trajectories $\omega(t)$ up to time t. The relation of Markov times to Markov processes is discussed later. Thus, for $y > 0$ the random time $\tau_y(\omega) = \inf\{t \geq 0 \,|\, x(t,\omega) > y\}$ is the first passage time of $x(t,\omega)$ through the value y. To see that τ_y is a Markov time, we proceed in a manner similar to that of Example 2.4 above. The event $\{\omega \in \Omega \,|\, \tau_y(\omega) > t\}$ consists of all Brownian trajectories ω for which $x(t,\omega)$ stays below the level y for all times prior to t. Thus, due to the continuity of the paths of $x(t,\omega)$,

$$\{\omega \in \Omega \,|\, \tau_y(\omega) \leq t\} = \bigcup_{r \in \mathbb{Q}_t} \{\omega \in \Omega \,|\, x(r,\omega) \geq y\},$$

which is in $\mathcal{F}_t$ because $x(t,\omega)$ is an adapted process. □

Example 2.8 (Last passage time). The last passage time of $x(t,\omega)$ to a given point y before time T, denoted $LPT(y,T,\omega)$, is not a Markov time relative to the Brownian filtration, because at time t it is not a Brownian event that depends on the Brownian trajectories up to time t. As mentioned in Example 2.5, it depends on events after time t, because the last passage may occur after that; that is, $\{\omega \in \Omega \,|\, LPT(y,T,\omega) \leq t\} \notin \mathcal{F}_t$ for all $0 \leq t < T$. □

Example 2.9 (First exit time). The *first exit time* from an interval is a Markov time for the Brownian motion. It is defined for $a < 0 < b$ by

$$\tau_{[a,b]} = \inf\{t \geq 0 \,|\, w(t) < a \text{ or } w(t) > b\}.$$

□

Exercise 2.2 (First exit time: continuation). Prove the claim of Example 2.9: show that $\tau_{[a,b]}$ is a Markov time for the Brownian motion. □

There are two slightly different concepts of a *signed measure*, depending on whether one allows it to take infinite values (see, e.g., Wikipedia).

Definition 2.2.10 (Signed measure). *A* signed real-valued measure *in the space $(\Omega,\mathcal{F})$ is a function*

$$\mu : \mathcal{F} \to \mathbb{R} \tag{2.9}$$

such that for any sequence of disjoint set $\{A_n\}_{n=1}^{\infty}$ in $\mathcal{F}$,

$$\mu\left\{\bigcup_{n=1}^{\infty} A_n\right\} = \sum_{n=1}^{\infty} \mu\{A_n\}.$$

Definition 2.2.11 (Real measure). *If $\mu(A) \geq 0$ for all $A \in \mathcal{F}$, we say that it is a* measure. *The triple $(\Omega,\mathcal{F},\mu)$ is called a* measure space.

Definition 2.2.12 (Null sets). *A measurable set A such that $\mu(A) = 0$ is a μ-*null set. *Any property that holds for all $\omega \in \Omega$, except on a μ-null set is said to hold μ-almost everywhere (μ-*a.e.*) or for μ-*almost all $\omega \in \Omega$.

We assume that all subsets of null sets are measurable (in measure theory this means that the measure space is *complete*). A measure is a monotone set function in the sense that if A and B are measurable sets such that $A \subset B$, then $\mu(A) \leq \mu(B)$. If $\mu(\Omega) < \infty$, we say that $\mu(A)$ is a *finite measure.* If Ω is a countable union of sets of finite measure, we say that Ω is a *σ-finite measure.*

Definition 2.2.13 (Integration with respect to a measure). *A measure $\mu(A)$ defines an integral of a nonnegative measurable function $f(\omega)$ by*

$$\int_{\Omega} f(\omega)\, d\mu(\omega) = \lim_{h\to 0}\lim_{N\to\infty}\sum_{n=0}^{N} nh\mu\{\omega : nh \leq f(\omega) \leq (n+1)h\}, \tag{2.10}$$

whenever the limit exists. In this case, we say that $f(\omega)$ is an integrable function.

For every measurable function $f(\omega)$ the nonnegative functions $f^+(\omega)$ and $f^-(\omega)$ are nonnegative measurable functions and $f(\omega) = f^+(\omega) - f^-(\omega)$. We say that $f(\omega)$ is an integrable function if both $f^+(\omega)$ and $f^-(\omega)$ are integrable and

$$\int_{\Omega} f(\omega)\, d\mu(\omega) = \int_{\Omega} f^+(\omega)\, d\mu(\omega) - \int_{\Omega} f^-(\omega)\, d\mu(\omega).$$

The function $f(\omega)$ is integrable if and only if $|f(\omega)|$ is integrable, because $|f(\omega)| = f^+(\omega) + f^-(\omega)$. For any set $A \in \mathcal{F}$ the indicator function $\mathbf{1}_A(\omega)$ (see Example 2.4) is integrable and $\int \mathbf{1}_A(\omega)\, d\mu(\omega) = \mu(A)$. We define an integral over a measurable set A by $\int_A f(\omega)\, d\mu(\omega) = \int_\Omega \mathbf{1}_A(\omega) f(\omega)\, d\mu(\omega)$. If $\int_A f(\omega)\, d\mu(\omega)$ exists, we say that $f(\omega)$ is *integrable in* A. In that case $f(\omega)$ is integrable in every measurable subset of A. If $\mu(A) = 0$ then $\int_A f(\omega)\, d\mu(\omega) = 0$. If $f(\omega)$ is an integrable function with respect to a measure $\mu(A)$, then the integral

$$\nu(A) = \int_A f(\omega)\, d\mu(\omega) \tag{2.11}$$

defines $\nu(A)$ as a signed measure in $\mathcal{F}$. Obviously, if $\mu(A) = 0$ then $\nu(A) = 0$.

Definition 2.2.14 (Differentiation of measures). *If the measures ν and μ satisfy eq.* (2.11)*, the function $f(\omega)$ is the Radon–Nikodym derivative of the measure ν with respect to the measure μ at the point ω and is denoted*

$$f(\omega) = \frac{d\nu(\omega)}{d\mu(\omega)}. \tag{2.12}$$

Definition 2.2.15 (Absolute continuity). *A signed measure ν is* absolutely continuous *with respect to the measure μ if μ-null sets are ν-null sets.*

Thus the measure ν in (2.11) is absolutely continuous with respect to μ. If two measures are absolutely continuous with respect to each other, they are said to be *equivalent*.

Theorem 2.2.1 (Radon–Nikodym [58]). *If ν is a finite signed measure, absolutely continuous with respect to a σ-finite measure μ, then there exists a μ integrable function $f(\omega)$, uniquely defined up to μ-null sets, such that* (2.11) *holds for all $A \in \mathcal{F}$. For a constant c the inequality $\nu(A) \geq c\mu(A)$ for all $A \in \mathcal{F}$ implies $f(\omega) \geq c$ for μ-almost all $\omega \in \Omega$.*

The function $f(\omega)$ in the theorem is called the Radon–Nikodym derivative and is denoted as in (2.12).

Definition 2.2.16 (Probability measure). *A positive measure* $\Pr$ *such that* $\Pr\{\Omega\} = 1$ *is called* a probability measure *and the probability of an event $A \in \mathcal{F}$ is denoted* $\Pr\{A\}$.

Thus the probability of an event (a measurable set) is a number between 0 and 1. An event whose probability is 1 is called a *sure event*. The event Ω is a sure event. There are many ways for assigning probabilities to events, depending on the degree of uncertainty we have about a given event; different persons may assign different probabilities to the same events. We may think of the probability of an event as a measure of our uncertainty about it [53]. Recall that a measurable function on a probability space $(\Omega, \mathcal{F}, \Pr)$ is called a random variable, denoted $X(\omega)$.

Definition 2.2.17 (Expectations). *Integrals of random variables with respect to the probability measure* $\Pr\{A\}$ *are called* expectations *and are denoted*

$$\mathbb{E}X(\omega) = \int_{\Omega} X(\omega)\, d\Pr\{\omega\}. \tag{2.13}$$

For any set A in $\mathcal{F}$, we define

$$\begin{aligned}\mathbb{E}\{X, A\} &= \int_{A} X(\omega)\, d\Pr(\omega) \\ &= \int_{-\infty}^{\infty} x \Pr\{\omega \in A : x \leq X(\omega) \leq x + dx\}.\end{aligned} \tag{2.14}$$

Applications of integration with respect to Pr are given in the next section.

Definition 2.2.18 (PDF and pdf). *For an integrable random variable $X(\omega)$ the function*

$$F_X(x) = \Pr\{\omega : X(\omega) \leq x\}$$

is called the probability distribution function (PDF) *of $X(\omega)$. The function (or generalized function* [152]*)*

$$f_X(x) = \frac{d}{dx} F_X(x)$$

is called the probability density function (pdf) *of $X(\omega)$. The expectation $\mathbb{E}X(\omega)$ can be written as*

$$\mathbb{E}X(\omega) = \int x\, dF_X(x) = \int x f_X(x)\, dx. \tag{2.15}$$

Note that the PDF $F_X(x)$ need not be differentiable, so the pdf $f_X(x)$ need not be a function, but rather a generalized function (a distribution).

Exercise 2.3 (Coin tossing). Construct a probability space on $\mathbb{R}$, a random variable, PDF, and pdf for the experiment of tossing a fair coin. □

Definition 2.2.19 (Conditional expectations). *For a sub σ-field $\mathcal{F}_1$ of $\mathcal{F}$ and a random variable $X(\omega)$ with finite variance on $\mathcal{F}$, the* conditional expectation $\mathbb{E}(X \mid \mathcal{F}_1)$ *is a random variable measurable with respect to $\mathcal{F}_1$ such that for every random variable $Y(\omega)$ measurable with respect to $\mathcal{F}_1$, whose variance is finite*

$$\mathbb{E}X(\omega)Y(\omega) = \mathbb{E}[\mathbb{E}(X \mid \mathcal{F}_1)Y(\omega)]. \tag{2.16}$$

If we confine the functions $Y(\omega)$ to indicators of events $A \in \mathcal{F}_1$, then (2.16) implies that $\mathbb{E}(X \mid \mathcal{F}_1)$ is a random variable that satisfies

$$\int\limits_A X(\omega)\, d\Pr(\omega) = \int\limits_A \mathbb{E}(X \mid \mathcal{F}_1)\, d\Pr(\omega) \tag{2.17}$$

for all $A \in \mathcal{F}_1$. To determine the existence of the conditional expectation, we denote the left-hand side of eq. (2.17) $\nu(A)$ and recall that $\nu(A)$ is a signed measure on $\mathcal{F}_1$. It follows from the Radon–Nikodym theorem that the desired function is simply

$$\mathbb{E}(X \mid \mathcal{F}_1) = \frac{d\nu(\omega)}{d\mu(\omega)}. \tag{2.18}$$

Definition 2.2.20 (Conditional probabilities, distributions, and densities). *For any event $M \in \mathcal{F}$ the* conditional probability *of M, given $\mathcal{F}_1$, is defined as*

$$\Pr\{M \mid \mathcal{F}_1\} = \mathbb{E}[\mathbf{1}_M \mid \mathcal{F}_1].$$

The conditional probability distribution function (CPDF) *of the random variable $X(\omega)$, given $\mathcal{F}_1$, is defined as*

$$F_{X|\mathcal{F}_1}(x,\omega) = \Pr\{\omega :\, X(\omega) \le x \mid \mathcal{F}_1\}.$$

Note that for every ω the CPDF $F_{X|\mathcal{F}_1}(x,\omega)$ is a probability distribution function. Its density,

$$f_{X|\mathcal{F}_1}(x,\omega) = \frac{d}{dx} F_{X|\mathcal{F}_1}(x,\omega),$$

is also a random function.

2.2.1 The Wiener measure of Brownian trajectories

Having constructed the set $\mathcal{F}$ of events for the Brownian trajectories, we proceed to construct a probability measure of these events. The probability measure is used to construct a mathematical theory of the Brownian motion that can describe experiments.

A probability measure Pr can be defined on Ω (i.e., on the events $\mathcal{F}$ in Ω) to conform with the Einstein–Langevin description of the Brownian motion. It is enough to define the probability measure $\Pr\{\cdot\}$ on cylinder sets and then to extend it to all events in $\mathcal{F}$ by the elementary properties of a probability measure (see [106] for a more detailed exposition of this topic). The following probability measure in $\mathcal{F}$ is called *the Wiener measure* [248]. Consider the cylinder $C(t; I)$, where $t \geq 0$ and $I = (a, b)$ and set

$$\Pr\{C(t; I)\} = \frac{1}{\sqrt{2\pi t}} \int_a^b e^{-x^2/2t}\, dx. \tag{2.19}$$

If $0 = t_0 < t_1 < t_2 < \cdots < t_n$ and I_k $(k = 1, 2, \ldots, n)$ are real intervals, set

$$\begin{aligned} &\Pr\{C(t_1, t_2, \ldots, t_n; I_1, I_2, \ldots, I_n)\} \\ &= \int_{I_1}\int_{I_2}\cdots\int_{I_n} \prod_{k=1}^{n} \frac{dx_k}{\sqrt{2\pi(t_k - t_{k-1})}} \exp\left\{-\frac{(x_k - x_{k-1})^2}{2(t_k - t_{k-1})}\right\}, \end{aligned} \tag{2.20}$$

where $x_0 = 0$ (the extension of the Wiener measure from cylinders to $\mathcal{F}$ is described in [106, 208]). The integral (2.20) is called *Wiener's discrete path integral*. The obvious features of the Wiener measure that follow from eqs. (2.4) and (2.20) are

$$\Pr\{\Omega\} = \frac{1}{\sqrt{2\pi t}} \int_{-\infty}^{\infty} e^{-x^2/2t}\, dx = 1$$

and for $t_1 < t$,

$$\begin{aligned} &\Pr\{C(t_1, t; I_1, \mathbb{R})\\ &= \frac{1}{2\pi\sqrt{(t - t_1)t_1}} \int_{I_1}\int_{-\infty}^{\infty} \exp\left\{-\frac{(x - x_1)^2}{2(t - t_1)}\right\} \exp\left\{-\frac{x_1^2}{2t_1}\right\} dx\, dx_1 \\ &= \frac{1}{\sqrt{2\pi t_1}} \int_{I_1} \exp\left\{-\frac{x_1^2}{2t_1}\right\} dx_1 = \Pr\{C(t_1; I_1)\}. \end{aligned}$$

The Wiener measure (2.20) of a cylinder is the probability of sampling points of a trajectory in the cylinder by the simulation

$$x(t_k) = x(t_{k-1}) + \Delta w(t_k), \quad k = 1, \ldots, n, \tag{2.21}$$

where t_k are ordered as above, and $\Delta w(t_k) \sim \mathcal{N}(0, t_k - t_{k-1})$ are independent normal variables. The vertices of the trajectories in Figures 2.1 and 2.2 were sampled according to (2.21) and interpolated linearly.

The axiomatic definition of the Brownian motion, consistent with the Einstein–Langevin theory is as follows

Definition 2.2.21 (The MBM). *A real-valued stochastic process $w(t,\omega)$ defined on $\mathbb{R}_+ \times \Omega$ is a* mathematical Brownian motion *if*
(1) $w(0,\omega) = 0$ w.p. 1
(2) $w(t,\omega)$ is almost surely a continuous function of t
(3) For every $t, s \geq 0$ the increment $\Delta w(s,\omega) = w(t+s,\omega) - w(t,\omega)$ is independent of $\mathcal{F}_t$, and is a zero mean Gaussian random variable with variance

$$\mathbb{E}\left|\Delta w(s)\right|^2 = s. \tag{2.22}$$

The mathematical Brownian motion, defined by axioms (1)–(3), and its velocity process are mathematical idealizations of physical processes. This idealization may lead to unexpected and counterintuitive consequences.

There are other equivalent definitions of the MBM (see Wikipedia). According to Definition 2.2.21, the cylinders (2.3) are identical to the cylinders

$$C(t_1,\ldots,t_n;\, I_1,\ldots,I_n) = \{\omega \in \Omega \,|\, w(t_k,\omega) \in I_k, \text{ for all } 1 \leq k \leq n\}. \tag{2.23}$$

To understand the conceptual difference between the definitions (2.3) and (2.23), we note that in (2.3) the cylinder is defined directly in terms of elementary events whereas in (2.23) the cylinder is defined in terms of a stochastic process. It is coincidental that such two different definitions produce the same cylinder. In Section 5.6 below cylinders are defined in terms of other stochastic processes, as in (2.23). It should be borne in mind, however, that the extension of the Wiener measure from cylinders to $\mathcal{F}$ is not straightforward [106].

The expectation (2.22) is meant in the sense of the definition (2.13). Properties (1)–(3) are axioms that define the Brownian motion as a mathematical entity. It has to be shown that a stochastic process satisfying these axioms actually exists. Before showing constructions of the MBM (see Section 2.3), we can derive some of its properties in a straightforward manner.

First, we note that by eq. (2.19) the PDF of the MBM is

$$\begin{aligned} F_w(x,t) &= \Pr\{\omega \in \Omega \,|\, w(t,\omega) \leq x\} = \Pr\{C(t, I^x)\} \\ &= \frac{1}{\sqrt{2\pi t}} \int_{-\infty}^{x} e^{-y^2/2t}\, dy \end{aligned} \tag{2.24}$$

and the pdf is

$$f_w(x,t) = \frac{\partial}{\partial x} F_w(x,t) = \frac{1}{\sqrt{2\pi t}} e^{-x^2/2t}. \tag{2.25}$$

It is well-known (and easily verified) that $f_w(x,t)$ is the solution of the initial value problem for the diffusion equation

$$\frac{\partial f_w(x,t)}{\partial t} = \frac{1}{2}\frac{\partial^2 f_w(x,t)}{\partial x^2}, \qquad \lim_{t\downarrow 0} f_w(x,t) = \delta(x). \tag{2.26}$$

Second, we note that (1) and (2) are not contradictory, despite the fact that not all continuous functions vanish at time $t = 0$. Property (1) asserts that all trajectories of the Brownian motion that do not start at the origin are assigned probability 0. More specifically,

Theorem 2.2.2. *The Wiener measure has property (1).*

Proof. The set $\{\omega \in \Omega \,|\, w(0,\omega) = 0\}$ is in $\mathcal{F}$, because it can be represented as a countable intersection of cylinders. Indeed, consider two sequences t_k and ε_n that decrease to zero and define the cylinders

$$C(t_k;[-\varepsilon_n,\varepsilon_n]) = \left\{\omega \in \Omega \,\middle|\, |w(t_k,\omega)| < \varepsilon_n\right\}.$$

Then

$$\{\omega \in \Omega \,|\, w(0,\omega) = 0\} = \bigcap_{n=1}^{\infty} \bigcup_{m=1}^{\infty} \bigcap_{k=m}^{\infty} C(t_k;[-\varepsilon_n,\varepsilon_n]). \tag{2.27}$$

It follows from probability theory that

$$\begin{aligned}
\Pr\{\omega \in \Omega \,|\, w(0,\omega) = 0\Big\} &= \Pr\left\{\bigcap_{n=1}^{\infty} \bigcup_{m=1}^{\infty} \bigcap_{k=m}^{\infty} C(t_k;[-\varepsilon_n,\varepsilon_n])\right\}\\
&= \lim_{n\to\infty} \Pr\left\{\bigcup_{m=1}^{\infty} \bigcap_{k=m}^{\infty} C(t_k;[-\varepsilon_n,\varepsilon_n])\right\}\\
&\geq \lim_{n\to\infty} \lim_{m\to\infty} \Pr\left\{\bigcap_{k=m}^{\infty} C(t_k;[-\varepsilon_n,\varepsilon_n])\right\}\\
&= \lim_{n\to\infty} \lim_{k\to\infty} \Pr\{C(t_k;[-\varepsilon_n,\varepsilon_n])\}.
\end{aligned} \tag{2.28}$$

Now, the Wiener measure (2.20) of a cylinder $C(t;[-\varepsilon,\varepsilon])$ is

$$\Pr\{C(t;[-\varepsilon,\varepsilon])\} = \int_{-\varepsilon}^{\varepsilon} \frac{1}{\sqrt{2\pi t}} e^{-x^2/2t}\,dx = \frac{2}{\sqrt{2\pi}} \int_0^{\varepsilon/\sqrt{t}} e^{-x^2/2}\,dx, \tag{2.29}$$

so that eqs.(2.28) and (2.29) give

$$\Pr\{\omega \in \Omega \,|\, w(0,\omega) = 0\} = \lim_{n\to\infty} \lim_{k\to\infty} \frac{2}{\sqrt{2\pi}} \int_0^{\varepsilon_n/\sqrt{t_k}} e^{-x^2/2}\,dx = 1. \tag{2.30}$$

This completes the proof of the assertion (see Figure 2.2). □

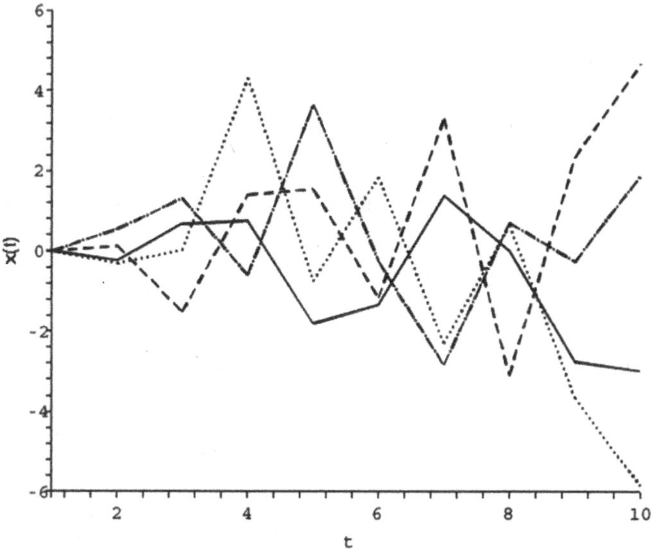

Figure 2.2. Three Brownian trajectories sampled at discrete times according to the Wiener measure $\Pr_0\{\cdot\}$ by the scheme (2.21).

The initial point

In view of the above, $x_0 = 0$ in the definition (2.20) of the Wiener measure of a cylinder means that the Brownian paths are those continuous functions that take the value 0 at time 0. That is, the Brownian paths are conditioned on starting at time $t = 0$ at the point $x_0 = w(0, \omega) = 0$. To emphasize this point, we modify the notation of the Wiener measure to $\Pr_0\{\cdot\}$. If, in eq. (2.20), this condition is replaced with $x_0 = x$, the above proof of Theorem 2.2.2 shows that $\Pr_x\{w(0, \omega) = x\} = 1$ under the modified Wiener measure, now denoted $\Pr_x\{\cdot\}$,

Thus conditioning reassigns probabilities to the Brownian paths; the set of trajectories $\{w(0, \omega) = x\}$, which was assigned the probability 0 under the measure $\Pr_0\{\cdot\}$, is now assigned the probability 1 under the measure $\Pr_x\{\cdot\}$.

Similarly, replacing the condition $t_0 = 0$ with $t_0 = s$ and setting $x_0 = x$ in eq. (2.20) shifts the Wiener measure, now denoted $\Pr_{x,s}$, so that

$$\Pr_{x,s}\{C(t; [a, b])\} = \Pr_0\{C(t - s; [a - x, b - x])\}. \tag{2.31}$$

This means that for all positive t the increment $\Delta w(s, \omega) = w(t + s, \omega) - w(t, \omega)$, as a function of s, is a MBM so that the probabilities of any Brownian event of $\Delta w(s, \omega)$ are independent of t. This property is stated in the form

(4) *The increments of the MBM are stationary.*

The above argument shows that equation (2.22) in property (3) of the MBM follows from the definition of the Wiener measure.

Theorem 2.2.3. *The Wiener measure has property* (3) *of Definition* 2.2.21 *(i.e., the increment* $\Delta w(s,\omega) = w(t+s,\omega) - w(t,\omega)$ *is independent of the Brownian events* $\mathcal{F}_t$*) and is a zero mean Gaussian variable with variance given by* (2.22).

Proof. We have to show that the joint PDF of $\Delta w(s,\omega) = w(t+s,\omega) - w(t,\omega)$ and any event $A \in \mathcal{F}_t$ is a product of the PDF of $\Delta w(s,\omega)$ and $\Pr\{A\}$. The same is true for any event $A \in \mathcal{F}_t$, because any cylinder $C \in \mathcal{F}_t$ is generated by cylinders of the form $C(s, I^x) = \{\omega \in \Omega \,|\, w(s,\omega) \leq x\}$ with $0 \leq s \leq t$. It suffices therefore to show the independence of the increment $\Delta w(s,\omega)$ and $w(u,\omega)$ for all $s \geq 0$ and $u \leq t$.

To show this, recall that by definition the joint PDF of $w(t_1,\omega), w(t_2,\omega)$, and $w(t_3,\omega)$ for $0 < t_1 < t_2 < t_3$ is the Wiener measure of the cylinder $C = C(t_1,t_2,t_3; I^x, I^y, I^z)$. According to the definition (2.20) of the Wiener measure,

$$
\begin{aligned}
F_{w(t_1),w(t_2),w(t_3)}(x,y,z) &= \Pr_0(C(t_1,t_2,t_3;I^x,I^y,I^z)) \\
&= \int_{-\infty}^{x} d\xi \int_{-\infty}^{y} d\eta \int_{-\infty}^{z} d\zeta \, \frac{1}{\sqrt{(2\pi)^3 t_1(t_2-t_1)(t_3-t_2)}} \\
&\quad \times \exp\left\{-\frac{\xi^2}{2t_1} - \frac{(\eta-\xi)^2}{2(t_2-t_1)} - \frac{(\zeta-\eta)^2}{2(t_3-t_2)}\right\}.
\end{aligned} \tag{2.32}
$$

It follows that the joint pdf of $w(t_1,\omega), w(t_2,\omega)$, and $w(t_3,\omega)$ is given by

$$
\begin{aligned}
f_{w(t_1),w(t_2),w(t_3)}(\xi,\eta,\zeta) &= \frac{\partial^3}{\partial\xi\partial\eta\partial\zeta} F_{w(t_1),w(t_2),w(t_3)}(\xi,\eta,\zeta) \\
&= \frac{1}{\sqrt{(2\pi)^3 t_1(t_2-t_1)(t_3-t_2)}} \\
&\quad \times \exp\left\{-\frac{\xi^2}{2t_1} - \frac{(\eta-\xi)^2}{2(t_2-t_1)} - \frac{(\zeta-\eta)^2}{2(t_3-t_2)}\right\}.
\end{aligned}
$$

Now, for any x and y

$$
\begin{aligned}
&\Pr\{\omega \in \Omega \,|\, w(t_3,\omega) - w(t_2,\omega) < x,\; w(t_1,\omega) < y\} \\
&= \int_{-\infty}^{y} d\xi \int_{-\infty}^{\infty} d\eta \int_{-\infty}^{x+\eta} d\zeta \, f_{w(t_1),w(t_2),w(t_3)}(\xi,\eta,\zeta) \\
&= \int_{-\infty}^{y} d\xi \int_{\zeta-x}^{\infty} d\eta \int_{-\infty}^{\infty} d\zeta \, \frac{1}{\sqrt{(2\pi)^3 t_1(t_2-t_1)(t_3-t_2)}} \\
&\quad \times \exp\left\{-\frac{\xi^2}{2t_1} - \frac{(\eta-\xi)^2}{2(t_2-t_1)} - \frac{(\zeta-\eta)^2}{2(t_3-t_2)}\right\}.
\end{aligned} \tag{2.33}
$$

Substituting $\eta = \zeta - z$ and noting that

$$\int_{-\infty}^{\infty} \frac{1}{\sqrt{2\pi(t_2 - t_1)}} \exp\left\{-\frac{(\zeta - z - \xi)^2}{2(t_2 - t_1)}\right\} d\zeta = 1, \tag{2.34}$$

we obtain from eq. (2.33) that

$$\begin{aligned} &\Pr\{\omega \in \Omega \,|\, w(t_3, \omega) - w(t_2, \omega) < x,\ w(t_1, \omega) < y\} \qquad (2.35)\\ &= \frac{1}{\sqrt{2\pi t_1}} \int_{-\infty}^{y} \exp\left\{-\frac{\xi^2}{2t_1}\right\} d\xi \frac{1}{\sqrt{2\pi(t_3 - t_2)}} \int_{-\infty}^{x} \exp\left\{-\frac{z^2}{2(t_3 - t_2)}\right\} dz\\ &= \Pr\{\omega \in \Omega \,|\, w(t_1, \omega) < y\}\Pr\{\omega \in \Omega \,|\, w(t_3, \omega) - w(t_2, \omega) < x\}. \end{aligned}$$

Equation (2.35) means that $w(t_3, \omega) - w(t_2, \omega)$ and $w(t_1, \omega)$ are independent, as stated in property (3) of the MBM.

Next, we calculate the moments of the MBM according to the definition (2.13). Using (2.14), we find from (2.25) that

$$\begin{aligned} \mathbb{E}w(t, \omega) &= \int_{-\infty}^{\infty} x \Pr\{\omega \in \Omega \,|\, x \le w(t, \omega) \le x + dx\}\\ &= \int_{-\infty}^{\infty} \frac{x}{\sqrt{2\pi t}} e^{-x^2/2t}\, dx = 0. \end{aligned}$$

Similarly, $Ew^2(t, \omega) = (2\pi t)^{-1/2} \int_{-\infty}^{\infty} x^2 e^{-x^2/2t}\, dx = t$. Now, property eq. (2.22) follows from the independence of the increments of the MBM. □

We recall that the autocorrelation function of a stochastic process $x(t, \omega)$ is defined as the expectation $R_x(t, s) = \mathbb{E}x(t, \omega)x(s, \omega)$. Using the notation $t \wedge s = \min\{t, s\}$, we have the following

Theorem 2.2.4 (Property (5)). *The autocorrelation function of $w(t, \omega)$ is*

$$Ew(t, \omega)w(s, \omega) = t \wedge s. \tag{2.36}$$

Proof. Assuming that $t \ge s \ge 0$ and using property (3), we find that

$$\begin{aligned} Ew(t, \omega)w(s, \omega) =& E\left[w(t, \omega) - w(s, \omega)\right]\left[w(s, \omega) - w(0, \omega)\right] + Ew(s, \omega)w(s, \omega)\\ =& s = t \wedge s. \end{aligned}$$

□

2.2.2 The MBM in $\mathbb{R}^d$

If $w_1(t), w_2, (t), \dots, w_d(t)$ are independent Brownian motions, the vector process

$$\boldsymbol{w}(t) = (w_1(t), w_2(t), \dots, w_d(t))^T$$

is defined as the *d-dimensional Brownian motion*. The probability space $\boldsymbol{\Omega}$ for the d-dimensional Brownian motion consists of all $\mathbb{R}^d$-valued continuous functions of t. Thus

$$\boldsymbol{\omega}(t) = (\omega_1(t), \omega_2(t), \dots, \omega_d(t))^T,$$

where $\omega_j(t) \in \Omega$. Cylinder sets are defined by the following

Definition 2.2.22 (Cylinder sets in $\mathbb{R}^d$). *A* cylinder set *of d-dimensional Brownian trajectories is defined by times $0 \le t_1 < t_2 < \cdots < t_k$ and open sets $\boldsymbol{I}_k$, $(k = 1, 2, \dots, k)$ as*

$$C(t_1, \dots, t_k; \boldsymbol{I}_1, \dots, \boldsymbol{I}_k) = \{\boldsymbol{\omega} \in \boldsymbol{\Omega} \,|\, \boldsymbol{\omega}(t_j) \in \boldsymbol{I}_j \text{ for } j = 1, \dots, k\}. \tag{2.37}$$

The open sets $\boldsymbol{I}_j$ can be, for example, open boxes or balls in $\mathbb{R}^d$. In particular, we write $\boldsymbol{I}_{\boldsymbol{x}} = \{\boldsymbol{\omega} \le \boldsymbol{x}\} = \{\omega_1 \le x_1, \dots, \omega_d \le x_d\}$. The appropriate σ-algebra $\boldsymbol{\mathcal{F}}$ and the filtration $\boldsymbol{\mathcal{F}}_t$ are constructed as in Section 2.2.

Definition 2.2.23 (The Wiener measure for the d-dimensional MBM). *The d-dimensional* Wiener measure *of a cylinder is defined as*

$$\Pr\{C(t_1, t_2, \dots, t_k; \boldsymbol{I}_1, \boldsymbol{I}_2, \dots, \boldsymbol{I}_k)\}$$

$$= \int\limits_{\boldsymbol{I}_1}\int\limits_{\boldsymbol{I}_2}\cdots\int\limits_{\boldsymbol{I}_k} \prod_{j=1}^{k} \frac{d\boldsymbol{x}_j}{[2\pi(t_j - t_{j-1})]^{n/2}} \exp\left\{-\frac{|\boldsymbol{x}_j - \boldsymbol{x}_{j-1}|^2}{2(t_j - t_{j-1})}\right\}. \tag{2.38}$$

The PDF of the d-dimensional MBM is

$$F_{\boldsymbol{w}}(\boldsymbol{x}, t) = \Pr\{\boldsymbol{\omega} \in \boldsymbol{\Omega} \,|\, \boldsymbol{w}(t, \boldsymbol{\omega}) \le \boldsymbol{x}\}$$

$$= \frac{1}{(2\pi t)^{n/2}} \int\limits_{-\infty}^{x_1} \cdots \int\limits_{-\infty}^{x_d} e^{-|\boldsymbol{y}|^2/2t}\, dy_1 \cdots dy_d \tag{2.39}$$

and the pdf is

$$f_{\boldsymbol{w}}(\boldsymbol{x}, t) = \frac{\partial^d F_{\boldsymbol{w}}(\boldsymbol{x}, t)}{\partial x_1 \partial x_2 \cdots \partial x_d} = \frac{1}{(2\pi t)^{n/2}} e^{-|\boldsymbol{x}|^2/2t}. \tag{2.40}$$

Equations (2.26) imply that $f_{\boldsymbol{w}}(\boldsymbol{x},t)$ satisfies the d-dimensional diffusion equation and the initial condition

$$\frac{\partial f_{\boldsymbol{w}}(\boldsymbol{x},t)}{\partial t} = \frac{1}{2}\Delta f_{\boldsymbol{w}}(\boldsymbol{x},t), \quad \lim_{t\downarrow 0} f_{\boldsymbol{w}}(\boldsymbol{x},t) = \delta(\boldsymbol{x}). \tag{2.41}$$

It can be seen from eq. (2.38) that any rotation of the d-dimensional Brownian motion is d-dimensional Brownian motion.

Higher-dimensional stochastic processes are defined by the following

Definition 2.2.24 (Vector valued processes). *A vector-valued function $\boldsymbol{x}(t,\omega)$: $\mathbb{R}_+ \times \boldsymbol{\Omega} \mapsto \mathbb{R}^d$ is called a* stochastic process *in $(\boldsymbol{\Omega}, \boldsymbol{\mathcal{F}})$ with continuous trajectories if*

(i) $\boldsymbol{x}(t,\omega)$ is a continuous function of t for every $\omega \in \boldsymbol{\Omega}$,

(ii) for every $t \geq 0$ and $\boldsymbol{x} \in \mathbb{R}^d$ the sets $\{\omega \in \boldsymbol{\Omega} \,|\, \boldsymbol{x}(t,\omega) \leq \boldsymbol{x}\}$ are Brownian events; that is, if $\{\omega \in \boldsymbol{\Omega} \,|\, \boldsymbol{x}(t,\omega) \leq \boldsymbol{x}\} \in \boldsymbol{\mathcal{F}}$.

Note that the dimension of the elementary events $\omega(\cdot)$: $\mathbb{R}_+ \mapsto \mathbb{R}^n$ and the dimension of the space in which the trajectories $\boldsymbol{x}(t,\omega)$ move, d, are not necessarily the same. As above, when $\{\omega \in \boldsymbol{\Omega} \,|\, \boldsymbol{x}(t,\omega) \leq \boldsymbol{x}\} \in \boldsymbol{\mathcal{F}}$, we say that the process $\boldsymbol{x}(t,\omega)$ is $\boldsymbol{\mathcal{F}}$-measurable. The PDF of $\boldsymbol{x}(t,\omega)$ is defined as

$$F_{\boldsymbol{x}}(\boldsymbol{y},t) = \Pr\{\omega \in \boldsymbol{\Omega} \,|\, \boldsymbol{x}(t,\omega) \leq \boldsymbol{y}\} \tag{2.42}$$

and the pdf is defined as

$$f_{\boldsymbol{x}}(\boldsymbol{y},t) = \frac{\partial^d F_{\boldsymbol{x}}(\boldsymbol{y},t)}{\partial y^1 \partial y^2 \cdots \partial y^d}. \tag{2.43}$$

The expectation of a matrix-valued function $\boldsymbol{g}(\boldsymbol{x})$ of a vector-valued process $\boldsymbol{x}(t,\omega)$ is the matrix

$$\mathbb{E}\boldsymbol{g}(\boldsymbol{x}(t,\omega)) = \int \boldsymbol{g}(\boldsymbol{y}) f_{\boldsymbol{x}}(\boldsymbol{y},t)\, d\boldsymbol{y}. \tag{2.44}$$

Definition 2.2.25 (Autocorrelation and autocovariance). *The autocorrelation matrix of $\boldsymbol{x}(t,\omega)$ is defined as the $n \times n$ matrix*

$$\boldsymbol{R}_{\boldsymbol{x}}(t,s) = \mathbb{E}\boldsymbol{x}(t)\boldsymbol{x}^T(s) \tag{2.45}$$

and the autocovariance matrix is defined as

$$\boldsymbol{Cov}_{\boldsymbol{x}}(t,s) = \mathbb{E}\left[\boldsymbol{x}(t) - \mathbb{E}\boldsymbol{x}(t)\right]\left[\boldsymbol{x} - \mathbb{E}\boldsymbol{x}(s)\right]^T. \tag{2.46}$$

The autocovariance matrix of the d-dimensional Brownian motion is found from (2.36) as

$$\boldsymbol{Cov}_{\boldsymbol{w}}(t,s) = \boldsymbol{I}(t \wedge s), \tag{2.47}$$

where $\boldsymbol{I}$ is the identity matrix.

Exercise 2.4 (Transformations preserving the MBM). Show, by verifying properties (1)–(3), that the following processes are Brownian motions:
(i) $w_1(t) = w(t+s) - w(s)$
(ii) $w_2(t) = cw(t/c^2)$, where c is any positive constant
(iii) $w_3(t) = tw(1/t)$,
(iv) $w_4(t) = w(T) - w(T-t)$ for $0 \le t \le T$,
(v) $w_5(t) = -w(t)$. □

Exercise 2.5 (Changing scale). Give necessary and sufficient conditions on the functions $f(t)$ and $g(t)$ such that the process $w_4(t) = f(t)w(g(t))$ is MBM. □

Exercise 2.6 (The joint pdf of the increments). Define

$$\Delta \boldsymbol{w} = (\Delta w(t_1), \Delta w(t_2), \ldots, \Delta w(t_n))^T.$$

Find the joint pdf of $\Delta \boldsymbol{w}$. □

Exercise 2.7 (The radial MBM). Define the *radial MBM* by $y(t) = |\boldsymbol{w}(t)|$, where $\boldsymbol{w}(t)$ is the d-dimensional MBM. Find the pdf of $y(t)$, the partial differential equation, and the initial condition it satisfies. □

2.3 Constructions of the MBM

Two mathematical problems arise with the axiomatic definition of the Brownian motion. One is the question of existence, or construction of such a process and the other is of computer simulations of Brownian trajectories with different refinements of time discretization. The first proof of existence and a mathematical construction of the Brownian motion is due to Paley and Wiener [197] and is presented in Section 2.3.1. The second construction, due to P. Lévy [150], and the method of refinement of computer simulations of the Brownian paths are presented in Section 2.3.2.

2.3.1 The Paley–Wiener construction of the Brownian motion

Assume that X_k, Y_k $(k = 0, \pm 1, \pm 2, \ldots)$ are zero mean and unit variance Gaussian i.i.d. random variables defined in a probability space $\tilde{\Omega}$. The probability space $\tilde{\Omega}$ can be chosen, for example, as the interval $[0,1]$ or $\mathbb{R}$ (see [106]). The variable $Z_k = (X_k + iY_k)/\sqrt{2}$ is called a *complex Gaussian variable.* The simplest properties of Z_k are $\mathbb{E} Z_k Z_l = 0$ and $\mathbb{E} Z_k \bar{Z}_l = \delta_{kl}$. The series

$$\tilde{Z}_1(t) = tZ_0 + \sum_{n \neq 0} \frac{Z_n \left(e^{int} - 1\right)}{in} \tag{2.48}$$

converges in $L^2([0, 2\pi] \times \Omega)$, because the coefficients are $O(n^{-1})$. Each trajectory of $\tilde{Z}_1(t)$ is obtained from an infinite sequence of the numbers Z_k that are drawn independently from a standard Gaussian distribution. That is, every trajectory of $\tilde{Z}_1(t)$

is obtained from a realization of the infinite sequence of the random variables Z_k. Every continuous function has a unique Fourier series representation in $L^2[0,2\pi]$, however not all Fourier series that converge in the $L^2[0,2\pi]$ sense are continuous functions. Thus the space Ω of continuous functions is a subset of all the possible realizations of $\tilde{Z}_1(t)$ obtained from realizations of infinite sequences of the independent Gaussian random variables Z_k. It follows that $\tilde{Z}_1(t)$ may be a stochastic process on $L^2[0,2\pi]$ rather than on Ω. Note that (2.48) embeds infinite-dimensional Gaussian vectors $\{Z_k\}_{k=1}^{\infty}$ in $L^2[0,2\pi]$.

We set $Z(t) = \tilde{Z}_1(t)/\sqrt{2\pi}$ and note that $Z(t)$ satisfies the properties (1), (3)–(5) of the Brownian motion. Indeed, $Z(t)$ is obviously a zero mean Gaussian process with independent increments on the probability space $\tilde{\Omega}$. The autocorrelation function is calculated from

$$\mathbb{E}\tilde{Z}_1(t)\overline{\tilde{Z}_1(s)} = ts + \sum_{n\neq 0} \frac{\left(e^{int}-1\right)\left(e^{-ins}-1\right)}{n^2}.$$

It is well-known from Fourier series theory that

$$\sum_{n\neq 0} \frac{\left(e^{int}-1\right)\left(e^{-ins}-1\right)}{n^2} = \begin{cases} s(2\pi - t) & \text{for } 0 \le s \le t \le 2\pi \\ t(2\pi - s) & \text{for } 0 \le t \le s \le 2\pi. \end{cases} \tag{2.49}$$

Exercise 2.8 (Proof of (2.49)). Prove eq. (2.49). □

It follows that

$$\mathbb{E}\tilde{Z}_1(t)\bar{\tilde{Z}}_1(s) = \begin{cases} st + s(2\pi - t) = 2\pi s & \text{for } 0 \le s \le t \le 2\pi \\ ts + t(2\pi - s) = 2\pi t & \text{for } 0 \le t \le s \le 2\pi \end{cases} = 2\pi(t \wedge s).$$

Separating the complex process into its real and imaginary parts, we get two independent Brownian motions.

Now that $Z(t)$ has been shown to satisfy all the requirements of the definition of a Brownian motion, but the continuity property, it remains to show that almost all its paths are continuous. To show that the paths of $Z(t)$ are continuous, the almost sure (in $\tilde{\Omega}$) uniform convergence of the series (2.48) has to be demonstrated. Once this is done, the space of realizations of the infinite sequence Z_k can be identified with the space Ω of continuous functions through the one-to-one correspondence (2.48). Thus, we write $Z(t,\omega)$ to denote any realization of the path. For any ω denote

$$Z_{m,n}(t,\omega) = \sum_{k=m+1}^{n} \frac{Z_k e^{ikt}}{ik} \quad \text{for } n > m. \tag{2.50}$$

Theorem 2.3.1 (Paley–Wiener). *The sum* $\sum_{n\neq 0} Z_{2^n,2^{n+1}}(t)$ *converges uniformly for* $t \in \mathbb{R}$ *to a Brownian motion, except possibly on a set of probability 0 in* $\tilde{\Omega}$.

Proof. According to eq. (2.50),

$$
\begin{aligned}
|Z_{m,n}(t,\omega)|^2 &= \sum_{k=m+1}^{n} \frac{|Z_k(\omega)|^2}{k^2} + 2\text{Re}\left\{\sum_{j=1}^{n-m-1} e^{ijt} \sum_{k=m+1+j}^{n} \frac{Z_k \bar{Z}_{k-j}}{k(k-j)}\right\} \\
&\leq \sum_{k=m+1}^{n} \frac{|Z_k(\omega)|^2}{k^2} + 2\sum_{j=1}^{n-m-1} \left| \sum_{k=m+1+j}^{n} \frac{Z_k \bar{Z}_{k-j}}{k(k-j)} \right|.
\end{aligned}
$$

Setting $T_{m,n}(\omega) = \max\limits_{0\leq t \leq 2\pi} |Z_{m,n}(t,\omega)|$, we get

$$
\mathbb{E}T^2_{m,n} \leq \sum_{k=m+1}^{n} \frac{1}{k^2} + 2\sum_{j=1}^{n-m-1} \mathbb{E}\left| \sum_{k=m+1+j}^{n} \frac{Z_k \bar{Z}_{k-j}}{k(k-j)} \right|.
$$

Using the Cauchy-Schwarz inequality $\mathbb{E}\left|\sum\right| \leq \left(\mathbb{E}\left|\sum\right|^2\right)^{1/2}$, we obtain the inequality

$$
\begin{aligned}
&\sum_{j=1}^{n-m-1} \left(\mathbb{E}\left| \sum_{k=m+1+j}^{n} \frac{Z_k \bar{Z}_{k-j}}{k(k-j)} \right|\right)^2 \\
\leq & \mathbb{E} \sum_{k=m+1+j}^{n} \frac{|Z_k|^2 |Z_{k-j}|^2}{k^2(k-j)^2} + 2\text{Re} \sum_{m+1+j<l<k\leq n} \mathbb{E}\frac{Z_k \bar{Z}_{k-j} Z_l \bar{Z}_{l-j}}{k(k-j)l(l-j)} \\
= & \sum_{k=m+1+j}^{n} \frac{1}{k^2(k-j)^2}.
\end{aligned}
$$

It follows that

$$
\begin{aligned}
\mathbb{E}T^2_{m,n} &\leq \sum_{k=m+1}^{n} \frac{1}{k^2} + 2\sum_{j=1}^{n-m-1} \left\{ \sum_{k=m+1+j}^{n} \frac{1}{k^2(k-j)^2} \right\}^{1/2} \\
&\leq \frac{n-m}{m^2} + 2(n-m)\left(\frac{n-m}{m^4}\right)^{1/2}.
\end{aligned}
$$

Now, we choose $n = 2m$ and apply the Cauchy–Schwarz inequality again to get

$$
\mathbb{E}\left(\max_t |Z_{m,2m}(t,\omega)|\right) \leq \left(\mathbb{E}T^2_{m,2m}\right)^{1/2} \leq \sqrt{\frac{1}{m} + \frac{2}{\sqrt{m}}} \leq 2m^{-1/4}.
$$

It follows that

$$
\sum_{n=1}^{\infty} \mathbb{E}\left(\max_t \left|Z_{2^n,2^{n+1}}(t,\omega)\right|\right) \leq 2\sum_{n=1}^{\infty} 2^{-n/4} < \infty.
$$

Lebesgue's monotone convergence theorem [183] asserts that if the nonnegative functions f_n are integrable such that $\int \sum_{n=1}^{m} f_n(x)\,dx \to$ limit as $m \to \infty$, then $\sum_{n=1}^{m} f_n(x) \to$ limit a.e. It follows that the sum $\sum_{n\neq 0} Z_{2^n,2^{n+1}}(t)$ converges uniformly in t for almost all ω. $\square$

2.3.2 P. Lévy's method and refinements

P. Lévy [150] proposed a construction of the Brownian motion that is particularly useful in computer simulations of Brownian trajectories. The Brownian paths are constructed by a process that mimics the sampling of the Brownian path on different time scales, beginning with a coarse scale through consecutive refinements.

Consider a sequence of standard Gaussian i.i.d. random variables $\{Y_k\}$, for $k = 0, 1, \ldots$ defined in $\tilde{\Omega}$. We denote by ω any realization of the infinite sequence $\{Y_k\}$ and construct a continuous path corresponding to this realization. We consider a sequence of binary partitions of the unit interval,

$$T_1 = \{0, 1\}, \quad T_2 = \left\{0, \frac{1}{2}, 1\right\}, \quad T_3 = \left\{0, \frac{1}{4}, \frac{1}{2}, \frac{3}{4}, 1\right\} \ldots,$$
$$T_{n+1} = \left\{\frac{k}{2^n},\ k = 0, 1, \ldots, 2^n\right\}.$$

The set $T_0 = \bigcup_{n=1}^{\infty} T_n$ contains all the binary numbers in the unit interval. The binary numbers are dense in the unit interval in the sense that for every $0 \leq x \leq 1$ there is a sequence of binary numbers $x_j = k_j 2^{n_j}$ with $0 \leq k_j \leq 2^{n_j}$ such that $x_j \to x$ as $j \to \infty$.

Define $X_1(\omega) = tY_1(\omega)$ for $0 \leq t \leq 1$. Keeping in mind that $T_2 = \{0, \frac{1}{2}, 1\}$ and $T_1 \setminus T_1 = \{\frac{1}{2}\}$, we refine by keeping the "old" points; that is, by setting $X_2(t,\omega) = X_1(t,\omega)$ for $t \in T_1$ and in the "new" point, $T_2 \setminus T_1 = \{\frac{1}{2}\}$, we set $X_2\left(\frac{1}{2},\omega\right) = \frac{1}{2}[X_1(0,\omega) + X_1(1,\omega)] + \frac{1}{2}Y_2(\omega)$. The process $X_2(t,\omega)$ is defined in the interval by linear interpolation between the points of T_2. We proceed by induction,

$$X_{n+1}(t,\omega) = \begin{cases} X_n(t,\omega) \quad \text{for } t \in T_n \text{ (old points)} \\ \dfrac{1}{2}\left\{X_n\left(t + \dfrac{1}{2^n}, \omega\right) + X_n\left(t - \dfrac{1}{2^n}, \omega\right)\right\} + \dfrac{1}{2^{(n+1)/2}} Y_k(\omega) \\ \text{for } t \in T_{n+1} \setminus T_n,\ k = 2^{n-1} + \frac{1}{2}\left(2^n t - 1\right) \text{ (new points)} \\ \text{connect linearly between consecutive points} \end{cases}$$

(see Figure 2.3). A Brownian trajectory sampled at 1024 points is shown in Figure 2.4.

Thus $X_{n+1}(t)$ is a refinement of $X_n(t)$. Old points stay put! So far, for every realization ω, we constructed an infinite sequence of continuous functions. We show below that for almost all (in the sense of $\tilde{\Omega}$) realizations ω the sequence $X_n(t)$ converges uniformly to a continuous function, thus establishing a correspondence

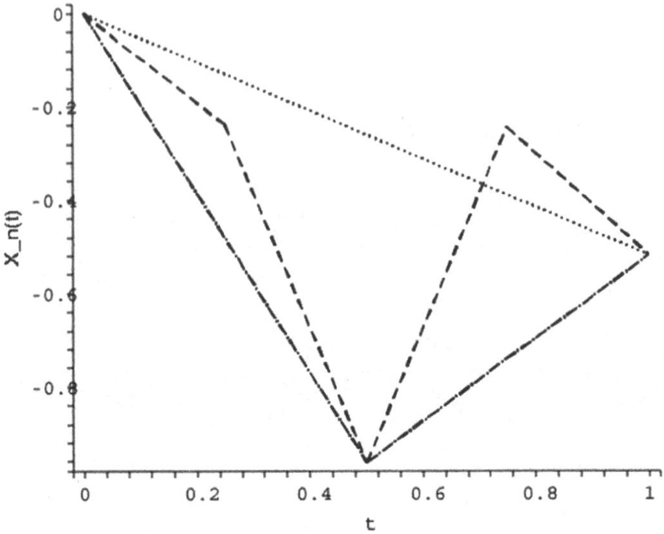

Figure 2.3. The graphs of $X_1(t)$ (dots), its first refinement $X_2(t)$ (dot-dash), and second refinement $X_3(t)$ (dash).

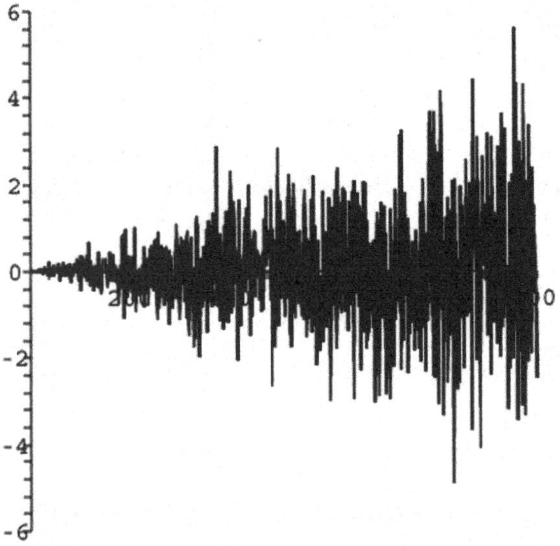

Figure 2.4. A Brownian trajectory sampled at 1024 points.

between ω and a continuous function. Obviously, the correspondence can be reversed in this construction.

Exercise 2.9 (MBM at binary points). Show that at binary points, $t_{k,n} = k2^{-n}, 0 \le k \le 2^n$, the process $X_n(t,\omega)$ has the properties of the Brownian motion $w(t)$. □

Exercise 2.10 (Refinements). If a Brownian trajectory is sampled at points $0 = t_0 < t_1 < \cdots < t_n = T$ according to the scheme (2.21) or otherwise, how should the sampling be refined by introducing an additional sampling point $\tilde{t}_i$ such that $t_i < \tilde{t}_i < t_{i+1}$? □

Exercise 2.11 (L^2 convergence*). Show that $X_n(t,\omega) \xrightarrow{L^2} X(t,\omega)$, where $X(t,\omega)$ has continuous paths [101]. □

Exercise 2.12 (A_n i.o.). Consider an infinite sequence of events (sets) $\{A_n\}$. The set A_n i.o. (infinitely often) is the set of all elementary events (points) that occur in infinitely many of the events A_n (that belong to infinitely many sets A_n). Thus $x \in A_n$ i.o. if and only if there is an infinite sequence of indices n_k, $(k = 1, 2, \dots)$, such that $x \in A_{n_k}$ for all $k = 1, 2, \dots$. Show that A_n i.o. $= \bigcap_{m=1}^\infty \bigcup_{n=m}^\infty A_n$. □

We need the following

Lemma 2.3.1 (Borel–Cantelli). *If* $\sum_{n=1}^\infty \Pr\{A_n\} < \infty$, *then* $\Pr\{A_n \text{ i.o.}\} = 0$.

Proof. From Exercise 2.12 it follows that A_n i.o. $\subset \bigcup_{n=m}^\infty A_n$ for all $m > 0$. The convergence of the series $\sum_{n=1}^\infty \Pr\{A_n\}$ implies the inequalities $\Pr\{A_n \text{ i.o.}\} \le \Pr\left\{\bigcup_{n=m}^\infty A_n\right\} \le \sum_{n=m}^\infty \Pr\{A_n\} \to 0$ as $m \to \infty$. □

The lemma means that if the sum of the probabilities converges, then, with probability one, only a finite number of the events occur.

Theorem 2.3.2 (P. Lévy). $X_n(t,\omega) \to X(t,\omega)$ almost surely in $\tilde{\Omega}$, where $X(t,\omega)$ is continuous for $0 \le t \le 1$.

Proof. Set $Z_n = X_{n+1} - X_n$; then $Z_n(t) = 0$ for $t \in T_n$ and

$$\max_{0\le t\le 1} |Z_n(t,\omega)| = \max_{2^{n-1}\le k\le 2^n} 2^{-n+1}|Y_k|.$$

It follows that for every $\lambda_n > 0$,

$$P_n = \Pr\left\{\max_{0\le t\le 1} |Z_n(t,\omega)| > \lambda_n\right\} \le 2^{n-1}\Pr\left\{|Y_k| \ge 2^{\frac{n+1}{2}}\lambda_n\right\}$$
$$\le 2^{n-1}\frac{2}{\sqrt{2\pi}} \int\limits_{2^{(n+1)/2}\lambda_n}^\infty e^{-x^2/2}\,dx \le \frac{2^{n-1}}{\sqrt{2^n\pi}}\frac{1}{\lambda_n}\exp\left\{-\frac{1}{2}\left(2^{(n+1)/2}\lambda_n\right)^2\right\}$$

because (show!) $(2\pi)^{-1/2}\int_a^\infty e^{-x^2/2}\,dx < (2\pi)^{-1/2}a^{-1}e^{-a^2/2}$. Choosing $\lambda_n = 2^{-(n+1)/2}\sqrt{2cn\log 2}$, $(c>1)$, we get

$$\sum_n P_n < \frac{1}{\sqrt{2\pi}}\sum_n \frac{2^{(n-1)/2}2^{(n+1)/2}}{\sqrt{2cn\log 2}}\exp\left\{-\left(2^{(n+1)/2}\frac{\sqrt{2cn\log 2}}{2^{(n+1)/2}}\right)^2\right\}$$

$$= \sum_n \frac{2^n 2^{-(cn+1/2)}}{2\sqrt{\pi cn\log 2}} = C_1\sum_n 2^{(1-c)n-1/2} < \infty,$$

where C_1 is a constant. It follows from the Borel–Cantelli lemma that with probability one, only a finite number of the events $\{\max_{0\leq t\leq 1}|Z_n(t,\omega)| > \lambda_n\}$ occur. Thus, for n sufficiently large,

$$\max_{0\leq t\leq 1}|Z_n(t,\omega)| \leq \lambda_n 2^{-(n+1)/2}\sqrt{2cn\log 2} \quad \text{w.p. 1.}$$

It follows from the Weierstrass M-test that for almost all trajectories ω the series $\sum_{n=1}^\infty Z_n(t,\omega)$ converges uniformly for all $t\in[0,2\pi]$. Cauchy's theorem asserts that the sum of a uniformly convergent series of continuous functions is continuous. The trajectories of the process $X(t,\omega)=\sum_{n=1}^\infty Z_n(t,\omega)$ are continuous with probability one, because, with probability one, each function $Z_n(t,\omega)$ is continuous.

□

Exercise 2.13 (Lévy's construction gives a MBM). Show that if $X_1(t)$ and $X_2(t)$ are independent Brownian motions on the interval $[0,1]$, then the process

$$X(t) = \begin{cases} X_1(t) & \text{for } 0\leq t\leq 1\\ X_1(1)+tX_2\left(\dfrac{1}{t}\right)-X_2(1) & \text{for } t>1\end{cases}$$

is a Brownian motion on $\mathbb{R}_+$. □

2.4 Analytical and statistical properties of Brownian paths

The Wiener measure assigns probability 0 to several important classes of Brownian paths. These classes include all differentiable paths, all paths that satisfy the Lipschitz condition at any point, all continuous paths with bounded variation on any interval, and so on. The Brownian paths have many interesting properties, whose investigation exceeds the scope of this book. For a more detailed description of the Brownian paths see, for example, [106], [101], [208]. Here we list only a few of the most prominent features of the Brownian paths. As shown below, although continuous, the Brownian paths are nondifferentiable at any point with probability 1. This means that the Wiener measure assigns probability 0 to all differentiable paths. This fact implies that the white noise process $\dot{w}(t)$ does not exist, so that strictly speaking, none of the calculations carried out under the assumption that $\dot{w}(t)$ exists are valid. This means that the velocity process of the MBM (white noise) should be interpreted as the overdamped limit [214] of the Brownian velocity process described in Chapter 1. In 1933 Paley, Wiener, and Zygmund [198] proved the following

Theorem 2.4.1 (Paley, Wiener, Zygmund). *The Brownian paths are nondifferentiable at any point with probability* 1.

Proof. **(Dvoretzky, Erdös, Kakutani [60])** We construct a set B of trajectories that contains all the trajectories such that the derivative $\dot{w}(t,\omega)$ exists for some $t \in [0,1]$ and show that $\Pr\{B\} = 0$. The set B is constructed in the form $B = \bigcup_{l=1}^{\infty}\bigcup_{m=1}^{\infty} B_{l,m}$, such that $\Pr\{B_{l,m}\} = 0$ for all l, m. Consider a trajectory ω such that $\dot{w}(t,\omega)$ exists for some $t \in [0,1]$. This means that $\lim_{s\to t}[w(t,\omega) - w(s,\omega)]/(t-s) = \dot{w}(t,\omega) \neq \pm\infty$, or, equivalently, for every $\varepsilon > 0$ there exists $\delta > 0$ such that $|[w(t,\omega) - w(s,\omega)]/(t-s) - \dot{w}(t,\omega)| < \varepsilon$ if $|t-s| < \delta$. It follows that $|[w(t,\omega) - w(s,\omega)]/(t-s)| \leq |\dot{w}(t,\omega)| + \varepsilon < l$, for some $l \geq 1$. Define $i = [nt] + 1$, then

$$\left|t - \frac{i}{n}\right| < \frac{1}{n} \to 0 \text{ as } n \to \infty. \tag{2.51}$$

Now choose $s_{i,j} = (i+j)/n$ for $j = 0, 1, 2, 3$, then

$$|w(s_{i,0},\omega) - w(s_{i,1},\omega)| \tag{2.52}$$

$$\leq |w(s_{i,0},\omega) - w(t,\omega)| + |w(t,\omega) - w(s_{i,1},\omega)| < \frac{3l}{n}.$$

Similarly, $|w(s_{i,j},\omega) - w(s_{i,j+1},\omega)| \leq 7l/n$ for $j = 1, 2$. Thus, if $\dot{w}(t,\omega)$ exists for some ω and t, then there exists $l \in \{1, 2, \ldots, n\}$ such that for all sufficiently large n there exists an i such that inequalities (2.51) and (2.52) hold for $j = 0, 1, 2$. It follows that the set of ω such that the derivative exists at some point, $\{\omega \in \Omega \,|\, \dot{w}(t,\omega) \text{ exists for some } t \in [0,1]\}$, is contained in the set

$$\left\{\omega \in \Omega \,\middle|\, \bigcup_{l=1}^{\infty}\bigcup_{m=1}^{\infty}\bigcap_{n=m}^{\infty}\bigcup_{i=1}^{n}\bigcap_{j=0}^{2}\left\{|w(s_{i,j},\omega) - w(s_{i,j+1},\omega)| \leq \frac{7l}{n}\right\}\right\}.$$

We denote $B_{l,m} = \bigcap_{n=m}^{\infty}\bigcup_{i=1}^{n}\bigcap_{j=0}^{2}\{|w(s_{i,j},\omega) - w(s_{i,j+1},\omega)| \leq 7l/n\}$ and show $\Pr\{B_{l,m}\} = 0$. Indeed, the increments $w(s_{i,j}) - w(s_{i,j+1})$ are independent random zero mean Gaussian variables with variance $1/n$. It follows that the probability that the three events $\{|w(s_{i,j},\omega) - w(s_{i,j+1},\omega)| \leq 7l/n\}$ for $j = 0, 1, 2$ occur simultaneously is the product of the probabilities of each event occurring separately; that is,

$$\Pr\left\{\bigcap_{j=0}^{2}\left\{|w(s_{i,j},\omega) - w(s_{i,j+1},\omega)| \leq \frac{7l}{n}\right\}\right\} = \left(\sqrt{\frac{n}{2\pi}}\int_{-7l/n}^{7l/n} e^{-nx^2/2}\,dx\right)^3.$$

The probability of a union of n such events does not exceed the sum of the probabilities; that is,

$$\Pr\left\{\bigcup_{i=1}^{n}\bigcap_{j=0}^{2}\left\{|w(s_{i,j},\omega)-w(s_{i,j+1},\omega)|\leq\frac{7l}{n}\right\}\right\}$$
$$\leq n\left(\sqrt{\frac{n}{2\pi}}\int_{-7l/n}^{7l/n}e^{-nx^2/2}\,dx\right)^3.$$

The probability of the intersection $\bigcap_{n\geq m}$ does not exceed the probability of any of the intersected sets, thus, changing the variable of integration to $y=\sqrt{n}x$, we obtain that for all $n\geq m$,

$$\Pr\{B_{l,m}\}\leq\frac{8n}{\sqrt{8\pi^3}}\left(\int_0^{7l/\sqrt{n}}e^{-y^2/2}\,dy\right)^3\leq\frac{8n}{\sqrt{8\pi^3}}\left(\frac{7l}{\sqrt{n}}\right)^3=const.n^{-1/2}\to 0$$

as $n\to\infty$. That is, $\Pr\{B_{l,m}\}=0$. □

The proofs of the following theorems are given, for example, in [106], [101], [208].

Theorem 2.4.2 (The Khinchine–Lévy law of the iterated logarithm).

$$\limsup_{t\to\infty}\frac{w(t)}{\sqrt{2t\log\log t}}=1,\quad \liminf_{t\to\infty}\frac{w(t)}{\sqrt{2t\log\log t}}=-1. \tag{2.53}$$

Theorem 2.4.3 (Modulus of continuity).

$$\limsup_{h\to 0}\frac{w(t+h)-w(t)}{\sqrt{2|h|\log\log|h^{-1}|}}=1,\quad \liminf_{h\to 0}\frac{w(t+h)-w(t)}{\sqrt{2|h|\log\log|h^{-1}|}}=-1. \tag{2.54}$$

In particular

$$\limsup_{h\to 0}\frac{|\Delta w|}{|h|^\alpha}=\begin{cases}\infty & \text{if } \alpha\geq\frac{1}{2}\\[2ex] 0 & \text{if } \alpha<\frac{1}{2}.\end{cases} \tag{2.55}$$

Theorem 2.4.4 (The level-crossing property). *For any level a the times t such that $w(t)=a$ form a perfect set (i.e., every point of this set is a limit of points in this set).*

Thus, when a Brownian path reaches a given level at time t is recrosses it infinitely many times in every interval $[t,t+\Delta t]$.

Exercise 2.14 (Properties of Brownian trajectories). Compare plots of simulated trajectories $w(t)$ with $\sqrt{2t\log\log t}$. □

2.4.1 The Markov property of the MBM

Definition 2.4.1 (Markov process). *A stochastic process $\zeta(t)$ on $[0,T]$ is called a* Markov process *if for any sequences $0 \le t_0 < \cdots < t_n \le T$ and $x_0, x_1, \ldots, x_n$, its transition probability distribution function has the property*

$$\Pr\Big\{\zeta(t_n) < x_n \,|\, \zeta(t_{n-1}) < x_{n-1}, \zeta(t_{n-2}) < x_{n-2}, \ldots, \zeta(t_0) < x_0\Big\}$$
$$= \Pr\Big\{\zeta(t_n) < x_n \,|\, \zeta(t_{n-1}) < x_{n-1}\Big\}. \tag{2.56}$$

The transition probability density function, defined by

$$p\left(x_n, t_n \,|\, x_{n-1}, t_{n-1}, \ldots, x_1, t_1\right)$$
$$= \frac{\partial}{\partial x_n} \Pr\Big\{\zeta(t_n) < x_n \,|\, \zeta(t_{n-1}) = x_{n-1}, \zeta(t_{n-2}) = x_{n-2}, \ldots, \zeta(t_0) = x_0\Big\},$$

then satisfies

$$p\left(x_n, t_n \,|\, x_{n-1}, t_{n-1}, \ldots, x_1, t_1\right) = p\left(x_n, t_n \,|\, x_{n-1}, t_{n-1}\right). \tag{2.57}$$

The Markov property eq. (2.56) means that the process "forgets" the past in the sense that if the process is observed at times $t_0, t_1, \ldots, t_{n-1}$ such that $0 \le t_0 < \cdots < t_{n-1} \le T$, its "future" evolution (at times $t > t_{n-1}$) depends only on the "latest" observation (at time t_{n-1}).

For any three times $t > \tau > s$ and any points x, y, z, we can write the identities

$$\begin{aligned} p\left(y, t, z, \tau \,|\, x, s\right) &= p\left(y, t \,|\, z, \tau, x, s\right) p\left(z, \tau \,|\, x, s\right) \\ &= p\left(y, t \,|\, z, \tau\right) p\left(z, \tau \,|\, x, s\right), \end{aligned} \tag{2.58}$$

the last equation being a consequence of the Markov property. Now, using the identities (2.58) and writing $p\left(y, t \,|\, x, s\right)$ as a marginal density of $p\left(y, t, z, \tau \,|\, x, s\right)$, we obtain

$$\begin{aligned} p\left(y, t \,|\, x, s\right) &= \int p\left(y, t, z, \tau \,|\, x, s\right) dz = \int p\left(y, t \,|\, z, \tau, x, s\right) p\left(z, \tau \,|\, x, s\right) dz \\ &= \int p\left(y, t \,|\, z, \tau\right) p\left(z, \tau \,|\, x, s\right) dz. \end{aligned} \tag{2.59}$$

Equation (2.59) is called the *Chapman–Kolmogorov equation* (CKE). More general properties of Markov processes are described in Chapter 9.

Theorem 2.4.5. *The MBM is a Markov process.*

Proof. To determine the Markov property of the Brownian motion, consider any sequences $0 = t_0 < t_1 < \cdots < t_n$ and $x_0 = 0, x_1, \ldots, x_n$. The joint pdf of the vector

$$\boldsymbol{w} = \left(w(t_1), w(t_2), \ldots, w(t_n)\right)^T \tag{2.60}$$

is given by (see eq. (2.20))

$$p\left(x_1,t_1;x_2,t_2;\ldots;x_n,t_n\right)=\Pr\{w(t_1)=x_1,w(t_2)=x_2,\ldots,w(t_n)=x_n\}$$
$$=\prod_{k=1}^{n}\left[\left\{2\pi(t_k-t_{k-1})\right\}^{-1/2}\exp\left\{-\frac{(x_k-x_{k-1})^2}{2(t_k-t_{k-1})}\right\}\right], \tag{2.61}$$

so that for $0=t_0<t_1<\cdots<t_n<t=t_{n+1}$ and $0=x_0,x_1,\ldots,x_n,x=x_{n+1}$,

$$\Pr\{w(t)=x\,|\,w(t_n)=x_n,\ldots,w(t_1)=x_1\}$$
$$=\frac{\Pr\{w(t_{n+1})=x_{n+1},w(t_n)=x_n,\ldots,w(t_1)=x_1\}}{\Pr\{w(t_n)=x_n,\ldots,w(t_1)=x_1\}}$$
$$=\frac{\displaystyle\prod_{k=1}^{n+1}\left[\left\{2\pi(t_k-t_{k-1})\right\}^{-1/2}\exp\left\{-\frac{(x_k-x_{k-1})^2}{2(t_k-t_{k-1})}\right\}\right]}{\displaystyle\prod_{k=1}^{n}\left[\left\{2\pi(t_k-t_{k-1})\right\}^{-1/2}\exp\left\{-\frac{(x_k-x_{k-1})^2}{2(t_k-t_{k-1})}\right\}\right]}$$
$$=\frac{1}{\sqrt{2\pi(t-t_n)}}\exp\left\{-\frac{(x_{n+1}-x_n)^2}{2(t_{n+1}-t_n)}\right\}=\Pr\{w(t)=x\,|\,w(t_n)=x_n\};$$

that is, the Brownian motion is a Markov process. □

It follows that it suffices to know the two-point transition pdf of the Brownian motion, $p\left(y,t\,|\,x,s\right)=\Pr\{w(t)=y\,|\,w(s)=x\}$ for $t>s$, to calculate the joint and conditional probability densities of the vector (2.60); that is, $p\left(x_1,t_1;x_2,t_2;\ldots;x_n,t_n\right)=\prod_{k=1}^{n}p\left(x_k,t_k\,|\,x_{k-1},t_{k-1}\right)$.

Theorem 2.4.6 (The strong Markov property of the MBM). *If τ is a Markov time for the Brownian motion, then the process $\tilde{w}(t)=w(t+\tau)-w(\tau)$ is a Brownian motion.*

Exercise 2.15. (Strong Markov property of MBM)
(i) Verify the Chapman–Kolmogorov equation for the MBM.
(ii) Prove Theorem 2.4.6 [106], [101]. □

Exercise 2.16 (The velocity process). Consider the velocity process $y(t)$ in Definition 1.4.1, $y(t)=w(t)-\int_0^t e^{-(t-s)}\,w(s)\,ds$, and define the displacement process $x(t)=\int_0^t y(s)\,ds$.
(i) Prove that $y(t)$ is a Markov process.
(ii) Prove that $x(t)$ is not a Markov process.
(iii) Prove that the two-dimensional process $\boldsymbol{z}(t)=(x(t),y(t))$ is Markovian. □

2.4.2 Reflecting and absorbing walls

The variants of the Brownian motion defined below appear in many applications of diffusion theory. The calculation of the transition probability density functions of the variants can be done directly from the definition (see, e.g., [199]) and also by solving boundary value problems for partial differential equations that the transition probability density functions satisfy. The partial differential equations and the boundary conditions that these transition probability density functions satisfy are derived in Section 3.4. With these equations the calculations become straightforward. In this chapter the calculations based on the definition of the variants of the Brownian motion are presented as exercises.

Definition 2.4.2 (Reflected MBM). *The process* $w^+(t) = |w(t)|$ *is called the* reflected Brownian motion.

The reflected Brownian motion is obtained from the Brownian motion by observing its trajectories in the negative x-axis in a mirror placed at the origin. The reflected Brownian motion is used to describe the motion of freely diffusing particles in the presence of an impermeable wall.

Exercise 2.17. (The transition pdf of the reflected MBM).

(i) Find the transition probability density function of the reflected Brownian motion.

(ii) Prove that the transition pdf of the reflected Brownian motion, $f_{|w|}(x,t)$, satisfies the diffusion equation and the initial condition (2.26) on the positive ray, and the boundary condition $f'_{|w|}(0,t) = 0$ for $t > 0$.

(iii) Prove that (i) and (ii) imply that if $x_0 > 0$, then $\int_0^\infty f_{|w|}(x,t)\,dx = 1$ for all $t \geq 0$. □

Exercise 2.18 (The reflected MBM is Markovian). Show that the reflected MBM is a Markov process. □

Exercise 2.19 (The reflection principle). Prove the following reflection principle. Let τ_a be the first passage time to a; then for every Brownian path $w(t,\omega_1),\ t \geq \tau_a$, there is another path, $w(t,\omega_2),\ t \geq \tau_a$, which is the mirror image of $w(t,\omega_1)$ about the line $L_a : w = a$ [199]. □

Exercise 2.20 (The joint PDF of the FPT and the maximum of the MBM motion). Setting $M(t) = \max_{0\leq s\leq t} w(s)$, we find, by definition,

$$\Pr\{M(t) \leq a\} = \Pr\{\tau_a \geq t\}. \tag{2.62}$$

Thus the PDFs of $M(t)$ and τ_a are related through eq. (2.62). Prove that if $x \leq a$, then

$$\Pr\{w(t) \leq x,\, M(t) \geq a\} = \frac{1}{\sqrt{2\pi t}} \int_{2a-x}^{\infty} e^{-y^2/2t}\,dy$$

$$\Pr\{w(t) \leq x,\, M(t) \leq a\} = \frac{1}{\sqrt{2\pi t}} \int_{-\infty}^{x} e^{-y^2/2t}\,dy - \frac{1}{\sqrt{2\pi t}} \int_{2a-x}^{\infty} e^{-y^2/2t}\,dy.$$

□

Exercise 2.21. (The PDF of the FPT).

(i) Prove that $\Pr\{\tau_a \le t\} = 2(2\pi t)^{-1/2}\int_a^\infty e^{-y^2/2t}\,dy$. Conclude that the first passage time to a given point is finite with probability 1 but its mean is infinite. This is the continuous time version of the Gambler's Ruin "Paradox": gambling with even odds against an infinitely rich adversary leads to sure ruin in a finite number of games, but on the average, the gambler can play forever (see [72, Ch. XIV.3]).

(ii) Use the above result to conclude that the one-dimensional MBM is *recurrent* in the sense that $\Pr\{w(t,\omega) = x \text{ for some } t > T\} = 1$ for every x and every T. This means that the MBM returns to every point infinitely many times for arbitrary large times.

(iii) Consider two independent Brownian motions, $w_1(t)$ and $w_2(t)$ that start at x_1 and x_2 on the positive axis and denote by τ_1 and τ_2 their first passage times to the origin, respectively. Define $\tau = \tau_1 \wedge \tau_2$, the first passage time of the first Brownian motion to reach the origin. Find the PDF and mean value of τ. □

Exercise 2.22 (Absorbed MBM). If the Brownian motion is stopped at the moment it reaches a for the first time, the process $y(t) = w(t)$ for $t \le \tau_a$ and $y(t) = a$ for $t \ge \tau_a$ is called the absorbed Brownian motion.

(i) Prove

$$\Pr\{y(t) \le y\} = \begin{cases} \dfrac{1}{\sqrt{2\pi t}}\displaystyle\int_{-\infty}^{y} e^{-z^2/2t}\,dz - \dfrac{1}{\sqrt{2\pi t}}\displaystyle\int_{2a-y}^{\infty} e^{-z^2/2t}\,dz & \text{for } y < a \\ 1 & \text{for } y \ge a. \end{cases}$$

Prove that the pdf of $y(t)$, denoted $f_{y(\cdot)}(x,t)$, satisfies for $x < a$ the diffusion equation and the initial condition (2.26) and the boundary condition $f_{y(\cdot)}(x,t) = 0$ for $x \ge a$.

(ii) Assume that the trajectories of the MBM begin at time t_0 at a point $x_0 < a$. Find the pdf of the absorbed MBM for this case.

(iii) Prove that the pdf of the absorbed MBM in (ii), denoted $f_{y(\cdot)}(x,t\,|\,x_0,t_0)$, is the solution of the initial and boundary value problem for the diffusion equation

$$\begin{aligned} \frac{\partial f_{y(\cdot)}}{\partial t} &= \frac{1}{2}\frac{\partial^2 f_{y(\cdot)}}{\partial x^2} \quad \text{for } x < a \\ \lim_{t\downarrow t_0} f_{y(\cdot)} &= \delta(x - x_0), \quad f_{y(\cdot)} = 0 \ \text{ for } x \ge a,\ t > t_0. \end{aligned} \tag{2.63}$$

(iv) Find the partial differential equation and the terminal and boundary conditions that the function $f_{y(\cdot)}(x,t\,|\,x_0,t_0)$ satisfies with respect to the initial point and time (x_0,t_0).

(v) Verify that eq. (2.62) holds.

(vi) It is known from the theory of parabolic partial differential equations [78] that the partial differential equations and boundary conditions in (iii) and (iv) have a unique solutions. Use this information to derive and solve partial differential equations and boundary conditions for the PDF of the FPT from a point x to a point y.

(vii) Prove that $\Pr\{\tau_a > t \,|\, x_0, t_0) = \int_{-\infty}^{a} f_{y(\cdot)}(x, t \,|\, x_0, t_0)\, dx$.
(viii) Prove in the presence of an absorbing boundary at a the population of Brownian trajectories in $(-\infty, a)$ decays in time; that is,

$$\lim_{t\to\infty} \int_{-\infty}^{a} f_{y(\cdot)}(x, t \,|\, x_0, t_0)\, dx = 0,$$

which is equivalent to the decay of $\Pr\{\tau_a > t \,|\, x_0, t_0\}$ as $t \to \infty$. Reconcile this with (i).

(ix) Prove that the function $u(t \,|\, x_0, t_0) = \Pr\{\tau_a \le t \,|\, x_0, t_0\}$ is the solution of the terminal and boundary value problem

$$\frac{\partial u}{\partial t_0} + \frac{1}{2}\frac{\partial^2 u}{\partial x_0^2} = -1 \text{ for } x_0 < a,\ t_0 < t$$
$$\lim_{t_0 \uparrow t} u = 0 \text{ for } x_0 < a, \quad u(t \,|\, a, t_0) = 1.$$

(x*) Consider the d-dimensional MBM that starts at a distance r from the origin and assume that an absorbing sphere of radius a is centered at the origin. The FPT to the sphere $|\boldsymbol{w}| = a$ is defined by $\tau_a = \inf\{t \,|\, |\boldsymbol{w}| = a\}$. □

Exercise 2.23 (Absorbed MBM in $\mathbb{R}^d$). The d-dimensional Brownian motion with absorption at the sphere is defined by

$$\boldsymbol{y}(t) = \begin{cases} \boldsymbol{w}(t) & \text{for } t \le \tau_a \\ \boldsymbol{w}(\tau_a) & \text{for } t \ge \tau_a. \end{cases}$$

Denote $y(t) = |\boldsymbol{y}(t)|$.

(i) Formulate and prove a reflection principle for the process $y(t)$ (use reflection of the d-dimensional MBM in a sphere of radius a, as defined in eq. (2.64) below).

(ii) Formulate and solve initial and boundary value problems for the pdf and the FPT of $y(t)$, analogous to (i)–(ix) above. □

Exercise 2.24 (The absorbed MBM is Markovian). Show that the absorbed MBM is a Markov process. □

Exercise 2.25 (Reflecting wall). If upon hitting the line $L_a : w = a$ for the first time, the Brownian path is reflected in L_a, the resulting process is defined by

$$x(t) = \begin{cases} w(t) & \text{for } w(t) \le a \\ 2a - w(t) & \text{for } w(t) > a. \end{cases}$$

Thus $x(t) \le a$ for all t.

(i) Prove that for $x \le a$,

$$\Pr\{x(t) \le x\} = 1 - \frac{1}{\sqrt{2\pi t}} \int_{x}^{2a-x} e^{-y^2/2t}\, dy$$

(see [199]).

(ii) For the d-dimensional MBM $\boldsymbol{w}(t)$, define a reflected d-dimensional MBM in a sphere of radius a centered at the origin by

$$\boldsymbol{w}^*(t) = \begin{cases} \boldsymbol{w}(t) & \text{for } |\boldsymbol{w}(t)| \leq a \\ a^2 \dfrac{\boldsymbol{w}(t)}{|\boldsymbol{w}(t)|^2} & \text{for } |\boldsymbol{w}(t)| \geq a. \end{cases} \tag{2.64}$$

Obviously, $|\boldsymbol{w}^*(t)| \leq a$. Find the pdf of $\boldsymbol{w}^*(t)$, the partial differential equation, the initial conditions, and boundary conditions it satisfies on the sphere. □

Exercise 2.26 (The Brownian bridge). Start the Brownian motion at the point x; that is, set $x(t) = w(t) + x$, and consider only the trajectories that pass through the point y at time t_0. That is, condition $x(t)$ on $w(t_0) + x = y$. Thus the paths of the Brownian bridge $x(t)$ are those paths of the Brownian motion that satisfy the given condition.

(i) Show that $x(t) = w(t) - (t/t_0)\,[w(t_0) - y + x] + x$ for $0 \leq t \leq t_0$.

(ii) Show that $x(t)$ is a Gaussian process.

(iii) Calculate the mean and the autocorrelation function of the Brownian bridge.

(iv) Show that $x(t)$ and $x(t_0 - t)$ have the same PDF for $0 \leq t \leq t_0$.

(v) Show that $x(t)$ is a Markov process [117]. □

Exercise 2.27 (Maximum of the Brownian bridge). Find the distribution of the maximum of the Brownian bridge in the interval $[0, t]$ for $0 < t < t_0$ (see also Exercise 6.11 below). □

2.4.3 MBM and martingales

Definition 2.4.3 (Martingales). *A* martingale *is a stochastic process $x(t)$ such that $\mathbb{E}|x(t)| < \infty$ for all t and for every $t_1 < t_2 < \cdots < t_n < t$ and $x_1, x_2, \ldots, x_n$,*

$$\mathbb{E}\left[x(t) \,|\, x(t_1) = x_1, x(t_2) = x_2, \ldots, x(t_n) = x_n\right] = x_n \tag{2.65}$$

(see [208], [115], [178]). In gambling theory $x(t)$ often represents the capital of a player at time t. The martingale property (2.65) means that the game is *fair*; that is, not biased.

Theorem 2.4.7. *The MBM is a martingale.*

Proof. Indeed,

$$\mathbb{E}|w(t)| = \frac{1}{\sqrt{2\pi t}} \int_{-\infty}^{\infty} |x| e^{-x^2/2t}\, dx = \sqrt{\frac{2t}{\pi}} < \infty$$

and due to the Markov property of the MBM (see Section 2.4.1)

$$\begin{aligned}&\mathbb{E}\left[w(t)\,|\,w(t_1)=x_1,w(t_2)=x_2,\ldots,w(t_n)=x_n\right]\\&=\int_{-\infty}^{\infty}xp\left(x,t\,|\,x_1,t_1,x_2,t_2,\ldots,x_n,t_n\right)dx\\&=\int_{-\infty}^{\infty}xp\left(x,t\,|\,x_n,t_n\right)dx=\frac{1}{\sqrt{2\pi\left(t-t_n\right)}}\int_{-\infty}^{\infty}xe^{-(x-x_n)^2/2(t-t_n)}\,dx=x_n.\end{aligned}$$

□

Theorem 2.4.8. *For every α, the process $x(t)=\exp\{\alpha w(t)-\alpha^2t/2\}$ is a martingale.*

Proof. The property $\mathbb{E}|x(t)|<\infty$ is obtained from

$$\begin{aligned}\mathbb{E}|x(t)|=\mathbb{E}x(t)&=\frac{1}{\sqrt{2\pi t}}\int_{-\infty}^{\infty}\exp\left\{\alpha x-\frac{\alpha^2}{2}t\right\}\exp\left\{-\frac{x^2}{2t}\right\}dx\\&=\frac{1}{\sqrt{2\pi t}}\int_{-\infty}^{\infty}\exp\left\{-\frac{(x-\alpha t)^2}{2t}\right\}dx=1.\end{aligned}\tag{2.66}$$

To verify the martingale property (2.65), we use the identity

$$x(t)=\exp\left\{\alpha\left[w(t)-w(s)\right]-\frac{\alpha^2}{2}(t-s)\right\}x(s).\tag{2.67}$$

Setting $\tilde{w}(t-s)=w(t)-w(s)$, we rewrite (2.67) as $x(t)=\tilde{x}(t-s)x(s)$ and recall that the increment $\tilde{w}(t-s)$ is a MBM independent of $x(s)$. It follows that the process

$$\tilde{x}(t-s)=\exp\left\{\alpha\tilde{w}(t-s)-\frac{\alpha^2}{2}(t-s)\right\}$$

has the same probability law as $x(t-s)$, but is independent of $x(\tau)$ for all $\tau\leq s$. Hence, using (2.66), we obtain

$$\begin{aligned}&\mathbb{E}\left[x(t)\,|\,x(t_1)=x_1,x(t_2)=x_2,\ldots,x(t_n)=x_n\right]\\&=\mathbb{E}\left[\tilde{x}(t-t_n)x(t_n)\,|\,x(t_1)=x_1,x(t_2)=x_2,\ldots,x(t_n)=x_n\right]\\&=x_n\mathbb{E}\tilde{x}(t-t_n)=x_n.\end{aligned}\tag{2.68}$$

□

Definition 2.4.4 (Submartingales). *If instead of (2.65) $x(t)$ satisfies the inequality*

$$\mathbb{E}\left[x(t)\,|\,x(t_1)=x_1,x(t_2)=x_2,\ldots,x(t_n)=x_n\right]\geq x_n\tag{2.69}$$

(in addition to $\mathbb{E}|x(t)|<\infty$), then $x(t)$ is said to be a submartingale.

For example, if $x(t)$ is a martingale, then $y(t) = |x(t)|$ is a *submartingale*, because $y(t) \geq \pm x(t)$ and

$$\begin{aligned} &\mathbb{E}\left[y(t) \mid y(t_1) = y_1, y(t_2) = y_2, \ldots, y(t_n) = y_n\right] \\ \geq &\mathbb{E}\left[\pm x(t) \mid x(t_1) = \pm y_1, x(t_2) = \pm y_2, \ldots, x(t_n) = \pm y_n\right] = \pm y_1 \end{aligned} \tag{2.70}$$

for all possible combinations of + and –, which is the submartingale condition (2.69). Thus the reflected Brownian motion $|w(t)|$ is a submartingale.

Exercise 2.28 (Martingales). Show that the following are martingales.

(i) $w^2(t) - t$

(ii) $\exp\left\{-\alpha^2 t \cosh\left[\sqrt{2}\alpha\, w(t)\right]\right\}$. □

Exercise 2.29 (Martingales: continued). Assume that α in Theorem 2.4.8 is a random variable in Ω such that $\alpha(\omega)$ is $\mathcal{F}_t$-measurable for $a \leq t \leq b$ (it is independent of the Brownian increments $w(s+\Delta s, \omega) - w(s, \omega)$ for all $s \geq b$ and $\Delta s > 0$).

(i) Show that if $\alpha(\omega)$ is bounded, then $x(t, \omega)$ in Theorem 2.4.8 is a martingale in the interval $[a, b]$.

(ii) Can the boundedness condition be relaxed? How?

(iii) Consider the following generalization: Let $C_1(\omega)$ and $C_2(\omega)$ be random variables in Ω and let $0 < t_1 < T$. Assume that $C_1(\omega)$ is independent of $w(t, \omega)$ for all t and $C_2(\omega)$ is $\mathcal{F}_{t_1}$-measurable (independent of $w(t, \omega) - w(t_1, \omega)$ for all $t > t_1$). Consider the following process in the interval $[0, T]$,

$$x(t, \omega) = \begin{cases} \exp\left\{C_1(\omega) w(t, \omega)) - \dfrac{C_1^2(\omega)}{2} t\right\} & \text{if } 0 \leq t \leq t_1 \\ \exp\{C_1(\omega) w(t_1, \omega)) + C_2(\omega)\left[w(t, \omega) - w(t_1, \omega)\right]\} & \\ \times \exp\left\{-\dfrac{1}{2}\left[C_1^2(\omega) t_1 + C_2^2(\omega)(t - t_1)\right]\right\} & \text{if } t_1 < t \leq T. \end{cases}$$

Show that if $C_1(\omega)$ and $C_2(\omega)$ are bounded, then $x(t, \omega)$ is a martingale in $[0, T]$.

(iv) Find a more general condition than boundedness that ensures the same result [115]. □

Chapter 3

Itô Integration and Calculus

3.1 Integration of white noise

The Brownian motion is nowhere differentiable with probability 1 (see the Paley–Wiener–Zygmund theorem in Chapter 2); it therefore makes not much sense to calculate integrals of the type that appear in the solution of Langevin's equation (1.27). For example, in Langevin's equation for a free Brownian particle,

$$\dot{v} + \gamma v = \sqrt{\frac{2\gamma kT}{m}}\,\dot{w}, \tag{3.1}$$

the symbol $\dot{w}$ is not well-defined. The equation makes more sense if written in the integral form

$$v(t) = v(0) - \gamma \int_0^t v(s)\,ds + \sqrt{\frac{2\gamma kT}{m}}\,w(t), \tag{3.2}$$

because $w(t)$ is a well-defined continuous function for almost all trajectories of the Brownian motion. For the same reason the solution of the Langevin equation (3.1), written as

$$v(t) = v(0)e^{-\gamma t} + \sqrt{\frac{2\gamma kT}{m}} \int_0^t e^{-\gamma(t-s)}\,dw(s),$$

makes no mathematical sense. However, the form obtained by formally integrating by parts,

$$v(t) = v(0)e^{-\gamma t} + \sqrt{\frac{2\gamma kT}{m}} \left[w(t) - \gamma \int_0^t w(s)e^{-\gamma(t-s)}\,ds \right],$$

makes sense and actually solves the integral equation (3.2). It can be defined as the solution of the Langevin equation (3.1) in some weak sense. This procedure,

Z. Schuss, *Theory and Applications of Stochastic Processes: An Analytical Approach*, Applied Mathematical Sciences 170, DOI 10.1007/978-1-4419-1605-1_3,

however, fails when we try to make sense of integrals of the form

$$I = \int_a^b w(s)\,dw(s), \tag{3.3}$$

which appear in stochastic equations.

Exercise 3.1 (The integral (3.3)). Convert the differential equation $dx = 2\sqrt{x}\,dw$ to an integral equation and integrate by parts. □

If all functions involved in the integration (3.3) obeyed the rules of the differential and integral calculus, we would expect that $I = \frac{1}{2}[w^2(b) - w^2(a)]$ and that this result would be independent of the sequence of partitions of the interval $[a, b]$ and the choice of points in the partition intervals for the evaluation of the integral sums that define I. To gain some insight into the difficulties integrals of the type (3.3) pose, we compare three ways to calculate the integral by evaluating the limits of the integral sums that define I with different choices for the points inside the partition intervals.

We consider three methods for the calculation of the integral (3.3). First, for every partition $a = t_0 < t_1 < \cdots < t_n = b$, and choose the left endpoint of the partition interval for the evaluation of the integrand $w(t)$ in the integral sum

$$\begin{aligned}
\sigma_n^{(1)} &= \sum_{i=1}^n w(t_{i-1})\,[w(t_i) - w(t_{i-1})]\\
&= \frac{1}{2}\sum_{i=1}^n \left\{w^2(t_i) - w^2(t_{i-1}) - [w(t_i) - w(t_{i-1})]^2\right\}\\
&= \frac{1}{2}\left[w^2(t_n) - w^2(t_0)\right] - \frac{1}{2}\sum_{i=1}^n [w(t_i) - w(t_{i-1})]^2\\
&= \frac{1}{2}[w^2(b) - w^2(a)] - \eta_n.
\end{aligned} \tag{3.4}$$

The expectation is

$$\mathbb{E}\eta_n = \frac{1}{2}\sum_{i=1}^n \mathbb{E}\,(\Delta_i w)^2 = \frac{1}{2}\sum_{i=1}^n \Delta t_i = \frac{1}{2}\,(b-a)\,.$$

Next, we show that

$$\lim_{n\to\infty} \eta_n \overset{\Pr}{=} \mathbb{E}\eta_n = \frac{1}{2}\,(b-a) \tag{3.5}$$

in the sense that for every $\varepsilon > 0$

$$\lim_{n\to\infty} \Pr\{|\eta_n - \mathbb{E}\eta_n| > \varepsilon\} = 0. \tag{3.6}$$

This type of convergence is called *convergence in probability*. We recall the following

Theorem 3.1.1 (Chebyshev's inequality). *If X is a random variable with finite mean and variance, $\mathbb{E}X = m_1$, $\text{Var}X = \mathbb{E}\left(X - \mathbb{E}X\right)^2 = \sigma^2$, then*

$$\Pr\{|X - m_1| > \varepsilon\} \leq \frac{\sigma^2}{\varepsilon^2}$$

for every $\varepsilon > 0$.

Exercise 3.2 (Chebyshev's inequality). Prove Chebyshev's inequality. □

To prove (3.6), we calculate $\text{Var}\,\eta_n$ and use it in Chebyshev's inequality. For simplicity, we choose a partition of the interval $[a, b]$ into equal segments $\Delta t_i = (b - a)/n$. We have

$$\begin{aligned}\text{Var}\,\eta_n &= \frac{1}{4}\sum_{i=1}^{n} \text{Var}\,\left(\Delta_i w\right)^2 = \frac{1}{4}\sum_{i=1}^{n}\left[\mathbb{E}\left(\Delta_i w\right)^4 - \left(\mathbb{E}\left(\Delta_i w\right)^2\right)^2\right] \\ &\leq \frac{1}{4}\sum_{i=1}^{n}\mathbb{E}\left(\Delta_i w\right)^4 = \frac{n}{4\sqrt{2\pi\Delta t}}\int_{-\infty}^{\infty} x^4 e^{-x^2/2\Delta t}\,dx = \frac{3n}{4}\left(\Delta t\right)^2 \\ &= \frac{3(b-a)^2}{4n} \to 0 \text{ as } n \to \infty,\end{aligned}$$

because $\left(\Delta_i w\right)^2$ are independent random variables. Chebyshev's inequality gives $\Pr\{\left|\eta_n - \mathbb{E}\eta_n\right| > \varepsilon\} \leq \text{Var}\,\eta_n/\varepsilon^2 \to 0$ as $n \to \infty$. It follows that (3.5) holds. Now, it follows from (3.4) that the integral sum $\sigma_n^{(1)}$ in eq. (3.4) gives I the value

$$I = \int_a^b w(s)\,dw(s) = \frac{1}{2}[w^2(b) - w^2(a)] - \frac{1}{2}(b - a). \tag{3.7}$$

In particular, for $a = 0, b = t$, we have $I = \frac{1}{2}\left[w^2(t) - t\right]$.

Second, with the same notation as above, we choose the *middle point* of the partition interval, denoted $t_{i/2} = \frac{1}{2}(t_i + t_{i-1})$, for the evaluation of the integrand. The corresponding integral sum is now

$$\begin{aligned}\sigma_n^{(2)} &= \sum_{i=1}^{n} w\left(t_{i/2}\right)\left[w(t_i) - w(t_{i-1})\right] \qquad (3.8)\\ &= \sum_{i=1}^{n} w\left(t_{i/2}\right)\left[w(t_i) - w\left(t_{i/2}\right)\right] + \sum_{i=1}^{n} w\left(t_{i-1}\right)\left[w\left(t_{i/2}\right) - w(t_{i-1})\right] \\ &\quad + \sum_{i=1}^{n}\left[w\left(t_{i/2}\right) - w(t_{i-1})\right]^2.\end{aligned}$$

The first two sums combine to a sum that corresponds to the calculation by Method 1 with a partition refined by the midpoints, so they converge to the same limit as

$\sigma_n^{(1)}$ in (3.4) above. The last sum converges to one half of the sum in (3.5), so that

$$\lim_{n\to\infty} \sigma_n^{(2)} \overset{\Pr}{=} \frac{1}{2}[w^2(b) - w^2(a)]. \tag{3.9}$$

Third, we use the same partition as above, however, we choose the right end-point of the partition interval for the evaluation of the integrand. The corresponding integral sum is now

$$\sigma_n^{(3)} = \sum_{i=1}^{n} w(t_i)\,[w(t_i) - w(t_{i-1})] = \sigma_n^{(1)} + \sum_{i=1}^{n} [w(t_i) - w(t_{i-1})]^2. \tag{3.10}$$

It was shown above that

$$\lim_{n\to\infty} \sigma_n^{(1)} \overset{\Pr}{=} \frac{1}{2}[w^2(b) - w^2(a)] - \frac{1}{2}(b-a), \quad \lim_{n\to\infty} \sum_{i=1}^{n} [w(t_i) - w(t_{i-1})]^2 \overset{\Pr}{=} b-a,$$

so that

$$\lim_{n\to\infty} \sigma_n^{(3)} \overset{\Pr}{=} \frac{1}{2}[w^2(b) - w^2(a)] + \frac{1}{2}(b-a), \tag{3.11}$$

which is not the same as (3.7).

Obviously, different choices of the points in the partition intervals for the evaluation of the integrand lead to different results. It is apparent from these calculations that the traditional differential and integral calculus is unsuitable to handle stochastic integrals in a straightforward manner.

3.2 The Itô, Stratonovich, and other integrals

According to Definition 2.2.9, a stochastic process $f(t,\omega)$ that is adapted to the Brownian filtration $\mathcal{F}_t$, is independent of the increments of the Brownian motion $w(t,\omega)$ "in the future;" that is, $f(t,\omega)$ is independent of $w(t+s,\omega) - w(t,\omega)$ for all $s > 0$. For example, if $f(x)$ is an integrable deterministic function, then the functions $f(w(t,\omega))$ and $\int_0^t f(w(s,\omega))\,ds$ are $\mathcal{F}_t$-adapted.

Definition 3.2.1 (The class $H_2[0,T]$). *We denote by $H_2[0,T]$ the class of $\mathcal{F}_t$-adapted stochastic processes $f(t,\omega)$ on an interval $[0,T]$ such that*

$$\int_0^T \mathbb{E} f^2(s,\omega)\,ds < \infty.$$

3.2.1 The Itô integral

Integration with respect to white noise is defined in this class of stochastic processes. Itô's construction of the integral of a function $f(t)$ of the class $H_2[0,T]$ is similar to

Method 1 above. For any partition $0 \leq t_0 < t_1 < \cdots < t_n = t \leq T$, form the sum

$$\sigma_n(t,\omega) = \sum_{i=1}^{n} f(t_{i-1},\omega)\left[w(t_i,\omega) - w(t_{i-1},\omega)\right]. \tag{3.12}$$

Note that the increment $\Delta_i w = w(t_i,\omega) - w(t_{i-1},\omega)$ is independent of $f(t_{i-1},\omega)$, because $f(t,\omega)$ is $\mathcal{F}_t$-adapted. It can be shown (see Section 3.3) that for any sequence of partitions of the interval, such that $\max_i(t_i - t_{i-1}) \to 0$, the sequence $\{\sigma_n(t,\omega)\}$ converges (in some sense, as explained in Section 3.3) to the same limit, denoted

$$(I)\int_0^t f(s,\omega)\,dw(s,\omega) \stackrel{\text{Pr}}{=} \lim_{\max_i(t_i - t_{i-1})\to 0} \sigma_n(t,\omega), \tag{3.13}$$

and called *the Itô integral of* $f(t,\omega)$. It can also be shown (see Section 3.3) that the convergence in (3.13) is uniform in t with probability one; that is, on almost every trajectory of the Brownian motion $w(t,\omega)$. The Itô integral is a $\mathcal{F}_t$-adapted stochastic process in Ω. It takes different values on different realizations ω of the Brownian trajectories.

Next, we investigate some properties of the Itô integral. We use the abbreviated notation $(I)\int_0^t f(s)\,dw(s) = \int_0^t f(s)\,dw(s)$. The following properties are analogous to those of the classical Riemann integral [171], [101], [115], [204], [57].

1. The Itô integral is a linear functional in the sense that if $f(t), g(t) \in H_2[0,T]$ and α, β are real numbers, then $\alpha f(t) + \beta g(t) \in H_2[0,T]$ and

$$\int_0^t \left[\alpha f(s) + \beta g(s)\right]\,dw(s) = \alpha \int_0^t f(s)\,dw(s) + \beta \int_0^t g(s)\,dw(s).$$

2. The Itô integral is an additive function of the interval in the sense that if $f(t) \in H_2[0,T]$ and $0 < T_1 < T$, then $f(t) \in H_2[0,T_1]$ and $f(t) \in H_2[T_1,T]$ and

$$\int_0^T f(s)\,dw(s) = \int_0^{T_1} f(s)\,dw(s) + \int_{T_1}^T f(s)\,dw(s).$$

3. If $f(t)$ is an integrable deterministic function, then

$$\int_0^t f(s)\,dw(s) \sim \mathcal{N}\left(0, \int_0^t f^2(s)\,ds\right).$$

4. For $f(t) \in H_2[0,T]$, and any $0 \leq \tau \leq t \leq T$,

$$\mathbb{E}\int_0^t f(s)\,dw(s) = 0$$

$$\mathbb{E}\left[\int_0^t f(s)\,dw(s) \mid \int_0^\tau f(s)\,dw(s) = x\right] = x \tag{3.14}$$

$$\mathbb{E}\left[\int_0^T f(s)\,dw(s)\right]^2 = \int_0^T \mathbb{E}f^2(s)\,ds. \tag{3.15}$$

5. For $f(t), g(t) \in H_2[0,T]$,

$$\mathbb{E}\left[\int_0^T f(s)\,dw(s) \int_0^T g(s)\,dw(s)\right] = \int_0^T \mathbb{E}\left[f(s)g(s)\right]\,ds. \tag{3.16}$$

Property (3.14) follows from the construction of the Itô integral and the independence of $f(t)$ from the increments $w(t'') - w(t')$ for all $t \leq t' \leq t''$. It is easy to see that properties (3.15) and (3.16) are equivalent.

Exercise 3.3. (Method 1).

(i) Show that eq. (3.7) represents the Itô integral

$$(I)\int_a^b w(s)\,dw(s) = \frac{1}{2}[w^2(b) - w^2(a)] - \frac{1}{2}(b-a). \tag{3.17}$$

(ii) Derive an equation analogous to (3.14) for the conditional expectation

$$\mathbb{E}\left[\left(\int_0^t f(s)\,dw(s)\right)^2 \,\middle|\, \int_0^\tau f(s)\,dw(s) = x\right].$$

(iii) Derive an equation analogous to (3.16) for conditional expectations and for conditional expectations of the product $\int_0^{T_1} f(s)\,dw(s) \int_0^{T_2} g(s)\,dw(s)$, when $T_1 \neq T_2$. □

Exercise 3.4 (The Itô integral is a martingale). Show that the Itô integral is a martingale (see Section 2.4.3). □

3.2.2 The Stratonovich integral

A different construction of a stochastic integral is based on Method 2 with a choice of a "midpoint" in the integral sum. Specifically, keeping the notation of the previous section, for functions $f(w(t),t)$ of class $H_2[0,T]$ define the sum

$$\sigma_n = \sum_{i=1}^{n} f\left(w\left(t_{i/2}\right), t_{i-1}\right) \Delta_i w, \tag{3.18}$$

where, as above, $t_{i/2} = \frac{1}{2}(t_i + t_{i-1})$, and define the Stratonovich integral as the limit

$$\int_a^b f(w(t),t)\, d_S w(t) \stackrel{\Pr}{=} \lim_{\max_i (t_i - t_{i-1}) \to 0} \sigma_n. \tag{3.19}$$

Exercise 3.5 (Another Stratonovich sum). Use instead of (3.18) the sums

$$\sigma_n = \sum_{i=1}^{n} f\left(\frac{w(t_i) + w(t_{i-1})}{2}, t_{i-1}\right) \Delta_i w \tag{3.20}$$

to define the Stratonovich integral. Show that the definitions (3.20) and (3.18) are equivalent in $H_2[0,T]$. □

It is obvious from the definition of the Stratonovich integral that eq. (3.9) represents the Stratonovich integral

$$\int_a^b w(s)\, d_S w(s) = \frac{1}{2}[w^2(b) - w^2(a)]. \tag{3.21}$$

The Stratonovich integral is related to the Itô integral by the following correction term.

Theorem 3.2.1 (The Wong–Zakai correction). *If $f(x,t)$ has a continuous derivative of second order such that $|f_{xx}(x,t)| < A(t)e^{\alpha(t)|x|}$ for some positive continuous functions $\alpha(t)$ and $A(t)$ for all $a \le t \le b$, then*

$$\int_a^b f(w(t),t)\, d_S w(t) = \int_a^b f(w(t),t)\, dw(t) + \frac{1}{2}\int_a^b \frac{\partial}{\partial x} f(w(t),t)\, dt \tag{3.22}$$

in the sense that the left-hand side of (3.22) exists if and only if the right-hand side exists and they are equal.

Proof. First, we assume that the second-order partial derivative $f_{x,x}(x,t)$ is a continuous function of both variables. We expand the function in eq. (3.20) in powers of $\Delta_i w = w(t_i,\omega) - w(t_{i-1},\omega)$ about the point $(w(t_{i-1}), t_{i-1})$ according to Taylor's formula with a remainder,

$$f\left(\frac{1}{2}[w(t_i) + w(t_{i-1})], t_{i-1}\right) = f\left(w(t_{i-1}), t_{i-1}\right) + \frac{1}{2} f_x\left(w(t_{i-1}), t_{i-1}\right) \Delta_i w$$
$$+ \frac{1}{8} f_{x,x}\left(\xi_i, t_{i-1}\right) \left(\Delta_i w\right)^2, \tag{3.23}$$

where ξ_i is some point between $\frac{1}{2}[w(t_i) + w(t_{i-1})]$ and $w(t_{i-1})$. Using eq. (3.23) in eq. (3.20), we obtain

$$\sigma_n = \sum_{i=1}^{n} \left[f\left(w(t_{i-1}), t_{i-1}\right) \Delta_i w + \frac{1}{2} \frac{\partial f\left(w(t_{i-1}), t_{i-1}\right)}{\partial x} \left(\Delta_i w\right)^2 \right.$$
$$\left. + \frac{1}{8} \frac{\partial^2 f\left(\xi_i, t_{i-1}\right)}{\partial x^2} \left(\Delta_i w\right)^3 \right].$$

We denote the three sums

$$\sum_{i=1}^{n} f \Delta_i w = I_1, \quad \frac{1}{2} \sum_{i=1}^{n} f_x \left(\Delta_i w\right)^2 = I_2, \quad \frac{1}{8} \sum_{i=1}^{n} f_{x,x}\left(\xi_i, t_{i-1}\right) \left(\Delta_i w\right)^3 = I_3.$$

Obviously, $I_1 \stackrel{\Pr}{\to} \int_a^b f(w(t), t)\, dw(t)$ as $\max_i(t_i - t_{i-1}) \to 0$ (see eq. (3.13)). To show that I_2 converges to the second integral in eq. (3.22), we recall its definition as the limit of the Riemann sums, $\eta_n = \frac{1}{2} \sum_{i=1}^{n} f_x\left(w(t_{i-1}), t_{i-1}\right) \Delta_i t$; that is,

$$\frac{1}{2} \int_a^b f_x(w(t), t)\, dt \stackrel{\Pr}{=} \lim_{\max_i(t_i - t_{i-1}) \to 0} \eta_n.$$

Now, $I_2 - \eta_n = \frac{1}{2} \sum_{i=1}^{n} f_x\left(w(t_{i-1}), t_{i-1}\right) \left[\left(\Delta_i w\right)^2 - \Delta_i t \right]$. It follows that

$$\mathbb{E}\left[I_2 - \eta_n\right] = \frac{1}{2} \sum_{i=1}^{n} \mathbb{E}\left[f_x\left(w(t_{i-1}), t_{i-1}\right)\right] \mathbb{E}\left[\left(\Delta_i w\right)^2 - \Delta_i t \right] = 0,$$

because $f_x\left(w(t_{i-1}), t_{i-1}\right)$ is independent of $\left(\Delta_i w\right)^2$ and $\mathbb{E}\left(\Delta_i w\right)^2 = \Delta_i t$. We show below that

$$\lim_{\max_i(t_i - t_{i-1}) \to 0} \mathbb{E}\left[I_2 - \eta_n\right]^2 = 0, \tag{3.24}$$

hence $I_2 \stackrel{\Pr}{\to} \frac{1}{2} \int_a^b f_x(w(t), t)\, dt$ as $\max_i(t_i - t_{i-1}) \to 0$.

First, note that from the Gaussian distribution of $\Delta_i w$, we have

$$\mathbb{E}\left[\left(\Delta_i w\right)^2 - \Delta_i t \right]^2 = 3\left(\Delta_i t\right)^2 - 2\left(\Delta_i t\right)^2 + \left(\Delta_i t\right)^2 = 2\left(\Delta_i t\right)^2. \tag{3.25}$$

Again, due to independence, we have from eq. (3.25)

$$\mathbb{E}\left[\left(\Delta_i w\right)^2 - \Delta_i t \right]^2 = \frac{1}{4} \sum_{i=1}^{n} \mathbb{E}\left[f_x\left(w(t_{i-1}), t_{i-1}\right)\right]^2 \mathbb{E}\left[\left(\Delta_i w\right)^2 - \Delta_i t \right]^2$$
$$= \frac{1}{2} \sum_{i=1}^{n} \mathbb{E}\left[f_x\left(w(t_{i-1}), t_{i-1}\right)\right]^2 \left(\Delta_i t\right)^2. \tag{3.26}$$

The continuous function $\mathbb{E}\left[f_x(w(t),t)\right]^2$ is (Riemann) integrable, so that

$$\lim_{\max_i(t_i-t_{i-1})\to 0}\frac{1}{2}\sum_{i=1}^{n}\mathbb{E}\left[f_x\left(w(t_{i-1}),t_{i-1}\right)\right]^2\Delta_i t=\frac{1}{2}\int_a^b\mathbb{E}\left[f_x(w(t),t)\right]^2\,dt.$$

It follows that the extra factor of $\Delta_i t$ in eq. (3.26), which converges to zero, implies that eq. (3.24) holds. Finally, using the inequalities $|\xi_i|\leq|w(t_{i_1})|+|w(t_i)|$ and $\mathbb{E}\left(\Delta_i w\right)^6=3\cdot 5\left(\Delta_i t\right)^3$, we obtain, as above,

$$\begin{aligned}E|I_3|^2&=\frac{1}{64}\mathbb{E}\left|\sum_{i=1}^{n}f_{x,x}\left(\xi_i,t_{i-1}\right)\left(\Delta_i w\right)^3\right|\leq\frac{1}{64}\sum_{i=1}^{n}\mathbb{E}\left|f_{x,x}\left(\xi_i,t_{i-1}\right)\right|^2\mathbb{E}\left(\Delta_i w\right)^6\\&\leq\frac{3\cdot 5}{64}\sum_{i=1}^{n}A^2(t_{i-1})\mathbb{E}e^{2\alpha(t_{i-1})[|w(t_{i-1})|+|w(t_i)|]}\left(\Delta_i t\right)^3\to 0.\end{aligned}$$

□

Exercise 3.6 (Convergence). Prove the last convergence above. □

As an example of the Wong–Zakai correction, we calculate the integral $I=\int_a^b w(s)\,dw_S(s)$. The function $f(x,t)$ in this case is $f(x,t)=x$, so that $f_x(x,t)=1$. Converting from the Stratonovich to the Itô integral according to the Wong–Zakai formula eq. (3.22) and using the previously derived result (eq. (3.17)), we obtain

$$\begin{aligned}I&=\int_a^b w(s)\,dw_S(s)=\int_a^b w(s)\,dw(s)+\frac{1}{2}\int_a^b 1\,ds\\&=\frac{1}{2}[w^2(b)-w^2(a)]-\frac{1}{2}(b-a)+\frac{1}{2}(b-a)=\frac{1}{2}[w^2(b)-w^2(a)],\end{aligned}$$

as in the ordinary integral calculus! This is not a fluke, actually, we have

Theorem 3.2.2 (Stratonovich calculus). *The Stratonovich integral satisfies the rules of the integral calculus:*
(1) The fundamental theorem of the integral calculus: if $F'(x)=f(x)$ is a continuous function and $|f''(x)|<e^{\alpha|x|}$ for some $\alpha>0$, then

$$\int_a^b f\left(w(s)\right)\,dw_S(s)=F\left(w(b)\right)-F\left(w(a)\right).\tag{3.27}$$

(2) The integration by parts formula: if $F'(x)=f(x)$ and $G'(x)=g(x)$ are continuous functions, then

$$\begin{aligned}&\int_a^b\left[F(w(t))G'(w(t))+F'(w(t))G(w(t))\right]\,d_S w(t)\\&=F(w(b))G(w(b))-F(w(a))G(w(a)).\end{aligned}$$

(3) The change of variables rule: if $f'(x)$ and $\varphi'(x)$ are continuous functions on the real line, then

$$\int_a^b f'\left(\varphi(w(t))\,\varphi'(w(t))\,d_S w(t) = f\left(\varphi(w(b))\right) - f\left(\varphi(w(a))\right).$$

Proof. We prove (1) and leave (2) and (3) as an exercise. Setting $w_i = w(t_i)$ and assuming differentiability, as above, we have the Taylor expansions

$$F(w_i) = F\left(\frac{w_{i-1}+w_i}{2}\right) + f\left(\frac{w_{i-1}+w_i}{2}\right)\frac{\Delta_i w}{2} + f'\left(\frac{w_{i-1}+w_i}{2}\right)\frac{(\Delta_i w)^2}{8}$$
$$+ f''(\xi_i)\frac{(\Delta_i w)^3}{24}$$

and

$$F(w_{i-1}) = F\left(\frac{w_{i-1}+w_i}{2}\right) - f\left(\frac{w_{i-1}+w_i}{2}\right)\frac{\Delta_i w}{2}$$
$$+ f'\left(\frac{w_{i-1}+w_i}{2}\right)\frac{(\Delta_i w)^2}{8} - f''(\xi_i')\frac{(\Delta_i w)^3}{24}.$$

It follows that

$$F(w_i) - F(w_{i-1}) = f\left(\frac{w_{i-1}+w_i}{2}\right)\Delta_i w + \left[f''(\xi_i) + f''(\xi_i')\right]\frac{(\Delta_i w)^3}{24}.$$

Summing the above equations from $i = 1$ to $i = n$, we obtain a telescopic sum that gives

$$F(w(b)) - F(w(a)) = \sum_{i=1}^n f\left(\frac{w(t_i)+w(t_{i-1})}{2}\right)\Delta_i w + \frac{1}{24}\sum_{i=1}^n \left[f''(\xi_i)\right.$$
$$\left. + f''(\xi_i')\right](\Delta_i w)^3.$$

The first sum converges to the Stratonovich integral and the second sum converges to zero, as in Exercise 3.6. Hence eq. (3.27). □

As an important consequence of the Wong–Zakai formula and the properties (3.15) of the Itô integral is that the mean value of the Stratonovich integral is not zero. Rather, eq. (3.22) gives

$$\mathbb{E}\left[\int_a^b f(w(t),t)\,d_S w(t)\right] = \mathbb{E}\left[\int_a^b f(w(t),t)\,dw(t)\right] + \frac{1}{2}\mathbb{E}\left[\int_a^b \frac{\partial f(w(t),t)}{\partial x}\,dt\right]$$
$$= \frac{1}{2}\int_a^b \mathbb{E}\left[\frac{\partial f(w(t),t)}{\partial x}\,dt\right]. \tag{3.28}$$

It follows that the Stratonovich integral is not a martingale.

3.2.3 The backward integral

The *backward integral* is defined, as in Method 3, by integral sums, in which the function $f(t)$ is evaluated at the right endpoint of the partition interval as

$$\int_a^b f(t)\,d_B w(t) \stackrel{\Pr}{=} \lim_{\max_i(t_i-t_{i-1})\to 0} \sum_{i=1}^n f(t_i)\left[w(t_i)-w(t_{i-1})\right]. \tag{3.29}$$

The Wong–Zakai formula is now

$$\int_a^b f(w(t),t)\,d_B w(t) = \int_a^b f(w(t),t)\,dw(t) + \int_a^b \frac{\partial}{\partial x} f(w(t),t)\,dt. \tag{3.30}$$

It is obvious from the definition of the backward integral that eq. (3.11) represents the backward integral

$$\int_a^b w(s)\,d_B w(s) = \frac{1}{2}[w^2(b)-w^2(a)] + \frac{1}{2}(b-a). \tag{3.31}$$

It is evident that the above-mentioned Methods 1, 2, and 3 for the calculation of the integral (3.3) correspond to its Itô, Stratonovich, and the backward integral definitions, respectively.

Exercise 3.7. (The Stratonovich and backward integrals)

(i) Use the Wong–Zakai corrections (3.22) and (3.30) to derive the relationship between the Stratonovich and the backward integrals.

(ii) Verify your calculation by applying it to the integrals $\int_a^b w_j(t)\,dw(t)$ $(j = I, S, B)$. □

Exercise 3.8 (Second-order integral). The integral sums

$$\sigma_n(t) = \sum_{i=1}^n f(\tilde{t}_i)\left[w(t_i)-w(t_{i-1})\right]^2,$$

where $t_{i-1} \leq \tilde{t}_i \leq t_i$ and set $m = \max_i(t_i - t_{i-1})$. Define a second-order integral in $H_2[a,b]$ by

$$\int_a^b f(t)\,[\,dw(t)]^2 \stackrel{\Pr}{=} \lim_{m\to 0} \sigma_n(t). \tag{3.32}$$

Use the Cauchy–Schwarz inequality and the method used in the proof of the Wong–Zakai correction to show that

(i) $\int_a^b f(t)\,[\,dw(t)]^2 = \int_a^b f(t)\,dt$,

(ii) an analogous construction of an nth order integral for $n > 2$ gives

$$\int_a^b f(t)\,[dw(t)]^n = 0.$$

□

Exercise 3.9 (The λ-integral). For any $0 \leq \lambda \leq 1$ define a λ-integral by the sums

$$\sigma_n(t) = \sum_{i=1}^{n} f(\lambda w(t_{i-1}) + (1-\lambda) w(t_i), t_{i-1})\,\Delta_i w.$$

Show that for $\lambda = 0$ it leads to the backward integral, for $\lambda = \frac{1}{2}$ to the Stratonovich integral, and for $\lambda = 1$ to the Itô integral. Find the relationship between the λ-integral and the Itô integral. □

The different stochastic integrals lead to different definitions of stochastic differential equations. The question of which integral is the "correct" one to use in modeling physical phenomena is of fundamental significance. This question is discussed later in the text.

3.3 The construction of the Itô integral

In this section we show that for every $f \in H_2[0,T]$ and on almost every trajectory $\omega \in \Omega$ (i.e., except for a set of trajectories of probability 0) there is a sequence of Itô sums (3.12) that converges uniformly to a limit and that the limit is independent of the sequence of partitions. To this end, we make use of the martingale property of certain functions of the MBM [171].

We begin with the observation that according to Itô's definition (3.12), the integral of a (random) constant function $f(\omega) \in H_2[0,T]$ is

$$\int_0^T f(\omega)\,dw(t,\omega) = f(\omega)\left[w(T,\omega) - w(0,\omega)\right]. \tag{3.33}$$

It follows from Section 2.4.3 that for a bounded function $f(\omega) \in H_2[0,T]$ and every constant α the process

$$x(t,\omega) = \exp\left\{\alpha \int_0^t f(\omega)\,dw(s,\omega) - \frac{\alpha^2}{2}\int_0^t f^2(\omega)\,ds\right\} \tag{3.34}$$

is a martingale and that for all $t \in [0,T]$

$$\mathbb{E}x(t,\omega) = 1. \tag{3.35}$$

The next step is to construct and investigate the properties of Itô integrals of step functions $f(t,\omega)$ in $H_2[a,b]$. A step function takes (random constant) values $C_j(\omega)$

in the (non random, i.e., independent of ω) intervals (t_{j-1}, t_j), $(j = 1, \dots, N)$, where the points $a = t_0 < t_1 < \cdots < t_n = b$ form a partition of the interval $[a, b]$. For $a \le t \le b$, if $a < t \le t_1$, the function is constant in the interval (a, t) so its Itô integral is given by (3.33) as

$$I(t,\omega) = \int_a^t f(s,\omega)\, dw(s,\omega) = C_1(\omega)\left[w(t,\omega) - w(a,\omega)\right].$$

If $t_{j-1} < t \le t_j$ for $j > 1$, the Itô integral of such a step function is given by

$$\begin{aligned} I(t,\omega) &= \int_a^t f(s,\omega)\, dw(s,\omega) &(3.36)\\ &= \sum_{k=1}^{j-1} C_k(\omega)\left[w(t_k,\omega) - w(t_{k-1},\omega)\right] + C_j(\omega)\left[w(t,\omega) - w(t_{j-1},\omega)\right]. \end{aligned}$$

Obviously, $I(t,\omega)$ is an $\mathcal{F}_t$-adapted continuous martingale in $[a, b]$. Note that $I(t,\omega)$ in eq. (3.36) does not change if the partition of the interval is refined. Furthermore, the condition $f(t,\omega)$ in $H_2[a,b]$ means that the constant $C_j(\omega)$ is $\mathcal{F}_{t_{j-1}}$-measurable and has a finite second moment.

Lemma 3.3.1. *For a bounded step function $f(t,\omega)$ in $H_2[a,b]$ the process*

$$x(t,\omega) = \exp\left\{\int_0^t \alpha f(s,\omega)\, dw(s,\omega) - \frac{\alpha^2}{2}\int_0^t f^2(s,\omega)\, ds\right\} \tag{3.37}$$

is a martingale.

Proof. We proceed by induction on the number n of intervals in the partition of the interval. In Section 2.4.3 and in Exercise 2.28 it was shown that $x(t,\omega)$ is a martingale and that (3.35) holds for the case of one and two partition intervals. Consider now $a \le s < t \le b$ and $n > 2$. If the number of partition intervals between s and t is less than n, then the conclusion of the lemma holds by assumption. If it is n, choose t_1 in a partition interval between s and t and write

$$\mathbb{E}\left[x(t,\omega) \,|\, x(s,\omega) = x\right] = \mathbb{E}\left\{\mathbb{E}\left[x(t,\omega) \,|\, x(t_1,\omega)\right] \,|\, x(s,\omega) = x\right\}. \tag{3.38}$$

By the inductive assumption, $x(\cdot,\omega)$ is a martingale in both intervals $[s, t_1]$ and $[t_1, t]$. Thus

$$\mathbb{E}\left[x(t,\omega) \,|\, x(t_1,\omega)\right] = x(t_1,\omega) \tag{3.39}$$

$$\mathbb{E}\left[x(t_1,\omega) \,|\, x(s,\omega) = x\right] = x. \tag{3.40}$$

Using (3.39) and (3.40) in (3.38), we obtain

$$\mathbb{E}\left[x(t,\omega) \,|\, x(s,\omega) = x\right] = E\left[x(t_1,\omega) \,|\, x(s,\omega) = x\right] = x.$$

We have $\mathbb{E}x(t,\omega) = \mathbb{E}\left[x(t,\omega) \,|\, x(a,\omega) = 1\right] = 1$, because $x(a,\omega) = 1$ with probability 1. This proves (3.35). □

The boundedness condition in Lemma 3.3.1 can be relaxed by imposing boundedness conditions on exponential moments (see, e.g., [80], [115]), for example, by the following

Definition 3.3.1 (Novikov's condition). *A function* $f(t,\omega) \in H_2[0,T]$ *is said to satisfy* Novikov's condition, *if*

$$\mathbb{E}\exp\left\{\frac{1}{2}\int_0^T f^2(t,\omega)\,dt\right\} < \infty. \tag{3.41}$$

We consider henceforward to the end of this section only functions that satisfy the condition (3.41). The key to the proof of convergence is the following *submartingale inequality* [106, 115, 57], adapted to the needs of this chapter.

Theorem 3.3.1 (Submartingale inequality). *Let* $x(t,\omega)$ *be a continuous nonnegative* $\mathcal{F}_t$*-adapted submartingale in* $[0,T]$, *then for any* $\alpha > 0$

$$\Pr\left\{\omega \in \Omega \,\middle|\, \max_{0\leq t\leq T} x(t,\omega) > \alpha\right\} \leq \frac{\mathbb{E}x\,(T,\omega)}{\alpha}. \tag{3.42}$$

Proof. For any points $0 \leq t_1 < t_2 < \cdots < t_n \leq T$ define

$$B = \left\{\omega \in \Omega \,\middle|\, \max_{1\leq i\leq n} x(t_i,\omega) > \alpha\right\}.$$

If

$$\Pr\{B\} \leq \frac{\mathbb{E}x\,(T,\omega)}{\alpha}, \tag{3.43}$$

then (3.42) follows by continuity of $x(t,\omega)$. To prove (3.43), we define the sets

$$B_i = \{\omega \in \Omega \,|\, x(t_j,\omega) \leq \alpha \text{ for all } j < i,\ x(t_i,\omega) > \alpha\}.$$

For each $\omega \in B_i$ the sequence $x(t_j,\omega)$ crosses the level α at time t_i and not sooner. The sets B_i are mutually disjoint and $B = \bigcup_{i=1}^n B_i$, and furthermore, $B_i \in \mathcal{F}_{t_i}$. Recall that from the definition (2.14)

$$\begin{aligned}
\mathbb{E}x(T,\omega) &= \int x\Pr\{\omega \in \Omega \,|\, x \leq x(T,\omega) \leq x + dx\} \\
&\geq \int x\Pr\{\omega \in B \,|\, x \leq x(T,\omega) \leq x + dx\} \\
&= \sum_{i=1}^n \int x\Pr\{\omega \in B_i \,|\, x \leq x(T,\omega) \leq x + dx\} \\
&= \sum_{i=1}^n \mathbb{E}\left\{x(T,\omega), B_i\right\} = \sum_{i=1}^n \mathbb{E}\{\mathbb{E}\left[x(T,\omega) \,|\, x(t_i,\omega)\right], B_i\},
\end{aligned} \tag{3.44}$$

because $B_i \in \mathcal{F}_{t_i}$. The submartingale property (2.69) implies that

$$\mathbb{E}\left[x(T,\omega)\,|\,x(t_i,\omega)\right] \geq x(t_i,\omega),$$

so that (3.44) gives

$$\begin{aligned}
\mathbb{E}x(T,\omega) &= \sum_{i=1}^{n} \mathbb{E}\{x(t_i,\omega), B_i\} = \sum_{i=1}^{n} \int x\Pr\{\omega \in B_i\,|\,x \leq x(t_i,\omega) \leq x+dx\} \\
&> \sum_{i=1}^{n} \int \alpha \Pr\{\omega \in B_i\,|\,x \leq x(t_i,\omega) \leq x+dx\} \\
&= \sum_{i=1}^{n} \alpha \Pr\{B_i\} = \alpha \Pr\{B\},
\end{aligned}$$

which is (3.43). It follows that (3.42) holds. □

Lemma 3.3.1 and Theorem 3.3.1 imply the following

Lemma 3.3.2. *For any step function $f(t,\omega) \in H_2[0,T]$ and any $\alpha > 0$ and β,*

$$\Pr\left\{\omega \in \Omega \,\middle|\, \max_{0\leq t\leq T} \left[\int_0^t f(s,\omega)\,dw(s,\omega) - \frac{\alpha}{2}\int_0^t f^2(s,\omega)\,ds\right] > \beta\right\} \leq e^{-\alpha\beta}. \tag{3.45}$$

Proof. By changing f to αf and exponentiating the inequality inside the braces in (3.45), we find that (3.45) is equivalent to the inequality

$$\begin{aligned}
&\Pr\left\{\omega \in \Omega \,\middle|\, \max_{0\leq t\leq T} \left[\int_0^t \left[\alpha f(s,\omega)\right]\,dw(s,\omega) - \frac{1}{2}\int_0^t \left[\alpha f(s,\omega)\right]^2\,ds\right] > \alpha\beta\right\} \\
&= \Pr\left\{\omega \in \Omega \,\middle|\, \max_{0\leq t\leq T} \exp\left[\int_0^t \alpha f(s,\omega)\,dw(s,\omega) - \frac{1}{2}\int_0^t \left[\alpha f(s,\omega)\right]^2\,ds\right] > e^{\alpha\beta}\right\} \\
&\leq e^{-\alpha\beta}.
\end{aligned} \tag{3.46}$$

To prove (3.46), we note that the maximized function in (3.46),

$$x(t,\omega) = \exp\left[\int_0^t \alpha f(s,\omega)\,dw(s,\omega) - \frac{1}{2}\int_0^t \left[\alpha f(s,\omega)\right]^2\,ds\right],$$

is a martingale, as shown in Lemma 3.3.1. It follows from the martingale inequality (3.42) that

$$\Pr\left\{\omega \in \Omega \,\middle|\, \max_{0\leq t\leq T} x(t,\omega) > e^{\alpha\beta}\right\} \leq e^{-\alpha\beta}\mathbb{E}x(T,\omega) = e^{-\alpha\beta}, \tag{3.47}$$

because $\mathbb{E}x(T,\omega) = 1$, as shown in (3.35). Combining the inequalities (3.46) and (3.47), we obtain (3.45). □

The next lemma asserts that if a sequence of step functions in $H_2[0,T]$ converges to zero sufficiently fast in $L^2\{[0,T]\}$, so does the sequence of their Itô integrals. Specifically,

Lemma 3.3.3. *If* $f_n(t,\omega) \in H_2[0,T]$ *are step functions such that*

$$\Pr\left\{\omega \in \Omega \,\middle|\, \int_0^T f_n^2(s,\omega)\,ds > 2^{-n} \text{ i.o.}\right\} = 0, \tag{3.48}$$

then, for any $\nu > 1$,

$$\Pr\left\{\omega \in \Omega \,\middle|\, \max_{0\leq t\leq T} \left|\int_0^t f_n(s,\omega)\,dw\right| \geq \nu\left(2^{-n+1}\log n\right)^{1/2} \text{ i.o.}\right\} = 0. \tag{3.49}$$

The condition (3.48) means that on every trajectory $\omega \in \Omega$, except for a set of probability zero, there is an index $n(\omega)$ such that $\int_0^T f_n^2(s,\omega)\,ds \leq 2^{-n}$ for all $n > n(\omega)$. The result (3.49) means that on every trajectory $\omega \in \Omega$, except for a set of probability zero, there is an index $n(\nu,\omega)$ such that

$$\max_{0\leq t\leq T} \left|\int_0^t f_n(s,\omega)\,dw(s,\omega)\right| < \nu\left(2^{-n+1}\log n\right)^{1/2}$$

for all $n > n(\nu,\omega)$.

Proof. We choose $\alpha = \left(2^{n+1}\log n\right)^{1/2}$ and $\beta = \nu\left(2^{-n-1}\log n\right)^{1/2}$ and observe that condition (3.48) implies that for almost every $\omega \in \Omega$ the inequality

$$\frac{\alpha}{2}\int_0^T f_n^2(s,\omega)\,ds \leq 2^{-n-1}\alpha$$

holds for all but a finite number of indices n and that $2\beta - 2^{-n-1}\alpha = (2\nu - 1)\left(2^{-n-1}\log n\right)^{1/2}$. It follows that for almost every $\omega \in \Omega$,

$$2\beta - \frac{\alpha}{2}\int_0^t f_n^2(s,\omega)\,ds \geq (2\nu-1)\left(2^{-n-1}\log n\right)^{1/2}, \tag{3.50}$$

for all but a finite number of indices n. We also use the obvious inequality

$$\max_{0\leq t\leq T} a(t) - b(t) \leq \max_{0\leq t\leq T}[a(t) - b(t)]$$

for any continuous functions $a(t)$ and $b(t)$ in the interval $[0,T]$. Keeping the above inequalities in mind, we use (3.45) to obtain that for all but a finite number of indices

n,

$$\Pr\left\{\max_{0\le t\le T}\int_0^t f_n(s,\omega)\,dw(s,\omega) \ge \left(2^{-n+1}\log n\right)^{1/2}\right\}$$

$$= \Pr\left\{\max_{0\le t\le T}\int_0^t f_n(s,\omega)\,dw(s,\omega) - \frac{\alpha}{2}\int_0^t f_n^2(s,\omega)\,ds \ge 2\beta\right. \tag{3.51}$$

$$\left. - \frac{\alpha}{2}\int_0^t f_n^2(s,\omega)\,ds\right\}$$

$$\le \Pr\left\{\max_{0\le t\le T}\left[\int_0^t f_n(s,\omega)\,dw(s,\omega) - \frac{\alpha}{2}\int_0^t f_n^2(s,\omega)\,ds\right] \ge (2\nu-1)\right. \tag{3.52}$$

$$\left.\times\left(2^{-n-1}\log n\right)^{1/2}\right\}$$

$$\le e^{-(2\nu-1)\log n} = n^{-(2\nu-1)}. \tag{3.53}$$

The same argument is valid if the functions $f_n(t)$ are replaced with the functions $-f_n(t)$. It follows that

$$\Pr\left\{\omega\in\Omega \;\middle|\; \max_{0\le t\le T}\left|\int_0^t f_n(s,\omega)\,dw\right| \ge \nu\left(2^{-n+1}\log n\right)^{1/2}\right\} \le n^{-(2\nu-1)}$$

for all but a finite number of indices n. Now, because $2\nu - 1 > \nu > 1$, the series $\sum_{n=1}^{\infty} n^{-(2\nu-1)}$ converges and it follows from the Borel–Cantelli lemma that (3.49) holds. □

Next, we prove the uniform convergence of the Itô sums (3.12) in $[0,T]$ on almost every trajectory $\omega\in\Omega$.

Theorem 3.3.2. *For every $f \in H_2[0,T]$ there is a sequence of step functions $f_n(t)\in H_2[0,T]$ such that*

$$\lim_{n\to\infty}\int_0^T |f(t,\omega) - f_n(t,\omega)|^2\,dt = 0 \tag{3.54}$$

for almost all $\omega\in\Omega$, the limit

$$I(t,\omega) = \lim_{n\to\infty}\int_0^t f_n(s,\omega)\,dw(s) \tag{3.55}$$

is uniform in $[0,T]$ for almost all $\omega\in\Omega$ and is independent of the sequence $f_n(t,\omega)\in H_2[0,T]$ that satisfies (3.54).

Proof. First, we show that for every $f(t,\omega) \in H_2[0,T]$, there exists a sequence of step functions $f_n(t,\omega) \in H_2[0,T]$ that converges to $f(t,\omega)$ in the sense of (3.54) such that (3.55) holds uniformly in $[0,T]$ for almost all $\omega \in \Omega$. Then, we show that the same conclusion is valid for every sequence of functions $f_n(t,\omega) \in H_2[0,T]$ that converges to $f(t,\omega)$ in the sense of (3.54). In particular, for any sequence of partitions $0 = t_{0,n} < t_{1,n} < t_{2,n} < \cdots < t_{n,n} = T$, such that $\lim_{n\to\infty} \max_{0\leq i\leq n}(t_{i,n} - t_{i-1,n}) = 0$ as $n \to \infty$, and for any function $f(t,\omega) \in H_2[0,T]$, the step functions in the Itô sums (3.12), defined by

$$f_n(t,\omega) = f(t_{i-1,n},\omega) \text{ for } t_{i-1,n} \leq t < t_{i,n}, \tag{3.56}$$

satisfy the condition (3.54). It follows, therefore, that the Itô sums converge to $I(t,\omega)$ uniformly in $[0,T]$ on almost every $\omega \in \Omega$.

To carry out this plan, we note that because for almost every $\omega \in \Omega$ the function $f(t,\omega)$ is integrable with respect to t, it is equal to the derivative of its integral for almost all $t \in [0,T]$. Hence, defining $f(s,\omega) = 0$ for $s < 0$ and setting $\tilde{f}_k(t,\omega) = 2^k \int_{t-2^{-k}}^t f(s,\omega)\,ds$, we obtain $\lim_{k\to\infty} \tilde{f}_k(t,\omega) = f(t,\omega)$ for almost all $t \in [0,T]$. The function $\tilde{f}_k(t,\omega)$ is absolutely continuous so that the approximating step functions $\tilde{f}_{m,k}(t,\omega) = \tilde{f}_k\left(2^{-m}\left[2^m t\right]\right)$, where $[x]$ is the greatest integer that does not exceed x, converge to $\tilde{f}_k(t,\omega)$ uniformly as $m \to \infty$. It follows that $\lim_{m,k\to\infty} \int_0^T |f(t,\omega) - \tilde{f}_{m,k}(t,\omega)|\,dt = 0$ for almost every $\omega \in \Omega$. Thus, for every n there is a pair of indices m_n, k_n such that

$$\Pr\left\{\omega \in \Omega \;\middle|\; \int_0^T \left|f(t,\omega) - \tilde{f}_{m_n,k_n}(t,\omega)\right|^2 dt > 2^{-n+1}\right\} < 2^{-n}.$$

Defining $f_n(t,\omega) = \tilde{f}_{m_n,k_n}(t,\omega)$, we get from the Borel–Cantelli lemma that

$$\Pr\left\{\int_0^T |f(t,\omega) - f_n(t,\omega)|^2\,dt > 2^{-n+1} \text{ i.o}\right\} = 0;$$

hence

$$\Pr\left\{\int_0^T |f_n(t,\omega) - f_{n-1}(t,\omega)|^2\,dt > 2^{-n} \text{ i.o}\right\} = 0. \tag{3.57}$$

Now, it follows from Lemma 3.3.3 that for any $\nu > 1$,

$$\Pr\left\{\max_{0\leq t\leq T}\left|\int_0^t [f_n(s,\omega) - f_{n-1}(s,\omega)]\,dw(s,\omega)\right| \geq \nu\left(2^{-n+1}\log n\right)^{1/2} \text{ i.o.}\right\}$$
$$= 0. \tag{3.58}$$

This, in turn, means that the series of functions

$$\sum_{n=1}^{\infty} \left| \int_0^t [f_n(s,\omega) - f_{n-1}(s,\omega)]\, dw(s,\omega) \right|$$

is dominated by the convergent series of constants $\sum_{n=1}^{\infty} \nu \left(2^{-n+1} \log n\right)^{1/2}$, for almost all $\omega \in \Omega$. It follows from the Weierstrass M-test that the telescopic series

$$\sum_{n=1}^{\infty} \int_0^t [f_n(s,\omega) - f_{n-1}(s,\omega)]\, dw(s,\omega)$$
$$= \lim_{n\to\infty} \int_0^t f_n(s,\omega)\, dw(s,\omega) - \int_0^t f_0(s,\omega)\, dw(s)$$

converges uniformly in $[0,T]$ for almost every $\omega \in \Omega$. We denote

$$\int_0^t f(s,\omega)\, dw(s,\omega) = \lim_{n\to\infty} \int_0^t f_n(s,\omega)\, dw(s,\omega).$$

It remains to show that the same conclusion is true for every sequence of step functions that satisfies (3.49). The argument used above to construct the series f_n that satisfies (3.57) and (3.58) shows that every sequence that satisfies (3.49) has a subsequence that satisfies (3.57) and (3.58). It follows in a straightforward manner that (3.49) implies the uniform convergence (3.55). □

Corollary 3.3.1. *The Itô integral $I(t,\omega)$ is continuous in $[0,T]$ for almost all $\omega \in \Omega$.*

Indeed, because $\sigma_n(t,\omega)$ are continuous functions of t, the uniform limit of such sums is also a continuous function in $[0,T]$.

Remark 3.3.1. In general, if $f \in H_2[0,T]$, the process $x(t,\omega)$ defined in eq. (3.37) is not a martingale, but rather a *supermartingale*; that is,

$$\mathbb{E}[x(t,\omega) \,|\, x(s,\omega)] \leq x(s,\omega).$$

If, however, f satisfies the Novikov condition (3.41), $x(t,\omega)$ is a martingale [115].

3.4 The Itô calculus

The deviation of Itô's stochastic integration from the usual rules of the differential and integral calculus, as demonstrated, for example, in the integral (3.3), can be expected to give rise to a different calculus. We begin with the usual definition of a differential of an integral and immediately face the deviations from the usual rules of the differential calculus. These are exhibited in the simplest rules, such as differentiation of a product of two functions, the differentiation of a composite function, and so on.

Consider two processes, $a(t), b(t)$, of class $H_2[0,T]$ and define the stochastic process

$$x(t) = x_0 + \int_0^t a(s)\,ds + \int_0^t b(s)\,dw(s), \tag{3.59}$$

where x_0 is a random variable independent of $w(t)$ for all $t > 0$. Then, for $0 \leq t_1 < t_2 \leq T$,

$$x(t_2) - x(t_1) = \int_{t_1}^{t_2} a(s)\,ds + \int_{t_1}^{t_2} b(s)\,dw(s). \tag{3.60}$$

We abbreviate this notation as

$$dx(t) = a(t)\,dt + b(t)\,dw(t). \tag{3.61}$$

If instead of the Itô integral the Stratonovich integral is used in the definition of the stochastic differential (3.60), we use the abbreviation

$$d_S x(t) = a(t)\,dt + b(t)\,d_S w(t), \tag{3.62}$$

and if the backward integral is used, we abbreviate

$$d_B x(t) = a(t)\,dt + b(t)\,d_B w(t). \tag{3.63}$$

Example 3.1 (The differential of $w^2(t)$). Equation (3.17) gives $w^2(t_2) - w^2(t_1) = 2\int_{t_1}^{t_2} w(t)\,dw(t) + \int_{t_1}^{t_2} 1\,dt$ for the process $x(t) = w^2(t)$. According to eqs. (3.60) and (3.61), this can be written as $dw^2(t) = 1\,dt + 2w(t)\,dw(t)$; that is, $a(t) = 1$ and $b(t) = 2w(t)$. If, however, the Itô integral in the definition (3.60) is replaced with the Stratonovich integral, then (3.21) gives $d_S w^2(t) = 2w(t)\,d_S w(t)$. Thus the Itô differential (3.61) does not satisfy the usual rule $dx^2 = 2x\,dx$, however the Stratonovich differential (3.62) does. □

Example 3.2 (The differential of $f(t)w(t)$). If $f(t)$ is a smooth deterministic function, then integration by parts is possible so that

$$\int_{t_1}^{t_2} f(t)\,dw(t) = f(t_2)w(t_2) - f(t_1)w(t_1) - \int_{t_1}^{t_2} f'(t)w(t)\,dt.$$

Thus, setting $x(t) = f(t)w(t)$, we obtain

$$dx(t) = f'(t)w(t)\,dt + f(t)dw(t) = w(t)\,df(t) + f(t)\,dw(t),$$

as in the classical calculus. In this case, $a(t) = f'(t)w(t)$ and $b(t) = f(t)$. □

Exercise 3.10 (Backward differentials). Find the backward differential $d_B w^2(t)$. □

Before establishing the general rules of the stochastic calculus, we further illustrate the discrepancy between the rules of the classical calculus and the stochastic Itô calculus with the rule for differentiating a product of two processes. The classical rule for differentiation of a product of two differentiable functions is $dx_1(t)x_2(t) = x_1(t)\,dx_2(t) + x_2(t)\,dx_1(t)$. This is not the case, however, for the Itô differentials (3.61), as shown below.

Example 3.3 (Differential of a product). If $x_1(t)$ and $x_2(t)$ have the Itô differentials

$$dx_1(t) = a_1(t)\,dt + b_1(t)\,dw(t), \quad dx_2(t) = a_2(t)\,dt + b_2(t)\,dw(t),$$

where $a_1, a_2, b_1, b_2 \in H_2[0,T]$, then

$$d\left[x_1(t)x_2(t)\right] = x_1(t)\,dx_2(t) + x_2(t)\,dx_1(t) + b_1(t)b_2(t)\,dt. \tag{3.64}$$

Indeed, assume first that a_i, b_i are constants; then $x_i(t) = x_i(0) + a_i t + b_i w(t)$ for $i = 1,2$, so that

$$\begin{aligned} x_1(t)x_2(t) =& x_1(0)x_2(0) + \left[x_1(0)a_2 + x_2(0)a_1\right]t + \left[x_1(0)b_2 + x_2(0)b_1\right]w(t) \\ &+ a_1a_2t^2 + \left[a_1b_2 + a_2b_1\right]tw(t) + b_1b_2w^2(t). \end{aligned}$$

It follows that

$$\begin{aligned} &d\left[x_1(t)x_2(t)\right]) = \left[x_1(0)a_2 + x_2(0)a_1\right]\,dt + \left[x_1(0)b_2 + x_2(0)b_1\right]\,dw(t) \\ &+ 2a_1a_2t\,dt + \left[a_1b_2 + a_2b_1\right]\left[t\,dw(t) + w(t)\,dt\right] + 2b_1b_2w(t)\,dw(t) + b_1b_2\,dt \\ =& \left[x_1(0) + a_1t + b_1w(t)\right]\left[a_2\,dt + b_2\,dw(t)\right] + \left[x_2(0) + a_2t + b_2w(t)\right] \\ &\times \left[a_1\,dt + b_1\,dw(t)\right] + b_1b_2\,dt \\ =& x_1(t)\,dx_2(t) + x_2(t)\,dx_1(t) + b_1(t)b_2(t)\,dt. \end{aligned}$$

If a_i, b_i are step functions in $H_2[0,T]$, $(i = 1,2)$, then eq. (3.64) holds as well. From the construction of the Itô integral (see Section 3.3) it is clear that it is enough to prove (3.64) for coefficients that are step functions in $H_2[0,T]$. Note that (3.64) differs from the classical differential of a product by the additional term $b_1(t)b_2(t)\,dt$, which is often called *Itô's correction.* □

Exercise 3.11. (Stratonovich differentials).

(i) Use Wong–Zakai corrections (3.89) to show that the Stratonovich differential (3.62) satisfies that classical rule

$$d_S x_1(t)x_2(t) = x_1(t)\,d_S x_2(t) + x_2(t)\,d_S x_1(t).$$

(ii) Determine the rule for backward differentiation of a product. □

Exercise 3.12. (Differentials of polynomials).

(i) Use the product rule (3.64) and mathematical induction to show that for positive integers n,

$$dw^n = nw^{n-1}\,dw + \frac{n(n-1)}{2}w^{n-2}\,dt. \tag{3.65}$$

(ii) What are the corresponding expressions for the Stratonovich and backward differentials? □

Exercise 3.13. (Differentials of twice differentiable functions).

(i) Setting $P(x) = x^n$, show that eq. (3.65) can be written as

$$dP(w) = P'(w)\,dw + \frac{1}{2}P''(w)\,dt.$$

(ii) Obtain the latter for every polynomial $P(x) = \sum_{m=0}^{n} a_m x^m$.

(iii) Use polynomial approximations to prove the same for any twice continuously differentiable function $P(x)$.

(iv) Derive the analogous expressions for Stratonovich and backward differentiation. □

The essence of the differentiation rules is captured in the chain rule for differentiating composite functions. For example, given a chain rule for expressing the differential of a composite function in terms of the differentials of its components, the rule (3.64) is simply an application of the chain rule to the composite function $f(x_1, x_2) = x_1 x_2$.

Consider n Itô-differentiable processes

$$dx^i = a^i\,dt + \sum_{j=1}^{m} b^{ij}\,dw^j, \quad i = 1, 2, \ldots, n,$$

where $a^i, b^{ij} \in H_2[0,T]$ for $i = 1, 2, \ldots, n$, $j = 1, 2, \ldots, m$ and w^j are independent Brownian motions and a function $f(x^1, x^2, \ldots, x^n, t)$ that has continuous partial derivatives of second order in $x^1, x^2, \ldots, x^n$ and a continuous partial derivative with respect to t. For a d-dimensional process $\boldsymbol{x}(t)$ that is differentiable in the ordinary sense, the classical chain rule is

$$\begin{aligned} df(\boldsymbol{x}(t),t) &= \frac{\partial f(\boldsymbol{x}(t),t)}{\partial t}\,dt + \nabla_{\boldsymbol{x}} f(\boldsymbol{x}(t),t)\cdot d\boldsymbol{x}(t) \\ &= \left(\frac{\partial f(\boldsymbol{x}(t),t)}{\partial t} + \sum_{i=1}^{n} a^i(t)\frac{\partial f(\boldsymbol{x}(t),t)}{\partial x^i}\right)dt \\ &\quad + \sum_{i=1}^{n}\sum_{j=1}^{m} b^{ij}(t)\,\frac{\partial f(\boldsymbol{x}(t),t)}{\partial x^i}\,dw^j. \end{aligned} \tag{3.66}$$

For processes differentiable in the Itô sense, but not in the ordinary sense, eq. (3.66) does not hold. Rather, we have the following

Theorem 3.4.1 (Itô's formula). *Assume $a^i(t)$ and $b^{ij}(t)$ are continuous functions and that $f(\boldsymbol{x},t)$ is a twice continuously differentiable function such that $|f_{xx}(x,t)| \leq A(t)e^{\alpha(t)|x|}$ for some positive continuous functions $A(t)$ and $\alpha(t)$. Then*

$$df(\boldsymbol{x}(t),t) = \left[\frac{\partial f(\boldsymbol{x}(t),t)}{\partial t} + L^*_{\boldsymbol{x}} f(\boldsymbol{x},t)\right] dt + \sum_{i=1}^{n}\sum_{j=1}^{m} b^{ij}(t) \frac{\partial f(\boldsymbol{x}(t),t)}{\partial x^i}\, dw^j, \tag{3.67}$$

where

$$L^*_{\boldsymbol{x}} f(\boldsymbol{x},t) = \sum_{i=1}^{n}\sum_{j=1}^{n} \sigma^{ij}(t) \frac{\partial^2 f(\boldsymbol{x},t)}{\partial x^i \partial x^j} + \sum_{i=1}^{n} a^i(t) \frac{\partial f(\boldsymbol{x},t)}{\partial x^i} \tag{3.68}$$

and

$$\sigma^{ij}(t) = \frac{1}{2}\sum_{k=1}^{m} b^{ik}(t)\, b^{jk}(t). \tag{3.69}$$

The $n \times n$ matrix $\left\{\sigma^{ij}(t)\right\}$ is called the *diffusion matrix*. In matrix notation,

$$\boldsymbol{B}(t) = \left\{b^{ij}(t)\right\}_{n \times m} \tag{3.70}$$

is the *noise matrix* and the diffusion matrix $\boldsymbol{\sigma}(t)$ is given by

$$\boldsymbol{\sigma}(t) = \frac{1}{2}\boldsymbol{B}(t)\,\boldsymbol{B}^T(t).$$

The operator $L^*_{\boldsymbol{x}}$ in (3.68) is called the *backward Kolmogorov operator.*

Proof. For each N, we partition the interval by points $0 = t_0 < t_1 < \cdots < t_N = T$ such that $\max_{1 \leq i \leq N}(t_i - t_{i-1}) \to 0$ as $N \to \infty$. Using Taylor's formula, we can write the identity

$$\begin{aligned} & f(\boldsymbol{x}(t),t) - f(\boldsymbol{x}(0),0) \\ &= \sum_{k=1}^{N} \left[f(\boldsymbol{x}(t_k),t_k) - f(\boldsymbol{x}(t_{k-1}),t_{k-1})\right] \\ &= \sum_{k=1}^{N} f_t(\boldsymbol{x}(t_{k-1}),t_{k-1})\Delta t_k && (3.71) \\ &+ \sum_{k=1}^{N}\sum_{i=1}^{n} f_{x^i}(\boldsymbol{x}(t_{k-1}),t_{k-1})\Delta x^i_k && (3.72) \\ &+ \frac{1}{2}\sum_{k=1}^{N}\sum_{i=1}^{n}\sum_{j=1}^{n} f_{x^i,x^j}(\tilde{\boldsymbol{x}}(t_{k-1}),\tilde{t}_{k-1})\Delta x^i_k \Delta x^j_k, && (3.73) \end{aligned}$$

where $(\tilde{\boldsymbol{x}}(\tilde{t}_{k-1}), \tilde{t}_{k-1})$ is a point on the line connecting the points $(\boldsymbol{x}(t_{k-1}), t_{k-1})$ and $(\boldsymbol{x}(t_k), t_k)$ and $f_{x^i}(\cdot,\cdot) = \partial f(\cdot,\cdot)/\partial x^i$ and so on. For $i = 1, 2, \ldots, n$,

$$\begin{aligned}\Delta x_k^i &= \int_{t_{k-1}}^{t_k} a^i(s)\,ds + \sum_{j=1}^m \int_{t_{k-1}}^{t_k} b^{ij}(s)\,dw^j(s)\\ &= a^i(t'_{i,k})\Delta t_k + \sum_{j=1}^m b^{ij}(t''_{i,k})\Delta w_k^j,\end{aligned} \tag{3.74}$$

where $t'_{i,k}, t''_{i,k} \in (t_{k-1}, t_k)$. We consider each sum separately. Obviously, the sum (3.71) gives

$$\lim_{N\to\infty} \sum_{k=1}^N f_t(\boldsymbol{x}(t_{k-1}), t_{k-1})\Delta t_k = \int_0^t f_t(\boldsymbol{x}(s), s)\,ds. \tag{3.75}$$

To evaluate the sum (3.72), we substitute (3.74) for Δx_k^i and write

$$\begin{aligned}&\sum_{k=1}^N f_{x^i}(\boldsymbol{x}(t_{k-1}), t_{k-1})\Delta x_k^i \qquad (3.76)\\ &= \sum_{k=1}^N f_{x^i}(\boldsymbol{x}(t_{k-1}), t_{k-1})a^i(t'_{i,k})\Delta t_k + \sum_{k=1}^N f_{x^i}(\boldsymbol{x}(t_{k-1}), t_{k-1}) \sum_{j=1}^m b^{ij}(t''_{i,k})\Delta w_k^j.\end{aligned}$$

As above, for each i, the first sum on the right-hand side of (3.76) gives

$$\lim_{N\to\infty} \sum_{k=1}^N f_{x^i}(\boldsymbol{x}(t_{k-1}), t_{k-1})a^i(t'_{i,k})\Delta t_k = \int_0^t f_{x^i}(\boldsymbol{x}(s), s)a^i(s)\,ds. \tag{3.77}$$

For each i, j the second sum on the right-hand side of (3.76) gives

$$\sum_{k=1}^N f_{x^i} b^{ij}(t''_{i,k})\Delta w_k^j = \sum_{k=1}^N f_{x^i}\left[b^{ij}(t_{i,k-1}) + \Delta_k b^{ij}\right]\Delta w_k^j, \tag{3.78}$$

where $\Delta_k b^{ij} = b^{ij}(t''_{i,k}) - b^{ij}(t_{i,k-1})$. By definition,

$$\begin{aligned}&\lim_{N\to\infty} \sum_{k=1}^N f_{x^i}(\boldsymbol{x}(t_{k-1}), t_{k-1})b^{ij}(t_{i,k-1})\Delta w_k^j\\ &\stackrel{\Pr}{=} \int_0^t f_{x^i}(\boldsymbol{x}(s), s)b^{ij}(s)\,dw^j(s).\end{aligned} \tag{3.79}$$

To estimate the variance of the sum $I_N = \sum_{k=1}^N f_{x^i}(\boldsymbol{x}(t_{k-1}), t_{k-1})\Delta_k b^{ij} \Delta w_k^j$, we use the Cauchy–Schwarz inequality,

$$\mathbb{E} I_N^2 \leq \sum_{k=1}^N \mathbb{E} \left| f_{x^i}(\boldsymbol{x}(t_{k-1}), t_{k-1}) \right|^2 \Delta t_k \sum_{k=1}^N \mathbb{E} \left| \Delta_k b^{ij} \right|^2 \Delta t_k,$$

because $\mathbb{E} \left| \Delta w_k^j \right|^2 = \Delta t_k$. We get

$$\lim_{N \to \infty} \sum_{k=1}^N \mathbb{E} \left| \Delta_k b^{ij} \right|^2 \Delta t_k = \int_0^T \mathbb{E} \left| b^{ij}(s) - b^{ij}(s) \right|^2 ds = 0,$$

because the integrals $\int_0^T \mathbb{E} \left| f_{x^i}(\boldsymbol{x}(s), s) \right|^2 ds$ and $\int_0^T \mathbb{E} \left| b^{ij}(s) \right|^2 ds$ ar assumed to exist. It follows that (3.78) gives in the limit

$$\lim_{N \to \infty} \sum_{k=1}^N f_{x^i}(\boldsymbol{x}(t_{k-1}), t_{k-1}) b^{ij}(t''_{i,k}) \Delta w_k^j = \int_0^t f_{x^i}(\boldsymbol{x}(s), s) b^{ij}(s)\, dw^j(s).$$

Finally, using (3.74) for each i, j in the triple sum (3.73), we get

$$\frac{1}{2} \sum_{k=1}^N f_{x^i,x^j}(\tilde{\boldsymbol{x}}(t_{k-1}), \tilde{t}_{k-1}) \Delta x_k^i \Delta x_k^j = \frac{1}{2} \sum_{k=1}^N f_{x^i,x^j}(\tilde{\boldsymbol{x}}(t_{k-1}), \tilde{t}_{k-1}) \tag{3.80}$$
$$\times \left[a^i(t'_{i,k}) \Delta t_k + \sum_{l=1}^m b^{il}(t''_{i,k}) \Delta w_k^l \right] \left[a^j(t'_{j,k}) \Delta t_k + \sum_{l=1}^m b^{jl}(t''_{j,k}) \Delta w_k^l \right].$$

Expanding the terms in brackets on the right-hand side of eq. (3.80), we obtain sums with factors $(\Delta t_k)^2$, $\Delta t_k \Delta w_k^l$, and $\Delta w_k^{l'} \Delta w_k^{l''}$. It can be shown in a straightforward manner that all sums containing the factors $\Delta t_k \Delta w_k^l$, and $\Delta w_k^{l'} \Delta w_k^{l''}$ converge to zero as $N \to \infty$. Similarly, for $l' \neq l''$ sums containing $\Delta w_k^{l'} \Delta w_k^{l''}$ converge to zero as $N \to \infty$. For $l' = l''$, we use the result of Exercise 3.16 to show that for each i, j the sums containing the factors $[\Delta w_k^l]^2$ converge to the integrals

$$\int_0^t \frac{1}{2} f_{x^i,x^j}(\boldsymbol{x}(s), s) \sum_{k=1}^m b^{ik}(s) b^{jk}(s)\, ds. \tag{3.81}$$

It follows that in the limit $N \to \infty$ the triple sum (3.73) gives

$$\lim_{N \to \infty} \frac{1}{2} \sum_{k=1}^N \sum_{i=1}^n \sum_{j=1}^n f_{x^i,x^j}(\tilde{\boldsymbol{x}}(t_{k-1}), \tilde{t}_{k-1}) \Delta x_k^i \Delta x_k^j \tag{3.82}$$
$$= \frac{1}{2} \int_0^t \sum_{i=1}^n \sum_{j=1}^n f_{x^i,x^j}(\boldsymbol{x}(s), s) \sum_{k=1}^m b^{ik}(s) b^{jk}(s)\, ds.$$

Using (3.75), (3.79), and (3.82) in (3.71)–(3.73), we obtain

$$
\begin{aligned}
f(\boldsymbol{x}(t),t) = f(\boldsymbol{x}(0),0) + \int_0^t \Bigg[& \frac{\partial f(\boldsymbol{x}(s),s)}{\partial t} + \sum_{i=1}^n a^i(s) \frac{\partial f(\boldsymbol{x}(s),s)}{\partial x^i} \\
& + \frac{1}{2}\sum_{i=1}^n \sum_{j=1}^n \sigma^{ij}(s) \frac{\partial^2 f(\boldsymbol{x}(s),s)}{\partial x^i \partial x^j} \Bigg] ds \\
& + \int_0^t \sum_{i=1}^n \sum_{j=1}^m b^{ij}(s) \frac{\partial f(\boldsymbol{x}(s),s)}{\partial x^i} \, dw^j(s),
\end{aligned}
$$

hence (3.67). □

Remark 3.4.1. Rewriting Itô's formula (3.67) in the form

$$
\begin{aligned}
df(\boldsymbol{x}(t),t) = & \left[\frac{\partial f(\boldsymbol{x}(t),t)}{\partial t} + \sum_{i=1}^n a^i(t) \frac{\partial f(\boldsymbol{x}(t),t)}{\partial x^i} \right] dt \\
& + \sum_{i=1}^n \sum_{j=1}^m b^{ij}(\boldsymbol{x}(t),t) \frac{\partial f(\boldsymbol{x}(t),t)}{\partial x^i} \, dw^j \\
& + \frac{1}{2} \sum_{i=1}^n \sum_{j=1}^n \sigma^{ij}(\boldsymbol{x}(t),t) \frac{\partial^2 f(\boldsymbol{x}(t),t)}{\partial x^i \partial x^j},
\end{aligned}
\tag{3.83}
$$

we see that without the last sum the formula is the ordinary differential (3.66), so that the correction due to Itô's integration is the last term in eq. (3.83).

Exercise 3.14 (Itô's formula in 1-D). Specialize Itô's formula (3.67) to the one-dimensional case: for a process $x(t)$ with differential $dx = a(t)\,dt + b(t)\,dw$, where $a(t), b(t) \in H_2[0,T]$ and a twice continuously differentiable function $f(x,t)$,

$$
\begin{aligned}
df(x(t),t) = & \left[\frac{\partial f(x(t),t)}{\partial t} + a(t) \frac{\partial f(x(t),t)}{\partial x} + \frac{1}{2} b^2(t) \frac{\partial^2 f(x(t),t)}{\partial x^2} \right] dt \\
& + b(t) \frac{\partial f(x(t),t)}{\partial x} \, dw(t).
\end{aligned}
\tag{3.84}
$$

□

Exercise 3.15. (Itô's formula as the chain rule).

(i) Apply Itô's formula (3.67) to the function $f(x_1, x_2) = x_1 x_2$ and obtain the rule (3.64) for differentiating a product.

(ii) Apply Itô's one-dimensional formula of Exercise 3.14 to the function $f(x) = e^x$. Obtain a differential equation for the function $y(t) = e^{\alpha w(t)}$.

(iii) Use the transformation $y = \log x$ to solve the linear stochastic differential equation

$$dx(t) = ax(t)\,dt + bx(t)\,dw(t), \quad x(0) = x_0. \tag{3.85}$$

Show that the solution cannot change sign. □

Exercise 3.16. (Applications to moments).

(i) Use the one-dimensional Itô formula to prove

$$\mathbb{E}e^{w(t)} = 1 + \frac{1}{2}\int_0^t \mathbb{E}e^{w(s)}\,ds = e^{t/2}.$$

(ii) Calculate the first and the second moments of $e^{aw(t)}$, $e^{iw(t)}$, $\sin aw(t)$, and $\cos aw(t)$, where a is a real constant. □

Exercise 3.17 (Rotation of white noise). Given two independent Brownian motions $w_1(t), w_2(t)$ and a process $x(t) \in H_2[0,T]$, define the processes $u_1(t), u_2(t)$ by their differentials

$$\begin{aligned} du_1(t) &= -\sin x(t)\,dw_1(t) + \cos x(t)\,dw_2(t) \\ du_2(t) &= \cos x(t)\,dw_1(t) + \sin x(t)\,dw_2(t). \end{aligned}$$

Show that $u_1(t)$ and $u_2(t)$ are independent Brownian motions. □

Exercise 3.18 (A stochastic equation for a martingale). Consider an adapted process $x(t)$ that satisfies the Novikov condition (3.41). Define the processes

$$y(t) = \int_0^t x(s)\,dw(s) - \frac{1}{2}\int_0^t x^2(s)\,ds, \quad z(t) = e^{y(t)}. \tag{3.86}$$

(i) Apply Itô's formula to the function $f(y) = e^y$ to show that

$$dz(t) = z(t)x(t)\,dw(t).$$

(ii) Show that $\mathbb{E}z(t) = 1$.

(iii) Show that $z(t)$ is a martingale. □

For processes defined by the Stratonovich integral,

$$x(t) = x_0 + \int_0^t a(s)\,ds + \int_0^t b(s)\,dw_S(s),$$

where $a, b \in H_2[0,T]$, we write

$$d_S x(t) = a(t)\,dt + b(t)\,d_S w(t). \tag{3.87}$$

Theorem 3.4.2 (The Stratonovich chain rule). *The chain rule for Stratonovich differentials is the usual rule* (3.66)*; that is,*

$$d_S f(x,t) = \frac{\partial f}{\partial x}\,d_S x + \frac{\partial f}{\partial t}\,dt = \left[\frac{\partial f}{\partial t} + a\,\frac{\partial f}{\partial x}\right] dt + b\frac{\partial f}{\partial x}\,d_S w. \tag{3.88}$$

Proof. First, we convert (3.87) to Itô's form by introducing the Wong–Zakai corrections,

$$dx(t) = \left(a(t) + \frac{1}{2}\frac{\partial b}{\partial w}\right) dt + b(t)\,dw(t). \tag{3.89}$$

If the dependence of $b(t)$ on $w(t)$ is expressed as

$$b(t) = B\left(x(t), t\right),$$

where $B(x,t)$ is a differentiable function in both variables, then the Wong–Zakai correction is found as follows.

$$\frac{\Delta b(t)}{\Delta w(t)} = \frac{\Delta B\left(x(t),t\right)}{\Delta x(t)}\frac{\Delta x(t)}{\Delta w(t)} = \frac{\Delta B\left(x(t),t\right)}{\Delta x(t)}\frac{a(t)\,\Delta t + b(t)\,\Delta w(t) + o(\Delta t)}{\Delta w(t)}$$

$$= \frac{\partial B\left(x(t),t\right)}{\partial x}\left[b(t) + a(t)O\left(\frac{\Delta t}{\Delta w(t)}\right)\right].$$

Note that

$$\Pr\left\{\left|\frac{\Delta t}{\Delta w}\right| > \varepsilon\right\} = \Pr\left\{|\Delta w| < \frac{\Delta t}{\varepsilon}\right\} = \frac{1}{\sqrt{2\pi\Delta t}}\int\limits_{-\Delta t/\varepsilon}^{\Delta t/\varepsilon} e^{-x^2/2\Delta t}\,dx$$

$$= \frac{1}{\sqrt{2\pi}}\int\limits_{-\sqrt{\Delta t}/\varepsilon}^{\sqrt{\Delta t}/\varepsilon} e^{-z^2/2}\,dz \to 0 \text{ as } \Delta t \to 0,$$

so that $\lim_{\Delta t\to 0} \Delta t/\Delta w(t) \stackrel{\Pr}{=} 0$. It follows that in this case the Wong–Zakai correction is

$$\frac{1}{2}\frac{\partial b}{\partial w} = \frac{1}{2}B\left(x,t\right)\frac{\partial B\left(x,t\right)}{\partial x}. \tag{3.90}$$

Next, from Itô's formula and eq. (3.89), we have

$$f(x(t),t) - f(x(t_0),t_0)$$

$$= \int\limits_{t_0}^{t}\left\{\frac{\partial f(x(s),s)}{\partial t} + \left[a(s) + \frac{1}{2}\frac{\partial b(s)}{\partial w(s)}\right]\frac{\partial f(x(s),s)}{\partial x} + \frac{1}{2}b^2(s)\frac{\partial^2 f(x(s),s)}{\partial x^2}\right\} ds$$

$$+ \int\limits_{t_0}^{t} b(s)\frac{\partial f(x(s),s)}{\partial x}\,dw(s).$$

Now, we convert the Itô integral into a Stratonovich integral using the Wong–Zakai correction,

$$\int_{t_0}^{t} b(s)\frac{\partial f(x(s),s)}{\partial x}\,dw(s) = \int_{t_0}^{t} b(s)\frac{\partial f(x(s),s)}{\partial x}\,dw_S(s) - \frac{1}{2}\int_{t_0}^{t} \frac{\partial}{\partial w(s)}\left[b(s)\frac{\partial f(x(s),s)}{\partial x}\right] ds. \tag{3.91}$$

Using the differentiation rule (3.90), we find that

$$\frac{\partial}{\partial w(t)}\left[b(t)\frac{\partial f(x(t),t)}{\partial x}\right] = \frac{\partial b(t)}{\partial w(t)}\frac{\partial f(x(t),t)}{\partial x} + \frac{\partial^2 f(x(t),t)}{\partial x^2}b^2(t), \tag{3.92}$$

so (3.91) gives

$$\begin{aligned}
&f(x(t),t) - f(x(t_0),t_0)\\
&= \int_{t_0}^{t}\left\{\frac{\partial f}{\partial t} + \left[a(s) + \frac{1}{2}\frac{\partial b(s)}{\partial w(s)}\right]\frac{\partial f}{\partial x} + \frac{1}{2}b^2(s)\frac{\partial^2 f}{\partial x^2}\right\} ds\\
&\quad + \int_{t_0}^{t} b(s)\frac{\partial f}{\partial x}\,dw_S(s) - \frac{1}{2}\int_{t_0}^{t}\left[\frac{\partial b(s)}{\partial w(s)}\frac{\partial f}{\partial x} + \frac{\partial^2 f}{\partial x^2}b^2(s)\right] ds\\
&= \int_{t_0}^{t}\left[f_t + a(s)f_x\right] ds + \int_{t_0}^{t} b(s)f_x\,dw_S(s),
\end{aligned}$$

where $f = f(x(s),s)$, as asserted. In differential form this is identical to (3.87). □

Exercise 3.19 (The differential of $\exp\{w(t)\}$). Set $x(t) = e^{w(t)}$. Show that $d_S x(t) = x(t)\,d_S w(t)$ and $dx(t) = x(t)\,d_I w(t) + \frac{1}{2}x(t)\,dt$. This can be done by power series expansion or by using Itô's formula. □

Exercise 3.20 (The stability of Itô and Stratonovich linear equations). Investigate the stability of the origin for solutions of the stochastic linear differential equation (3.85) if the differentials are interpreted in the Itô sense or in the Stratonovich sense, show that different stability criteria are obtained depending on the interpretation of the differentials in the equation. □

Chapter 4

Stochastic Differential Equations

Dynamics driven by white noise, often written as

$$d\boldsymbol{x} = \boldsymbol{a}(\boldsymbol{x}, t)\, dt + \boldsymbol{B}(\boldsymbol{x}, t)\, d\boldsymbol{w}, \quad \boldsymbol{x}(0) = \boldsymbol{x}_0, \tag{4.1}$$

or

$$\dot{\boldsymbol{x}} = \boldsymbol{a}(\boldsymbol{x}, t) + \boldsymbol{B}(\boldsymbol{x}, t)\, \dot{\boldsymbol{w}}, \quad \boldsymbol{x}(0) = \boldsymbol{x}_0,$$

is usually understood as the integral equation

$$\boldsymbol{x}(t) = \boldsymbol{x}(0) + \int_0^t \boldsymbol{a}(\boldsymbol{x}(s), s)\, ds + \int_0^t \boldsymbol{B}(\boldsymbol{x}(s), s)\, d\boldsymbol{w}(s), \tag{4.2}$$

where $\boldsymbol{a}(\boldsymbol{x}, t)$ and $\boldsymbol{B}(\boldsymbol{x}, t)$ are random coefficients, which can be interpreted in several different ways, depending on the interpretation of the stochastic integral in (4.2) as Itô, Stratonovich, backward, or otherwise. Different interpretations lead to very different solutions and to qualitative differences in the behavior of the solution. For example, a noisy dynamical system of the form (4.1) may be stable if the Itô integral is used in (4.2), but unstable if the Stratonovich or the backward integral is used instead. Different interpretations lead to different numerical schemes for the computer simulation of the equation. A different approach, based on path integrals, is given in Chapter 5.

In modeling stochastic dynamics with equations of the form (4.2), a key question arises of which of the possible interpretations is the right one to use. This question is particularly relevant if the noise is state-dependent; that is, if the coefficients $\boldsymbol{B}(\boldsymbol{x}, t)$ depend on $\boldsymbol{x}$. This situation is encountered in many different applications, for example, when the friction coefficient or the temperature in Langevin's equation are not constant. The answer to this question depends on the origin of the noise.

Z. Schuss, *Theory and Applications of Stochastic Processes: An Analytical Approach*, Applied Mathematical Sciences 170, DOI 10.1007/978-1-4419-1605-1_4,

The correlation-free white noise (or the nondifferentiable MBM), is an idealization of a physical process that may have finite though short correlation time (or differentiable trajectories). This is illustrated in Section 1.2, where the correlated velocity process of a free Brownian particle becomes white noise in the limit of large friction (or Section 1.3, where the displacement process becomes Brownian motion in that limit). The white noise approximation may originate in a model with discontinuous paths in the limit of small or large frequent jumps, and so on.

Thus, the choice of the integral in (4.2) is not arbitrary, but rather derives from the underlying more microscopic model and from the passage to the white noise limit. In certain situations this procedure leads to an Itô interpretation and in others to Stratonovich or other interpretations. The limiting procedures are described in Section 7.3. In this chapter, we consider the Itô and Stratonovich interpretations and their interrelationship. The backward interpretation is left as an exercise.

4.1 Itô and Stratonovich SDEs

First, we consider the one-dimensional version of eq. (4.1) and interpret it in the Itô sense as the output of an Euler numerical scheme of the form

$$x_E(t+\Delta t,\omega) = x_E(t,\omega) + a(x_E(t,\omega),\omega)\Delta t + b(x_E(t,\omega),\omega)\Delta w(t,\omega) \quad (4.3)$$

in the limit $\Delta t \to 0$. To each realization of the MBM, $w(t,\omega)$, constructed numerically, for example, by any of the methods of Section 2.3, eq. (4.3) assigns a realization $x_E(t,\omega)$ of the solution at grid points. The right-hand side of eq. (4.3) can assume any value in $\mathbb{R}$, so that $x_E(t,\omega)$ can assume any value at every time t, because $\Delta w(t,\omega) = w(t+\Delta t,\omega) - w(t,\omega)$ is a Gaussian random variable. This implies that $a(x,t,\omega)$ and $b(x,t,\omega)$ have to be defined for all $x \in \mathbb{R}$. If for each $x \in \mathbb{R}$ the random coefficients $a(x,t,\omega)$ and $b(x,t,\omega)$ are adapted processes, say of class $H_2[0,T]$ for all $T > 0$, the output process $x_E(t,\omega)$ is also an adapted process.

The output process at grid times $t_j = j\Delta t$, given by

$$x_E(t_j,\omega) = x_0 + \sum_{k=0}^{j-1} \left[a(x_E(t_k),t_k,\omega)\Delta t + b(x_E(t_k),t_k,\omega)\Delta w(t_k,\omega)\right], \quad (4.4)$$

has the form of two integral sums, one for the integral $\int_0^t a(x(s,\omega),s,\omega)\,ds$ and the other of the Itô integral $\int_0^t b(x(s,\omega),s,\omega)\,dw(s,\omega)$, where $x(t,\omega) = \lim_{\Delta t\to 0} x(t_j,\omega)$ for $t_j \to t$, if the limit exists in some sense.

If the coefficients $a(x,t,\omega)$ and $b(x,t,\omega)$ are adapted processes, (of class $H_2[0,T]$ for all $T > 0$), equation (4.1) is written in the Itô form

$$dx = a(x,t,\omega)\,dt + b(x,t,\omega)\,dw(t,\omega), \quad x(0,\omega) = x_0, \quad (4.5)$$

or as an equivalent Itô integral equation

$$x(t,\omega) = x_0 + \int_0^t a(x(s,\omega),s,\omega)\,ds + \int_0^t b(x(s,\omega),s,\omega)\,dw(s,\omega). \tag{4.6}$$

The initial condition x_0 is assumed independent of $w(t)$.

There are several different definitions of a solution to the stochastic differential equation (SDE) (4.5), including strong, weak, a solution to the martingale problem, path integral interpretation (see Chapter 5), and so on. Similarly, there are several different notions of uniqueness, including uniqueness in the strong sense, pathwise uniqueness, and uniqueness in probability law. For the definitions and relationship among the different definitions see [153], [115]. We consider here only strong solutions (abbreviated as *solutions*) of (4.5).

Definition 4.1.1 (Solution of a SDE). *A stochastic process $x(t,\omega)$ is a* solution *of the initial value problem* (4.5) *in the Itô sense if*
I. $x(t,\omega) \in H_2[0,T]$ for all $T > 0$.
II. Equation (4.6) *holds for almost all $\omega \in \Omega$.*

We assume that the coefficients $a(x,t,\omega)$ and $b(x,t,\omega)$ satisfy the *uniform Lipschitz condition*; that is, there exists a constant K such that

$$|a(x,t,\omega) - a(y,t,\omega)| + |b(x,t,\omega) - b(y,t,\omega)| \le K|x-y| \tag{4.7}$$

for all $x, y \in \mathbb{R},\ t \ge 0$, and $\omega \in \Omega$.

Theorem 4.1.1 (Existence and uniqueness). *If $a(x,t,\omega)$ and $b(x,t,\omega)$ satisfy the Lipschitz condition* (4.7)*, uniformly for all x, t, and for almost all $\omega \in \Omega$, then there exists a unique solution to the initial value problem* (4.5)*. Its trajectories are continuous with probability* 1.

The proof of existence and uniqueness, as well as analytical properties of solutions, are discussed in Section 4.7 (see also the proof of Skorokhod's theorem 5.1.1). For any domain $D \subset \mathbb{R}^n$ such that $x_0 \in D$, we denote the first exit time from D by

$$\tau_D(\omega) = \inf\{t > 0 \,|\, x(t,\omega) \notin D\}. \tag{4.8}$$

It can be shown from the proof of the existence theorem that $\tau_D(\omega)$ is a Markov time.

If the coefficients $a(x,t,\omega)$ and $b(x,t,\omega)$ are not defined for all x, but only in some domain D, the definition of the solution has to be modified. First, we need the following. [84]

Theorem 4.1.2 (Localization principle). *Assume that for $i = 1,2$ the coefficients $a_i(x,t,\omega)$ and $b_i(x,t,\omega)$ satisfy the Lipschitz condition uniformly for all x, t,*

$$a_1(x,t,\omega) = a_2(x,t,\omega), \quad b_1(x,t,\omega) = b_2(x,t,\omega)$$

for all $x \in D$, $\omega \in D$, *and* $x_0 \in D$. *Let* $x_1(t,\omega)$ *and* $x_2(t,\omega)$ *be the solutions of*

$$dx(t,\omega) = a_i(x,t,\omega)\,dt + b_i(x,t,\omega)\,dw(t,\omega), \quad x_i(0,\omega) = x_0, \quad i = 1,2,$$

respectively, and $\tau_1(\omega)$, $\tau_2(\omega)$ *be their first exit times from* D. *Then* $\tau_1(\omega) = \tau_2(\omega)$ *with probability* 1, *and* $x_1(t,\omega) = x_2(t,\omega)$ *for all* $t < \tau_1(\omega)$ *and almost all* $\omega \in D$.

The localization theorem can be used to define solutions to Itô equations in finite domains. Assume that $a(x,t,\omega)$ and $b(x,t,\omega)$ are defined only for $x \in D$ and satisfy there the Lipschitz condition and can be extended to all x as uniformly Lipschitz functions. Then solutions are defined for the extended equations. The localization principle ensures that all solutions, corresponding to different extensions, are the same for all $t < \tau_D(\omega)$.

Exercise 4.1 (Proof of localization). Prove the localization principle (see the proof of uniqueness in Section 4.7). □

Exercise 4.2 (Growth estimate). Show that if $\mathbb{E}x_0^{2m} < \infty$, then the solution of (4.6) satisfies the inequality

$$\mathbb{E}x^{2m}(t,\omega) \leq \mathbb{E}\left(1 + x_0^{2m}\right)e^{Ct},$$

where C is a constant. (HINT: Use Itô's formula). □

Exercise 4.3 (Modulus of continuity). Show that

$$\mathbb{E}|x(t,\omega) - x(0)|^{2m} \leq C_1\mathbb{E}\left(1 + |x_0|^{2m}\right)e^{C_2 t}t^m. \tag{4.9}$$

where C_1 is another constant. □

Exercise 4.4 (Test of uniqueness). For which values of α has the equation $dx = |x|^\alpha\,dw$ a unique solution satisfying the initial condition $x(0) = 0$? □

Exercise 4.5 (Example of nonuniqueness). For any $T \geq 0$, denote by $\tau_T(\omega)$ the first passage time of the MBM to the origin after time T; that is, $\tau_T(\omega) = \inf\{s \geq T \,|\, w(s,\omega) = 0\}$. Show that the stochastic equation $dx = 3x^{1/3}\,dt + 3x^{2/3}\,dw$, with the initial condition $x(0) = 0$, has infinitely (uncountably) many solutions of the form

$$x_T(t,\omega) = \begin{cases} 0 & \text{for } 0 \leq t < \tau_T(\omega) \\ w^3(t,\omega) & \text{for } t \geq \tau_T(\omega). \end{cases}$$

This example is due to Itô and Watanabe. □

Next, we consider a system of Itô equations of the form

$$dx^i = a^i(\boldsymbol{x},t)\,dt + \sum_{j=1}^{m} b^{ij}(\boldsymbol{x},t)\,dw^j, \quad x^i(0) = x_0^i, \quad i = 1,2,\,\ldots,\,n, \tag{4.10}$$

where $w^j(t)$ are independent MBMs and $\boldsymbol{x} = (x^1, x^2, \,\ldots,\, x^n)$. If the coefficients satisfy a uniform Lipschitz condition, the proofs of the existence and uniqueness theorem and of the localization principle are generalized in a straightforward manner to include the case of systems of the form (4.10).

Exercise 4.6 (Existence and uniqueness for (4.10)). Generalize the existence and uniqueness theorem and the localization principle, given in Section 4.7, for the system (4.10). □

The stochastic differential equation (4.1) can be interpreted in the Stratonovich sense by making the integral in (4.2) a Stratonovich integral; that is, by writing it in the form

$$x(t,\omega)=x_0+\int_0^t a(x(s,\omega),s,\omega)\,ds+\int_0^t b(x(s,\omega),s,\omega)\,d_Sw(s,\omega). \tag{4.11}$$

The corresponding differential notation is

$$d_Sx=a(x,t)\,dt+b(x,t)\,d_Sw,\quad x(0)=x_0. \tag{4.12}$$

Existence and uniqueness theorems for (4.12) are proved in much the same way as for the Itô version. To make the proofs go through the Stratonovich integrals are converted into the equivalent Itô form by means of the Wong–Zakai corrections, as described in Chapter 3. More specifically, the Itô equation equivalent to (4.12) is obtained by applying eq. (3.90) to $b(x(t),t)$. The resulting Itô equation is given by

$$dx=\left[a(x,t)+\frac{1}{2}\,b(x,t)\frac{\partial}{\partial x}b(x,t)\right]\,dt+b(x,t)\,dw. \tag{4.13}$$

To convert in the other direction; that is, from Itô form to Stratonovich form, the Wong–Zakai correction is subtracted. Thus, the Itô equation $dx=a(x,t)\,dt+b(x,t)\,dw$ is converted to the equivalent Stratonovich form

$$d_Sx=\left[a(x,t)-\frac{1}{2}b(x,t)b_x(x,t)\right]\,dt+b(x,t)\,d_Sw.$$

In d dimensions the Stratonovich system

$$\boldsymbol{x}(t)=\boldsymbol{x}(0)+\int_0^t \boldsymbol{a}(\boldsymbol{x}(s),s)\,ds+\int_0^t \boldsymbol{b}(\boldsymbol{x}(s),s)\,d\boldsymbol{w}_S(s) \tag{4.14}$$

is converted to Itô form by the Wong–Zakai corrections

$$dx^i(t)=\left[a^i(\boldsymbol{x}(t),t)+\frac{1}{2}\sum_{k=1}^n\sum_{j=1}^m b^{i,j}(\boldsymbol{x}(t),t)\frac{\partial}{\partial x^k}b^{ij}(\boldsymbol{x}(t),t)\right]\,dt$$
$$+\sum_{j=1}^m b^{kj}(\boldsymbol{x}(t),t)\,dw^j. \tag{4.15}$$

Exercise 4.7 (Backward stochastic differential equations). Develop a theory of backward stochastic differential equations. □

4.2 Transformations of Itô equations

Differential equations are often transformed by changing the independent variable or the dependent variable, or both. Transformations of differential equations are used for many different purposes, for example, to change the scales of time and space so comparisons of different orders of magnitude can be done, to simplify the equations, to change coordinates, and so on. In transforming stochastic differential equations Itô's formula plays a central role, as shown below.

In changing the independent variable in Itô equations the Brownian scaling laws of Exercise 2.4 have to be borne in mind. Thus, changing the time scale $t = \alpha s$, where α is a constant, transforms the Brownian motion and its differential as follows.

$$w(t) = w(\alpha s) = \sqrt{\alpha}\left[\frac{1}{\sqrt{\alpha}}w(\alpha s)\right] = \sqrt{\alpha}w_\alpha(s), \tag{4.16}$$

where $w_\alpha(s)$ is Brownian motion. The differential $dw(t)$ is expressed in terms of the differential $dw_\alpha(s)$ as

$$d_t w(t) = d_t w(\alpha s) = \sqrt{\alpha}\, d_s w_\alpha(s). \tag{4.17}$$

Setting $x(t) = x_\alpha(s)$ the integral equation (4.6) becomes

$$x_\alpha(s) = x_0 + \alpha \int_0^s a(x_\alpha(u), \alpha u)\, du + \sqrt{\alpha} \int_0^s b(x_\alpha(u), \alpha u)\, dw_\alpha(u). \tag{4.18}$$

The Itô differential equation (4.5) is therefore transformed into

$$dx_\alpha(s) = \alpha\, a(x_\alpha(s), \alpha s)\, ds + \sqrt{\alpha}\, b(x_\alpha(s), \alpha s)\, dw_\alpha(s). \tag{4.19}$$

Exercise 4.8 (Change of Wiener measures). The space Ω of Brownian trajectories remains unchanged under the change of scale, however, the same sets of trajectories acquire different probabilities under the Wiener measure corresponding to $w_\alpha(s)$. What is the relationship between the two Wiener measures? □

If the dependent variable in an Itô equation, $x(t)$, is transformed to $y(t) = f(x(t), t)$, the differential equation for $y(t)$ is found from Itô's formula. We consider here two applications of this principle, first, to reducing equations to explicitly solvable form and second, to change coordinates.

Equations of the form

$$dx(t) = a(t)\, dt + b(t)\, dw(t), \tag{4.20}$$

where $a, b, \in H_2[0, T]$ can be solved explicitly as

$$x(t) = x(s) + \int_s^t a(u)\, du + \int_s^t b(u)\, dw(u). \tag{4.21}$$

If $x(s) = x$, we denote the solution $x_{x,s}(t)$. When the coefficients $a(t)$ and $b(t)$ are deterministic functions, the solution is a Gaussian process with conditional mean

$$\bar{x}_{x,s}(t) = \mathbb{E}[x(t) \,|\, x(s) = x] = x + \int\limits_s^t a(u)\,du \tag{4.22}$$

and conditional variance

$$\begin{aligned} \operatorname{Var} x_{x,s}(t) &= \mathbb{E}\left\{[x(t) - \bar{x}_{x,s}(t)]^2 \,|\, x(s) = x\right\} \\ &= \mathbb{E}\left[\left(\int\limits_s^t b(u)\,dw(u)\right)^2 \,|\, x(s) = x\right] = \int\limits_s^t b^2(u)\,du. \end{aligned} \tag{4.23}$$

The transition pdf is therefore given by

$$\begin{aligned} &p\,(y, t \,|\, x, s) \\ &= \left(2\pi \int\limits_s^t b^2(u)\,du\right)^{-1/2} \exp\left\{-\left[y - x - \int\limits_s^t a(u)\,du\right]^2 \Bigg/ 2\int\limits_s^t b^2(u)\,du\right\}. \end{aligned} \tag{4.24}$$

It follows that equations that can be transformed into the form (4.20) can be solved explicitly. It should be borne in mind, however, that the usefulness of explicit solutions is quite limited. For example, what can be learned from the explicit solution (4.21) and its explicit conditional transition probability density function (4.24) about the first passage time of the solution (4.21) to a given point?

Next, we consider equations that can be reduced to the form (4.20). We begin with general a linear equation of the form

$$dx(t) = [\alpha(t) + \beta(t)x]\,dt + [\gamma(t) + \delta(t)x]\,dw(t), \tag{4.25}$$

where the coefficients $\alpha(t), \beta(t), \gamma(t)$, and $\delta(t)$ are in $H_2[0, T]$. First, we consider the homogeneous case $\alpha(t) = \gamma(t) = 0$; that is,

$$dx(t) = \beta(t)x\,dt + \delta(t)x\,dw(t). \tag{4.26}$$

The transformation $y = \log x$ and Itô's formula give

$$dy(t) = \left[\beta(t) - \frac{1}{2}\delta^2(t)\right]\,dt + \delta(t)\,dw(t), \tag{4.27}$$

which is of the form (4.20) and can be solved explicitly. If the coefficients are deterministic and $\delta(t) \neq 0$, then $x(t)$ is log-normal. We denote the solution of the homogeneous equation (4.26) by $x_h(t)$. The inhomogeneous case reduced to the homogeneous case by the substitution $x(t) = x_h(t)\xi(t)$.

The method of changing variables can be applied to reduce equation (4.5) to the form (4.20). A sufficient condition for reducibility can be found as follows. The change of variables $y(t) = f(x(t), t)$ and Itô's formula transforms the equation into

$$dy(t) = \left[f_t(x(t), t) + a(x(t), t) f_x(x(t), t) + \frac{1}{2} b^2(x(t), t) f_{xx}(x(t), t) \right] dt$$
$$+ b(x(t), t) f_x(x(t), t)\, dw(t), \tag{4.28}$$

where subscripts denote partial derivatives. Equation (4.28) is of the form (4.20) if

$$f_t(x, t) + a(x, t) f_x(x, t) + \frac{1}{2} b^2(x, t) f_{xx}(x, t) = A(t) \tag{4.29}$$
$$b(x, t) f_x(x, t) = B(t), \tag{4.30}$$

where $A(t), B(t)$ are independent of x. Equation (4.30) gives

$$f_x(x, t) = \frac{B(t)}{b(x, t)} \tag{4.31}$$
$$f_{xx}(x, t) = -\, B(t) \frac{b_x(x, t)}{b^2(x, t)} \tag{4.32}$$
$$f_{tx}(x, t) = \frac{\dot{B}(t) b(x, t) - B(t) b_t(x, t)}{b^2(x, t)}. \tag{4.33}$$

Differentiating eq. (4.29) with respect to x gives

$$f_{tx}(x, t) = -\left[a(x, t) f_x(x, t) + \frac{1}{2} b^2(x, t) f_{xx}(x, t) \right]_x,$$

which together with (4.31)–(4.33) gives

$$\frac{\dot{B}(t) b(x, t) - B(t) b_t(x, t)}{b^2(x, t)} = -\left[a(x, t) \frac{B(t)}{b(x, t)} - \frac{1}{2} B(t) b_x(x, t) \right]_x.$$

Dividing out $B(t)$ and multiplying through by $b(x, t)$ gives

$$\frac{\dot{B}(t)}{B(t)} = b(x, t) \left\{ \frac{b_t(x, t)}{b^2(x, t)} - \left[\frac{a(x, t)}{b(x, t)} \right]_x + \frac{1}{2} b_{xx}(x, t) \right\}. \tag{4.34}$$

Thus the reducibility condition is

$$\left\{ b(x, t) \left[\frac{b_t(x, t)}{b^2(x, t)} - \left(\frac{a(x, t)}{b(x, t)} \right)_x + \frac{1}{2} b_{xx}(x, t) \right] \right\}_x = 0. \tag{4.35}$$

If (4.35) is satisfied, the right-hand side of (4.34) is independent of x and $B(t)$ can be recovered by integration. The function $f(x, t)$ is then recovered from (4.31). Now, because eq. (4.34) is equivalent to

$$\left[f_t(x, t) + a(x, t) f_x(x, t) + \frac{1}{2} b^2(x, t) f_{xx}(x, t) \right]_x = 0,$$

the expression in brackets is independent of x and eq. (4.29) defines $A(t)$.

For example, if the coefficients are independent of t, then the reducibility condition (4.35) is

$$\left\{b(x,t)\left[-\left(\frac{a(x)}{b(x)}\right)' + \frac{1}{2}b_{xx}(x)\right]\right\}_x = 0$$

and eq. (4.34) becomes $\dot{B}(t)/B(t) = C$, where C is a constant, so that $B(t) = e^{Ct}$ and eq. (4.31) gives $f(x,t) = e^{Ct}\int^x [b(x)]^{-1}\,dx$. Finally, eq. (4.29) gives

$$A(t) = e^{Ct}\left[C\int^x \frac{dx}{b(x)} + \frac{a(x)}{b(x)} - \frac{1}{2}b'(x)\right].$$

Exercise 4.9. (Linear SDEs)

(i) Reduce the inhomogeneous linear equation (4.25) to a homogeneous equation by the indicated substitution and write down the explicit solution.

(ii) Calculate the transition pdf for the case that $\alpha(t), \beta(t), \gamma(t)$, and $\delta(t)$ are deterministic functions.

(iii) Consider the case $\delta(t) = 0$ separately (this is the case of the Ornstein–Uhlenbeck process). □

Exercise 4.10 (Continuations). Find and solve all Itô equations with coefficients that are independent of t and $b(x) = x$, which can be transformed into a linear equation. □

The change of coordinates by a transformation of the form $\boldsymbol{y} = \boldsymbol{f}(\boldsymbol{x})$ has to be done by applying Itô's formula to each component of $\boldsymbol{f}(\boldsymbol{x})$. This is illustrated by the conversion of a system of equations in the plane from Cartesian to polar coordinates. We consider the example of the *Rayleigh process*, which is often used as a model of a random electric field. The two components of the field are assumed independent identically distributed Ornstein–Uhlenbeck processes (see Section 2.4). Specifically, consider a two-dimensional field $\boldsymbol{E}(t) = E_1(t) + iE_2(t)$ and assume that each component satisfies the stochastic differential equation $dE_j(t) = -\gamma E_j(t)dt + \varepsilon\, dw_j(t)$, where $j = 1,2$ and $w_j(t)$ are independent Brownian motions. Converting to polar coordinates, $E_1(t) = r(t)\cos\theta(t)$ and $E_2(t) = r(t)\sin\theta(t)$, or $r(t) = \sqrt{E_1^2(t) + E_2^2(t)}$ and $\theta(t) = \arctan(E_2(t)/E_1(t))$, we obtain from Itô's formula that

$$dr(t) = \left(-\gamma r(t) + \frac{\varepsilon^2}{2r(t)}\right)dt + \varepsilon\left[\cos\theta(t)\,dw_1(t) + \sin\theta(t)\,dw_2(t)\right]. \quad (4.36)$$

Recalling from Exercise 3.17 that the solutions of the equations

$$\begin{aligned} dw_r(t) &= \cos\theta(t)\,dw_1(t) + \sin\theta(t)\,dw_2(t) \\ dw_\theta(t) &= -\sin\theta(t)\,dw_1(t) + \cos\theta(t)\,dw_2(t) \end{aligned}$$

are independent Brownian motions, we write eq. (4.36) in the form

$$dr(t) = \left[-\gamma r(t) + \frac{\varepsilon^2}{2r(t)}\right]dt + \varepsilon\,dw_r(t).$$

The equation for the phase $\theta(t)$ is found from Itô's formula for the differential of the transformation $d\theta(t) = \arctan(E_2(t)/E_1(t))$. The resulting equation is $d\theta(t) = [\varepsilon/r(t)]\, dw_\theta(t)$. Note that the radial equation is independent of $\theta(t)$ so that it can be solved first and then used in the phase equation.

Exercise 4.11 (Conversion to an integral equation). Assume that $b(x,t,\omega)$ is a nonnegative continuously differentiable function of x and t and that $1/b(x,t,\omega)$ is an integrable function of x on every finite interval.
(i) Use Itô's formula for the function $f(x,t,\omega) = \int_0^x [b(s,t,\omega)]^{-1}\, ds$ to show that the integral equation (4.6) can be converted to an equation of the form $y(t,\omega) = y(0,\omega) + \int_0^t A(y,s,\omega)\, dt + w(t,\omega)$.
(ii) solve explicitly the equation $dx = \left(\sqrt{1+x^2} + \frac{1}{2}x\right)\, dt + \sqrt{1+x^2}\, dw$. □

Exercise 4.12 (Rayleigh process). Derive the three-dimensional version of the Rayleigh process in spherical coordinates. □

4.3 Solutions of SDEs are Markovian

The Markov property of the Brownian motion was discussed in Section 2.4.1. It was shown that the transition probability density function of a Markov process, $p\,(y,t\,|\,x,s)$, can be expressed in terms of the transition probabilities at intermediate times by the Chapman–Kolmogorov equation (2.59); that is,

$$
\begin{aligned}
p\,(y,t\,|\,x,s) &= \int p\,(y,t,z,\tau\,|\,x,s)\, dz = \int p\,(y,t\,|\,z,\tau\,|\,x,s)p\,(z,\tau\,|\,x,s)\, dz \\
&= \int p\,(y,t\,|\,z,\tau)p\,(z,\tau\,|\,x,s)\, dz. \qquad (4.37)
\end{aligned}
$$

To show that the solution of the Itô SDE (4.5) is a Markov process, we note that for $t > s$,

$$
x(t) = x(s) + \int_s^t a(x(u),u)\, du + \int_s^t b(x(u),u)\, dw(u), \qquad (4.38)
$$

the existence and uniqueness theorem asserts that the initial condition $x(s)$ determines the solution of the Itô integral equation (4.38) uniquely. The solution in the interval $[s,t]$ depends only on $x(s)$ and on a, b, and the increments of w in this interval, because $a, b \in H[0,T]$ and dw is a forward difference of the Brownian motion. It follows from eq. (4.38) that for $t > s > s_1 > \cdots > s_n$,

$$
\Pr\{x(t) < x \mid x(s) = x_0,\ \ldots,\ x(s_n) = x_n\} = \Pr\{x(t) < x \mid x(s) = x_0\},
$$

which means that $x(t)$ is a Markov process.

Definition 4.3.1 (Diffusion process in $\mathbb{R}$). *A one-dimensional Markov process $x(t)$is called a* diffusion process with (deterministic) drift $a(x,t)$ *and* (deterministic) diffusion coefficient $b^2(x,t)$ *if it has continuous trajectories,*

$$\lim_{\Delta t\to 0}\frac{1}{\Delta t}\mathbb{E}\left\{x(t+\Delta t)-x(t)\mid x(t)=x\right\}=a(x,t) \tag{4.39}$$

$$\lim_{\Delta t\to 0}\frac{1}{\Delta t}\mathbb{E}\left\{[x(t+\Delta t)-x(t)]^{\,2}\mid x(t)=x\right\}=b^2(x,t), \tag{4.40}$$

and for some $\delta>0$

$$\lim_{\Delta t\to 0}\frac{1}{\Delta t}\mathbb{E}\left\{[x(t+\Delta t)-x(t)]^{\,2+\delta}\mid x(t)=x\right\}=0. \tag{4.41}$$

Theorem 4.3.1 (SDEs and diffusions). *Solutions of the Itô SDE* (4.5) *are diffusion processes.*

Proof. We show that solutions of the Itô SDE (4.5) satisfy the conditions of Definitions 4.3.1 and 4.3.2, respectively. Indeed, according to eq. (4.38),

$$x(t+\Delta t)=x(t)+\int\limits_t^{t+\Delta t}a(x(u),u)\,du+\int\limits_t^{t+\Delta t}b(x(u),u)\,dw(u)$$

hence, using the fact that the mean value of an Itô integral vanishes, we obtain

$$\begin{aligned}
&\lim_{\Delta t\to 0}\frac{1}{\Delta t}\mathbb{E}\left\{x(t+\Delta t)-x(t)\mid x(t)=x\right\}\\
=&\lim_{\Delta t\to 0}\frac{1}{\Delta t}\mathbb{E}\left\{\int\limits_t^{t+\Delta t}a(x(u),u)\,du+\int\limits_t^{t+\Delta t}b(x(u),u)\,dw(u)\mid x(t)=x\right\}\\
=&\lim_{\Delta t\to 0}\frac{1}{\Delta t}\mathbb{E}\left\{\int\limits_t^{t+\Delta t}a(x(u),u)\,du\mid x(t)=x\right\}=a(x,t),
\end{aligned}$$

due to the continuity of the trajectory and of $a(x,t)$. Thus eq. (4.39) is satisfied. Next, we have the identity

$$\begin{aligned}
\mathbb{E}\left\{[x(t+\Delta t)-x(t)]^{\,2}\mid x(t)=x\right\}=&\,\mathbb{E}\left[x^2(t+\Delta t)\,|\,x(t)=x\right]\\
&-2x\mathbb{E}\left[x(t+\Delta t)\,|\,x(t)=x\right]+x^2.
\end{aligned}\tag{4.42}$$

We have obtained so far that $\mathbb{E}\left[x(t+\Delta t)\,|\,x(t)=x\right]=x+a(x,t)\Delta t+o(\Delta t)$, so that

$$-2x\mathbb{E}\left[x(t+\Delta t)\,|\,x(t)=x\right]+x^2=x^2-2x^2-2xa(x,t)\Delta t+o(\Delta t). \tag{4.43}$$

Using Itô's formula with the function $f(x)=x^2$, we write

$$dx^2(t)=\left[2xa(x,t)+\frac{1}{2}(2b^2(x,t))\right]dt+2xb(x,t)\,dw,$$

or

$$x^2(t+\Delta t) = x^2(t) + \int_t^{t+\Delta t} \left[2x(u)a(x(u),u) + b^2(x(u),u)\right] du + \int_t^{t+\Delta t} 2x(u)b(x(u),u)\, dw(u).$$

Thus, using again the fact that the mean value of an Itô integral vanishes, we obtain

$$\begin{aligned} &\mathbb{E}\left[x^2(t+\Delta t) \,|\, x(t) = x\right] \\ &= x^2 + \mathbb{E}\left\{\int_t^{t+\Delta t} \left[2x(u)a(x(u),u) + b^2(x(u),u)\right] du \,|\, x(t) = x\right\} \\ &+ \mathbb{E}\left\{\int_t^{t+\Delta t} 2x(u)b(x(u),u)\, dw(u) |\, x(t) = x\right\} \\ &= x^2 + 2xa(x,t)\Delta t + b^2(x,t)\Delta t + o(\Delta t). \end{aligned} \tag{4.44}$$

Now, using eqs. (4.42) and (4.43) in eq. (4.44), we obtain

$$\mathbb{E}\left\{\left[x(t+\Delta t) - x(t)\right]^2 \mid x(t) = x\right\} = b^2(x,t)\Delta t + o(\Delta t),$$

hence eq. (4.40) is satisfied. Finally, using Itô's equation for the function $f(x) = x^4$, we find that eq. (4.41) is satisfied with $\delta = 2$.

Exercise 4.13 (Detail of the proof). Use inequality (4.9) in Exercise 4.3 to prove the last statement. □

Definition 4.3.2 (Diffusion process in $\mathbb{R}^d$). *A d-dimensional Markov process $\boldsymbol{x}(t)$ is called a* diffusion process with (deterministic) drift vector $\boldsymbol{a}(\boldsymbol{x},t)$ *and* (deterministic) diffusion matrix $\boldsymbol{\sigma}(\boldsymbol{x},t)$ *if it has continuous trajectories,*

$$\begin{aligned} &\lim_{\Delta t \to 0} \frac{1}{\Delta t} \mathbb{E}\left\{\boldsymbol{x}(t+\Delta t) - \boldsymbol{x}(t) \,|\, \boldsymbol{x}(t) = \boldsymbol{x}\right\} = \boldsymbol{a}(\boldsymbol{x},t) \\ &\lim_{\Delta t \to 0} \frac{1}{\Delta t} \mathbb{E}\left\{\left[x^i(t+\Delta t) - x^i(t)\right]\left[x^j(t+\Delta t) - x^j(t)\right] \,|\, \boldsymbol{x}(t) = \boldsymbol{x}\right\} \\ &= \sigma^{ij}(\boldsymbol{x},t) \end{aligned} \tag{4.45}$$

for $i, j = 1, 2, \ldots, d$, and for some $\delta > 0$

$$\lim_{\Delta t \to 0} \frac{1}{\Delta t} \mathbb{E}\left\{|\boldsymbol{x}(t+\Delta t) - \boldsymbol{x}(t)|^{\,2+\delta} \mid \boldsymbol{x}(t) = \boldsymbol{x}\right\} = 0.$$

Now, consider a system of Itô SDEs

$$d\boldsymbol{x}(t) = \boldsymbol{a}(\boldsymbol{x}, t)\, dt + \boldsymbol{B}(\boldsymbol{x}, t)\, d\boldsymbol{w}, \quad \boldsymbol{x}(0) = \boldsymbol{x}_0, \tag{4.46}$$

where

$$\begin{aligned}
\boldsymbol{x}(t) &= \left(x^1(t), \ldots, x^d(t)\right)^T \\
\boldsymbol{a}(\boldsymbol{x}, t) &= \left[a^1\left(x^1(t), \ldots, x^d(t)\right), \ldots, a^d\left(x^1(t), \ldots, x^d(t)\right)\right]^T \\
\boldsymbol{B}(\boldsymbol{x}, t) &= \left\{b^{ij}\left(x^1(t), \ldots, x^d(t)\right)\right\}_{i \le d, j \le m} \\
\boldsymbol{w}(t) &= \left(w^1(t), \ldots, w^m(t)\right)^T,
\end{aligned}$$

$w^i(t)$ are independent Brownian motions, $a^i, b^{ij} \in H[0, T]$, and $\boldsymbol{x}_0$ is independent of $\boldsymbol{w}(t)$ and a^i, b^{ij}. Also in this case the existence and uniqueness theorem implies that the solution is a d-dimensional Markov process with continuous trajectories and that it is a diffusion process with drift vector $\boldsymbol{a}(\boldsymbol{x}, t)$ and diffusion matrix $\boldsymbol{\sigma}(\boldsymbol{x}, t) = \frac{1}{2}\boldsymbol{B}(\boldsymbol{x}, t)\boldsymbol{B}^T(\boldsymbol{x}, t)$. □

Exercise 4.14 (Detail of the proof). Prove the last statement. □

Also a partial converse is true: assume that $\boldsymbol{x}(t)$ is a diffusion process with (deterministic) drift vector $\boldsymbol{a}(\boldsymbol{x}, t)$ and (deterministic) diffusion matrix $\boldsymbol{\sigma}(\boldsymbol{x}, t)$. If $\boldsymbol{a}(\boldsymbol{x}, t)$ is a uniformly Lipschitz continuous vector and $\boldsymbol{\sigma}(\boldsymbol{x}, t)$ is a uniformly Lipschitz continuous strictly positive definite matrix, then there exists a matrix uniformly Lipschitz continuous $\boldsymbol{B}(\boldsymbol{x}, t)$ and a Brownian motion $\boldsymbol{w}(t)$ such that $\boldsymbol{x}(t)$ is a solution of eq. (4.46) (see, e.g., [115]).

4.4 Stochastic and partial differential equations

Many useful functionals of solutions of stochastic differential equations, such as the transition probability density function, conditional and weighted expectations, functionals of the first passage times, escape probabilities from a given domain, and others, can be found by solving *deterministic* partial differential equations. These include Kolmogorov's representation formulas, the Andronov–Vitt–Pontryagin equation for the expected first passage time, the Feynman–Kac formula for the transition pdf when trajectories can be terminated at random times, and so on. These partial differential equations reflect the continuum macroscopic properties of the underlying stochastic dynamics of the individual trajectories.

Throughout this section $\boldsymbol{x}_{\boldsymbol{x},\, s}(t)$ with $t > s$ denotes the solution of the Itô system

$$d\boldsymbol{x}(t) = \boldsymbol{a}(\boldsymbol{x}(t), t)\, dt + \boldsymbol{B}(\boldsymbol{x}(t), t)\, d\boldsymbol{w}(t), \quad \boldsymbol{x}(s) = \boldsymbol{x}, \tag{4.47}$$

where $\boldsymbol{a}(\boldsymbol{x}, t) : \mathbb{R}^d \times [0, T] \mapsto \mathbb{R}^d$, $\boldsymbol{B}(\boldsymbol{x}, t) : \mathbb{R}^d \times [0, T] \mapsto \mathbb{M}_{n,m}$, and $\boldsymbol{w}(t)$ is m-dimensional Brownian motion. We assume that $\boldsymbol{a}(\boldsymbol{x}, t)$ and $\boldsymbol{B}(\boldsymbol{x}, t)$ satisfy the conditions of the existence and uniqueness theorem. The notation (1.56) and (3.70) is used.

Consider the Itô system (4.47) and the corresponding backward Kolmogorov operator (3.68)

$$L^*_{\boldsymbol{x}} u(\boldsymbol{x}, s) = \boldsymbol{a}(\boldsymbol{x}, s) \cdot \nabla u(\boldsymbol{x}, s) + \sum_{i=1}^{d}\sum_{j=1}^{d} \sigma^{ij}(\boldsymbol{x}, s) \frac{\partial^2 u(\boldsymbol{x}, s)}{\partial x^i x^j}. \tag{4.48}$$

Assume the function $u(\boldsymbol{x}, s, t)$ satisfies the backward parabolic equation

$$\frac{\partial u(\boldsymbol{x}, s, t)}{\partial s} + L^*_{\boldsymbol{x}} u(\boldsymbol{x}, s, t) = 0 \text{ for } t > s \tag{4.49}$$

with the terminal value

$$\lim_{s \uparrow t} u(\boldsymbol{x}, s, t) = f(\boldsymbol{x}) \tag{4.50}$$

for a sufficiently regular function $f(\boldsymbol{x})$ (e.g., a test function). We denote by $\boldsymbol{x}(t)$ any solution of eq. (4.47) and by $\boldsymbol{x}_{\boldsymbol{x},s}(t)$ the solution of eq. (4.47) for $t > s$, satisfying the initial condition

$$\boldsymbol{x}_{\boldsymbol{x},s}(s) = \boldsymbol{x} \tag{4.51}$$

Theorem 4.4.1 (Kolmogorov's representation formula). *The solution of the terminal value problem* (4.49), (4.50) *has the representation*

$$u(\boldsymbol{x}, s, t) = \mathbb{E}\left[f(\boldsymbol{x}(t)) \,|\, \boldsymbol{x}(s) = \boldsymbol{x}\right], \tag{4.52}$$

where $\boldsymbol{x}(t) = \boldsymbol{x}_{\boldsymbol{x},s}(t)$.

Proof. Substituting $\boldsymbol{x}_{\boldsymbol{x},s}(t)$ into $u(\boldsymbol{x}, s, t)$, we obtain from Itô's formula that for $t > \tau > s$

$$\begin{aligned} u(\boldsymbol{x}_{\boldsymbol{x},s}(\tau), \tau, t) =& u(\boldsymbol{x}_{\boldsymbol{x},s}(s), s, t) \\ &+ \int_s^\tau \left[\frac{\partial u(\boldsymbol{x}_{\boldsymbol{x},s}(t'), t', t)}{\partial s} + L^*_{\boldsymbol{x}} u(\boldsymbol{x}_{\boldsymbol{x},s}(t'), t', t)\right] dt' \\ &+ \int_s^\tau \sum_{i=1}^{d}\sum_{j=1}^{m} b^{ij}(\boldsymbol{x}_{\boldsymbol{x},s}(t'), t') \frac{\partial u(\boldsymbol{x}_{\boldsymbol{x},s}(t'), t', t)}{\partial x^i} \, dw^j(t'), \end{aligned} \tag{4.53}$$

where $L^*_{\boldsymbol{x}}$ is the backward Kolmogorov operator (3.68). The first integral in eq. (4.53) vanishes due to eq. (4.49). Now, we set $\tau = t$ and take the expectation of both sides of eq. (4.53), conditioned on the initial value (4.51). We obtain that for $t > s$ eq. (4.52) is satisfied, because the expectation of the stochastic integral vanishes. □

Equation (4.52) represents the solution of the terminal value problem eqs. (4.49), (4.50) as an expectation of the terminal value at the terminal point of the solution of the stochastic differential equation (4.47) with the initial condition (4.51). Kolmogorov's representation formula can be used for the calculation of any moment of the solution to the stochastic dynamics (4.47), given any initial point or distribution of points. A moment of the solution at time t, given the initial point $\boldsymbol{x}(s) = \boldsymbol{x}$, is the conditional expectation

$$\mathbb{E}\left[f(\boldsymbol{x}(t)) \,|\, \boldsymbol{x}(s) = \boldsymbol{x}\right]. \tag{4.54}$$

For example, choosing $f(\boldsymbol{x}) = \boldsymbol{x}$ in eq. (4.54) gives the conditional mean value of the solution, and choosing $f(\boldsymbol{x}) = x^i x^j$ gives the second moments, and so on. According to Kolmogorov's representation formula, the conditional moment (4.54) of the solution is the solution of the terminal value problem (4.49), (4.50).

One simple consequence of the representation formula is the *maximum principle*, which asserts that the solution of the terminal value problem (4.49), (4.50) satisfies the inequality $u(\boldsymbol{x}, s, t) \leq \sup_{\boldsymbol{x}} f(\boldsymbol{x})$. This is due to the fact that the average of a function does not exceed its upper bound. Another consequence of the formula is a numerical method for evaluating the solution of the terminal value problem eqs. (4.49), (4.50). Simply, run trajectories of the stochastic differential equation (4.47) with the initial condition (4.51) on the time interval $[s, t]$ with some ODE integrator.[1]

Note that if the coefficients $\boldsymbol{a}$ and $\boldsymbol{B}$ are independent of t, then the solution is a function of $t - s$ so that $\partial u/\partial s = -\partial u/\partial t$ and the backward parabolic equation (4.49) with the terminal value (4.50) becomes the forward parabolic equation

$$\frac{\partial u(\boldsymbol{x}, \tau)}{\partial \tau} = L^*_{\boldsymbol{x}} u(\boldsymbol{x}, \tau) \tag{4.55}$$

with the initial condition $\lim_{\tau \to 0} u(\boldsymbol{x}, \tau) = f(\boldsymbol{x})$, where $\tau = t - s$. Kolmogorov's representation formula for this case takes the form

$$u(\boldsymbol{x}, \tau) = \mathbb{E}\left[f(\boldsymbol{x}(\tau)) \,|\, \boldsymbol{x}(0) = \boldsymbol{x}\right].$$

The Feynman–Kac formula provides a representation of the solution to a backward parabolic terminal value problem of the form

$$\frac{\partial v(\boldsymbol{x}, t)}{\partial t} + L^*_{\boldsymbol{x}} v(\boldsymbol{x}, t) + g(\boldsymbol{x}, t) v(\boldsymbol{x}, t) = 0, \quad t < T \tag{4.56}$$

$$\lim_{t \uparrow T} v(\boldsymbol{x}, t) = f(\boldsymbol{x}), \tag{4.57}$$

where $L^*_{\boldsymbol{x}}$ is the backward Kolmogorov operator (3.68), and $g(\boldsymbol{x}, t)$ and $f(\boldsymbol{x})$ are given sufficiently smooth functions, as a conditional expectation of a certain functional of the solution to the Itô system (4.47).

[1] Make sure the integrator is "forward" with respect to $d\boldsymbol{w}$. Thus Euler and Runge–Kutta type schemes preserve the class $H_2[s, t]$, whereas predictor–correctors do not. In the Runge–Kutta schemes the noise has to be added at the end of the time step, not in intermediate steps, and average the initial function $f(\boldsymbol{x})$, at the terminal points of the trajectories [131], [132].

Theorem 4.4.2 (The Feynman–Kac formula). *If the initial value problem* (4.47) *and the terminal value problem* (4.56), (4.57) *have unique solutions, then*

$$v(\boldsymbol{x},t) = \mathbb{E}\left[f(\boldsymbol{x}(T))\exp\left\{\int_t^T g(\boldsymbol{x}(s),s)\,ds\right\} \mid \boldsymbol{x}(t)=\boldsymbol{x}\right], \tag{4.58}$$

where $\boldsymbol{x}(s)$ is the solution of the Itô system (4.47) *for $s>t$ with the initial condition $\boldsymbol{x}(t)=\boldsymbol{x}$.*

Proof. We apply Itô's formula to the function

$$F(s) = v(\boldsymbol{x}(s),s)\exp\left\{\int_t^s g(\boldsymbol{x}(\tau),\tau)\,d\tau\right\}$$

to get

$$d\left[v(\boldsymbol{x}(s),s)\exp\left\{\int_t^s g(\boldsymbol{x}(\tau),\tau)\,d\tau\right\}\right] = \exp\left\{\int_t^s g(\boldsymbol{x}(\tau),\tau)\,d\tau\right\} dv(\boldsymbol{x}(s),s)$$

$$+\ g(\boldsymbol{x}(s),s)v(\boldsymbol{x}(s),s)\exp\left\{\int_t^s g(\boldsymbol{x}(\tau),\tau)\,d\tau\right\} ds. \tag{4.59}$$

Itô's formula and eq. (4.56) give

$$\begin{aligned} dv(\boldsymbol{x}(s),s) &= \left[\frac{\partial v(\boldsymbol{x}(s),s)}{\partial s} + L^*_{\boldsymbol{x}} v(\boldsymbol{x}(s),s)\right] ds \\ &\quad + \sum_{i=1}^d \sum_{j=1}^m b^{ij}(\boldsymbol{x}(s),s)\frac{\partial v(\boldsymbol{x}(s),s)}{\partial x^i}\,dw^j(s) \\ &= -\ g(\boldsymbol{x}(s),s)v(\boldsymbol{x}(s),s)\,ds \\ &\quad + \sum_{i=1}^d \sum_{j=1}^m b^{ij}(\boldsymbol{x}(s),s)\frac{\partial v(\boldsymbol{x}(s),s)}{\partial x^i}\,dw^j(s), \end{aligned} \tag{4.60}$$

so using (4.60) in (4.59) and taking the expectation of both sides, conditioned on the initial value $\boldsymbol{x}(t)=\boldsymbol{x}$, we obtain

$$d\mathbb{E}\left[v(\boldsymbol{x}(s),s)\exp\left\{\int_t^s g(\boldsymbol{x}(\tau),\tau)\,d\tau\right\} \mid \boldsymbol{x}(t)=\boldsymbol{x}\right] = 0.$$

This means that the conditional expectation is independent of s in the interval $[t,T]$. Therefore choosing $s=t$, we obtain that

$$\begin{aligned} \mathbb{E}\left[v(\boldsymbol{x}(s),s)\exp\left\{\int_t^s g(\boldsymbol{x}(\tau),\tau)\,d\tau\right\} \mid \boldsymbol{x}(t)=\boldsymbol{x}\right] &= \mathbb{E}\left[v(\boldsymbol{x}(t),t) \mid \boldsymbol{x}(t)=\boldsymbol{x}\right] \\ &= v(\boldsymbol{x},t), \end{aligned}$$

and choosing $s = T$, we obtain from the above

$$\begin{aligned} v(\boldsymbol{x},t) &= \mathbb{E}\left[v(\boldsymbol{x}(s),s)\exp\left\{\int_t^s g(\boldsymbol{x}(\tau),\tau)\,d\tau\right\} \mid \boldsymbol{x}(t)=\boldsymbol{x}\right] \\ &= \mathbb{E}\left[v(\boldsymbol{x}(T),T)\exp\left\{\int_t^T g(\boldsymbol{x}(\tau),\tau)\,d\tau\right\} \mid \boldsymbol{x}(t)=\boldsymbol{x}\right]. \end{aligned}$$

Hence, using (4.57), we obtain (4.58). □

The Feynman–Kac formula can be interpreted as the expectation of the function $f(\boldsymbol{x}(T))$, where $\boldsymbol{x}(t)$ is a solution of the stochastic dynamics (4.47), whose trajectories can terminate at any point and at any time with a certain probability. Such dynamics are referred to as *stochastic dynamics with killing*. The *killing rate* $-g(\boldsymbol{x},t)$ is defined as follows. Assume that at each point $\boldsymbol{x}$ and time t there is a probability $-g(\boldsymbol{x},t) \geq 0$ per unit time that the trajectory of the solution $\boldsymbol{x}(t)$ terminates there and then, independently of the past; that is, of $\mathcal{F}_t$. Partition the time interval $[t,T]$ into N small intervals of length Δt, $t = t_0 < t_1 < \cdots < T$. Then the probability at time t that the solution $\boldsymbol{x}(t)$ survives by time T is the product of the probabilities that it survives each one of preceding N time intervals,

$$\Pr_N\{\text{killing time} > T\} = \prod_{i=1}^{N}\left[1 + g(\boldsymbol{x}(t_i),t_i)\,\Delta t\right] + o(\Delta t). \tag{4.61}$$

In the limit $N \to \infty$ the product (4.61) converges to the integral

$$\begin{aligned} \Pr\{\text{killing time} > T\} &= \lim_{N\to\infty} \Pr_N\{\text{killing time} > T\} \\ &= \exp\left\{\int_t^T g(\boldsymbol{x}(t'),t')\,dt'\right\}. \end{aligned} \tag{4.62}$$

Hence,

$$\begin{aligned} &\mathbb{E}\left[f(\boldsymbol{x}(T)), \text{killing time} > T \mid \boldsymbol{x}(t)=\boldsymbol{x}\right] \\ &= \mathbb{E}\left[f(\boldsymbol{x}(T))\exp\left\{\int_t^T g(\boldsymbol{x}(s),s)\,ds\right\} \mid \boldsymbol{x}(t)=\boldsymbol{x}\right], \end{aligned}$$

which is (4.58).

Exercise 4.15 (Representation for a nonhomogeneous terminal value problem). Use Itô's formula to derive the representation

$$v(\boldsymbol{x},t) = \mathbb{E}\left[\int_t^T f(\boldsymbol{x}(s),s)\,ds \mid \boldsymbol{x}(t)=\boldsymbol{x}\right]$$

for the solution of the terminal value problem

$$\frac{\partial v(\boldsymbol{x},t)}{\partial t} + L^*_{\boldsymbol{x}} v(\boldsymbol{x},t) + f(\boldsymbol{x},t) = 0 \text{ for } t < T,\ \boldsymbol{x} \in \mathbb{R}^d, \quad \lim_{t \uparrow T} v(\boldsymbol{x},t) = 0.$$

□

Exercise 4.16 (Representation for a complex field). Derive a representation formula for the solution of the terminal value problem

$$\frac{\partial v(\boldsymbol{x},t)}{\partial t} + L^*_{\boldsymbol{x}} v(\boldsymbol{x},t) + i\boldsymbol{h}(\boldsymbol{x},t) \cdot \nabla v(\boldsymbol{x},t) + g(\boldsymbol{x},t)v(\boldsymbol{x},t) = 0, \ t < T,\ \boldsymbol{x} \in \mathbb{R}^d$$

$$\lim_{t \uparrow T} v(\boldsymbol{x},t) = f(\boldsymbol{x}),$$

where $\boldsymbol{h}(\boldsymbol{x},t)$ is a given field [84]. □

4.4.1 The Andronov–Vitt–Pontryagin equation

It is shown below that the backward Kolmogorov equation (BKE) in a bounded domain $D \subset \mathbb{R}^d$, with sufficiently regular boundary ∂D,

$$\frac{\partial p(\boldsymbol{y},t\,|\,\boldsymbol{x},s)}{\partial s} + L^*_{\boldsymbol{x}} p(\boldsymbol{y},t\,|\,\boldsymbol{x},s) = 0 \text{ for } \boldsymbol{x},\boldsymbol{y} \in D, \quad s < t, \tag{4.63}$$

with the boundary condition

$$p(\boldsymbol{y},t\,|\,\boldsymbol{x},s) = 0 \text{ for } \boldsymbol{x} \in \partial D, \boldsymbol{y} \in D, \quad s < t$$

and the terminal condition

$$p(\boldsymbol{y},t\,|\,\boldsymbol{x},t) = \delta(\boldsymbol{y}-\boldsymbol{x}) \text{ for } \boldsymbol{x},\boldsymbol{y} \in D, \quad t \in \mathbb{R},$$

is the transition probability density function $p(\boldsymbol{y},s\,|\,\boldsymbol{x},t)\,d\boldsymbol{y} = \Pr\{\boldsymbol{x}(s) \in \boldsymbol{y} + d\boldsymbol{y}\,|\,\boldsymbol{x}(t) = \boldsymbol{x}\}$ of the process (4.47) in D, whose trajectories are terminated (absorbed) when they hit ∂D for the first time. It is shown in Chapter 5 below that this terminal and boundary value problem has a unique solution. Under mild regularity conditions the solution decays exponentially fast as $t \to \infty$ (see [78]).

Lemma 4.4.1 (Backward parabolic equations). *For every integrable function $f(\boldsymbol{x},s)$ in $D \times [s,\infty]$ and all $s \in \mathbb{R}$ the boundary value problem*

$$\frac{\partial u(\boldsymbol{x},s)}{\partial s} + L^*_{\boldsymbol{x}} u(\boldsymbol{x},s) = -\ f(\boldsymbol{x},s) \quad \text{for} \quad (\boldsymbol{x},s) \in D \times \mathbb{R} \tag{4.64}$$

$$u(\boldsymbol{x},s) = 0 \quad \text{for} \quad \boldsymbol{x} \in \partial D, \tag{4.65}$$

has a unique solution, given by

$$u(\boldsymbol{x},s) = \int_s^\infty \int_D f(\boldsymbol{y},t) p(\boldsymbol{y},t\,|\,\boldsymbol{x},s)\,d\boldsymbol{y}\,dt. \tag{4.66}$$

The lemma is a straightforward consequence of the above assumptions.

Theorem 4.4.3 (The Andronov–Vitt–Pontryagin formula). *Assume the boundary value problem*

$$\frac{\partial u(\boldsymbol{x},s)}{\partial s} + L^*_{\boldsymbol{x}} u(\boldsymbol{x},s) = -1 \quad \text{for } \boldsymbol{x} \in D, \quad \text{for all } s \in \mathbb{R} \tag{4.67}$$

$$u(\boldsymbol{x},s) = 0 \qquad \text{for } \boldsymbol{x} \in \partial D, \tag{4.68}$$

where $L^*_{\boldsymbol{x}}$ *is the backward Kolmogorov operator* (3.68), *has a unique bounded solution. Then the mean first passage time* $\mathbb{E}\left[\tau_D \,|\, \boldsymbol{x}(s) = \boldsymbol{x}\right]$ *of the solution* $\boldsymbol{x}(t)$ *of* (4.47) *from every point* $\boldsymbol{x}$ *in a bounded domain* D *to the boundary* ∂D *is finite and*

$$\mathbb{E}\left[\tau_D \,|\, \boldsymbol{x}(s) = \boldsymbol{x}\right] = s + u(\boldsymbol{x},s). \tag{4.69}$$

The assumption of the theorem is satisfied if the coefficients are continuously differentiable functions and $\boldsymbol{\sigma}(\boldsymbol{x},t)$ is a uniformly positive definite matrix in the domain.

Proof. We can apply Itô's formula to $u(\boldsymbol{x}(t),t)$ for all $s < t < \tau_D$, because the formula is valid for Markov times. Substituting $t = \tau_D$ in Itô's formula and taking the conditional expectation, as above, we obtain Dynkin's equation

$$\begin{aligned}&\mathbb{E}\left[u(\boldsymbol{x}(\tau_D),\tau_D) \,|\, \boldsymbol{x}(s) = \boldsymbol{x}\right] \\ &= u(\boldsymbol{x},s) + \mathbb{E}\left[\int_s^{\tau_D} \left(\frac{\partial u(\boldsymbol{x}(t),t)}{\partial s} + L^*_{\boldsymbol{x}} u(\boldsymbol{x}(t),t)\right) dt | \boldsymbol{x}(s) = \boldsymbol{x}\right].\end{aligned}$$

The boundary condition (4.68) implies that $u(\boldsymbol{x}(\tau_D),\tau_D) = 0$, because $\boldsymbol{x}(\tau_D)$ is on the boundary ∂D, so that the left-hand side of Dynkin's equation vanishes. Furthermore, $\boldsymbol{x}(t) \in D$ for all $s < t < \tau_D$, so that eq. (4.67) implies that the integrand equals $- - 1$. It follow that the integral equals $-(\tau_D - s)$ and therefore the expectation of the integral on the right-hand side of Dynkin's equation is simply $-\mathbb{E}\left[\tau_D - s \,|\, \boldsymbol{x}(s) = \boldsymbol{x}\right]$. Thus Dynkin's equation reduces to $0 = u(\boldsymbol{x},s) - \mathbb{E}\left[\tau_D - s \,|\, \boldsymbol{x}(s) = \boldsymbol{x}\right]$, hence the representation formula (4.69) follows. □

Corollary 4.4.1 (The Andronov–Vitt-Pontryagin boundary value problem for the autonomous case). *Under the assumptions of Theorem* 4.4.3, *if the coefficients* $\boldsymbol{a}$ *and* $\boldsymbol{B}$ *are independent of* t, *the solution of eq.* (4.67) *is independent of* s *so that the backward parabolic boundary value problem eqs.* (4.67), (4.68) *reduces to the elliptic boundary value problem of Andronov, Vitt, and Pontryagin*

$$L^*_{\boldsymbol{x}} u(\boldsymbol{x}) = -1 \ \text{ for } \boldsymbol{x} \in D, \quad u(\boldsymbol{x}) = 0 \ \text{ for } \boldsymbol{x} \in \partial D. \tag{4.70}$$

The representation formula (4.69) *simplifies to* $\mathbb{E}\left[\tau_D \,|\, \boldsymbol{x}(0) = \boldsymbol{x}\right] = u(\boldsymbol{x})$.

Another proof is given in Section 6.2 below. It can be shown that if the boundary value problem (4.70) has a finite solution, the MFPT is finite [84].

Example 4.1 (The mean exit time of the MBM). To find the mean exit time of Brownian motion from an interval $[a,b]$, given that it starts at a point x in the interval, we have to solve the equation (4.70) $\frac{1}{2}u''(x) = -1$ for $a < x < b$ and the boundary conditions $u(a) = u(b) = 0$. The solution is given by

$$\mathbb{E}\left[\tau_{[a,b]}\,|\,w(0)=x\right] = u(x) = (b-x)(x-a).$$

In particular, $\lim_{b\to\infty} u(x) = \infty$; that is, the mean time to exit the half-line $[0,\infty)$ is infinite. This is the well-known "Gambler's Paradox," mentioned in Exercise *2.21*, of the random walker that starts at x and is sure to fall off a cliff in finite time, but on the average it will take an infinite number of steps to do so. This means that in a simulation of random walks almost every trajectory will reach the endpoint of the half-line in a finite number of steps. However, if the number of steps to get there is averaged over a sample of N trajectories, the average will grow indefinitely as $N \to \infty$. □

Example 4.2 (The mean exit time of the Ornstein–Uhlenbeck process). To solve the same problem for the Ornstein–Uhlenbeck process, recall that it is defined by the SDE $dx = -\alpha x\,dt + \beta\,dw$. Equation (4.70) is now $\frac{1}{2}\beta^2 u''(x) - \alpha x u'(x) = -1$ for $a < x < b$ and the boundary conditions are $u(a) = u(b) = 0$. The solution is given by

$$u(x) = C\int\limits_a^x e^{-\alpha y^2/\beta^2}\,dy - \frac{2}{\beta^2}\int\limits_a^x\int\limits_a^y e^{-\alpha(y^2-z^2)/\beta^2}\,dz\,dy,$$

where

$$C = 2\int\limits_a^b\int\limits_a^y e^{-\alpha(y^2-z^2)/\beta^2}\,dz\,dy\Big/\beta^2\int\limits_a^b e^{-\alpha y^2/\beta^2}\,dy.$$

Does $\lim_{b\to\infty} u(x) = \infty$ hold in this case as well? □

Exercise 4.17 (Higher moments of the FPT). Derive boundary value problems similar to eqs. (4.67) and (4.70) for higher moments of the FPT. (HINT: Replace -1 on the right-hand side of the equation with an appropriate power of t). □

4.4.2 The exit distribution

Consider again the system (4.47) in a domain D and assume that if $\boldsymbol{x}(s) \in D$, then the first exit time τ_D of the solution from D after time s is finite with probability 1.

Theorem 4.4.4 (Kolmogorov's representation of the exit distribution). *The conditional probability density function of the exit points $\boldsymbol{x}(\tau_D)$, for $\tau_D > s$, of trajectories of (4.47) with $\boldsymbol{x}(s) = \boldsymbol{x}$ is Green's function for the boundary value problem*

$$\frac{\partial u(\boldsymbol{x},t)}{\partial t} + L^*_{\boldsymbol{x}} u(\boldsymbol{x},t) = 0 \quad \text{for } \boldsymbol{x} \in D, \quad -\infty < t < \infty \tag{4.71}$$

$$u(\boldsymbol{x},t) = f(\boldsymbol{x}) \quad \text{for } \boldsymbol{x} \in \partial D, \quad -\infty < t < \infty, \tag{4.72}$$

where $L^*_{\boldsymbol{x}}$ is the backward Kolmogorov operator (3.68), *and*

$$u(\boldsymbol{x},s) = \mathbb{E}\left[f(\boldsymbol{x}(\tau_D))\,|\,\boldsymbol{x}(s)=\boldsymbol{x}\right]. \tag{4.73}$$

First, we prove the following

Lemma 4.4.2 (Backward boundary value problems). *Assume that the backward Kolmogorov operator $L^*_{\boldsymbol{x}}$ defined in* (3.68) *is uniformly elliptic in a bounded domain $D \subset \mathbb{R}^d$, whose boundary ∂D is smooth and for all $t \in \mathbb{R}$ and has continuously differentiable and bounded coefficients in $D \times \mathbb{R}$. Then the boundary value problem* (4.71), (4.72) *has a unique solution for every smooth function $f(\boldsymbol{x})$ on ∂D.*

Proof. For each $t \in \mathbb{R}$ the elliptic boundary value problem

$$L^*_{\boldsymbol{x}} v(\boldsymbol{x},t) = 0 \quad \text{for } \boldsymbol{x} \in D, \quad -\infty < t < \infty \tag{4.74}$$

$$v(\boldsymbol{x},t) = f(\boldsymbol{x}) \quad \text{for } \boldsymbol{x} \in \partial D, \quad -\infty < t < \infty, \tag{4.75}$$

has a unique solution $v(\boldsymbol{x},t)$, whose derivative $\partial v(\boldsymbol{x},t)/\partial t$ is a continuous and bounded function in $D \times \mathbb{R}$. Setting $w(\boldsymbol{x},t) = u(\boldsymbol{x},t) - v(\boldsymbol{x},t)$, we obtain for $w(\boldsymbol{x},t)$ the boundary value problem

$$\frac{\partial w(\boldsymbol{x},t)}{\partial t} + L^*_{\boldsymbol{x}} u(\boldsymbol{x},t) = -\frac{\partial v(\boldsymbol{x},t)}{\partial t} \quad \text{for } \boldsymbol{x} \in D, \quad -\infty < t < \infty \tag{4.76}$$

$$w(\boldsymbol{x},t) = 0 \quad \text{for } \boldsymbol{x} \in \partial D, \quad -\infty < t < \infty, \tag{4.77}$$

whose solution is given by

$$w(\boldsymbol{x},t) = \int_t^\infty \int_D p\,(\boldsymbol{y},s\,|\,\boldsymbol{x},t)\frac{\partial v(\boldsymbol{y},s)}{\partial t}\,d\boldsymbol{y}\,ds, \tag{4.78}$$

according to Lemma 4.4.1. □

We can now prove Theorem 4.4.4.

Proof. Applying Itô's formula to the process $u(\boldsymbol{x}(t),t)$, we obtain

$$u(\boldsymbol{x}(t),t) = u(\boldsymbol{x},s) + \int_s^t \left[\frac{\partial u(\boldsymbol{x}(t'),t')}{\partial t'} + L^*_{\boldsymbol{x}} u(\boldsymbol{x}(t'),t')\right] dt'$$

$$+ \int_s^t \sum_{i,j} b^{ij}(\boldsymbol{x}(t'),t')\frac{\partial u(\boldsymbol{x}(t'),t')}{\partial x^i}\,dw^j.$$

Denoting by τ_D the first passage time of $\boldsymbol{x}(t)$ to ∂D after s, we set $t = \tau_D$ and take expectation, conditioned on the initial point. As long as $t < \tau_D$ the expression in brackets vanishes, because $\boldsymbol{x}(t') \in D$ for $t' < t < \tau_D$ and in D eq. (4.71) is satisfied. Furthermore, due to the boundary condition (4.72), $u(\boldsymbol{x}(\tau_D),\tau_D) =$

$f(\boldsymbol{x}(\tau_D))$. It follows that $\mathbb{E}\left[u(\boldsymbol{x}(\tau_D),\tau_D)\,|\,\boldsymbol{x}(s)=\boldsymbol{x}\right] = u(\boldsymbol{x},s)$, hence (4.73). Finally, the solution to the boundary value problem is expressed in terms of the Green's function $G(\boldsymbol{x},\boldsymbol{y},t)$ as

$$u(\boldsymbol{x},s) = \oint\limits_{\partial D} G(\boldsymbol{x},\boldsymbol{y},s)f(\boldsymbol{y})\,dS_{\boldsymbol{y}} \tag{4.79}$$

and the expectation in (4.73) is, by definition,

$$\mathbb{E}\left[f(\boldsymbol{x}(\tau_D))\,|\,\boldsymbol{x}(s)=\boldsymbol{x}\right] = \oint\limits_{\partial D} f(\boldsymbol{y})\Pr\left\{\boldsymbol{x}(\tau_D)\in\boldsymbol{y}+dS_{\boldsymbol{y}}\,|\,\boldsymbol{x}(s)=\boldsymbol{x}\right\}. \tag{4.80}$$

Equation (4.73) means that (4.79) and (4.80) represent the same function $u(\boldsymbol{x},s)$ for all sufficiently regular boundary functions $f(\boldsymbol{x})$. It follows that

$$\Pr\left\{\boldsymbol{x}(\tau_D)\in\boldsymbol{y}+dS_{\boldsymbol{y}}\,|\,\boldsymbol{x}(s)=\boldsymbol{x}\right\} = G(\boldsymbol{x},\boldsymbol{y},s)\,dS_{\boldsymbol{y}} \tag{4.81}$$

for almost all $\boldsymbol{x}\in\partial D$. Thus $G(\boldsymbol{x},\boldsymbol{y},s)$ is the surface probability density of exit points $\boldsymbol{y}\in\partial D$ of trajectories that start at $\boldsymbol{x}\in D$ at time s. □

Exercise 4.18 (Green's function). Express Green's functions for the inhomogeneous problem (4.76) with the homogeneous boundary conditions (4.77) and for the homogeneous problem (4.71) with inhomogeneous boundary conditions (4.72) in terms of $p\left(\boldsymbol{y},t\,|\,\boldsymbol{x},s\right)$ (see Section 6.3). □

If $\boldsymbol{a}(\boldsymbol{x},t)$ and $\boldsymbol{B}(\boldsymbol{x},t)$ are independent of t, the boundary value problem (4.71) becomes the elliptic boundary value problem

$$L^*_{\boldsymbol{x}}u(\boldsymbol{x}) = 0 \text{ for } \boldsymbol{x}\in D, \quad u(\boldsymbol{x}) = f(\boldsymbol{x}) \text{ for } \boldsymbol{x}\in\partial D \tag{4.82}$$

and Kolmogorov's formula becomes

$$u(\boldsymbol{x}) = \mathbb{E}\left[f(\boldsymbol{x}(\tau_D))\,|\,\boldsymbol{x}(0)=\boldsymbol{x}\right]. \tag{4.83}$$

Kolmogorov's formula (4.83) indicates that the solution of the boundary value problem can be constructed by running trajectories of the SDE that start at $\boldsymbol{x}$ until they hit ∂D and averaging the boundary function at the points where the trajectories hit ∂D.

Equation (4.83) leads to an important interpretation of Green's function for the elliptic boundary value problem (4.82). By definition, Green's function, $G(\boldsymbol{x},\boldsymbol{y})$, is characterized by the relation

$$u(\boldsymbol{x}) = \oint\limits_{\partial D} f(\boldsymbol{y})G(\boldsymbol{x},\boldsymbol{y})\,dS_{\boldsymbol{y}}, \tag{4.84}$$

where $dS_{\boldsymbol{y}}$ is a surface area element on ∂D. On the other hand, eq. (4.83) can be written as

$$u(\boldsymbol{x}) = \oint\limits_{\partial D} f(\boldsymbol{y})p\left(\boldsymbol{x}(\tau_D)=\boldsymbol{y}\,|\,\boldsymbol{x}(0)=\boldsymbol{x}\right)dS_{\boldsymbol{y}}. \tag{4.85}$$

We must have

$$G(\boldsymbol{x},\boldsymbol{y}) = \Pr\{\boldsymbol{x}(\tau_D) = \boldsymbol{y} \,|\, \boldsymbol{x}(0) = \boldsymbol{x}\},$$

because eqs. (4.84) and (4.85) hold for all smooth functions $f(\boldsymbol{y})$ on ∂D. This means that Green's function is the pdf of the exit points on ∂D of trajectories of (4.47) that start at $\boldsymbol{x}$. In a simulation it counts the fraction of trajectories that, starting at $\boldsymbol{x}$, hit the boundary at $\boldsymbol{y}$.

Exercise 4.19. Exit distribution of MBM from a half-space and the Cauchy process

(i) The MBM $\boldsymbol{w}(t) = (w_1(t), w_2(t), \ldots, w_d(t))$ in $\mathbb{R}^d$ starts in the upper half space, at $\boldsymbol{w}(0) = (0, 0, \ldots, 0, z)$ with $z > 0$. Find the distribution of its exit points in the plane $z = 0$.

(ii) Let τ_z be the FPT to the line $z = 0$ in $\mathbb{R}^2$. Show that $x(z) = w_1(\tau_z)$ is the Cauchy process, defined by the transition probability density function

$$p\,(y, z \,|\, x, 0) = \frac{z}{\pi} \frac{1}{(x-y)^2 + z^2}. \tag{4.86}$$

See Exercise 7.2 and [231] for more details. □

4.4.3 The PDF of the FPT

The results of this section are derived from Itô's formula. In Section 6.1 they are derived again directly from the solution of the Fokker–Planck equation. The derivation here is based on Itô's formula. We consider again the solution $\boldsymbol{x}(t)$ of the Itô system (4.47) that starts at time s in a domain D and we denote by τ_D the first passage time of the solution to the boundary ∂D of D. That is, we assume that $\boldsymbol{x}(s) \in D$ at some time s and $\tau_D = \inf\{t > s \,|\, \boldsymbol{x}(t) \in \partial D\}$. In particular, $\boldsymbol{x}(\tau_D) \in \partial D$. The PDF of τ_D, conditioned on $\boldsymbol{x}(s) = \boldsymbol{x} \in D$ is the conditional probability $P(T \,|\, \boldsymbol{x}, s) = \Pr\{\tau_D < T \,|\, \boldsymbol{x}(s) = \boldsymbol{x}\}$ for every $T > s$. Obviously, if $\boldsymbol{x} \in \partial D$, then $P(T \,|\, \boldsymbol{x}, s) = 1$, because in this case the trajectories of $\boldsymbol{x}(t)$ start out on the boundary so that surely $\tau_D = s < T$. Similarly, $P(T \,|\, \boldsymbol{x}, T) = 0$ for all $\boldsymbol{x} \in D$, because the trajectories of the solution $\boldsymbol{x}(t)$ cannot be at the same time T both inside D and on its boundary ∂D.

Theorem 4.4.5 (A boundary value problem for the PDF of the FPT).

$$\Pr\{\tau_D < T \,|\, \boldsymbol{x}(s) = \boldsymbol{x}\} = u(\boldsymbol{x}, s, T), \tag{4.87}$$

where $u(\boldsymbol{x}, t, T)$ is the solution of the backward parabolic terminal boundary value problem

$$\frac{\partial u(\boldsymbol{x},t,T)}{\partial t} + L^*_{\boldsymbol{x}} u(\boldsymbol{x},t,T) = 0 \quad \text{for } \boldsymbol{x} \in D,\ t < T \tag{4.88}$$

$$u(\boldsymbol{x},t,T) = 1 \quad \text{for } \boldsymbol{x} \in \partial D,\ t < T \tag{4.89}$$

$$u(\boldsymbol{x},T,T) = 0 \quad \text{for } \boldsymbol{x} \in D. \tag{4.90}$$

Proof. Assuming, as we may, that (4.88)–(4.90) has a unique solution, we get from Itô's formula

$$u(\boldsymbol{x}(t),t,T) = u(\boldsymbol{x}(s),s,T) + \int_s^t \left[\frac{\partial u(\boldsymbol{x}(t'),t',T)}{\partial t} + L^*_{\boldsymbol{x}} u(\boldsymbol{x}(t'),t',T)\right] dt'$$
$$+ \int_s^t \sum_{i,j} b^{ij}(\boldsymbol{x}(t'),t') \frac{\partial u(\boldsymbol{x}(t'),t',T)}{\partial x^i}\, dw^j. \tag{4.91}$$

We have from eq. (4.88) that $u_t(\boldsymbol{x}(t'),t',T) + L^*_{\boldsymbol{x}} u(\boldsymbol{x}(t'),t',T) = 0$, because $\boldsymbol{x}(t') \in D$ at times $t' < \tau_D \wedge T = \min\{\tau_D, T\}$, which are prior to exit. Now, setting $t = \tau_D \wedge T$ in eq. (4.91), we get

$$u(\boldsymbol{x}(\tau_D \wedge T), \tau_D \wedge T, T) = u(\boldsymbol{x}(s),s,T) \tag{4.92}$$
$$+ \int_s^{\tau_D \wedge T} \sum_{i,j} b^{ij}(\boldsymbol{x}(t),t) \frac{\partial u(\boldsymbol{x}(t),t,T)}{\partial x^i}\, dw^j(t).$$

The expectation of both sides of eq. (4.92), conditioned on $\boldsymbol{x}(s) = \boldsymbol{x}$, gives

$$\mathbb{E}\left[u(\boldsymbol{x}(\tau_D \wedge T), \tau_D \wedge T, T) \,|\, \boldsymbol{x}(s) = \boldsymbol{x}\right] = u(\boldsymbol{x}, s, T). \tag{4.93}$$

On the other hand,

$$\begin{aligned}
&\mathbb{E}\left[u(\boldsymbol{x}(\tau_D \wedge T), \tau_D \wedge T, T) \,|\, \boldsymbol{x}(s) = \boldsymbol{x}\right]\\
&= \mathbb{E}\left[u(\boldsymbol{x}(\tau_D \wedge T), \tau_D \wedge T, T) \,|\, \boldsymbol{x}(s) = \boldsymbol{x}, \tau_D \wedge T = T\right]\\
&\times \Pr\{\tau_D \wedge T = T \,|\, \boldsymbol{x}(s) = \boldsymbol{x}\}\\
&+ \mathbb{E}\left[u(\boldsymbol{x}(\tau_D \wedge T), \tau_D \wedge T, T) \,|\, \boldsymbol{x}(s) = \boldsymbol{x}, \tau_D \wedge T = \tau_D\right]\\
&\times \Pr\{\tau_D \wedge T = \tau_D \,|\, \boldsymbol{x}(s) = \boldsymbol{x}\}.
\end{aligned} \tag{4.94}$$

When $\tau_D \wedge T = T$ the trajectory has not reached ∂D by time T and therefore $\boldsymbol{x}(\tau_D \wedge T) = \boldsymbol{x}(T) \in D$. It follows from the terminal condition (4.90) that $u(\boldsymbol{x}(\tau_D \wedge T), \tau_D \wedge T, T) = u(\boldsymbol{x}(T), T, T) = 0$. When $\tau_D \wedge T = \tau_D$, we have $\boldsymbol{x}(\tau_D \wedge T) = \boldsymbol{x}(\tau_D) \in \partial D$ so that the boundary condition (4.89) gives $u(\boldsymbol{x}(\tau_D \wedge T), \tau_D \wedge T, T) = u(\boldsymbol{x}(\tau_D), \tau_D, T) = 1$. It follows that the first term on the right-hand side of eq. (4.94) vanishes and that the conditional expectation in the second term equals 1. In addition,

$$\Pr\{(\tau_D \wedge T) = \tau_D \,|\, \boldsymbol{x}(s) = \boldsymbol{x}\} = \Pr\{\tau_D < T \,|\, \boldsymbol{x}(s) = \boldsymbol{x}\}.$$

Now, equations (4.93) and (4.94) reduce to

$$u(\boldsymbol{x}, s, T) = \mathbb{E}\left[u(\boldsymbol{x}(\tau_D \wedge T), \tau_D \wedge T, T) \,|\, \boldsymbol{x}(s) = \boldsymbol{x}\right] = \Pr\{\tau_D < T \,|\, \boldsymbol{x}(s) = \boldsymbol{x}\},$$

which means that $u(\boldsymbol{x}, s, T)$ is the conditional PDF of the FPT,

$$\Pr\{\tau_D < T \,|\, \boldsymbol{x}(s) = \boldsymbol{x}\} = u(\boldsymbol{x}, s, T). \tag{4.95}$$

□

In the autonomous case, where the coefficients in the stochastic system (4.47) are independent of t, the solution of eq. (4.88) is a function of the difference $T-t$, so the change of the time variable $\tau = T-t$ and the substitution $v(\boldsymbol{x},\tau) = 1-u(\boldsymbol{x},\tau)$ transform the terminal boundary value problem eqs. (4.88)–(4.90) into the forward homogeneous initial boundary value problem

$$\begin{aligned}\frac{\partial v(\boldsymbol{x},\tau)}{\partial \tau} &= L^*_{\boldsymbol{x}} v(\boldsymbol{x},\tau) \;\text{ for } \boldsymbol{x} \in D,\ \tau > 0 \\ v(\boldsymbol{x},\tau) &= 0 \;\text{ for } \boldsymbol{x} \in \partial D,\ \tau > 0, \quad v(\boldsymbol{x},0) = 1 \;\text{ for } \boldsymbol{x} \in D.\end{aligned} \tag{4.96}$$

We may assume that $s = 0$, because the above problem is invariant to time shifts, and then

$$v(\boldsymbol{x},\tau) = \Pr\left\{\tau_D > \tau \,|\, \boldsymbol{x}(0) = \boldsymbol{x}\right\}. \tag{4.97}$$

Example 4.3 (Recurrence). The solution $\boldsymbol{x}(t)$ of the autonomous stochastic dynamics (4.47) in $\mathbb{R}^d$ is said to be *recurrent* if it returns into any sphere at a sequence of times that increases to infinity. To show that given dynamics are recurrent, it is sufficient to show that the FPT from any point $\boldsymbol{x}$ to any sphere is finite with probability one. This means that the process is recurrent if $\lim_{t\to\infty} v(\boldsymbol{x},t) = 0$ for all $\boldsymbol{x}$ outside the sphere, where D denotes the domain $|\boldsymbol{x}| > a$, outside the sphere of radius a centered at the origin. In the autonomous case, the operator $L^*_{\boldsymbol{x}}$ in eq. (4.96) is independent of τ so that the Laplace transform with respect to τ results in the elliptic boundary value problem

$$\lambda\hat{v}(\boldsymbol{x},\lambda) - 1 = L^*_{\boldsymbol{x}}\hat{v}(\boldsymbol{x},\lambda) \text{ for } \boldsymbol{x} \in D, \quad \hat{v}(\boldsymbol{x},\lambda) = 0 \;\text{ for } \boldsymbol{x} \in \partial D,$$

where $\hat{v}(\boldsymbol{x},\lambda) = \int_0^\infty e^{-\lambda t} v(\boldsymbol{x},t)\,dt$. The recurrence condition in the time domain, $\lim_{t\to\infty} v(\boldsymbol{x},t) = 0$, can be written as $\lim_{\lambda\to 0} \lambda\hat{v}(\boldsymbol{x},\lambda) = 0$ for all $\boldsymbol{x}$ outside the sphere in the Laplace domain. □

Example 4.4 (Solution to Exercise 2.21). Exercise 2.21 is to find the pdf of the FPT of a Brownian motion to a level a. The Brownian motion $w(t)$ is the solution of the stochastic dynamics $dx(t) = dw(t)$ with $x(0) = 0$. For $a > 0$ the domain D in the boundary value problem (4.96) is the ray $(-\infty, a]$ and the initial point is $x = 0$. Thus the problem (4.96) is

$$\frac{\partial v(x,t)}{\partial t} = \frac{1}{2}\frac{\partial^2 v(x,t)}{\partial x^2} \quad \text{for } x < a,\ t > 0 \tag{4.98}$$

$$v(a,t) = 0, \;\text{ for } t > 0, \quad v(x,0) = 1 \;\text{ for } x < a. \tag{4.99}$$

The initial boundary value problem (4.98)–(4.99) is solved by first shifting the origin to a by the change of variables $y = x - a$ and by the method of images. Setting $v(x,t) = R(y,t)$, we find that (4.98)–(4.99) is transformed into

$$\frac{\partial R(y,t)}{\partial t} = \frac{1}{2}\frac{\partial^2 R(y,t)}{\partial y^2} \quad \text{for } y < 0,\ t > 0 \tag{4.100}$$

$$R(0,t) = 0 \;\text{ for } t > 0, \quad R(y,0) = 1 \;\text{ for } y < 0. \tag{4.101}$$

We extend the system (4.100)–(4.101) to the entire line by reflecting the initial condition (4.101) anti-symmetrically to

$$R(y,0)=1 \;\text{ for } y<0, \quad R(y,0)=-1 \;\text{ for } y>0. \tag{4.102}$$

The solution of the diffusion equation (4.100) with the initial condition (4.102) is given by

$$\begin{aligned} R(y,t) &= \frac{1}{\sqrt{2\pi t}}\int_{-\infty}^{0} e^{-(x-y)^2/2t}\,dx - \frac{1}{\sqrt{2\pi t}}\int_{0}^{\infty} e^{-(x-y)^2/2t}\,dx \\ &= \frac{1}{\sqrt{2\pi t}}\int_{y}^{-y} e^{-x^2/2t}\,dx = \frac{2}{\sqrt{2\pi t}}\int_{0}^{-y} e^{-x^2/2t}\,dx. \end{aligned}$$

For $y=-a$, we obtain $R(-a,t)=2(2\pi t)^{-1/2}\int_0^a e^{-x^2/2t}\,dx$ and

$$\Pr\{\tau_a<t\}=u(0,t)=1-R(-a,t)=\frac{2}{\sqrt{2\pi t}}\int_{a}^{\infty} e^{-x^2/2t}\,dx, \tag{4.103}$$

as in Exercise 2.21(i). To see that the Brownian motion is recurrent, we have to show that the FPT is finite with probability 1. To this end, we rewrite eq. (4.103) in the form $\Pr\{\tau_a<t\}=2(2\pi)^{-1/2}\int_{a/\sqrt{t}}^{\infty} e^{-x^2/2}\,dx$ and obtain that

$$\begin{aligned} \Pr\{\tau_a<\infty\} &= \lim_{t\to\infty}\Pr\{\tau_a<t\} = \lim_{t\to\infty}\frac{2}{\sqrt{2\pi}}\int_{a/\sqrt{t}}^{\infty} e^{-x^2/2}\,dx \\ &= \frac{2}{\sqrt{2\pi}}\int_{0}^{\infty} e^{-x^2/2}\,dx = 1. \end{aligned}$$

□

Example 4.5 (Recurrence in $\mathbb{R}^d$). To examine the recurrence of Brownian motion in higher dimensions, we have to determine if its FPT from any point to any sphere is finite. Due to the homogeneity properties of the relevant partial differential equations, it suffices to consider the FPT τ from any point $\boldsymbol{x}$ to the unit sphere. Due to the spherical symmetry of the problem, the probability for all points $\boldsymbol{x}$ that are equidistant from the sphere is the same. We therefore determine the recurrence probability from a sphere $|\boldsymbol{x}|=r>1$ to the unit sphere. The initial boundary value problem for $v(\boldsymbol{x},t)=\Pr\{\tau_D>t\,|\,\boldsymbol{x}(0)=\boldsymbol{x}\}$, (4.96), is

$$\frac{\partial v(\boldsymbol{x},t)}{\partial t} = \frac{1}{2}\Delta v(\boldsymbol{x},t) \;\text{ for } |\boldsymbol{x}|>1,\; t>0 \tag{4.104}$$

$$v(\boldsymbol{x},t) = 0 \;\text{ for } |\boldsymbol{x}|=1,\; t>0 \tag{4.105}$$

$$v(\boldsymbol{x},0) = 1 \;\text{ for } |\boldsymbol{x}|>1. \tag{4.106}$$

The initial boundary value problem (4.71)–(4.106) has a radial solution, $v(r,t)$ that satisfies

$$\frac{\partial v(r,t)}{\partial t} = \frac{1}{2}\left[\frac{\partial^2 v(r,t)}{\partial r^2} + \frac{d-1}{r}\frac{\partial v(r,t)}{\partial r}\right] \quad \text{for } r>1,\ t>0 \tag{4.107}$$

$$v(1,t) = 0, \quad \text{for } t>0, \quad v(r,0) = 1 \quad \text{for } r>1. \tag{4.108}$$

Obviously, the limit

$$v(r) = \lim_{t\to\infty} v(r,t) = \Pr\{\tau = \infty \mid |\boldsymbol{x}(0)| = r\} \tag{4.109}$$

is a nonnegative, nondecreasing bounded function of r that satisfies the stationary equation

$$\frac{\partial^2 v(r)}{\partial r^2} + \frac{d-1}{r}\frac{\partial v(r)}{\partial r} = 0 \quad \text{for } r>1 \tag{4.110}$$

$$v(1) = 0. \tag{4.111}$$

Because in one dimension ($d=1$) the only solution of eqs. (4.110), (4.111) that satisfies these conditions is a linear function, it must vanish identically, so that

$$\Pr\{\tau = \infty \mid |\boldsymbol{x}(0)| = r\} = 0,$$

or

$$\Pr\{\tau < \infty \mid |\boldsymbol{x}(0)| = r\} = 1 - \Pr\{\tau = \infty \mid |\boldsymbol{x}(0)| = r\} = 1,$$

as shown in the calculation above. We conclude, as above, that the one-dimensional Brownian motion is recurrent.

In two dimensions ($d=2$), the general solution of eq. (4.110) is $v(r) = a\log r + b$, where a and b are constants. The function $\log r$ is unbounded for large r, therefore the only solution of eqs. (4.110), (4.111) that satisfies these conditions must vanish identically as well. Thus the Brownian motion in the plane is also recurrent.

In higher dimensions ($d>2$) there are two solutions that satisfy all the required conditions. They are $v_0(r) = 0$ and the nontrivial solution $v_1(r) = 1 - r^{-d+2}$. To find out which one is the limit (4.109), consider the difference $u(r,t) = v(r,t) - v_1(r)$. Obviously, $u(r,t)$ is the solution of the equation (4.71) with the boundary condition (4.105) and the initial condition $u(r,0) = r^{-d+2} > 0$. Because eq. (4.71) is the diffusion equation in the domain $r>1$ with absorption at the boundary $r=1$, its Green's function is positive [78]. It follows that the solution of (4.71), with the boundary condition (4.105) is positive for every positive initial condition, $u(r,0) = r^{-d+2}$ being no exception. Thus $u(r,t) = v(r,t) - v_1(r) > 0$ for all $t>0$. It follows that $\lim_{t\to\infty} v(r,t) = v_1(r)$, so that $\Pr\{\tau = \infty \mid |\boldsymbol{x}(0)| = r\} = 1 - r^{-d+2}$. This means that the Brownian motion in dimension $d>2$ is nonrecurrent. Thus, there is a set of Brownian trajectories that never reach the unit sphere and whose probability is positive. □

Exercise 4.20 (BM on a sphere). Find the probability of the Brownian motion to reach a sphere of radius a from a concentric sphere of radius $b > a$ in any dimension $d \geq 2$. □

Exercise 4.21. (Recurrence of the Ornstein–Uhlenbeck process)

(i) Determine the recurrence of the d-dimensional Ornstein–Uhlenbeck process, defined by the stochastic dynamics $d\boldsymbol{x} = -\alpha\boldsymbol{x}\,dt + d\boldsymbol{w}(t)$, where α is a positive constant.

(ii) Is the Ornstein–Uhlenbeck process recurrent for α positive and negative?

(iii) Consider the case that α is a matrix. □

Exercise 4.22 (Lévy's arcsine law[117]). Define the occupation time $T(t)$ = the time a BM spent on the positive axis up to time t.

(i) Show that

$$\Pr\{T(t) < \tau \,|\, w(0) = 0\} = \frac{2}{\pi}\arcsin\sqrt{\frac{\tau}{t}} \quad \text{for } 0 \leq \tau \leq t. \tag{4.112}$$

(HINT: Obviously, $T(t) = \int_0^t \mathbf{1}_{[0,\infty)}(z)\,dz$. Use the Feynman–Kac formula for the function $u(x,t) = \mathbb{E}\left[\exp\{-sT(t)\} \,|\, w(0) = x\right]$. The initial condition for $u(x,t)$ is $u(x, 0+) = 1$. Assume $u(x,t)$ and its x derivative are continuous at the origin. Use the Laplace transform with respect to t to convert the partial differential equation of $u(x,t)$ to an ordinary differential equation. Show that

$$u(0,t) = \frac{1}{\pi}\int_0^t \frac{e^{-sz}}{\sqrt{z(t-z)}}\,dz \quad \text{for } s > 0. \tag{4.113}$$

Use eq. (4.112) to identify $1/\pi\sqrt{z(t-z)}$ as the pdf of $T(t)$.)

(ii) Find the Laplace transform of the pdf of $\Pr\{T(t) < \tau \,|\, w(0) = x\}$ for positive and negative x. (HINT: Condition on arriving at the origin before and after t.)

(iii) Find the Laplace transform of the pdf of the time spent in a finite interval.

(iv) Find the Laplace transform of the pdf of the time spent in a set that consists of a periodic arrangement of finite intervals.

(v) Find the Laplace transform of the pdf of the time spent outside a finite interval.

(vi) Use the results of (i)–(v) to determine the decay rate of the survival probability of a BM with unit killing rate on the line, the positive axis, on an interval, on the periodic set of (iv), and outside a finite interval. Show that in the first and last cases the decay rate is exponential, whereas in the other cases it is algebraic with the same power law for all cases. Can a set be found so that the survival probability decays at a different algebraic rate? □

4.5 The Fokker–Planck equation

The transition probability density function of the solution $\boldsymbol{x}_{\boldsymbol{x},s}(t)$ of the stochastic differential equation (4.47), denoted $p(\boldsymbol{y}, t \,|\, \boldsymbol{x}, s)$, satisfies two different partial differential equations, one with respect to the "forward variables" $(\boldsymbol{y}, t)$ and one with respect to the "backward variables" $(\boldsymbol{x}, s)$. The former is called the *Fokker–Planck equation* (FPE) or the *forward Kolmogorov equation*, and is the subject of this section, and the latter is called *the backward Kolmogorov equation* (see (4.63)) and is derived in Section 4.6.

Definition 4.5.1 (The Fokker–Planck operator). *The operator*

$$L_{\boldsymbol{y}} p = \sum_{i=1}^{d} \frac{\partial}{\partial y^i} \left\{ \sum_{j=1}^{d} \frac{\partial}{\partial y^j} \sigma^{ij}(\boldsymbol{y}, t) p - a^i(\boldsymbol{y}, t) p \right\} \tag{4.114}$$

is called the Fokker–Planck operator, or the forward Kolmogorov operator.

Note that the forward operator $L_{\boldsymbol{y}}$ is the formal adjoint, with respect to the $L^2(\mathbb{R}^d)$ inner product $\langle \cdot, \cdot \rangle_{L^2}$, of the operator $L^*_{\boldsymbol{x}}$, defined by (3.68), that appears in Itô's formula (3.67), in the sense that for all sufficiently smooth functions $f(\boldsymbol{x}), g(\boldsymbol{x})$ in $\mathbb{R}^d$ that vanish sufficiently fast at infinity

$$\int_{\mathbb{R}^d} g(\boldsymbol{y}) L_{\boldsymbol{y}} f(\boldsymbol{y})\, d\boldsymbol{y} = \langle L_{\boldsymbol{y}} f, g \rangle_{L^2} = \langle f, L^*_{\boldsymbol{y}} g \rangle_{L^2} = \int_{\mathbb{R}^d} f(\boldsymbol{y}) L^*_{\boldsymbol{y}} g(\boldsymbol{y})\, d\boldsymbol{y}. \tag{4.115}$$

Theorem 4.5.1 (The FPE). *The pdf* $p(\boldsymbol{y}, t \mid \boldsymbol{x}, s)$ *satisfies the initial value problem*

$$\frac{\partial p(\boldsymbol{y}, t \mid \boldsymbol{x}, s)}{\partial t} = L_{\boldsymbol{y}} p(\boldsymbol{y}, t \mid \boldsymbol{x}, s) \quad \text{for } \boldsymbol{x}, \boldsymbol{y} \in \mathbb{R}^d, \quad t > s \tag{4.116}$$

$$\lim_{t \to s} p(\boldsymbol{y}, t \mid \boldsymbol{x}, s) = \delta(\boldsymbol{x} - \boldsymbol{y}). \tag{4.117}$$

Proof. To derive the forward equation, we take the expectation in Itô's formula,

$$\begin{aligned} &f(\boldsymbol{x}(t)) - f(\boldsymbol{x}(s)) \\ &= \int_s^t \left[\sum_{i=1}^d a^i(\boldsymbol{x}(\tau), \tau) \frac{\partial f(\boldsymbol{x}(\tau))}{\partial x^i} + \sum_{i=1}^d \sum_{j=1}^d \sigma^{ij}(\boldsymbol{x}(\tau), \tau) \frac{\partial^2 f(\boldsymbol{x}(\tau))}{\partial x^i \partial x^j} \right] d\tau \\ &\quad + \sum_{j=1}^m \int_s^t \left[\sum_{i=1}^d b^{ij}(\boldsymbol{x}(\tau), \tau) \frac{\partial f(\boldsymbol{x}(\tau))}{\partial x^i} \right] dw^j(\tau), \end{aligned}$$

conditioned on $\boldsymbol{x}(s) = \boldsymbol{x}$, where $f(\boldsymbol{x})$ is a smooth function that vanishes outside some bounded set. We obtain

$$\begin{aligned} \mathbb{E}\left[f(\boldsymbol{x}(t)) \mid \boldsymbol{x}(s) = \boldsymbol{x}\right] = f(\boldsymbol{x}) + \mathbb{E}\Bigg\{ \int_s^t \Bigg[\sum_{i=1}^d a^i(\boldsymbol{x}(\tau), \tau) \frac{\partial f(\boldsymbol{x}(\tau))}{\partial x^i} \\ + \sum_{i=1}^d \sum_{j=1}^d \sigma^{ij}(\boldsymbol{x}(\tau), \tau) \frac{\partial^2 f(\boldsymbol{x}(\tau))}{\partial x^i \partial x^j} \Bigg] d\tau \,\Big|\, \boldsymbol{x}(s) = \boldsymbol{x} \Bigg\}. \end{aligned}$$

This means that

$$\int f(\boldsymbol{y})\,p(\boldsymbol{y},t\,|\,\boldsymbol{x},s) = f(\boldsymbol{x}) + \int d\boldsymbol{y} \int_s^t \left[\sum_{i=1}^d a^i(\boldsymbol{y},\tau) \frac{\partial f(\boldsymbol{y})}{\partial y^i} d\boldsymbol{y} \right. \tag{4.118}$$

$$\left. + \sum_{i=1}^d \sum_{j=1}^d \sigma^{ij}(\boldsymbol{y},\tau) \frac{\partial^2 f(\boldsymbol{y})}{\partial y^i \partial y^j} \right] p(\boldsymbol{y},\tau\,|\,\boldsymbol{x},s)\,d\tau.$$

Equation (4.118) is a *weak* form of a partial differential equation that the transition pdf $p(\boldsymbol{y},t\,|\,\boldsymbol{x},s)$ satisfies. A *weak solution* is a function $p(\boldsymbol{y},t\,|\,\boldsymbol{x},s)$ that satisfies eq. (4.118) for all test functions $f(\boldsymbol{y})$; that is, for all infinitely differentiable functions $f(\boldsymbol{y})$ that vanish outside a bounded domain [152], [78].

A partial differential equation for $p(\boldsymbol{y},t\,|\,\boldsymbol{x},s)$ is obtained by integrating by parts in eq. (4.118) and changing the order of integration,

$$\int f(\boldsymbol{y})\,p(\boldsymbol{y},t\,|\,\boldsymbol{x},s)\,d\boldsymbol{y} = \int f(\boldsymbol{y})\,\delta(\boldsymbol{x}-\boldsymbol{y})\,d\boldsymbol{y} + \int f(\boldsymbol{y}) \tag{4.119}$$

$$\times \int_s^t \sum_{i=1}^d \frac{\partial}{\partial y^i} \left[\sum_{j=1}^d \frac{\partial \sigma^{ij}(\boldsymbol{y},\tau) p(\boldsymbol{y},\tau\,|\,\boldsymbol{x},s)}{\partial y^j} - a^i(\boldsymbol{y},\tau)\,p(\boldsymbol{y},\tau\,|\,\boldsymbol{x},s) \right] d\tau d\boldsymbol{y}.$$

Equation (4.119) means that $p(\boldsymbol{y},t\,|\,\boldsymbol{x},s)$ satisfies the equation

$$p(\boldsymbol{y},t\,|\,\boldsymbol{x},s) = \delta(\boldsymbol{x}-\boldsymbol{y}) \tag{4.120}$$

$$+ \int_s^t \sum_{i=1}^d \frac{\partial}{\partial y^i} \left[\sum_{j=1}^d \frac{\partial \sigma^{ij}(\boldsymbol{y},\tau) p(\boldsymbol{y},\tau\,|\,\boldsymbol{x},s)}{\partial y^j} - a^i(\boldsymbol{y},\tau)\,p(\boldsymbol{y},\tau\,|\,\boldsymbol{x},s) \right] d\tau$$

in the sense of distributions. Equation (4.120) can be written in differential form as (4.116) with the initial condition (4.117). □

A classical solution of eq. (4.116) is a function that has all the derivatives that appear in the equation and the equation is satisfied at all points. It is known from the theory of parabolic partial differential equations [78] that under a mild regularity assumption, if $\boldsymbol{\sigma}(\boldsymbol{y},\tau)$ is a strictly positive definite matrix, the initial value problem (4.116), (4.117) has a unique classical solution,

The one-dimensional Fokker–Planck equation has the form

$$\frac{\partial p(y,t\,|\,x,s)}{\partial t} = \frac{1}{2} \frac{\partial^2 \left[b^2(y,t)p(y,t\,|\,x,s)\right]}{\partial y^2} - \frac{\partial \left[a(y,t)p(y,t\,|\,x,s)\right]}{\partial y} \tag{4.121}$$

with the initial condition

$$\lim_{t \downarrow s} p(y,t\,|\,x,s) = \delta(y-x). \tag{4.122}$$

Exercise 4.23. (The solution of the FPE satisfies the CKE and (4.39)–(4.41))

(i) Use the existence and uniqueness theorem for linear parabolic initial value problems to show that the solution $p\,(\boldsymbol{y},t\,|\,\boldsymbol{x},s)$ of (4.116), (4.117) satisfies the Chapman–Kolmogorov equation (2.59).

(ii) Prove that if $\boldsymbol{a}(\boldsymbol{x},t)$ and $\boldsymbol{\sigma}(\boldsymbol{x},t)$ are sufficiently regular, then the pdf $p\,(\boldsymbol{y},t\,|\,\boldsymbol{x},s)$ of (4.116), (4.117) satisfies (4.39)–(4.41) in the sense that

$$\lim_{\Delta t\to 0}\frac{1}{\Delta t}\int_{\mathbb{R}^d}(\boldsymbol{y}-\boldsymbol{x})p\,(\boldsymbol{y},t+\Delta t\,|\,\boldsymbol{x},t)\,d\boldsymbol{y}=\boldsymbol{a}(\boldsymbol{x},t) \tag{4.123}$$

$$\lim_{\Delta t\to 0}\frac{1}{\Delta t}\int_{\mathbb{R}^d}(\boldsymbol{y}-\boldsymbol{x})(\boldsymbol{y}-\boldsymbol{x})^T p\,(\boldsymbol{y},t+\Delta t\,|\,\boldsymbol{x},t)\,d\boldsymbol{y}=\boldsymbol{\sigma}(\boldsymbol{x},t) \tag{4.124}$$

$$\lim_{\Delta t\to 0}\frac{1}{\Delta t}\int_{\mathbb{R}^d}|\boldsymbol{y}-\boldsymbol{x})|^{2+\delta}p\,(\boldsymbol{y},t+\Delta t\,|\,\boldsymbol{x},t)\,d\boldsymbol{y}=0\quad\text{for }\delta>0. \tag{4.125}$$

□

The transition pdf of the solution of the Stratonovich equation (4.12) does not satisfy the Fokker–Planck equation (4.116), however, the partial differential equation it satisfies can be found by converting the Stratonovich dynamics (4.12) into the equivalent Itô dynamics (4.13). The Fokker–Planck equation corresponding to (4.13) is found by replacing $a(y,t)$ in (4.121) with the corrected drift $\tilde{a}(y,t) = a(y,t)+\frac{1}{2}b(y,t)b_y(y,t)$. This results in the Fokker–Planck equation

$$\begin{aligned}\frac{\partial p\,(y,t\,|\,x,s)}{\partial t}&=\frac{1}{2}\frac{\partial^2\left[b^2(y,t)p\,(y,t\,|\,x,s)\right]}{\partial y^2}\\&-\frac{\partial}{\partial y}\left\{\left[a(y,t)+\frac{1}{2}b(y,t)\frac{\partial}{\partial y}b(y,t)\right]p\,(y,t\,|\,x,s)\right\}.\end{aligned} \tag{4.126}$$

Now, using the identity

$$\begin{aligned}\frac{1}{2}\frac{\partial}{\partial y}\left\{b(y,t)\frac{\partial}{\partial y}\left[b(y,t)p\,(y,t\,|\,x,s)\right]\right\}&=\frac{\partial^2}{\partial y^2}\left[b^2(y,t)p\,(y,t\,|\,x,s)\right]\\&-\frac{\partial}{\partial y}\left\{\left[b(y,t)\frac{\partial}{\partial y}b(y,t)\right]p\,(y,t\,|\,x,s)\right\},\end{aligned}$$

we rewrite eq. (4.126) in the form

$$\begin{aligned}\frac{\partial p\,(y,t\,|\,x,s)}{\partial t}&=\frac{1}{2}\frac{\partial}{\partial y}\left\{b(y,t)\frac{\partial\left[b(y,t)p\,(y,t\,|\,x,s)\right]}{\partial y}\right\}\\&-\frac{\partial\left[a(y,t)p\,(y,t\,|\,x,s)\right]}{\partial y}.\end{aligned} \tag{4.127}$$

Equation (4.127) is called the *Fokker–Planck–Stratonovich equation*.

Note that neither the Fokker–Planck equation (4.121) nor the Fokker–Planck–Stratonovich equation (4.127) are Fickian. Although Fick's laws (1.1) give the driftless (i.e., $a(y,t) = 0$) diffusion equation $p_t(y,t) = [D(y,t)p_y(y,t)]_y$, the Fokker–Planck equation (4.121) gives the driftless diffusion equation in the form $p_t(y,t) = [D(y,t)p\,(y,t)]_{yy}$, where $D(y,t) = b^2(y,t)/2$ and the Fokker–Planck–Stratonovich equation (4.127) gives the driftless diffusion equation in the form $p_t(y,t\,|\,x,s) = [\sqrt{D(y,t)}(\sqrt{D(y,t)}p\,(y,t\,|\,x,s))_y]_y$. Obviously, for constant diffusion coefficients all three diffusion equations are the same. However, if the diffusion coefficient is variable, they are all different. State-dependent diffusion coefficients arise, for example, in the diffusion of ions in liquid (e.g., water) when the ambient temperature is not uniform.

The simplest example of the Fokker–Planck equation corresponds to the case $a(x,t) = 0$ and $b(x,t) = 1$; that is, $x(t)$ is the Brownian motion $w(t)$. In this case, the Fokker–Planck equation (4.121) and the initial condition (4.122) reduce to the diffusion equation and the initial condition (2.26), moved from the origin to the point x.

The Fokker–Planck equation corresponding to the Ornstein–Uhlenbeck process (or colored noise), defined by the stochastic dynamics (1.60); that is, by

$$dx(t) = -ax(t)\,dt + b\,dw(t), \quad x(s) = x, \tag{4.128}$$

is

$$\frac{\partial p\,(y,t\,|\,x,s)}{\partial t} = \frac{b^2}{2}\frac{\partial^2 p\,(y,t\,|\,x,s)}{\partial y^2} + a\frac{\partial y p\,(y,t\,|\,x,s)}{\partial y} \tag{4.129}$$

$$p\,(y,t\,|\,x,s) \to \delta(y-x) \ \text{ as } t \downarrow s \quad \text{(converges from above).} \tag{4.130}$$

Exercise 4.24 (Explicit solution of the FPE (4.129), (4.130)). Use the explicit solution of eq. (4.128) (see Exercise 4.9(iii)) to find the explicit solution of the Fokker–Planck equation (4.129), (4.130). □

Exercise 4.25 (Explicit solution of the FPE for exactly solvable SDEs). Write down the Fokker–Planck equation s and initial conditions corresponding to the exactly solvable stochastic differential equations of Section 4.2 and use the explicit solutions of the stochastic equations to construct the solutions of the corresponding Fokker–Planck initial value problems. □

Exercise 4.26 (Fokker–Planck–Stratonovich equation for (4.14)). Write down the Fokker–Planck–Stratonovich equation corresponding to the Stratonovich dynamics (4.14) in $\mathbb{R}^d$ by using (4.15) in (4.116). □

Exercise 4.27 (FPE for SDEs defined with the backward integral). Write down the Fokker–Planck equation corresponding to stochastic differential equations defined with the backward integral. □

4.6 The backward Kolmogorov equation

Theorem 4.6.1 (The backward Kolmogorov equation). *The transition probability density function $p(\boldsymbol{y},t\,|\,\boldsymbol{x},s)$ of the solution $\boldsymbol{x}_{\boldsymbol{x},s}(t)$ of the stochastic differential equation* (4.47) *satisfies with respect to the backward variables $(\boldsymbol{x},s)$ the backward Kolmogorov equation*

$$\begin{aligned}
&\frac{\partial p(\boldsymbol{y},t\,|\,\boldsymbol{x},s)}{\partial s}\\
&= -\sum_{i=1}^{d} a^i(\boldsymbol{x},s)\frac{\partial p(\boldsymbol{y},t\,|\,\boldsymbol{x},s)}{\partial x^i} - \sum_{i=1}^{d}\sum_{j=1}^{d}\sigma^{ij}(\boldsymbol{x},s)\frac{\partial^2 p(\boldsymbol{y},t\,|\,\boldsymbol{x},s)}{\partial x^i \partial x^j}\\
&= -L^*_{\boldsymbol{x}}p(\boldsymbol{y},t\,|\,\boldsymbol{x},s)
\end{aligned} \tag{4.131}$$

with the terminal condition

$$\lim_{s\to t} p(\boldsymbol{y},t\,|\,\boldsymbol{x},s) = \delta(\boldsymbol{x}-\boldsymbol{y}). \tag{4.132}$$

Proof. Due to the Chapman–Kolmogorov equation (2.59), if $s < s+\Delta s < t$, then we can write

$$p(\boldsymbol{y},t\,|\,\boldsymbol{x},s) = \int p(\boldsymbol{y},t\,|\,\boldsymbol{z},s+\Delta s)p(\boldsymbol{z},s+\Delta s\,|\,\boldsymbol{x},s)\,d\boldsymbol{z}. \tag{4.133}$$

Using this, and the fact that

$$\int p(\boldsymbol{z},s+\Delta s\,|\,\boldsymbol{x},s)\,d\boldsymbol{z} = 1, \tag{4.134}$$

we obtain

$$\begin{aligned}
&p(\boldsymbol{y},t\,|\,\boldsymbol{x},s+\Delta s) - p(\boldsymbol{y},t\,|\,\boldsymbol{x},s)\\
&= \int p(\boldsymbol{z},s+\Delta s\,|\,\boldsymbol{x},s)\left[p(\boldsymbol{y},t\,|\,\boldsymbol{x},s+\Delta s) - p(\boldsymbol{y},t\,|\,\boldsymbol{z},s+\Delta s)\right]d\boldsymbol{z}.
\end{aligned} \tag{4.135}$$

Next, we expand

$$\begin{aligned}
p(\boldsymbol{y},t\,|\,\boldsymbol{z},s+\Delta s) &= p(\boldsymbol{y},t\,|\,\boldsymbol{x},s+\Delta s) + \sum_{i=1}^{d}(z^i-x^i)\frac{\partial p(\boldsymbol{y},t\,|\,\boldsymbol{x},s+\Delta s)}{\partial x^i}\\
&+ \frac{1}{2}\sum_{i=1}^{d}\sum_{j=1}^{d}(z^i-x^i)(z^j-x^j)\frac{\partial^2 p(\boldsymbol{y},t\,|\,\boldsymbol{x},s+\Delta s)}{\partial x^i\partial x^j} + O\left(|\boldsymbol{x}-\boldsymbol{z}|^3\right)
\end{aligned} \tag{4.136}$$

and substitute in eq. (4.135). From eq. (4.134), we obtain

$$\int p(\boldsymbol{z},s+\Delta s\,|\,\boldsymbol{x},s)p(\boldsymbol{y},t\,|\,\boldsymbol{x},s+\Delta s)d\boldsymbol{z} = p(\boldsymbol{y},t\,|\,\boldsymbol{x},s+\Delta s),$$

and from the smoothness of $p(\boldsymbol{y},t\,|\,\boldsymbol{x},s)$ and from eqs. (4.45), we find that

$$\begin{aligned}
&\int p(\boldsymbol{z},s+\Delta s\,|\,\boldsymbol{x},s)(z^i-x^i)\frac{\partial p(\boldsymbol{y},t\,|\,\boldsymbol{x},s+\Delta s)}{\partial x^i}\,d\boldsymbol{z}\\
&=a^i(\boldsymbol{x},t)\frac{\partial p(\boldsymbol{y},t\,|\,\boldsymbol{x},s)}{\partial x^i}\,\Delta s+o(\Delta s)
\end{aligned}$$

and

$$\begin{aligned}
&\int p(\boldsymbol{z},s+\Delta s\,|\,\boldsymbol{x},s)(z^i-x^i)(z^j-x^j)\frac{\partial^2 p(\boldsymbol{y},t\,|\,\boldsymbol{x},s+\Delta s)}{\partial x^i\partial x^j}\,d\boldsymbol{z}\\
&=2\sigma^{ij}(\boldsymbol{x}(\tau),\tau)\frac{\partial^2 p(\boldsymbol{y},t\,|\,\boldsymbol{x},s)}{\partial x^i\partial x^j}\Delta s+o(\Delta s).
\end{aligned}\tag{4.137}$$

Finally, from eq. (4.9) and the Cauchy–Schwarz inequality, we find that

$$\int O\left(|\boldsymbol{x}-\boldsymbol{z}|^3\right)p(\boldsymbol{z},s+\Delta s\,|\,\boldsymbol{x},s)\,d\boldsymbol{z}=O\left(\Delta s^{3/2}\right).\tag{4.138}$$

Using eqs. (4.137) and (4.138) in eq. (4.136), dividing by Δs, and taking the limit $\Delta s\to 0$, we obtain (4.131). □

Exercise 4.28 (The solution of the FPE solves the BKE). Use Exercise (4.23) and the Chapman–Kolmogorov equation (2.59), as in the proof of Theorem (4.131), to prove that if $\boldsymbol{a}(\boldsymbol{x},t)$ and $\boldsymbol{\sigma}(\boldsymbol{x},t)$ are sufficiently regular, then the solution $p(\boldsymbol{y},t\,|\,\boldsymbol{x},s)$ of the FPE (4.116), (4.117) satisfies the BKE (4.131), (4.132). □

4.7 Appendix: Proof of Theorem 4.1.1

We construct the solution to (4.5) by the usual method of successive approximations which are defined by

$$\begin{aligned}
x_0(t,\omega)&=x_0\\
x_{n+1}(t,\omega)&=x_0+\int_0^t a(x_{n-1}(s,\omega),s,\omega)\,ds\\
&\quad+\int_0^t b(x_{n-1}(s,\omega),s,\omega)\,dw(s,\omega),\quad (n\geq 1).
\end{aligned}\tag{4.139}$$

Next, we will show that the approximating sequence converges uniformly in $[0,T]$ for positive, but sufficiently small T for almost all $\omega\in\Omega$. First, it can be easily shown by induction that the Lipschitz condition ensures that $x_n(t,\omega)\in H_2[0,T]$ for each $T>0$ and $n>0$. Subtracting two consecutive equations (4.139), we obtain the inequalities

$$\begin{aligned}
&\mathbb{E}\left[x_{n+1}(t,\omega)-x_n(t,\omega)\right]^2\\
&\leq 2\mathbb{E}\left|\int_0^t \left[a(x_n(s,\omega),s,\omega)-a(x_{n-1}(s,\omega),s,\omega)\right]ds\right|^2\\
&\quad+2\mathbb{E}\left|\int_0^t \left[b(x_n(s,\omega),s,\omega)-b(x_{n-1}(s,\omega),s,\omega)\right]dw(s,\omega)\right|^2\\
&\leq 2t\int_0^t \mathbb{E}\left|a(x_n(s,\omega),s,\omega)-a(x_{n-1}(s,\omega),s,\omega)\right|^2 ds\\
&\quad+2\int_0^t \mathbb{E}\left|b(x_n(s,\omega),s,\omega)-b(x_{n-1}(s,\omega),s,\omega)\right|^2 ds. \qquad (4.140)
\end{aligned}$$

We used the inequality $(A+B)^2 \leq 2(A^2+B^2)$ in the first inequality of (4.140), the Cauchy–Schwarz inequality in the first integral of the second inequality and the property (3.15) of the variance of the Itô integral in the second. Applying the uniform Lipschitz condition (4.7) to the integrands in (4.140), we obtain the inequality

$$\begin{aligned}
&\mathbb{E}\left[x_{n+1}(t,\omega)-x_n(t,\omega)\right]^2\\
&\leq 4K^2(T+1)\int_0^t \mathbb{E}\left[x_n(s,\omega)-x_{n-1}(s,\omega)\right]^2 ds. \qquad (4.141)
\end{aligned}$$

Iteration of (4.141) gives

$$\mathbb{E}\left[x_{n+1}(t,\omega)-x_n(t,\omega)\right]^2 \leq \frac{M\left[4K^2(T+1)t\right]^{n+1}}{(n+1)!}, \qquad (4.142)$$

where $M = \int_0^T \mathbb{E}\left[a^2(x_0,s,\omega)+b^2(x_0,s,\omega)\right] ds$. In particular, (4.142) implies that

$$\begin{aligned}
m_n(t) =& \mathbb{E}\left|\int_0^t \left[a(x_n(s,\omega),s,\omega)-a(x_{n-1}(s,\omega),s,\omega)\right]ds\right| \qquad (4.143)\\
&\leq K\left\{T\int_0^T \mathbb{E}\left[x_n(s,\omega)-x_{n-1}(s,\omega)\right]^2 ds\right\}^{1/2} \leq K\sqrt{MT\frac{C^n}{n!}},
\end{aligned}$$

where

$$C = 4K^2(T+1)T < 1, \qquad (4.144)$$

if T is sufficiently small. Denoting

$$X_n(t,\omega) = \left| \int_0^t [a(x_n(s,\omega),s,\omega) - a(x_{n-1}(s,\omega),s,\omega)]\, ds \right|,$$

we have $\mathbb{E}X_n(t,\omega) = m_n(t)$ and

$$\sigma_n^2(t) = \mathbb{E}\left[X_n(t,\omega) - m_n(t)\right]^2 \le \mathbb{E}[X_n(t,\omega)]^2 \le K^2MT^2\frac{C^n}{n!}. \tag{4.145}$$

Choosing $\varepsilon_n = 1/\sqrt{(n-2)!}$, we find that for sufficiently large n

$$\varepsilon_n - m_n(t) \ge \frac{\varepsilon_n}{2}.$$

Now, Chebyshev's inequality gives

$$\begin{aligned} Pr\{|X_n(t,\omega)| \ge \varepsilon_n\} &\le \Pr\{|X_n(t,\omega) - m_n(t)| \ge \varepsilon_n - m_n(t)\} \\ &\le \Pr\left\{|X_n(t,\omega) - m_n(t)| \ge \frac{\varepsilon_n}{2}\right\} \\ &\le \frac{4\sigma_n^2(t)}{\varepsilon_n^2} \le \frac{4K^2MT^2C^n}{n(n-1)}. \end{aligned} \tag{4.146}$$

It follows that

$$\Pr\left\{ \max_{0\le t\le T} \left| \int_0^t [a(x_n(s,\omega),s,\omega) - a(x_{n-1}(s,\omega),s,\omega)]\, ds \right| \ge \frac{1}{\sqrt{(n-2)!}} \right\}$$
$$\le \frac{4K^2MT^2C^n}{n(n-1)},$$

because the inequality (4.146) holds for all $t \in [0,T]$. Given (4.144), we see that

$$\sum_{n=2}^{\infty} \frac{4K^2MT^2C^n}{n(n-1)} < \infty,$$

so that by the Borel–Cantelli lemma

$$\Pr\left\{ \max_{0\le t\le T} \left| \int_0^t [a(x_n(s,\omega),s,\omega) - a(x_{n-1}(s,\omega),s,\omega)]\, ds \right| \right.$$
$$\left. \ge \frac{1}{\sqrt{(n-2)!}} \text{ i.o.} \right\} = 0.$$

This means that for almost all $\omega \in \Omega$

$$\begin{aligned} \sum_{n=N(\omega)}^{\infty} \left| \int_0^t [a(x_n(s,\omega),s,\omega) - a(x_{n-1}(s,\omega),s,\omega)]\, ds \right| &\le \sum_{n=N(\omega)}^{\infty} \frac{1}{\sqrt{(n-2)!}} \\ &< \infty, \end{aligned} \tag{4.147}$$

which, in turn, means that for almost all $\omega \in \Omega$ the series

$$\sum_{n=1}^{\infty} \int_0^t [a(x_n(s,\omega), s, \omega) - a(x_{n-1}(s,\omega), s, \omega)]\, ds$$

converges uniformly in the interval $[0, T]$.

Now, we derive a similar estimate for the difference

$$Y_n(t,\omega) = \int_0^t [b(x_n(s,\omega), s, \omega) - b(x_{n-1}(s,\omega), s, \omega)]\, dw(s,\omega).$$

First, we show that $Y_n^2(t,\omega)$ is a submartingale. To see this, we write

$$dY_n(t,\omega) = B(t,\omega)\, dw(t,\omega),$$

where

$$B(t,\omega) = b(x_n(t,\omega), t, \omega) - b(x_{n-1}(t,\omega), t, \omega),$$

and apply Itô's formula (3.67) to the function $f(y) = y^2$. We obtain

$$dY_n^2(t,\omega) = B^2(t,\omega)\, dt + 2B(t,\omega)Y_n(t,\omega)\, dw(t,\omega),$$

or in integral form,

$$Y_n^2(t,\omega) = Y_n^2(s,\omega) + \int_s^t B^2(s,\omega)\, ds + \int_s^t 2B(s,\omega)Y_n(s,\omega)\, dw(t,\omega).$$

Taking conditional expectation and remembering that the conditional expectation of the Itô integral vanishes, we obtain

$$\mathbb{E}\left\{Y_n^2(t,\omega) \,|\, Y_n^2(s,\omega) = z\right\} = z + \int_s^t \mathbb{E}B^2(s,\omega)\, ds \geq z,$$

which means that $Y_n^2(t,\omega)$ is a submartingale.

It follows from the submartingale inequality (2.69), the Lipschitz condition, and

(4.142) that

$$
\begin{aligned}
&\Pr\left\{\max_{0\le t\le T}|Y_n(t,\omega)|\ge\varepsilon_n\right\}\\
=&\Pr\left\{\max_{0\le t\le T}|Y_n(t,\omega)|^2\ge\varepsilon_n^2\right\}\le\frac{\mathbb{E}\,|Y_n(T,\omega)|^2}{\varepsilon_n^2}\\
=&\frac{1}{\varepsilon_n^2}\int_0^T\mathbb{E}\,[b(x_n(s,\omega),s,\omega)-b(x_{n-1}(s,\omega),s,\omega)]^2\,ds\\
\le&\left(\frac{K}{\varepsilon_n}\right)^2E\int_0^T[x_n(s,\omega)-x_{n-1}(s,\omega)]^2\,ds\\
\le&8MK^2\frac{C^n}{n!\varepsilon_n^2}\le\frac{8K^2M^2C^n}{n(n-1)}.
\end{aligned}
\tag{4.148}
$$

We conclude from the Borel–Cantelli lemma that the series $\sum_{n=1}^{\infty}Y_n(t,\omega)$ also converges uniformly for almost all $\omega\in\Omega$.

It follows now from eq. (4.140) that the series $\sum_{n=1}^{\infty}[x_{n+1}(t,\omega)-x_n(t,\omega)]$ converges uniformly for almost all $\omega\in\Omega$. Setting $x(t,\omega)=\lim_{n\to\infty}x_n(t,\omega)$, we see that $x(t,\omega)$ is continuous, adapted, and satisfies the Itô integral equation (4.6), as required by conditions I and II of the definition of a solution. This proves existence of a solution.

To prove uniqueness of solutions, assume that $x_1(t,\omega)$ and $x_2(t,\omega)$ are two solutions of (4.5), denote by $\tau_1(\omega)$ and $\tau_2(\omega)$ their first exit times from a sphere of radius N; that is, $\tau_i(\omega)=\inf\{t>0\,|\,|x_i(t,\omega)|\ge N\}$ $(i=1,2)$, and denote the minimum between the two $T_N(\omega)=\tau_1(\omega)\wedge\tau_2(\omega)$. Note that $T_N\to\infty$ as $N\to\infty$ on almost every trajectory $\omega\in\Omega$. From eq. (4.6), we obtain for $0\le t\le T$,

$$
\begin{aligned}
&x_1(t\wedge T_N(\omega),\omega)-x_2(t\wedge T_N(\omega),\omega)\\
=&\int_0^{t\wedge T_N(\omega)}[a(x_1(s,\omega),s,\omega)-a(x_2(s,\omega),s,\omega)]\,ds\\
+&\int_0^{t\wedge T_N(\omega)}[b(x_1(s,\omega),s,\omega)-b(x_2(s,\omega),s,\omega)]\,dw(s,\omega)
\end{aligned}
$$

Using the previously derived inequalities, we get

$$\begin{aligned}
&\mathbb{E}\left|x_1\left(t\wedge T_N(\omega),\omega\right)-x_2\left(t\wedge T_N(\omega),\omega\right)\right|^2\\
&\leq 4(T+1)K^2\mathbb{E}\int_0^{t\wedge T_N(\omega)}\left|x_1(s,\omega)-x_2(s,\omega)\right|^2\,ds\\
&=4(T+1)K^2\int_0^t\mathbb{E}\left|x_1\left(s\wedge T_N(\omega),\omega\right)-x_2\left(s\wedge T_N(\omega),\omega\right)\right|^2\,ds.
\end{aligned}$$

It follows from Gronwall's inequality (4.150) that

$$\mathbb{E}\left|x_1(t\wedge T_N(\omega),\omega)-x_2(t\wedge T_N(\omega),\omega)\right|^2=0$$

for all $0\leq t\leq T$ and all T; that is, for all $t\geq 0$. Thus

$$x_1(t\wedge T_N(\omega),\omega)=x_2(t\wedge T_N(\omega),\omega)$$

for almost all $\omega\in\Omega$ and for all N. Taking the limit $N\to\infty$ gives that

$$x_1(t,\omega)=x_2(t,\omega)$$

for all $t>0$, which proves uniqueness.

Theorem 4.7.1 (Gronwall's inequality). *If a continuous function $f(t)$ satisfies the inequality*

$$0\leq f(t)\leq A(t)+\int_0^t B(s)f(s)\,ds \tag{4.149}$$

in some interval $[0,T]$ for an integrable function $A(t)$ and a nonnegative integrable function $B(t)$, then

$$f(t)\leq A(t)+\int_0^t A(s)B(s)\exp\left\{\int_s^t B(u)\,du\right\}ds. \tag{4.150}$$

Proof. If $B(t)$ is continuous, we denote

$$y(t)=\int_0^t B(s)f(s)\,ds$$

and note that $y(t)$ is differentiable and $y(0)=0$. Multiplying both sides of the inequality (4.149) by the nonnegative function $B(t)$, we write the inequality in the form

$$y'(t)\leq A(t)B(t)+B(t)y(t),$$

which, in turn, can be written as

$$\frac{d}{dt}\exp\left\{-\int_0^t B(s)\,ds\right\}y(t) \le A(t)B(t)\exp\left\{-\int_0^t B(s)\,ds\right\}.$$

Integrating between 0 and t, we obtain

$$y(t) \le \int_0^t A(t')B(t')\exp\left\{\int_{t'}^t B(s)\,ds\right\}dt'.$$

Finally, adding $A(t)$ to both sides and using (4.149), we obtain (4.150). □

Corollary 4.7.1. *If $A(t) = 0$, then $f(t) = 0$.*

4.7.1 Continuous dependence on parameters

Assume the coefficients $\boldsymbol{a}_k(\boldsymbol{x}, t, \omega)$, the $n \times m$ diffusion matrices $\boldsymbol{B}_k(\boldsymbol{x}, t, \omega)$, and the initial conditions $\boldsymbol{x}_k(0, \omega)$ $(k = 1, 2, \dots)$ satisfy the condition of the existence and uniqueness theorem with the same Lipschitz constant for all k. We consider the stochastic integral equations

$$\boldsymbol{x}_k(t,\omega) = \boldsymbol{x}_k(0,\omega) + \int_0^t \boldsymbol{a}_k(\boldsymbol{x}_k(s,\omega), s, \omega)\,ds + \int_0^t \boldsymbol{B}_k(\boldsymbol{x}_k(s,\omega), s, \omega)\,d\boldsymbol{w}(s,\omega), \tag{4.151}$$

where $\boldsymbol{w}(t, \omega)$ is an m-dimensional Brownian motion. If the coefficients and the initial conditions converge to limits, in an appropriate sense, then so do the solutions. Specifically,

Theorem 4.7.2 (Continuous dependence on parameters). *If for every $N > 0$, $t \in [0, T]$, and $\varepsilon > 0$*

$$\lim_{k\to\infty} \Pr\left\{\sup_{|\boldsymbol{x}|\le N}\left[|\boldsymbol{a}_k(\boldsymbol{x},t,\omega) - \boldsymbol{a}_0(\boldsymbol{x},t,\omega)| + |\boldsymbol{B}_k(\boldsymbol{x},t,\omega) - \boldsymbol{B}_0(\boldsymbol{x},t,\omega)|\right] > \varepsilon\right\} = 0$$

and

$$\lim_{k\to\infty} \mathbb{E}\left[\boldsymbol{x}_k(0,\omega) - \boldsymbol{x}_0(0,\omega)\right]^2 = 0,$$

then

$$\lim_{k\to\infty}\sup_{0\le t\le T} \mathbb{E}\left[\boldsymbol{x}_k(t,\omega) - \boldsymbol{x}_0(t,\omega)\right]^2 = 0.$$

Corollary 4.7.2. *If k is a parameter varying in some set such that the limit $k \to k_0$ is defined, then the continuous dependence on parameters theorem holds with the limit $k \to \infty$ replaced by $k \to k_0$.*

The solution $\boldsymbol{x}_k(t,\omega)$ is differentiable with respect to k and its derivative satisfies a linear stochastic differential equation if the coefficients and the initial conditions satisfy certain differentiability and integrability conditions [84].

Theorem 4.7.3 (Differentiability with respect to parameters). *Assume the following conditions.*
1. $\mathbb{E}\left|\boldsymbol{x}_k(0,\omega)\right|^2 < \infty$, the derivative

$$\boldsymbol{x}'_k(0,\omega) = \frac{\partial \boldsymbol{x}_k(0,\omega)}{\partial k}$$

exists, and

$$\lim_{\Delta k\to 0} \mathbb{E}\left|\boldsymbol{x}'_k(0,\omega) - \frac{\Delta \boldsymbol{x}_k(0,\omega)}{\Delta k}\right|^2 = 0.$$

2. The derivatives $\boldsymbol{a}'_k(\boldsymbol{x},t,\omega)$ and $\boldsymbol{B}'_k(\boldsymbol{x},t,\omega)$ exist and

$$\lim_{\Delta k\to 0} \mathbb{E}\int_0^T \left\{\left[\frac{\Delta \boldsymbol{a}_k(\boldsymbol{x},s,\omega)}{\Delta k} - \boldsymbol{a}'_k(\boldsymbol{x},t,\omega)\right]^2 \right.$$
$$\left. + \left[\frac{\Delta \boldsymbol{B}_k(\boldsymbol{x},s,\omega)}{\Delta k} - \boldsymbol{B}'_k(\boldsymbol{x},t,\omega)\right]^2\right\}_{\boldsymbol{x}=\boldsymbol{x}_k(s,\omega)} ds = 0.$$

3. The coefficients are continuously differentiable functions of $\boldsymbol{x}$ and their derivatives are uniformly bounded by the same constant K for almost all $\omega \in \Omega$. Then $\boldsymbol{x}_k(t,\omega)$ is differentiable with respect to k and

$$\boldsymbol{x}'_k(t,\omega) \tag{4.152}$$
$$= \boldsymbol{x}'_k(0,\omega) + \int_0^t \left[\boldsymbol{a}'_k(\boldsymbol{x}_k(s,\omega),s,\omega) + \boldsymbol{a}_k(\boldsymbol{x}_k(s,\omega),s,\omega)\cdot \boldsymbol{x}'_k(s,\omega)\right] ds$$
$$+ \int_0^t \left[\boldsymbol{B}'_k(\boldsymbol{x}_k(s,\omega),s,\omega) + \nabla \boldsymbol{B}_k(\boldsymbol{x}_k(s,\omega),s,\omega)\cdot \boldsymbol{x}'_k(s,\omega)\right] d\boldsymbol{w}(s,\omega).$$

Note that the integral equation (4.152) is obtained from (4.151) by taking the formal total derivative with respect to k of both sides of the equation. If higher-order derivatives of the coefficients with respect to k exist and satisfy similar conditions, higher-order derivatives of the solution exist and satisfy linear equations. These equations are found repeatedly differentiating eq. (4.151) with respect to k.

Chapter 5

The Discrete Approach and Boundary Behavior

The path integral or the Wiener measure interpretation of stochastic differential equations is useful for both the conceptual understanding of stochastic differential equations and for deriving differential equations that govern the evolution of the pdfs of their solutions. A simple illustration of the computational usefulness of the Wiener measure is the easy derivation of the explicit expression (2.25) for the pdf of the MBM. Unfortunately, no explicit expressions exist in general for the pdf of the solution to (5.1). The second best to such an explicit expression is a (deterministic) differential equation for the pdf, whose solution can be studied both analytically and numerically directly from the differential equation. A case in point is the diffusion equation and the initial condition (2.26) that the pdf of the MBM satisfies.

The discrete approach to SDEs provides insight into the behavior of the random trajectories of the SDE that is not contained in the FPE. Thus, for example, the flux in the FPE is net flux and cannot be separated into its unidirectional components. The need for such a separation arises in connecting discrete simulations to the continuum. Also the boundary behavior of the random trajectories is not easily expressed in terms of boundary conditions for the FPE. These problems are handled in a natural way by the discrete simulation and by its limit.

5.1 The Euler simulation scheme and its convergence

Itô's definition of the stochastic integral on the lattice $t_k = t_0 + k\Delta t$, with $\Delta t = T/N$ and $\Delta w(t) = w(t + \Delta t) - w(t)$, defines the solution of the SDE

$$dx = a(x,t)\,dt + b(x,t)\,dw, \quad x(0) = x_0, \tag{5.1}$$

Z. Schuss, *Theory and Applications of Stochastic Processes: An Analytical Approach*,
Applied Mathematical Sciences 170, DOI 10.1007/978-1-4419-1605-1_5,

or equivalently, of the Itô integral equation

$$x(t) = x_0 + \int_0^t a(x(s), s)\, ds + \int_0^t b(x(s), s)\, dw(s), \tag{5.2}$$

as the limit of the solution of the Euler scheme

$$\begin{aligned} x_N(t + \Delta t) &= x_N(t) + a(x_N(t), t)\Delta t + b(x_N(t), t)\, \Delta w(t) \\ x_N(0) &= x_0 \end{aligned} \tag{5.3}$$

as $\Delta t \to 0$. The increments $\Delta w(t)$ are independent random variables that can be constructed by Lévy's method (see Section 2.3.2) as $\Delta w(t) = n(t)\sqrt{\Delta t}$, where the random variables $n(t)$, for each t on the numerical mesh, are independent standard Gaussian variables $\mathcal{N}(0,1)$. According to the recursive scheme (5.3), at any time t (on the numerical mesh) the process $x_N(t)$ depends on the sampled trajectory $w(s)$ for $s \leq t$, so it is $\mathcal{F}_t$-adapted. The existence of the limit $x(t) = \lim_{N\to\infty} x_N(t)$ is proved in the following

Theorem 5.1.1 (Skorokhod [224]). *If $a(x,t)$ and $b(x,t)$ are uniformly Lipschitz continuous functions in $x \in \mathbb{R}$, $t \in [t_0, T]$, then the limit $x(t) \stackrel{\Pr}{=} \lim_{N\to\infty} x_N(t)$ (convergence in probability) exists and is the solution of* (5.2).

Proof. We consider a refinement $t_0 = t'_0 < t'_1 < \cdots < t'_{N'} = T$ of the given partition $t_0 < t_1 < \cdots < t_N = T$ into N' intervals of length $\Delta t'_l = t'_{l+1} - t'_l$, and the Brownian increments $\Delta w_k = w(t_{k+1}) - w(t_k)$ and $\Delta w'_l = w(t'_{l+1}) - w(t'_l)$, respectively. The two Euler schemes are

$$\begin{aligned} x_k &= x_{k-1} + a(x_{k-1}, t_{k-1})\Delta t + b(x_{k-1}, t_{k-1})\Delta w_k, \\ x'_l &= x'_{l-1} + a(x'_{l-1}, t'_{l-1})\Delta t' + b(x'_{l-1}, t'_{l-1})\Delta w'_l. \end{aligned}$$

We define $x_N(t) = x_k$ for $t \in [t_k, t_{k+1})$ and $x'_{N'}(t) = x'_l$ for $t \in [t'_l, t'_{l+1})$. To show that $x_N(t)$ converges in probability as $N \to \infty$, we have to show that $x_N(t) - x'_{N'}(t) \stackrel{\Pr}{\to} 0$ as $N, N' \to \infty$. To this end, we define

$$\Delta^+(t) = \begin{cases} x_N(t) - x'_{N'}(t) & \text{if } x_N(t) > x'_{N'}(t) \\ 0 & \text{if } x_N(t) \leq x'_{N'}(t) \end{cases}$$

$$\Delta^-(t) = \begin{cases} x'_{N'}(t) - x_N(t) & \text{if } x'_{N'}(t) > x_N(t) \\ 0 & \text{if } x'_{N'}(t) \leq x_N(t). \end{cases}$$

First, we estimate $\Delta^+(t)$ when it is positive. Assuming that $\Delta^+(s) > 0$ for all

$s \in [t_0, t]$, we have

$$\begin{aligned}\Delta^+(t) &= x_0 - x_0' + \sum_{t_k \le t} [a(x_{k-1}, t_{k-1})\Delta t + b(x_{k-1}, t_{k-1})\Delta w_k] \\ &\quad - \sum_{t_l' \le t} [a(x_{l-1}', t_{l-1}')\Delta t' + b(x_{l-1}', t_{l-1}')\Delta w_l'] \\ &\le |x_0 - x_0'| + \sup_{t_0 \le s \le t} \left| \sum_{s < t_{k-1} \le t} [a(x_{k-1}, t_{k-1})\Delta t + b(x_{k-1}, t_{k-1})\Delta w_k] \right. \\ &\quad \left. - \sum_{s < t_l' \le t} [a(x_{l-1}', t_{l-1}')\Delta t' + b(x_{l-1}', t_{l-1}')\Delta w_l'] \right|. \end{aligned} \tag{5.4}$$

Now, we assume that $\Delta^+(t) > 0$, but $\Delta^+(s)$ vanishes for some $s < t$. If τ is the maximal point in $[t_0, t]$, for which $\Delta^+(\tau-) = 0$ and $\Delta^+(\tau+) > 0$, then clearly $|x_N(\tau) - x_{N'}'(\tau)| \le \max_k |x_k - x_{k-1}| + \max_l |x_l' - x_{l-1}'|$. Obviously, $\Delta^+(s) > 0$ for $\tau < s \le t$, so inequality (5.4) is valid for the interval $[\tau, t]$. Therefore

$$\begin{aligned}\Delta^+(t) &\le |x_0 - x_0'| + \max_k |x_k - x_{k-1}| + \max_l |x_l' - x_{l-1}'| \\ &\quad + \sup_{t_0 \le s \le t} \left| \sum_{s < t_{k-1} \le t} [a(x_{k-1}, t_{k-1})\Delta t + b(x_{k-1}, t_{k-1})\Delta w_k] \right. \\ &\quad \left. - \sum_{s < t_l' \le t} [a(x_{l-1}', t_{l-1}')\Delta t' + b(x_{l-1}', t_{l-1}')\Delta w_l'] \right|. \end{aligned}$$

Using analogous estimates for $\Delta^-(t)$, we obtain

$$\begin{aligned}|x_N(t) - x_{N'}'(t)| &= \max[\Delta^+(t), \Delta^-(t)] \\ &\le |x_0 - x_0'| + \max_k |x_k - x_{k-1}| + \max_l |x_l' - x_{l-1}'| \\ &\quad + \sup_{t_0 \le s \le t} \left| \sum_{s < t_{k-1} \le t} [a(x_{k-1}, t_{k-1})\Delta t + b(x_{k-1}, t_{k-1})\Delta w_k] \right. \\ &\quad \left. - \sum_{s < t_{l-1}' \le t} [a(x_{l-1}', t_{l-1}')\Delta t' + b(x_{l-1}', t_{l-1}')\Delta w_l'] \right|. \end{aligned} \tag{5.5}$$

Keeping in mind the fact that one partition is a refinement of the other, we can transform the sum inside the absolute value as follows. Assume that $t_r \le s < t_{r+1}$

and $t'_g \le s < t'_{g+1}$, then

$$
\begin{aligned}
&\sum_{t_k\le s} b(x_k,t_k)\Delta w_k - \sum_{t'_l\le s} b(x'_l,t'_l)\Delta w'_l\\
&= \sum_{t_k\le s}\ \sum_{t_{k-1}\le t'_l<t_k} [b(x_{k-1},t'_l) - b(x'_l,t'_l)]\Delta w'_l\\
&\quad + \sum_{t_r\le t'_l<g} [b(x_r,t'_l) - b(x'_l,t'_l)]\Delta w'_l\\
&\quad + \sum_{t_k\le s}\ \sum_{t_{k-1}\le t'_l<t_k} [b(x_{k-1},t_{k-1}) - b(x_{k-1},t'_l)]\Delta w'_l\\
&\quad + \sum_{t_r\le t'_l\le s} [b(x_r,t_{r-1}) - b(x_r,t'_l)]\Delta w'_l + b(x_r,t_r)[w(t_{r+1}) - w(t'_{g+1})].
\end{aligned}
$$

Transforming the difference $\sum_{t_k\le s} a(x_k,t_k)\Delta t_k - \sum_{t'_l\le s} a(x'_l,t'_l)\Delta t'_l$ in an analogous manner and using the inequality

$$
\sup_{1\le k\le N}\left|\sum_{i=k}^{N} a_i\right| \le 2\sup_{1\le k\le N}\left|\sum_{i=1}^{k} a_i\right|,
$$

we obtain the estimate

$$
\begin{aligned}
&|x_N(t) - x'_{N'}(t)| \qquad (5.6)\\
&\le |x_0 - x'_0| + \max_k |x_k - x_{k-1}| + \max_l |x'_l - x'_{l-1}|\\
&\quad + 2\sup_{1\le j\le N, t\in[t_0,T]} |b(x_k,t)| \max_{|s_1+s_2|\le\max\Delta t_k} |w(s_2) - w(s_1)|\\
&\quad + 2\sup_{1\le j\le N, t\in[t_0,T]} |a(x_k,t)| \max_{1\le k\le N} \Delta t_k\\
&\quad + 2\sup_{s\le t}\left|\sum_{t'_l<s} [b(x_N(t'_l),t'_l) - b(x'_{N'}(t'_l),t'_l)]\Delta w'_l\right|\\
&\quad + 2\sup_{t_0\le s\le t}\left|\sum_{t'_l<s} [a(x_N(t'_l),t'_l) - a(x'_{N'}(t'_l),t'_l)]\Delta t'_l\right|\\
&\quad + 2\sup_{t_0\le s\le t}\left|\sum_{t_k\le t'_l<t_{k+1}\le s} [b(x_k,t_k) - b(x_k,t'_l)]\Delta w'_l\right|\\
&\quad + 2\sup_{t_0\le s\le t}\left|\sum_{t_k\le t'_l<t_{k+1}\le s} [a(x_k,t_k) - a(x_k,t'_l)]\Delta t'_l\right|.
\end{aligned}
$$

The first three rows of (5.6) converge to zero in probability and in the L^2 norm as $N\to\infty$. Setting $x_N(t) - x'_{N'}(t) = \Delta x_{N,N'}(t)$ and using the properties of the Itô integral for step functions

$$\sum_{t'_l<s}[b(x_N(t'_l),t'_l)-b(x'_{N'}(t'_l),t'_l)]\Delta w'_l$$
$$=\int_0^s\left[\frac{b(x_N(t'),t')-b(x_{N'}(t'),t')}{\Delta x_{N,N'}(t')}\right]\Delta x_{N,N'}(t')\,dw(t'),$$

we obtain

$$\mathbb{E}\left[\sup_{s\le t}\left|\sum_{t'_l<s}[b(x_N(t'_l),t'_l)-b(x'_{N'}(t'_l),t'_l)]\Delta w'_l\right|\right]^2$$
$$=\mathbb{E}\left[\sup_{s\le t}\left|\int_0^s\left[\frac{b(x_N(t'),t')-b(x_{N'}(t'),t')}{\Delta x_{N,N'}(t')}\right]\Delta x_{N,N'}(t')\,dw(t')\right|\right]^2.$$

The integral is a continuous function of $s\in[t_0,T]$, so the maximum is achieved at some $s_0\in[t_0,T]$, for which we have, due to the Lipschitz continuity of the coefficients, the inequality

$$\mathbb{E}\left|\int_0^{s_0}\left[\frac{b(x_N(t'),t')-b(x_{N'}(t'),t')}{\Delta x_{N,N'}(t')}\right]\Delta x_{N,N'}(t')\,dw(t')\right|^2$$
$$=\mathbb{E}\int_0^{s_0}\left[\frac{b(x_N(t'),t')-b(x_{N'}(t'),t')}{\Delta x_{N,N'}(t')}\right]^2\left|\Delta x_{N,N'}(t')\right|^2dt'$$
$$\le L^2\int_0^t\left|\Delta x_{N,N'}(t')\right|^2dt',$$

where L is the Lipschitz constant. The remaining sums are estimated in a similar manner. Gronwall's inequality (4.150) implies that $\mathbb{E}\left|\Delta x_{N,N'}(t')\right|^2\to 0$ as $N,N'\to\infty$ and Chebyshev's inequality (3.1.1) implies convergence in probability. □

Exercise 5.1 (Itô's formula). Use the proof of Theorem 5.1.1 to prove Itô's formula (3.67). □

5.2 The pdf of Euler's scheme in $\mathbb{R}$ and the FPE

We assume that the coefficients $a(x,t)$ and $b(x,t)$ are smooth functions in $\mathbb{R}\times\mathbb{R}^+$, with $b(x,t)>\delta>0$ for some constant δ. The coefficients can be allowed to be random in a way such that for each $x\in\mathbb{R}$ the stochastic processes $a(x,t,\omega)$ and $b(x,t,\omega)$ are adapted to $\mathcal{F}_t$ in the sense of Definition 2.2.9. We assume for now that $a(x,t)$ and $b(x,t)$ are deterministic.

Theorem 5.2.1. *The pdf* $p_N(x,t\,|\,x_0)$ *of the solution* $x_N(t,\omega)$ *of* (5.3) *converges to the solution* $p\,(x,t\,|\,x_0)$ *of the initial value problem* (4.121), (4.122) *as* $N \to \infty$.

Proof. First, we extend $x_N(t)$ off the lattice as follows. If $0 < t \le \Delta t$, we define

$$\tilde{x}_N(t) = x_0 + a(x_0,0)\,t + b(x_0,0)\,w(t), \quad x_N(0) = x_0. \tag{5.7}$$

If $k\Delta t < t \le (k+1)\Delta t$, where $k = 1,2,\dots$, we define $\tilde{t} = k\Delta t$, $\Delta\tilde{t} = t - \tilde{t}$, $\Delta\tilde{w}(t) = w(t) - w(\tilde{t})$, and

$$\tilde{x}_N(t) = \tilde{x}_N(\tilde{t}) + a(\tilde{x}_N(\tilde{t}),\tilde{t})\Delta\tilde{t} + b(\tilde{x}_N(\tilde{t}),\tilde{t})\,\Delta\tilde{w}(t). \tag{5.8}$$

Obviously, for each realization of $w(t)$, we have $\tilde{x}_N(t) = x_N(t)$ at lattice points t. The pdf $\tilde{p}_N(x,t\,|\,x_0)$ of $\tilde{x}_N(t)$ is identical with the pdf $p_N(x,t\,|\,x_0)$ of $x_N(t)$ on lattice points t and it satisfies on the lattice the recursion relation

$$\begin{aligned}&\tilde{p}_N(x,t+\Delta t\,|\,x_0)\\ &= \int_{\mathbb{R}} \frac{1}{\sqrt{2\pi\Delta t}\,b(y,t)} \exp\left\{-\frac{[x-y-a(y,t)\Delta t]^2}{2b^2(y,t)\Delta t}\right\} \tilde{p}_N(y,t\,|\,x_0)\,dy.\end{aligned} \tag{5.9}$$

Off the lattice we have the recursion

$$\begin{aligned}&\tilde{p}_N(x,t\,|\,x_0)\\ &= \int_{\mathbb{R}} \frac{1}{\sqrt{2\pi\Delta\tilde{t}}\,b(y,\tilde{t})} \exp\left\{-\frac{\left[x-y-a(y,\tilde{t})\Delta\tilde{t}\right]^2}{2b^2(y,\tilde{t})\Delta\tilde{t}}\right\} \tilde{p}_N(y,\tilde{t}\,|\,x_0)\,dy,\end{aligned} \tag{5.10}$$

where $\tilde{p}_N(x,\tilde{t}\,|\,x_0) = p_N(x,\tilde{t}\,|\,x_0)$. Note that $\tilde{p}_N(x,t\,|\,x_0)$ is differentiable with respect to t and twice differentiable with respect to x. Therefore the analysis of (5.10) applies to (5.9) as well. We observe that integrating $\tilde{p}_N(x,t\,|\,x_0)$ with respect to x_0 against a bounded sufficiently smooth initial function $p_0(x_0)$ results in a sequence of bounded and twice continuously differentiable functions

$$\tilde{p}_N(x,t) = \int_{\mathbb{R}} \tilde{p}_N(x,t\,|\,x_0)p_0(x_0)\,dx_0, \tag{5.11}$$

that satisfy the recursion (5.10), the initial condition

$$\lim_{t\to 0} \tilde{p}_N(x,t) = p_0(x), \tag{5.12}$$

uniformly on finite intervals, and whose partial derivatives up to second order are uniformly bounded (see Exercise 5.2 below).

Differentiation with respect to t at off-lattice points is equivalent to differentiation with respect to $\Delta\tilde{t}$. Differentiating and expanding all functions in powers of $\sqrt{\Delta\tilde{t}}$, we obtain (see Exercises 5.3–5.6 below)

$$\frac{\tilde{p}_N(x,t)}{\partial t} = \frac{1}{2}\frac{\partial^2\left[b^2(x,t)\tilde{p}_N(x,t)\right]}{\partial x^2} - \frac{\partial\left[a(x,t)\tilde{p}_N(x,t)\right]}{\partial x} + O(\Delta t)\,, \tag{5.13}$$

uniformly for $x \in \mathbb{R}$ and $t > 0$. At lattice points, we use the change of variables

$$y = x - a(y,t)\Delta t + \eta b(y,t)\sqrt{\Delta t} \tag{5.14}$$

in (5.9), and expanding in powers of Δt, we obtain (5.13) again. If $p(x,t)$ is the (unique) solution of the initial value problem (5.12) for the FPE (4.121), then $p(x,t) - p_N(x,t)$ satisfies the inhomogeneous FPE with homogeneous initial value and right-hand side that is uniformly $O(\Delta t)$. It follows from the maximum principle for parabolic initial value problems [78], [203] that the difference converges uniformly to zero. □

Remark 5.2.1. There are many types of convergence of the Euler scheme [132], [131]. Theorems 5.1.1 and 5.2.1 concern convergence in probability and of probability density and therefore cannot be used as measures for the error of the Euler numerical scheme in a given simulation. Such estimates depend on the sample size and are the subject of numerical analysis of stochastic differential equations.

Exercise 5.2 (Regularity of $p_N(x,t)$). Use the recursion (5.9) to prove that the functions $p_N(x,t)$ and their partial derivatives with respect to x, up to second order, are uniformly bounded and the convergence (5.12) is uniform on finite intervals. □

Exercise 5.3 (The differential). Prove that the differential of the transformation (5.14) is given by

$$\begin{aligned}
dy &= \frac{\sqrt{\Delta t}\, b(y,t)}{1 - \sqrt{\Delta t}\,\eta b_y(y,t) + a_y(y,t)\Delta t}\, d\eta \qquad (5.15)\\
&= \left\{1 + \eta\sqrt{\Delta t}\, b_x(x,t) - a_x(x,t)\Delta t + \eta^2 \Delta t \left[(b_x(x,t))^2 + b_{xx}(x,t)b(x,t)\right]\right\}\\
&\quad \times \sqrt{\Delta t}\, b(x,t)\left[1 + O\left(\sqrt{\Delta t}\right)\right]\, d\eta,
\end{aligned}$$

where subscripts denote partial derivatives. □

Exercise 5.4. (The exponent).

(i) Prove that the exponent in (5.9) is expanded as

$$\begin{aligned}
&\frac{\left[\eta b(y,t)\sqrt{\Delta t} + a\left(x + \eta b(y,t)\sqrt{\Delta t}, t\right)\Delta t\right]^2}{2b^2(y,t)\Delta t}\\
&= \eta^2 \left(\frac{a(x,t)}{b(x,t)}\right)_x b(x,t) + \frac{a^2(x,t)}{2b^2(x,t)} + O\left(\Delta t^{3/2}\right).
\end{aligned}$$

(ii) Show that the exponential function in (5.9) can be expanded as

$$\begin{aligned}
&\exp\left\{-\frac{[x - y - a(y,t)\Delta t]^2}{2b^2(y,t)\Delta t}\right\} = \exp\left\{-\frac{\eta^2}{2}\right\}\\
&\times \left\{1 - \eta\sqrt{\Delta t}\,\frac{a(x,t)}{b(x,t)} - \Delta t\left[\eta^2\left(\frac{a(x,t)}{b(x,t)}\right)_x b(x,t) + \left(1 - \eta^2\right)\frac{a^2(x,t)}{2b^2(x,t)}\right]\right\}\\
&+ O\left(\Delta t^{3/2}\right).
\end{aligned}$$

□

Exercise 5.5. (The density $p_N(y,t)$).
(i) Expand

$$\begin{aligned}p_N(y,t) =& p_N(x+\eta b(y,t)\sqrt{\Delta t},t)\\ =& p_N(x,t)+p_{N,x}(x,t)\eta\left[b(x,t)+b_x(x,t)\eta b(x,t)\sqrt{\Delta t}\right]\sqrt{\Delta t}\\ &+\frac{1}{2}p_{N,xx}(x,t)\eta^2 b^2(x,t)\Delta t+O\left(\Delta t^{3/2}\right).\end{aligned}$$

(ii) Show that the pre-exponential factor in (5.9), up to $O\left(\Delta t^{3/2}\right)$, has the form

$$\begin{aligned}&\Big\{p_N(x,t)+p_{N,x}(x,t)b(x,t)\eta\sqrt{\Delta t}\\ &+\frac{1}{2}\left[p_{N,xx}(x,t)b^2(x,t)+\left(b^2(x,t)\right)_x p_{N,x}(x,t)\right]\eta^2\Delta t\Big\}\\ &\times\left\{\left[1-\frac{a^2(x,t)}{2b^2(x,t)}\Delta t\right]-\eta\sqrt{\Delta t}\,\frac{a(x,t)}{b(x,t)}\right.\\ &\left.-\eta^2\Delta t\left[\left(\frac{a(x,t)}{b(x,t)}\right)_x b(x,t)-\frac{a^2(x,t)}{2b^2(x,t)}\right]\right\}\\ =&\,p_N(x,t)\left[1-\frac{a^2(x,t)}{2b^2(x,t)}\Delta t\right]+\eta\sqrt{\Delta t}\left\{\left[1-\frac{a^2(x,t)}{2b^2(x,t)}\Delta t\right]p_{N,x}(x,t)b(x,t)\right.\\ &\left.-\frac{a(x,t)}{b(x,t)}p_N(x,t)\right\}+\eta^2\Delta t\left\{\frac{1}{2}\left[p_{N,xx}(x,t)b^2(x,t)+\left(b^2(x,t)\right)_x p_{N,x}(x,t)\right]\right.\\ &\times\left[1-\frac{a^2(x,t)}{2b^2(x,t)}\Delta t\right]-a(x,t)p_{N,x}(x,t)-\left[\left(\frac{a(x,t)}{b(x,t)}\right)_x b(x,t)\right.\\ &\left.\left.-\frac{a^2(x,t)}{2b^2(x,t)}\right]p\left(x\right)\right\}.\end{aligned}\tag{5.16}$$

(iii) Using eqs. (5.15)–(5.16) in eq. (5.9), obtain

$$\begin{aligned}&p_N(x,t+\Delta t)\\ &=\frac{1}{\sqrt{2\pi}}\int_{\mathbb{R}}\left\{\left\{1-a_x(x,t)\Delta t+\eta^2\Delta t\left[(b_x)^2(x,t)+b_{xx}(x,t)b(x,t)\right]\right\}\right.\\ &\quad\times\left[1-\frac{a^2(x,t)}{2b^2(x,t)}\Delta t\right]p_N(x,t)\\ &\quad+\eta^2\Delta t b_x(x,t)\left[p_{N,x}(x,t)b(x,t)-\frac{a(x,t)}{b(x,t)}p_N(x,t)\right]\\ &\quad+\eta^2\Delta t\left\{\frac{1}{2}\left[p_{N,xx}(x,t)b^2(x,t)+\left(b^2(x,t)\right)_x p_{N,x}(x,t)\right]\right.\\ &\quad\left.\left.-\left[\left(\frac{a(x,t)}{b(x,t)}\right)_x b(x,t)-\frac{a^2(x,t)}{2b^2(x,t)}\right]p_N(x,t)\right\}\right\}\exp\left\{-\frac{\eta^2}{2}\right\}d\eta.\end{aligned}\tag{5.17}$$

□

Exercise 5.6 (The FPE). Evaluate the Gaussian integrals in (5.17) and show that

$$\begin{aligned}&\frac{p_N(x,t+\Delta t)-p_N(x,t)}{\Delta t}\\ =&p_N(x,t)\left[\left(b_x\right)^2(x,t)+b_{xx}(x,t)b(x,t)\right]\\ &+\frac{1}{2}p_N(x,t)b^2(x,t)+\left(b^2(x,t)\right)_x p_{N,x}(x,t)-\left[a(x,t)p_N(x,t)\right]_x+O\left(\Delta t^{1/2}\right)\\ =&\frac{1}{2}\left[b^2(x,t)p_N(x,t)\right]_{xx}-\left[a(x,t)p_N(x,t)\right]_x+O\left(\Delta t^{1/2}\right),\end{aligned}$$

and hence (5.13). Can the last estimate be improved? □

Exercise 5.7 (The initial condition). Prove that $p(x,t)-p_N(x,t)\to 0$ uniformly for finite intervals in x and $0<t<T$. □

Exercise 5.8 (The Feynman–Kac formula). Prove that if the recursion (5.10) is modified to

$$\begin{aligned}&\tilde{p}_N(x,t+\Delta t\,|\,x_0)\\ &=\int_{\mathbb{R}}\frac{\tilde{p}_N(y,t\,|\,x_0)}{\sqrt{2\pi\Delta t}\,b(y,t)}\exp\left\{-\frac{[x-y-a(y,t)\Delta t]^2}{2b^2(y,t)\Delta t}+g(y,t)\Delta t\right\}dy,\end{aligned}\tag{5.18}$$

where $g(y,t)$ is a sufficiently regular function, then $\lim_{N\to\infty}p_N(x,t\,|\,x_0,s)=p(x,t\,|\,x_0,s)$, where $p(x,t\,|\,x_0,s)$ is the solution of the initial value problem

$$p_t=\frac{1}{2}(b^2p)_{xx}-(ap)_x+gp,\quad \lim_{t\downarrow s}p=\delta(x-x_0).$$

□

Exercise 5.9 (Simulation of the Feynman–Kac formula). How should the Euler scheme (5.3) be modified so that the corresponding pdf satisfies the recursion (5.18), in the case where $g(x,t)$ is nonpositive? What is the interpretation of $p_N(x,t\,|\,x_0,s)$ and $p(x,t\,|\,x_0,s)$ if $g(x,t)$ can be positive? How should the Euler scheme (5.3) be modified for this case? □

Exercise 5.10. (The backward Kolmogorov equation)

(i) Prove that

$$p_N(y,t\,|\,x,s)=\int_{\mathbb{R}}\frac{p_N(y,t\,|\,z,s+\Delta s)}{\sqrt{2\pi\Delta s}b(x,s)}\exp\left\{-\frac{(z-x-a(x,s)\Delta s)^2}{2b^2(x,s)\Delta s}\right\}dz.$$

(ii) Prove that the transition pdf $p(y,t\,|\,x,s)=\lim_{N\to\infty}p_N(y,t\,|\,x,s)$ satisfies in $\mathbb{R}$ with respect to the backward variables (x,s) the backward Kolmogorov equation $p_s+a(x,s)p_x+\frac{1}{2}b^2(x,s)p_{xx}=0$ and the terminal condition $\lim_{t\uparrow s}p=\delta(y-x)$. (HINT: Change the variable of integration to $z=b(x,s)\xi\sqrt{\Delta s}+x+a(x,s)\Delta s$ and expand everything in sight in powers of $\sqrt{\Delta s}$, as above. Finally, prove convergence by using the maximum principle.) □

We consider now the d-dimensional stochastic dynamics

$$d\boldsymbol{x} = \boldsymbol{a}(\boldsymbol{x},t)\,dt + \sqrt{2}\boldsymbol{B}(\boldsymbol{x},t)\,d\boldsymbol{w}, \quad \boldsymbol{x}(0) = \boldsymbol{x}_0, \tag{5.19}$$

where $\boldsymbol{a}(\boldsymbol{x},t) : \mathbb{R}^d \times [0,\infty) \mapsto \mathbb{R}^d$ is a vector of smooth functions for all $\boldsymbol{x} \in \mathbb{R}$, $t \geq 0$, $\boldsymbol{B}(\boldsymbol{x},t) : \mathbb{R}^d \times [0,\infty) \mapsto \mathbb{M}_{n\times m}$ is a smooth $n \times m$ matrix of smooth functions, and $\boldsymbol{w}(t) : [0,\infty) \mapsto \mathbb{R}^m$ is a vector of m independent MBMs. We assume that the diffusion tensor $\boldsymbol{\sigma}(\boldsymbol{x},t) = \boldsymbol{B}(\boldsymbol{x},t)\boldsymbol{B}^T(\boldsymbol{x},t)$ is uniformly positive definite in $\mathbb{R}^d$. The Euler scheme for (5.19) is

$$\begin{aligned}\boldsymbol{x}(t+\Delta t) &= \boldsymbol{x}(t) + \boldsymbol{a}(\boldsymbol{x}(t),t)\Delta t + \sqrt{2}\boldsymbol{B}(\boldsymbol{x}(t),t)\,\Delta\boldsymbol{w}(t,\Delta t) \\ \boldsymbol{x}_N(0) &= \boldsymbol{x}_0.\end{aligned} \tag{5.20}$$

Exercise 5.11 (Convergence of trajectories). Generalize the proof of Skorokhod's theorem 5.1.1 to the d-dimensional case. □

The proof of convergence of the pdf of the trajectories of (5.20) is similar to that in one dimension. Setting

$$\mathcal{B}(\boldsymbol{x},\boldsymbol{y},t) = [\boldsymbol{y} - \boldsymbol{x} - \boldsymbol{a}(\boldsymbol{x},t)\Delta t]^T \boldsymbol{\sigma}^{-1}(\boldsymbol{x},t)\,[\boldsymbol{y} - \boldsymbol{x} - \boldsymbol{a}(\boldsymbol{x},t)\Delta t], \tag{5.21}$$

we see that the pdf of the trajectories of (5.20) satisfies the d-dimensional version of the recursion relation (5.9),

$$p_N(\boldsymbol{y},t+\Delta t) = \int_{\mathbb{R}^d} \frac{p_N(\boldsymbol{x},t)\,d\boldsymbol{x}}{(2\pi\Delta t)^{d/2}\sqrt{\det\boldsymbol{\sigma}(\boldsymbol{x},t)}} \exp\left\{-\frac{\mathcal{B}(\boldsymbol{x},\boldsymbol{y},t)}{2\Delta t}\right\}. \tag{5.22}$$

Theorem 5.2.2. *Under the above assumptions, if the initial point $\boldsymbol{x}_0$ is chosen from a smooth bounded density $p_0(\boldsymbol{x}_0)$, then the pdf $p_N(\boldsymbol{y},t)$ of the solution $\boldsymbol{x}_N(t)$ of* (5.20) *converges as $N \to \infty$ to the solution $p(\boldsymbol{y},t)$ of the initial value problem*

$$\frac{\partial p(\boldsymbol{y},t)}{\partial t} = \sum_{i=1}^d\sum_{j=1}^d \frac{\partial^2 \sigma^{ij}(\boldsymbol{y},t)p(\boldsymbol{y},t)}{\partial y^i \partial y^j} - \sum_{i=1}^d \frac{\partial a^i(\boldsymbol{y},t)p(\boldsymbol{y},t)}{\partial y^i} \tag{5.23}$$

$$\lim_{t\downarrow 0} p(\boldsymbol{y},t) = p_0(\boldsymbol{y}). \tag{5.24}$$

Proof. As above, we change variables in (5.22) to $\boldsymbol{z} = \boldsymbol{\sigma}^{-1/2}(\boldsymbol{x},t)(\boldsymbol{x} - \boldsymbol{y} + \boldsymbol{a}(\boldsymbol{x},t)\Delta t)/\sqrt{\Delta t}$ and expand the integrand in powers of $\sqrt{\Delta t}$. First, we need to expand the Jacobian of the transformation. Differentiating the identity $\boldsymbol{\sigma}^{1/2}\boldsymbol{\sigma}^{-1/2} = \boldsymbol{I}$, we write

$$\boldsymbol{\sigma}^{1/2}\nabla\left(\boldsymbol{\sigma}^{-1/2}\right) + \nabla\left(\boldsymbol{\sigma}^{1/2}\right)\boldsymbol{\sigma}^{-1/2} = \boldsymbol{0}, \tag{5.25}$$

or

$$\boldsymbol{\sigma}^{1/2}\nabla\left(\boldsymbol{\sigma}^{-1/2}\right)\boldsymbol{\sigma}^{1/2} = -\nabla(\boldsymbol{\sigma}^{1/2}),$$

from which it follows that the Jacobian matrix is

$$\frac{\partial \boldsymbol{z}}{\partial \boldsymbol{x}} = \frac{\boldsymbol{\sigma}^{-1/2}}{\sqrt{\Delta t}}\left[\boldsymbol{I} - \sqrt{\Delta t}\,\nabla\left(\boldsymbol{\sigma}^{1/2}\right)\cdot \boldsymbol{z} + O(\Delta t)\right]$$

and that the Jacobian of the transformation is

$$\begin{aligned}\mathcal{J} &= \left|\det\left(\frac{\partial \boldsymbol{z}}{\partial \boldsymbol{x}}\right)\right| \\ &= \frac{1}{(\Delta t)^{d/2}\sqrt{\det \boldsymbol{\sigma}}}\left\{1 - \sqrt{\Delta t}\ \mathrm{tr}\left[\nabla\left(\boldsymbol{\sigma}^{1/2}\right)\cdot \boldsymbol{z}\right] + O(\Delta t)\right\}.\end{aligned} \tag{5.26}$$

Expanding the transformed integrand about $\boldsymbol{y}$ in powers of $\sqrt{\Delta t}$, we note that terms linear in $\boldsymbol{z}$ vanish, because they give rise to Gaussian integrals with an odd integrand. We end up with the approximate Fokker–Plank equation for $p_N(\boldsymbol{y}, t)$:

$$\begin{aligned}&\frac{\partial p_N(\boldsymbol{y}, t)}{\partial t} \\ &= \sum_{i=1}^{d}\sum_{j=1}^{d} \frac{\partial^2 \sigma^{ij}(\boldsymbol{y}, t) p_N(\boldsymbol{y}, t)}{\partial y^i \partial y^j} - \sum_{i=1}^{d} \frac{\partial a^i(\boldsymbol{y}, t) p_N(\boldsymbol{y}, t)}{\partial y^i} + O(\sqrt{\Delta t}).\end{aligned} \tag{5.27}$$

The uniform convergence of $p_N(\boldsymbol{y}, t)$ to the solution $p(\boldsymbol{y}, t)$ of the initial value problem (5.23), (5.24) is proved as in the one-dimensional case above. □

Exercise 5.12. (The FPE for Langevin's equation).

(i) Write the Langevin equation

$$\ddot{x} + \gamma \dot{x} = f(x) + \sqrt{2\varepsilon\gamma}\,\dot{w}. \tag{5.28}$$

as the phase space system

$$\dot{x} = v, \quad \dot{v} = -\gamma v + f(x) + \sqrt{2\varepsilon\gamma}\,\dot{w} \tag{5.29}$$

and the Euler scheme

$$\begin{aligned}x_\Delta(t + \Delta t) &= x_\Delta(t) + v_\Delta(t)\Delta t + o(\Delta t) \\ v_\Delta(t + \Delta t) &= v_\Delta(t) + [-\gamma v_\Delta(t) + f(x_\Delta(t))]\Delta t + \sqrt{2\varepsilon\gamma}\,\Delta w + o(\Delta t).\end{aligned}$$

(ii) Derive the equation

$$\begin{aligned}p_\Delta(x, v, t + \Delta t) &= \frac{1}{\sqrt{2\varepsilon\gamma\pi\Delta t}} \int_{\mathbb{R}}\int_{\mathbb{R}} p_\Delta(\xi, \eta, t)\delta(x - \xi - \eta\Delta t) \\ &\times \exp\left\{-\frac{[v - \eta - [-\gamma\eta + f(\xi)]\Delta t]^2}{2\varepsilon\gamma\Delta t}\right\} d\xi\, d\eta + o(\Delta t)\end{aligned}$$

for the pdf $p_\Delta(x, v, t)$ of the phase plane pair $(x_\Delta(t), v_\Delta(t))$.

(iii) Change variables in the above integral to

$$-u = v - \eta - \frac{[-\gamma\eta + f(x - \eta\Delta t)]\Delta t}{\sqrt{\varepsilon\gamma\Delta t}},$$

expand in powers of Δt, and write the integral in the form

$$p_\Delta(x, v, t + \Delta t) = \frac{1}{\sqrt{\pi}(1 - \gamma\Delta t + o(\Delta t))} \int_{-\infty}^{\infty} e^{-u^2/2}\, du \tag{5.30}$$
$$\times\, p_\Delta\Big(x - v(1 + \gamma\Delta t)\Delta t + o(\Delta t), v(1 + \gamma\Delta t) + u\sqrt{\varepsilon\gamma\Delta t}$$
$$- f(x)\Delta t(1 + \gamma\Delta t) + o(\Delta t), t\Big).$$

(iv) Re-expand in powers of Δt to get

$$p_\Delta\Big(x - v(1 + \gamma\Delta t)\Delta t + o(\Delta t), v(1 + \gamma\Delta t) + u\sqrt{\varepsilon\gamma\Delta t}$$
$$- f(x)\Delta t(1 + \gamma\Delta t) + o(\Delta t), t\Big)$$
$$= p_\Delta(x, v, t) - v\Delta t \frac{\partial p_\Delta(x, v, t)}{\partial x}$$
$$+ \frac{\partial p_\Delta(x, v, t)}{\partial v}\Big(v\gamma\Delta t + u\sqrt{\varepsilon\gamma\Delta t} - f(x)\Delta t + o(\Delta t)\Big)$$
$$+ \varepsilon\gamma u^2 \Delta t \frac{\partial^2 p_\Delta(x, v, t)}{\partial v^2} + o(\Delta t).$$

(v) Show that (5.30) gives

$$p_\Delta(x, v, t + \Delta t) - \frac{p_\Delta(x, v, t)}{1 - \gamma\Delta t}$$
$$= -\frac{1}{1 - \gamma\Delta t} v\Delta t \frac{\partial p_\Delta(x, v, t)}{\partial x} + \frac{\Delta t}{1 - \gamma\Delta t} \frac{\partial p_\Delta(x, v, t)}{\partial v}$$
$$\times\, (v\gamma - f(x)) + \frac{\varepsilon\gamma\Delta t}{1 - \gamma\Delta t} \frac{\partial^2 p_\Delta(x, v, t)}{\partial v^2} + O\left(\Delta t^{3/2}\right).$$

(vi) Prove that $\lim_{\Delta t \to 0} p_\Delta(x, v, t) = p(x, v, t)$ exists and satisfies the Fokker–Planck equation

$$\frac{\partial p(x, v, t)}{\partial t} \tag{5.31}$$
$$= -v \frac{\partial p(x, v, t)}{\partial x} + \frac{\partial}{\partial v}\left[(\gamma v - f(x))\, p(x, v, t)\right] + \varepsilon\gamma \frac{\partial^2 p(x, v, t)}{\partial v^2}.$$

(vii) Write the above FPE in the conservation law form

$$p_t = -\nabla \cdot \boldsymbol{J}, \tag{5.32}$$

where the components of the net probability flux density vector $\boldsymbol{J}$ are

$$J_x(x,v,t) = vp\,(x,v,t)$$
$$J_v(x,v,t) = -\,(\gamma v - f(x))\,p\,(x,v,t) - \varepsilon\gamma\frac{\partial p\,(x,v,t)}{v} \tag{5.33}$$

(see [222]). □

5.2.1 Unidirectional and net probability flux density

The flux density in continuum diffusion theory (Fick's laws (1.1)) is the net flux through a given point (or surface, in higher dimensions). Unidirectional fluxes are not defined in the diffusion or Fokker–Planck equations, because velocity is not a state variable, so the equations cannot separate between probability flux densities. However, it is often necessary to evaluate the probability flux density across a given interface in simulations of diffusive trajectories of particles. This is the case, for example, if a simulation of diffusing particles is connected to a region, where only a coarse-grained continuum description of the particles is used. In this case, the exchange of trajectories between the two regions, across the interface, requires the calculation of the unidirectional diffusion flux from the continuum region into the simulated region. This situation is encountered in simulations of ionic motion through the protein channel of biological membranes, where the number of ions in the salt solution away from the channel is too large to simulate. This issue is discussed further in [214]. In this section we keep the notation of the previous one.

Definition 5.2.1 (Unidirectional flux). *The* unidirectional probability current (flux) density *at a point x_1 is the probability per unit time of trajectories that propagate from the ray $x < x_1$ into the ray $x > x_1$. It is given by*

$$J_{LR}(x_1,t) = \lim_{\Delta t\to 0} J_{LR}(x_1,t,\Delta t), \tag{5.34}$$

where

$$J_{LR}(x_1,t,\Delta t) \tag{5.35}$$
$$= \frac{1}{\Delta t}\int\limits_{x_1}^{\infty} dx \int\limits_{-\infty}^{x_1} \frac{dy}{\sqrt{2\pi\Delta t\sigma(y,t)}} \exp\left\{-\frac{[x-y-a(y,t)\Delta t]^2}{2\sigma(y,t)\Delta t}\right\} p_N(y,t).$$

Remark 5.2.2. Note that the dependence of p_N on the initial point has been suppressed in (5.35).

Theorem 5.2.3 (Unidirectional and net fluxes in one dimension). *The discrete probability flux densities at a point x_1 are given by*

$$J_{LR,RL}(x_1,t,\Delta t) = \sqrt{\frac{\sigma(x_1,t)}{2\pi\Delta t}}\,p_N(x_1,t) \pm \frac{1}{2}J(x_1,t) + O(\sqrt{\Delta t}), \tag{5.36}$$

where the net flux is

$$J(x_1,t) = \lim_{\Delta t\to 0}\left[J_{LR}(x_1,t,\Delta t) - J_{RL}(x_1,t,\Delta t)\right] \tag{5.37}$$
$$= \left\{-\frac{\partial\left[\sigma(x,t)p\,(x,t)\right]}{\partial x} + a(x,t)p\,(x,t)\right\}_{x=x_1}.$$

Remark 5.2.3. It is clear from (5.36) that the probability flux densities in Definition 5.2.1 are infinite, but the net flux is finite.

Proof. The integral (5.35) can be calculated by the Laplace method [20] at the saddle point $x = y = x_1$. First, we change variables in (5.35) to $x = x_1 + \xi\sqrt{\Delta t}$ and $y = x_1 - \eta\sqrt{\Delta t}$ to obtain

$$J_{LR}(x_1,t,\Delta t) = \int_0^\infty d\xi \int_0^\infty \frac{d\eta}{\sqrt{2\pi\Delta t\sigma(x_1-\eta\sqrt{\Delta t},t)}}$$
$$\times \exp\left\{-\frac{\left[\xi+\eta-a(x_1-\eta\sqrt{\Delta t},t)\sqrt{\Delta t}\right]^2}{2\sigma(x_1-\eta\sqrt{\Delta t},t)}\right\} p_N\left(x_1-\eta\sqrt{\Delta t},t\right),$$

and changing the variable in the inner integral to $\eta = \zeta - \xi$, we get

$$J_{LR}(x_1,t,\Delta t) = \int_0^\infty d\xi \int_\xi^\infty \frac{d\zeta}{\sqrt{2\pi\Delta t\sigma(x_1-(\zeta-\xi)\sqrt{\Delta t},t)}} \tag{5.38}$$
$$\times \exp\left\{-\frac{\left[\zeta-a(x_1-(\zeta-\xi)\sqrt{\Delta t},t)\sqrt{\Delta t}\right]^2}{2\sigma(x_1-(\zeta-\xi)\sqrt{\Delta t},t)}\right\} p_N\left(x_1-(\zeta-\xi)\sqrt{\Delta t},t\right).$$

Next, we expand the exponent in powers of $\sqrt{\Delta t}$ to obtain

$$\frac{\left[\zeta-a(x_1-(\zeta-\xi)\sqrt{\Delta t},t)\sqrt{\Delta t}\right]^2}{2\sigma(x_1-(\zeta-\xi)\sqrt{\Delta t},t)} \tag{5.39}$$
$$= \frac{\zeta^2}{2\sigma(x_1,t)} + \left[\frac{\zeta^2(\zeta-\xi)\left[\sigma^2(x_1,t)\right]_x}{8\sigma^2(x_1,t)} - \frac{\zeta a(x_1,t)}{2\sigma(x_1,t)}\right]\sqrt{\Delta t} + O(\Delta t),$$

the pre-exponential factor,

$$\frac{1}{\sqrt{\sigma\left(x_1-(\zeta-\xi)\sqrt{\Delta t},t\right)}} = \frac{\left[1+\dfrac{\sigma_x(x_1,t)}{2\sigma(x_1,t)}(\zeta-\xi)\sqrt{\Delta t}+O(\Delta t)\right]}{\sqrt{\sigma(x_1,t)}},$$

and the pdf

$$\begin{aligned} & p_N\left(x_1-(\zeta-\xi)\sqrt{\Delta t},t\right) \\ &= p_N(x_1,t)-\frac{\partial p_N(x_1,t)}{\partial x}(\zeta-\xi)\sqrt{\Delta t}+O(\Delta t). \end{aligned} \tag{5.40}$$

Using the expansions (5.39)–(5.40) in (5.38), we obtain

$$J_{LR}(x_1,t,\Delta t) \tag{5.41}$$

$$\begin{aligned} &= \int_0^\infty d\xi \int_\xi^\infty \frac{d\zeta}{\sqrt{2\pi\Delta t\sigma(x_1,t)}} \exp\left\{-\frac{\zeta^2}{2\sigma(x_1,t)}\right\} \left\{p_N(x_1,t)\right. \\ &\quad - \sqrt{\Delta t}\left[\left(\frac{\zeta^2(\zeta-\xi)\sigma_x(x_1,t)}{\sigma^2(x_1,t)}-\frac{2\zeta a(x_1,t)}{\sigma(x_1,t)}-\frac{\sigma_x(x_1,t)(\zeta-\xi)}{2\sigma(x_1,t)}\right)p_N(x_1,t)\right. \\ &\quad \left.\left. + (\zeta-\xi)p_{N,x}(x_1,t)\right]+O(\Delta t)\right\}. \end{aligned} \tag{5.42}$$

Similarly, $J_{RL}(x_1,t)=\lim_{\Delta t\to 0} J_{RL}(x_1,t,\Delta t)$, where

$$J_{RL}(x_1,t,\Delta t) \tag{5.43}$$

$$= \frac{1}{\Delta t}\int_{-\infty}^{x_1} dx \int_{x_1}^{\infty} \frac{dy}{\sqrt{2\pi\Delta t\sigma(y,t)}} \exp\left\{-\frac{[x-y-a(y,t)\Delta t]^2}{2\sigma(y,t)\Delta t}\right\} p_N(y,t).$$

The change of variables $x=x_1-\xi\sqrt{\Delta t},\ y=x_1+\eta\sqrt{\Delta t}$ in (5.43) gives

$$J_{RL}(x_1,t,\Delta t)$$

$$\begin{aligned} &= \int_0^\infty d\xi \int_\xi^\infty \frac{d\zeta}{\sqrt{2\pi\Delta t\sigma(x_1,t)}} \exp\left\{-\frac{\zeta^2}{2\sigma(x_1,t)}\right\} \left\{p_N(x_1,t)\right. \\ &\quad + \sqrt{\Delta t}\left[\left(\frac{\zeta^2(\zeta-\xi)\sigma_x(x_1,t)}{\sigma^2(x_1,t)}-\frac{2\zeta a(x_1,t)}{\sigma(x_1,t)}-\frac{\sigma_x(x_1,t)(\zeta-\xi)}{2\sigma(x_1,t)}\right)p_N(x_1,t)\right. \\ &\quad \left.\left. - (\zeta-\xi)p_{N,x}(x_1,t)\right]+O(\Delta t)\right\}. \end{aligned} \tag{5.44}$$

Both $J_{LR}(x_1,t)$ and $J_{RL}(x_1,t)$ are infinite, because $p_N(x_1,t)>0$. Using the identities of Exercise 5.13 below, we find that the net flux density is, however, finite and is given by

$$\begin{aligned}
J_{\text{net}}(x_1,t) &= \lim_{\Delta t\to 0}\{J_{LR}(x_1,t,\Delta t) - J_{RL}(x_1,t,\Delta t)\}\\
&= -2\int_0^\infty d\xi \int_\xi^\infty \frac{d\zeta}{\sqrt{2\pi\Delta t\sigma(x_1,t)}}\exp\left\{-\frac{\zeta^2}{2\sigma(x_1,t)}\right\}\\
&\times \left[\left(\frac{\zeta^2(\zeta-\xi)\,\sigma_x(x_1,t)}{\sigma^2(x_1,t)} - \frac{2\zeta a(x_1,t)}{\sigma(x_1,t)} - \frac{\sigma_x(x_1,t)\,(\zeta-\xi)}{2\sigma(x_1,t)}\right)\right.\\
&\left.\times\, p_N(x_1,t) + (\zeta-\xi)\,p_{N,x}(x_1,t)\right]\\
&= \left\{-\frac{\partial\left[\sigma(x,t)p(x,t)\right]}{\partial x} + a(x,t)p(x,t)\right\}_{x=x_1},
\end{aligned} \tag{5.45}$$

as asserted. □

Exercise 5.13 (Identities). Prove the following identities are obtained by changing the order of integration,

$$\begin{aligned}
\int_0^\infty d\xi \int_\xi^\infty \frac{\zeta^2(\zeta-\xi)\,d\zeta}{\sqrt{2\pi\sigma}}\exp\left\{-\frac{\zeta^2}{2\sigma}\right\} &= \int_0^\infty \frac{\zeta^4\,d\zeta}{2\sqrt{2\pi\sigma}}\exp\left\{-\frac{\zeta^2}{2\sigma}\right\} = 3\sigma^2\\
\int_0^\infty d\xi \int_\xi^\infty \frac{\zeta\,d\zeta}{\sqrt{2\pi\sigma}}\exp\left\{-\frac{\zeta^2}{2\sigma}\right\} &= \int_0^\infty \frac{\zeta^2\,d\zeta}{\sqrt{2\pi\sigma}}\exp\left\{-\frac{\zeta^2}{2\sigma}\right\} = \sigma\\
\int_0^\infty d\xi \int_\xi^\infty \frac{(\zeta-\xi)\,d\zeta}{\sqrt{2\pi\sigma}}\exp\left\{-\frac{\zeta^2}{2\sigma}\right\} &= \frac{1}{2}\int_0^\infty \frac{\zeta^2\,d\zeta}{\sqrt{2\pi\sigma}}\exp\left\{-\frac{\zeta^2}{2\sigma}\right\} = \frac{\sigma}{2}.
\end{aligned}$$

□

Equation (5.37) is the classical expression for the probability (or heat) current in diffusion theory [82]. The FPE (5.129) can be written in terms of the flux density function $J(x,t)$ in the conservation law form

$$\frac{\partial p(x,t)}{\partial t} = -\frac{\partial J(x,t)}{\partial x}. \tag{5.46}$$

In $\mathbb{R}^d$ the probability flux density is the probability density of trajectories that propagate per unit time from a domain D across its boundary, ∂D, into the complementary part of space, D^c. It is given by $J_{\text{out}}(\partial D,t) = \lim_{\Delta t\to 0} J_{\text{out}}(\partial D,t,\Delta t)$,

where

$$J_{\text{out}}(\partial D, t, \Delta t)$$
$$= \frac{1}{\Delta t} \int_{D^c} d\boldsymbol{x} \int_{D} \frac{p_N(\boldsymbol{y}, t)\, d\boldsymbol{y}}{(2\pi \Delta t)^{d/2} \sqrt{\det \boldsymbol{\sigma}(\boldsymbol{y}, t)}}$$
$$\times \exp\left\{ -\frac{(\boldsymbol{x} - \boldsymbol{y} - \boldsymbol{a}(\boldsymbol{y}, t)\Delta t)^T \boldsymbol{\sigma}^{-1}(\boldsymbol{y}, t)(\boldsymbol{x} - \boldsymbol{y} - \boldsymbol{a}(\boldsymbol{y}, t)\Delta t)}{2\Delta t} \right\}. \tag{5.47}$$

Similarly, the probability flux density into the domain is defined as the limit of

$$J_{\text{in}}(\partial D, t, \Delta t)$$
$$= \frac{1}{\Delta t} \int_{D} d\boldsymbol{x} \int_{D^c} \frac{p_N(\boldsymbol{y}, t)\, d\boldsymbol{y}}{(2\pi \Delta t)^{d/2} \sqrt{\det \boldsymbol{\sigma}(\boldsymbol{y}, t)}}$$
$$\times \exp\left\{ -\frac{(\boldsymbol{x} - \boldsymbol{y} - \boldsymbol{a}(\boldsymbol{y}, t)\Delta t)^T \boldsymbol{\sigma}^{-1}(\boldsymbol{y}, t)(\boldsymbol{x} - \boldsymbol{y} - \boldsymbol{a}(\boldsymbol{y}, t)\Delta t)}{2\Delta t} \right\}. \tag{5.48}$$

The net flux from the domain is defined as the limit

$$J_{\text{net}}(\partial D, t) = \lim_{\Delta t \to 0} J_{\text{net}}(\partial D, t, \Delta t),$$

where

$$J_{\text{net}}(\partial D, t, \Delta t) = J_{\text{out}}(\partial D, t) - J_{\text{in}}(\partial D, t, \Delta t).$$

Theorem 5.2.4 (Unidirectional and net fluxes in $\mathbb{R}^d$). *The discrete probability flux density densities at a boundary point $\boldsymbol{x}_B$ are given by*

$$\boldsymbol{J}_{\text{out,in}}(\boldsymbol{x}_B, t) \cdot \boldsymbol{n}(\boldsymbol{x}_B) = \sqrt{\frac{\sigma_n(\boldsymbol{x}_B, t)}{2\pi \Delta t}}\, p\,(\boldsymbol{x}_B, t) \pm \frac{1}{2} \boldsymbol{J}_{\text{net}}(\boldsymbol{x}_B, t) \cdot \boldsymbol{n}(\boldsymbol{x}_B)$$
$$+ O(\sqrt{\Delta t}), \tag{5.49}$$

where $\boldsymbol{n}(\boldsymbol{x})$ is the unit outer normal at a boundary point $\boldsymbol{x}$,

$$\sigma_n(\boldsymbol{x}_B, t) = \boldsymbol{n}(\boldsymbol{x}_B)^T \boldsymbol{\sigma}(\boldsymbol{x}_B, t) \boldsymbol{n}(\boldsymbol{x}_B),$$

and the net flux density vector is

$$J^i_{\text{net}}(\boldsymbol{x}_B, t) = -\left\{ \sum_{j=1}^{d} \frac{\partial \sigma^{ij}(\boldsymbol{x}, t) p\,(\boldsymbol{x}, t)}{\partial x^j} + a^i(\boldsymbol{x}, t) p\,(\boldsymbol{x}, t) \right\}_{\boldsymbol{x} = \boldsymbol{x}_B}, \tag{5.50}$$

for $i = 1, 2, \ldots, d$. The net flux is

$$J_{\text{net}}(\partial D, t) = \oint_{D} \boldsymbol{J}_{\text{net}}(\boldsymbol{x}, t) \cdot \boldsymbol{n}(\boldsymbol{x})\, dS_{\boldsymbol{x}}. \tag{5.51}$$

Proof. To evaluate the unidirectional and net fluxes, we define near a boundary point $\boldsymbol{x}_B$ the vector $\boldsymbol{v}(\boldsymbol{x}_B) = \boldsymbol{\sigma}^{1/2}(\boldsymbol{x}_B, t)\boldsymbol{n}(\boldsymbol{x}_B)$, where $\boldsymbol{n}(\boldsymbol{x}_B)$ is the unit outer normal at $\boldsymbol{x}_B$, and map a two-sided neighborhood $\mathcal{N}$ of the boundary by the transformation

$$\boldsymbol{x} = \boldsymbol{x}_B + \boldsymbol{\sigma}^{1/2}(\boldsymbol{x}_B, t)\left[\boldsymbol{x}^{\perp} - \xi \boldsymbol{v}(\boldsymbol{x}_B)\right]\sqrt{\Delta t}, \tag{5.52}$$

where $\boldsymbol{x}^{\perp}$ are $d-1$ variables orthogonal to $\xi\boldsymbol{v}(\boldsymbol{x}_B)$. Here $\xi < 0$ for $\boldsymbol{x} \in D$ and $\xi > 0$ for $\boldsymbol{x} \in D^c$ (this applies to both $\boldsymbol{x}$ and $\boldsymbol{y}$ in the integrals (5.47) and (5.48)). The boundary is then mapped into the hyperplane $\xi = 0$. We may confine the domain of integration in the double integral (5.47) to $\mathcal{N}$, because the contribution of integration outside $\mathcal{N}$ decays exponentially fast as $\Delta t \to 0$. We partition the boundary into patches $\mathcal{P}_B$ about a finite set of boundary points $\boldsymbol{x}_B$ and freeze the coefficients at $\boldsymbol{x}_B$ inside the slice $\left\{(\boldsymbol{x}^{\perp}, \xi) \in \mathcal{N} : \boldsymbol{x}^{\perp} \in \mathcal{P}_B\right\}$. We expand first

$$\begin{aligned}&\frac{(\boldsymbol{x} - \boldsymbol{y} - \boldsymbol{a}(\boldsymbol{y}, t)\Delta t)^T \boldsymbol{\sigma}^{-1}(\boldsymbol{x}_B, t)(\boldsymbol{x} - \boldsymbol{y} - \boldsymbol{a}(\boldsymbol{y}, t)\Delta t)}{\Delta t}\\ &= (\boldsymbol{y}^{\perp} - \boldsymbol{x}^{\perp})^T \boldsymbol{\sigma}^{-1}(\boldsymbol{x}, t)(\boldsymbol{y} - \boldsymbol{x}) - 2\boldsymbol{a}^T(\boldsymbol{x}, t)\boldsymbol{\sigma}^{-1}(\boldsymbol{x}, t)(\boldsymbol{y} - \boldsymbol{x})\Delta t + O(\Delta t^2),\end{aligned}$$

and then about $\boldsymbol{x}_B$ in the variables $(\boldsymbol{x}^{\perp}, \xi)$. The transformation (5.52) maps each side of the slice onto a half space. The variables $\boldsymbol{x}^{\perp}$ integrate out in the double integrals (5.47), (5.48), expressed in the variables $\boldsymbol{x}^{\perp}, \xi$ (in both integrals) and the calculation of the probability flux density reduces to that in the one-dimensional case. We obtain the probability flux density in the form (5.49)–(5.51), as asserted. □

Exercise 5.14 (Details of the proof). Fill in the missing details of the proof. □

Exercise 5.15 (The FPE is a conservation law). Prove that in analogy with (5.46), the FPE in $\mathbb{R}^d$ can also be written in a conservation law form. □

5.3 Boundary behavior of diffusions

Diffusion processes often model particles confined to a given domain in space, for example ions in biological cells. The behavior of the diffusion paths at the boundary of the domain is often determined by physical laws, for example, ions cannot penetrate biological cell membranes due to the much lower dielectric constant of the lipid cell membrane (about $\varepsilon = 2$) than that of the intracellular salt solution (about $\varepsilon = 80$). Sometimes diffusing trajectories that cross the boundary of a domain cannot return for a long time and can be considered instantaneously terminated then and there. This can happen, for example, in modeling the diffusive motion of an atom inside a molecule that collides thermally with other molecules. Due to the collisions the atom, held by the chemical bond, can be displaced to a distance at which the chemical bond is broken, thus dissociating from the molecule permanently. In other situations the diffusing paths can be terminated at the boundary with a given probability; for example, a diffusing protein can stick to a receptor on the cell membrane, or continue its diffusive motion inside the cell. There are many more modes of boundary behavior of diffusion processes inside bounded domains (see, e.g., [71], [161], [117]), so a theory of diffusion inside bounded domains with different boundary behavior is needed.

The easiest way to define a diffusion process inside a given domain, with a prescribed boundary behavior, is to run discrete computer simulations. The relevant mathematical problems are the question of convergence, of the partial differential equations that the transition probabilities and their functionals satisfy, of boundary conditions for the partial differential equations, and the probability measures defined in function space by the confined diffusions. The imposed boundary conditions on the simulated trajectories are reflected in the pdf, in boundary conditions for the FPE, but sometimes more complicated connections show up. And conversely, often boundary conditions imposed on the FPE to express physical processes that occur at the boundary (e.g., a reactive boundary condition that expresses a possible binding of a molecule) can be expressed in terms of the boundary behavior of simulated trajectories of a SDE. The Wiener path integral is a convenient tool for the study of the duality between the boundary behavior of trajectories and boundary (and other) conditions for the FPE, as discussed below.

5.4 Absorbing boundaries

The simplest simulation of the Itô dynamics

$$dx = a(x,t)\,dt + b(x,t)\,dw \quad \text{for } t > s, \quad x(s) = x_0, \tag{5.53}$$

is the Euler scheme

$$\begin{aligned} x_N(t+\Delta t) &= x_N(t) + a(x_N(t),t)\Delta t + b(x_N(t),t)\,\Delta w(t) \\ x_N(s) &= x_0. \end{aligned} \tag{5.54}$$

Equation (5.54) defines $x_N(t)$ as a Markov chain (see Section 7.1). If the trajectories of $x_N(t,\omega)$ that start at $x_0 > 0$ (and are determined by (5.54)), are truncated at the first time they cross the origin, we say that the origin is an *absorbing boundary*.

Exercise 5.16 (Convergence of trajectories). Generalize the proof of Skorokhod's theorem 5.1.1 to Euler's scheme with an absorbing boundary (see [224]). □

The path integral corresponding to this situation is defined on the subset of trajectories that never cross a from left to right. Thus the integration in the definition (5.9) of the pdf does not extend over $\mathbb{R}$, but rather is confined to the ray $[0,\infty)$. That is, the pdf is given by

$$p_N(x,t\,|\,x_0,s) = \underbrace{\int_0^\infty dy_1 \int_0^\infty dy_2 \cdots \int_0^\infty}_{N-1} dy_{N-1} \prod_{j=1}^N \frac{1}{\sqrt{2\pi\Delta t}\, b(y_{j-1},t_{j-1})} \times \exp\left\{ -\frac{[y_j - y_{j-1} - a(y_{j-1},t_{j-1})\Delta t]^2}{2b^2(y_{j-1},t_{j-1})\Delta t} \right\}, \tag{5.55}$$

where $t_0 = s$, $y_0 = x_0$ and $t_N = t$, $y_N = x$. As in Chapter 5, we denote $p_N(x,t) = \int_0^\infty p_N(x,t\,|\,x_0,s)p_0(x_0)\,dx_0$, where $p_0(x_0)$ is a sufficiently smooth test density with compact support on the positive axis.

Theorem 5.4.1. *For every $T > 0$, the Wiener integral $p_N(x,t)$ converges to the solution $p\,(x,t)$ of the boundary value problem* (5.129), (5.130) *and*

$$p\,(0,t) = 0 \quad \text{for} \quad t > 0, \tag{5.56}$$

uniformly for all $x > 0$, $s < t < T$.

Proof. If $x > 0$, then the change of variables $y = x - a(y,t)\Delta t + \eta b(y,t)\sqrt{\Delta t}$ maps the domain of integration from $0 < y < \infty$ unto the ray

$$-\frac{x - a(y,t)\Delta t}{b(y,t)\sqrt{\Delta t}} < \eta < \infty,$$

so integration can be extended to $\mathbb{R}$ with exponentially decaying error as $\Delta t \to 0$. The proof of Theorem 5.2.1 then shows that the limit function $p\,(x,t)$ satisfies (5.129), (5.130). If, however, we set $x = 0$ in the expansion of the path integral (5.55) that leads to (5.17), the change of variables maps the domain of integration onto only the half-line $0 \leq \eta < \infty$, rather than onto the entire line. The value of the Gaussian integral over this domain is $\frac{1}{2}$, so assuming that the limit of $p_N(x,t) \to p\,(x,t)$ as $N \to \infty$ exists, we obtain the identity $p\,(0,t) = \frac{1}{2}p\,(0,t)$, which apparently implies that $p\,(y,t)$ satisfies the boundary condition (5.56).

The pdf $p_N(y,t)$, however, does not necessarily converge to the solution $p\,(y,t)$ of (5.129), (5.130) with the boundary condition (5.56), uniformly up to the boundary. More specifically, it is not clear that

$$\lim_{y\to 0}\lim_{N\to\infty} p_N(y,t) = \lim_{N\to\infty}\lim_{y\to 0} p_N(y,t), \tag{5.57}$$

because, as is typical for diffusion approximations of Markovian jump processes that jump over the boundary [133], [137], [135], [69], the convergence is not necessarily uniform and typically, a boundary layer is formed. A boundary layer expansion is needed to capture the boundary phenomena. To examine the convergence of $p_N(y,t)$ near $y = 0$, we rewrite (5.55) as the integral equation

$$p_N(y,t+\Delta t\,|\,x_0) = \int_0^\infty \frac{p_N(x,t)}{\sqrt{4\pi\sigma(x,t)\Delta t}}\exp\left\{-\frac{(y-x-a(x,t)\Delta t)^2}{4\sigma(x,t)\Delta t}\right\}dx, \tag{5.58}$$

where $\sigma(x,t) = \frac{1}{2}b^2(x,t)$, and introduce the local variable $y = \eta\sqrt{\Delta t}$ and the boundary layer solution $p_{\text{bl}}(\eta,t) = p_N(\eta\sqrt{\Delta t},t\,|\,x_0)$. Changing variable of integration $x = \xi\sqrt{\Delta t}$ in (5.58) gives

$$p_{\text{bl}}(\eta, t + \Delta t) = \int_0^\infty \frac{p_{\text{bl}}(\xi, t)}{\sqrt{4\pi\sigma(\xi\sqrt{\Delta t}, t)}} \exp\left\{ -\frac{\left[\eta - \xi - a(\xi\sqrt{\Delta t}, t)\sqrt{\Delta t}\right]^2}{4\sigma(\xi\sqrt{\Delta t}, t)} \right\} d\xi. \tag{5.59}$$

The boundary layer solution has an asymptotic expansion in powers of $\sqrt{\Delta t}$

$$p_{\text{bl}}(\eta, t) \sim p_{\text{bl}}^{(0)}(\eta, t) + \sqrt{\Delta t}\, p_{\text{bl}}^{(1)}(\eta, t) + \Delta t\, p_{\text{bl}}^{(2)}(\eta, t) + \cdots. \tag{5.60}$$

Expanding all functions in (5.59) in powers of $\sqrt{\Delta t}$ and equating similar orders, we obtain integral equations that the asymptotic terms of (5.60) must satisfy. The leading-order $O(1)$ term gives the Wiener–Hopf equation on the half-line [187]

$$p_{\text{bl}}^{(0)}(\eta, t) = \int_0^\infty \frac{p_{\text{bl}}^{(0)}(\xi, t)}{\sqrt{4\pi\sigma(0, t)}} \exp\left\{ -\frac{(\eta - \xi)^2}{4\sigma(0, t)} \right\} d\xi, \quad \text{for } \eta > 0. \tag{5.61}$$

Integrating (5.61) with respect to η over $\mathbb{R}^+$, changing the order of integration, and changing variables to $\eta = \xi + z$ on the right-hand side, we obtain

$$\begin{aligned} \int_0^\infty p_{\text{bl}}^{(0)}(\eta, t)\, d\eta &= \int_0^\infty \frac{p_{\text{bl}}^{(0)}(\xi, t)}{\sqrt{4\pi\sigma(0, t)}} \int_{-\xi}^\infty \exp\left\{ -\frac{z^2}{4\sigma(0, t)} \right\} dz\, d\xi \\ &= \int_0^\infty p_{\text{bl}}^{(0)}(\xi, t) \left[1 - \frac{1}{\sqrt{4\pi\sigma(0, t)}} \int_\xi^\infty \exp\left\{ -\frac{z^2}{4\sigma(0, t)} \right\} dz \right] d\xi, \end{aligned}$$

hence

$$\int_0^\infty \frac{p_{\text{bl}}^{(0)}(\xi, t)}{\sqrt{4\pi\sigma(0, t)}} \int_\xi^\infty \exp\left\{ -\frac{z^2}{4\sigma(0, t)} \right\} dz\, d\xi = 0. \tag{5.62}$$

It follows that $p_{\text{bl}}^{(0)}(\xi, t) = 0$, because all functions in (5.62) are continuous and nonnegative.

Away from the boundary layer, the solution admits an outer expansion

$$p_{\text{out}}(y, t) \sim p_{\text{out}}^{(0)}(y, t) + \sqrt{\Delta t} p_{\text{out}}^{(1)}(y, t) + \cdots, \tag{5.63}$$

where $p_{\text{out}}^{(0)}(y, t)$ is yet an undetermined function that satisfies (5.13). The leading-order matching condition of the boundary layer and the outer solutions is

$$\lim_{\eta \to \infty} p_{\text{bl}}^{(0)}(\eta, t) = p_{\text{out}}^{(0)}(0, t),$$

so that $p_{\text{out}}^{(0)}(0,t)=0$. The limits are interchangeable and (5.57) holds, because

$$\lim_{y\to 0}\lim_{N\to\infty} p_N(y,t)=p_{\text{out}}^{(0)}(0,t)=0,\qquad \lim_{N\to\infty}\lim_{y\to 0} p_N(y,t)=p_{\text{bl}}^{(0)}(0,t)=0,$$

and so does the boundary condition (5.56).

The remainder of the proof follows that of Theorem 5.2.1. We extend $p_N(x,t)$ to t off the lattice by an interpolation $\tilde{p}_N(x,t)$, as in (5.7) and (5.8). The boundary layer expansion of $\tilde{p}_N(x,t)$ is similar to that of $p_N(x,t)$ and implies that for every $\varepsilon>0$ and $T>0$ there is $\delta>0$, such that if $0\leq x<\delta$ and $t<T$, then $\tilde{p}_N(x,t)<\varepsilon$, $p(x,t)<\varepsilon$, $|\tilde{p}_N(\delta,t)-p(\delta,t)|<2\varepsilon$, and $\tilde{p}_N(x,0)-p(x,0)=p_0(x)-p_0(x)=0$. The maximum principle implies that $|\tilde{p}_N(x,t)-p(x,t)|<2\varepsilon$ for all $x>\delta$, $0<t<T$. The convergence is uniform, because δ is arbitrarily small. □

Exercise 5.17 (Diffusion in an interval with absorbing boundaries). Generalize Theorem 5.4.1 to diffusion in a finite interval with absorption at both boundaries. Generalize Exercises 5.2–5.9 to this case. □

Exercise 5.18 (Convergence of trajectories in d dimensions). Generalize the proof of Skorokhod's theorem 5.1.1 to Euler's scheme in a domain $D\subset\mathbb{R}^d$ with an absorbing boundary. □

Theorem 5.4.2. *For every $T>s\geq 0$ the pdf $p_{\Delta t}(\boldsymbol{y},t\,|\,\boldsymbol{x},s)$ of the Euler scheme*

$$\begin{aligned}\boldsymbol{x}(t+\Delta t)&=\boldsymbol{x}(t)+\boldsymbol{a}(\boldsymbol{x}(t),t)\Delta t+\sqrt{2}\boldsymbol{B}(\boldsymbol{x}(t),t)\,\Delta\boldsymbol{w}(t,\Delta t)\\ \boldsymbol{x}_N(s)&=\boldsymbol{x}_0,\end{aligned}\tag{5.64}$$

where all trajectories are instantaneously terminated when they exit D, converges in the limit $\Delta t\to 0$ to the solution $p(\boldsymbol{y},t\,|\,\boldsymbol{x},s)$ of the initial value problem for the FPE, (5.23), (5.24), *with the absorbing (Dirichlet) boundary condition*

$$p(\boldsymbol{y},t\,|\,\boldsymbol{x},s)=0\ \text{ for } \boldsymbol{y}\in\partial D,\ \boldsymbol{x}\in D.\tag{5.65}$$

Exercise 5.19 (Proof of Theorem 5.4.2). Prove Theorem 5.4.2 by following the steps

(i) Derive the Chapman–Kolmogorov equation

$$\begin{aligned}&p_{\Delta t}(\boldsymbol{y},t+\Delta t\,\boldsymbol{x}_0,s)\\ &=\int_D \frac{p_{\Delta t}(\boldsymbol{x},t\,|\,\boldsymbol{x}_0,s)\,d\boldsymbol{x}}{(2\pi\Delta t)^{d/2}\sqrt{\det\boldsymbol{\sigma}(\boldsymbol{x},t)}}\exp\left\{-\frac{\mathcal{B}(\boldsymbol{x},\boldsymbol{y},t)}{2\Delta t}\right\},\end{aligned}\tag{5.66}$$

where $\mathcal{B}(\boldsymbol{x},\boldsymbol{y},t)=[\boldsymbol{y}-\boldsymbol{x}-\boldsymbol{a}(\boldsymbol{x},t)\Delta t]^T\boldsymbol{\sigma}^{-1}(\boldsymbol{x},t)\,[\boldsymbol{y}-\boldsymbol{x}-\boldsymbol{a}(\boldsymbol{x},t)\Delta t]$, as in eq. (5.21).

(ii) Generalize Exercises 5.2–5.9 to the integral (5.66).

(iii) Show that there is no boundary layer.

(iv) Use the maximum principle to prove convergence. □

5.4.1 Unidirectional flux and the survival probability

The trajectories absorbed at the boundary give rise to a unidirectional probability flux from the domain into the boundary. The absorbing boundary condition (5.56) implies that the pdf vanishes for all $x \geq 0$ so that its right derivatives at the origin vanish. It follows from eq. (5.43) that $J_{RL}(0,t) = 0$. On the other hand, eqs. (5.35) and (5.37) give

$$J(0,t) = J_{LR}(0,t) = -\left.\frac{\partial \sigma(x,t)p\,(x,t)}{\partial x}\right|_{x=0}.$$

It follows that $J(0,t) > 0$, because $\sigma(x,t) > 0$ and $p\,(x,t) > 0$ for $x < 0$, but $p\,(0,t) = 0$. This means that there is positive flux into the absorbing boundary so that the probability of trajectories that survive in the region to the left of the absorbing boundary, $\int_{-\infty}^{0} p\,(x,t)\,dx$, must be a decreasing function of time. This can be seen directly from eq. (5.46) by integrating it with respect to x over the ray $(-\infty, 0)$ and using the fact that $\lim_{x\to-\infty} J(x,t) = 0$,

$$\frac{d}{dt}\int\limits_{-\infty}^{0} p\,(x,t)\,dx = -J(0,t) < 0. \tag{5.67}$$

Equation (5.67) means that the total population of trajectories in the domain $x < 0$ decreases with time.

Definition 5.4.1 (The survival probability). *The* survival probability $S(t \,|\, \boldsymbol{x}, s)$ *of trajectories of*

$$d\boldsymbol{x} = \boldsymbol{a}(\boldsymbol{x},t)\,dt + \sqrt{2}\boldsymbol{B}(\boldsymbol{x},t)\,d\boldsymbol{w} \tag{5.68}$$

in a domain D at time t, that started at time $s < t$ at a point $\boldsymbol{x} \in D$, is the conditional probability that the first passage time τ to the boundary ∂D of the domain does not exceed t,

$$S(t \,|\, \boldsymbol{x}, s) = \Pr\{\tau > t \,|\, \boldsymbol{x}, s) = \int\limits_{D} p\,(\boldsymbol{y}, t \,|\, \boldsymbol{x}, s)\,d\boldsymbol{y}, \tag{5.69}$$

where the transition pdf $p\,(\boldsymbol{y}, t \,|\, \boldsymbol{x}, s)$ of the process (5.68), *with absorption in ∂D, is the solution of the initial boundary value problem for the Fokker–Planck equation* (5.23), (5.24), (5.65).

The flux density vector $\boldsymbol{J}(\boldsymbol{y}, t \,|\, \boldsymbol{x}, s)$ in (5.50) reduces to

$$J^i(\boldsymbol{y}, t \,|\, \boldsymbol{x}, s) = -\sum_{j=1}^{n} \frac{\partial\left[\sigma^{ij}(\boldsymbol{y},t)p\,(\boldsymbol{y}, t \,|\, \boldsymbol{x}, s)\right]}{\partial y^j}, \tag{5.70}$$

where $\boldsymbol{\sigma}(\boldsymbol{x},t) = \boldsymbol{B}^T(\boldsymbol{x},t)\boldsymbol{B}(\boldsymbol{x},t)$. The probability per unit time of trajectories that are absorbed into a given surface $S \subset \partial D$ is given by $F = \int_S \boldsymbol{J}(\boldsymbol{y}, t \,|\, \boldsymbol{x}, s) \cdot \boldsymbol{n}(\boldsymbol{y})\,dS_{\boldsymbol{y}}$, which can be interpreted as the following

Theorem 5.4.3 (Normal flux density at an absorbing boundary). *The normal flux density $\boldsymbol{J}(\boldsymbol{y},t\,|\,\boldsymbol{x},s)\cdot\boldsymbol{n}(\boldsymbol{y})$ at point $\boldsymbol{y}$ of the absorbing boundary is the conditional probability per unit surface area and per unit time, that passes through the surface at the boundary point $\boldsymbol{y}$ at time t. Thus it is the conditional probability density (per unit area) of stochastic trajectories absorbed at $\boldsymbol{y}$ at a given instance of time $t > s$, given that they started at $\boldsymbol{x}$ at time s.*

The survival probability and the probability distribution function of the first passage time τ to the boundary ∂D are related by the following

Theorem 5.4.4 (Survival probability and the first passage time). *The MFPT to the boundary after time s is*

$$\mathbb{E}[\tau\,|\,\boldsymbol{x},s] = \int_s^\infty S(t\,|\boldsymbol{x},s)\,dt = \int_s^\infty\int_D p\,(\boldsymbol{y},t\,|\,\boldsymbol{x},s)\,d\boldsymbol{y}\,dt. \tag{5.71}$$

Example 5.1 (Flux in 1-D). The one-dimensional Fokker–Planck equation has the form $p_t = -J_y(y,t\,|\,x,s)$, where the one-dimensional flux is given by

$$J(y,t\,|\,x,s) = a(y,t)p\,(y,t\,|\,x,s) - \left[\sigma(y,t)p\,(y,t\,|\,x,s)\right]_y.$$

At an absorbing boundary

$$J(y,t\,|\,x,s) = -\left[\sigma(y,t)p\,(y,t\,|\,x,s)\right]_y$$

for $x \in D$ and $y \in \partial D$, because $p\,(y,t)|_{y\in\partial D} = 0$. □

Exercise 5.20 (The probability flux density of the Langevin equation). The instantaneous unidirectional probability flux from left to right in the Langevin equation (5.28), $J_{LR}(x_1,t)$, is usually defined in statistical mechanics as the integral of $J_x(x_1,v,t)$ in (5.33) over the positive velocities, $J_{LR}(x_1,t) = \int_0^\infty vp\,(x_1,v,t)\,dv$. Show by the following steps that this integral actually represents the probability of the trajectories that move from left to right across x_1 per unit time.

(i) Prove that $J_x(x,v,t)$ and $J_v(x,v,t)$, given in (5.33), and are the probability flux density densities in the phase plane.

(ii) Prove that the instantaneous unidirectional probability flux from left to right in the Langevin equation (5.28) (or equivalently, (5.29)), at a point x_1, is

$$\begin{aligned}
&J_{LR}(x_1,t)\\
&= \lim_{\Delta t\to 0}\frac{1}{\Delta t}\int_{-\infty}^{x_1} d\xi\int_{x_1}^{\infty} dx\int_{-\infty}^{\infty} d\eta\int_{-\infty}^{\infty} dv\frac{1}{\sqrt{4\varepsilon\gamma\pi\Delta t}}p\,(\xi,\eta,t)\delta(x-\xi-\eta\Delta t)\\
&\times\exp\left\{-\frac{\left[v-\eta-[-\gamma\eta+f(\xi)]\Delta t\right]^2}{4\varepsilon\gamma\Delta t}\right\}.
\end{aligned} \tag{5.72}$$

(iii) Integrate with respect to v to get (5.33),

$$\begin{aligned} J_{LR}(x_1,t) &= \lim_{\Delta t\to 0}\frac{1}{\Delta t}\iint\limits_{x-\eta\Delta t<x_1} p(x-\eta\Delta t,\eta,t)\,d\eta\,dx \\ &= \lim_{\Delta t\to 0}\frac{1}{\Delta t}\int_0^\infty d\eta \int_{x_1-\eta\Delta t}^{x_1} p(u,\eta,t)\,du = \int_0^\infty \eta p(x_1,\eta,t)\,d\eta \end{aligned} \tag{5.73}$$

(see [222]). □

5.5 Reflecting and partially reflecting boundaries

The Fokker–Planck equation (5.23) in a domain D, written as

$$\frac{\partial p(\boldsymbol{y},t\,|\,\boldsymbol{x},s)}{\partial t} = -\nabla_{\boldsymbol{y}}\cdot \boldsymbol{J}(\boldsymbol{y},t\,|\,\boldsymbol{x},s) \quad \text{for all} \quad \boldsymbol{y},\boldsymbol{x}\in D, \tag{5.74}$$

$$\lim_{t\downarrow s} p(\boldsymbol{y},t\,|\,\boldsymbol{x},s) = \delta(\boldsymbol{y}-\boldsymbol{x}), \tag{5.75}$$

where the components of the flux vector $\boldsymbol{J}(\boldsymbol{y},t\,|\,\boldsymbol{x},s)$ are defined by

$$\begin{aligned} &J^k(\boldsymbol{y},t\,|\,\boldsymbol{x},s) \\ &= -a^k(\boldsymbol{y},t)p(\boldsymbol{y},t\,|\,\boldsymbol{x},s) + \sum_{j=1}^d \frac{\partial}{\partial y_j}\left[\sigma^{j,k}(\boldsymbol{y},t)p(\boldsymbol{y},t\,|\,\boldsymbol{x},s)\right], \end{aligned} \tag{5.76}$$

with the *partially absorbing boundary condition (radiation, reaction, Robin)*

$$-\boldsymbol{J}(\boldsymbol{y},t\,|\,\boldsymbol{x},s)\cdot\boldsymbol{n} = \kappa(\boldsymbol{y},t)p(\boldsymbol{y},t\,|\,\boldsymbol{x},s), \quad \text{for } \boldsymbol{y}\in\partial D,\ \boldsymbol{x}\in D, \tag{5.77}$$

is widely used in chemical and biological applications to express reactive boundaries. The underlying trajectories of the diffusing particles are believed to be partially absorbed and partially reflected at the reactive boundary, however, the relation between the reaction parameter $\kappa(\boldsymbol{y},t)$ in the Robin boundary condition (5.77) and the reflection probability is not well defined. In this section we define the partially reflected process as a limit of the Markovian jump process generated by the Euler scheme for the underlying Itô dynamics (5.68) with partial boundary reflection. Trajectories that cross the boundary at a point $(\boldsymbol{x},t)$ are terminated with probability $P(\boldsymbol{x},t)\sqrt{\Delta t}$ and are otherwise reflected in a normal or oblique direction. We consider here only the one-dimensional problem. Partial co-normal reflection in higher dimensions is discussed in [214].

5.5.1 Total and partial reflection in one dimension

The one-dimensional Robin problem for $p = p\,(y,t\,|\,x,s)$ in $\mathbb{R}^+$ is

$$\frac{\partial p}{\partial t} = -\frac{\partial[a(y,t)p]}{\partial y} + \frac{\partial^2[\sigma(y,t)p]}{\partial y^2} \quad \text{for } x,y>0, \tag{5.78}$$

$$\lim_{t\downarrow s} p = \delta(y-x) \quad \text{for } x,y>0 \tag{5.79}$$

$$-J(0,t\,|\,x,s) = \kappa(t)p\,(0,t\,|\,x,s) \quad \text{for } x>0,\ t>s, \tag{5.80}$$

where

$$J(y,t\,|\,x,s) = a(y,t)p\,(y,t\,|\,x,s) - \frac{\partial[\sigma(y,t)p\,(y,t\,|\,x,s)]}{\partial y}, \tag{5.81}$$

We abbreviate henceforward $\kappa(t) = \kappa$. The Dirichlet boundary condition (5.56) is recovered from (5.80) if $\kappa = \infty$, and a *no flux (reflecting)* boundary condition

$$-J(0,t\,|\,x,s) = 0 \ \text{ for } x>0\,t>s, \tag{5.82}$$

is obtained if $\kappa = 0$ in (5.80), respectively.

Next, we construct an Itô equation

$$dx = a(x,t)\,dt + b(x,t)\,dw \tag{5.83}$$

in $\mathbb{R}^+$ with a totally or partially absorbing boundary at $x = 0$, whose pdf satisfies (5.78)–(5.80). It is defined as the limit of the Markovian jump processes generated by the Euler scheme for $t > s$

$$x_{\Delta t}(t+\Delta t) = x_{\Delta t}(t) + a(x_{\Delta t}(t),t)\Delta t + b(x_{\Delta t}(t),t)\,\Delta w(t,\Delta t) \tag{5.84}$$

$$x_{\Delta t}(s) = x \tag{5.85}$$

for $x_{\Delta t}(t) \in \mathbb{R}^+$, for $0 \leq t-s \leq T$, with $\Delta t = T/N$, $t-s = iT/N$ ($i = 0,1,\ldots,N$), where for each t the random variables $\Delta w(t,\Delta t)$ are normally distributed and independent with zero mean and variance Δt. The partially absorbing boundary condition for (5.84) has to be chosen so that the pdf $p_{\Delta t}(x,t)$ of $x_{\Delta t}(t)$ converges to the solution of (5.78)–(5.80). At a partially reflecting boundary for (5.84), the trajectories that cross the origin are instantaneously reflected with probability (w.p.) $R(t)$, and are otherwise terminated (absorbed).

Exercise 5.21 (Convergence of trajectories). Generalize the proof of Skorokhod's theorem 5.1.1 to Euler's scheme with a reflecting or partially reflecting boundary. □

It is shown below that keeping $R(t)$ independent of Δt (e.g., $R = \frac{1}{2}$) leads in the limit $\Delta t \to 0$ to the convergence of the pdf $p_{\Delta t}(x,t)$ to the solution of the FPE with an absorbing rather than the Robin boundary condition. Thus the reflection probability $R(t)$ must increase to 1 as $\Delta t \to 0$ in order to yield the Robin condition (5.80). Moreover, the reactive "constant" $\kappa(t)$ is related to the limit

$$\lim_{\Delta t\to 0} \frac{1-R(t)}{\sqrt{\Delta t}} = P(t). \tag{5.86}$$

The absorbing boundary condition (5.56) is obtained for $P = 1/\sqrt{\Delta t} \to \infty$, and the no-flux (totally reflecting) boundary condition (5.82) is recovered for $P(t) = 0$. These considerations lead to the following simple boundary behavior for the simulated trajectories that cross the boundary, identified by

$$x' = x_{\Delta t}(t) + a(x_{\Delta t}(t), t)\Delta t + b(x_{\Delta t}(t), t)\,\Delta w < 0,$$

namely,

$$x_{\Delta t}(t + \Delta t) \tag{5.87}$$
$$= \begin{cases} -(x_{\Delta t}(t) + a(x_{\Delta t}(t), t)\Delta t + b(x_{\Delta t}(t), t)\,\Delta w) & \text{w.p. } 1 - P(t)\sqrt{\Delta t} \\ \text{terminate trajectory otherwise.} \end{cases}$$

We abbreviate notation henceforward by suppressing the dependence of $P(t)$ on t. The exiting trajectory is normally reflected with probability

$$R = 1 - P\sqrt{\Delta t} \tag{5.88}$$

and is otherwise terminated (absorbed). Choosing $R = 0$ recovers the absorbing Euler scheme of Section 5.4 and choosing $R = 1$ gives total reflection at the boundary. The scaling of the termination probability with $\sqrt{\Delta t}$ reflects the fact that the discrete unidirectional diffusion current at any point, including the boundary, is $O(1/\sqrt{\Delta t})$ (see eq. (5.36)). This means that the number of discrete trajectories hitting or crossing the boundary in any finite time interval Δt increases as $O(1/\sqrt{\Delta t})$. Therefore, to keep the efflux of trajectories finite as $\Delta t \to 0$, the termination probability $1 - R$ of a crossing trajectory has to be $O(\sqrt{\Delta t})$. The derivative of $p_{\Delta t}(x, t)$, however, does not converge to that of the solution $p\,(x, t)$ of (5.78)–(5.80) on the boundary, as shown in the proof of Theorem 5.5.1 below. This is due to the formation of a boundary layer, as is typical for diffusion approximations of Markovian jump processes that jump over the boundary, as mentioned in Section 5.4.

Note that when using (5.84) and other schemes, the pdf of the solution of (5.84), (5.85) converges to the solution of the FPE (5.78) and the initial condition (5.79), however, it does not satisfy the boundary condition (5.80), not even approximately. For a general diffusion coefficient and drift term, the boundary condition is not satisfied even for the case of a reflecting boundary condition. This problem plagues other schemes as well. This apparent paradox is due to the nonuniform convergence of $p_{\Delta t}(y, t \,|\, x, s)$ to the solution $p\,(y, t \,|\, x, s)$ of the Fokker–Planck equation, caused by a boundary layer that $p_{\Delta t}(y, t \,|\, x, s)$ develops for small Δt. The limit $p\,(y, t \,|\, x, s)$, however, satisfies the boundary condition (5.80). To derive the Chapman–Kolmogorov equation (2.59) corresponding to the scheme (5.87), we note that a trajectory of (5.87) can reach a point $y \in \mathbb{R}^+$ at time $t + \Delta t$ in one time step Δt in two ways. One is to reach it from a point $x \in \mathbb{R}^+$ and the other is to reach it from the point $x' = x_{\Delta t}(t) + a(x_{\Delta t}(t), t)\Delta t + b(x_{\Delta t}(t), t)\,\Delta w < 0$. In the latter case the jump to y is contingent on the trajectory not being terminated at x'. Thus, using abbreviated notation, the pdf $p_{\Delta t}(y, t \,|\, x, s) = p_{\Delta t}(y, t)$ satisfies the

the Chapman–Kolmogorov equation

$$p_{\Delta t}(y,t+\Delta t) = \int_0^\infty \frac{p_{\Delta t}(x,t)}{\sqrt{4\pi\sigma(x,t)\Delta t}} \left[\exp\left\{-\frac{(y-x-a(x,t)\Delta t)^2}{4\sigma(x,t)\Delta t}\right\} + (1-P\sqrt{\Delta t})\exp\left\{-\frac{(y+x+a(x,t)\Delta t)^2}{4\sigma(x,t)\Delta t}\right\}\right] dx, \quad (5.89)$$

where $\sigma(x,t) = \frac{1}{2}b^2(x,t)$. For $P = 0$ the pdf $p_{\Delta t}(y,t)$ satisfies the boundary condition

$$\frac{\partial p_{\Delta t}(0,t)}{\partial y} = 0, \quad (5.90)$$

which is obtained by differentiation of (5.89) with respect to y at $y = 0$. If $P \neq 0$, we obtain

$$\frac{\partial p_{\Delta t}(0,t+\Delta t)}{\partial y} = \frac{p_{\Delta t}(0,t)P}{\sqrt{4\pi\sigma(0,t)}} + O(\sqrt{\Delta t}), \quad (5.91)$$

which in the limit $\Delta t \to 0$ gives

$$\frac{\partial p\,(0,t)}{\partial y} = \frac{p\,(0,t)P}{\sqrt{4\pi\sigma(0,t)}}. \quad (5.92)$$

However, this is not the boundary condition that the limit function

$$p\,(y,t) = \lim_{\Delta t \to 0} p_{\Delta t}(y,t)$$

satisfies. To find the boundary condition of $p\,(y,t)$, in either case, we note that $p\,(y,t)$ satisfies the FPE (5.78) and the initial condition (5.79). We must have

$$0 = \frac{d}{dt}\int_0^\infty p\,(x,t)\,dx = -\frac{\partial[\sigma(0,t)p\,(0,t)]}{\partial y} + a(0,t)p\,(0,t) = J(0,t), \quad (5.93)$$

because for $P = 0$ the simulation preserves probability (the population of trajectories), which contradicts (5.90). The discrepancy between (5.93) and (5.90) is due to the nonuniform convergence of $p_{\Delta t}(y,t)$ to its limit $p\,(y,t)$ in the interval. There is a boundary layer of width $O(\sqrt{\Delta t})$, in which the boundary condition (5.90) for $p_{\Delta t}(y,t)$ changes into the boundary condition (5.93) that $p\,(y,t)$ satisfies. Equation (5.93) is the correct no-flux boundary condition (5.82).

Theorem 5.5.1 (Partially reflected diffusion). *For every $T > 0$, the pdf $p_{\Delta t}(x,t)$ of the Markovian jump process $x_{\Delta t}(t)$ generated by* (5.84), (5.85), (5.87) *converges to the solution $p\,(x,t)$ of the initial and boundary value problem* (5.78)–(5.81), *uniformly for all $x > 0$, $s < t < T$.*

Proof. We construct a uniform asymptotic expansion of the solution $p_{\Delta t}(y,t\,|\,x,s)$ of (5.89) to examine its convergence to the solution $p\,(y,t\,|\,x,s)$ of (5.78)–(5.80), and to find the relation between the parameter P of (5.87) and the reactive "constant" $\kappa(t)$ in (5.80). To analyze the discrepancy between (5.90) and (5.93), we proceed as in Section 5.4; that is, we introduce the local variable $y=\eta\sqrt{\Delta t}$ and the boundary layer solution

$$p_{\rm bl}(\eta,t)=p_{\Delta t}(\eta\sqrt{\Delta t},t). \tag{5.94}$$

Changing variables $x=\xi\sqrt{\Delta t}$ in the integral (5.89) gives

$$\begin{aligned} &p_{\rm bl}(\eta,t+\Delta t)\\ &=\int_0^\infty \frac{p_{\rm bl}(\xi,t)}{\sqrt{4\pi\sigma(\xi\sqrt{\Delta t},t)}}\left[\exp\left\{-\frac{\left(\eta-\xi-a(\xi\sqrt{\Delta t},t)\sqrt{\Delta t}\right)^2}{4\sigma(\xi\sqrt{\Delta t},t)}\right\}\right.\\ &\left.+\ (1-P\sqrt{\Delta t})\exp\left\{-\frac{\left(\eta+\xi+a(\xi\sqrt{\Delta t},t)\sqrt{\Delta t}\right)^2}{4\sigma(\xi\sqrt{\Delta t},t)}\right\}\right]d\xi. \end{aligned} \tag{5.95}$$

The boundary layer solution has an asymptotic expansion in powers of $\sqrt{\Delta t}$

$$p_{\rm bl}(\eta,t)\sim p_{\rm bl}^{(0)}(\eta,t)+\sqrt{\Delta t}\,p_{\rm bl}^{(1)}(\eta,t)+\Delta t\,p_{\rm bl}^{(2)}(\eta,t)+\cdots. \tag{5.96}$$

Expanding all functions in (5.95) in powers of $\sqrt{\Delta t}$ and equating similar orders, we obtain integral equations that the asymptotic terms of (5.96) must satisfy. The leading-order $O(1)$ term gives on the half-line $\eta>0$ the Wiener–Hopf equation

$$\begin{aligned} &p_{\rm bl}^{(0)}(\eta,t)\\ &=\int_0^\infty \frac{p_{\rm bl}^{(0)}(\xi,t)}{\sqrt{4\pi\sigma(0,t)}}\left[\exp\left\{-\frac{(\eta-\xi)^2}{4\sigma(0,t)}\right\}+\exp\left\{-\frac{(\eta+\xi)^2}{4\sigma(0,t)}\right\}\right]d\xi. \end{aligned} \tag{5.97}$$

To solve (5.97), we note first that the kernel

$$K(\eta,\xi)=\exp\left\{-\frac{(\eta-\xi)^2}{4\sigma(0,t)}\right\}+\exp\left\{-\frac{(\eta+\xi)^2}{4\sigma(0,t)}\right\} \tag{5.98}$$

is an even function of η and ξ; that is, $K(\eta,\xi)=K(-\eta,\xi)=K(\eta,-\xi)=K(-\eta,-\xi)$. Therefore, we extend $p_{\rm bl}^{(0)}(\xi,t)$ to the entire line as an even function ($p_{\rm bl}^{(0)}(\xi,t)=p_{\rm bl}^{(0)}(-\xi,t)$), and rewrite (5.97) as the convolution equation on the entire line $-\infty<\eta<\infty$,

$$p_{\rm bl}^{(0)}(\eta,t)=\int_{-\infty}^\infty \frac{p_{\rm bl}^{(0)}(\xi,t)}{\sqrt{4\pi\sigma(0,t)}}\exp\left\{-\frac{(\eta-\xi)^2}{4\sigma(0,t)}\right\}d\xi. \tag{5.99}$$

The Fourier transform of (5.99) shows that the only solution is the constant function; that is, $p_{\text{bl}}^{(0)}(\eta,t) = f(t)$, independent of η.

Away from the boundary layer the solution has the regular (outer) expansion

$$p_{\text{out}}(y,t) \sim p_{\text{out}}^{(0)}(y,t) + \sqrt{\Delta t}\, p_{\text{out}}^{(1)}(y,t) + \cdots, \tag{5.100}$$

where $p_{\text{out}}^{(0)}$ satisfies the Fokker–Planck equation (5.78) with the radiation boundary condition (5.80). The matching condition of the boundary layer and the outer solution is at leading order $\lim_{\eta\to\infty} p_{\text{bl}}^{(0)}(\eta,t) = p_{\text{out}}^{(0)}(0,t)$, hence

$$p_{\text{bl}}^{(0)}(\eta,t) = p_{\text{out}}^{(0)}(0,t). \tag{5.101}$$

The first-order matching condition gives

$$\lim_{\eta\to\infty} \frac{\partial p_{\text{bl}}^{(1)}(\eta,t)}{\partial\eta} = \frac{\partial p_{\text{out}}^{(0)}(0,t)}{\partial y}, \tag{5.102}$$

where the first-order boundary layer term is the solution of the integral equation

$$p_{\text{bl}}^{(1)}(\eta,t) = \sum_{j=1}^{5} I_j, \tag{5.103}$$

where

$$I_1 = \int_0^\infty \frac{p_{\text{bl}}^{(1)}(\xi,t)}{\sqrt{4\pi\sigma(0,t)}} \left[\exp\left\{-\frac{(\eta-\xi)^2}{4\sigma(0,t)}\right\} + \exp\left\{-\frac{(\eta+\xi)^2}{4\sigma(0,t)}\right\}\right] d\xi$$

$$I_2 = -P\int_0^\infty \frac{p_{\text{bl}}^{(0)}(\xi,t)}{\sqrt{4\pi\sigma(0,t)}} \exp\left\{-\frac{(\eta+\xi)^2}{4\sigma(0,t)}\right\} d\xi$$

$$I_3 = -\frac{\sigma_y(0,t)}{2\sigma(0,t)}\int_0^\infty \frac{p_{\text{bl}}^{(0)}(\xi,t)\,\xi}{\sqrt{4\pi\sigma(0,t)}} \left[\exp\left\{-\frac{(\eta-\xi)^2}{4\sigma(0,t)}\right\} + \exp\left\{-\frac{(\eta+\xi)^2}{4\sigma(0,t)}\right\}\right] d\xi$$

and

$$\begin{aligned} I_4 = \frac{\sigma_y(0,t)}{4\sigma(0,t)^2}\int_0^\infty \frac{p_{\text{bl}}^{(0)}(\xi,t)\,\xi}{\sqrt{4\pi\sigma(0,t)}} &\left[(\eta-\xi)^2 \exp\left\{-\frac{(\eta-\xi)^2}{4\sigma(0,t)}\right\}\right. \\ &\left.+(\eta+\xi)^2 \exp\left\{-\frac{(\eta+\xi)^2}{4\sigma(0,t)}\right\}\right] d\xi \\ I_5 = \frac{2a(0,t)}{4\sigma(0,t)}\int_0^\infty \frac{p_{\text{bl}}^{(0)}(\xi,t)}{\sqrt{4\pi\sigma(0,t)}} &\left[(\eta-\xi)\exp\left\{-\frac{(\eta-\xi)^2}{4\sigma(0,t)}\right\}\right. \\ &\left.-(\eta+\xi)\exp\left\{-\frac{(\eta+\xi)^2}{4\sigma(0,t)}\right\}\right] d\xi. \end{aligned}$$

The last four integrals in (5.103) can be evaluated explicitly and together with (5.101) give

$$
\begin{aligned}
&p_{\text{bl}}^{(1)}(\eta,t)\\
&=\int_0^\infty \frac{p_{\text{bl}}^{(1)}(\xi,t)}{\sqrt{4\pi\sigma(0,t)}}\left[\exp\left\{-\frac{(\eta-\xi)^2}{4\sigma(0,t)}\right\}+\exp\left\{-\frac{(\eta+\xi)^2}{4\sigma(0,t)}\right\}\right]d\xi \qquad (5.104)\\
&\quad -\frac{P}{2}p_{\text{out}}^{(0)}(0,t)\,\text{erfc}\left(\frac{\eta}{2\sqrt{\sigma(0,t)}}\right)+\frac{\sigma_y(0,t)-a(0,t)}{\sqrt{\pi\sigma(0,t)}}p_{\text{out}}^{(0)}(0,t)\\
&\quad\times\exp\left\{-\frac{\eta^2}{4\sigma(0,t)}\right\}.
\end{aligned}
$$

Differentiating (5.104) with respect to η and integrating by parts, we obtain

$$
\begin{aligned}
&\frac{\partial p_{\text{bl}}^{(1)}(\eta,t)}{\partial\eta} \qquad (5.105)\\
&=\frac{1}{\sqrt{4\pi\sigma(0,t)}}\int_0^\infty \frac{\partial p_{\text{bl}}^{(1)}(\xi,t)}{\partial\eta}\left[\exp\left\{-\frac{(\eta-\xi)^2}{4\sigma(0,t)}\right\}-\exp\left\{-\frac{(\eta+\xi)^2}{4\sigma(0,t)}\right\}\right]d\xi\\
&\quad+\frac{P}{2\sqrt{\pi\sigma(0,t)}}p_{\text{out}}^{(0)}(0,t)\exp\left\{-\frac{\eta^2}{4\sigma(0,t)}\right\}-\frac{\sigma_y(0,t)-a(0,t)}{2\sqrt{\pi}\,\sigma(0,t)^{3/2}}p_{\text{out}}^{(0)}(0,t)\\
&\quad\times\eta\exp\left\{-\frac{\eta^2}{4\sigma(0,t)}\right\},
\end{aligned}
$$

which can be simplified by the substitution

$$
g(\eta,t)=\frac{\partial p_{\text{bl}}^{(1)}(\eta,t)}{\partial\eta}-\frac{P}{2\sqrt{\pi\sigma(0,t)}}p_{\text{out}}^{(0)}(0,t)\exp\left\{-\frac{\eta^2}{4\sigma(0,t)}\right\} \qquad (5.106)
$$

to the more compact form

$$
\begin{aligned}
&g(\eta,t) \qquad (5.107)\\
&=\phi(\eta,t)+\frac{1}{\sqrt{4\pi\sigma(0,t)}}\int_0^\infty g(\xi,t)\left[\exp\left\{-\frac{(\eta-\xi)^2}{4\sigma(0,t)}\right\}-\exp\left\{-\frac{(\eta+\xi)^2}{4\sigma(0,t)}\right\}\right]d\xi,
\end{aligned}
$$

where the odd function $\phi(\eta,t)$ is defined as

$$
\begin{aligned}
\phi(\eta,t)&=\frac{P}{\sqrt{8\pi\sigma(0,t)}}p_{\text{out}}^{(0)}(0,t)\exp\left\{-\frac{\eta^2}{8\sigma(0,t)}\right\}\text{erf}\left(\frac{\eta}{\sqrt{8\sigma(0,t)}}\right)\\
&\quad-\frac{\sigma_y(0,t)-a(0,t)}{2\sqrt{\pi}\,\sigma(0,t)^{3/2}}p_{\text{out}}^{(0)}(0,t)\,\eta\,\exp\left\{-\frac{\eta^2}{4\sigma(0,t)}\right\}. \qquad (5.108)
\end{aligned}
$$

The kernel in the integral equation (5.108) is also an odd function of η, so setting $g(\eta,t) = -g(-\eta,t)$ for $\eta < 0$ extends $g(\eta,t)$ to the entire line as an odd function. Thus (5.107) can be rewritten as the convolution equation

$$g(\eta,t) = \phi(\eta,t) + \frac{1}{\sqrt{4\pi\sigma(0,t)}} \int_{-\infty}^{\infty} g(\xi,t)\exp\left\{-\frac{(\eta-\xi)^2}{4\sigma(0,t)}\right\} d\xi, \tag{5.109}$$

which is solved in Fourier space as

$$\hat{g}(k,t) = \frac{\hat{\phi}(k,t)}{1-\exp[-\sigma(0,t)k^2]}. \tag{5.110}$$

Using the Wiener–Hopf method, as in Section (5.4), we decompose

$$\hat{g}(k,t) = \hat{g}_+(k,t) + \hat{g}_-(k,t), \tag{5.111}$$

where $g_+(\eta) = g(\eta)\chi_{[0,\infty)}(\eta)$, $g_-(\eta) = g(\eta)\chi_{(-\infty,0]}(\eta)$. The Fourier transform $\hat{g}(k,t)$ exists in the sense of distributions, and $\hat{g}_\pm(k,t)$ are analytic in the upper and lower halves of the complex plane, respectively. Taylor's expansion of $\hat{\phi}(k,t)$ in eq. (5.108) gives

$$\hat{\phi}(k,t) = 2ip_{\text{out}}^{(0)}(0,t)\left\{\frac{P\sqrt{\sigma(0,t)}}{\sqrt{\pi}} - [\sigma_y(0,t) - a(0,t)]\right\}k + O(k^3) \text{ as } k \to 0.$$

The nonzero poles of (5.110) split evenly between $\hat{g}_+(k,t)$ and $\hat{g}_-(k,t)$, and using $\hat{g}_+(k,t) = -\hat{g}_-(-k,t)$, the pole at the origin gives

$$\hat{g}_+(k,t) = ip_{\text{out}}^{(0)}(0,t)\left\{\frac{P}{\sqrt{\pi\sigma(0,t)}} - \frac{\sigma_y(0,t) - a(0,t)}{\sigma(0,t)}\right\}\frac{1}{k} + O(k) \text{ as } k \to 0.$$

Inverting the Fourier transform $\hat{g}_+(k,t)$, by closing the contour of integration around the lower half plane, we obtain

$$\lim_{\eta\to\infty} \frac{\partial p_{\text{bl}}^{(1)}(\eta,t)}{\partial\eta} = p_{\text{out}}^{(0)}(0,t)\left\{\frac{P}{\sqrt{\pi\sigma(0,t)}} - \frac{\sigma_y(0,t) - a(0,t)}{\sigma(0,t)}\right\}. \tag{5.112}$$

The matching condition (5.102) implies

$$\frac{\partial p_{\text{out}}^{(0)}(0,t)}{\partial y} = p_{\text{out}}^{(0)}(0,t)\left\{\frac{P}{\sqrt{\pi\sigma(0,t)}} - \frac{\sigma_y(0,t) - a(0,t)}{\sigma(0,t)}\right\}. \tag{5.113}$$

Multiplying by $\sigma(0,t)$ and rearranging, we obtain the radiation boundary condition

$$-J(0,t) = \frac{\partial}{\partial y}\left[\sigma(0,t)p_{\text{out}}^{(0)}(0,t)\right] - a(0,t)p_{\text{out}}^{(0)}(0,t) = \frac{P\sqrt{\sigma(0,t)}}{\sqrt{\pi}}p_{\text{out}}^{(0)}(0,t).$$

The reactive "constant" in (5.80) is

$$\kappa(t) = \frac{P(t)\sqrt{\sigma(0,t)}}{\sqrt{\pi}}, \tag{5.114}$$

because $p\,(y,t) = p^{(0)}_{\text{out}}(y,t)$. The proof of the convergence of $p_{\Delta t}(y,t\,|\,x,s)$ to $p\,(y,t\,|\,x,s)$ follows along the lines of the proof of Theorem 5.4.1. □

5.5.2 Partially reflected diffusion in higher dimensions

Partially reflected diffusions in higher dimensions involve the geometry of the boundary of the domain, the anisotropy of the diffusion tensor, and its spatial dependence. The main result (see [214]) is as follows. Consider the Itô system

$$d\boldsymbol{x} = \boldsymbol{a}(\boldsymbol{x},t)\,dt + \sqrt{2}\boldsymbol{B}(\boldsymbol{x},t)\,d\boldsymbol{w} \tag{5.115}$$

is in the half-space

$$D = \{\boldsymbol{x} \in \mathbb{R}^d : x_1 > 0\}, \tag{5.116}$$

with the unit inner normal $\boldsymbol{n} = (1,0,\ldots,0)^T$, and diffusion tensor $\boldsymbol{\sigma}(\boldsymbol{x},t) = \boldsymbol{B}(\boldsymbol{x},t)\boldsymbol{B}^T(\boldsymbol{x},t)$, which is a positive definite differentiable matrix in D. The Euler scheme for (5.115) is

$$\begin{aligned} \boldsymbol{x}(t+\Delta t) &= \boldsymbol{x}(t) + \boldsymbol{a}(\boldsymbol{x}(t),t)\Delta t + \sqrt{2}\boldsymbol{B}(\boldsymbol{x}(t),t)\,\Delta\boldsymbol{w}(t,\Delta t) \\ \boldsymbol{x}_N(0) &= \boldsymbol{x}_0. \end{aligned} \tag{5.117}$$

If $\boldsymbol{x} \in D$, but $\boldsymbol{x}' = \boldsymbol{x} + \boldsymbol{a}(\boldsymbol{x},t)\Delta t + \sqrt{2}\boldsymbol{B}(\boldsymbol{x},t)\,\Delta\boldsymbol{w}(t,\Delta t) \notin D$, the Euler scheme for (5.115) with oblique reflection in ∂D reflects the point $\boldsymbol{x}'$ with respect to the boundary in the (variable) direction of the unit vector $\boldsymbol{v}(\boldsymbol{x}_B,t) = (v_1(\boldsymbol{x}_B,),\ldots,v_n(\boldsymbol{x}_B,t))^T$, such that $\boldsymbol{v}(\boldsymbol{x}_B,t)\cdot\boldsymbol{n}(\boldsymbol{x}_B,t) \neq 0$ (i.e. $v_1(\boldsymbol{x}_B,t) \neq 0$), as

$$\boldsymbol{x}'' = \boldsymbol{x}' - \frac{2x_1'\,\boldsymbol{v}(\boldsymbol{x}_B,t)}{v_1(\boldsymbol{x}_B,t)} = \boldsymbol{x}' - x_1'\boldsymbol{\theta} \tag{5.118}$$

with

$$\boldsymbol{\theta} = \boldsymbol{\theta}\,(\boldsymbol{x}_B,t) = (\theta_1,\ldots,\theta_n)^T = 2\left(1, \frac{v_2(\boldsymbol{x}_B,t)}{v_1(\boldsymbol{x}_B,t)},\ldots,\frac{v_n(\boldsymbol{x}_B,t)}{v_1(\boldsymbol{x}_B,t)}\right)^T. \tag{5.119}$$

With this notation, define the Euler–Markov chain scheme with partial oblique reflection as follows. Setting $\boldsymbol{x} = \boldsymbol{x}(t)$ and $\boldsymbol{x}' = \boldsymbol{x} + \boldsymbol{a}(\boldsymbol{x},t)\Delta t + \sqrt{2}\boldsymbol{B}(\boldsymbol{x},t)\,\Delta\boldsymbol{w}(t)$, define

$$\begin{aligned} &\boldsymbol{x}(t+\Delta t) \\ &= \begin{cases} \boldsymbol{x}' & \text{if } \boldsymbol{x} \in D,\ \boldsymbol{x}' \in D \\ \boldsymbol{x}'' \quad \text{w.p.} 1 - P\,(\boldsymbol{x}_B)\sqrt{\Delta t} & \text{if } \boldsymbol{x} \in D,\ \boldsymbol{x}' \notin D \\ \text{terminate trajectory w.p. } P\,(\boldsymbol{x}_B)\sqrt{\Delta t} & \text{for } \boldsymbol{x}' \notin D. \end{cases} \end{aligned} \tag{5.120}$$

Here $\boldsymbol{x}''$ is the oblique reflection of $\boldsymbol{x}'$ in the direction of $\boldsymbol{v}\,(\boldsymbol{x}_B)$, given by (5.118). The main result in this case is as follows (see [214]).

Theorem 5.5.2 (Partially reflected diffusion). *For given finite functions $P(\boldsymbol{x})$ and $\kappa(\boldsymbol{x},t)$ defined on the ∂D, the pdf of* (5.120) *in the half-space D* (5.116) *converges to the solution of the initial and boundary value problem* (5.74)–(5.77) *if and only if*

$$P(\boldsymbol{x}_B,t) = \sqrt{\frac{\pi}{\sigma_{1,1}(\boldsymbol{x}_B,t)}}\,\kappa(\boldsymbol{x}_B,t) \tag{5.121}$$

and

$$\boldsymbol{v}(\boldsymbol{x}_B,t) = \frac{\boldsymbol{\sigma}(\boldsymbol{x}_B,t)\,\boldsymbol{n}(\boldsymbol{x}_B)}{|\boldsymbol{\sigma}(\boldsymbol{x}_B,t)\,\boldsymbol{n}(\boldsymbol{x}_B)|}. \tag{5.122}$$

The application to general domains with curved boundaries is discussed in [214]. The particular case of totally reflecting boundary is described by the following theorem.

Theorem 5.5.3 (Reflected diffusion). *The pdf $p_{\Delta t}(\boldsymbol{y},t\,|\,\boldsymbol{x},s)$ of the Markovian jump process $\boldsymbol{x}_{\Delta t}(t)$ generated by* (5.120) *with $P(\boldsymbol{x},t)=0$ converges to the solution $p(\boldsymbol{x},t)$ of the initial and boundary value problem* (5.74)–(5.75) *with the no-flux (reflecting) boundary condition*

$$\boldsymbol{J}(\boldsymbol{y},t\,|\,\boldsymbol{x},s)\cdot\boldsymbol{n}(\boldsymbol{y}) = 0 \ \text{ for } \boldsymbol{y}\in\partial D\ \ \boldsymbol{x}\in D, \tag{5.123}$$

uniformly for all $\boldsymbol{x}\in D,\ s<t<T$ for every $T>0$.

The transition pdf $p(\boldsymbol{y},t\,|\,\boldsymbol{x},s)$ of the solution $\boldsymbol{x}(t)$ of the Itô equation (5.19) in $\mathbb{R}^d$ satisfies with respect to the backward variables $(\boldsymbol{x},s)$ the backward Kolmogorov equation (4.131). In the case of absorbing, partially, or totally reflecting boundaries the pdf satisfies the following boundary conditions [214].

Theorem 5.5.4 (Absorbing boundary condition for the backward equation). *The transition pdf $p(\boldsymbol{y},t\,|\,\boldsymbol{x},s) = \lim_{\Delta t\to 0} p_{\Delta t}(\boldsymbol{y},t\,|\,\boldsymbol{x},s)$ of the solution $\boldsymbol{x}(t)$ of the Itô equation* (5.19) *in a sufficiently smooth domain D with absorbing, partially, or totally reflecting boundary, satisfies with respect to the backward variables $(\boldsymbol{x},s)$ the backward Kolmogorov equation (4.131) in D and the boundary conditions for $\boldsymbol{y}\in\partial D,\ \boldsymbol{x}\in D$,*

$$p(\boldsymbol{y},t\,|\,\boldsymbol{x},s) = 0 \tag{5.124}$$

$$-\frac{\partial p(\boldsymbol{y},t\,|\,\boldsymbol{x},s)}{\partial\tilde{n}(\boldsymbol{x},s)} = \kappa(\boldsymbol{x},t)p(\boldsymbol{y},t\,|\,\boldsymbol{x},s) \tag{5.125}$$

$$\frac{\partial p(\boldsymbol{y},t\,|\,\boldsymbol{x},s)}{\partial\tilde{n}(\boldsymbol{x},s)} = 0, \tag{5.126}$$

respectively, where $\tilde{\boldsymbol{n}}(\boldsymbol{x})$ is the co-normal vector $\tilde{\boldsymbol{n}}(\boldsymbol{x},s)=\boldsymbol{\sigma}(\boldsymbol{x},s)\boldsymbol{n}(\boldsymbol{x})$.

The co-normal derivative in (5.126) is spelled out as

$$\frac{\partial p(\boldsymbol{y},t\,|\,\boldsymbol{x},s)}{\partial\tilde{n}(\boldsymbol{x},s)} = \sum_{i,j=1}^{d} n^i(\boldsymbol{x})\sigma^{ij}(\boldsymbol{x},s)\frac{\partial p(\boldsymbol{y},t\,|\,\boldsymbol{x},s)}{\partial x^j}.$$

Exercise 5.22 (Mixed boundary conditions). Design an Euler simulation of a stochastic differential equation in a domain D, whose boundary consists of an absorbing part ∂D_a and a reflecting part ∂D_r. Prove convergence and find the boundary conditions for the (defective) transition probability density function on ∂D_a and ∂D_r. □

Exercise 5.23 (Mixed boundary conditions for the backward Kolmogorov equation). Find the boundary conditions the transition probability density of the mixed boundary problem of Exercise 5.22 satisfies with respect to the backward variables on ∂D_a and ∂D_r. □

The following elementary theorem concerns the generalization of the definition (4.115) of adjoint partial differential operators to include boundary conditions.

Theorem 5.5.5 (Adjoint boundary conditions). *The Fokker–Planck operator* $L_{\boldsymbol{y}}$ *defined in* (4.114)*, with the absorbing, partially reflecting, or reflecting boundary conditions* (5.65), (5.77), *or* (5.123), *respectively, is adjoint to the backward Kolmogorov operator* $L^*_{\boldsymbol{y}}$ *defined in (3.68), with the boundary conditions* (5.124), (5.125), *or* (5.126), *respectively, in the sense that*

$$\int_D g(\boldsymbol{y}) L_{\boldsymbol{y}} f(\boldsymbol{y})\, d\boldsymbol{y} = \langle L_{\boldsymbol{y}} f, g\rangle_{L^2} = \langle f, L^*_{\boldsymbol{y}} g\rangle_{L^2} = \int_D f(\boldsymbol{y}) L^*_{\boldsymbol{y}} g(\boldsymbol{y})\, d\boldsymbol{y} \tag{5.127}$$

for all sufficiently smooth functions $f(\boldsymbol{y})$ *in* D *that satisfy the boundary conditions* (5.65), (5.77), *or* (5.123), *and all sufficiently smooth functions* $g(\boldsymbol{y})$ *in* D *that satisfy the boundary conditions (5.124), (5.125), or* (5.126), *respectively.*

Proof. We consider, for simplicity, the case of time-independent coefficients and reflecting boundary conditions (5.123) and (5.126). Writing

$$\int_D g(\boldsymbol{y}) L_{\boldsymbol{y}} f(\boldsymbol{y})\, d\boldsymbol{y} = -\int_D g(\boldsymbol{y}) \nabla\cdot \boldsymbol{J}(\boldsymbol{y})\, d\boldsymbol{y}$$

$$J^i(\boldsymbol{y}) = a^i(\boldsymbol{y}) f(\boldsymbol{y}) - \sum_{j=1}^d \frac{\partial\left[\sigma^{ij}(\boldsymbol{y}) f(\boldsymbol{y})\right]}{\partial y^j}$$

and applying the divergence theorem, we find that

$$\begin{aligned}\int_D g(\boldsymbol{y}) L_{\boldsymbol{y}} f(\boldsymbol{y})\, d\boldsymbol{y} &= -\oint_{\partial D} g(\boldsymbol{y}) \boldsymbol{J}(\boldsymbol{y})\cdot \boldsymbol{n}\, dS_{\boldsymbol{y}} + \int_D \boldsymbol{J}(\boldsymbol{y})\cdot \nabla g(\boldsymbol{y})\, d\boldsymbol{y}\\ &= \int_D \boldsymbol{J}(\boldsymbol{y})\cdot\nabla g(\boldsymbol{y})\, d\boldsymbol{y},\end{aligned}$$

because of (5.123). Applying the divergence theorem once again, we obtain

$$\int_D g(\boldsymbol{y})L_{\boldsymbol{y}}f(\boldsymbol{y})\,d\boldsymbol{y} = \oint_{\partial D} f(\boldsymbol{y})\sum_{i=1}^d\sum_{j=1}^d \boldsymbol{n}^i(\boldsymbol{y})\sigma^{ij}(\boldsymbol{y})\frac{\partial g(\boldsymbol{y})}{\partial y^j}\,dS_{\boldsymbol{y}} + \int_D f(\boldsymbol{y})L^*_{\boldsymbol{y}}g(\boldsymbol{y})\,d\boldsymbol{y}.$$

It follows that (5.127) holds for all smooth functions $f(\boldsymbol{y})$ in D that satisfy (5.123) if and only if the functions $g(\boldsymbol{y})$ satisfy the boundary condition (5.126). □

Exercise 5.24. (Derivation of total and partial reflection from a potential).
(i) Can a no flux boundary condition be derived as a limit of diffusion in a potential field?
(ii) Can a partially reflecting boundary condition be derived from diffusion in a potential field with killing? (See discussion after the proof of the Feynman–Kac formula Theorem 4.4.2.) □

5.5.3 Discontinuous coefficients

If the coefficients in the Fokker–Planck equation suffer a jump discontinuity across a smooth surface S, the solution is determined from the continuity condition $p|_{S_+} = p|_{S_-}$ and the flux conservation law $\boldsymbol{J}\cdot\boldsymbol{\nu}|_{S_+} = \boldsymbol{J}\cdot\boldsymbol{\nu}|_{S_-}$. Note that stochastic differential equations with jump discontinuities with respect to the spatial variables in the coefficients are not well defined. The theorem of existence and uniqueness requires the Lipschitz condition. Thus this is a problem in continuum theory, not in stochastic differential equations.

5.5.4 Diffusion on a sphere

If the solution of a two-dimensional stochastic differential equation stays on a circle; that is, $x^2(t)+y^2(t) = r^2$, where $r = const.$, a transformation to polar coordinates, $x(t) = r\cos\theta(t)$, $y(t) = r\sin\theta(t)$, defines $\theta(t)$ as a diffusion process (through Itô's formula). The pdf of $\theta(t)$ is a 2π-periodic function of θ. The solution of the corresponding Fokker–Planck equation is then periodic and so is the flux. Similarly, if the solution stays on a sphere, the solution of the Fokker–Planck equation and the corresponding flux are defined on the sphere and are required to be periodic functions of the spherical coordinates [193].

Exercise 5.25 (The diffusion equation on a Riemannian manifold [234], [171], [14]). Write the diffusion equation on a Riemannian manifold using the Laplace–Beltrami operator. □

5.6 The Wiener measure induced by SDEs

The solution of $x(t)$ of the SDE

$$dx = a(x,t)\,dt + b(x,t)\,dw, \quad x(0) = x_0 \tag{5.128}$$

is a Markov process, so its multidimensional density is determined uniquely by the transition probability density function $p\,(y,t\,|\,x,s)$, which is the solution of the FPE

$$\frac{\partial p\,(y,t\,|\,x,s)}{\partial t} = \frac{1}{2}\frac{\partial^2\left[b^2(y,t)p\,(y,t\,|\,x,s)\right]}{\partial y^2} - \frac{\partial\left[a(y,t)p\,(y,t\,|\,x,s)\right]}{\partial y} \tag{5.129}$$

with the initial condition

$$\lim_{t\downarrow s} p\,(y,t\,|\,x,s) = \delta(y-x). \tag{5.130}$$

We can use it to construct a Wiener measure on the space of continuous functions (trajectories) in analogy to that constructed in Section 2.2.1. The cylinder sets are defined as

$$C(t_1,\,\ldots,\,t_K; I_1,\,\ldots,\,I_K) = \{\omega \in \Omega\,|\,x(t_1,\omega) \in I_1,\ldots,x(t_K,\omega) \in I_K\}.$$

These are the same cylinder sets as in Section 2.2.1, but they are assigned different probabilities. Specifically, we define the measure of a cylinder as the discrete path integral

$$\Pr\{C(t_1,\,\ldots,t_K; I_1,\,\ldots,I_K)\} = \int_{I_1}\cdots\int_{I_K}\prod_{j=1}^{K} p\,(y_j,t_j\,|\,y_{j-1},t_{j-1})\,dy_j. \tag{5.131}$$

The transition probability density function $p\,(y,t\,|\,x,s)$ satisfies the CKE (2.59), so the consistency condition

$$\begin{aligned} &C\left(t_1,\ldots,t_K; I_1,I_2,\ldots,I_j=\mathbb{R},\ldots,I_K\right)\\ =&C\left(t_1,\ldots t_{j-1},t_{j+1},\ldots,t_K; I_1,I_2,\ldots,I_{j-1},I_{j+1},\ldots,I_K\right)\end{aligned}$$

is satisfied.

Using Theorem 5.2.1, we can write each factor $p\,(y_j,t_j\,|\,y_{j-1},t_{j-1})$ in (5.131) as a limit. More specifically, we partition each interval $[t_{j-1},t_j]$ $(j=1,2,\ldots,K)$ by the points $t_{j-1} = t_j^{(0)} < t_j^{(1)} < \cdots < t_j^{(N_j)} = t_j$ such that $\Delta t_{k,N_j} = t_j^{(k)} - t_j^{(k-1)} = (t_j - t_{j-1})/N_j \to 0$ as $N_j \to \infty$ and write each pdf as the path integral

$$\begin{aligned} &p\,(y_j,t_j\,|\,y_{j-1},t_{j-1})\\ =&\lim_{N_j\to\infty}\underbrace{\int\!\!\int\cdots\int}_{N_j-1}\prod_{l=1}^{N_j-1}dz_l\prod_{k=1}^{N_j}\frac{1}{\sqrt{2\pi\Delta t_{k,N_j}}\,b\left(z_{k-1},t_{k-1,N_j}\right)}\\ &\times\exp\left\{\frac{-\left[z_k - z_{k-1} - a\left(z_{k-1},t_{k-1,N_j}\right)\Delta t_{k-1,N_j}\right]^2}{2b^2\left(z_{k-1},t_{k-1,N_j}\right)\Delta t_{k,N_j}}\right\},\end{aligned} \tag{5.132}$$

with $z_{N_j} = y_j$, $z_0 = y_{j-1}$, which can be used in (5.131). We denote by $\Pr_{a,b}\{A\}$ the extension of this probability measure from cylinders to any set A in $\mathcal{F}$. The case where $a(x,t)$ and $b(x,t)$ are adapted stochastic processes is handled in a similar manner [84], [76].

For example, consider a stochastic process $\xi(t,\omega)$ adapted to $\mathcal{F}_t$ and define the Itô integral $y(t,\omega) = \int_0^t \xi(s,\omega)\,dw(s,\omega)$ by the Itô dynamics $dy(t,\omega) = \xi(t,\omega)\,dw(t,\omega)$ in the sense of the corresponding Wiener measure $\Pr_{0,\xi}\{A\}$ defined above. The Wiener measure defined in Chapter 2 is $\Pr\{A\} = \Pr_{0,1}\{A\}$. Integrals with respect to $d\Pr_{a,b}\{\omega\}$ of measurable functions on Ω can now be defined, as described in Chapter 2.

The driftless case $a(x,t) = 0$ in the stochastic dynamics (5.128) has important applications in the theory of games of chance, including the stock market and the theory of pricing financial derivatives [178]. Here, this particular case is presented as an application of the Wiener measure.

In the case $a(x,t) = 0$, the stochastic dynamics (5.128) is

$$dx = b(x,t)\,dw; \tag{5.133}$$

that is, $\dot{x}(t,\omega)$ is noise, whose intensity $b(x,t)$ is state- and time-dependent.

Lemma 5.6.1. *Driftless diffusions are martingales.*

Proof. Obviously, $x(t) = x(t_0) + \int_{t_0}^t b(x(s),s)\,dw(s)$ and thus $x(t)$ is an Itô integral. □

The Wiener measure $\Pr_{0,b}\{A\}$ generated by a driftless diffusion is called *a martingale measure* in Ω. If $x(t,\omega)$ represents the possible outcomes of a game of chance, the martingale property expresses its "fairness" in the sense that the game is not biased in favor of any of the players. For example, a well-known fact is that the game of tossing a fair coin is a martingale. Changing the Wiener measure into an equivalent martingale measure has several important applications.

The constructions of different Wiener measures in Ω, be it for the Brownian motion or for the stochastic dynamics (5.128), raise the question of the relationship between different measures in Ω. An important result is Girsanov's lemma [86], which asserts that the measures in Ω corresponding to different drift coefficients, but the same (strictly positive) diffusion coefficient, are equivalent in the sense that the density of one can be obtained from that of the other by multiplication by a positive random variable. In particular, it asserts that the drift term in eq. (5.128) can be eliminated by such a change of measure. This means that for every stochastic process defined by (5.128) there exists an equivalent measure with respect to which the process is a martingale. A well-known fact in the continuous theory of pricing financial derivatives is that there exists a martingale measure on the space Ω of possible paths of the discounted stock price (see [178] for a full exposition). The Nobel Prize in economics was awarded in 1997 to Scholes and Merton for making this fact known and taking advantage of it in developing the Black–Scholes formula and its applications.

We restrict below the space Ω to continuous functions on a given interval $0 \leq t \leq T$.

Theorem 5.6.1 (Girsanov). *Assume $a(x,t)$ and $b(x,t)$ are sufficiently smooth functions that satisfy the conditions of the existence and uniqueness theorems for* (5.128), *$b(x,t) > 0$, and $x(t,\omega)$ is the solution of* (5.128) *with $a(x,t) = 0$. Assume that Novikov's condition 3.3.1 is satisfied for $a(x(t,\omega))/b^2(x(t,\omega))$. Then the Wiener measures $\Pr_{a,b}\{A\}$, induced in the function space Ω by* (5.128), *and $\Pr_{0,b}\{A\}$, induced by* (5.128) *with $a(x,t) = 0$, are equivalent and*

$$d\Pr_{a,b}\{\omega\} = X(T,\omega)\, d\Pr_{0,b}\{\omega\}, \tag{5.134}$$

where

$$\begin{aligned} &X(T,\omega) \\ &= \exp\left\{ -\int_0^T \xi(s,x(s,\omega))\, dw(s,\omega) - \frac{1}{2}\int_0^T b^2(x(s,\omega),s)\xi^2(s,x(s,\omega))\, ds \right\} \end{aligned} \tag{5.135}$$

and

$$b^2(x,t)\xi(t,x) = -a(x,t). \tag{5.136}$$

The random variable $X(T,\omega)$ is positive and

$$\mathbb{E}X(T,\omega) = \int_\Omega X(T,\omega)\, d\Pr_{0,b}\{\omega\} = 1. \tag{5.137}$$

Proof. Condition (5.137) means that $\int_\Omega d\Pr_{a,b}\{\omega\} = 1$, as required from a probability density function. We consider random variables of the form (5.135) and denote by $\Pr_{a,b}\{\omega\}$ the measure defined by the right-hand side of eq. (5.134) with the random variable (5.135) (in this notation we suppress the dependence of the measures on T). Note that $\xi(t,x(t,\omega))$ and $b^2(x(t,\omega),t)$ are functions of the trajectory ω. Thus, for any realization of ω such that $\omega(t_k) = y_k$ for $k = 0,1,\ldots,K$, the realization of the approximation $x_N(t,\omega)$ depends on the values of y_k such that $t_k \leq t$.

First, we have to verify that condition (5.137) is satisfied. To this end, we define $0 = t_0 < t_1 < t_2 < \cdots < t_{N+1} = T$ such that $\Delta t_k = t_{k+1} - t_k \to 0$ as $N \to \infty$ and calculate the expectation by means of the path integral

$$\begin{aligned} &\mathbb{E}X(T,\omega) \\ &= \int_\Omega X(T,\omega)\, d\Pr_{0,b}\{\omega\} = \lim_{N\to\infty} \underbrace{\int\int\cdots\int}_{N} \prod_{k=1}^{N} \frac{dy_k}{\sqrt{2\pi\Delta t_k}\, b(y_{k-1},t_{k-1})} \\ &\times \exp\left\{ -\xi(t_{k-1},y_{k-1})(y_k - y_{k-1}) - \frac{b^2(y_{k-1},t_{k-1})\xi^2(t_{k-1},y_{k-1})}{2}\Delta t_k \right\} \\ &\times \exp\left\{ -\frac{(y_k - y_{k-1})^2}{2b^2(y_{k-1},t_{k-1})\Delta t_k} \right\}. \end{aligned} \tag{5.138}$$

Completing the exponent to a square, we get from (5.138) that

$$\mathbb{E}X(T,\omega) = \lim_{N\to\infty} \underbrace{\int\int\cdots\int}_{N} \prod_{k=1}^{N} \frac{dy_k}{\sqrt{2\pi\Delta t_k}\, b(y_{k-1},t_{k-1})}$$

$$\times \exp\left\{\frac{-\left[y_k - y_{k-1} + \Delta t_k b^2(y_{k-1},t_{k-1})\xi(t_{k-1},y_{k-1})\right]^2}{2b^2(y_{k-1},t_{k-1})\Delta t_k}\right\} = 1,$$

as asserted.

According to (5.131) and (5.132), the measure assigned to a cylinder by the right-hand side of (5.134) is the path integral

$$\Pr\{C(t_1,\ldots,t_K;I_1,I_2,\ldots,I_K)\} \tag{5.139}$$

$$= \int_{I_1}\int_{I_2}\cdots\int_{I_K} \prod_{j=1}^{K} \lim_{\min_j\{N_j\}\to\infty} \underbrace{\int\int\cdots\int}_{N_j-1} \prod_{m=1}^{N_j-1} dz_m$$

$$\times \frac{dy_j}{\sqrt{2\pi\Delta t_j}\, b\left(y_{j-1},t_{j-1}\right)} \frac{1}{\sqrt{2\pi\Delta t_{m,N_j}}\, b\left(z_{m-1},t_{m-1,N_j}\right)}$$

$$\times \exp\left\{\frac{-\left[z_m - z_{m-1} - a\left(z_{m-1},t_{m-1,N_j}\right)\Delta t_{m-1,N_j}\right]^2}{2b^2\left(z_{m-1},t_{m-1,N_j}\right)\Delta t_{m,N_j}}\right\}$$

$$\times \exp\left\{\frac{-\left[y_j - y_{j-1} + \Delta t_j b^2\left(y_{j-1},t_{j-1}\right)\xi\left(t_{j-1},y_{j-1}\right)\right]^2}{2b^2\left(y_{j-1},t_{j-1}\right)\Delta t_j}\right\}.$$

If ξ in (5.139) is chosen as in (5.136), then for $0 \le t \le T$,

$$\Pr\{C(t_1,\ldots,t_K;I_1,I_2,\ldots,I_K)\}$$

$$= \int_{I_1}\int_{I_2}\cdots\int_{I_K} \lim_{\min_j\{N_j\}\to\infty} \underbrace{\int\int\cdots\int}_{N_j-1} \prod_{m=1}^{N_j-1} dz_m$$

$$\times \prod_{j=1}^{K} \frac{dy_j}{\sqrt{2\pi\Delta t_j}\, b\left(y_{j-1},t_{j-1}\right)} \frac{1}{\sqrt{2\pi\Delta t_{m,N_j}}\, b\left(z_{m-1},t_{m-1,N_j}\right)}$$

$$\times \exp\left\{\frac{-\left[z_m - z_{m-1} - a\left(z_{m-1},t_{m-1,N_j}\right)\Delta t_{m-1,N_j}\right]^2}{2b^2\left(z_{m-1},t_{m-1,N_j}\right)\Delta t_{m,N_j}}\right\}$$

$$\times \exp\left\{\frac{-\left[y_j - y_{j-1} - \Delta t_j a\left(y_{j-1},t_{j-1}\right)\right]^2}{2b^2\left(y_{j-1},t_{j-1}\right)\Delta t_j}\right\}. \tag{5.140}$$

The probability (5.140) is identical to (5.131), which is the measure assigned to the cylinder
$C(t_1,\ldots,t_K;I_1,I_2,\ldots,I_K)$ by the measure $\Pr_{a,b}\{\cdot\}$, whenever $0 \le t_1 < \cdots < t_K \le T$. □

Obviously, the inverse substitution to (5.134),

$$d\mathrm{Pr}_{0,b}\{\omega\} = \frac{d\mathrm{Pr}_{a,b}\{\omega\}}{X(T,\omega)}, \tag{5.141}$$

converts the dynamics (5.128) to the driftless (5.133) in the sense that the Wiener measure for (5.128) is transformed to that for (5.133) on the interval $[0, T]$.

Exercise 5.26 (Change of MBM). Show that the measure $\mathrm{Pr}_{a,1}\{\omega\}\{\cdot\}$, corresponding to the case $b = 1$ in (5.128), assigns to the process

$$\tilde{w}(t,\omega) = w(t,\omega) - \int_0^t \xi(s, x(s,\omega))\, ds \tag{5.142}$$

the Wiener path integral

$$\begin{aligned}&\mathrm{Pr}\{C(t_1,\ldots,t_n; I_1,\ldots,I_n)\}\\ &= \int_{I_1}\cdots\int_{I_n}\prod_{k=1}^{n}\frac{dx_k}{\sqrt{2\pi(t_k - t_{k-1})}}\exp\left\{-\frac{(x_k - x_{k-1})^2}{2(t_k - t_{k-1})}\right\};\end{aligned} \tag{5.143}$$

that is, for each fixed T, the process $\tilde{w}(t,\omega)$ is a MBM in the probability space $(\Omega, F_T, \mathrm{Pr}_{a,1,T}\{\cdot\})$ in the interval $[0, T]$ in the sense that the measure $\mathrm{Pr}_{a,1,T}\{\cdot\}$ assigns to the cylinders $\{\omega \in \Omega \mid \tilde{w}(t_1,\omega) \in I_1, \ldots, \tilde{w}(t_K,\omega) \in I_K\}$ the same probability that the Wiener measure (5.143) assigns to the cylinders $\{\omega \in \Omega \mid w(t_1,\omega) \in I_1, \ldots, w(t_K,\omega) \in I_K\}$. □

Exercise 5.27 (The Radon–Nikodym derivative). Use Itô's formula to show that the Radon–Nikodym derivative $X(T,\omega) = dP_{a,b}(\omega)/dP_{0,b}(\omega)$ in (5.135) satisfies the Itô equation

$$dX(t,\omega) = -X(t,\omega)\xi(t, x(t))\, dw(t,\omega). \tag{5.144}$$

□

5.7 Annotations

This chapter is based on [222], [165], and [223]. It is concerned with the convergence of the partially reflecting Markovian jump process generated by (5.84), (5.87) in one and higher dimensions. It shows that the partially reflecting Euler scheme, with the additional requirement that the pdf converges to the solution of the FPE with a given Robin boundary condition, defines a unique diffusion process with partial reflection at the boundary. In contrast to the Collins and Kimball discrete scheme [43], this definition is not restricted to lattice points and to constant drift and diffusion coefficients. From the theoretical point of view, (5.87) serves as a physical interpretation for the behavior of diffusive trajectories near a reactive boundary.

The definition of probability flux densities for diffusions and the decomposition of the net flux into its two unidirectional components, as presented in Section 5.2.3, was done in [165], [222]. The recursion (5.9) was used in [120]. The definition of the Itô stochastic dynamics (5.1) on the positive axis with total or partial reflection at the origin was given first by Feller [72] for the one-dimensional case with $a(x,t)$ and $b(x,t)$ independent of t, as a limit of processes, which are terminated when they reach the boundary or moved instantaneously to a point $x = \rho_j > 0$ with probability p_j. When $p_j \to 1$ and $\rho_j \to 0$ with $\lim_{j\to\infty}(1-p_j)/\rho_j = c$, where c is a constant, the partially reflected process converges to a limit. The pdf of the limit process was shown to satisfy the initial value problem for the Fokker–Planck equation with the radiation boundary condition (5.80), where k is a constant related to the constant c and to the values of the coefficients at the boundary. The no flux and Dirichlet boundary conditions are recovered if $c = 0$ or $c = \infty$, respectively. Feller's method does not translate into a Brownian dynamics simulation of the limit process.

The FPE with radiation (also called reactive or Robin) boundary conditions is widely used to describe diffusion in biological cells with chemical reactions on the membrane [145], [253], [239], [10], [21], [176], [230]. The derivation of the radiation condition has a long history. Collins and Kimball [43] (see also [89]) derived the radiation boundary condition (5.80) for the limit $p(x,t) = \lim_{N\to\infty} p_N(x,t)$ from an underlying discrete random walk model on a semi-infinite one-dimensional lattice with partial absorption at the endpoint. This model assumes constant diffusion coefficient and vanishing drift, for which a reactive constant is found and expressed in terms of the absorption probability and the diffusion coefficient. The present scheme, in contrast, allows the jumps not to be restricted to lattice points and the drift and diffusion coefficient to vary. Moreover, a different relation is found between the reactive constant and the absorption probability. Other simulation schemes that recover the Robin boundary condition [145], [83] make use of the explicit solution to the half space FPE with linear drift term and constant diffusion coefficient with a Robin condition. The advantage of the scheme (5.87) is its simplicity, which is both easily and efficiently implemented and amenable to analysis. There is no need to make any assumptions on the structure of the diffusion coefficient or the drift. From the theoretical point of view, it serves as a physical interpretation for the behavior of diffusive trajectories near a reactive boundary. Moreover, (5.87) is a natural generalization of the specular reflection method near a reflecting boundary, which has been shown to be superior to other methods, such as rejection, multiple rejection, and interruption [238, and references therein]. The scheme (5.87) is generalized in [214] to diffusion in higher dimensions with partial co-normal reflection at the boundary, as summarized in Section 5.5.1.

The analysis of this chapter can be extended to other schemes in a pretty straightforward way. The appearance of a boundary layer should come as no surprise. It is well known that the Euler scheme produces an $O(\sqrt{\Delta t})$ error in estimating the mean first passage time to reach an absorbing boundary. There are several recipes to reduce the discretization error to $O(\Delta t)$ [11], [103], [163], [164]. Another manifestation of the boundary layer is that the approximation error of the pdf near absorbing or reflecting boundaries is $O(\sqrt{\Delta t})$, and improved methods, including [145], [201], reduce this error to $O(\Delta t)$. Thus, we expect the formation of a boundary layer of size $O(\sqrt{\Delta t})$ for the Euler scheme (5.84) with the boundary behavior (5.87).

The formation of spurious boundary layers in semi-continuous simulations is discussed in [222] and [214], where a special treatment of the simulation near the boundary is proposed to avoid this phenomenon. In contrast, a fully discrete simulation, for example, a random walk on a semi-infinite or finite one-dimensional lattice does not require special boundary treatment [43], [212]. The disadvantage of the fully discrete random walk simulations is their low computational efficiency. The relation $\Delta t = (\Delta x)^2/2b^2$ for the diffusion limit, which the random walk must satisfy, enforces the adoption of a small time step Δt for a given space resolution Δx. The same space resolution can be achieved with the Euler scheme with a larger time step, because the jumps are arbitrary in size rather than confined to the lattiee. Indeed, the need for many iterations of the discrete random walk to approximate a single normally distributed jump of the Euler scheme is a consequence of the law of large numbers (see Exercise 7.6). Also the connection between the termination probability and the radiation constant is not the same in the two schemes.

The stochastic model (5.1) appears in many applications and improved discretization schemes have been proposed to ensure a higher-order convergence rate than that of the Euler scheme (5.84) (see [174], [175], [132]). We note that replacing the Euler scheme with such a higher-order scheme reproduces the Robin boundary condition as long as the boundary behavior (5.87) is preserved.

Skorokhod [224] defines the reflection process inside the boundary. Several numerical schemes have been proposed for evaluating this process (see, e.g., [224], [7], [149], [44]). The main issue there is to approximate the local time spent on the boundary. The convergence of the pdf of an Euler scheme has been studied in [88], [28] for the higher-dimensional problem with oblique reflection. Bounds on the integral norm of the approximation error are given for the solution of the backward Kolmogorov equation. These, however, do not resolve the boundary layer of the pdf of the numerical solution so that a boundary layer expansion is needed to capture the boundary phenomena.

The explicit solution was first given by Bryan in 1891 [31] (see [37, §14.2, p. 358]) by the method of images. The Laplace transform method was later employed [145], [3] to obtain the explicit solution, which reduces for $k = 0$ to Smoluchowski's [227] explicit analytical solution for a reflecting boundary with a constant drift term, whereas setting $a = 0$ reduces it to Bryan's solution.

Note that the transition pdf $p\,(\boldsymbol{y}, t \,|\, \boldsymbol{x}, s)$ of a problem with total or partial absorption at the boundary is not normalized to 1 and actually decays to zero with time. This is due to the fact that more trajectories get absorbed in the boundary as t increases. In contrast, the conditional transition pdf of the surviving trajectories is normalized to 1 at all times.

Path integrals and measures in function space are discussed in several books, such as [211], [76], [129], [241]. The present approach is taken from [165]. Cameron and Martin [34] have established results similar to Girsanov's theorem for nonlinear transformations of Brownian motion.

Chapter 6

The First Passage Time of Diffusions

6.1 The FPT and escape from a domain

The basic statistical properties of the FPT from a point to the boundary of a given domain were investigated in Chapter 4.4 by means of equations derived from Itô's formula and Kolmogorov's representation formulas. The PDF of the FPT was discussed in Section 4.4.3, its moments were discussed in Section 4.4.1, and the exit distribution in Section 4.4.2.

We consider, again, a system of Itô stochastic differential equations in a domain D,

$$d\boldsymbol{x} = \boldsymbol{a}(\boldsymbol{x},t)\,dt + \sqrt{2}\boldsymbol{B}(\boldsymbol{x},t)\,d\boldsymbol{w} \text{ for } t > s, \quad \boldsymbol{x}(s) = \boldsymbol{x}, \tag{6.1}$$

but assume in this section that $\boldsymbol{a}(\boldsymbol{x},t)$ and $\boldsymbol{B}(\boldsymbol{x},t)$ are independent of t. We adopt here a different, more direct approach to the investigation of the statistical properties of the FPT than that in Chapter 4.4. We assume in this chapter that $p\,(\boldsymbol{y},t\,|\,\boldsymbol{x},s)$ is the solution of the initial and boundary value problem for the FPE

$$\begin{aligned}
\frac{\partial p\,(\boldsymbol{y},t\,|\,\boldsymbol{x},s)}{\partial t} &= L_{\boldsymbol{y}}\,p\,(\boldsymbol{y},t\,|\,\boldsymbol{x},s) \;\text{ for } \boldsymbol{x},\boldsymbol{y} \in D \\
\lim_{t\downarrow s} p\,(\boldsymbol{y},t\,|\,\boldsymbol{x},s) &= \delta(\boldsymbol{y}-\boldsymbol{x}) \;\text{ for } \boldsymbol{x},\boldsymbol{y} \in D \\
p\,(\boldsymbol{y},t\,|\,\boldsymbol{x},s) &= 0 \;\text{ for } \boldsymbol{x} \in D,\ \boldsymbol{y} \in \partial D.
\end{aligned} \tag{6.2}$$

We recall that it is also the solution of the terminal and boundary value problem for

Z. Schuss, *Theory and Applications of Stochastic Processes: An Analytical Approach*, Applied Mathematical Sciences 170, DOI 10.1007/978-1-4419-1605-1_6,

the BKE (see Section 4.6)

$$\frac{\partial p\,(\boldsymbol{y},t\,|\,\boldsymbol{x},s)}{\partial s} = -\,L^*_{\boldsymbol{x}} p\,(\boldsymbol{y},t\,|\,\boldsymbol{x},s) \text{ for } \boldsymbol{x},\boldsymbol{y} \in D,\ s<t \tag{6.3}$$
$$\lim_{s\uparrow t} p\,(\boldsymbol{y},t\,|\,\boldsymbol{x},s) = \delta(\boldsymbol{x}-\boldsymbol{y}) \text{ for } \boldsymbol{x},\boldsymbol{y} \in D$$
$$p\,(\boldsymbol{y},t\,|\,\boldsymbol{x},s) = 0 \text{ for } \boldsymbol{y} \in D,\ \boldsymbol{x} \in \partial D.$$

As mentioned in Section 5.4.1 (Theorem 5.4.4), the solution of the Fokker–Planck equation with absorbing boundary conditions is the joint transition probability density and the probability that the FPT to the boundary, τ, exceeds t, as given in (5.69),

$$p\,(\boldsymbol{y},t\,|\,\boldsymbol{x},s)\,d\boldsymbol{y} = \Pr\left\{\boldsymbol{x}(t) = \boldsymbol{y}+d\boldsymbol{y},\ \tau>t\,|\,\boldsymbol{x}(s)=\boldsymbol{x}\right\}.$$

It follows that the complementary PDF of the FPT is the marginal distribution

$$\Pr\left\{\tau>t\,|\,\boldsymbol{x}(0)=\boldsymbol{x}\right\} = \int_D \Pr\left\{\boldsymbol{x}(t)=\boldsymbol{y}+d\boldsymbol{y},\ \tau>t\,|\,\boldsymbol{x}(0)=\boldsymbol{x}\right\}$$
$$= \int_D p\,(\boldsymbol{y},t\,|\,\boldsymbol{x},0)\,d\boldsymbol{y}. \tag{6.4}$$

When trajectories of (6.1) are absorbed at the boundary, the total population of trajectories in the domain D decreases in time. To make this notion mathematically meaningful, we define

Definition 6.1.1 (Total population). *The* total population *in D at time t of trajectories the dynamics* (6.1) *with absorption at ∂D, that started at $\boldsymbol{x} \in D$ at and earlier time s is defined as*

$$N(\boldsymbol{x},t\,|\,s) = \int_D p\,(\boldsymbol{y},t\,|\,\boldsymbol{x},s)\,d\boldsymbol{y}. \tag{6.5}$$

The function $N(\boldsymbol{x},t\,|\,s)$ is identical to the conditional survival probability $S(t\,|\boldsymbol{x},s)$.

Thus (5.69) means that

$$N(\boldsymbol{x},t\,|\,s) = \int_D p\,(\boldsymbol{y},t\,|\,\boldsymbol{x},s)\,d\boldsymbol{y} = \Pr\left\{\tau>t\,|\,\boldsymbol{x}(s)=\boldsymbol{x}\right\}; \tag{6.6}$$

that is, $N(\boldsymbol{x},t\,|\,s)$ is the probability of a trajectory that starts at $\boldsymbol{x}$ to survive beyond time t. Recall that the pdf $p\,(\boldsymbol{y},t\,|\,\boldsymbol{x},s)$ is the solution of the forward (Fokker–Planck) and backward Kolmogorov equations (4.116), (4.131) with the initial and terminal conditions $p\,(\boldsymbol{y},s\,|\,\boldsymbol{x},s) = 0$ for $\boldsymbol{y} \in \partial D$, $\boldsymbol{x} \in D$ and $\boldsymbol{x} \in \partial D$, $\boldsymbol{y} \in D$, respectively. To simplify notation, we assume $s=0$ and drop this variable.

Definition 6.1.2 (Instantaneous absorption rate). *The* instantaneous relative rate of change *of the population of trajectories in D that started at $\boldsymbol{x}$ is defined as*

$$\kappa(\boldsymbol{x},t) = -\frac{\dot{N}(\boldsymbol{x},t)}{N(\boldsymbol{x},t)}. \tag{6.7}$$

Theorem 6.1.1 (Rate is the ratio of flux to population).

$$\kappa(\boldsymbol{x},t)=\frac{F(\boldsymbol{x},t)}{N(\boldsymbol{x},t)}. \tag{6.8}$$

Proof. Using the definition (6.5) and the Fokker–Planck equation (6.2), we find that

$$\begin{aligned}\dot{N}(\boldsymbol{x},t) &= \int_D \frac{\partial}{\partial t}p(\boldsymbol{x},t)\,d\boldsymbol{y} = -\int_D \nabla\cdot\boldsymbol{J}(\boldsymbol{x},t)\,d\boldsymbol{y} = -\oint_{\partial D}\boldsymbol{J}(\boldsymbol{x},t)\cdot\boldsymbol{\nu}(\boldsymbol{x},t)\,dS_{\boldsymbol{y}}\\ &= -F(\boldsymbol{x},t),\end{aligned}$$

where $F(\boldsymbol{x},t)$ is the total probability flux out of D. Hence (6.8) follows. □

Next, we relate the rate to the MFPT.

Theorem 6.1.2 (MFPT and total population).

$$\mathbb{E}\left[\tau\,|\,\boldsymbol{x}(0)=\boldsymbol{x}\right]=\int_0^\infty N(\boldsymbol{x},t)\,dt. \tag{6.9}$$

Proof. Integrating by parts, we obtain

$$\begin{aligned}\mathbb{E}\left[\tau\,|\,\boldsymbol{x}(0)=\boldsymbol{x}\right] &= \int_0^\infty t d_t\left[\Pr\{\tau<t\,|\,\boldsymbol{x}(0)=\boldsymbol{x}\}-1\right]\\ &= \int_0^\infty \Pr\{\tau>t\,|\,\boldsymbol{x}(0)=\boldsymbol{x}\}\,dt\\ &= \int_0^\infty dt\int_D p(\boldsymbol{y},t\,|\,\boldsymbol{x},0)\,d\boldsymbol{y}=\int_0^\infty N(\boldsymbol{x},t)\,dt.\end{aligned}$$

□

Theorem 6.1.3 (Quasi steady-state rate and eigenvalues). *If D is a bounded domain, ∂D has a piecewise continuous normal, and L^* (in eq. (4.131)) is a uniformly elliptic operator with sufficiently smooth coefficients in D, then the "steady-state" rate is the principal eigenvalue of the Fokker–Planck operator with absorbing boundary conditions.*

Proof. The Fokker–Planck equation can be solved by separation of variables,

$$p(\boldsymbol{y},t\,|\,\boldsymbol{x})=\sum_{n=1}^\infty \bar{\psi}_n(\boldsymbol{x})\phi_n(\boldsymbol{y})e^{-\lambda_n t},$$

where λ_n, $\phi_n(\boldsymbol{y})$ are the eigenvalues and eigenfunctions of the forward Kolmogorov operator (Fokker–Planck), defined in eq. (4.116), respectively,

$$L\phi_n(\boldsymbol{y})=-\lambda_n\phi_n(\boldsymbol{y})\text{ for }\boldsymbol{y}\in D,\quad \phi_n(\boldsymbol{y})=0\text{ for }\boldsymbol{y}\in\partial D$$

and those of the backward Kolmogorov operator are

$$L^*\psi_n(\boldsymbol{x}) = -\bar{\lambda}_n\psi_n(\boldsymbol{x}) \text{ for } \boldsymbol{x} \in D, \quad \psi_n(\boldsymbol{x}) = 0 \text{ for } \boldsymbol{x} \in \partial D.$$

The eigenfunctions form a biorthogonal set

$$\int_D \bar{\psi}_n(\boldsymbol{y})\phi_m(\boldsymbol{y})\,d\boldsymbol{y} = \delta_{m,n}.$$

It is known that $\lambda_1 > 0$ and $\phi_1(\boldsymbol{y}) > 0$ in D and λ_1 is a simple eigenvalue. It follows that

$$N(\boldsymbol{x},t) = \int_D p(\boldsymbol{y},t\,|\,\boldsymbol{x})\,d\boldsymbol{y} = \sum_{n=1}^{\infty} \bar{\psi}_n(\boldsymbol{x}) \int_D \phi_n(\boldsymbol{y})\,d\boldsymbol{y}\, e^{-\lambda_n t}, \tag{6.10}$$

hence

$$\dot{N}(\boldsymbol{x},t) = -\sum_{n=1}^{\infty} \lambda_n\bar{\psi}_n(\boldsymbol{x}) \int_D \phi_n(\boldsymbol{y})\,d\boldsymbol{y}\, e^{-\lambda_n t}$$

and

$$\kappa(\boldsymbol{x},t) = -\frac{\dot{N}(\boldsymbol{x},t)}{N(\boldsymbol{x},t)} = \frac{\displaystyle\sum_{n=1}^{\infty} \lambda_n\bar{\psi}_n(\boldsymbol{x}) \int_D \phi_n(\boldsymbol{y})\,d\boldsymbol{y}\, e^{-\lambda_n t}}{\displaystyle\sum_{n=1}^{\infty} \bar{\psi}_n(\boldsymbol{x}) \int_D \phi_n(\boldsymbol{y})\,d\boldsymbol{y}\, e^{-\lambda_n t}}.$$

It follows that

$$\kappa = \lim_{t\to\infty} \kappa(\boldsymbol{x},t) = \lambda_1. \tag{6.11}$$

□

Theorem 6.1.4 (The quasi-steady-state). *The steady-state pdf of the surviving trajectories is the (normalized) first eigenfunction $\phi_1(\boldsymbol{y})$ and the steady-state MFPT is λ_1^{-1}.*

Proof. Equation (6.10) gives

$$\mathbb{E}\left[\tau\,|\,\boldsymbol{x}(0) = \boldsymbol{x}\right] = \int_0^{\infty} N(\boldsymbol{x},t)\,dt = \sum_{n=1}^{\infty} \lambda_n^{-1}\bar{\psi}_n(\boldsymbol{x}) \int_D \phi_n(\boldsymbol{y})\,d\boldsymbol{y}. \tag{6.12}$$

To interpret this result, we note that the probability that $\boldsymbol{x}(t)$ is observed at the point $\boldsymbol{y}$ at time $t_0 >> 1$ is $p\,(\boldsymbol{y},t_0\,|\,\boldsymbol{x})$. From eq. (6.4), we find that the probability that this occurs, given that the trajectory has survived, is

$$p_C(\boldsymbol{y},t_0\,|\,\boldsymbol{x}) = \frac{p\,(\boldsymbol{y},t_0\,|\,\boldsymbol{x})}{\displaystyle\int_D p\,(\boldsymbol{y},t_0\,|\,\boldsymbol{x})\,d\boldsymbol{y}},$$

hence

$$\lim_{t_0\to\infty} p_C(\boldsymbol{y}, t_0 \,|\, \boldsymbol{x}) = \frac{\phi_1(\boldsymbol{y})}{\displaystyle\int_D \phi_1(\boldsymbol{y})\, d\boldsymbol{y}}. \tag{6.13}$$

The pdf of $\boldsymbol{x}(t)$ at time $t_0 + t$, given that the trajectory survived beyond time t_0, where t_0 is very large (such that $\lambda_1 t_0 >> 1$) is therefore the solution of the Fokker–Planck equation with the initial condition (6.13) given at time t_0. Thus

$$p_C(\boldsymbol{y}, t) = \lim_{t_0\to\infty} p\,(\boldsymbol{y}, t_0 + t \,|\, \tau > t_0) = \frac{\phi_1(\boldsymbol{y})}{\displaystyle\int_D \phi_1(\boldsymbol{y})\, d\boldsymbol{y}} e^{-\lambda_1 t}.$$

It follows that the conditional MFPT after a long time t_0 is given by

$$\lim_{t_0\to\infty} \mathbb{E}\,[\tau - t_0 \,|\, \tau > t_0] = \int_0^\infty \int_D p_C(\boldsymbol{y}, t)\, d\boldsymbol{y}\, dt = \int_0^\infty \int_D \frac{\phi_1(\boldsymbol{y})}{\displaystyle\int_D \phi_1(\boldsymbol{y})\, d\boldsymbol{y}} e^{-\lambda_1 t}\, d\boldsymbol{y}\, dt$$

$$= \frac{1}{\lambda_1} = \frac{1}{\kappa}, \tag{6.14}$$

where κ is the steady-state rate (6.11). □

6.2 The PDF of the FPT

An alternative proof of Theorem 4.4.5, based on the representation (6.6), is given below.

Proof. Assume first that the problem is autonomous; then

$$v(\boldsymbol{x}, t) = \Pr\{\tau > t \,|\, \boldsymbol{x}(0) = \boldsymbol{x}\}$$

and differentiation with respect to t leads to

$$\frac{\partial v(\boldsymbol{x}, t)}{\partial t} = \int_D \frac{\partial p\,(\boldsymbol{y}, t \,|\, \boldsymbol{x})}{\partial t}\, d\boldsymbol{y} = \int_D L^*_{\boldsymbol{x}} p\,(\boldsymbol{y}, t \,|\, \boldsymbol{x})\, d\boldsymbol{y} = L^*_{\boldsymbol{x}} v(\boldsymbol{x}, t),$$

where $L^*_{\boldsymbol{x}}$ is the backward Kolmogorov operator defined in eq. (3.68). The initial condition is obviously $v(\boldsymbol{x},0) = 1$, because the first passage time from an interior point to the boundary is positive with probability 1. Similarly, the boundary condition is $v(\boldsymbol{x},t) = 0$ for $\boldsymbol{x} \in \partial D,\ t > 0$, because the time to reach the boundary from a point on the boundary is zero with probability 1. The PDF $u(\boldsymbol{x}, t) = 1 - v(\boldsymbol{x}, t) = \Pr\{\tau \le t \,|\, \boldsymbol{x}(0) = \boldsymbol{x}\}$ satisfies the equation $u_t(\boldsymbol{x}, t) = L^*_{\boldsymbol{x}} u(\boldsymbol{x}, t)$ for $\boldsymbol{x} \in D$, the initial condition $u(\boldsymbol{x}, 0) = 0$ for $\boldsymbol{x} \in D$, and the boundary condition $u(\boldsymbol{x}, t) = 1$ for $\boldsymbol{x} \in \partial D,\ t > 0$. □

Exercise 6.1 (The time-dependent case). Find the equation for the PDF of the FPT for the case of time-dependent coefficients. □

Similarly, Corollary 4.4.1 (the boundary value problem (4.70) for the MFPT) can be derived directly from the representation (6.9), the backward Kolmogorov equation (4.49) and the terminal condition (4.50).

Proof. **(of Corollary 4.4.1)** Applying the backward Kolmogorov operator to eq. (6.9), using (4.49) with $f(\boldsymbol{y}) = \delta(\boldsymbol{y} - \boldsymbol{x})$, and changing the order of integration, we obtain

$$L^*_{\boldsymbol{x}}\mathbb{E}\left[\tau \,|\, \boldsymbol{x}(0) = \boldsymbol{x}\right] = \int_D d\boldsymbol{y} \int_0^\infty L^*_{\boldsymbol{x}} p\left(\boldsymbol{y}, t \,|\, \boldsymbol{x}\right) dt = \int_D d\boldsymbol{y} \int_0^\infty \frac{\partial p\left(\boldsymbol{y}, t \,|\, \boldsymbol{x}\right)}{\partial t}\, dt$$
$$= -\int_D \delta(\boldsymbol{y} - \boldsymbol{x})\, d\boldsymbol{y} = -1. \tag{6.15}$$

□

Theorem 6.2.1 (The density of the mean time spent at a point). *The function $p\left(\boldsymbol{y} \,|\, \boldsymbol{x}\right) = \int_0^\infty p\left(\boldsymbol{y}, t \,|\, \boldsymbol{x}\right) dt$ is the density of the mean time a trajectory that starts at $\boldsymbol{x} \in D$ spends at $\boldsymbol{y} \in D$ prior to absorption in ∂D.*

Proof. Equation (6.9) can be written in the form

$$\mathbb{E}\left[\tau \,|\, \boldsymbol{x}(0) = \boldsymbol{x}\right] = \int_D \int_0^\infty p\left(\boldsymbol{y}, t \,|\, \boldsymbol{x}\right) dt\, d\boldsymbol{y} = \int_D p\left(\boldsymbol{y} \,|\, \boldsymbol{x}\right) d\boldsymbol{y}.$$

The function $p\left(\boldsymbol{y} \,|\, \boldsymbol{x}\right)$ has several interesting interpretations. To understand the meaning of $p\left(\boldsymbol{y} \,|\, \boldsymbol{x}\right)$, consider an open set $A \subset D$ and define the characteristic function of A as

$$\chi_A(\boldsymbol{x}) = \begin{cases} 1 & \text{for } \boldsymbol{x} \in A \\ 0 & \text{otherwise} \end{cases}.$$

Then the integral $\int_0^\tau \chi_A(\boldsymbol{x}(t))\, dt$ is the time the trajectory $\boldsymbol{x}(t)$ spends in the set A prior to absorption (at time τ). We have

$$\mathbb{E}\left[\int_0^\tau \chi_A(\boldsymbol{x}(t))\, dt \,\middle|\, \boldsymbol{x}(0) = \boldsymbol{x}\right] = \int_0^\infty dt \int_A p\left(\boldsymbol{y}, t \,|\, \boldsymbol{x}\right) d\boldsymbol{y} = \int_A p\left(\boldsymbol{y} \,|\, \boldsymbol{x}\right) d\boldsymbol{y},$$

because $p\left(\boldsymbol{y}, t \,|\, \boldsymbol{x}\right) d\boldsymbol{y} = \Pr\left\{\boldsymbol{x}(t) \in \boldsymbol{y} + d\boldsymbol{y},\ \tau > t \,|\, \boldsymbol{x}(0) = \boldsymbol{x}\right\}$. It follows that the pdf $p\left(\boldsymbol{y} \,|\, \boldsymbol{x}\right)$ is the density of the mean time a trajectory that starts at $\boldsymbol{x}$ spent at $\boldsymbol{y}$ prior to absorption . □

Another interpretation of the function $p(\boldsymbol{y} \mid \boldsymbol{x})$ is given in the following theorem.

Theorem 6.2.2 (Steady-state density with a source). *If a source is placed at $\boldsymbol{x} \in D$ and trajectories are absorbed at ∂D, then $p(\boldsymbol{y} \mid \boldsymbol{x})$ is the steady-state density of the trajectories in D.*

Proof. Integrating the Fokker–Planck equation with respect to t, we obtain

$$\begin{aligned} p(\boldsymbol{y},\infty \mid \boldsymbol{x}) - p(\boldsymbol{y},0 \mid \boldsymbol{x}) &= -p(\boldsymbol{y},0 \mid \boldsymbol{x}) = -\delta(\boldsymbol{y}-\boldsymbol{x}) = L_{\boldsymbol{y}} \int_0^\infty p(\boldsymbol{y},t \mid \boldsymbol{x})\,dt \\ &= L_{\boldsymbol{y}} p(\boldsymbol{y} \mid \boldsymbol{x}); \end{aligned}$$

that is,

$$L_{\boldsymbol{y}} p(\boldsymbol{y} \mid \boldsymbol{x}) = -\delta(\boldsymbol{y}-\boldsymbol{x}), \tag{6.16}$$

which can be written in the equivalent divergence form $\nabla_{\boldsymbol{y}} \cdot \boldsymbol{J}(\boldsymbol{y} \mid \boldsymbol{x}) = \delta(\boldsymbol{y}-\boldsymbol{x})$ with the boundary condition $p(\boldsymbol{y} \mid \boldsymbol{x}) = 0$ for $\boldsymbol{y} \in \partial D$. Integrating over the domain and using the divergence theorem, we obtain

$$F(\boldsymbol{x}) = \oint_{\partial D} \boldsymbol{J}(\boldsymbol{y} \mid \boldsymbol{x}) \cdot \boldsymbol{\nu}(\boldsymbol{y})\, dS_{\boldsymbol{y}} = 1. \tag{6.17}$$

This means that the total flux out of the domain equals the total output of the source. This leads to the following interpretation of $p(\boldsymbol{y} \mid \boldsymbol{x})$. If a source is placed at $\boldsymbol{x}$ and all trajectories are absorbed at the boundary, then $p(\boldsymbol{y} \mid \boldsymbol{x})$ is the steady-state density of trajectories in D that started at $\boldsymbol{x}$. □

This interpretation describes the situation where all absorbed trajectories are instantaneously re-injected at $\boldsymbol{x}$. The total population of trajectories that started at $\boldsymbol{x}$ is then

$$N(\boldsymbol{x}) = \int_D p(\boldsymbol{y} \mid \boldsymbol{x})\, d\boldsymbol{y} = \mathbb{E}\left[\tau \mid \boldsymbol{x}(0) = \boldsymbol{x}\right].$$

This equation can be written in the "population over flux" form

$$\begin{aligned} \mathbb{E}\left[\tau \mid \boldsymbol{x}(0) = \boldsymbol{x}\right] &= \int_D p(\boldsymbol{y} \mid \boldsymbol{x})\, d\boldsymbol{y} = \frac{N(\boldsymbol{x})}{1} = \frac{N(\boldsymbol{x})}{\displaystyle\oint_{\partial D} \boldsymbol{J}(\boldsymbol{y} \mid \boldsymbol{x}) \cdot \boldsymbol{\nu}(\boldsymbol{y})\, dS_{\boldsymbol{y}}} \\ &= \frac{N(\boldsymbol{x})}{F(\boldsymbol{x})}. \end{aligned} \tag{6.18}$$

The identity (6.18) holds even if the normalization (6.17) is changed. The steady-state absorption rate in the boundary of trajectories that start at $\boldsymbol{x}$ is

$$\kappa(\boldsymbol{x}) = \frac{F(\boldsymbol{x})}{N(\boldsymbol{x})}.$$

Corollary 6.2.1 (Averaged rate). *If trajectories are started with an initial density $f(\boldsymbol{x})$ in D, then the averaged absorption rate is*

$$\langle\kappa\rangle = \frac{F}{N} = \frac{1}{\mathbb{E}\tau}, \tag{6.19}$$

where $\mathbb{E}\tau = \int_D \mathbb{E}\left[\tau \,|\, \boldsymbol{x}(0) = \boldsymbol{x}\right] f(\boldsymbol{x})\, d\boldsymbol{x}$.

Note that this is the correct way to average rates!

Proof. When an initial density is given, eq. (6.16) is replaced by

$$L_{\boldsymbol{y}} p\,(\boldsymbol{y}) = -f(\boldsymbol{y}) \tag{6.20}$$

or $p\,(\boldsymbol{y}) = \int_D f(\boldsymbol{x}) p\,(\boldsymbol{y} \,|\, \boldsymbol{x})\, d\boldsymbol{x}$; that is, $p\,(\boldsymbol{y} \,|\, \boldsymbol{x})$ is Green's function for the elliptic boundary value problem (6.20) in D with the boundary condition $p\,(\boldsymbol{y}) = 0$ for $\boldsymbol{y} \in \partial D$. Also in this case the total flux is

$$\oint_{\partial D} \boldsymbol{J}(\boldsymbol{y} \,|\, \boldsymbol{x}) \cdot \boldsymbol{\nu}(\boldsymbol{y})\, dS_{\boldsymbol{y}} = \int_D f(\boldsymbol{x})\, d\boldsymbol{x} = 1.$$

It follows that the MFPT from $\boldsymbol{x}$, averaged with respect to the initial density $f(\boldsymbol{x})$, is

$$\begin{aligned}\mathbb{E}\tau = \frac{N}{F} &= \int_D p\,(\boldsymbol{x})\, d\boldsymbol{x} = \int_D \int_D p\,(\boldsymbol{y} \,|\, \boldsymbol{x})\, d\boldsymbol{y} f(\boldsymbol{x})\, d\boldsymbol{x} \\ &= \int_D \mathbb{E}\left[\tau \,|\, \boldsymbol{x}(0) = \boldsymbol{x}\right] f(\boldsymbol{x})\, d\boldsymbol{x}.\end{aligned}$$

Thus the average rate of absorption is (6.19). □

Still another interpretation comes from the description of a source and absorbing boundary as reinjection.

Theorem 6.2.3 (Reinjection). *Assume that the strength of the source is the total flux at the boundary, then*

$$\mathbb{E}\left[\tau \,|\, \boldsymbol{x}(0) = \boldsymbol{x}\right] = \frac{\displaystyle\int_D p\,(\boldsymbol{y} \,|\, \boldsymbol{x}) d\boldsymbol{y}}{\displaystyle\oint_{\partial D} \boldsymbol{J}(\boldsymbol{y} \,|\, \boldsymbol{x}) \cdot \boldsymbol{\nu}(\boldsymbol{y})\, dS_{\boldsymbol{y}}} \tag{6.21}$$

and

$$\mathbb{E}\tau = \frac{\displaystyle\int_D \left[\int_D p\,(\boldsymbol{y} \,|\, \boldsymbol{x}) d\boldsymbol{y}\right] f(\boldsymbol{x})\, d\boldsymbol{x}}{\displaystyle\int_D \left[\oint_{\partial D} \boldsymbol{J}(\boldsymbol{y} \,|\, \boldsymbol{x}) \cdot \boldsymbol{\nu}(\boldsymbol{y})\, dS_{\boldsymbol{y}}\right] f(\boldsymbol{x})\, d\boldsymbol{x}}. \tag{6.22}$$

Proof. The Fokker–Planck equation has the form

$$\frac{\partial p\,(\boldsymbol{y},t\,|\,\boldsymbol{x})}{\partial t} = -\nabla_{\boldsymbol{y}}\cdot\boldsymbol{J}(\boldsymbol{y},t\,|\,\boldsymbol{x})+F(\boldsymbol{x},t)\delta(\boldsymbol{y}-\boldsymbol{x}),$$

where

$$F(\boldsymbol{x},t)=\oint\limits_{\partial D}\boldsymbol{J}(\boldsymbol{x},t\,|\,\boldsymbol{x}')\cdot\boldsymbol{\nu}(\boldsymbol{x}')\,dS_{\boldsymbol{x}'}.$$

In the steady-state, $\nabla_{\boldsymbol{y}}\cdot\boldsymbol{J}(\boldsymbol{y}\,|\,\boldsymbol{x}) = F(\boldsymbol{x})\delta(\boldsymbol{y}-\boldsymbol{x})$, where the steady-state flux density vector is $\boldsymbol{J}(\boldsymbol{y}\,|\,\boldsymbol{x}) = \lim_{t\to\infty}\boldsymbol{J}(\boldsymbol{y},t\,|\,\boldsymbol{x})$ and $F(\boldsymbol{x}) = \lim_{t\to\infty}F(\boldsymbol{x},t)$. Hence (6.21) and (6.22) follow. □

6.3 The exit density and probability flux density

It is intuitively obvious that the flux density on the boundary is the density of the points on the boundary, where trajectories are absorbed. To formalize this relationship, we prove the following

Theorem 6.3.1 (The exit flux density is the exit pdf on ∂D). *The normal absorption flux density on ∂D is the pdf of the points on ∂D where trajectories are absorbed.*

Proof. Recall that (4.81) means that Green's function for the homogeneous boundary value problem (4.82) is the exit density on ∂D. Thus, we have to show that Green's function for the homogeneous boundary value problem

$$L^*u(\boldsymbol{x})=0\ \text{ for }\boldsymbol{x}\in D,\quad u(\boldsymbol{x})=g(\boldsymbol{x})\ \text{ for }\boldsymbol{x}\in\partial D, \tag{6.23}$$

denoted $G(\boldsymbol{x},\boldsymbol{y})$, and the flux density $\boldsymbol{J}(\boldsymbol{y}\,|\,\boldsymbol{x})$ of Green's function $\Gamma(\boldsymbol{x},\boldsymbol{y})$ for the inhomogeneous boundary value problem,

$$Lp\,(\boldsymbol{x})=-f(\boldsymbol{x})\ \text{ for }\boldsymbol{x}\in D,\quad p\,(\boldsymbol{x})=0\ \text{ for }\boldsymbol{x}\in\partial D, \tag{6.24}$$

are related by

$$G(\boldsymbol{x},\boldsymbol{y})=\boldsymbol{J}(\boldsymbol{y}\,|\,\boldsymbol{x})\cdot\boldsymbol{\nu}(\boldsymbol{y}),\quad \boldsymbol{x}\in D,\ \boldsymbol{y}\in\partial D, \tag{6.25}$$

where $\boldsymbol{\nu}(\boldsymbol{y})$ is the unit outer normal at $\boldsymbol{y}\in\partial D$. To this end, we recall that the solution of (6.23) is given by

$$u(\boldsymbol{y})=\oint\limits_{\partial D}g(\boldsymbol{x})G(\boldsymbol{x},\boldsymbol{y})\,dS_{\boldsymbol{x}} \tag{6.26}$$

and the solution of (6.24) is given by $p\,(\boldsymbol{y})=\int_D f(\boldsymbol{x})\Gamma(\boldsymbol{x},\boldsymbol{y})\,d\boldsymbol{x}$. Note that

$$\begin{aligned} L_{\boldsymbol{y}}\Gamma(\boldsymbol{x},\boldsymbol{y}) &= -\,\delta(\boldsymbol{x}-\boldsymbol{y})\ \text{ for }\boldsymbol{x},\boldsymbol{y}\in D\\ \Gamma(\boldsymbol{x},\boldsymbol{y}) &= 0\ \text{ for }\boldsymbol{x}\in\partial D,\ \boldsymbol{y}\in D.\end{aligned} \tag{6.27}$$

First, we multiply (6.23) by $\Gamma(\boldsymbol{y},\boldsymbol{x})$ and integrate over the domain, to obtain

$$
\begin{aligned}
0 =& \int_D \Gamma(\boldsymbol{y},\boldsymbol{x}) L^*_{\boldsymbol{x}} u(\boldsymbol{x})\,d\boldsymbol{x} = \int_D u(\boldsymbol{x}) L_{\boldsymbol{x}} \Gamma(\boldsymbol{y},\boldsymbol{x})\,d\boldsymbol{x} \\
&+ \oint_{\partial D} \left\{ \Gamma(\boldsymbol{y},\boldsymbol{x}) \left[\sum_{i,j=1}^n \sigma^{i,j}(\boldsymbol{x}) \frac{\partial u(\boldsymbol{x})}{\partial x^i} \nu^j(\boldsymbol{x}) + \sum_{i=1}^n a^i(\boldsymbol{x}) \nu^i(\boldsymbol{x}) u(\boldsymbol{x}) \right] \right. \\
&\left. - u(\boldsymbol{x}) \sum_{i,j=1}^n \frac{\partial \sigma^{i,j}(\boldsymbol{x}) \Gamma(\boldsymbol{y},\boldsymbol{x})}{\partial x^i} \nu^j(\boldsymbol{x}) \right\} dS_{\boldsymbol{x}}. \qquad (6.28)
\end{aligned}
$$

The first equation in (6.27) gives $u(\boldsymbol{y}) = -\int_D u(\boldsymbol{x}) L_{\boldsymbol{x}} \Gamma(\boldsymbol{y},\boldsymbol{x})\,d\boldsymbol{x}$ and the second implies that the integrand in the second line of (6.28) vanishes. Finally, we note that in the third line of (6.28), we have $u(\boldsymbol{x}) = g(\boldsymbol{x})$ on ∂D and the sum is the normal component of the flux of $\Gamma(\boldsymbol{x},\boldsymbol{y})$. Hence $u(\boldsymbol{z}) = \oint_{\partial D} g(\boldsymbol{x}) \boldsymbol{J}(\boldsymbol{z}\,|\,\boldsymbol{x}) \cdot \boldsymbol{\nu}(\boldsymbol{x})\,dS_{\boldsymbol{x}}$. Together with (6.26), this implies that (6.25) holds. □

Corollary 6.3.1. *For any subset $A \subset \partial D$ the exit probability at A is given by*

$$
\Pr\{\boldsymbol{x}(\tau) \in A \,|\, \boldsymbol{x}\} = \int_A G(\boldsymbol{x},\boldsymbol{y})\,dS_{\boldsymbol{y}} = \int_A \boldsymbol{J}(\boldsymbol{x}\,|\,\boldsymbol{y}) \cdot \boldsymbol{\nu}(\boldsymbol{y})\,dS_{\boldsymbol{y}} \qquad (6.29)
$$

and the function $u(\boldsymbol{x}) = \Pr\{\boldsymbol{x}(\tau) \in A \,|\, \boldsymbol{x}\}$ is the solution of the boundary value problem

$$
L^*_{\boldsymbol{x}} u(\boldsymbol{x}) = 0 \text{ for } \boldsymbol{x} \in D, \quad u(\boldsymbol{x}) = 1 \text{ for } \boldsymbol{x} \in A, \quad u(\boldsymbol{x}) = 0 \text{ for } \boldsymbol{x} \in \partial D - A.
$$

6.4 The exit problem in one dimension

The exit problem is to calculate the MFPT of random trajectories to the boundary of a domain and the probability distribution of their exit points on the boundary. Calculations of the MFPT and of the exit probability in one dimension are considerably simplified if the diffusion coefficient is small. First, we note that small noise in a dynamical system is not a regular perturbation, in the sense that the behavior of the noiseless dynamics is close to that of the noisy system, but rather a singular perturbation, in the sense that it can cause large deviations from the noiseless behavior. For example, the origin is a locally stable attractor in the interval $-1 < x < 1$ for the one-dimensional dynamics

$$
\dot{x} = -U'(x), \quad x(0) = x_0, \qquad (6.30)
$$

where the potential is $U(x) = x^2/2 - x^4/4$ (see Figure 6.1). That is, if $-1 < x_0 < 1$, then $x(t) \to 0$ as $t \to \infty$ and if $|x_0| > 1$, then $|x(t)| \to \infty$ in finite time. Indeed, the explicit solution of (6.30) is given by

$$
x(t) = \frac{x_0}{\sqrt{x_0^2(1 - e^{2t}) + e^{2t}}}. \qquad (6.31)
$$

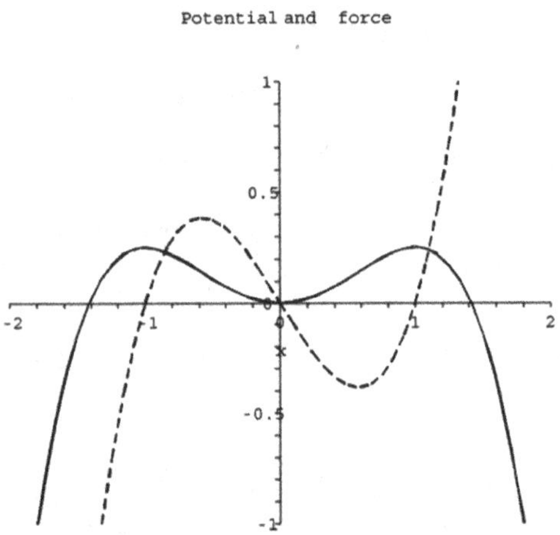

Figure 6.1. The potential $U(x) = x^2/2 - x^4/4$ (solid line) and the force $F(x) = -U'(x)$ (dashed line). The domain of attraction of the stable equilibrium $x = 0$ is the interval $D = (-1, 1)$.

If $|x_0| < 1$, the denominator is positive for all $t > 0$ and $x(t) \to 0$ as $t \to \infty$. On the other hand, if $|x_0| > 1$, then the denominator vanishes for $t = \log\sqrt{x_0^2/(x_0^2 - 1)}$, so the solution blows up in finite time. Therefore the solution that starts in the interval $D = (-1, 1)$, which is the domain of attraction of the origin, never leaves it (see Figure 6.2).

In contrast, if small white noise is added to (6.30), it becomes the SDE

$$dx = (-x + x^3)\,dt + \sqrt{2\varepsilon}\,dw, \quad x(0) = x_0, \tag{6.32}$$

where ε is a small parameter. The MFPT of $x(t)$ from $x = x_0 \in D$ to the boundary of D is finite, because, according to the Andronov,–Vitt–Pontryagin Theorem 4.4.3, the MFPT $\bar{\tau}(x_0) = \mathbb{E}[\tau_D \,|\, x(0) = x_0]$ is the solution of the boundary value problem

$$\varepsilon\bar{\tau}''(x) - U'(x)\bar{\tau}'(x) = -1 \text{ for } x \in D, \quad \bar{\tau}(x) = 0 \text{ for } x \in \partial D, \tag{6.33}$$

whose solution is given by

$$\bar{\tau}(x) = \frac{\displaystyle\int_{-1}^{x} \exp\left\{\frac{U(y)}{\varepsilon}\right\} \int_{-1}^{y} \exp\left\{\frac{U(y) - U(z)}{\varepsilon}\right\} dz\,dy}{\displaystyle\varepsilon \int_{-1}^{1} \exp\left\{\frac{U(y)}{\varepsilon}\right\} dy}, \tag{6.34}$$

which is finite for all $x \in D$. Thus almost all trajectories exit the domain of attraction D in finite time with probability one. In this section we explore analytically

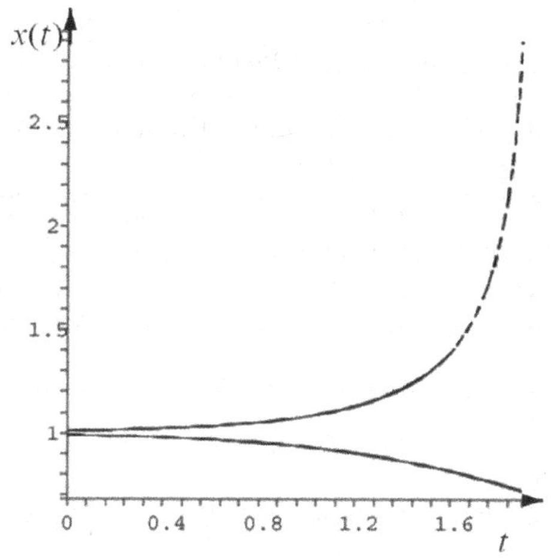

Figure 6.2. The solution (6.31) of the noiseless dynamics (6.30) with $x(0) = 0.99$ inside (bottom line) and $x(0) = 1.01$ outside (top line) the domain of attraction D.

large and small deviations caused by small noise in locally stable and unstable dynamical systems.

In many models of stochastic dynamics the diffusion matrix is "small" in some sense so that it can be scaled with a small parameter ε and written as

$$dx = a\,(x)\;dt + \sqrt{2\varepsilon}\,b(x)\,dw, \quad x\,(0) = x. \tag{6.35}$$

The exit problem for the stochastic dynamics (6.35) is to find an asymptotic expression (for small ε) for the probabilities that the trajectories of (6.35) hit the boundaries of an interval (α, β) for the first time at α or at β, as well as for the PDF of the FPT. The partial differential equations for these probability distributions were given in Section 6.3. For small ε the boundary value problem eq. (6.24) is now

$$L^{\varepsilon}p\,(x) = -f(x) \;\text{ for } \alpha < x < \beta, \quad p\,(\alpha) = 0, \quad p\,(\beta) = 0, \tag{6.36}$$

where the Fokker–Planck operator is defined by

$$L^{\varepsilon}p\,(x) = \varepsilon\frac{\partial^2[\sigma(x)p\,(x)]}{\partial x^2} - \frac{\partial[a\,(x)\,p\,(x)]}{\partial x} \tag{6.37}$$

and $\sigma(x) = b^2\,(x)$. The boundary value problem (6.36) is a singular perturbation problem, because the reduced boundary value problem, corresponding to $\varepsilon = 0$,

$$L^{0}p\,(x) = -f(x) \;\text{ for } \alpha < x < \beta, \quad p\,(\alpha) = 0, \quad p\,(\beta) = 0, \tag{6.38}$$

where $L^0 p\,(x) = -\partial a\,(x)\,p\,(x)\,/\partial x$, involves only a first-order differential equation. The boundary value problem (6.38) is in general overdetermined, because the solution cannot satisfy, in general, both boundary conditions (6.38), in contrast to the solution of the second-order equation (6.36). The disappearance of a boundary condition in the limit $\varepsilon \to 0$ gives rise to a singular behavior of the solution to the boundary value problem (6.36). Boundary layers, which are regions of large gradients of the solution, appear near one or both boundary points. Other singularities of the solution appear as well.

To understand intuitively the nature of the singularities, we consider first the probability of exit at one boundary point or the other. When ε is small the trajectories of (6.35) can be expected to stay close to the trajectory of the reduced (noiseless) dynamics

$$\dot{x} = a\,(x)\,, \quad x(0) = x_0. \tag{6.39}$$

If the trajectory of (6.39) hits α, we may expect the exit probability to be concentrated about α, due to the small noise perturbation in the stochastic dynamics (6.35). Not all trajectories of (6.39), however, reach the boundary of the interval, as is the case in the example (6.30), whose trajectories in D stay there forever. The noisy dynamics (6.35), however, exit the interval in finite time with probability 1, as in the example (6.31). In this case it is not intuitively obvious what the probability of exit at α looks like.

The FPT to the boundary is expected to be close to the FPT of the noiseless system (6.35) when ε is small. The FPT from a point x on a trajectory of (6.35) to the boundary, denoted $T_0\,(x)$, can be calculated directly from the reduced dynamics (6.39). Assuming that the trajectory of (6.39) that starts at x hits the boundary at α, we rewrite eq. (6.39) as $dt = dx/a(x)$ and find by direct integration that $T_0\,(x) = \int_x^\alpha [a\,(s)]^{-1}\,ds$. Obviously, $T\,(x)$ is the solution of the first-order equation

$$a\,(x)\,T_0'\,(x) = -1, \quad T_0\,(\alpha) = 0. \tag{6.40}$$

The FPT $T_0\,(x)$ is continuous in the interval up to the boundary point where the trajectory of (6.39) that starts at x hits the boundary. For example, (6.30) in an interval (α, β) such that $0 < \alpha < 1$ and $\beta > 1$

$$T_0(x_0) = \frac{1}{2}\log\frac{x_0^2(1-\alpha^2)}{\alpha^2(1-x_0^2)} \quad \text{for } \alpha < x_0 < 1. \tag{6.41}$$

The time $T_0(x_0)$ to reach α from x_0 blows up as $x_0 \uparrow 1$. In the general case (6.39), the time suffers a discontinuity as x approaches the boundary point where trajectories of (6.39) enter the interval. At any point x in the interval such that the trajectory of (6.39) that starts at x never reaches the boundary, $T_0\,(x) = \infty$. In particular, if all trajectories of (6.39) stay forever in the interval, then $T_0\,(x) = \infty$ for all x.

In contrast, the FPT of the stochastic dynamics (6.35) is finite if the MFPT

$$\tau^\varepsilon(x) = \mathbb{E}\,[\tau \mid x\,(0) = x]$$

is finite. This is the case, for example, if the Andronov–Vitt–Pontryagin boundary value problem (4.70), which is now

$$L^{*\varepsilon}\bar{\tau}^\varepsilon\,(x) = -1 \ \text{ for } \alpha < x < \beta, \quad \bar{\tau}^\varepsilon(\alpha) = 0, \quad \bar{\tau}^\varepsilon(\beta) = 0, \tag{6.42}$$

has a finite solution. Here $L^{*\varepsilon}$ is the backward Kolmogorov operator defined in (3.68), which for the stochastic dynamics (6.35) has the form

$$L^{*\varepsilon}\bar{\tau}^{\varepsilon}(x) = \varepsilon\sigma(x)\bar{\tau}^{\varepsilon\prime\prime}(x) + a(x)\bar{\tau}^{\varepsilon\prime}(x). \tag{6.43}$$

Note that the differential equation (6.40) is obtained from the boundary value problem (6.42) by setting $\varepsilon = 0$ and dropping one of the boundary conditions.

If, for example, the diffusion coefficient $\sigma(x)$ is bounded below by a positive constant in the interval (i.e., if the backward Kolmogorov operator $L^{*\varepsilon}$ is not degenerate in the interval), the Andronov–Vitt–Pontryagin boundary value problem (6.42) has a unique finite solution for all $\varepsilon > 0$, regardless of the nature of the trajectories of the reduced dynamics (6.39). It follows that the FPT to the boundary of almost all trajectories of the stochastic system (6.35) that start at any point in the interval, is finite. We conclude that in the case where the noiseless dynamics (6.39) persists in the interval forever (and therefore the $T(x) = \infty$ for all $\alpha < x < \beta$), the MFPT $\bar{\tau}^{\varepsilon}(x)$ becomes infinite in the limit $\varepsilon \to 0$. The exit problem in this case is to find an asymptotic expression for the MFPT and for the rate κ^{ε} as defined in Sections 6.1 and 6.4.

In one dimension the boundary value problem (6.23) becomes the ordinary differential equation

$$\begin{aligned} \varepsilon\sigma(x)u_{\varepsilon}''(x) + a(x)u_{\varepsilon}'(x) &= 0 \quad \text{for } \alpha < x < \beta \\ u_{\varepsilon}(\alpha) = g_{\alpha}, \quad u_{\varepsilon}(\beta) &= g_{\beta}, \end{aligned} \tag{6.44}$$

and g_{α}, g_{β} are given numbers. We assume that $a(x)$ and $\sigma(x)$ are uniformly Lipschitz continuous functions in $[\alpha, \beta]$ and that $\sigma(x)$ has a positive minimum in the interval.

The exit distribution on the boundary of the domain; that is, at the points α, β, is the Green's function for the boundary value problem (6.44), evaluated at these points. Under the given assumptions the solution of (6.44) is given by

$$u_{\varepsilon}(x) = \frac{g_{\alpha}\displaystyle\int_x^{\beta}\exp\left\{\frac{\Phi(s)}{\varepsilon}\right\}ds + g_{\beta}\int_{\alpha}^{x}\exp\left\{\frac{\Phi(s)}{\varepsilon}\right\}ds}{\displaystyle\int_{\alpha}^{\beta}\exp\left\{\frac{\Phi(s)}{\varepsilon}\right\}ds},$$

where the potential is

$$\Phi(x) = -\int_{\alpha}^{x}\frac{a(y)}{\sigma(y)}\,dy. \tag{6.45}$$

According to eq. (6.29), the exit distribution is

$$\Pr\{x(\tau)=\alpha\,|\,x(0)=x\}=\lim_{\zeta\to 0}\int\limits_{\alpha-\zeta}^{\alpha+\zeta}G(x,y)\,dy=\frac{\displaystyle\int\limits_x^\beta \exp\left\{\frac{\Phi(s)}{\varepsilon}\right\}ds}{\displaystyle\int\limits_\alpha^\beta \exp\left\{\frac{\Phi(s)}{\varepsilon}\right\}ds} \tag{6.46}$$

$$\Pr\{x(\tau)=\beta\,|\,x(0)=x\}=\lim_{\zeta\to 0}\int\limits_{\beta-\zeta}^{\beta+\zeta}G(x,y)\,dy=\frac{\displaystyle\int\limits_\alpha^x \exp\left\{\frac{\Phi(s)}{\varepsilon}\right\}ds}{\displaystyle\int\limits_\alpha^\beta \exp\left\{\frac{\Phi(s)}{\varepsilon}\right\}ds}.$$

Denoting by τ_y the FPT to y, we can write

$$\Pr\{x(\tau)=\alpha\,|\,x(0)=x\}=\Pr\{\tau_\alpha<\tau_\beta\,|\,x(0)=x\}.$$

An alternative derivation of eq. (6.46) is obtained from the identities of Section 6.3. It was shown there that for trajectories that start at x, the probability of exit at β is the flux density at β, calculated from the Fokker–Planck equation with a source at x and absorbing boundaries at α and β. Specifically,

$$\Pr\{\tau_\beta<\tau_\alpha\,|\,x(0)=x\}=-\frac{\partial\varepsilon\sigma(y)\,p_\varepsilon(\beta\,|\,x)}{\partial y}, \tag{6.47}$$

where the function $p_\varepsilon(y\,|\,x)$ is the solution of the Fokker–Planck equation corresponding to the system (6.35),

$$\frac{\partial^2\varepsilon\sigma(y)\,p_\varepsilon(y\,|\,x)}{\partial y^2}-\frac{\partial a(y)\,p_\varepsilon(y\,|\,x)}{\partial y}=-\,\delta(y-x)\quad\text{for } \alpha<x,y<\beta \tag{6.48}$$

$$p_\varepsilon(\alpha\,|\,x)=p_\varepsilon(\beta\,|\,x)=0. \tag{6.49}$$

The solution of (6.48), (6.49) is given by

$$p_\varepsilon(y\,|\,x)=\frac{\exp\left\{-\dfrac{\Phi(y)}{\varepsilon}\right\}}{\varepsilon\sigma(y)} \tag{6.50}$$

$$\times\left[\frac{\displaystyle\int\limits_x^\beta \exp\left\{\frac{\Phi(s)}{\varepsilon}\right\}ds}{\displaystyle\int\limits_\alpha^\beta \exp\left\{\frac{\Phi(s)}{\varepsilon}\right\}ds}\int\limits_\alpha^y \exp\left\{\frac{\Phi(s)}{\varepsilon}\right\}ds-\int\limits_\alpha^y H(s-x)\exp\left\{\frac{\Phi(s)}{\varepsilon}\right\}ds\right],$$

where $H(s-x)$ is the Heaviside step function. A detailed analysis of the asymptotic form of the function $p_\varepsilon(y\,|\,x)$ for small ε is given in Section 10.1.2. It is shown that eq. (6.47) gives eq. (6.46).

6.4.1 The exit time

It was shown in Section 6.2 that the complementary PDF of the FPT to the boundary, $v(x,t) = \Pr\{\tau > t\,|\,x(0)=x\}$, is the solution of the initial boundary value problem for $\alpha < x < \beta$ and $t > 0$,

$$\frac{\partial v(x,t)}{\partial t} = L^{*\varepsilon} v(x,t) = \varepsilon\sigma(x)\frac{\partial^2 v(x,t)}{\partial x^2} - a(x)\frac{\partial v(x,t)}{\partial x} \tag{6.51}$$

with the boundary and the initial conditions

$$v(\beta,t) = v(\beta,t) = 0 \text{ for } t>0, \quad v(x,0)=1 \text{ for } \alpha < x < b.$$

The solution is found by the method of separation of variables and is given by

$$v(x,t) = \sum_{n=0}^{\infty} v_n \Phi_n(x)\, e^{-\lambda_n t},$$

where λ_n and $\Phi_n(x)$ are the eigenvalues and eigenfunctions, respectively, of the self-adjoint boundary value problem

$$\left(e^{\Phi(x)/\varepsilon}\Phi_n'(x)\right)' + \frac{\lambda_n \sigma(x)}{\varepsilon} e^{\Phi(x)/\varepsilon}\Phi_n(x) = 0 \text{ for } \alpha < x < \beta$$
$$\Phi_n(\alpha) = \Phi_n(\beta) = 0.$$

As is well-known from the theory of ordinary differential equations [41], the eigenvalues λ_n are real positive numbers. The numbers v_n are the coefficients in the Fourier expansion of the initial function $v(x,0)=1$ with respect to the eigenfunctions $\{\Phi_n(x)\}_{n=0}^{\infty}$, which are orthonormal with the weight $\sigma(x)\,e^{\Phi(x)/\varepsilon}$. That is,

$$v_n = \frac{\displaystyle\int_\alpha^\beta \sigma(x)\, e^{\Phi(x)/\varepsilon}\Phi_n(x)\,dx}{\displaystyle\int_\alpha^\beta \sigma(x)\, e^{\Phi(x)/\varepsilon}\Phi_n^2(x)\,dx}.$$

Thus the PDF of the FPT from x to the boundary of the interval $[\alpha,\beta]$ is given by

$$\Pr\{\tau \le t\,|\,x(0)=x\} = 1 - \sum_{n=0}^{\infty} v_n \Phi_n(x)\, e^{-\lambda_n t}. \tag{6.52}$$

The MFPT is given by

$$\mathbb{E}\left[\tau \mid x(0)=x\right]=\int_0^\infty \Pr\left\{\tau>t \mid x(0)=x\right\}\,dt=\sum_{n=0}^\infty \frac{v_n\Phi_n(x)}{\lambda_n}. \tag{6.53}$$

If the initial point is distributed with density $\psi(x)$, the MFPT is

$$\mathbb{E}\tau=\sum_{n=0}^\infty \frac{v_n}{\lambda_n}\int_\alpha^\beta \Phi_n(x)\,\psi(x)\,dx.$$

In particular, if we begin to observe the dynamics a long time after it started, then, according to Section 6.1, the density of the surviving trajectories (the conditional pdf, given $\tau>t$) at the moment observation begins is approximately $\psi(x)=\psi_0(x)$, where $\psi_0(x)$ is the principal eigenfunction of the adjoint problem

$$\varepsilon\frac{\partial^2\sigma(x)\,\psi_n(x)}{\partial x^2}+\frac{\partial a(x)\,\psi_n(x)}{\partial x}=-\,\lambda_n\psi_n(x)\quad\text{for }\alpha<x<\beta$$
$$\psi_n(\alpha)=\psi_n(\beta)=0,$$

which is normalized by $\int_\alpha^\beta \psi_0(x)\,dx=1$. Then, due to the biorthogonality of the eigenfunctions $\{\Phi_n(x)\}$ and $\{\psi_n(x)\}$, the MFPT is

$$\mathbb{E}\tau=\int_\alpha^\beta \mathbb{E}\left[\tau \mid x(0)=x\right]\psi_0(x)\,dx=\sum_{n=0}^\infty \frac{v_n}{\lambda_n}\int_\alpha^\beta \psi_0(x)\,\Phi_n(x)$$
$$=\frac{v_0}{\lambda_0}\int_\alpha^\beta \psi_0(x)\,\Phi_0(x)\,dx.$$

An alternative calculation of the MFPT is based on the Andronov–Vitt–Pontryagin equation (see Corollary 4.4.1). The MFPT $\bar\tau(x)=\mathbb{E}\left[\tau \mid x(0)=x\right]$ is the solution of the boundary value problem

$$L^{*\varepsilon}\bar\tau(x)=-1\quad\text{for }\alpha<x<\beta,\quad \bar\tau(\alpha)=\bar\tau(\beta)=0,$$

where the operator $L^{*\varepsilon}$ is defined by the left-hand side of eq. (6.51). The solution is given by

$$\bar\tau(x)=\int_\alpha^x \exp\left\{-\frac{\Phi(s)}{\varepsilon}\right\}ds\,\frac{\displaystyle\int_\alpha^\beta \exp\left\{-\frac{\Phi(y)}{\varepsilon}\right\}\int_\alpha^y \exp\left\{\frac{\Phi(s)}{\varepsilon}\right\}ds\,dy}{\displaystyle\int_\alpha^\beta \exp\left\{-\frac{\Phi(s)}{\varepsilon}\right\}}$$
$$-\int_\alpha^x \exp\left\{-\frac{\Phi(y)}{\varepsilon}\right\}\int_\alpha^y \exp\left\{\frac{\Phi(s)}{\varepsilon}\right\}ds\,dy. \tag{6.54}$$

Exercise 6.2 (The PDF of the FPT for the MBM). Calculate the PDF of the FPT for the Brownian motion. □

Example 6.1. (MBM with constant drift and the Ornstein–Uhlenbeck process). (a) Find the explicit expression for the pdf $p_\varepsilon(y,t\,|\,x)$ of the trajectories $x(t)$ for the MBM with constant drift, defined by

$$dx = -1\,dt + \sqrt{2\varepsilon}\,dw$$

on the positive axis with absorption at the origin. Plot $p_\varepsilon(y,t\,|\,x)$ versus (y,t) for $x=1$ and $\varepsilon = 0.01$ in the rectangle $0<y<2$, $0<t<2$.
(b) Use the expression

$$\Pr\{\tau > t\,|\,x\} = \int_0^\infty p_\varepsilon(y,t\,|\,x)\,dy$$

to express $\Pr\{\tau > t\,|\,x\}$ in terms of error functions and plot the pdf of the FPT from $x=1$ to the origin for the values $\varepsilon = 0.01$ and $\varepsilon = 0.001$.
(c) Do the same for the Ornstein–Uhlenbeck process

$$dx = -x\,dt + \sqrt{2\varepsilon}\,dw.$$

Solution:
(a) The FPE for $p_\varepsilon(y,t\,|\,x)$ is given by

$$p_t = p_x + \varepsilon p_{xx} \quad \text{for } t,x,y>0, \tag{6.55}$$

with the initial and boundary conditions

$$p_\varepsilon(y,0\,|\,x) = \delta(y-x), \quad p_\varepsilon(0,t\,|\,x) = 0.$$

The substitution

$$p = q\exp\left\{-\frac{x+\frac{1}{2}t}{2\varepsilon}\right\} \tag{6.56}$$

converts the FPE (6.55) to the diffusion equation $q_t = \varepsilon q_{xx}$ with the initial condition $q(y,0\,|\,x) = \delta(y-x)\exp\{x/2\varepsilon\}$ and the boundary condition $q(0,t\,|\,x) = 0$. The solution is found by solving the diffusion equation on the entire line with the initial condition $q(y,0\,|\,x) = \delta(y-x)\exp\{x/2\varepsilon\} - \delta(y+x)\exp\{x/2\varepsilon\}$. The solution is an odd function, due to the antisymmetric initial condition, and therefore vanishes at $y=0$ (this is the method of images). Thus the solution of the initial boundary value problem (6.55), (6.56) is given by

$$p_\varepsilon(y,t\,|\,x) = \frac{1}{2\sqrt{\pi\varepsilon t}}\exp\left\{-\frac{y-x+\frac{1}{2}t}{2\varepsilon}\right\}\left[\exp\left\{-\frac{(y-x)^2}{4\varepsilon t}\right\} - \exp\left\{-\frac{(y+x)^2}{4\varepsilon t}\right\}\right].$$

(b) We have

$$\Pr\{\tau > t \mid x\} = \int_0^\infty p_\varepsilon(y,t \mid x)\,dy$$
$$= \frac{1}{2}\operatorname{erfc}\left(\frac{t-x}{2\sqrt{\varepsilon t}}\right) - \frac{1}{2}\left[\exp\left\{\frac{x}{\varepsilon}\right\}\operatorname{erfc}\left(\frac{t+x}{2\sqrt{\varepsilon t}}\right)\right].$$

It follows that the density of the FPT is given by

$$\Pr\{\tau = t \mid x\} = \frac{x}{2\sqrt{\pi\varepsilon t^3}}\exp\left\{-\frac{(t-x)^2}{4\varepsilon t}\right\}.$$

(c) Answer:

$$p_\varepsilon(y,t \mid x)$$
$$= \frac{1}{\sqrt{2\pi\varepsilon\left(1-e^{-2t}\right)}}\left[\exp\left\{-\frac{1}{2}\frac{\left(-y+xe^{-t}\right)^2}{\varepsilon\left(1-e^{-2t}\right)}\right\} - \exp\left\{-\frac{1}{2}\frac{\left(y+xe^{-t}\right)^2}{\varepsilon\left(1-e^{-2t}\right)}\right\}\right],$$

$$\Pr\{\tau > t \mid x\} = \operatorname{erf}\left(\frac{xe^{-t}}{\sqrt{2\varepsilon\left(1-e^{-2t}\right)}}\right),$$

and

$$\Pr\{\tau = t \mid x\} = \frac{2xe^{-t}\exp\left\{-\frac{1}{2}\frac{x^2e^{-2t}}{\varepsilon\left(1-e^{-2t}\right)}\right\}}{\sqrt{\pi\left(1-e^{-2t}\right)^3}}.$$

Note that although τ is finite with probability 1, and even has all moments, its pdf peaks at

$$t = -\frac{1}{2}\log\varepsilon\left(1+O(\varepsilon)\right) \to \infty \quad as\ \varepsilon \to 0,$$

which reflects the fact that the FPT of the noiseless system from x to the origin is infinite. □

6.4.2 Application of the Laplace method

The Laplace method for evaluating the integrals in the previous section is based on the observation that the function $\exp\{\Phi(x)/\varepsilon\}$ is sharply peaked at the maximum of $\Phi(x)$ so that the main contribution to the integrals comes from the absolute maxima of $\Phi(x)$ in the interval $[\alpha,\beta]$. We consider the two cases that $\Phi(x)$ achieves its maximum at the boundary and that $\Phi(x)$ achieves its maximum at an interior point.

If $\Phi(x)$ achieves its maximum at the boundary, at $x = \alpha$, say, then $\Phi'(\alpha) \leq 0$; that is, $a(\alpha) \geq 0$ and the main contribution to the integral comes from the point $x = \alpha$. If $a(\alpha) > 0$, the function $\Phi(x)$ can be approximated near α by

$$\Phi(x) = \Phi'(\alpha)(x-\alpha) + o(|x-\alpha|) = -\frac{a(\alpha)}{\sigma(\alpha)}(x-\alpha) + o(|x-\alpha|) \quad (6.57)$$

and the integrals are evaluated by changing the variable of integration to

$$\frac{a(\alpha)}{\sigma(\alpha)}\left(\frac{s-\alpha}{\varepsilon}\right) = z. \quad (6.58)$$

We obtain

$$\int_\alpha^x \exp\left\{\frac{\Phi(s)}{\varepsilon}\right\} ds = \varepsilon\frac{\sigma(\alpha)}{a(\alpha)}\int_0^u e^{-z}(1+O(\varepsilon|z|))\,dz \quad (6.59)$$
$$= \varepsilon\frac{\sigma(\alpha)}{a(\alpha)}\left(1-e^{-u}\right)(1+O(\varepsilon)).$$

where

$$u = \frac{a(\alpha)}{\sigma(\alpha)}\frac{x-\alpha}{\varepsilon}.$$

It follows that for all $\alpha \leq x \leq \beta$

$$\Pr\{\tau_\beta < \tau_\alpha \mid x(0) = x\} = \left[1 - \exp\left\{-\frac{a(\alpha)}{\varepsilon\sigma(\alpha)}(x-\alpha)\right\}\right](1+O(\varepsilon)). \quad (6.60)$$

(note that here $O(\varepsilon) < 0$). In this case practically all trajectories that start away from the left boundary exit the interval at the right boundary.

Exercise 6.3. (Flatter extrema).

(i) Assume $a(\alpha) = a'(\alpha) = \cdots = a^{(n-1)}(\alpha) = 0$, but $a^{(n)}(\alpha) > 0$. Replace the local expansion (6.57) with

$$\Phi(x) = -\frac{a^{(n)}(\alpha)}{(n+1)!\sigma(\alpha)}(x-\alpha)^{n+1} + O\left(|x-\alpha|^{n+2}\right)$$

to approximate the Laplace-type integral $\int_\alpha^x \exp\{\Phi(s)/\varepsilon\}\,ds$ by the incomplete Euler Gamma function $\Gamma(x,y) = \int_0^y t^{x-1}e^{-t}\,dt$.

(ii) Show that for $0 < y << 1$ and all $x > 0$ the asymptotic behavior of the incomplete Euler Gamma function is

$$\Gamma(x,y) = \frac{y^x}{x}(1+o(1)) \quad (6.61)$$

whereas for and $y >> 1$ and $x > 0$, it is

$$\Gamma(x,y) = \Gamma(x) - \frac{y^x}{x}e^{-y}(1+o(1)), \quad (6.62)$$

where $\Gamma(x)$ is Euler's Gamma function

$$\Gamma(x) = \Gamma(x, \infty).$$

(iii) Conclude that if $\Phi(x)$ attains its maximum at $x = \alpha$, then eq. (6.60) is replaced with

$$\Pr\left\{\tau_\beta < \tau_\alpha \,|\, x(0) = x\right\} = \frac{\Gamma\left(\dfrac{1}{n+1}, \dfrac{a^{(n)}(\alpha)}{\varepsilon(n+1)!\sigma(\alpha)}(x-\alpha)^{n+1}\right)}{\Gamma\left(\dfrac{1}{n+1}\right)} \times \left(1 + O\left(\varepsilon^{1/(n+1)}\right)\right). \tag{6.63}$$

(iv) If the maximum of $\Phi(x)$ is achieved at both endpoints, assume $a(\alpha) = a'(\alpha) = \cdots = a^{(n-1)}(\alpha) = 0$, but $a^{(n)}(\alpha) > 0$, and $a(\beta) = a'(\beta) = \cdots = a^{(m-1)}(\beta) = 0$, but $a^{(m)}(\beta) < 0$. Find for this case $\Pr\left\{\tau_\beta < \tau_\alpha \,|\, x(0) = x\right\}$.
(v) If $n > m$, then $\varepsilon^{1/(m+1)} << \varepsilon^{1/(n+1)}$ for $\varepsilon << 1$ and $x - \alpha >> \varepsilon^{1/(n+1)}$. Conclude that

$$\Pr\left\{\tau_\beta < \tau_\alpha \,|\, x(0) = x\right\} = 1 + O\left(\varepsilon^{(n-m)/(n+m+2)}\right).$$

For $0 < x - \alpha << \varepsilon^{1/(n+1)}$,

$$\Pr\left\{\tau_\beta < \tau_\alpha \,|\, x(0) = x\right\} = O\left(\frac{(x-\alpha)^{n+1}}{\varepsilon}\right).$$

(vi) If $n = m$, then $\beta - x >> \varepsilon^{1/n+1}$ for $\varepsilon << 1$ and $x - \alpha >> \varepsilon^{1/(n+1)}$. Conclude that

$$\Pr\left\{\tau_\beta < \tau_\alpha \,|\, x(0) = x\right\} \sim \frac{\left(\dfrac{\sigma(\alpha)}{a^{(n)}(\alpha)}\right)^{1/(n+1)}}{\left(\dfrac{\sigma(\alpha)}{a^{(n)}(\alpha)}\right)^{1/(n+1)} + \left(\dfrac{\sigma(\beta)}{-a^{(n)}(\beta)}\right)^{1/(n+1)}}.$$

(vii) Give the result (6.60) the following physical interpretation. Under the given conditions, the noiseless dynamics,

$$dx = a(x)\, dt,$$

drifts (deterministically) to the right with nonnegative velocity $a(x)$. Thus small noise has but negligible chance of pushing the trajectories upstream toward the left endpoint. If the dynamics (6.35) is interpreted as the motion of an overdamped Brownian particle in a field of force (see Chapter 1), the potential is normalized by the height of the potential barrier,

$$\Delta E = \Phi(\beta) - \Phi(\alpha), \tag{6.64}$$

and ε represents the absolute temperature measured in units of the potential barrier height. To push the particle from an interior point x to exit at the left endpoint of the interval, the noise has to supply kinetic energy to the particle of at least $\Phi(x) - \Phi(\alpha)$. According to Maxwell's distribution of velocities, the probability for this is

$$P = \exp\left\{-\frac{\Phi(x) - \Phi(\alpha)}{\varepsilon}\right\}. \tag{6.65}$$

Taylor's expansion reduces eq. (6.65) to eq. (6.60). This explanation is an intuitive interpretation of the result, not its alternative derivation. □

Exercise 6.4 (Kramers' formula for the MFPT). Assume that $\alpha = -\infty$, and that the function $\Phi(x)$ in eq. (6.54) has the following properties: $\Phi(x) \to \infty$ as $x \to -\infty$, $\Phi(x)$ has a single local minimum at $x_A < \beta$, a single local maximum at $x_C = \beta$, and $\Phi'(x_C) = 0$. Assume that $\Phi''(x_A) = \omega_A^2 > 0$ and $\Phi''(x_C) = -\omega_C^2 < 0$ (see Figure 8.1 below). Expand (6.54) by the Laplace method to prove that for $\varepsilon \ll \Phi(x_C) - \Phi(x_A) = \Delta E$ the MFPT is given by Kramers' formula [140]

$$\tau_\varepsilon(x_A) \sim \frac{\pi}{\omega_A\omega_C}e^{\Delta E/\varepsilon}. \tag{6.66}$$

In dimensional variables the MFPT is

$$\tau_{\text{overdamped}}(x_A) \sim \frac{\pi\gamma}{\omega_A\omega_C}e^{\Delta E/\varepsilon}. \tag{6.67}$$

and Kramers' rate is

$$\kappa_{\text{overdamped}} = \frac{1}{2\tau_{\text{overdamped}}} = \frac{\omega_A\omega_C}{2\pi\gamma}e^{-\Delta E/\varepsilon}. \tag{6.68}$$

□

Exercise 6.5 (The full Laplace expansion). Use the full Laplace expansion in eq. (6.59) (see [20]) to construct a full asymptotic series approximation to the solutions $u_\varepsilon(x)$ and $p(y\,|\,x)$ of eqs. (6.44) and (6.48). □

Exercise 6.6 (The MFPT). Repeat Exercise 6.5 for the MFPT. □

6.5 Conditioning

Conditioning a stochastic process on a given event amounts to selecting only a subset of the trajectories and assigning to them a new probability measure. For example, if in a simulation of Brownian particles diffusing in a field of force in the presence of a membrane and a channel, we want to produce only the trajectories that traverse the channel from left to right, we condition the trajectories that entered the channel on the left, on reaching the right end before reaching left end of the channel. The conditional probability of trajectories that do not traverse the channel is then zero. The conditional probabilities assigned to trajectories that do traverse the channel differ from the unconditional probabilities they have. Thus the conditioning in effect changes the stochastic dynamics of the trajectories so that nontraversing trajectories are not produced. The process can also be conditioned on the future, for example, on reaching a certain point at a given future time (see Exercises 6.10 and the Brownian bridge 2.26 [117]).

6.5.1 Conditioning on trajectories that reach A before B

Consider the Itô system

$$d\boldsymbol{x} = \boldsymbol{a}(\boldsymbol{x})\,dt + \sqrt{2}\boldsymbol{B}(\boldsymbol{x})\,d\boldsymbol{w} \tag{6.69}$$

in a domain D, whose boundary consists of two parts, A and B. If the trajectories $\boldsymbol{x}(t)$ that reach B before A are eliminated, the remaining trajectories form a process conditioned on reaching A before B. We denote this process by $\boldsymbol{x}^*(t)$ and the first passage times to A and to B by τ_A and τ_B, respectively. Thus $\boldsymbol{x}^*(t)$ is obtained from $\boldsymbol{x}(t)$ by conditioning on the event $\{\tau_A < \tau_B\}$. We set, as usual, $\boldsymbol{\sigma}(\boldsymbol{x}) = \boldsymbol{B}(\boldsymbol{x})\boldsymbol{B}^T(\boldsymbol{x})$.

Theorem 6.5.1 (Conditioned diffusion). *Conditioning the solution of* (6.69) *reaching a set A before a set B results in a diffusion process $\boldsymbol{x}^*(t)$, whose drift vector and diffusion matrix are given by*

$$\boldsymbol{a}^*(\boldsymbol{x}) = \boldsymbol{a}(\boldsymbol{x}) + \boldsymbol{\sigma}(\boldsymbol{x})\frac{\nabla P(\boldsymbol{x})}{P(\boldsymbol{x})}, \quad \boldsymbol{\sigma}^*(\boldsymbol{x}) = \boldsymbol{\sigma}(\boldsymbol{x}), \tag{6.70}$$

respectively, where $P(\boldsymbol{x})$ is determined from the boundary value problem, provided boundary conditions can be posed there[1]

$$\begin{aligned} L^*P(\boldsymbol{x}) &= 0 \ \text{ for } \boldsymbol{x} \in D \\ P(\boldsymbol{x}) &= 1 \ \text{ for } \boldsymbol{x} \in A, \quad P(\boldsymbol{x}) = 0 \ \text{ for } \boldsymbol{x} \in B. \end{aligned} \tag{6.71}$$

Proof. Obviously, the trajectories of $\boldsymbol{x}^*(t)$ are continuous. The pdf of $\boldsymbol{x}^*(t)$, denoted $p^*(\boldsymbol{y}, t \,|\, \boldsymbol{x})$, is given by

$$p^*(\boldsymbol{y}, t \,|\, \boldsymbol{x})\,d\boldsymbol{y} = \Pr\{\boldsymbol{x}(t) \in \boldsymbol{y} + d\boldsymbol{y}, t \,|\, \boldsymbol{x}(0) = \boldsymbol{x},\, \tau_A < \tau_B\}.$$

From Bayes' rule, we obtain

$$\begin{aligned} &\Pr\{\boldsymbol{x}(t) \in \boldsymbol{x} + \Delta\boldsymbol{y}, t \,|\, \boldsymbol{x}(0) = \boldsymbol{x},\, \tau_A < \tau_B\} \\ &= \Pr\{\boldsymbol{x}(t) \in \boldsymbol{y} + \Delta\boldsymbol{y}, t \,|\, \boldsymbol{x}(0) = \boldsymbol{x}\}\, \frac{\Pr\{\tau_A < \tau_B \,|\, \boldsymbol{x}(0) = \boldsymbol{x},\, \boldsymbol{x}(t) = \boldsymbol{y}\}}{\Pr\{\tau_A < \tau_B \,|\, \boldsymbol{x}(0) = \boldsymbol{x}\}}. \end{aligned}$$

[1] It is known in partial differential equations theory in higher dimensions that at boundary points where $\sum_{i,j} \sigma^{ij}(\boldsymbol{x})\nu^i(\boldsymbol{x})\nu^j(\boldsymbol{x}) = 0$, boundary conditions can be imposed only at points where $\boldsymbol{a}(\boldsymbol{x})\cdot\boldsymbol{\nu}(\boldsymbol{x}) < 0$ [73].

We obtain $\Pr\{\tau_A < \tau_B \,|\, \boldsymbol{x}(0) = \boldsymbol{x},\, \boldsymbol{x}(t) = \boldsymbol{y}\} = \Pr\{\tau_A < \tau_B \,|\, \boldsymbol{x}(0) = \boldsymbol{y}\}$ from the Markov property and time homogeneity, hence

$$p^*(\boldsymbol{y},t\,|\,\boldsymbol{x}) = p(\boldsymbol{y},t\,|\,\boldsymbol{x}) \frac{\Pr\{\tau_A < \tau_B|\,\boldsymbol{x}(0) = \boldsymbol{y}\}}{\Pr\{\,\tau_A < \tau_B\,|\,\boldsymbol{x}(0) = \boldsymbol{x}\}}. \tag{6.72}$$

It is evident from eq. (6.72) that $p^*(\boldsymbol{y},t\,|\,\boldsymbol{x})$ is a probability density function and that it satisfies the properties of a pdf of a diffusion process. It remains to calculate its infinitesimal drift vector and diffusion matrix. Note that the function $P(\boldsymbol{x}) = \Pr\{\tau_A < \tau_B\,|\,\boldsymbol{x}(0) = \boldsymbol{x}\}$ can be determined from the boundary value problem (6.71).

Next, we calculate the infinitesimal drift vector of $\boldsymbol{x}^*(t)$. By definition,

$$\begin{aligned} \boldsymbol{a}^*(\boldsymbol{x}) &= \lim_{h\downarrow 0} \frac{1}{h} \int p^*(\boldsymbol{y},h\,|\,\boldsymbol{x})(\boldsymbol{y}-\boldsymbol{x})\,d\boldsymbol{y} \\ &= \lim_{h\downarrow 0} \frac{1}{h} \int p(\boldsymbol{y},h\,|\,\boldsymbol{x}) \frac{P(\boldsymbol{y})}{P(\boldsymbol{x})}(\boldsymbol{y}-\boldsymbol{x})\,d\boldsymbol{x}. \end{aligned} \tag{6.73}$$

We expand $P(\boldsymbol{y})$ about $\boldsymbol{y} = \boldsymbol{x}$ in Taylor's series,

$$P(\boldsymbol{y}) = P(\boldsymbol{x}) + (\boldsymbol{y}-\boldsymbol{x})\cdot\nabla P(\boldsymbol{x}) + \frac{1}{2}(\boldsymbol{y}-\boldsymbol{x})^T\mathcal{H}\left(P(\boldsymbol{x})\right)(\boldsymbol{y}-\boldsymbol{x}) + o\left(|\boldsymbol{y}-\boldsymbol{x}|^2\right),$$

where $\mathcal{H}\left(P(\boldsymbol{x})\right)$ is the Hessian matrix of $P(\boldsymbol{x})$. Substituting the expansion in eq. (6.73), we obtain

$$\begin{aligned} \boldsymbol{a}^*(\boldsymbol{x}) = \lim_{h\downarrow 0} \frac{1}{h} \int p(\boldsymbol{y},h\,|\,\boldsymbol{x}) &\Bigg[(\boldsymbol{y}-\boldsymbol{x}) + (\boldsymbol{y}-\boldsymbol{x}) \cdot \frac{\nabla P(\boldsymbol{x})}{P(\boldsymbol{x})}(\boldsymbol{y}-\boldsymbol{x}) \\ &+ o\left(|\boldsymbol{y}-\boldsymbol{x}|^2\right) \Bigg]\,d\boldsymbol{y} = \boldsymbol{a}(\boldsymbol{x}) + \boldsymbol{\sigma}(\boldsymbol{x}) \frac{\nabla P(\boldsymbol{x})}{P(\boldsymbol{x})}, \end{aligned}$$

which is (6.70). Similarly,

$$\begin{aligned} \boldsymbol{\sigma}^*(\boldsymbol{x}) &= \lim_{h\downarrow 0} \frac{1}{h} \int p^*(\boldsymbol{y},h\,|\,\boldsymbol{x})(\boldsymbol{y}-\boldsymbol{x})(\boldsymbol{y}-\boldsymbol{x})^T\,d\boldsymbol{y} \\ &= \lim_{h\downarrow 0} \frac{1}{h} \int p(\boldsymbol{y},h\,|\,\boldsymbol{x})(\boldsymbol{y}-\boldsymbol{x})(\boldsymbol{y}-\boldsymbol{x})^T \\ &\quad \times \left[1 + \frac{\nabla P(\boldsymbol{x})}{P(\boldsymbol{x})} \cdot (\boldsymbol{y}-\boldsymbol{x}) + O\left(|\boldsymbol{y}-\boldsymbol{x}|^2\right)\right] d\boldsymbol{y} = \boldsymbol{\sigma}(\boldsymbol{x}). \end{aligned}$$

□

Note that eq. (6.71) implies that the second term in eq. (6.70) becomes infinite as $\boldsymbol{x}$ approaches the part B of the boundary. The direction of $\boldsymbol{a}^*(\boldsymbol{x})$ is into the domain, away from the boundary. Indeed, assume that B is an open subset of the boundary ∂D. We have $\nabla P(\boldsymbol{x})/|\nabla P(\boldsymbol{x})| = -\boldsymbol{\nu}(\boldsymbol{x})$, where $\boldsymbol{\nu}(\boldsymbol{x})$ is the unit outer normal at the boundary point $\boldsymbol{x}$, because $P(\boldsymbol{x}) = 0$ at all points $\boldsymbol{x} \in B$ and $P(\boldsymbol{x}) > 0$ for $\boldsymbol{x} \in D$. It follows that for points $\boldsymbol{x} \in D$ near B,

$$\begin{aligned} \boldsymbol{a}^*(\boldsymbol{x}) \cdot \boldsymbol{\nu}(\boldsymbol{x}) &= \boldsymbol{a}(\boldsymbol{x}) \cdot \boldsymbol{\nu}(\boldsymbol{x}) + \boldsymbol{\sigma}(\boldsymbol{x}) \frac{\nabla P(\boldsymbol{x})}{P(\boldsymbol{x})} \cdot \boldsymbol{\nu}(\boldsymbol{x}) + o(1) \\ &= \boldsymbol{a}(\boldsymbol{x}) \cdot \boldsymbol{\nu}(\boldsymbol{x}) - \frac{|\nabla P(\boldsymbol{x})|}{P(\boldsymbol{x})} \boldsymbol{\nu}^T(\boldsymbol{x}) \boldsymbol{\sigma}(\boldsymbol{x}) \boldsymbol{\nu}(\boldsymbol{x}) + o(1). \end{aligned} \tag{6.74}$$

If $\boldsymbol{\sigma}(\boldsymbol{x})$ is a positive definite matrix, then

$$\boldsymbol{a}^*(\boldsymbol{x}) \cdot \boldsymbol{\nu}(\boldsymbol{x}) \to -\infty \text{ as } \boldsymbol{x} \to B, \tag{6.75}$$

because $P(\boldsymbol{x}) \to 0$ as $\boldsymbol{x} \to B$. If $\boldsymbol{\nu}^T(\boldsymbol{x})\boldsymbol{\sigma}(\boldsymbol{x})\boldsymbol{\nu}(\boldsymbol{x}) = 0$ on B, then necessarily $\boldsymbol{a}(\boldsymbol{x}) \cdot \boldsymbol{\nu}(\boldsymbol{x}) < 0$, whenever boundary conditions can be imposed at B (see preceding footnote). Equation(6.75) also means that $\boldsymbol{a}^*(\boldsymbol{x}) \cdot \boldsymbol{\nu}(\boldsymbol{x}) < 0$ near B. This means that the angle between $\boldsymbol{a}^*(\boldsymbol{x})$ and $\boldsymbol{\nu}(\boldsymbol{x})$ is obtuse; that is, $\boldsymbol{a}^*(\boldsymbol{x})$ points into D. It follows that the trajectories $\boldsymbol{x}^*(t)$ cannot exit D at B. When the diffusion in the normal direction vanishes at the boundary, and the drift vector pushes the trajectories $\boldsymbol{x}(t)$ away from B, a trajectory $\boldsymbol{x}^*(t)$ cannot leave D through B, either.

The effect of conditioning on reaching A before reaching B is that the drift vector $\boldsymbol{a}(\boldsymbol{x})$ is replaced by the drift vector $\boldsymbol{a}^*(\boldsymbol{x})$, and the diffusion matrix remains unchanged. The dynamics (6.69) changes so that the dynamics of the conditioned process becomes $d\boldsymbol{x}^* = \boldsymbol{a}^*(\boldsymbol{x}^*)\,dt + \sqrt{2}\boldsymbol{B}(\boldsymbol{x}^*)\,d\boldsymbol{w}$. To simulate only those trajectories that satisfy the condition, the function $P(\boldsymbol{x})$ has to be known. Finding this function from the simulation may be as costly as running the unconditioned simulation.

Exercise 6.7 (Conditioning in 1-D). Find the one-dimensional version of eq. (6.74) [117]. □

Example 6.2 (Last passage time). Consider the SDE $dx = -U'(x)\,dt + \sqrt{2\varepsilon}\,dw$ in the ray $[-\infty, A]$, where $U(x)$ is a potential that forms a single well with the bottom at $x = B$ and ε is a small parameter. Find the *mean last passage time* (MLPT) of the trajectory through B when it reaches A. On its way to A the trajectory passes through B for the last time. The part of the trajectory that reaches A without returning to B is the conditional process $x^*(t)$. The mean first passage time of $x^*(t)$ from B to A is the MLPT. It also represents the escape time, once the trajectory starts the escape (i.e., it no longer returns to B before reaching A). To calculate the MLPT, we first calculate the function $P(x) = \Pr\{\tau_A < \tau_B \,|\, x(0) = x\}$, where $B < x < A$. We have to solve the boundary value problem (6.71); that is, $\varepsilon P''(x) - U'(x)P'(x) = 0$ for $B < x < A$ with the boundary conditions $P(A) = 1$ and $P(B) = 0$. The solution is given by $P(x) = \int_B^x e^{U(y)/\varepsilon}\,dy / \int_B^A e^{U(y)/\varepsilon}\,dy$. The modified drift is

$$a^*(x) = -U'(x) + \frac{2\varepsilon e^{U(x)/\varepsilon}}{\displaystyle\int_B^x e^{U(y)/\varepsilon}\,dy}. \tag{6.76}$$

Note that $U(x)$ is an increasing function in the interval $B < x < A$. It follows that for small ε the main contribution to the integrals comes from the upper limit of integration. That is, writing $U(y) = U(x) + U'(x)(y-x) + O\left((y-x)^2\right)$, we evaluate the integrals as

$$\begin{aligned}
\int_B^x e^{U(y)/\varepsilon}\,dy &= \int_B^x e^{\left[U(x)+U'(x)(y-x)+O\left((y-x)^2\right)\right]/\varepsilon}\,dy \\
&= e^{U(x)/\varepsilon}\int_B^x e^{\left[U'(x)(y-x)+O\left((y-x)^2\right)\right]/\varepsilon}\,dy \\
&= \varepsilon e^{U(x)/\varepsilon}\int_{(B-x)/\varepsilon}^0 e^{\left[U'(x)z+\varepsilon O\left(z^2\right)\right]}\,dz \\
&= \frac{\varepsilon e^{U(x)/\varepsilon}}{U'(x)\left(1+O(\varepsilon)\right)}. \qquad (6.77)
\end{aligned}$$

Substituting (6.77) in (6.76), we see that $a^*(x) = U'(x)(1+O(\varepsilon))$. Thus the conditioned dynamics is approximately $dx^* = U'(x^*)\,dt + \sqrt{2\varepsilon}\,dw$; that is, the well is turned upside down. For small noise the MLPT is approximately the time it takes to slide down from A to B. It can be found by solving the boundary value problem

$$\varepsilon\bar{\tau}''(x) + a^*(x)\bar{\tau}'(x) = -1 \text{ for } B < x < A, \quad \bar{\tau}(A) = 0, \quad \bar{\tau}(B) < \infty. \qquad (6.78)$$

Then the MLPT is $\bar{\tau}(B)$. □

Exercise 6.8 (A boundary condition). Explain the boundary condition (6.78). □

Exercise 6.9 (Conditioned MBM). Find the dynamics of the Brownian motion on the positive axis conditioned on reaching the point $x = 1$ before reaching the origin, given that $w(0) = x$, where $0 < x < 1$. □

Exercise 6.10 (Conditioning on a future interval [117]). Consider a diffusion process defined by the Itô equation $dx = a(x)\,dt + \sqrt{2\sigma(x)}\,dw$. Reduce the sample space of a diffusion process $x(t)$ to the cylinder $\alpha < x(1) < \beta$; that is, condition the trajectories of $x(t)$ on reaching the interval (α, β) at the future time $t = 1$. Denote the resulting diffusion process $x^*(t)$, for $0 < t < 1$. Define the conditional probability of the cylinder

$$P(x,t) = \Pr\left\{\alpha < x(1) < \beta \,|\, x(t) = x\right\}.$$

(i) Use Bayes' rule to show that for $0 < t < s < 1$,

$$\begin{aligned}
p^*(y,s\,|\,x,t)\,dy &= \Pr\left\{y \le x^*(s) < y+dy \,|\, x^*(t) = x\right\} \\
&= \frac{p(y,s\,|\,x,t)P(y,s)}{P(x,t)}\,dy.
\end{aligned}$$

(ii) Expand $P(y, t+h)$ in Taylor's series in the variables h and $y-x$ up to first-order terms and use it to obtain

$$a^*(x,t) = a(x) + \frac{\partial P(x,t)/\partial x}{P(x,t)}\sigma(x), \quad \sigma^*(x,t) = \sigma(x).$$

□

Exercise 6.11 (The Brownian bridge). Assume that $x(t)$ in Exercise 6.10 is the Brownian motion.

(i) Show that

$$P(x,t) = \frac{1}{\sqrt{2\pi(1-t)}} \int\limits_{\alpha}^{\beta} e^{-(y-x)^2/2(1-t)}\, dy.$$

(ii) Condition on $w(1) = \alpha$, by taking the limit $\beta - \alpha = \varepsilon \to 0$. Show that

$$\lim_{\varepsilon \downarrow 0} \frac{\partial P(x,t)/\partial x}{P(x,t)} = \frac{\alpha - x}{1-t}, \quad a^*(x,t) = \frac{\alpha - x}{1-t}, \quad \sigma^*(x,t) = 1.$$

For $\alpha = 0$, the resulting process is the Brownian bridge 2.26 [117]. □

6.6 Killing measure and the survival probability

One-dimensional Brownian motion (particle) with killing and an absorbing boundary can be considered as an approximation to free diffusion in a uniform long and narrow cylinder, whose lateral boundary contains many small absorbing holes, with one reflecting and one absorbing base. The small absorbers on the boundary may represent, for example, binding sites for diffusing ions or molecules on the membrane of the narrow neck of a dendritic spine of a neuronal synapse. The strength of the killing measure is related to the absorption flux of the three-dimensional Brownian motion through the small holes on the boundary of the cylinder. The killing measure $k(x,t)$ is the probability per unit time and unit length that the Brownian trajectory is terminated at point x and time t. According to Section 5.4, the joint (defective) pdf–PDF of finding a trajectory at x at time t before it is terminated (i.e., $\tau > t$), $p(x, t \,|\, y)\, dx = \Pr\{x(t) \in x + dx,\, \tau > t \,|\, x(0) = y\}$, is the solution of the boundary value problem $p_t = Dp_{xx} - k(x,t)\,p$ for $x \in \mathbb{R}$, $t > 0$ and the initial condition $p(x,0) = \delta(x-y)$. In the case that $k(x,t) = V_0$, the solution is given by $p(x,t \,|\, y) = (4\pi D t)^{-1/2} \exp\{-V_0 t - (x-y)^2/4Dt\}$. The probability per unit time of being killed (absorbed) inside the interval $[a,b]$ at time t is

$$\Pr\{x(\tau) \in [a,b],\, \tau = t \,|\, x(0) = y\} = \int\limits_a^b k(x,t)\, p(x,t \,|\, y)\, dx,$$

whereas the probability of being killed in the interval before time t is

$$\Pr\{x(\tau) \in [a,b],\, \tau < t \,|\, x(0) = y\} = \int\limits_0^t \int\limits_a^b k(x,s) p(x,s \,|\, y)\, dx\, ds.$$

The probability of ever being killed in the interval is

$$\Pr\{x(\tau) \in [a,b] \mid x(0) = y\} = \int_0^\infty \int_a^b k(x,t)\, p(x,t \mid y)\, dx\, dt,$$

and the density of ever being killed at x is therefore

$$\Pr\{x(\tau) = x \mid x(0) = y\} = \int_0^\infty k(x,t) p(x,t \mid y)\, dt. \tag{6.79}$$

The survival probability (5.69) in one dimension is $S(t \mid y) = \Pr\{\tau > t \mid x(0) = y\} = \int_{\mathbb{R}} p(x,t \mid y)\, dx$, which for the case of a constant killing rate $k(x,t) = V_0$ is

$$\Pr\{\tau > t \mid x(0) = y\} = \int_{\mathbb{R}} p(x,t \mid y)\, dx = e^{-V_0 t}.$$

Thus V_0 is the rate at which trajectories (particles) disappear from the domain (medium). Out of N_0 initial independent Brownian particles in $\mathbb{R}$ the expected number of particles that have disappeared by time t is $N_0(1 - e^{-V_0 t})$. The probability density of being killed at point x, given by eq. (6.79), is

$$\begin{aligned} P(x \mid y) &= V_0 \int_0^\infty \frac{1}{2\sqrt{\pi D t}} \exp\left\{-V_0 t - \frac{(x-y)^2}{4Dt}\right\} dt \\ &= \frac{1}{2}\sqrt{\frac{V_0}{D}} \exp\left\{-\sqrt{\frac{V_0}{D}}|x-y|\right\}. \end{aligned}$$

We assume henceforward that the killing measure is time-independent.

We consider now a particle diffusing in a domain $D \subset \mathbb{R}^d$ with a killing measure $k(\boldsymbol{x})$ and an absorbing part $\partial D_a \subset \partial D$ of the boundary ∂D. Thus the trajectory of the particle can terminate in two ways, it can either be killed inside D or absorbed in ∂D_a. The difference between the killing and the absorbing processes is that although the trajectory has a finite probability of not being terminated at points $\boldsymbol{x}$, where $k(\boldsymbol{x}) > 0$, it is terminated with probability 1 the first time it hits ∂D_a. The trajectory may often traverse killing regions, where $k(\boldsymbol{x}) > 0$, but it cannot emerge from the absorbing part of the boundary. Two random termination times are defined on the trajectories of the diffusion process: the time to killing, denoted T, and the time to absorption in ∂D_a, denoted τ, which is the first passage time to ∂D_a. We calculate below the probability $\Pr\{T < \tau \mid y\}$, and the conditional distribution $\Pr\{\tau < t \mid \tau < T, y\}$.

More generally, we consider the trajectories of the stochastic differential equation $d\boldsymbol{x} = \boldsymbol{a}(\boldsymbol{x})\, dt + \sqrt{2}\boldsymbol{B}(\boldsymbol{x})\, d\boldsymbol{w}(t)$ for $\boldsymbol{x}(t) \in D$ with killing measure $k(\boldsymbol{x})$. The transition pdf of $\boldsymbol{x}(t)$ satisfies the Fokker–Planck equation

$$\frac{\partial p(\boldsymbol{x},t \mid \boldsymbol{y})}{\partial t} = -\nabla \cdot \boldsymbol{J}(\boldsymbol{x},t \mid \boldsymbol{y}) - k(\boldsymbol{x}) p(\boldsymbol{x},t \mid \boldsymbol{y}) \ \text{for} \boldsymbol{x}, \boldsymbol{y} \in D, \tag{6.80}$$

where

$$J^i(\boldsymbol{x},t\,|\,\boldsymbol{y}) = -\sum_{j=1}^{d} \frac{\partial \sigma^{i,j}(\boldsymbol{x})p\,(\boldsymbol{x},t\,|\,\boldsymbol{y})}{\partial x^i} + a^i(\boldsymbol{x})p\,(\boldsymbol{x},t\,|\,\boldsymbol{y}) \tag{6.81}$$

and the initial and boundary conditions

$$p\,(\boldsymbol{x},0\,|\,\boldsymbol{y}) = \delta(\boldsymbol{x}-\boldsymbol{y}) \quad \text{for } \boldsymbol{x},\boldsymbol{y}\in D \tag{6.82}$$

$$p\,(\boldsymbol{x},t\,|\,\boldsymbol{y}) = 0 \quad \text{for } t>0,\ \boldsymbol{x}\in\partial D,\ \boldsymbol{y}\in D_a \tag{6.83}$$

$$\boldsymbol{J}(\boldsymbol{x},t\,|\,\boldsymbol{y})\cdot\boldsymbol{n}(\boldsymbol{x}) = 0 \quad \text{for } t>0,\ \boldsymbol{x}\in\partial D-\partial D_a,\ \boldsymbol{y}\in D. \tag{6.84}$$

The transition pdf $p\,(\boldsymbol{x},t\,|\,\boldsymbol{y})$ is actually the joint pdf

$$p\,(\boldsymbol{x},t\,|\,\boldsymbol{y})\,d\boldsymbol{x} = \Pr\{\boldsymbol{x}(t)\in\boldsymbol{x}+d\boldsymbol{x},\ T>t,\ \tau>t\,|\,\boldsymbol{y}\}; \tag{6.85}$$

that is, $p\,(\boldsymbol{x},t\,|\,\boldsymbol{y})$ is the probability density that the trajectory survived to time t (was neither killed nor absorbed in ∂D_a) and reaches $\boldsymbol{x}$ at time t.

Theorem 6.6.1 (Killing before absorption). *The probability that a diffusion process in a domain D, with killing $k(\boldsymbol{x})$ and absorption at $\partial D_a\subset\partial D$ is killed before it is absorbed is given by*

$$\Pr\{T<\tau\,|\,\boldsymbol{y}\} = \int_0^\infty\int_D k(\boldsymbol{x})p\,(\boldsymbol{x},t\,|\,\boldsymbol{y})\,d\boldsymbol{x}\,dt. \tag{6.86}$$

Proof. First, assume that the entire boundary is absorbing; that is, $\partial D_a = \partial D$. Then the probability density of surviving up to time t and being killed at time t at point $\boldsymbol{x}$ can be represented by the limit as $N\to\infty$ of

$$\Pr\{\boldsymbol{x}_N(t_{1,N})\in D,\ldots,\boldsymbol{x}_N(t)=\boldsymbol{x}, t\le T\le t+\Delta t\,|\,\boldsymbol{x}(0)=\boldsymbol{y}\} \tag{6.87}$$

$$= \int_D\cdots\int_D \prod_{j=1}^{N} \frac{d\boldsymbol{y}_j}{\sqrt{(2\pi\Delta t)^d \det\boldsymbol{\sigma}(\boldsymbol{x})(t_{j-1,N}))}}$$

$$\times\exp\left\{-\frac{1}{2\Delta t}\left[\boldsymbol{y}_j-\boldsymbol{x}(t_{j-1,N})-\boldsymbol{a}(\boldsymbol{x}(t_{j-1,N}))\Delta t\right]^T\boldsymbol{\sigma}^{-1}(\boldsymbol{x}(t_{j-1,N}))\right.$$

$$\left.\times\left[\boldsymbol{y}_j-\boldsymbol{x}(t_{j-1,N})-\boldsymbol{a}(\boldsymbol{x}(t_{j-1,N}))\Delta t\right]\right\}\left[1-k(\boldsymbol{x}(t_{j,N})\Delta t\right]k(\boldsymbol{x})\Delta t,$$

where $\Delta t = t/N$ $t_{j,N} = j\Delta t$, and $\boldsymbol{x}(t_{0,N}) = \boldsymbol{y}$ in the product. The limit is the Wiener integral defined by the stochastic differential equation, with the killing measure $k(\boldsymbol{x})$ and the absorbing boundary condition (see Chapter 5.3). In the limit $N\to\infty$ the integral (6.87) converges to the solution of the Fokker–Planck equation (6.80) in D with the initial and boundary conditions (6.82) and (6.83). Integrating over D with respect to $\boldsymbol{x}$ and from 0 to ∞ with respect to t, we obtain, in view of (6.85), the representation (6.86).

A second proof begins with the integration of the Fokker–Planck equation (6.80),

$$1=\int_0^\infty \oint_{\partial D} \boldsymbol{J}(\boldsymbol{x},t\,|\,\boldsymbol{y})\cdot\boldsymbol{n}(\boldsymbol{x})\,dS_{\boldsymbol{x}}\,dt+\int_0^\infty\int_D k(\boldsymbol{x})p\,(\boldsymbol{x},t\,|\,\boldsymbol{y})\,d\boldsymbol{x}\,dt. \tag{6.88}$$

We write $J(t\,|\,\boldsymbol{y}) = \oint_{\partial D} \boldsymbol{J}(\boldsymbol{x},t\,|\,\boldsymbol{y})\cdot\boldsymbol{n}(\boldsymbol{x})\,dS_{\boldsymbol{x}}$ and note that this is the absorption probability current on ∂D. Therefore, in view of the boundary conditions (6.83), (6.84), $\int_0^\infty J(t\,|\,\boldsymbol{y})\,dt$ is the total probability that has ever been absorbed at the boundary ∂D_a. This is the probability of trajectories that have not been killed before reaching ∂D_a. Writing eq. (6.88) as

$$\int_0^\infty J(t\,|\,\boldsymbol{y})\,dt=1-\int_0^\infty\int_D k(\boldsymbol{x})p\,(\boldsymbol{x},t\,|\,\boldsymbol{y})\,d\boldsymbol{x}\,dt,$$

we obtain (6.86). □

The probability distribution function of T for trajectories that have not been absorbed in ∂D_a is found by integrating the Fokker–Planck equation with respect to $\boldsymbol{x}$ over D and with respect to t from 0 to t. It is given by

$$\Pr\{T<t\,|\,\tau>T,\boldsymbol{y}\}=\frac{\Pr\{T<t,\tau>T\,|\,\boldsymbol{y}\}}{\Pr\{\tau>T\,|\,\boldsymbol{y}\}}=\frac{\displaystyle\int_0^t\int_D k(\boldsymbol{x})p\,(\boldsymbol{x},s\,|\,\boldsymbol{y})\,d\boldsymbol{x}\,ds}{\displaystyle\int_0^\infty\int_D k(\boldsymbol{x})p\,(\boldsymbol{x},s\,|\,\boldsymbol{y})\,d\boldsymbol{x}\,ds}.$$

Hence

$$\mathbb{E}[T\,|\,T<\tau,\boldsymbol{y}]=\frac{\displaystyle\int_0^\infty\int_t^\infty\int_D k(\boldsymbol{x})p\,(\boldsymbol{x},s\,|\,\boldsymbol{y})\,d\boldsymbol{x}\,ds\,dt}{\displaystyle\int_0^\infty\int_D k(\boldsymbol{x})p\,(\boldsymbol{x},s\,|\,\boldsymbol{y})\,d\boldsymbol{x}\,ds}=\frac{\displaystyle\int_0^\infty s\int_D k(\boldsymbol{x})p\,(\boldsymbol{x},s\,|\,\boldsymbol{y})\,d\boldsymbol{x}\,ds\,dt}{\displaystyle\int_0^\infty\int_D k(\boldsymbol{x})p\,(\boldsymbol{x},s\,|\,\boldsymbol{y})\,d\boldsymbol{x}\,ds},$$

which can be expressed in terms of the Laplace transform

$$\hat{p}(\boldsymbol{x},q\,|\,\boldsymbol{y})=\int_0^\infty p\,(\boldsymbol{x},s\,|\,\boldsymbol{y})e^{-qs}ds$$

as

$$\mathbb{E}[T \mid T < \tau, \boldsymbol{y}] = -\frac{\displaystyle\int_D k(\boldsymbol{x}) \frac{d\hat{p}(\boldsymbol{x}, q \mid \boldsymbol{y})}{dq}\, d\boldsymbol{x}}{\displaystyle\int_D k(\boldsymbol{x})\hat{p}(\boldsymbol{x}, q \mid \boldsymbol{y}) d\boldsymbol{x}}$$

$$= -\frac{\partial}{\partial q}\left(\log\left\{\int_D k(\boldsymbol{x})\hat{p}(\boldsymbol{x}, q \mid \boldsymbol{y}) d\boldsymbol{x}\right\}\right)\Bigg|_{q=0}. \tag{6.89}$$

The conditional distribution of the first passage time to the boundary of trajectories, given they are absorbed, is

$$\Pr\{\tau < t \mid T > \tau, \boldsymbol{y}\} = \frac{\displaystyle\int_0^t J(s \mid \boldsymbol{y})\, ds}{1 - \displaystyle\int_0^\infty \int_D k(\boldsymbol{x}) p\,(\boldsymbol{x}, s \mid \boldsymbol{y})\, d\boldsymbol{x}\, ds}. \tag{6.90}$$

Thus the mean time to absorption in ∂D_a of trajectories that are absorbed is given by [182]

$$\mathbb{E}[\tau \mid T > \tau,\, \boldsymbol{y}] = \int_0^\infty \Pr\{\tau > t \mid T > \tau, \boldsymbol{y}\}\, dt = \frac{\displaystyle\int_0^\infty sJ(s \mid \boldsymbol{y})\, ds}{1 - \displaystyle\int_0^\infty \int_D k(\boldsymbol{x}) p\,(\boldsymbol{x}, s \mid \boldsymbol{y})\, d\boldsymbol{x}\, ds}.$$

The pdf $p\,(\boldsymbol{y}, t \mid \boldsymbol{x}, s)$ in Definition 5.4.1 of the survival probability (5.69) is now the solution of the Fokker–Planck equation (6.80).

Chapter 7

Markov Processes and their Diffusion Approximations

7.1 Markov processes

Recall that according to Definition 2.4.1, a stochastic process $\boldsymbol{x}(t)$ is a *Markov process* if for all times $\tau_1 \leq \tau_2 \leq \cdots \leq \tau_m \leq t_1 \leq t_2 \leq \cdots \leq t_n$ and all Borel sets $A_1, A_2, \ldots, A_m, B_1, B_2, \ldots, B_n$ in $\mathbb{R}^d$ its multidimensional conditional PDF satisfies the equation

$$\begin{aligned}&\Pr\{\boldsymbol{x}(t_1) \in B_1, \ldots, \boldsymbol{x}(t_n) \in B_n \mid \boldsymbol{x}(\tau_1) \in A_1, \ldots, \boldsymbol{x}(\tau_m) \in A_m\}\\ &= \Pr\{\boldsymbol{x}(t_1) \in B_1, \ldots, \boldsymbol{x}(t_n) \in B_n \mid \boldsymbol{x}(\tau_m) \in A_m\}. \end{aligned} \tag{7.1}$$

In terms of densities the Markov property (7.1) is expressed as

$$p(\boldsymbol{y}_1, t_1, \ldots, \boldsymbol{y}_n, t_n \mid \boldsymbol{x}_1, \tau_1, \ldots, \boldsymbol{x}_m, \tau_m) = p(\boldsymbol{y}_1, t_1, \ldots, \boldsymbol{y}_n, t_n \mid \boldsymbol{x}_m, \tau_m).$$

It was shown that the Markov property implies that for all $t_1 \leq t_2 \leq t_3$, the transition pdf satisfies the Chapman–Kolmogorov integral equation

$$p(\boldsymbol{x}_3, t_3 \mid \boldsymbol{x}_1, t_1) = \int p(\boldsymbol{x}_3, t_3 \mid \boldsymbol{x}_2, t_2) p(\boldsymbol{x}_2, t_2 \mid \boldsymbol{x}_1, t_1)\, d\boldsymbol{x}_2. \tag{7.2}$$

If the space is discrete, the integral is replaced by a sum. Markov processes in discrete time are called *Markov chains*. Specifically, the generalization of Definition 2.4.1 is as follows.

Definition 7.1.1 (Markov chain). *A* Markov chain *in a set $\boldsymbol{X}$ is a sequence of random variables $\boldsymbol{x}_1, \boldsymbol{x}_2, \boldsymbol{x}_3, \ldots$ in $\boldsymbol{X}$ such that for any n and any $\boldsymbol{y}, \boldsymbol{y}_1, \ldots, \boldsymbol{y}_n$ in $\boldsymbol{X}$*

$$\Pr\{\boldsymbol{x}_{n+1} = \boldsymbol{y} \mid \boldsymbol{x}_n = \boldsymbol{y}_n, \ldots, \boldsymbol{x}_1 = \boldsymbol{y}_1\} = \Pr\{\boldsymbol{x}_{n+1} = \boldsymbol{y} \mid \boldsymbol{x}_n = \boldsymbol{y}_n\}.$$

The set $\boldsymbol{X}$ is called the state space *of the chain.*

Z. Schuss, *Theory and Applications of Stochastic Processes: An Analytical Approach*, Applied Mathematical Sciences 170, DOI 10.1007/978-1-4419-1605-1_7,

The state space can be finite or infinite. The Chapman–Kolmogorov equation (7.2) is generalized in a straightforward manner to this case (see [46]).

Exercise 7.1 (Running a Markov process backwards). Show that if $\boldsymbol{x}(t)$ is a Markov process, then $\boldsymbol{x}_T(t) = \boldsymbol{x}(T-t)$ is also a Markov process. □

Exercise 7.2 (The Cauchy process). The Cauchy process defined in Exercise 4.19 is defined by the transition pdf (4.86),

$$p\,(y,t+\Delta t\,|\,x,t) = \frac{\Delta t}{\pi}\frac{1}{(x-y)^2+\Delta t^2}. \tag{7.3}$$

Show that the transition pdf of the Cauchy process satisfies the CKE (7.2). □

The quotient $\int_{|\boldsymbol{x}-\boldsymbol{y}|>\,\varepsilon} p\,(\boldsymbol{y},t+\Delta t\,|\,\boldsymbol{x},t)\,d\boldsymbol{y}/\Delta t$ is the probability per unit time of a jump of the discretized process from $\boldsymbol{x}$ to any point $\boldsymbol{y}$, more than ε away from $\boldsymbol{x}$. The principal value

$$\lim_{\varepsilon\,\to\,0}\frac{1}{\Delta t}\int\limits_{|\boldsymbol{x}-\boldsymbol{y}|>\,\varepsilon} p\,(\boldsymbol{y},t+\Delta t\,|\,\boldsymbol{x},t)\,d\boldsymbol{y} = \frac{1}{\Delta t}(\mathrm{PV})\int p\,(\boldsymbol{y},t+\Delta t\,|\,\boldsymbol{x},t)\,d\boldsymbol{y}$$

is the probability per unit time of any jump of the discretized process from $\boldsymbol{x}$.

Definition 7.1.2 (Jump rate). *The* rate of jumps *bigger than ε from $\boldsymbol{x}$ at time t is*

$$\lambda_\varepsilon(\boldsymbol{x},t) = \lim_{\Delta t\downarrow 0}\frac{1}{\Delta t}\int\limits_{|\boldsymbol{x}-\boldsymbol{y}|>\,\varepsilon} p\,(\boldsymbol{y},t+\Delta t\,|\,\boldsymbol{x},t)\,d\boldsymbol{y}. \tag{7.4}$$

The jump rate *from $\boldsymbol{x}$ at time t is*

$$\lambda(\boldsymbol{x},t) = \lim_{\varepsilon\to 0}\lambda_\varepsilon(\boldsymbol{x},t). \tag{7.5}$$

Therefore it is natural to expect that the continuity condition for trajectories of a Markov process is that $\lambda_\varepsilon(\boldsymbol{x},t) = 0$ for all $\boldsymbol{x}, t$, and $\varepsilon > 0$; that is, no jumps of any finite size occur in short time.

Theorem 7.1.1 (Lindeberg's condition for continuity of paths [84]). *The trajectories of a Markov process are continuous with probability* 1 *if*

$$\lambda_\varepsilon(\boldsymbol{x},t) = 0 \tag{7.6}$$

holds for all $\varepsilon > 0$, uniformly for all $\boldsymbol{x}$ and t.

The Lindeberg condition (7.6) means that the probability of moving more than ε in a short time Δt decays faster than linearly with Δt.

Example 7.1 (Lindeberg's condition for the MBM). For the MBM with diffusion coefficient D,

$$p\,(y,t+\Delta t\,|\,x,t) = \frac{1}{\sqrt{2\pi D\Delta t}}\exp\left\{-\frac{(x-y)^2}{2D\Delta t}\right\},$$

so the Lindeberg condition is

$$\begin{aligned}
\lambda_\varepsilon(x,t) &= \lim_{\Delta t\downarrow 0}\frac{1}{\Delta t}\int\limits_{|x-y|>\varepsilon}\frac{1}{\sqrt{2\pi D\Delta t}}\exp\left\{-\frac{(x-y)^2}{2D\Delta t}\right\}dy\\
&= \lim_{\Delta t\downarrow 0}\frac{1}{\Delta t}\int\limits_{|z|>\varepsilon}\frac{1}{\sqrt{2\pi D\Delta t}}\exp\left\{-\frac{z^2}{2D\Delta t}\right\}dz\\
&= \lim_{\Delta t\downarrow 0}\frac{2}{\Delta t}\int\limits_{\varepsilon}^{\infty}\frac{1}{\sqrt{2\pi D\Delta t}}\exp\left\{-\frac{z^2}{2D\Delta t}\right\}dz\\
&= \lim_{\Delta t\downarrow 0}\frac{2}{\Delta t}\int\limits_{\varepsilon/\sqrt{D\Delta t}}^{\infty}\frac{1}{\sqrt{2\pi}}\exp\left\{-\frac{u^2}{2}\right\}du\\
&= \lim_{\Delta t\downarrow 0}\frac{\varepsilon}{\sqrt{2\pi D\Delta t^3}}\exp\left\{-\frac{\varepsilon^2}{2D\Delta t}\right\}=0.
\end{aligned}$$

l'Hospital's rule was used in the last line above. It follows that the probability law of the MBM ensures the continuity of paths with probability 1. □

Example 7.2. (The Cauchy process).

(i) To examine Lindeberg's continuity condition for paths of the Cauchy process defined by eq. (7.3) in Exercise 7.2, we calculate the rate of jumps bigger than ε,

$$\begin{aligned}
\lambda_\varepsilon(x,t) &= \lim_{\Delta t\downarrow 0}\frac{1}{\pi}\int\limits_{|x-y|>\varepsilon}\frac{dy}{(x-y)^2+\Delta t^2}=\lim_{\Delta t\downarrow 0}\frac{2}{\pi\Delta t}\int\limits_{\varepsilon/\Delta t}^{\infty}\frac{\Delta t^2}{\Delta t^2\,(z^2+1)}\,dy\\
&= \lim_{\Delta t\downarrow 0}\frac{2}{\pi\Delta t}\left(\arctan(\infty)-\arctan\frac{\varepsilon}{\Delta t}\right)=\frac{2}{\varepsilon}\neq 0,
\end{aligned}$$

where l'Hospital's rule was used in the last line above. It follows that the continuity condition is not satisfied so that the trajectories of the Cauchy process cannot be expected to be continuous.

(ii) The jump rate of the Cauchy process is

$$\lambda(x,t)=\lim_{\Delta t\to 0}\frac{1}{\Delta t}(\mathrm{PV})\int p\,(y,t+\Delta t\,|\,x,t)\,dy=\int\frac{dy}{\pi(x-y)^2}=\infty.$$

Thus the process jumps at an infinite rate (all the time!). The jump distribution from a point x is determined from

$$\Pr\{\text{jump size }\geq\varepsilon\}=\frac{2}{\pi}\int\limits_{x+\varepsilon}^{\infty}\frac{\Delta t\,dy}{(x-y)^2+\Delta t^2}\to 0\ \text{ as }\Delta t\to 0.$$

It follows that in short times the jumps are small. □

Example 7.3 (The Poisson process). The Poisson process is defined for small Δt by the transition pdf

$$p\,(y,t+\Delta t\,|\,x,t) = (1-\lambda\Delta t)\delta(y-x) + \lambda\Delta t\delta(y-x-1) + o(\Delta t), \tag{7.7}$$

where $\lambda > 0$ is the *jump rate.* Its trajectories are step functions that increase by one with rate λ (see Examples 7.4 and 7.5). The Lindeberg condition gives

$$\lim_{\Delta t\downarrow 0}\frac{1}{\Delta t}\int\limits_{|x-y|>\varepsilon} p\,(y,t+\Delta t\,|\,x,t)\,dy = \lambda > 0, \tag{7.8}$$

so the trajectories cannot be expected to be continuous. It is shown below that the Poisson process jumps at exponentially distributed random times. □

Definition 7.1.3 (Markovian jump process). *A* Markovian jump processes *in the time interval* $[0,T]$ *is the limit in probability of the discrete simulation scheme with time step* $\Delta t = T/N$

$$\boldsymbol{x}_N(t+\Delta t) = \begin{cases} \boldsymbol{x}_N(t) & \text{w.p. } 1-\lambda_N(\boldsymbol{x}_N(t),t)\Delta t \\ \boldsymbol{x}_N(t)+\boldsymbol{\xi}_N(t) & \text{w.p. } \lambda_N(\boldsymbol{x}_N(t),t)\Delta t, \end{cases} \tag{7.9}$$

where $\lambda_N(\boldsymbol{x}_N(t),t)\Delta t$ *is the probability that* $x_N(t)$ *jumps in the time interval* $[t,t+\Delta t]$ *and* $\boldsymbol{\xi}_N(t)$ *is the jump size. The jump size* $\boldsymbol{\xi}_N(t)$ *is a random process such that for* $t \geq t_1 > t_2 > \cdots > t_n$ *its conditional probability density, given that* $\boldsymbol{x}_N(t)$ *jumped from the point* $\boldsymbol{x}$ *in the time interval* $[t,t+\Delta t]$*, is*

$$\Pr\{\boldsymbol{\xi}_N(t)\in \boldsymbol{y}+d\boldsymbol{y}\,|\,\boldsymbol{x}_N(t)=\boldsymbol{x},\ \boldsymbol{x}_N(t_1)=\boldsymbol{x}_1,\ \ldots,\ \boldsymbol{x}_N(t_n)=\boldsymbol{x}_n\} \\ = w_N(\boldsymbol{y}\,|\,\boldsymbol{x},t)\,d\boldsymbol{y}. \tag{7.10}$$

The transition pdf of $\boldsymbol{x}_N(t)$ satisfies the equation

$$p_N(\boldsymbol{y},t+\Delta t\,|\,\boldsymbol{x},s) = (1-\lambda_N(\boldsymbol{y},t)\Delta t)p_N(\boldsymbol{y},t\,|\,\boldsymbol{x},s) \\ + \Delta t\int \lambda_N(\boldsymbol{z},t)w_N(\boldsymbol{y}-\boldsymbol{z}\,|\,\boldsymbol{z},t)p_N(\boldsymbol{z},t\,|\,\boldsymbol{x},s)\,d\boldsymbol{z}, \tag{7.11}$$

which means that the pdf to reach $\boldsymbol{y}$ at time $t+\Delta t$ is the probability of getting to $\boldsymbol{y}$ at time t and staying there for the time interval $[t,t+\Delta t]$ (i.e., not jumping), plus the probability of getting at time t to any point $\boldsymbol{z}$ and jumping to the point $\boldsymbol{y}$ in the time interval $[t,t+\Delta t]$. Equation (7.11) implies that the processes $\boldsymbol{x}_N(t)$ are Markovian on the lattice $0,\Delta t, 2\Delta t,\ldots$. If the limits $\lambda_N(\boldsymbol{y},t)\to\lambda(\boldsymbol{y},t)$, $w_N(\boldsymbol{y}\,|\,\boldsymbol{x},t)\to w(\boldsymbol{y}\,|\,\boldsymbol{x},t)$, and $p_N(\boldsymbol{y},t\,|\,\boldsymbol{x},s)\to p\,(\boldsymbol{z},t\,|\,\boldsymbol{x},s)$ exist as $N\to\infty$, then the limit pdf satisfies the *master equation*

$$\frac{\partial p\,(\boldsymbol{y},t\,|\,\boldsymbol{x},s)}{\partial t} \\ = -\lambda(\boldsymbol{y},t)p\,(\boldsymbol{y},t\,|\,\boldsymbol{x},s) + \int \lambda(\boldsymbol{z},t)w(\boldsymbol{y}-\boldsymbol{z}\,|\,\boldsymbol{z},t)p\,(\boldsymbol{z},t\,|\,\boldsymbol{x},s)\,d\boldsymbol{z}. \tag{7.12}$$

We say then that $\boldsymbol{x}_N(t) \to \boldsymbol{x}(t)$ in distribution. The question of convergence is discussed in [62], [85], [208], [70], [233].

The PDF of the waiting time between jumps from a given point can be determined from eq. (7.9). To calculate the probability of no jump from $\boldsymbol{x}$ in the time interval $[s,t]$, we partition the interval into N parts, $s = t_0 < t_1 < \cdots < t_N = t$, set $\Delta t_i = t_i - t_{i-1}$, $(i = 1, 2, \ldots, N)$, and $\Delta t = \max_i \Delta t_i$. According to eq. (7.9),

$$\Pr\{\text{no jump from } \boldsymbol{x} \text{ in the interval } [t_{i-1}, t_i]\} = 1 - \Delta t_i \lambda(\boldsymbol{x}, t_{i-1}).$$

Due to the Markov property, the probability of a jump in the time interval $[t_i, t_{i+1}]$, given that the process is still at $\boldsymbol{x}$, is independent of the elapsed time prior to t_i. It follows that

$$\begin{aligned}\Pr\{\text{no jump from } \boldsymbol{x} \text{ in } [s,t]\}' &= \lim_{N\to\infty} \prod_{i=0}^{N} (1 - \Delta t_i \lambda(\boldsymbol{x}, t_{i-1})) \\ &= \exp\left\{ - \int_s^t \lambda(\boldsymbol{x}, \tau)\, d\tau \right\}.\end{aligned}$$

If the jump rate is independent of t, then

$$\Pr\{\text{no jump from } \boldsymbol{x} \text{ in } [s,t]\} = \exp\{-\lambda(\boldsymbol{x})(t-s)\}.$$

Example 7.4 (Poisson process). The *Poisson process* defined in Example 7.3 can be also defined by (7.9) with $\lambda(x,t) = \lambda = const.$ and $w(y \,|\, x,t) = \delta(y - x - 1)$. The master equation (7.12) is

$$\frac{\partial p\,(y,t \,|\, x,s)}{\partial t} = -\lambda p\,(y,t \,|\, x,s) + \lambda p\,(y-1,t \,|\, x,s). \tag{7.13}$$

For small $t - s = \Delta t$, eq. (7.13) can be written as (7.7). □

Exercise 7.3 (Exponential waiting times). Show that a random waiting time is independent of the elapsed time if and only if it is exponentially distributed. □

Example 7.5 (The jump rate of the Poisson process). According to (7.7) and (7.8) the jump rate of the Poisson process is λ. □

7.1.1 The general form of the master equation

Writing

$$W_N(\boldsymbol{y} \,|\, \boldsymbol{x},t) = \lambda_N(\boldsymbol{x},t) w_N(\boldsymbol{y} - \boldsymbol{x} \,|\, \boldsymbol{x},t)\Delta t \tag{7.14}$$

we find that

$$\lambda_N(\boldsymbol{x},t) = \frac{1}{\Delta t}(\text{PV}) \int W_N(\boldsymbol{y} \,|\, \boldsymbol{x},t)\, d\boldsymbol{y}, \tag{7.15}$$

and that $W_N(\boldsymbol{y}\,|\,\boldsymbol{x},t)$ is the joint probability of a jump in the interval $[t,t+\Delta t]$ and the jump $\boldsymbol{\xi}_N(t)$. The dynamics (7.9) can be written as

$$\boldsymbol{x}_N(t+\Delta t) \tag{7.16}$$
$$=\begin{cases} \boldsymbol{x}_N(t) & \text{w.p. } 1-(\text{PV})\displaystyle\int W_N(\boldsymbol{z}\,|\,\boldsymbol{x}_N(t),t)\,d\boldsymbol{z} \\ \boldsymbol{x}_N(t)+\boldsymbol{\xi}_N(t) & \text{w.p. } (\text{PV})\displaystyle\int W_N(\boldsymbol{z}\,|\,\boldsymbol{x}_N(t),t)\,d\boldsymbol{z}, \end{cases}$$

and (7.10) as

$$\Pr\{\boldsymbol{\xi}_N(t)\in \boldsymbol{y}+d\boldsymbol{y}\,|\,\boldsymbol{x}_N(t)=\boldsymbol{x},\,\boldsymbol{x}_N(t_1)=\boldsymbol{x}_1,\,\ldots,\,\boldsymbol{x}_N(t_n)=\boldsymbol{x}_n\}$$
$$=\frac{W_N(\boldsymbol{y}\,|\,\boldsymbol{x},t)\,d\boldsymbol{y}}{(\text{PV})\displaystyle\int W_N(\boldsymbol{z}\,|\,\boldsymbol{x},t)\,d\boldsymbol{z}}.$$

Equation (7.11) for the transition pdf of $\boldsymbol{x}_N(t)$ becomes

$$p_N(\boldsymbol{y},t+\Delta t\,|\,\boldsymbol{x},s)=\left(1-(\text{PV})\int W_N(\boldsymbol{z}\,|\,\boldsymbol{y},t)\,d\boldsymbol{z}\right)p_N(\boldsymbol{y},t\,|\,\boldsymbol{x},s)$$
$$+\,(\text{PV})\int W_N(\boldsymbol{y}\,|\,\boldsymbol{z},t)p_N(\boldsymbol{z},t\,|\,\boldsymbol{x},s)\,d\boldsymbol{z}. \tag{7.17}$$

As above, if the limits

$$\lim_{N\to\infty}\frac{W_N(\boldsymbol{z}\,|\,\boldsymbol{x},t)}{\Delta t}=W(\boldsymbol{z}\,|\,\boldsymbol{x},t)$$
$$\lim_{N\to\infty}\boldsymbol{\xi}_N(t)=\boldsymbol{\xi}(t),\qquad \lim_{N\to\infty}\boldsymbol{x}_N(t)=\boldsymbol{x}(t) \tag{7.18}$$

exist in some sense, the limit process $\boldsymbol{x}(t)$ is called a Markovian jump process. The master equation (7.12) becomes

$$\frac{\partial p\,(\boldsymbol{y},t\,|\,\boldsymbol{x},s)}{\partial t}$$
$$=(\text{PV})\int\left[W(\boldsymbol{y}\,|\,\boldsymbol{z},t)p\,(\boldsymbol{z},t\,|\,\boldsymbol{x},s)-W(\boldsymbol{z}\,|\,\boldsymbol{y},t)p\,(\boldsymbol{y},t\,|\,\boldsymbol{x},s)\right]d\boldsymbol{z}. \tag{7.19}$$

Note that eq. (7.15) implies that the jump rate (7.5) is

$$\lambda(\boldsymbol{y},t)=\lim_{\varepsilon\to 0}\int_{|\boldsymbol{z}-\boldsymbol{y}|>\varepsilon} W(\boldsymbol{z}\,|\,\boldsymbol{y},t)\,d\boldsymbol{z}=(\text{PV})\int W(\boldsymbol{z}\,|\,\boldsymbol{y},t)\,d\boldsymbol{z}. \tag{7.20}$$

Example 7.6 (The master equation for the Cauchy process). For the Cauchy process defined in Example 7.2 the joint probability $W_N(y\,|\,x,t)$ is defined in eq. (7.3) as $W_N(y\,|\,x,t)=\Delta t\left\{\pi\left[(x-y)^2+\Delta t^2\right]\right\}^{-1}$, so that $(\text{PV})\int W_N(\boldsymbol{z}\,|\,\boldsymbol{x},t)\,d\boldsymbol{z}=1$, which means that the probability of no jump in the interval $[t,t+\Delta t]$ is zero and the process jumps at each time step. (i.e., the jump rate is infinite). The dynamics (7.9) is then reduced to $x_N(t+\Delta t) = x_N(t) + \xi_N(t)$, where

$$\Pr\left\{\xi_N(t) \in y + dy \,|\, x(t) = x\right\} = \frac{\Delta t}{\pi\left[(x-y)^2 + \Delta t^2\right]}\, dy.$$

The result of Example (7.2) implies that $W(y\,|\,x,t) = [\pi(x-y)^2]^{-1}$, so the master equation (7.19) is

$$\frac{\partial p\,(y,t\,|\,x,s)}{\partial t} = (PV)\int \frac{[p\,(z,t\,|\,x,s) - p\,(y,t\,|\,x,s)]\,dz}{\pi(y-z)^2}.$$

□

Consider particles moving along the trajectories of the dynamical system in $\mathbb{R}^d$,

$$\dot{\boldsymbol{x}} = \boldsymbol{A}(\boldsymbol{x},t), \quad t > s, \quad \boldsymbol{x}(s) = \boldsymbol{x}, \tag{7.21}$$

where $\boldsymbol{x}_0$ is distributed with a given density $p_0(\boldsymbol{x})$. The measure in function space induced by (7.21) is concentrated on the trajectories of (7.21), parameterized by $\boldsymbol{x}_0 \in \mathbb{R}^d$. The pdf of a trajectory $\boldsymbol{x}(t,\boldsymbol{x}_0)$ is the pdf of its initial value $\boldsymbol{x}_0$. The process $\boldsymbol{x}(t,\boldsymbol{x}_0)$ is Markovian, because of the uniqueness of solutions of (7.21). The pdf $p_0(\boldsymbol{x})$ evolves in time to the density of particles $p\,(\boldsymbol{y},t\,|\,s)$.

Theorem 7.1.2 (Liouville's equation). *The pdf $p\,(\boldsymbol{y},t\,|\,s)$ is the solution of the initial value problem*

$$\begin{aligned}\frac{\partial p\,(\boldsymbol{y},t\,|\,s)}{\partial t} &= -\sum_{i=1}^{n}\frac{\partial\left[A^i\,(\boldsymbol{y},t)\,p\,(\boldsymbol{y},t\,|\,s)\right]}{\partial y^i} \quad \text{for} \quad t > s \\ p\,(\boldsymbol{y},s\,|\,s) &= p_0(\boldsymbol{y}).\end{aligned} \tag{7.22}$$

It is given by

$$p\,(\boldsymbol{y},t\,|\,s) = \int p\,(\boldsymbol{y},t\,|\,\boldsymbol{x},s)p_0(\boldsymbol{x})\,d\boldsymbol{x}$$

where the Green's function $p\,(\boldsymbol{y},t\,|\,\boldsymbol{x},s)$ is the solution of the initial value problem (7.21) with the initial condition $p\,(\boldsymbol{y},s\,|\,\boldsymbol{x},s) = \delta(\boldsymbol{y}-\boldsymbol{x})$.

Proof. In a given interval $s < t < T$ and for a given time step $\Delta t = T/N$, the Euler scheme for the solution of (7.21) is

$$\boldsymbol{x}_N(t+\Delta t) = \boldsymbol{x}_N(t) + \boldsymbol{A}(\boldsymbol{x}_N(t),t)\Delta t, \quad \boldsymbol{x}_N(s) = \boldsymbol{x}, \tag{7.23}$$

so that for every test function $f(\boldsymbol{y})$

$$\begin{aligned}\mathbb{E}f(\boldsymbol{x}_N(t+\Delta t)) &= \mathbb{E}f(\boldsymbol{x}_N(t) + \Delta\boldsymbol{x}) \\ &= \mathbb{E}f(\boldsymbol{x}_N(t)) + E\boldsymbol{A}(\boldsymbol{x}_N(t))\cdot\nabla f(\boldsymbol{x}_N(t))\Delta t + o(\Delta t).\end{aligned}$$

This means (integrate by parts)

$$\begin{aligned}
&\int f(\boldsymbol{y})p_N(\boldsymbol{y},t+\Delta t\,|\,s)\,d\boldsymbol{y}\\
&=\int p_N(\boldsymbol{y},t\,|\,s)\left[f(\boldsymbol{y})+\boldsymbol{A}(\boldsymbol{y},t)\cdot\nabla f(\boldsymbol{y})\Delta t+o(\Delta t)\right]\,d\boldsymbol{y}\\
&=\int f(\boldsymbol{y})\left[p_N(\boldsymbol{y},t\,|\,s)-\nabla\cdot\boldsymbol{A}(\boldsymbol{y},t)p_N(\boldsymbol{y},t\,|\,s)\Delta t+o(\Delta t)\right]\,d\boldsymbol{y},
\end{aligned}$$

which is equivalent to

$$\begin{aligned}
&\frac{\partial p_N(\boldsymbol{y},t\,|\,s)}{\partial t}=-\sum_{i=1}^{n}\frac{\partial\left[A^i(\boldsymbol{y},t)\,p_N(\boldsymbol{y},t\,|\,s)\right]}{\partial y^i}+o(1)\ \text{ for } N\to\infty,\quad t>s,\\
&p_N(\boldsymbol{y},s\,|\,s)=p_0(\boldsymbol{y}).
\end{aligned}$$

The convergence $p_N(\boldsymbol{y},t\,|\,s)\to p(\boldsymbol{y},t\,|\,s)$ gives (7.22). □

Exercise 7.4. (Solution of Liouville's equation).
(i) Prove under appropriate assumptions $p_N(\boldsymbol{y},t\,|\,s)\to p(\boldsymbol{y},t\,|\,s)$ as $N\to\infty$.
(ii) Solve Liouville's equation (7.22) by the methods of characteristics [45]. □

A Markovian drift and jump process corresponds to the simulated dynamics

$$\begin{aligned}
&\boldsymbol{x}_N(t+\Delta t) \qquad (7.24)\\
&=\begin{cases}\boldsymbol{x}_N(t)+\boldsymbol{A}(\boldsymbol{x}_N(t),t)\Delta t & \text{w.p. } 1-\lambda_N(\boldsymbol{x}_N(t),t)\Delta t\\ \boldsymbol{x}_N(t)+\boldsymbol{A}(\boldsymbol{x}_N(t),t)\Delta t+\boldsymbol{\xi}_N(t) & \text{w.p. } \lambda_N(\boldsymbol{x}_N(t),t)\Delta t,\end{cases}
\end{aligned}$$

where, as in (7.10),

$$\Pr\{(\boldsymbol{\xi}_N(t)\in\boldsymbol{z}+d\boldsymbol{z}\,|\,\boldsymbol{x}_N(t)=\boldsymbol{x},\,\ldots,\,\boldsymbol{x}_N(t_n)=\boldsymbol{x}_n\}=w_N(\boldsymbol{z}\,|\,\boldsymbol{x},t)\,d\boldsymbol{z}.$$

Obviously, $\boldsymbol{x}_N(t)$ is a discrete-time Markov process on the grid. The master equation is the combination of the Liouville equation (7.22) and the master equation (7.19) (or (7.12)),

$$\begin{aligned}
&\frac{\partial p(\boldsymbol{y},t\,|\,\boldsymbol{x},s)}{\partial t}\\
&=-\sum_{i=1}^{n}\frac{\partial\left[A^i(\boldsymbol{y},t)\,p(\boldsymbol{y},t\,|\,\boldsymbol{x},s)\right]}{\partial y^i}\\
&\quad-\lambda(\boldsymbol{y})p(\boldsymbol{y},t\,|\,\boldsymbol{x},s)+\int\lambda(\boldsymbol{z})w(\boldsymbol{y}-\boldsymbol{z}\,|\,\boldsymbol{y})p(\boldsymbol{z},t\,|\,\boldsymbol{x},s)\,d\boldsymbol{z}\\
&=-\sum_{i=1}^{n}\frac{\partial}{\partial y^i}\left[A^i(\boldsymbol{y},t)\,p(\boldsymbol{y},t\,|\,\boldsymbol{x},s)\right] \qquad (7.25)\\
&\quad+(\mathrm{PV})\int\left[W(\boldsymbol{y}\,|\,\boldsymbol{z},t)p(\boldsymbol{z},t\,|\,\boldsymbol{x},s)-W(\boldsymbol{z}\,|\,\boldsymbol{y},t)p(\boldsymbol{y},t\,|\,\boldsymbol{x},s)\right]\,d\boldsymbol{z}.
\end{aligned}$$

It is derived as above.

Example 7.7 (Virtual work in queueing theory). In queueing theory [130] a queue can be described by the *unfinished work in the system* or *virtual work* $U(t)$, which is the time it takes the queue to empty if the arrival of new customers is stopped at time t. If customers arrive at exponential waiting times with rate λ (that may depend on the unfinished work) and demand work x, whose pdf is $b(x)$ (which also may depend on $U(t)$), we denote the PDF of $U(t)$ by $F(w,t) = \Pr\{U(t) \leq w\}$ and the pdf by $f(w,t) = \partial F(w,t)/\partial w$. The server works at the fixed rate 1, because work is measured in units of time. The dynamics (7.24) of the unfinished work is given by

$$U(t+\Delta t) = \begin{cases} U(t) - \Delta t & \text{w.p. } 1 - \lambda(U)\Delta t + o(\Delta t) \\ U(t) + x & \text{w.p. } \lambda(U)\Delta t + o(\Delta t), \end{cases}$$

where $\Pr\{x = y \,|\, U(t) = w,\, U(t_1) = w_1, \ldots, U(t_n) = w_n\} = b(y,w)$. Thus the drift is $a = -1$, so that for $w > 0$ the master equation (7.12) is

$$\frac{\partial f(w,t)}{\partial t} = f_w(w,t) - \lambda(w) f(w,t) + \int_{0-}^{w} \lambda(y) b(w-y,y) f(y,t)\, dy. \tag{7.26}$$

This is the *generalized Takács equation* [130]. In general, $F(0+,t) \neq 0$, because there is a positive probability of an empty queue (i.e., of an *idle period*). On the other hand, $F(0-,t) = 0$. It follows that $F(w,t)$ is discontinuous at $w = 0$ so that

$$f(w,t) = \tilde{f}(w,t) + A(t)\delta(w), \tag{7.27}$$

where $\tilde{f}(w,t)$ is a regular function at $w = 0$ and $A(t)$ is the probability of an empty queue.

To derive equations for $\tilde{f}(w,t)$ and $A(t)$, we use (7.27) in the generalized Takacs equation (7.26). We obtain

$$\int_{0-}^{w} \lambda(y) b(w-y,y) f(y,t)\, dy = \int_{0-}^{w} \lambda(y) b(w-y,y) \left[\tilde{f}(y,t) + A(t)\delta(y)\right] dy;$$

that is, for $w > 0$,

$$\begin{aligned} \frac{\partial \tilde{f}(w,t)}{\partial t} &= \frac{\partial \tilde{f}(w,t)}{\partial w} - \lambda(w)\tilde{f}(w,t) \\ &\quad + \int_{0-}^{w} \lambda(y) b(w-y,y) \tilde{f}(y,t)\, dy + \lambda(0) b(w \,|\, 0) A(t). \end{aligned}$$

The equation for $A(t)$ is derived directly. We have

$$\begin{aligned} &A(t+\Delta t) \\ &= \Pr\{U(t+\Delta t) = 0\} = [1 - \lambda(0)\Delta t] \Pr\{U(t) = 0\} \\ &\quad + [1 - \lambda(0)\Delta t] \Pr\{0 < U(t) < \Delta t\} + \Delta t \int_{0-}^{\Delta t} \lambda(y) b(\Delta t - y, y) \tilde{f}(y,t)\, dy, \end{aligned}$$

because $A(t) = \Pr\{U(t) = 0\}$. In the limit $\Delta t \to 0$, we obtain

$$\dot{A}(t) = -\lambda(0)A(t) + \tilde{f}(0,t).$$

In addition, we have the normalization condition

$$\int_0^\infty \tilde{f}(w,t)\,dw + A(t) = 1$$

for all t [136], [138], [134]. □

Example 7.8 (The master equation for the point Josephson junction). The point Josephson junction is a sandwich of two superconductors separated by a thin layer of oxide (insulator) [18]. A current is driven across the junction by a voltage or current source. The capacity and resistance of the junction are represented in an equivalent circuit (the RSJ model) as a resistor (R) and a capacitor (C) in series connected in parallel to a nonlinear element. The voltage and the current across the junction, denoted $V(\tilde{t})$ and $\tilde{I}(\tilde{t})$, respectively, are functions of the dimensional time $\tilde{t}$.

The voltage in the junction is described by an order parameter θ. According to Josephson's law, the circuit equation (in the absence of normal resistance and noise) is given by

$$C\frac{dV(\tilde{t})}{d\tilde{t}} + I_J \sin\theta = \tilde{I} \tag{7.28}$$

$$\dot{\theta} = \frac{2e}{\hbar}V(\tilde{t}), \tag{7.29}$$

where I_J is a characteristic current of the junction and e is the electronic charge. Thus, for instance, if the current is driven by a constant voltage source, then $V = const.$ in eq. (7.29) gives $\theta = 2eV\tilde{t}/\hbar$, hence eq. (7.28) gives $\tilde{I} = I_J \sin 2eV\tilde{t}/\hbar$; that is, the constant voltage source produces an alternating current with frequency $2eV/\hbar$. Josephson, Esaki, and Giaever shared the 1973 Nobel Prize in physics for this discovery.

We introduce the notation $\omega_J^2 = 2eI_J/\hbar$, $I = \tilde{I}/I_J$, and $G = (\omega_J RC)^{-1}$, where ω_J is the *Josephson frequency*, and G is the conductance. We nondimensionalize time energy by setting $t = \omega_J\tilde{t}$ and $E_J = \hbar I_J/2e$. In these variables the circuit equation becomes $\ddot{\theta} + \sin\theta = I$. This is the equation of a pendulum driven by torque I. The circuit equation describes the current of so-called *Cooper pairs*, which flows without any voltage drop across the junction (super current).

In addition to the super current there is also a normal current that flows across the junction, with voltage V, resistance R, and noise. This current is due to tunneling of normal electrons across the oxide from left to right and from right to left when a voltage drop V from right to left is given. At a given temperature T, the energies of normal electrons on both sides of the junction are described by the *Fermi distribution*: the probabilities that a state E is occupied on the right or left side of the junction are given by

$$f_R(E) = \frac{1}{1+\exp\left\{\dfrac{E+eV-E_F}{k_BT}\right\}}, \quad f_L(E) = \frac{1}{1+\exp\left\{\dfrac{E-E_F}{k_BT}\right\}},$$

where E_F is the so-called *Fermi energy*. An electron can tunnel from right to left with energy E if the state E is occupied on the right and unoccupied on the left. Assuming tunneling at exponential waiting times, the rate of tunneling from right to left (and, similarly, from left to right) is

$$l(E) = cf_R(E)\,[1-f_L(E)]\,, \quad r(E) = cf_L(E)\,[1-f_R(E)]\,,$$

where c is some constant. It follows that

$$\frac{l(E)}{r(E)} = \exp\left\{-\frac{eV}{k_BT}\right\}, \quad \frac{r(E)+l(E)}{r(E)-l(E)} = \coth\frac{eV}{2k_BT},$$

or equivalently,

$$\begin{aligned} l(E) &= \exp\left\{-\frac{eV}{k_BT}\right\} r(E) \\ r(E)+l(E) &= [r(E)-l(E)]\coth\frac{eV}{2k_BT}. \end{aligned} \tag{7.30}$$

Denoting by $N(E)$ the density of states, we find that the net current across the junction (from left to right) is

$$I = \int N(E)e\,[r(E)-l(E)]\,dE = \int N(E)e\frac{r(E)+l(E)}{\coth\dfrac{eV}{2k_BT}}\,dE,$$

so that $I = V/R$ gives

$$\int N(E)\,[r(E)+l(E)]\,dE = \frac{V}{eR}\coth\frac{eV}{2k_BT}.$$

Setting $l = \int N(E)l(E)\,dE$ and $r = \int N(E)r(E)\,dE$, integration of eqs. (7.30) against $N(E)$ gives

$$l(V) = \frac{V}{eR\left[1-\exp\left\{-\dfrac{eV}{k_BT}\right\}\right]}, \quad r(V) = \frac{V}{eR\left[\exp\left\{\dfrac{eV}{k_BT}\right\}+1\right]}.$$

If the junction is driven by a constant DC current source I_{DC}, the stochastic dynamics of the voltage is now given by

$$V(t+\Delta t)$$
$$=\begin{cases} V(t)-\dfrac{I_J}{C}\sin\theta(t)\Delta t+\dfrac{I_{DC}}{C}\Delta t+o(\Delta t) & \text{w.p. } 1-(r+l)\Delta t+o(\Delta t)\\[2ex] V(t)-\dfrac{I_J}{C}\sin\theta(t)\Delta t+\dfrac{I_{DC}}{C}\Delta t+\dfrac{e}{C}+o(\Delta t) & \text{w.p. } r\Delta t+o(\Delta t)\\[2ex] V(t)-\dfrac{I_J}{C}\sin\theta(t)\Delta t+\dfrac{I_{DC}}{C}\Delta t-\dfrac{e}{C}+o(\Delta t) & \text{w.p. } l\Delta t+o(\Delta t). \end{cases}$$

It follows that the joint pdf of θ and V, denoted $p\,(\theta, V, t\,|\,\theta_0, V_0)$, satisfies the master equation

$$\frac{\partial p\,(V,\theta,t)}{\partial t}=\frac{2eV}{\hbar}\frac{\partial p\,(V,\theta,t)}{\partial\theta}-\left[\frac{I_{DC}}{C}\Delta t-\frac{I_J}{C}\sin\theta\right]\frac{\partial p\,(V,\theta,t)}{\partial V}$$
$$+\,r\left(V-\frac{e}{C}\right)p\left(V-\frac{e}{C},\theta,t\right)+l\left(V+\frac{e}{C}\right)p\left(V+\frac{e}{C},\theta,t\right)$$
$$-\,[r\,(V)+l\,(V)]\,p\,(V,\theta,t)\,.$$

Changing variables to

$$T\to\frac{2ek_BT}{\hbar I_J},\quad t\to\frac{t}{\omega_J},\quad E\to\frac{E}{E_J},$$

we write

$$eV=\frac{\hbar\omega_J}{2}\dot{\theta},\quad \frac{V}{RI_J}=G\dot{\theta},\quad q=\frac{e\omega_J}{I_J}=\frac{\hbar\omega_J}{2E_J},$$

so that $p\,(\theta, V, t)\to p\,(\theta,\dot{\theta},t)$. Then the master equation can be written as

$$\frac{\partial p\,(\theta,\dot{\theta},t)}{\partial t}=-\,\dot{\theta}\frac{\partial p\,(\theta,\dot{\theta},t)}{\partial\theta}-[I-\sin\theta]\,\frac{\partial p\,(\theta,\dot{\theta},t)}{\partial\dot{\theta}}$$
$$+\,r\left(\dot{\theta}-q\right)p\left(\theta,\dot{\theta}-q,t\right)+l\left(\dot{\theta}+q\right)p\left(\theta,\dot{\theta}+q,t\right)$$
$$-\left[r\left(\dot{\theta}\right)+l\left(\dot{\theta}\right)\right]p\left(\theta,\dot{\theta},t\right).$$

□

7.1.2 Jump-diffusion processes

The simulated dynamics of a diffusion process in $\mathbb{R}^d$ that jumps at random times is given by

$$\boldsymbol{x}_{\Delta t}(t+\Delta t) \tag{7.31}$$
$$=\begin{cases} \boldsymbol{x}_{\Delta t}(t)+\boldsymbol{A}(\boldsymbol{x}_{\Delta t}(t),t)\Delta t+\sqrt{2}\boldsymbol{B}(\boldsymbol{x}_{\Delta t}(t),t)\Delta\boldsymbol{w}(t)\\ \text{w.p. } 1-\lambda_{\Delta t}(\boldsymbol{x}_{\Delta t}(t),t)\Delta t\\[2ex] \boldsymbol{x}_{\Delta t}(t)+\boldsymbol{A}(\boldsymbol{x}_{\Delta t}(t),t)\Delta t+\sqrt{2}\boldsymbol{B}(\boldsymbol{x}_{\Delta t}(t),t)\Delta\boldsymbol{w}(t)+\boldsymbol{\xi}_{\Delta t}(t)\\ \text{w.p. } \lambda_{\Delta t}(\boldsymbol{x}_{\Delta t}(t),t)\Delta t, \end{cases}$$

where the drift vector $\boldsymbol{A}(\boldsymbol{x},t)$ and the diffusion matrix $\boldsymbol{B}(\boldsymbol{x},t)$ are sufficiently smooth functions (as required in the proof of the differential Chapman–Kolmogorov equation below) and $\boldsymbol{w}(t)$ is a MBM. As in (7.10),

$$\Pr\{\boldsymbol{\xi}_{\Delta t}(t) \in \boldsymbol{z} + d\boldsymbol{z} \,|\, \boldsymbol{x}_{\Delta t}(t) = \boldsymbol{x}, \ldots, \boldsymbol{x}_{\Delta t}(t_n) = \boldsymbol{x}_n\} = w_{\Delta t}(\boldsymbol{z} \,|\, \boldsymbol{x}, t)\, d\boldsymbol{z}.$$

The process $\boldsymbol{x}_{\Delta t}(t)$ is a discrete-time Markov process on the grid. With the notation (7.14) and (7.15), we assume that $\lambda_{\Delta t}(\boldsymbol{x},t) \to \lambda(\boldsymbol{x},t)$, $w_{\Delta t}(\boldsymbol{y} \,|\, \boldsymbol{x},t) \to w(\boldsymbol{y} \,|\, \boldsymbol{x},t)$, and $W_{\Delta t}(\boldsymbol{y} \,|\, \boldsymbol{x},t) \to W(\boldsymbol{y} \,|\boldsymbol{x},t)$ sufficiently strongly as $\Delta t \to 0$ to allow passage to the limit in integrals. Then the transition pdf $p_{\Delta t}(\boldsymbol{y},t \,|\, \boldsymbol{x},s)$ of the process $\boldsymbol{x}_{\Delta t}(t)$ converges in distribution to a Markov process $\boldsymbol{x}(t)$. Specifically, proceeding as in the proofs of the Fokker–Planck equation (Theorem 4.5.1) and Liouville's equation (Theorem 7.1.2), we obtain [62], [84], [82] the following theorem.

Theorem 7.1.3 (The differential Chapman–Kolmogorov equation). *The transition pdf of $\boldsymbol{x}(t)$ satisfies for $t > s$ the equation*

$$\begin{aligned}\frac{\partial p(\boldsymbol{y},t \,|\, \boldsymbol{x},s)}{\partial t} &= L_{\boldsymbol{y}} p(\boldsymbol{y},t \,|\, \boldsymbol{x},s) \qquad (7.32)\\ &= -\sum_{i=1}^{n} \frac{\partial \left[a^i(\boldsymbol{y},t)\, p(\boldsymbol{y},t \,|\, \boldsymbol{x},s)\right]}{\partial y^i} + \sum_{i=1}^{n}\sum_{j=1}^{n} \frac{\partial^2 \left[\sigma^{ij}(\boldsymbol{y},t) p(\boldsymbol{y},t \,|\, \boldsymbol{x},s)\right]}{\partial y^i \partial y^j}\\ &\quad + (\mathrm{PV}) \int \left[W(\boldsymbol{y} \,|\, \boldsymbol{z},t) p(\boldsymbol{z},t \,|\, \boldsymbol{x},s) - W(\boldsymbol{z} \,|\, \boldsymbol{y},t) p(\boldsymbol{y},t \,|\, \boldsymbol{x},s)\right] d\boldsymbol{z}\end{aligned}$$

and the initial condition

$$\lim_{t \downarrow s} p(\boldsymbol{y},t \,|\, \boldsymbol{x},s) = \delta(\boldsymbol{y} - \boldsymbol{x}). \qquad (7.33)$$

The pdf $p(\boldsymbol{y},t \,|\, \boldsymbol{x},s)$ satisfies with respect to the backward variables $(\boldsymbol{x},s)$ the backward Kolmogorov equation

$$\begin{aligned}\frac{\partial p(\boldsymbol{y},t \,|\, \boldsymbol{x},s)}{\partial s} &= L^*_{\boldsymbol{x}} p(\boldsymbol{y},t \,|\, \boldsymbol{x},s) \qquad (7.34)\\ &= -\sum_{i=1}^{n} a^i(\boldsymbol{x},s) \frac{\partial p(\boldsymbol{y},t \,|\, \boldsymbol{x},s)}{\partial x^i} - \sum_{i=1}^{n}\sum_{j=1}^{n} \sigma^{ij}(\boldsymbol{x},s) \frac{\partial^2 p(\boldsymbol{y},t \,|\, \boldsymbol{x},s)}{\partial x^i \partial x^j}\\ &\quad + (\mathrm{PV}) \int W(\boldsymbol{z} \,|\, \boldsymbol{x},s) \left[p(\boldsymbol{y},t \,|\, \boldsymbol{x},s) - p(\boldsymbol{y},t \,|\, \boldsymbol{z},s)\right] d\boldsymbol{z}\end{aligned}$$

and the terminal condition

$$\lim_{s \uparrow t} p(\boldsymbol{y},t \,|\, \boldsymbol{x},s) = \delta(\boldsymbol{y} - \boldsymbol{x}). \qquad (7.35)$$

The general definition of a jump diffusion process is as follows.

Definition 7.1.4 (The general definition of a jump-diffusion process). *A (multidimensional)* jump-diffusion process *is a Markov process $\boldsymbol{x}(t)$ in $\mathbb{R}^d$, whose transition pdf, $p(\boldsymbol{y},t \,|\, \boldsymbol{x},s)$, satisfies the following conditions.*

1. *For all* $\varepsilon > 0$,

$$\lim_{\Delta t \downarrow 0} \frac{1}{\Delta t} p\,(\boldsymbol{y}, t + \Delta t \,|\, \boldsymbol{x}, t) = W(\boldsymbol{y} \,|\, \boldsymbol{x}, t),$$

uniformly in $|\boldsymbol{x} - \boldsymbol{y}| > \varepsilon$ *and* t.

2. *For all* $1 \leq i \leq d$ *and small* $\varepsilon > 0$,

$$\lim_{\Delta t \downarrow 0} \frac{1}{\Delta t} \int\limits_{|\boldsymbol{x} - \boldsymbol{y}| < \varepsilon} (y^i - x^i) p\,(\boldsymbol{y}, t + \Delta t \,|\, \boldsymbol{x}, t)\, d\boldsymbol{y} = a^i(\boldsymbol{x}, t) + o(\varepsilon),$$

uniformly in $|\boldsymbol{x} - \boldsymbol{y}| > \varepsilon$ *and* t.

3. *For all* $1 \leq i, j \leq d$ *and small* $\varepsilon > 0$,

$$\lim_{\Delta t \downarrow 0} \frac{1}{\Delta t} \int\limits_{|\boldsymbol{x} - \boldsymbol{y}| < \varepsilon} (y^i - x^i)(y^j - x^j) p\,(\boldsymbol{y}, t + \Delta t \,|\, \boldsymbol{x}, t)\, d\boldsymbol{y} = \sigma^{i,j}(\boldsymbol{x}, t) + o(\varepsilon),$$

uniformly in $|\boldsymbol{x} - \boldsymbol{y}| > \varepsilon$ *and* t.

All higher-order differential moments vanish in the limit $\varepsilon \to 0$, because for any $\delta > 0$

$$\begin{aligned}
&\lim_{\Delta t \downarrow 0} \frac{1}{\Delta t} \int\limits_{|\boldsymbol{x} - \boldsymbol{y}| < \varepsilon} |\boldsymbol{x} - \boldsymbol{y}|^{2+\delta} p\,(\boldsymbol{y}, t + \Delta t \,|\, \boldsymbol{x}, t)\, d\boldsymbol{y} \\
&\leq \varepsilon^\delta \lim_{\Delta t \downarrow 0} \frac{1}{\Delta t} \int\limits_{|\boldsymbol{x} - \boldsymbol{y}| < \varepsilon} |\boldsymbol{x} - \boldsymbol{y}|^2 p\,(\boldsymbol{y}, t + \Delta t \,|\, \boldsymbol{x}, t)\, d\boldsymbol{y} \\
&= \varepsilon^\delta \left[\max_{i,j} \sigma^{i,j}(\boldsymbol{x}, t) + o(\varepsilon) \right].
\end{aligned}$$

It can be shown [82], as in the derivation of the Fokker–Planck equation, that the pdf of a general jump-diffusion process satisfies the differential Chapman–Kolmogorov equation (7.32). The meaning of the above conditions is that continuous paths are possible only if $W(\boldsymbol{y} \,|\, \boldsymbol{x}, t) = 0$ for all $\boldsymbol{y} \neq \boldsymbol{x}$. Thus $W(\boldsymbol{y} \,|\, \boldsymbol{x}, t)$ determines the probability distribution of the times between jumps and the probability of the jump size. The coefficients $a^i(\boldsymbol{x}, t)$ form the *infinitesimal drift vector* and the coefficients $\sigma^{i,j}(\boldsymbol{x}, t)$ form the *diffusion matrix*. Under the above assumptions Theorem 7.1.3 holds; that is, the transition pdf of $\boldsymbol{x}(t)$ satisfies the differential Chapman–Kolmogorov equation (7.32).

The pdf $p\,(\boldsymbol{y}, t \,|\, \boldsymbol{x}, s)$, in general, is not a pdf with respect to the backward variable $\boldsymbol{x}$. For example, the function $p\,(\boldsymbol{y}, t \,|\, \boldsymbol{x}, s) = 1$ is a solution of the backward equation, although it does not satisfy the terminal condition (7.35). If the coefficients $a^i\,(\boldsymbol{x}, s)$ and $\sigma^{ij}(\boldsymbol{x}, s)$, as well as the function $W(\boldsymbol{z} \,|\, \boldsymbol{x}, s)$ are independent of s, then

$$p\,(\boldsymbol{y}, t \,|\, \boldsymbol{x}, s) = p\,(\boldsymbol{y}, t - s \,|\, \boldsymbol{x}, 0) \tag{7.36}$$

and we can write $p(\boldsymbol{y},t\,|\,\boldsymbol{x},s)=p(\boldsymbol{y},\tau\,|\,\boldsymbol{x})$, where $\tau=t-s$, so that

$$\frac{\partial p(\boldsymbol{y},t\,|\,\boldsymbol{x},s)}{\partial t}=\frac{\partial p(\boldsymbol{y},\tau\,|\,\boldsymbol{x})}{\partial \tau}=-\frac{\partial p(\boldsymbol{y},t\,|\,\boldsymbol{x},s)}{\partial s}.$$

The forward and backward Kolmogorov equations can be written in the operator form, respectively, as

$$p_\tau(\boldsymbol{y},\tau\,|\,\boldsymbol{x})=L_{\boldsymbol{y}}p(\boldsymbol{y},\tau\,|\,\boldsymbol{x}),\quad p_\tau(\boldsymbol{y},\tau\,|\,\boldsymbol{x})=L^*_{\boldsymbol{x}}p(\boldsymbol{y},\tau\,|\,\boldsymbol{x})$$

where $L^*_{\boldsymbol{x}}$ is the formal adjoint to $L_{\boldsymbol{y}}$ in the sense that for all sufficiently smooth functions $U_1(\boldsymbol{y})$ and $\phi_2(\boldsymbol{y})$ that vanish outside a bounded set

$$\int U_1(\boldsymbol{y})L_{\boldsymbol{y}}U_2(\boldsymbol{y})\,d\boldsymbol{y}=\int U_2(\boldsymbol{y})L^*_{\boldsymbol{y}}U_1(\boldsymbol{y})\,d\boldsymbol{y}.$$

Definition 7.1.5 (Stationary process). *The process $\boldsymbol{x}(t)$ is said to be* stationary *if all its multidimensional densities are translation invariant in time.*

In particular, for a stationary process eq. (7.36) holds. If $\boldsymbol{x}(t)$ is a stationary process and its pdf converges to a limit $\lim_{\tau\to\infty}p(\boldsymbol{y},\tau\,|\,\boldsymbol{x})=p(\boldsymbol{y})$, independent of initial conditions, then $p(\boldsymbol{y})$ is the stationary pdf and satisfies the forward equation $L_{\boldsymbol{y}}p(\boldsymbol{y})=0$; that is, $p(\boldsymbol{y})$ is the eigenfunction of the forward operator $L_{\boldsymbol{y}}$ corresponding to the eigenvalue 0. The function $p^*(\boldsymbol{x})=1$ is then the eigenfunction of the backward Kolmogorov operator $L^*_{\boldsymbol{x}}$ corresponding to the same eigenvalue; that is, $L^*_{\boldsymbol{x}}1=0$. A general necessary and sufficient condition for the convergence of the solution of the Fokker–Planck equation to a stationary solution is unknown, although some sufficient conditions are given in [82], [206] and numerous journal articles.

If the trajectories of a Markovian jump-diffusion process are instantaneously truncated when they leave a domain $D\subset\mathbb{R}^d$, the pdf $p(\boldsymbol{y},t\,|\,\boldsymbol{x},s)$ satisfies the differential Chapman–Kolmogorov equation (7.32) in D with respect to the forward variable $\boldsymbol{y}$ and vanishes for $\boldsymbol{y}\notin D$, and the differential BKE (7.34) with respect to the backward variable $\boldsymbol{x}$ in D and also vanishes for $\boldsymbol{x}\notin D$. The pdf is not necessarily continuous at the boundary (see the proof of Theorem 7.4.1 below). Theorem 4.4.3 is generalized in a straightforward way to the following

Theorem 7.1.4. *The mean first passage time τ_D out of D time after s is the solution of the boundary value problem*

$$\frac{\partial \mathbb{E}[\tau_D\,|\,\boldsymbol{x},s]}{\partial s}+L^*_{\boldsymbol{x}}\mathbb{E}[\tau_D\,|\,\boldsymbol{x},s]=-1\quad\text{for }\boldsymbol{x}\in D,\ s<t \tag{7.37}$$

$$\mathbb{E}[\tau_D\,|\,\boldsymbol{x}(s)]=0\quad\text{for }\boldsymbol{x}\notin D, \tag{7.38}$$

where $L^*_{\boldsymbol{x}}$ is the backward differential Chapman–Kolmogorov operator (7.34).

Proof. The survival probability for $t>s$ is given by (5.69), where $p(\boldsymbol{y},t\,|\,\boldsymbol{x},s)$ is the pdf of the Markov process in D, whose trajectories are instantaneously truncated when it leaves D. Thus it is the solution of (7.34) with absorbing boundary conditions and the mean exit time after s is (5.71); that is,

$$\mathbb{E}\left[\tau_D \,|\, \boldsymbol{x}, s\right] = \int_s^\infty S(t \,|\, \boldsymbol{x}, s)\, dt = \int_s^\infty \int_D p\left(\boldsymbol{y}, t \,|\, \boldsymbol{x}, s\right) d\boldsymbol{y}\, dt. \tag{7.39}$$

Differentiating with respect to s, using the terminal condition (7.35) and the backward Chapman–Kolmogorov equation (7.34), we find that

$$\frac{\partial \mathbb{E}\left[\tau_D \,|\, \boldsymbol{x}, s\right]}{\partial s} = -1 + \int_s^\infty \int_D \frac{p\left(\boldsymbol{y}, t \,|\, \boldsymbol{x}, s\right)}{\partial s}\, d\boldsymbol{y}\, dt = -1 - L^*_{\boldsymbol{x}} \mathbb{E}\left[\tau_D \,|\, \boldsymbol{x}, s\right].$$

□

7.2 A semi-Markovian example: Renewal processes

Definition 7.2.1 (Semi-Markov process). *A continuous-time stochastic process $\boldsymbol{x}(t)$ is called a* continuous-time semi-Markov process *if it jumps at a strictly increasing sequence of times T_n and $\boldsymbol{x}(T_n)$ is a Markov chain (called the* embedded Markov chain*). The times between the consecutive jumps, $\tau_n = T_n - T_{n-1}$, are called* waiting times *or* holding times.

The pair $(\boldsymbol{x}(t), y(t))$, where $y(t)$ is the elapsed time because the last jump prior to time t, is Markovian. For example, the simplest semi-Markovian generalization of the Poisson counting process (Examples 7.3–7.5) is the *renewal (counting) process* $\{N(t),\, t \geq 0\}$, which counts the number of successive events in the time interval $(0, t]$, where the time durations between consecutive counts τ_i are *positive, independent, identically distributed* random variables. The formal definition is as follows

Definition 7.2.2 (Renewal process). *A continuous time semi-Markov process defined by $N(t) = \max_i \{i \,|\, T_i \leq t\}$, where τ_i are positive i.i.d. random waiting times, is called a* renewal process.

If $\Pr\{\tau_i = \infty\} > 0$, then $N(t)$ is called a *terminating* renewal process. This is the case, for example, of counting the number of consecutive returns of a diffusing particle to a given ball. Because diffusion is nonrecurrent in $\mathbb{R}^3$ (see Example 4.5), the sequence of returns will terminate [181], [180]. In contrast, the total charge $Q(t)$ carried by a current $I(t)$ of charges (electrons or ions) that arrive at i.i.d. random times at an electrode in the time interval $(0, t]$ is a renewal process, for example, the ionic current in an electrolytic solution [179].

The common PDF of the times τ_i is denoted

$$F_\tau(t) = \Pr\left\{\tau_i \leq t\right\}, \quad i = 1, 2, 3, \ldots \tag{7.40}$$

and their pdf by $f_\tau(t)$. If $F_\tau(t) = e^{-\lambda t}$, the renewal process $N(t)$ is Poissonian. The assumption $\tau_i > 0$ is expressed by the assumption $F_\tau(t) = 0$ for $t \leq 0$. We denote a generic waiting time τ_i by τ and its moments by

$$\mathbb{E}\tau^m = \int_0^\infty t^m f_\tau(t)\,dt.$$

Assuming that all moments exist, Taylor's expansion of the Laplace transform of $f_\tau(t)$ is given by

$$\hat{f}_\tau(s) = 1 + \sum_{n=1}^\infty \frac{(-1)^n}{n!} s^n \mathbb{E}\tau^n. \tag{7.41}$$

The random variable

$$T_n = \tau_1 + \tau_2 + \cdots + \tau_n, \quad n \geq 1,\ T_0 = 0$$

is the *waiting time* until the occurrence of the nth event. With this notation

$$N(t) = \max\{n \,|\, 0 < T_n \leq t\}.$$

The connection between $N(t)$ and T_n is also given by the identities

$$\{N(t) = n\} = \{T_n \leq t < T_{n+1}\}, \quad n = 0, 1, 2, \ldots. \tag{7.42}$$

The simplest example of a renewal process is the *Poisson process* with exponential waiting times between the counts. If λ is the exponential rate, the distribution of $N(t)$ is

$$\Pr\{N(t) = n\} = \frac{(\lambda t)^n}{n!} e^{-\lambda t}, \quad n = 0, 1, 2, \ldots.$$

The Laplace transform of the pdf of τ is

$$\hat{f}_\tau(s) = \frac{\lambda}{\lambda + s} = 1 + \sum_{n=1}^\infty \frac{(-1)^n s^n}{n!\lambda^n}, \quad 0 \leq s < \lambda.$$

The Poisson process is Markovian, because it jumps at exponential waiting times.

The pdf of T_n is the n-fold convolution of $f_\tau(t)$ with itself,

$$f_{T_n}(t) = \underbrace{f_\tau * f_\tau * \cdots * f_\tau}_{n}(t).$$

Taking the Laplace transform gives

$$\hat{f}_{T_n}(s) = \hat{f}_\tau^n(s).$$

The function

$$\varphi(t) = \sum_{n=1}^\infty f_{T_n}(t) \tag{7.43}$$

is called *the renewal function.* Its Laplace transform is

$$\hat{\varphi}(s) = \frac{\hat{f}_\tau(s)}{1 - \hat{f}_\tau(s)}.$$

For the Poisson process

$$\hat{\varphi}(s) = \frac{\lambda}{s}.$$

The simplest properties of $\varphi(t)$ are

$$\lim_{s\to 0} \hat{\varphi}(s) = \infty, \quad \lim_{s\to 0} s\hat{\varphi}(s) = \lim_{t\to\infty} \varphi(t) = \frac{1}{\mathbb{E}\tau}. \tag{7.44}$$

Exercise 7.5 (The renewal equation). The integral equation

$$u(t) = g(t) + \int_0^t u(t-s) f_\tau(s)\, ds = g(t) + u * f_\tau(t), \tag{7.45}$$

where $g(t)$ is a given function, is called a *renewal equation.* Show that its solution is given by

$$u(t) = g(t) + \int_0^t u(t-s)\varphi(s)\, ds = g(t) + u * \varphi(t), \tag{7.46}$$

where $\varphi(t)$ is the renewal function. □

In certain cases the time of the first event has a different distribution than that of the i.i.d. times τ_i for $i > 1$. This is the case, for example, if the counting process begins at some positive time t_0. Then the first count occurs after the *residual time* $\gamma_{t_0} = T_{i+1} - t_0$, where $T_i < t_0 \le T_{i+1}$. To determine the PDF of γ_{t_0}, we write

$$\begin{aligned}\Pr\{\gamma_{t_0} < x\} &= \sum_{n=0}^{\infty} \Pr\{\gamma_{t_0} < x,\, N(t_0) = n\} \\ &= \sum_{n=0}^{\infty} \Pr\{T_n \le t_0 < T_{n+1},\, T_{n+1} - t_0 < x\} \\ &= \sum_{n=0}^{\infty} \Pr\{T_n \le t_0 < T_{n+1} + \tau_{n+1} < t_0 + x\}.\end{aligned} \tag{7.47}$$

Because τ_{n+1} is independent of T_n, their joint pdf is $f_{T_n,\tau}(u,v) = f_{T_n}(u) f_\tau(v)$. It follows that

$$\Pr\{T_n \leq t_0 < T_{n+1} + \tau_{n+1} < t_0 + x\} \tag{7.48}$$
$$= \int_0^{t_0} du \int_{t_0-u}^{t_0+x-u} f_{T_n}(u) f_\tau(v)$$
$$= \int_0^{t_0} [F_\tau(t+x-u) - F_\tau(t-u)]\, f_{T_n}(u)\, du.$$

Using the definition (7.43) and eq. (7.48) in eq. (7.47), we obtain

$$\Pr\{\gamma_{t_0} < x\} = \int_0^{t_0} [F_\tau(t_0+x-u) - F_\tau(t_0-u)]\, \varphi(u)\, du.$$

Differentiating with respect to x, we obtain

$$f_{\gamma_{t_0}}(x) = \int_0^{t_0} f_\tau(t_0+x-u)\varphi(u)\, du = \int_x^{t_0+x} f_\tau(z)\varphi(t_0+x-z)\, dz.$$

In the limit $t_0 \to \infty$, we obtain

$$\lim_{t_0\to\infty} f_{\gamma_{t_0}}(x) = \frac{1}{\mathbb{E}\tau}\int_x^\infty f_\tau(z)\, dz = \frac{1-F_\tau(x)}{\mathbb{E}\tau}. \tag{7.49}$$

We denote a random variable with the pdf (7.49) by r and the pdf (7.49) by $f_r(x)$. The Laplace transform of $f_r(x)$ is given by

$$\hat{f}_r(s) = \frac{1-\hat{f}_\tau(s)}{s\mathbb{E}\tau}.$$

This means that if we begin to count at some time t_0 long after the arrival process has begun, the pdf of the time to the first count is $f_r(x)$. For the Poisson process the residual time is exponential with the same exponent as the arrival rate.

To find the distribution of $N(t)$, we assume that τ_1 is the residual time for the renewal process $N(t)$. We set $\tau_0 = 0$ so that its pdf is $\delta(t)$ and its Laplace transform is 1. Thus, the pdf of T_n becomes

$$f_{T_0}(t) = \delta(t), \quad f_{T_n}(t) = f_r * \underbrace{f_\tau * f_\tau * \cdots * f_\tau}_{n-1 \text{ times}}(t), \quad n = 1, 2, \ldots$$

and the Laplace transforms are

$$\hat{f}_{T_0}(s) = 1, \quad \hat{f}_{T_n}(s) = \frac{1-\hat{f}_\tau(s)}{s\mathbb{E}\tau}\hat{f}_\tau^{n-1}(s), \quad n = 1, 2, \ldots.$$

It follows from eq. (7.42) that

$$\Pr\{N(t)=0\} = \Pr\{r>t\} = 1-F_r(t)$$
$$\Pr\{N(t)=n\} = \Pr\{T_n \le t < T_{n+1}\}, \quad n=0,1,2,\dots.$$

Reasoning as above, we find that

$$\Pr\{T_n \le t < T_{n+1}\} = \int_0^t [1-F_\tau(t-u)]\, f_{T_n}(u)\,du = [1-F_\tau] * f_{T_n}(t),$$

so that, for $n>0$,

$$\Pr\{N(t)=n\} = [1-F_\tau] * f_{T_n}(t).$$

It follows that

$$\mathcal{L}\Pr\{N(t)=0\}(s) = \frac{1}{s}\left[1-\frac{1-\tilde{f}_\tau(s)}{s\mathbb{E}\tau}\right] \tag{7.50}$$

$$\mathcal{L}\Pr\{N(t)=n\}(s) = \frac{\left[1-\hat{f}_\tau(s)\right]^2}{s^2\mathbb{E}\tau}\hat{f}_\tau^{n-1}(s), \quad n=1,2,\dots.$$

Exercise 7.6 (Law of large numbers). Use the central limit theorem version of the large deviations theory (Section 9.3.2) to determine the asymptotic behavior of $\Pr\{N(t)=n\}$ for large n and large t such that $t=n\mathbb{E}\tau+y$. Conclude that

$$\frac{N(t)}{t} \xrightarrow{\text{a.s.}} \frac{1}{\mathbb{E}\tau} \quad \text{as } t\to\infty. \tag{7.51}$$

Equation (7.51) is the *(strong) law of large numbers.* □

Equations (7.50) can be used to calculate the moments of $N(t)$. Thus,

$$\mathbb{E}N(t) = \sum_{n=1}^{\infty} n\Pr\{N(t)=n\}$$

so that

$$\mathbb{E}\hat{N}(s) = \sum_{n=1}^{\infty} n\frac{\left[1-\hat{f}_\tau(s)\right]^2}{s^2\mathbb{E}\tau}\hat{f}_\tau^{n-1}(s) = \frac{1}{s^2\mathbb{E}\tau}.$$

It follows that

$$\mathbb{E}N(t) = \frac{t}{\mathbb{E}\tau}. \tag{7.52}$$

Similarly,

$$\begin{aligned}\mathcal{L}\mathbb{E}N^2(t)(s) &= \mathcal{L}\sum_{n=1}^{\infty} n^2\Pr\{N(t)=n\}(s)\sum_{n=1}^{\infty} n^2\frac{\left[1-\hat{f}_\tau(s)\right]^2}{s^2\mathbb{E}\tau}\hat{f}_\tau^{n-1}(s)\\ &= \frac{1+\hat{f}_\tau(s)}{s^2\mathbb{E}\tau\left[1-\hat{f}_\tau(s)\right]}.\end{aligned} \tag{7.53}$$

Classical *shot noise* is the noisy current due to the arrival of identical charges at an electrode at i.i.d. random times. The total charge registered by the electrode by time t is a renewal process. Thus shot noise is the derivative of a renewal process. The charges can be electrons or ions of various species. We denote the noisy current by $I(t)$. Denoting by q the charge of a single particle, we have $I(t) = q\dot{N}(t)$, and from eq. (7.52), we obtain

$$\langle I(t)\rangle = \frac{q}{\mathbb{E}\mathcal{T}}. \tag{7.54}$$

To calculate the autocorrelation and power spectral density of current fluctuations, we calculate first the autocorrelation function of the counting process $N(t)$,

$$R(t_1,t_2) = \mathbb{E}N(t_1)N(t_2).$$

Assuming $t_2 > t_1$, we have

$$R(t_1,t_2) = \sum_{m=1}^{\infty}\sum_{n=1}^{\infty} nm \Pr\{N(t_1)=n,\, N(t_2)=m\}. \tag{7.55}$$

Obviously, summation extends only over $m \geq n$. It follows that eq. (7.55) can be written as

$$\begin{aligned}
R(t_1,t_2) &= \sum_{n=1}^{\infty}\sum_{p=0}^{\infty} n(n+p)\Pr\{\,N(t_2)=n+p\,|\,N(t_1)=n\}\Pr\{N(t_1)=n\}\\
&= \sum_{n=1}^{\infty}\sum_{p=0}^{\infty} n(n+p)\Pr\{\,N(t_2-t_1)=p\,\}\Pr\{N(t_1)=n\}\\
&= \sum_{n=1}^{\infty} n^2\Pr\{N(t_1)=n\}\sum_{p=0}^{\infty}\Pr\{\,N(t_2-t_1)=p\,\}\\
&\quad+\sum_{n=1}^{\infty} n\Pr\{N(t_1)=n\}\sum_{p=1}^{\infty} p\Pr\{\,N(t_2-t_1)=p\,\}\\
&= \sum_{n=1}^{\infty} n^2\Pr\{N(t_1)=n\} + \mathbb{E}N(t_1)\mathbb{E}N(t_2-t_1)\\
&= \mathbb{E}N^2(t_1\wedge t_2) + \mathbb{E}N(t_1\wedge t_2)\mathbb{E}N(|t_2-t_1|),
\end{aligned} \tag{7.56}$$

where $t_1 \wedge t_2 = \min(t_1,t_2)$.

The autocorrelation of the current fluctuations is

$$\begin{aligned}
\left\langle\left(I(t_1)-\frac{q}{\mathbb{E}\mathcal{T}}\right)\left(I(t_2)-\frac{q}{\mathbb{E}\mathcal{T}}\right)\right\rangle &= \langle I(t_1)I(t_2)\rangle - \langle I\rangle^2 = \mathbb{E}\dot{N}(t_1)\dot{N}(t_2) - \langle I\rangle^2\\
&= \frac{\partial^2}{\partial t_1\partial t_2}R(t_1,t_2) - \langle I\rangle^2.
\end{aligned} \tag{7.57}$$

First, we note the following identities for $i \neq j,\ (i, j = 1, 2)$,

$$\frac{\partial t_i \wedge t_j}{\partial t_i} = H(t_j - t_i), \qquad \frac{\partial^2 t_1 \wedge t_2}{\partial t_2 \partial t_1} = \delta(t_2 - t_1)$$
$$\frac{\partial |t_i - t_j|}{\partial t_i} = 2H(t_i - t_j) - 1, \qquad \frac{\partial^2 |t_1 - t_2|}{\partial t_1 \partial t_2} = -2\delta(t_1 - t_2) \tag{7.58}$$

and

$$\frac{\partial t_1 \wedge t_2}{\partial t_1}\frac{\partial t_1 \wedge t_2}{\partial t_2} = H(t_2 - t_1)H(t_1 - t_2) = 0$$
$$\frac{\partial t_1 \wedge t_2}{\partial t_1}\frac{\partial |t_1 - t_2|}{\partial t_2} + \frac{\partial t_1 \wedge t_2}{\partial t_2}\frac{\partial |t_1 - t_2|}{\partial t_1} = 1. \tag{7.59}$$

Now, we have

$$\begin{aligned}\frac{\partial^2}{\partial t_1 \partial t_2}\mathbb{E}N^2(t_1 \wedge t_2) &= \frac{\partial}{\partial t_1}\frac{\partial t_1 \wedge t_2}{\partial t_2}\frac{d}{dt}\,\mathbb{E}N^2(t)\Big|_{t=t_1\wedge t_2}\\ &= \frac{\partial^2 t_1 \wedge t_2}{\partial t_2 \partial t_1}\frac{d}{dt}\,\mathbb{E}N^2(t)\Big|_{t=t_1\wedge t_2}\\ &\quad + \frac{\partial t_1 \wedge t_2}{\partial t_1}\frac{\partial t_1 \wedge t_2}{\partial t_2}\frac{d^2}{dt^2}\,\mathbb{E}N^2(t)\Big|_{t=t_1\wedge t_2}\\ &= \delta(t_2 - t_1)\frac{d}{dt}\,\mathbb{E}N^2(t)\Big|_{t=t_1\wedge t_2}\end{aligned} \tag{7.60}$$

and

$$\begin{aligned}&\frac{\partial^2}{\partial t_1 \partial t_2}\mathbb{E}N(t_1 \wedge t_2)\mathbb{E}N(|t_2 - t_1|)\\ &= \frac{\partial}{\partial t_1}\frac{\partial t_1 \wedge t_2}{\partial t_2}\mathbb{E}\dot{N}(t_1 \wedge t_2)\mathbb{E}N(|t_2 - t_1|)\\ &\quad + \frac{\partial}{\partial t_1}\frac{\partial |t_1 - t_2|}{\partial t_2}\mathbb{E}N(t_1 \wedge t_2)\mathbb{E}\dot{N}(|t_2 - t_1|)\\ &= \delta(t_2 - t_1)\mathbb{E}\dot{N}(t_1 \wedge t_2)\mathbb{E}N(|t_2 - t_1|) + \mathbb{E}\dot{N}(t_1 \wedge t_2)\mathbb{E}\dot{N}(|t_2 - t_1|)\\ &\quad - 2\delta(t_2 - t_1)\mathbb{E}N(t_1 \wedge t_2)\mathbb{E}\dot{N}(|t_2 - t_1|)\\ &= \frac{\delta(t_2 - t_1)}{(\mathbb{E}\tau)^2}|t_2 - t_1| + \frac{1}{(\mathbb{E}\tau)^2} - \frac{2\delta(t_2 - t_1)}{(\mathbb{E}\tau)^2}t_1 \wedge t_2\\ &= \frac{1}{(\mathbb{E}\tau)^2} - \frac{2\delta(t_2 - t_1)}{(\mathbb{E}\tau)^2}t_1 \wedge t_2.\end{aligned} \tag{7.61}$$

Using eq. (7.60) and (7.61) in eq. (7.57), we obtain

$$\begin{aligned}&\left\langle \left(I(t_1) - \frac{q}{\mathbb{E}\tau}\right)\left(I(t_2) - \frac{q}{\mathbb{E}\tau}\right)\right\rangle\\ &= \delta(t_2 - t_1)\left[\frac{d}{dt}\,\mathbb{E}N^2(t)\Big|_{t=t_1\wedge t_2} - \frac{2}{(\mathbb{E}\tau)^2}t_1 \wedge t_2\right].\end{aligned}$$

The large t asymptotics of $\mathbb{E}N^2(t)$ is found from the small Laplace variable s asymptotics in eq. (7.53). Using Taylor's expansion (7.41), we obtain for small s

$$\mathcal{L}\mathbb{E}N^2(t)(s) = \frac{1+\hat{f}_\tau(s)}{s^2\mathbb{E}\tau\left[1-\hat{f}_\tau(s)\right]} = \frac{2-s\mathbb{E}\tau+O(s^2)}{s\mathbb{E}\tau - s^2\dfrac{\mathbb{E}\tau^2}{2}+O\left(s^3\right)}\frac{1}{s^2\mathbb{E}\tau}$$
$$= \frac{2}{s^3\left(\mathbb{E}\tau\right)^2} + \frac{\mathbb{E}\tau^2-\left(\mathbb{E}\tau\right)^2}{s^2\left(\mathbb{E}\tau\right)^3} + O\left(\frac{1}{s}\right),$$

hence, inverting the Laplace transform, we obtain the long time asymptotic expansion

$$\mathbb{E}N^2(t) = \frac{t^2}{\left(\mathbb{E}\tau\right)^2} + \frac{\mathbb{E}\tau^2-\left(\mathbb{E}\tau\right)^2}{\left(\mathbb{E}\tau\right)^3}t + \cdots$$
$$\frac{d}{dt}\mathbb{E}N^2(t) = \frac{2t}{\left(\mathbb{E}\tau\right)^2} + \frac{\mathbb{E}\tau^2-\left(\mathbb{E}\tau\right)^2}{\left(\mathbb{E}\tau\right)^3} + \cdots.$$

It follows that for large $t_1 \wedge t_2$

$$\left\langle \left(I(t_1) - \frac{q}{\mathbb{E}\tau}\right)\left(I(t_2) - \frac{q}{\mathbb{E}\tau}\right)\right\rangle$$
$$= \delta(t_2-t_1)\left[\frac{d}{dt}\,\mathbb{E}N^2(t)\Big|_{t=t_1\wedge t_2} - \frac{2}{\left(\mathbb{E}\tau\right)^2}t_1\wedge t_2\right]$$
$$= \delta(t_2-t_1)\frac{\mathbb{E}\tau^2-\left(\mathbb{E}\tau\right)^2}{\left(\mathbb{E}\tau\right)^3} + o(1).$$

Thus the autocorrelation of the current fluctuations is given by

$$R_I(t) = \delta(t)q^2\frac{\mathbb{E}\tau^2-\left(\mathbb{E}\tau\right)^2}{\left(\mathbb{E}\tau\right)^3} = q\left\langle I\right\rangle\frac{\mathbb{E}\tau^2-\left(\mathbb{E}\tau\right)^2}{\left(\mathbb{E}\tau\right)^2}\delta(t), \tag{7.62}$$

where (7.54) was used. Equation (7.62) is sometime referred to as *Campbell's theorem.* If the arrivals are exponentially distributed, we obtain from eq. (7.62) Schottky's formula [232] (note the difference in the definition of the Fourier transform)

$$R_I(t) = q\left\langle I\right\rangle\delta(t). \tag{7.63}$$

The power spectral density function of the current fluctuations is given by

$$S_I(\omega) = q\left\langle I\right\rangle\frac{\mathbb{E}\tau^2-\left(\mathbb{E}\tau\right)^2}{\left(\mathbb{E}\tau\right)^2}. \tag{7.64}$$

For the exponential case, we have

$$S_I(\omega) = q\left\langle I\right\rangle. \tag{7.65}$$

7.3 Diffusion approximations of Markovian jump processes

The probability law of a Markov process can be often approximated by that of a diffusion process, which leads to a considerable mathematical simplification of the analysis of the process. The analytical machinery of diffusion processes, as described in the previous chapters, becomes available for the analysis of the approximated Markov process.

Modeling random phenomena as dynamical systems driven by white noise often leads to an oversimplification of the source of randomness. Thus, for instance, in modeling dynamics driven by state-dependent noise the resulting stochastic differential equations can be interpreted in the Itô, Stratonovich, or any other sense. It may not be obvious which interpretation is appropriate for a given application. Choosing the correct interpretation for the mathematical model may be crucial for the correct description of the system under consideration. For example, a model that uses the Itô interpretation can lead to a stable system whereas interpreting it in the Stratonovich sense may lead to instability of the model.

One way to choose the correct interpretation is to start with a model that involves the description of the source of randomness on a more fundamental level than using white noise. For example, white noise may be considered a limit, in some sense, of a jump process with frequent jumps, or of a correlated process with short correlation time. Thus diffusion approximations of various stochastic processes often involve separation of time and spatial scales and coarse-graining of the fast relative to the slow scale. The approximations may consist in the approximation of the random trajectories of the process by those of a diffusion process or in the approximation of the pdf of a process by that of a diffusion process. The latter often involves the expansion of the pdf in an asymptotic series and the coarse-grained equations are obtained as solvability conditions in the construction of the asymptotic series.

7.3.1 A refresher on solvability of linear equations

A basic fact from the theory of linear algebraic equations is that if the homogeneous vector equation

$$\boldsymbol{A}\boldsymbol{x} = \boldsymbol{0} \tag{7.66}$$

has a nontrivial solution; that is, if 0 is an eigenvalue of $\boldsymbol{A}$, then the inhomogeneous equation

$$\boldsymbol{A}\boldsymbol{x} = \boldsymbol{b} \tag{7.67}$$

has a solution if and only if $\boldsymbol{b}$ is orthogonal to all the eigenvectors of the adjoint matrix $\boldsymbol{A}^*$ corresponding to the eigenvalue 0. In a real space (with a real inner product $\langle\cdot,\cdot\rangle$) the matrix representation of the adjoint operator $\boldsymbol{A}^*$ is $\boldsymbol{A}^T$, in a complex space it is $\bar{\boldsymbol{A}}^T$. Thus, if (7.66) has a non-trivial solution, then (7.67) has a solution if and only if

$$\langle \boldsymbol{y}, \boldsymbol{b} \rangle = 0 \tag{7.68}$$

for all solutions $\boldsymbol{y}$ of the equation

$$\boldsymbol{A}^* \boldsymbol{y} = \boldsymbol{0}. \tag{7.69}$$

The generalization of the solvability condition to bounded linear operators between a Banach space $\boldsymbol{X}$ (e.g., $L^1(D)$) and its dual $\boldsymbol{X}^*$ (e.g., $L^\infty(D)$, where D is a bounded or unbounded domain in $\mathbb{R}^d$) is that if 0 is an eigenvalue of a bounded operator $\boldsymbol{A} : \boldsymbol{X} \mapsto \boldsymbol{X}^*$ (i.e., (7.66) has a nontrivial solution in $\boldsymbol{X}$), then a necessary condition for the solvability of (7.67) is that (7.68) holds for all solutions of the adjoint equation $\boldsymbol{A}^*\boldsymbol{y} = 0$ ($\boldsymbol{y} \in \boldsymbol{X}^*$). The "inner product" $\langle \cdot, \cdot \rangle$ in this case is the duality $\langle \boldsymbol{x}, \boldsymbol{y} \rangle = \boldsymbol{y}(\boldsymbol{x})$ [79].

The operator $\boldsymbol{A}$ in the context of stochastic processes is the inverse of the forward Kolmogorov operator, whose L^1 (or l^1) eigenfunction of the eigenvalue 0 is a probability density; that is, a nonnegative integrable function (nonnegative sequence), whose integral (sum) is 1. The dual space to L^1 is the space of essentially bounded functions $(L^1)^* = L^\infty$. The adjoint operator $\boldsymbol{A}^*$, with respect to the duality

$$\langle f, g \rangle = g^*(f) = \int f(\boldsymbol{x})\bar{g}(\boldsymbol{x})\, d\boldsymbol{x} \quad \text{for } f \in L^1,\ g \in L^\infty,$$

is the backward Kolmogorov operator. The eigenfunction of $\boldsymbol{A}^*$ in L^∞, corresponding to the eigenvalue 0 is a constant (see discussion below Definition 7.1.5). Thus a necessary condition for the inhomogeneous equation (7.67) to have a solution in L^1 is that the integral of $\boldsymbol{b}$ vanishes (i.e., $\boldsymbol{b}$ is orthogonal to 1). The solvability condition can be generalized to other linear spaces in a straightforward manner.

7.3.2 Dynamics with large and fast jumps

Consider a one-parameter family $\xi(x,t)_{x \in \boldsymbol{R}}$ of stationary zero mean Markov jump processes. Assume that the master equation for the pdf of $\xi(x,t)$ is

$$\begin{aligned} p_t(\xi, t \,|\, x) &= -\Lambda(\xi, x)p(\xi, t \,|\, x) + \int \Lambda(\xi - \eta, x)p(\eta, t \,|\, x)w(\xi - \eta \,|\, \eta, x)\, d\eta \\ &= L_\xi p_t(\xi, t \,|\, x). \end{aligned} \tag{7.70}$$

The stationary pdf of the process, $\Psi_0(\xi \,|\, x)$, is a nonnegative eigenfunction of the operator L_ξ with eigenvalue 0; that is, $L_\xi \Psi_0(\xi \,|\, x) = 0$, normalized by

$$\int \Psi_0(\xi \,|\, x)\, d\xi = 1. \tag{7.71}$$

We assume that this is the unique integrable eigenfunction of the operator; that is, we assume that it is the unique (up to a constant factor) eigenfunction in the function space $L^1(\mathbb{R})$. We also assume that the only bounded eigenfunction of the adjoint operator, L_ξ^*, is constant. In addition, we assume that the integrable eigenfunctions of L_ξ, denoted $\{\Psi_n(\xi\,|\,x)\}$, form a complete set and together with the eigenfunctions of the operator L_ξ^*, denoted $\{\Phi_n(\xi\,|\,x)\}$, form a bi-orthogonal set. That is, $\int \Psi_m(\xi\,|\,x)\bar{\Phi}_n(\xi\,|\,x)\,d\xi = \delta_{m,n}$. We denote the eigenvalues of L_ξ by $\lambda_n(x)$. These assumptions mean that all integrable functions $f(\xi)$ can be expanded in $L^1(\mathbb{R})$ in a series $f(\xi) = \sum_n a_n(x)\Psi_n(\xi\,|\,x)$ with

$$a_n(x) = \frac{\displaystyle\int f(\xi)\bar{\Phi}_n(\xi\,|\,x)\,d\xi}{\displaystyle\int \Psi_n(\xi\,|\,x)\bar{\Phi}_n(\xi\,|\,x)\,d\xi}.$$

Note that not all eigenvalues of L_ξ are necessarily real valued, because the operator is in general not self-adjoint.

The zero mean assumption means that

$$\int \xi\Psi_0(\xi\,|\,x)\,d\xi = 0. \tag{7.72}$$

Now, we consider a dynamical system driven by the noise process $\xi(x,t)$ in the form $\dot{x} = \alpha(x) + \xi(x,t)$. The jumps are speeded up by scaling the jump rate $\Lambda(\xi,x)$ with a small parameter $0 < \varepsilon < \varepsilon_0$ in the form $\Lambda(\xi,x) = \lambda(\xi,x)/\varepsilon$. The resulting jump process is denoted $\xi_\varepsilon(x,t)$. Its pdf satisfies the master equation $p_{\varepsilon,t}(\xi,t\,|\,x) = \varepsilon^{-1}L_\xi p_\varepsilon(\xi,t\,|\,x)$. Next, the jump size is scaled with $\varepsilon^{1/2}$ so that the scaled dynamics takes the form

$$\dot{x}_\varepsilon = \alpha(x_\varepsilon) + \varepsilon^{-1/2}\xi_\varepsilon(x_\varepsilon,t). \tag{7.73}$$

The pair $(x_\varepsilon(t), \xi_\varepsilon(x_\varepsilon(t),t))$ is a two-dimensional drift-jump Markov process, because the initial value problem for eq. (7.73) has a unique solution. It follows that the joint pdf of the pair satisfies the master equation

$$\frac{\partial p_\varepsilon(x,\xi,t)}{\partial t} = -\frac{\partial}{\partial x}\left[\alpha(x) + \frac{\xi}{\sqrt{\varepsilon}}\right]p_\varepsilon(x,\xi,t) + \frac{1}{\varepsilon}L_\xi p_\varepsilon(x,\xi,t). \tag{7.74}$$

To find the limit of $p_\varepsilon(x,\xi,t)$ as $\varepsilon \to 0$, we construct an asymptotic expansion of $p_\varepsilon(x,\xi,t)$ in the form

$$p_\varepsilon(x,\xi,t) = p^0(x,\xi,t) + \sqrt{\varepsilon}p^1(x,\xi,t) + \varepsilon p^2(x,\xi,t) + \cdots. \tag{7.75}$$

Inserting the expansion (7.75) into the master equation (7.74) and comparing the coefficients of like powers of ε on both sides, we obtain at the leading order ε^{-1} the equation $L_\xi p^0(x,\xi,t) = 0$. That is, $p^0(x,\xi,t)$ is proportional to the eigenfunction $\Psi_0(\xi\,|\,x)$ of the operator L_ξ with proportionality constant (with respect to ξ) that may depend on (x,t), which we denote $P^0(x,t)$. Hence $p^0(x,\xi,t) =$

$P^0(x,t)\Psi_0(\xi\,|\,x)$, where $P^0(x,t)$ is yet an undetermined function. At order $\varepsilon^{-1/2}$, we obtain the equation

$$L_\xi p^1(x,\xi,t) = -\xi\frac{\partial p^0(x,\xi,t)}{\partial x} = -\frac{\partial\left[\xi\Psi_0(\xi\,|\,x)P^0(x,t)\right]}{\partial x}. \tag{7.76}$$

The inhomogeneous equation (7.76) has a solution only if the right-hand side satisfies a solvability condition, because 0 is an eigenvalue of the operator L_ξ. The solvability condition, which is obtained by integration of both sides of the equation against an eigenfunction, is that the right-hand side be orthogonal to all eigenfunctions of the adjoint operator L_ξ^* corresponding to the same eigenvalues that belong to the function space $L^\infty(\mathbb{R})$ of essentially bounded functions, the dual space to $L^1(\mathbb{R})$. As mentioned above, this function is 1, so the solvability condition is simply found by integrating the equation with respect to ξ. The condition (7.72) ensures that the solvability condition is satisfied. The solution of the homogeneous equation, if it exists, is not unique and is defined up to an additive multiple of the eigenfunction $\Psi_0(\xi\,|\,x)$.

In the case at hand we show that the solution exists by constructing it. To construct the solution of eq. (7.76), we expand

$$\begin{aligned}
\xi\Psi_0(\xi\,|\,x) &= \sum_{n=0}^{\infty} a_n(x)\Psi_n(\xi\,|\,x) \qquad (7.77)\\
\xi\frac{\partial\Psi_0(\xi\,|\,x)}{\partial x} &= \sum_{n=0}^{\infty} b_n(x)\Psi_n(\xi\,|\,x)\\
\frac{\partial\Psi_0(\xi\,|\,x)}{\partial x} &= \sum_{n=0}^{\infty} c_n(x)\Psi_n(\xi\,|\,x)\\
p^1(x,\xi,t) &= \sum_{n=1}^{\infty} d_n(x,t)\Psi_n(\xi\,|\,x) + P^1(x,t)\Psi_0(\xi\,|\,x),
\end{aligned}$$

where $P^1(x,t)$ is yet an undetermined function,

$$a_n(x) = \frac{\displaystyle\int \xi\Psi_0(\xi\,|\,x)\bar{\Phi}_n(\xi\,|\,x)\,d\xi}{\displaystyle\int \Psi_n(\xi\,|\,x)\bar{\Phi}_n(\xi\,|\,x)\,d\xi},\quad b_n(x) = \frac{\displaystyle\int \xi\frac{\partial\Psi_0(\xi\,|\,x)}{\partial x}\bar{\Phi}_n(\xi\,|\,x)\,d\xi}{\displaystyle\int \Psi_n(\xi\,|\,x)\bar{\Phi}_n(\xi\,|\,x)\,d\xi}$$

$$c_n(x) = \frac{\displaystyle\int \frac{\partial\Psi_0(\xi\,|\,x)}{\partial x}\bar{\Phi}_n(\xi\,|\,x)\,d\xi}{\displaystyle\int \Psi_n(\xi\,|\,x)\bar{\Phi}_n(\xi\,|\,x)\,d\xi},$$

and $d_n(x,t)$ are yet undetermined coefficients. Eqation (7.76) gives

$$\lambda_n(x)d_n(x,t) = a_n(x)P_x^0(x,t) + b_n(x)P^0(x,t),$$

because

$$L_\xi p^1(x,\xi,t) = \sum_n \lambda_n(x) d_n(x,t) \Psi_n(\xi \,|\, x);$$

that is,

$$d_n(x,t) = \frac{a_n(x)}{\lambda_n(x)} \frac{\partial P^0(x,t)}{\partial x} + \frac{b_n(x)}{\lambda_n(x)} P^0(x,t).$$

It follows that

$$p^1(x,\xi,t) = A(\xi,x)\frac{\partial P^0(x,t)}{\partial x} + B(\xi,x)P^0(x,t) + P^1(x,t)\Psi_0(\xi \,|\, x),$$

where

$$A(\xi,x) = \sum_{n=1}^{\infty} \frac{a_n(x)}{\lambda_n(x)} \Psi_n(\xi \,|\, x), \quad B(\xi,x) = \sum_{n=1}^{\infty} \frac{b_n(x)}{\lambda_n(x)} \Psi_n(\xi \,|\, x).$$

Next, we compare terms that are $O(1)$. We obtain the equation

$$\begin{aligned}
&L_\xi p^2(x,\xi,t)\\
&= \frac{\partial p^0(x,\xi,t)}{\partial t} + \frac{\partial \alpha(x) p^0(x,\xi,t)}{\partial x} + \xi \frac{\partial p^1(x,\xi,t)}{\partial x}\\
&= \left[\frac{\partial P^0(x,t)}{\partial t} + \frac{\partial \alpha(x) P^0(x,t)}{\partial x}\right] \Psi_0(\xi \,|\, x) + C(\xi,x)\alpha(x)P^0(x,t)\\
&\quad - \xi \frac{\partial}{\partial x}\left[A(\xi,x)\frac{\partial P^0(x,t)}{\partial x} + B(\xi,x)P^0(x,t) + P^1(x,t)\Psi_0(\xi \,|\, x)\right],
\end{aligned}$$

where $C(\xi,x) = \sum_{n=1}^{\infty} c_n(x)\Psi_n(\xi \,|\, x)$. The solvability condition is obtained by integrating the right-hand side with respect to ξ. We have $\int C(\xi,x)\,d\xi = 0$, because $\Phi_0(\xi \,|\, x) = 1$ and the eigenfunctions form a bi-orthogonal system, $\int \Psi_n(\xi \,|\, x)\,d\xi = 0$ for $n > 0$. Conditions (7.71) and (7.72) give

$$\frac{\partial P^0(x,t)}{\partial t} = -\frac{\partial\left[\alpha(x) - B(x)\right] P^0(x,t)}{\partial x} + \frac{\partial}{\partial x} A(x) \frac{\partial P^0(x,t)}{\partial x}, \tag{7.78}$$

where $A(x) = \int \xi A(\xi,x)\,d\xi$ and $B(x) = \int \xi B(\xi,x)\,d\xi$.

If the dynamics (7.73) has the form

$$\dot{x}_\varepsilon = \alpha(x_\varepsilon) + \frac{\xi_\varepsilon(t)\beta(x_\varepsilon)}{\sqrt{\varepsilon}}$$

and the eigenfunctions of L_ξ are independent of x, then $A(\xi,x)$ and $B(\xi,x)$ are independent of x, so that $A(x)$ is constant, say D, and eq. (7.78) becomes

$$\frac{\partial P^0(x,t)}{\partial t} = -\frac{\partial \alpha(x) P^0(x,t)}{\partial x} + D\frac{\partial}{\partial x}\left\{\beta(x)\frac{\partial\left[\beta(x)P^0(x,t)\right]}{\partial x}\right\}. \tag{7.79}$$

This is the Fokker–Planck–Stratonovich equation for the pdf of the solution of the Stratonovich stochastic differential equation

$$d_S x = \alpha(x)\,dt + \sqrt{D}\beta(x)\,d_S w.$$

Example 7.9 (Goldstein's model of diffusion). Goldstein's model of diffusion in one dimension assumes that identical balls move on the line with velocities $\pm V$ and collide elastically at exponential waiting times with rate Λ. Upon collision they exchange velocities. As the velocities and the collision rate are increased the displacement of a ball satisfies the equation $\dot{x}_\varepsilon = \xi_\varepsilon(t)/\sqrt{\varepsilon}$, where $\xi_\varepsilon(t)$ is the speeded up random telegraph process. The operator L_ξ is given by $L_\xi p(\xi) = -\Lambda p(\xi) + \Lambda p(-\xi)$ and is self-adjoint, because $\xi_1(t)$ takes the values $\pm V$. The eigenvalues and eigenfunctions are

$$\begin{aligned}\mu_0 &= 0, \quad \Psi_0(\xi) = \frac{1}{2}\delta(\xi - V) + \frac{1}{2}\delta(\xi + V), \quad \Phi_0(\xi) = 1\\ \mu_1 &= 2\Lambda, \quad \Psi_1(\xi) = \delta(\xi - V) - \delta(\xi + V).\end{aligned}$$

Note that the density is not an integrable function, but rather a distribution. Obviously,

$$\int \Psi_0(\xi)\, d\xi = 1, \quad \int \xi\Psi_0(\xi)\, d\xi = 0, \quad \int \Psi_1(\xi)\Phi_0(\xi)\, d\xi = 0.$$

The expansion (7.77) is given by

$$\xi\Psi_0(\xi) = \frac{1}{2}\left[\xi\delta(\xi - V) + \xi\delta(\xi + V)\right] = \frac{1}{2}V\Psi_1(\xi);$$

that is, $a_1 = \frac{1}{2}V$. Now, we scale $\lambda = \Lambda/\varepsilon$ and $v = V/\sqrt{\varepsilon}$ to convert $\xi_1(t)$ to the speeded up process $\xi_\varepsilon(t)$.

Equation (7.74) for the joint pdf of x_ε and ξ_ε is given by

$$\frac{\partial p_\varepsilon(x, \xi, t)}{\partial t} = -\frac{\xi}{\sqrt{\varepsilon}}\frac{\partial p_\varepsilon(x, \xi, t)}{\partial x} - \frac{\lambda}{\varepsilon}\left[p_\varepsilon(x, \xi, t) - p_\varepsilon(x, -\xi, t)\right].$$

The expansion (7.75) leads to the leading term $p^0(x, \xi, t) = P^0(x, t)\Psi_0(\xi)$ and the Fokker–Planck–Stratonovich equation (7.79) is given by $P_t^0(x, t) = DP_{xx}^0(x, t)$ with the diffusion coefficient

$$D = \int \frac{a_1\xi\Psi_1(\xi)}{\mu_1}\, d\xi = \frac{V}{4\lambda}\int \xi\left[\delta(\xi - V) - \delta(\xi + V)\right]\, d\xi = \frac{V^2}{2\lambda}.$$

If there is additional drift, then

$$\dot{x}_\varepsilon = \alpha(x) + \frac{\xi_\varepsilon(t)}{\sqrt{\varepsilon}},$$

and the diffusion equation becomes

$$P_t^0(x, t) = -[\alpha(x)P^0(x, t)]_x + DP_{xx}^0(x, t),$$

where $D = V^2/2\lambda$. □

7.3.3 Small jumps and the Kramers–Moyal expansion

Consider the random walk on $\mathbb{R}$

$$x_{n+1} = x_n + \varepsilon\xi_n, \tag{7.80}$$

where $\Pr\{\xi_n = \xi \,|\, x_n = x,\, x_{n-1} = y,\, \ldots\} = w(\xi \,|\, x, \varepsilon)$ and x_0 is a random variable with a given pdf $p_0(x)$. The conditional moments of the jump process are denoted

$$m_n(x,\varepsilon) = \mathbb{E}\left[\xi_n \,|\, x_n = x\right] = \int_{\mathbb{R}} \xi^n w(\xi \,|\, x, \varepsilon)\, d\xi.$$

The master equation for the pdf $p_\varepsilon(x,n) = \Pr\{x_n = x\}$ of x_n is

$$\begin{aligned} &p_\varepsilon(x,n+1) - p_\varepsilon(x,n) \\ &= \int_{\mathbb{R}} \left[p_\varepsilon(x-\varepsilon\xi,n)w(\xi \,|\, x-\varepsilon\xi,\varepsilon) - p_\varepsilon(x,n)w(\xi \,|\, x,\varepsilon)\right] d\xi = L_\varepsilon p_\varepsilon(x,n) \end{aligned} \tag{7.81}$$

with a given initial condition

$$p_\varepsilon(x,0) = \phi_0(x). \tag{7.82}$$

Setting $n = t/\varepsilon^2$, $\Delta t = \varepsilon^2$, and $p_\varepsilon(x,t) = p_\varepsilon(x,n)$, we write the master equation (7.81) in the form

$$p_\varepsilon(x,t+\Delta t) = \int_{\mathbb{R}} p_\varepsilon(x-\varepsilon\xi,t)w(\xi \,|\, x-\varepsilon\xi,\varepsilon)\, d\xi. \tag{7.83}$$

We assume that the master equation (7.81) has a unique integrable solution that satisfies the given initial condition (7.82) and an a priori bound of the form

$$||p||_1 \leq C||f||_2 \tag{7.84}$$

on solutions of the inhomogeneous equation $p_\varepsilon(x,n+1) - L_\varepsilon p_\varepsilon(x,n) = f(x)$ with the homogeneous initial condition $\phi_0(x) = 0$, where C is a constant independent of ε, and $||\cdot||_1$, $||\cdot||_2$ are some norms. We assume furthermore that the conditional pdf of the jump size and its conditional moments have the asymptotic power series expansions for $\varepsilon \to 0$,

$$w(\xi \,|\, x,\varepsilon) \sim \sum_{i=0}^{\infty} \varepsilon^i w_i(\xi \,|\, x), \tag{7.85}$$

$$m_n(x,\varepsilon) \sim \sum_{i=0}^{\infty} \varepsilon^i m_{n,i}(x) \quad \text{for all} \quad n > 0, \tag{7.86}$$

where $w_i(\xi \,|\, x)$, $m_{n,i}(x)$ are regular functions.

The *Kramers–Moyal expansion* consists in expanding all functions in powers of the small parameter ε. We expand

$$p_\varepsilon(x,t) \sim \sum_0^\infty p_i(x,t), \tag{7.87}$$

and using (7.85) in (7.83), we write

$$p_\varepsilon(x,t+\Delta t) = \int_{\mathbb{R}} p_\varepsilon(x-\varepsilon\xi,t)\sum_{i=0}^\infty \varepsilon^i w_i(\xi\,|\,x-\varepsilon\xi)\,d\xi$$

and obtain

$$\begin{aligned}
&p_0(x,t) + \Delta t\frac{\partial p_0(x,t)}{\partial t} + \cdots\\
&= \int_{\mathbb{R}} p_0(x,t)w_0(\xi\,|\,x)\,d\xi + \sum_{n=1}^\infty \frac{(-\varepsilon)^n}{n!}\frac{\partial^n}{\partial x^n}\left[p_0(x,t)\int_{\mathbb{R}}\xi^n\sum_{i=0}^\infty \varepsilon^i w_i(\xi\,|\,x)\,d\xi\right].
\end{aligned}$$

Because normalization requires that $\int_{\mathbb{R}} p_0(x,t)w_0(\xi\,|\,x)\,d\xi = p_0(x,t)$, collecting terms up to order ε^2 gives

$$\begin{aligned}
&\Delta t\frac{\partial p_0(x,t)}{\partial t} + \cdots \\
&= -\varepsilon\frac{\partial[m_{1,0}(x)+\varepsilon m_{1,1}(x)][p_0(x,t)+\varepsilon p_1(x,t)]}{\partial x} + \frac{\varepsilon^2}{2}\frac{\partial^2 m_{2,0}(x)p_0(x,t)}{\partial x^2} + \cdots
\end{aligned} \tag{7.88}$$

Dividing by ε^2, we obtain

$$\begin{aligned}
&\frac{\partial p_0(x,t)}{\partial t}\\
&= -\frac{1}{\varepsilon}\frac{\partial[m_{1,0}(x)+\varepsilon m_{1,1}(x)][p_0(x,t)+\varepsilon p_1(x,t)]}{\partial x} + \frac{1}{2}\frac{\partial^2 m_2(x,\varepsilon)p_0(x,t)}{\partial x^2} + \cdots
\end{aligned}$$

Theorem 7.3.1 (Kramers–Moyal expansion). *Under the assumptions* (7.84)–(7.86), *if*

$$m_{1,0}(x,\varepsilon) = 0, \tag{7.89}$$

then for every $T > 0$, the pdf $p_\varepsilon(x,n)$ converges to the solution $p_0(x,t)$ of the initial value problem

$$\frac{\partial p_0(x,t)}{\partial t} = -\frac{\partial m_{1,1}(x)p_0(x,t)}{\partial x} + \frac{1}{2}\frac{\partial^2 m_{2,0}(x)p_0(x,t)}{\partial x^2} \quad \text{for } x > 0, \tag{7.90}$$

$$\lim_{t\to 0} p_0(x,t) = \phi_0(x), \tag{7.91}$$

uniformly for all $x > 0,\ 0 < t < T$.

Proof. Under the given regularity assumptions, $p_0(x,t)$ is a regular function, therefore the higher-order terms $p_i(x,t)$ in the expansion (7.87) are also regular functions. Setting

$$P^{(m)}(x,t) = \sum_{j=0}^{m} \varepsilon^j p_j(x,t),$$

there is a constant K (that depends on $\phi_0(x)$) such that for all $m \geq 0$

$$||p_\varepsilon(x,t) - P^{(m)}(x,t)||_1 \leq K\varepsilon^{m+1}. \tag{7.92}$$

Indeed, from the construction of the expansion terms $p_j(x,t)$ it follows that the difference $Q_n^m(x) = p_\varepsilon(x,t) - P^{(m)}(x,t)$ satisfies an equation of the form $Q_{n+1}^m(x) - L_\varepsilon Q_n^m(x) = O(\varepsilon^{m+1})$, and $Q_0^m(x) = 0$ for $m > 0$. Now, the *a priori* estimate (7.84) ensures that (7.92) holds. □

Note that (7.90) is the Fokker–Planck equation for the pdf of the solution of the Itô equation

$$dx = m_{1,1}(x)\,dt + \sqrt{m_{2,0}(x)}\,dw. \tag{7.93}$$

Exercise 7.7 (Convergence of trajectories). Prove that on any finite interval $[0,T]$, independent of ε, the piecewise constant trajectories $x_\varepsilon(t) = x_{[t/\varepsilon^2]}$ of the Markov process (7.80) converge in probability to the trajectories $x(t)$ of (7.93) as $\varepsilon \to 0$ (see [144]). □

Example 7.10 (A drift-jump process with small rapid jumps). The jumps of the continuous time Markov drift and jump process on $\mathbb{R}$, described by (7.24)–(7.25), are refined and sped up by introducing the temporal and spatial scaling with a small parameter ε,

$$x_\varepsilon(t+\Delta t) \tag{7.94}$$
$$= \begin{cases} x_\varepsilon(t) + A(x_\varepsilon(t))\Delta t + o(\Delta t) & \text{w.p. } 1 - \dfrac{\lambda(x_\varepsilon(t))}{\varepsilon^2}\Delta t + o(\Delta t) \\ x_\varepsilon(t) + A(x_\varepsilon(t))\Delta t + \varepsilon\xi(t) + o(\Delta t) & \text{w.p. } \dfrac{\lambda(x_\varepsilon(t))}{\varepsilon^2}\Delta t + o(\Delta t), \end{cases}$$

where

$$\Pr\{\xi(t) = \xi \,|\, x_\varepsilon(t) = x,\, x_\varepsilon(t_1) = x_1, \ldots\} = w(\xi \,|\, x)$$

for $t > t_1 > \cdots$. The master equation for the pdf of x is given by

$$\frac{\partial p_\varepsilon(x,t)}{\partial t} = -\frac{\partial A(x)p_\varepsilon(x,t)}{\partial x} + \frac{1}{\varepsilon^2}\left[\int_{\mathbb{R}} \lambda(x-\varepsilon\xi)p_\varepsilon(x-\varepsilon\xi,t)w(\xi \,|\, x-\varepsilon\xi)\,d\xi - \lambda(x)p_\varepsilon(x,t)\right].$$

The Kramers–Moyal expansion gives

$$\begin{aligned}\frac{\partial p_\varepsilon(x,t)}{\partial t} &= -\frac{\partial A(x)p_\varepsilon(x,t)}{\partial x} + \frac{\lambda(x)}{\varepsilon^2}\sum_{n=1}^{\infty}\frac{(-\varepsilon)^n}{n!}\frac{\partial^n m_n(x)p_\varepsilon(x,t)}{\partial x^n}\\ &= -\frac{\partial}{\partial x}\left[A(x)+\frac{\lambda(x)m_1(x)}{\varepsilon}\right]p_\varepsilon(x,t) + \frac{1}{2}\frac{\partial^2 m_2(x)p_\varepsilon(x,t)}{\partial x^2}\lambda(x)\\ &\quad + O(\varepsilon).\end{aligned}$$

If eq. (7.89) holds, we obtain the self-consistent diffusion approximation

$$\frac{\partial p_0(x,t)}{\partial t} = -\frac{\partial\left[A(x)+\lambda(x)m_{1,1}(x)\right]p_0(x,t)}{\partial x} + \frac{1}{2}\frac{\partial^2\lambda(x)m_{2,0}(x)p_0(x,t)}{\partial x^2}.$$

The diffusion limit of $x_\varepsilon(t)$ as $\varepsilon \to 0$ is the solution of the stochastic differential equation

$$\dot{x}(t) = A(x(t)) + \lambda(x(t))m_{1,1}(x(t)) + \sqrt{\lambda(x(t))m_{2,0}(x(t))}\,\dot{w}(t).$$

Note that the same results hold for time-dependent coefficients as well. □

The truncation of the Kramers–Moyal expansion (7.88) must be done with extreme care, because otherwise the resulting approximations may diverge from the solution of the master equation. The coarsest truncation gives the *fluid approximation*. If the scaling $n = t/\varepsilon^2$ is replaced with $n = t/\varepsilon$ and the expansion (7.88) is truncated after the leading-order term, the resulting equation is

$$\frac{\partial p_0(x,t)}{\partial t} = -\frac{\partial m_{1,0}(x)p_0(x,t)}{\partial x}.$$

This is the Liouville equation for the density of the trajectories of the deterministic ODE

$$\dot{x} = m_{1,0}(x), \quad x(0) = x_0 \tag{7.95}$$

with the initial density $p_0(x_0)$. The fluid approximation represents the averaged dynamics (7.80) in the sense that at each point the random term is replaced by its local mean. This approximation provides some information on the underlying deterministic motion of the system. For example, it can reveal stability properties of the dynamics, attractors, multistability, and so on.

The shortcomings of the fluid approximation are obvious. For example, if the dynamics (7.95) has an attractor with a finite domain of attraction D all trajectories of (7.95) that begin in D will stay there forever whereas the stochastic trajectories of (7.80) may leave D in finite time with probability 1.

A more refined diffusion approximation is obtained by replacing the scaling $n = t/\varepsilon^2$ with $n = t/\varepsilon$, as above, and truncating the expansion (7.88) after the *second*-order term. We obtain the Fokker–Planck equation

$$\frac{\partial p(x,t)}{\partial t} = -\frac{\partial m_{1,0}(x)p(x,t)}{\partial x} + \frac{\varepsilon}{2}\frac{\partial^2 m_{2,0}(x)p(x,t)}{\partial x^2}. \tag{7.96}$$

Its solution is the pdf of the diffusion process defined by the Itô equation

$$dx = m_{1,0}(x)\,dt + \sqrt{\varepsilon m_{2,0}(x)}\,dw. \tag{7.97}$$

Such an approximation assumes that the higher-order terms in the expansion (7.88) are smaller than the terms that have been retained in the Kramers–Moyal expansion; that is, that for small ε

$$\left|\frac{\varepsilon^{n+1}}{(n+1)!}\frac{\partial^{n+1} m_{n+1}(x)p\,(x,t)}{\partial x^{n+1}}\right| << \left|\frac{\varepsilon^n}{n!}\frac{\partial^n m_n(x)p\,(x,t)}{\partial x^n}\right|. \tag{7.98}$$

This, however, is inconsistent with the approximation (7.96) in general.

Example 7.11 (Random walk). Consider a random walk that jumps ε to the left and to the right at every point x with probabilities $l(x)$ and $r(x)$, respectively, where $l(x)+r(x)=1$. Assume that for $x>0$, we have $r(x)<l(x)$ (or $r(x)<\frac{1}{2}$) and for $x<0$, we have $l(x)<r(x)$ (or $r(x)>\frac{1}{2}$). The master equation for the pdf is $p\,(x,n+1)=p\,(x-\varepsilon,n)r(x-\varepsilon)+p\,(x+\varepsilon,n)l(x+\varepsilon)$. The conditional moments $m_n(x)$ are given by $m_n(x)=r(x)+(-1)^n l(x)$ so that $m_1(x)=r(x)-l(x)=-1+2r(x)$, $\quad m_2(x)=r(x)+l(x)=1$. This means that $xm_1(x)<0$; that is, at each point the drift points toward the origin. The fluid approximation (7.95) indicates that the origin is an attractor for the averaged dynamics.

The exact solution of the stationary master equation is given by [242]

$$p_\varepsilon(x)=p\,(n\varepsilon)=p_\varepsilon(0)\frac{l(0)}{r(n\varepsilon)}\prod_{j=1}^{n}\frac{r(j\varepsilon)}{l(j\varepsilon)}.$$

It follows that

$$\lim_{\varepsilon\to 0}\varepsilon\log p_\varepsilon(x)=\lim_{\varepsilon\to 0}\varepsilon\sum_{j=1}^{n}\log\frac{r(j\varepsilon)}{l(j\varepsilon)}=\int_0^x\log\frac{r(x)}{l(x)}\,dx$$

so that

$$\log p_\varepsilon(x)\sim\frac{1}{\varepsilon}\int_0^x\log\frac{r(x)}{l(x)}\,dx,$$

hence

$$\varepsilon^n\frac{\partial^n p_\varepsilon(x)}{\partial x^n}=O(1).$$

Thus the assumption (7.98) is not satisfied, because all terms in the Kramers–Moyal expansion are of the same order of magnitude.

The diffusion approximation (7.96) is

$$\frac{\partial p\,(x,t)}{\partial t}=-\frac{\partial\,[1-2r(x)]\,p\,(x,t)}{\partial x}+\frac{\varepsilon}{2}\frac{\partial^2 p\,(x,t)}{\partial x^2}.$$

The stationary solution is given by

$$p_\varepsilon(x) = C\exp\left\{-\frac{2}{\varepsilon}\int\limits_0^x [1-2r(x)]\,dx\right\} = C\exp\left\{-\frac{1}{\varepsilon}\psi(x)\right\},$$

where

$$\psi(x) = 2\int\limits_0^x [1-2r(x)]\,dx \neq -\int\limits_0^x \log\frac{r(x)}{1-r(x)}\,dx. \tag{7.99}$$

It follows that the approximation of the stochastic process x_n defined in (7.80) by the solution of the Itô equation, $x(\varepsilon n) = x(t)$, may lead to large errors in certain functionals of the process. The exact approximation of the pdf is the object of large deviations theory (see Chapter 9). □

Exercise 7.8 (Validity of the fluid approximation). Show that the leading term in Taylor's expansion of both sides of (7.99) for small x is the same. Conclude that the diffusion approximation can be valid for the description of only small fluctuations about the trajectories of the fluid approximation. □

7.3.4 An application to Brownian motion in a field of force

The following model of a Brownian particle in a field of force has been proposed in [122]. Consider a heavy particle of mass M moving in a one-dimensional potential $U(x)$ in a medium of light particles of mass m (per particle). Assume that the system is in thermal equilibrium at temperature T; that is, the velocity pdf of the small particles is Maxwellian,

$$f(v) = \sqrt{\frac{m}{2\pi kT}}\exp\left\{-\frac{mv^2}{2kT}\right\}.$$

The light particles collide elastically with the heavy particle at exponential waiting times with rate α. We denote the displacement of the heavy particle by x and its (random) velocity by V. Between collisions the heavy particle moves according to Newton's second law of motion; that is, $M\ddot{x} = -U'(x)$. The random dynamics of the heavy particle can be described by the system

$$x(t+\Delta t) = x(t) + V\Delta t + o(\Delta t) \tag{7.100}$$

$$V(t+\Delta t) = \begin{cases} V(t) - \dfrac{U'(x)}{M}\Delta t & \text{w.p. } 1-\alpha\Delta t + o(\Delta t) \\[2ex] V(t) - \dfrac{U'(x)}{M}\Delta t + \Delta V & \text{w.p. } \alpha\Delta t + o(\Delta t), \end{cases} \tag{7.101}$$

where ΔV is the change in the velocity of the heavy particle after a collision with a light particle. It is obtained from the laws of conservation of energy and momentum in an elastic collision,

$$\frac{MV^2}{2}+\frac{mv^2}{2}=\frac{M(V+\Delta V)^2}{2}+\frac{m(v+\Delta v)^2}{2}$$
$$MV+mv=M(V+\Delta V)+m(v+\Delta v),$$

as $\Delta V=2m(v-V)/(M+m)$. The master equation for the joint pdf of x and V is

$$\frac{\partial p\,(x,V,t)}{\partial t}=-\frac{\partial Vp\,(x,V,t)}{\partial x}+\frac{\partial}{\partial V}\frac{U'(x)}{M}p\,(x,V,t)$$
$$+\int \alpha p\,(x,V-z,t)w(z\,|\,V-z)\,dz-\alpha p\,(x,V,t),$$

where

$$\begin{aligned}w(z\,|\,V)\,dz &=\Pr\left\{\Delta V\in z+dz\,|\,V(t)=V\right\}\\&=\Pr\left\{\frac{2m}{M+m}(v-V)\in z+dz\right\},\\&=\Pr\left\{v\in V+\frac{M+m}{2m}(z+dz)\right\}\\&=f\left(V+\frac{M+m}{2m}z\right)\frac{M+m}{2m}\,dz.\end{aligned}$$

Hence

$$w(z\,|\,V-z)\,dz=f\left(V+\frac{M-m}{2m}z\right)\frac{m+M}{2m}\,dz.$$

Thus the explicit form of the master equation is

$$\frac{\partial p\,(x,V,t)}{\partial t}=-\frac{\partial Vp\,(x,V,t)}{\partial x}+\frac{\partial}{\partial V}\frac{U'(x)}{M}p\,(x,V,t)-\alpha p\,(x,V,t)$$
$$+\int \alpha p\,(x,V-z,t)f\left(V+\frac{M-m}{2m}z\right)\frac{m+M}{2m}\,dz.$$

Now, we change variables to $u=V+(M-m)z/2m$ and get

$$\frac{\partial p\,(x,V,t)}{\partial t}=-\frac{\partial Vp\,(x,V,t)}{\partial x}+\frac{\partial}{\partial V}\frac{U'(x)}{M}p\,(x,V,t)-\alpha p\,(x,V,t)$$
$$+\int \alpha p\left(x,\frac{M+m}{M-m}V-\frac{2m}{M-m}u,t\right)f\,(u)\frac{M+m}{M-m}\,du.$$

We introduce the small parameter $\varepsilon=2m/M$ and note that

$$\frac{2m}{M-m}=\varepsilon\left(1+\frac{\varepsilon}{2}+\frac{\varepsilon^4}{4}+\cdots\right),\quad \frac{M+m}{M-m}=\left(1+\varepsilon+\frac{\varepsilon^2}{2}+\cdots\right).$$

The master equation becomes

$$\begin{aligned}\frac{\partial p(x,V,t)}{\partial t} = &-\frac{\partial V p(x,V,t)}{\partial x} + \frac{\partial}{\partial V}\frac{U'(x)}{M}p(x,V,t) - \alpha p(x,V,t)\\ &+\int \alpha p\left(x,(1+\varepsilon+\cdots)V - \varepsilon\left(1+\frac{\varepsilon}{2}+\cdots\right)u,t\right)f(u)\,du\\ &\times(1+\varepsilon+\cdots).\end{aligned}$$

Now, we use the Kramers–Moyal expansion to obtain

$$\begin{aligned}\frac{\partial p(x,V,t)}{\partial t} = &-\frac{\partial V p(x,V,t)}{\partial x} + \frac{\partial}{\partial V}\frac{U'(x)}{M}p(x,V,t)\\ &+\frac{\partial}{\partial V}\varepsilon\alpha\left[\int (V-u)f(u)\,du\right]p(x,V,t)\\ &+\frac{\varepsilon^2\alpha}{2}\frac{\partial^2}{\partial V^2}\left[\int (V-u)^2 f(u)\,du\right]p(x,V,t) + O\left(\varepsilon^3\right).\end{aligned}$$

Next, using the identities

$$\int (V-u)f(u)\,du = V, \quad \int (V-u)^2 f(u)\,du = V^2 + \frac{kT}{m},$$

we obtain

$$\begin{aligned}\frac{\partial p(x,V,t)}{\partial t} = &-\frac{\partial V p(x,V,t)}{\partial x} + \frac{\partial}{\partial V}\frac{U'(x)}{M}p(x,V,t)\\ &+\varepsilon\alpha\frac{\partial V p(x,V,t)}{\partial V} + \frac{\varepsilon^2\alpha}{2}\frac{\partial^2}{\partial V^2}\left[V^2+\frac{kT}{m}\right]p(x,V,t) + O\left(\varepsilon^3\right).\end{aligned}$$

Finally, we set $\gamma = 2m\alpha/M$ so that

$$\frac{\varepsilon^2\alpha}{2} = \frac{\varepsilon\gamma}{2}, \quad \frac{\varepsilon^2\alpha}{2}\frac{kT}{m} = \frac{\gamma kT}{M}.$$

We assume that $\gamma = O(1)$ as $\varepsilon \to 0$, because reducing the mass of the light particles while keeping the temperature constant means that the mean kinetic energy per light particle is preserved, hence the velocity increases and so does the frequency of collisions. Now, the truncated Kramers–Moyal expansion has the form

$$\begin{aligned}\frac{\partial p(x,V,t)}{\partial t} = &-\frac{\partial V p(x,V,t)}{\partial x} + \frac{\partial}{\partial V}\left[\gamma V + \frac{U'(x)}{M}\right]p(x,V,t)\\ &+\frac{\gamma kT}{M}\frac{\partial^2 p(x,V,t)}{\partial V^2},\end{aligned} \tag{7.102}$$

which is the Fokker–Planck–Kramers equation for the joint pdf of the displacement and velocity in the Langevin equation

$$\ddot{x} + \gamma\dot{x} + U'(x) = \sqrt{\frac{2\gamma kT}{M}}\,\dot{w}. \tag{7.103}$$

Exercise 7.9 (Generalization to $\mathbb{R}^3$). Generalize the above derivation to $\mathbb{R}^3$. □

Exercise 7.10 (The BKE for BM in a field of force). Derive the backward Kolmogorov equation from the backward master equation in this model [122]. □

7.3.5 Dynamics driven by wideband noise

A model for wideband noise, $\xi_\alpha(t)$, is white noise passed through a wideband filter; that is,

$$\dot{\xi}_\alpha = -\alpha\xi_\alpha + \alpha\sqrt{2\varepsilon}\,\dot{w} \tag{7.104}$$

with $\alpha >> 1$. First, we note that

$$\lim_{\alpha\to\infty} \xi_\alpha \stackrel{F}{=} \dot{w} \tag{7.105}$$

in the sense that the pdf of $\int_0^t \xi_\alpha(s)\,ds$ converges to the pdf of $w(t)$ as $\alpha \to \infty$. The proof of this statement is contained in the theory below.

Consider the dynamics

$$\dot{x}_\alpha = A(x_\alpha) + B(x_\alpha)\xi_\alpha(t), \tag{7.106}$$

where $\xi_\alpha(t)$ is wideband noise as described above. In the limit $\alpha \to \infty$, we expect to have $x_\alpha(t) \to x(t)$, in some sense, where $x(t)$ is the solution of the stochastic differential equation

$$\dot{x} = A(x) + B(x)\dot{w}. \tag{7.107}$$

It is not a priori clear, however, if this equation has to be interpreted in the sense of Itô or Stratonovich. This distinction may be crucial in determining, for example, the stability of the system, and its many other properties. This is also an important question in modeling systems driven by noise. To answer this question, we have to carry out the limiting procedure and find out the type of the limiting equation (7.107). Note that the proof of (7.105) corresponds to the case $A(x) = 0$ and $B(x) = 1$.

To determine the limit, we note first that the pair $x_\alpha(t), \xi_\alpha(t)$ is the solution of the (Itô) system of stochastic differential equations (7.104) and (7.106). It follows that the joint pdf of $x_\alpha(t)$ and $\xi_\alpha(t)$ is the solution of the Fokker–Planck equation

$$\begin{aligned}\frac{\partial p_\alpha(x,\xi,t)}{\partial t} =& -\frac{\partial\left[A(x)+B(x)\xi\right]p_\alpha(x,\xi,t)}{\partial x} + \alpha\frac{\partial \xi p_\alpha(x,\xi,t)}{\partial \xi}\\ &+ \varepsilon\alpha^2\frac{\partial^2 p_\alpha(x,\xi,t)}{\partial \xi^2}.\end{aligned}$$

Scaling $\xi = \sqrt{\alpha}\eta$ and setting $p_\alpha(x,\xi,t) = P_\alpha(x,\eta,t)$, we obtain

$$\begin{aligned}\frac{\partial P_\alpha(x,\eta,t)}{\partial t} =& -\frac{\partial A(x)P_\alpha(x,\eta,t)}{\partial x} + \sqrt{\alpha}\frac{\partial B(x)\eta P_\alpha(x,\eta,t)}{\partial x}\\ &+ \alpha\left[\frac{\partial \eta P_\alpha(x,\eta,t)}{\partial \eta} + \varepsilon\frac{\partial^2 P_\alpha(x,\eta,t)}{\partial \eta^2}\right]\\ =& \,\alpha L_0 P_\alpha + \sqrt{\alpha}L_1 P_\alpha + L_2 P_\alpha,\end{aligned}$$

where

$$L_0 P_\alpha = \frac{\partial \eta P_\alpha}{\partial \eta} + \varepsilon \frac{\partial^2 P_\alpha}{\partial \eta^2}, \quad L_1 P_\alpha = \frac{\partial B(x) \eta P_\alpha}{\partial x}, \quad L_2 P_\alpha = -\frac{\partial A(x) P_\alpha}{\partial x}.$$

Expanding

$$P_\alpha = P_0 + \frac{1}{\sqrt{\alpha}} P_1 + \frac{1}{\alpha} P_2 + \cdots,$$

we obtain the hierarchy of equations

$$L_0 P_0 = 0 \tag{7.108}$$

$$L_0 P_1 = -L_1 P_0 \tag{7.109}$$

$$L_0 P_2 = \frac{\partial P_0}{\partial t} - L_1 P_1 - L_2 P_0, \tag{7.110}$$

and so on. Equation (7.108) gives $P_0(x, \eta, t) = P_0(x, t) e^{-\eta^2/2}$, where $P_0(x, t)$ is yet an undetermined function. It is easy to see that the solvability condition for eq. (7.109) is satisfied and the solution is given by

$$P_1 = -\eta e^{-\eta^2/2} \frac{\partial \left[B(x) P_0(x, t) \right]}{\partial x} + P_1(x, t) e^{-\eta^2/2},$$

where $P_1(x, t)$ is yet an undetermined function. The solvability condition for eq. (7.110) gives

$$\frac{\partial P_0(x, t)}{\partial t} = -\frac{\partial A(x) P_0(x, t)}{\partial x} - \varepsilon \frac{\partial}{\partial x} B(x) \frac{\partial B(x) P_0(x, t)}{\partial x}. \tag{7.111}$$

This is the Fokker–Planck–Stratonovich equation for the pdf of the Stratonovich equation (7.107). Thus the limiting equation should be interpreted as a Stratonovich equation. The special case $A(x) = 0$ and $B(x) = 1$ leads to the diffusion equation with diffusion coefficient ε. This proves the convergence (7.105) in the sense described above.

Exercise 7.11 (Overdamped harmonic oscillator with small wideband colored noise). Consider the case of the overdamped motion in a harmonic potential driven by a small wideband colored noise [124],

$$\dot{x} = -U'(x) + g(x) u, \quad \dot{u} = -\alpha u + \sqrt{2\varepsilon}\, \alpha\, \dot{w},$$

where

$$U(x) = \frac{\omega^2 x^2}{2}. \tag{7.112}$$

(i) Transform the problem to the variables x, $y = -U'(x) + u$.

(ii) Show that for the case $g(x) = 1$ the joint stationary pdf of x and y is given by

$$p = \exp\left\{ -\frac{1}{\varepsilon} \left[\frac{\omega^2}{2\alpha^2} + \frac{1}{\alpha} \left[\frac{\omega^4 x^2}{2} + \frac{y^2}{2} \right] + \frac{\omega^2 x^2}{2} \right] \right\}.$$

(iii) For an arbitrary potential for which a stationary density exists, construct an asymptotic solution to the stationary Fokker–Planck equation in the WKB (Wentzel, Kramers, Brillouin) form

$$p \sim K(x,y,\alpha)\exp\left\{-\frac{\psi(x,y,\alpha)}{\varepsilon}\right\} \quad \text{for } \alpha \gg 1,\ \varepsilon \ll 1.$$

Write the eikonal and transport equations for $\psi(x,y,\alpha)$ and $K(x,y,\alpha)$, respectively, and solve them by expanding

$$psi(x,y,\alpha) = \sum_{i=0}^{\infty} \alpha^{-i}\psi_i(x,y), \quad K(x,y,\alpha) = \sum_{i=0}^{\infty} \alpha^{-i}K_i(x,y).$$

ANSWER:

$$\psi = U(x) + \frac{1}{\alpha}\left(\frac{y^2}{2} + \frac{1}{2}\left[U'(x)\right]^2\right) + \frac{1}{\alpha^2}\left(U''(x)\frac{y^2}{2} - \frac{1}{2}\int^x U'^2 U'''\,dx\right)$$
$$+ O\left(\frac{1}{\alpha^3}\right)$$
$$K = 1 + \frac{3U''(x)}{2\alpha} + O\left(\frac{1}{\alpha^2}\right).$$

(iv) Show that for any $g(x)$ the result is

$$\psi = \int^x \frac{U'}{g}\,dx + \frac{1}{\alpha}\left[\frac{y^2}{2g^2} + \int^x \left(\frac{U'U''}{g^2} - \frac{g'U'^2}{g^3}\right)dx\right]$$
$$+ \frac{1}{\alpha^2}\left[\left(\frac{U''(x)}{g^2} - \frac{g'U'}{g^3}\right)\frac{y^2}{2}\right.$$
$$\left. - \frac{1}{2}\int^x U'^2\left(\frac{U'''}{g^2} - \frac{g'U'' + g''U'}{g^3} + \frac{g'^2U'}{g^4}\right)dx\right] + O\left(\frac{1}{\alpha^3}\right)$$
$$K = \frac{1}{g(x)} + O\left(\frac{1}{\alpha}\right).$$

This expansion corresponds to first expanding for small ε and then for large α. Show that the limit $\alpha \to \infty$ in the above expansion reproduces the result of the reverse order expansion (7.111). □

Exercise 7.12 (Dynamics driven by colored noise). Consider the damped dynamics

$$\ddot{x} + \beta x + U'(x) = u, \quad \dot{u} = -\alpha u + \sqrt{2\varepsilon\beta}\,\alpha\,\dot{w},$$

or equivalently, the three-dimensional problem

$$\dot{x} = y, \quad \dot{y} = -U'(x) - \beta y + u, \quad \dot{u} = -\alpha u + \sqrt{2\varepsilon\beta}\,\alpha\,\dot{w}. \tag{7.113}$$

In the limit $\alpha \to \infty$ the colored u becomes white noise and (7.113) reduces to the Langevin equation with friction β.
(i) show that in the limit $\beta \to \infty$ the system (7.113) reduces to

$$\dot{x} = -U'(x) + \sqrt{2\varepsilon}\,\dot{w}$$

(with scaled time).
(ii) Show that if $\alpha\beta = a$, where $a = const.$, (7.113) reduces to

$$\dot{x} = -U'(x) + u, \quad \dot{u} = -au + \sqrt{2\varepsilon}\,a\,\dot{w}.$$

(iii) Show that if the potential is harmonic (see (7.112)), the exact solution of the stationary Fokker–Planck equation corresponding to (7.113) is the Gaussian pdf $C\exp\{-\psi/\varepsilon\}$, where C is a normalization constant and ψ is given by

$$\psi = \left(\omega^2 + \frac{\omega^4}{\alpha^2}\right)\frac{x^2}{2} + \left[\frac{\omega^2}{\alpha^2} + \frac{\beta+\alpha)^2}{\alpha^2}\right]\frac{y^2}{2} + \left(\frac{\beta+\alpha}{\alpha^2\beta}\right)\frac{u^2}{2} + \frac{\beta\omega^2}{\alpha^2}xy - \frac{\omega^2}{\alpha^2}xu - \frac{\beta+\alpha}{\alpha^2}uy. \tag{7.114}$$

(iv) Show that in general, if $\varepsilon \ll 1,\ \alpha \gg 1$, the WKB expansion $p = \exp\{-\psi/\varepsilon\}$ gives

$$\psi = U(x) + \frac{y^2}{2} + \frac{1}{\alpha}\left[\frac{u^2}{2\beta} - uy + \beta y^2\right] + \frac{1}{\alpha^2}\left[\frac{u^2}{2} - \beta uy - uU'(x) + \frac{1}{2}\beta^2 y^2 + \beta U'(x)u + \frac{1}{2}\left(U'(x)\right)^2 + \int_0^t y\dot{y}U''\,dt\right] + O\left(\frac{1}{\alpha^3}\right),$$

where the integral is evaluated on the trajectory

$$\dot{x} = y, \quad \dot{y} = -U'(x) + \beta y$$

such that $x(t) = x,\ y(t) = y$. □

7.3.6 Boundary behavior of diffusion approximations

We consider the random walk (7.80) of Section 7.3.3,

$$x_{n+1} = x_n + \varepsilon\xi_n \tag{7.115}$$

on $\mathbb{R}^+$ with an absorbing boundary at $x = 0$, as in Section 5.4. Specifically, the random walk is defined by (7.115) as long as $x_n > 0$, but the trajectory is truncated when it exits into $\mathbb{R}^-$.

Theorem 7.3.2 (Kramers–Moyal expansion with an absorbing boundary). *Under the assumptions* (7.84)–(7.86), *if* (7.89) *holds and for some* $z > 0$

$$\int_{-\infty}^{0} w_0(\eta - z \,|\, 0)\, d\eta > 0, \tag{7.116}$$

then for every $T > 0$, *as* $\varepsilon \to 0$, *the pdf* $p_\varepsilon(x, n)$ *converges to the solution* $p_0(x, t)$ *of the initial value problem* (7.90), (7.91) *with the absorbing boundary condition*

$$p_0(0, t) = 0 \ \text{ for } t > 0, \tag{7.117}$$

uniformly for all $x > 0,\ s < t < T$.

The assumption (7.116) means that the probability of jumping from the boundary to the left is positive.

Proof. The master equation (7.81) has to be modified for the pdf of the truncated trajectories to

$$p_\varepsilon(y, t + \Delta t) = \int_{-\infty}^{y/\varepsilon} p_\varepsilon(y - \varepsilon z, t) w(x \,|\, y - \varepsilon z, \varepsilon)\, dz. \tag{7.118}$$

For $y \gg \varepsilon$ the upper limit of integration can be extended to infinity and, as in Theorem 7.3.1, we find that the leading order term $p_0(y, t)$ is the solution of the initial value problem (7.90), (7.91) for all $y > 0$. As in the proof of Theorem 5.4.1, a boundary layer expansion is needed for the proof of convergence of the Kramers–Moyal expansion up to the boundary.

We introduce the boundary layer variable $\eta = y/\varepsilon$, set $p_{\text{bl}}(\eta, t) = p_\varepsilon(y, t)$, and expand

$$p_{\text{bl}}(\eta, t) \sim p_{\text{bl}}^{(0)}(\eta, t) + \varepsilon p_{\text{bl}}^{(1)}(\eta, t) + \cdots. \tag{7.119}$$

Using (7.85) and (7.119) in (7.118), we obtain to leading order the Wiener–Hopf equation

$$p_{\text{bl}}^{(0)}(\eta, t) = \int_{-\infty}^{\eta} p_{\text{bl}}^{(0)}(\eta - z, t) w_0(z \,|\, 0)\, dz = \int_{0}^{\infty} p_{\text{bl}}^{(0)}(z, t) w_0(\eta - z \,|\, 0)\, dz. \tag{7.120}$$

Integrating (7.120 with respect to η over $\mathbb{R}^+$ and using the normalization

$$\int_{-\infty}^{\infty} w_0(z \,|\, 0)\, dz = 1,$$

we obtain

$$\int_0^\infty p_{\text{bl}}^{(0)}(\eta,t)\,d\eta = \int_0^\infty p_{\text{bl}}^{(0)}(z,t)\left[1-\int_{-\infty}^0 w_0(\eta-z\,|\,0)\,d\eta\right]dz,$$

hence

$$\int_0^\infty p_{\text{bl}}^{(0)}(z,t)\int_{-\infty}^0 w_0(\eta-z\,|\,0)\,d\eta\,dz=0. \tag{7.121}$$

Equations (7.121) and (7.116) imply that $p_{\text{bl}}^{(0)}(z,t)=0$. Matching the boundary layer with the outer Kramers–Moyal expansion gives the absorbing boundary condition (7.117). □

7.4 Diffusion approximation of the MFPT

The diffusion approximation of a Markov process with absorbing boundaries, as described in Theorem 7.3.2, is valid in general on fixed time intervals, independent of ε. It cannot be used, in general, for approximating the long-time behavior of the Markov process or its expected exit time from a given domain [143]. Specifically, the expected number of steps $n_\varepsilon(x)$, of the Markov process

$$x_{n+1}=x_n+\varepsilon\xi_n,\quad x_0=x\in D \tag{7.122}$$

to leave a domain D, is related to the MFPT $\mathbb{E}[\tau_D^\varepsilon\,|\,x]$ of the continuous time process $x_\varepsilon(t)=x_{[t/\varepsilon^2]}$ by $n_\varepsilon(x)=\varepsilon^{-2}\mathbb{E}[\tau_D^\varepsilon\,|\,x]$. The convergence of $\mathbb{E}[\tau_D^\varepsilon\,|\,x]$ to the MFPT $\mathbb{E}[\tau_D\,|\,x]$ of the Kramers–Moyal diffusion approximation (7.93), and the discrepancy between the mean number of steps $n_\varepsilon(x)$ to that predicted by the diffusion approximation,

$$n_{\text{d}}(x)=\varepsilon^{-2}\mathbb{E}[\tau_D\,|\,x], \tag{7.123}$$

are described in [135] and in Theorem 7.4.1 below.

We consider (7.122) in $D=\mathbb{R}^-$ and truncate trajectories that cross into $\mathbb{R}^+$, keeping the notation and assumptions of Section 7.3.3. In particular, we assume that (7.89) holds, that $m_{ij}(x)$ in (7.86) and $w_j(z\,|\,x)$ in (7.85) are analytic functions, that $n_\varepsilon(x)<\infty$, and that $n_\varepsilon(x)$ does not grow exponentially as $x\to-\infty$.

Theorem 7.4.1 (The MFPT of a Markov jump process). *Under the assumption of Theorem 7.3.1, the expected number of steps $n_\varepsilon(x)$ for the process x_n to leave D suffers a jump discontinuity at the boundary $x=0$ and*

$$n_\varepsilon(0-)-n_\varepsilon(0+)=O\left(\varepsilon^{-1}\right). \tag{7.124}$$

The relative error in estimating $n_\varepsilon(x)$ *by* $n_d(x)$ *is*

$$e_d(x) = \frac{n_\varepsilon(x) - n_d(x)}{n_\varepsilon(x)} \to 0 \text{ as } \varepsilon \to 0 \text{ for } x \ll \varepsilon, \tag{7.125}$$

but

$$e_d(0) \to 1 \text{ as } \varepsilon \to 0. \tag{7.126}$$

Proof. Equation (7.37) of Theorem 7.1.4 takes the form

$$\int_{-\infty}^{-x/\varepsilon} n_\varepsilon(x + \varepsilon z) w(z \mid x, \varepsilon)\, dz - n_\varepsilon(x) = -1 \text{ for } x < 0, \tag{7.127}$$

with the boundary condition (7.38), which is

$$n_\varepsilon(x) = 0 \text{ for } x \geq 0. \tag{7.128}$$

The MFPT $\mathbb{E}[\tau_D \mid x]$ of the diffusion approximation (7.93) is obtained from the outer regular expansion

$$n_\varepsilon(x) \sim \varepsilon^{-2} n_{-2}(x) + \varepsilon^{-1} n_{-1}(x) + n_0(x) + \cdots \text{ for } \varepsilon \to 0 \tag{7.129}$$

as

$$\mathbb{E}[\tau_D \mid x] = n_{-2}(x). \tag{7.130}$$

The proof of Theorem 7.1.4 shows that the expansion (7.129) is valid at each $x < 0$, independent of ε. Specifically, for $x \ll -\varepsilon$ we substitute the expansions (7.129) and (7.85) in (7.127), and extending the upper limit of integration to infinity, as we may, we equate the coefficient of each power of ε separately to zero to obtain a hierarchy of equations for the unknown coefficients $n_i(x)$. The leading-order equation is

$$L_0 n_{-2}(x) = m_{11}(x) \frac{dn_{-2}(x)}{dx} + \frac{1}{2} m_{20}(x) \frac{d^2 n_{-2}(x)}{dx^2} = -1 \tag{7.131}$$

and the next order equation is

$$\begin{aligned} L_0 n_{-1}(x) = & m_{12}(x) \frac{dn_{-2}(x)}{dx} - \frac{m_{21}(x)}{2} \frac{d^2 n_{-2}(x)}{dx^2} \\ & - \frac{m_{30}(x)}{6} \frac{d^3 n_{-2}(x)}{dx^3}. \end{aligned} \tag{7.132}$$

The relations (7.123) and (7.130) give

$$n_d(x) = \varepsilon^{-2} n_{-2}(x), \tag{7.133}$$

Note that (7.131) is the Andronov–Vitt–Pontryagin equation (4.67) for the MFPT $\mathbb{E}[\tau_D \mid x]$ of the Kramers–Moyal diffusion approximation (7.93). The boundary conditions for eqs. (7.131) and (7.132) are not obvious, because the functions $n_{-2}(x)$ and $n_{-1}(x)$ are not necessarily continuous at the boundary $x = 0$ and may suffer a discontinuity there. The boundary condition (7.128), which implies $n_\varepsilon(0) = 0$, does not necessarily imply that $n_\varepsilon(0-) = 0$, and as a matter of fact, $n_\varepsilon(0-) > 0$ in general. Thus the values of $n_i(0)$ have to be determined in a self-consistent way for all $i = -2, -1, 0, 1, \dots$.

We determine the boundary conditions by constructing a boundary layer expansion of $n_\varepsilon(x)$ and matching it with the outer expansion (7.129). In the boundary layer region $x = O(\varepsilon)$ the upper limit of integration in the master equation (7.127) cannot be extended to infinity, so the Kramers–Moyal expansion is not valid in this region.

As in previous sections, we introduce in the master equation the boundary layer variable and function

$$\eta = -x/\varepsilon > 0, \quad N(\eta) = n_\varepsilon(x) \tag{7.134}$$

and use the identity $n_\varepsilon(x + \varepsilon z) = N(\eta - z)$ to write (7.127) as

$$\int_{-\infty}^{\eta} N(\eta - z) w(z - \eta\varepsilon \,|\, \eta, \varepsilon)\, dz = -1 \quad \text{for } \eta > 0. \tag{7.135}$$

Expanding

$$N(\eta) \sim \varepsilon^{-2} N_{-2}(\eta) + \varepsilon^{-1} N_{-1}(\eta) + N_0(\eta) + \cdots, \tag{7.136}$$

we obtain the hierarchy of integral equations

$$\mathcal{L}_0 N_{-2}(\eta) = \int_{-\infty}^{\eta} N_{-2}(\eta - z) w_0(z \,|\, 0)\, dz - N_2(\eta) = 0 \tag{7.137}$$

$$\begin{aligned} \mathcal{L}_0 N_{-1}(\eta) &= \eta \int_{-\infty}^{\eta} N_{-2}(\eta - z) w_{0,x}(z \,|\, 0)\, dz - \int_{-\infty}^{\eta} N_{-2}(\eta - z) w_1(z \,|\, 0)\, dz \\ &= \mathcal{L}_1 N_{-2}(\eta) \end{aligned} \tag{7.138}$$

$$\begin{aligned} \mathcal{L}_0 N_0(\eta) &= \mathcal{L}_1 N_{-1}(\eta) - \frac{1}{2}\eta^2 \int_{-\infty}^{\eta} N_{-2}(\eta - z) w_{0,xx}(z \,|\, 0)\, dz \\ &\quad + \eta \int_{-\infty}^{\eta} N_{-2}(\eta - z) w_{1,x}(z \,|\, 0)\, dz - \int_{-\infty}^{\eta} N_{-2}(\eta - z) w_2(z \,|\, 0)\, dz \\ &\quad - 1, \end{aligned} \tag{7.139}$$

where $w_{i,x}(z \,|\, 0) = \partial w_i(z \,|\, x)/\partial x|_{x=0}$. The boundary condition (7.128) implies that $N_j(\eta) = 0$ for $\eta \le 0$ and all j. Next, we apply the Fourier transform to the

Wiener–Hopf eq. (7.137) and obtain

$$\hat{N}_{-2}(\alpha) = \int_{-\infty}^{\infty} e^{i\alpha\eta} N_{-2}(\eta)\, d\eta = \frac{\Phi(\alpha)}{1 - \hat{w}_0(\alpha \,|\, 0)}, \tag{7.140}$$

where

$$\Phi(\alpha) = \int_{-\infty}^{0} e^{i\alpha\eta} \int_{0}^{\infty} N_{-2}(z) w_0(\eta - z \,|\, 0)\, dz\, d\eta \tag{7.141}$$

is an analytic function in the lower half-plane. The function $\Phi(\alpha)$ is defined by equations (7.140) and (7.141) uniquely up to a multiplicative constant. The characteristic function of the process ξ_n, conditioned on $x = 0$,

$$\hat{w}_0(\alpha \,|\, 0) = \int_{-\infty}^{\infty} e^{i\alpha z} w_0(z \,|\, 0)\, dz, \tag{7.142}$$

generates the conditional moments $m_{i,0}(0)$.

The matching condition between the outer expansion (7.129) and the boundary layer expansion (7.136) is

$$n_\varepsilon(x) - N(\eta) \sim 0 \tag{7.143}$$

as $\varepsilon \to 0$, $x \to 0$, and $\eta \to \infty$, to all orders in ε. The behavior of $N_{-2}(\eta)$ as $\eta \to \infty$ is determined by the behavior of $\hat{N}_{-2}(\alpha)$ as $\alpha \to 0$. This in turn is determined by the poles of (7.140), which are the zeros of the denominator of (7.140). The assumptions

$$\hat{w}_0(\alpha \,|\, 0) = \int_{-\infty}^{\infty} w_0(z \,|\, 0)\, dz = 1, \quad \hat{w}_0'(\alpha \,|\, 0) = i m_{10}(0) = 0$$

imply that

$$1 - \hat{w}_0(\alpha \,|\, 0) = \frac{1}{2} m_{20}(0)\alpha^2 + O(\alpha^3)$$

and

$$\hat{N}_{-2}(\alpha) = \frac{\Phi(0) + \alpha\Phi'(0) + O(\alpha^2)}{\frac{1}{2}\alpha^2 m_{20}(0) + \frac{i}{6} m_{30}(0)\alpha^3 + O(\alpha^3)} = \frac{a_{-2}}{\alpha^2} + \frac{a_{-1}}{\alpha^1} + O(1),$$

where

$$a_{-2} = \frac{2\Phi(0)}{m_{20}(0)}, \quad a_{-1} = \frac{2\Phi'(0)}{m_{20}(0)} - \frac{2i\Phi(0) m_{30}(0)}{3 m_{20}^2(0)}, \quad \ldots. \tag{7.144}$$

Hence

$$N_{-2}(\eta) = -[ia_{-1} + \eta a_{-2} + o(1)] \text{ as } \eta \to \infty. \tag{7.145}$$

The values of $\Phi(0), \Phi'(0), \ldots$ have to be determined from the matching conditions for the outer and inner expansions (7.129) and (7.136), respectively, as $\varepsilon \to 0$, $x \to 0$, and $\eta \to \infty$, to all orders in ε. At order $O\left(\varepsilon^{-2}\right)$ we have the matching condition in the limit $x, \varepsilon \to 0$, $\eta \to \infty$,

$$n_{-2}(x) - N_{-2}(\eta) = n_{-2}(0) + ia_{-1} + \eta a_{-2} \to 0, \text{ as } x, \varepsilon \to 0,\ \eta \to \infty.$$

Hence $a_{-2} = 0$ and consequently $\Phi(0) = 0$. Setting $\alpha = 0$ in (7.141), we find that $N_{-2}(z) = 0$ for all z, because $N_{-2}(z) \geq 0$ and $w_0(z \,|\, 0) \geq 0$. Therefore $\Phi(\alpha) = 0$ and

$$n_{-2}(0) = 0. \tag{7.146}$$

Accordingly $n_{-2}(x)$ is continuous at the boundary $x = 0$.

To evaluate $n_{-1}(0)$ we have to solve eq. (7.138) for $N_{-1}(\eta)$. Setting $N_{-2}(\eta) = n_{-2}(x)$, we find that (7.138) is the same as the Wiener–Hopf equation (7.137) and therefore its solution is given again by (7.140) and (7.141). It follows, as above, that

$$N_{-1}(\eta) = -[ia_{-1} + \eta a_{-2} + o(1)] \text{ as } \eta \to \infty, \tag{7.147}$$

where $\Phi(0)$ is again an undetermined constant. In view of (7.146), Taylor's expansion of $n_{-2}(x)$ about $x = 0$ gives for $\varepsilon \ll 1$ and for all η

$$\varepsilon^{-2} n_{-2}(x) = \varepsilon^{-2} n'_{-2}(0) x + O\left(x^2\right) = \varepsilon^{-1} n'_{-2}(0)\eta + O(1).$$

Therefore the matching condition at order $O\left(\varepsilon^{-1}\right)$ is given by

$$n_{-1}(0) - n'_{-1}(0)\eta - N_{-1}(\eta) \to 0 \text{ as } \eta \to \infty$$

and hence, together with (7.147), we obtain

$$-ia_{-1} = n_{-1}(0), \quad a_{-2} = n'_{-2}(0). \tag{7.148}$$

Equations (7.148) and (7.144) give

$$\Phi(0) = \frac{1}{2} n'_{-2}(0) m_{20}(0), \tag{7.149}$$

which determines $N_{-1}(\eta)$ uniquely as the solution of the homogeneous Wiener–Hopf equation and consequently determines $\Phi(\alpha)$. Again, (7.147) gives

$$a_{-1} = \frac{2\Phi'(0)}{m_{20}(0)} - \frac{2i\Phi(0)m_{30}(0)}{3m_{20}^2(0)},$$

which together with (7.148) and (7.149) determines the boundary condition for $n_{-1}(x)$.

To measure the relative error of the diffusion approximation $n_2(x)$, we write

$$n_{-1}(0) = \frac{ia_{-1}}{a_{-2}} n'_{-2}(0) \tag{7.150}$$

and note that $n_\varepsilon(x)$ suffers a jump discontinuity at $x = 0$, because $n_\varepsilon(x) = 0$ for all $x \geq 0$. The jump is of order ε^{-1},

$$n_\varepsilon(0-) - n_\varepsilon(0+) = \varepsilon^{-1} N_{-1}(0) + O(1) \text{ as } \varepsilon \to 0,$$

as asserted in (7.124).

The relative error in using the diffusion approximation to estimate the expected number of steps to exit D is

$$e_{\mathrm{d}}(x) = \frac{n_\varepsilon(x) - n_{\mathrm{d}}(x)}{n_\varepsilon(x)} \to 0 \text{ as } \varepsilon \to 0 \text{ for } x \ll \varepsilon.$$

However, this is not the case near $x = 0$, because $e_d(0) \to 1$ as $\varepsilon \to 0$. We conclude that the diffusion approximation with absorbing conditions at the boundary correctly predicts the MFPT for initial conditions outside the boundary layer, but for initial conditions inside the boundary layer it leads to increasingly large errors. □

Corollary 7.4.1 (Uniform approximation). *The uniform approximation*

$$n_{\mathrm{db}}(x) = \varepsilon^{-2} n_{-2}(x) + \varepsilon^{-1} n_{-1}(x) + \varepsilon^{-1} \left[N_{-1}\left(-\frac{x}{\varepsilon}\right) - n_{-1}(0) - \frac{n'_{-2}(0)x}{\varepsilon} \right]$$

has relative error

$$e_{\mathrm{db}}(x) = O(\varepsilon) \text{ for all } \boldsymbol{x} \in D.$$

Example 7.12 (The MFPT of an asymmetric random walk). Consider a jump process on the lattice $\{n\varepsilon \,|\, n = 0, \pm 1, \pm 2, \ldots\}$, which jumps one lattice point to the left with probability $l(x,\varepsilon)$ and either one or two lattice points to the right with probability $r_1(x,\varepsilon)$ or $r_2(x,\varepsilon)$, respectively. The one step transition probability density is

$$\begin{aligned} w(z \,|\, x, \varepsilon) =& \, l(x,\varepsilon)\delta(z+1) + [1 - l(x,\varepsilon) - r_1(x,\varepsilon) - r_2(x,\varepsilon)]\delta(z) \\ &+ r_1(x,\varepsilon)\delta(z-1) + r_2(x,\varepsilon)\delta(z-2). \end{aligned}$$

Assume that there exists $x_0 < 0$ such that

$$m_1(x,\varepsilon) = -l(x,\varepsilon) + r_1(x,\varepsilon) + 2r_2(x,\varepsilon) = \varepsilon m_{11}(x) > 0 \text{ for } x < x_0,$$

so that that $n_\varepsilon(x) < \infty$ and does not grow exponentially as $x \to -\infty$, and (7.89) is satisfied. It follows that

$$\begin{aligned} m_2(x,\varepsilon) =& \, 2r_1(x,\varepsilon) + 6r_2(x,\varepsilon) - \varepsilon m_{11}(x) = m_{20}(x) + \varepsilon m_{21}(x) \\ m_3(x,\varepsilon) =& \, 6r_2(x,\varepsilon) + \varepsilon m_{11}(x) = m_{30}(x) + \varepsilon m_{31}(x). \end{aligned}$$

Scaling $t = \varepsilon^2 n$ and assuming all functions are sufficiently smooth in x, we obtain the Kramers–Moyal diffusion approximation

$$\dot{x} = m_{11}(x) + \sqrt{m_{20}(x)}\,\dot{w}. \tag{7.151}$$

The FPE for (7.151) is

$$\frac{\partial p}{\partial t} = -\frac{\partial[m_{11}(x)p]}{\partial x} + \frac{1}{2}\frac{\partial^2[m_{20}(x)p]}{\partial^2 x} = Lp.$$

The expected number of steps to exit $D = \mathbb{R}^-$, given $x_\varepsilon(0) = x < 0$, is given by $n_\varepsilon(x) \sim \varepsilon^{-2} n_{-2}(x) + \varepsilon^{-1} n_{-1}(x) + \cdots$, where

$$L^* n_{-2}(x) = m_{11}(x) n'_{-2}(x) + \frac{m_{20}(x)}{2} n''_{-2}(x) = -1 \quad \text{for } x < 0,$$

with the boundary condition $n_{-2}(0) = 0$ and the condition that $n_{-2}(x)$ does not grow exponentially as $x \to -\infty$. The correction $n_{-1}(x)$ is the solution of

$$L^* n_{-1}(x) = -\frac{1}{2} m_{21}(x) n''_{-2}(x) - \frac{m_{30}(x)}{6} n'''_{-2}(x) \quad \text{for } x < 0$$

with $n_{-1}(0) = A n'_{-2}(0)$ and the growth condition as $x \to -\infty$ imposed on $n_{-1}(x)$. Equation (7.150) implies that the constant A is given by

$$A = -\frac{[r_1(0,0) + 2r_2(0,0)]r_2(0,0)}{[r_1(0,0) + 3r_2(0,0)][3r_1(0,0) + r_2(0,0)]}.$$

The function $N_{-1}(\eta)$ is given by

$$N_{-1}(\eta) = -n'_{-2}(0)\left\{\eta - \frac{l(0,0)r_2(0,0)}{[l(0,0) + r_2(0,0)][r_1(0,0) + 2l(0,0)]}\left[1 - \left(\frac{-r_2(0,0)}{l(0,0)}\right)^{\eta}\right]\right\}.$$

Hence

$$n_{\text{db}}(x) = \varepsilon^{-2} n_{-2}(x) + \varepsilon^{-1} n_{-1}(x) - \varepsilon^{-1} n'_{-2}(0) \frac{l(0,0)r_2(0,0)}{[l(0,0) + r_2(0,0)][r_1(0,0) + 2l(0,0)]}\left(\frac{-r_2(0,0)}{l(0,0)}\right)^{-x/\varepsilon}.$$

Thus, for example, the mean number of steps to reach or exceed 0, starting one lattice point away from the boundary, is given by

$$n(-\varepsilon) \approx n_{\text{db}}(-\varepsilon) = -\frac{n'_{-2}(0)}{\varepsilon}\,\frac{r_1(0,0) + l(0,0) - r_2(0,0)}{r_1(0,0) + 2l(0,0)}[1 + o(1)]. \tag{7.152}$$

If the absorbing boundary conditions are employed with the diffusion approximation, the absolute error in the mean exit time will be $O\left(\varepsilon^{-1}\right)$, and the relative error will be $O(\varepsilon)$ if x is bounded away from $x = 0$, but will be $O(1)$ if x is within $O(\varepsilon)$ of the boundary point $x = 0$. In contrast, the relative error of the uniform approximation is $O(\varepsilon)$ for all $x \in \mathbb{R}^-$. □

Exercise 7.13 (The MFPT of a Markov drift-jump process). Formulate and prove a theorem analogous to 7.4.1 for the drift-jump process defined in Exercise 7.10 (see [135]). □

Exercise 7.14 (The MFPT of a Langevin drift-jump process [135]). Consider the drift-jump collision model of Section 7.3.4 with the potential $U(x) = -x^3/3 + x^2/2$.
(i) Write the Langevin equation (7.103) as the phase plane stochastic system

$$\dot{x} = V, \quad \dot{V} = -\gamma V - U'(x) + \sqrt{\frac{2\gamma T}{M}}\,\dot{w}. \tag{7.153}$$

and show that the noiseless system (with $T = 0$) has a stable attractor at the origin and a saddle point at $x = 1$, $V = 0$.
(ii) Find the domain of attraction D of the stable equilibrium point at the origin and its boundary ∂D, called the *separatrix*, by tracing (graphically) the unstable trajectories of the noiseless system (7.153), emanating from the saddle point $x = 1$, $V = 0$.
(iii) Write the boundary value problem (7.37), (7.38) for the MFPT $\mathbb{E}[\tau_D \mid x, V]$ for the model (7.100), (7.101) (use Exercise 7.10).
(iv) Write the Andronov–Vitt–Pontryagin boundary value problem (4.67), (4.68) in D for the MFPT $\tau_0(x, V)$ of the diffusion approximation (7.153) of the model (7.100), (7.101).
(v) Set $\varepsilon^2 = m/M$ and introduce near any point $(x, V_0) \in \partial D$ the boundary layer variable $\eta = (V_0 - V)/\varepsilon$. Use the boundary layer analysis of this section to show that the solution of the boundary value problem of (iii) converges to that of (iv) in D, outside a boundary layer near ∂D.
(vi) Show that $\mathbb{E}[\tau_D \mid x, V]$ suffers a jump discontinuity of size

$$-\sqrt{\frac{m}{M}}\frac{\partial \tau_0(x, V)}{\partial V} \quad \text{for } (x, V) \in \partial D.$$

(vii) Show that at boundary points, where $\partial \tau_0(x, V)/\partial V = 0$, the jump discontinuity of $\mathbb{E}[\tau_D \mid x, V]$ is of order ε^2. □

Chapter 8

Diffusion Approximations to Langevin's Equation

In 1940, H. Kramers [140] introduced a diffusion model for chemical reactions, based on the Langevin equation for a Brownian particle in a potential $U(x)$ (per unit mass) that forms a well (see Figure 8.1),

$$\ddot{x} + \gamma\dot{x} + U'(x) = \sqrt{2\varepsilon\gamma}\,\dot{w}, \tag{8.1}$$

where γ is the dissipation constant (here normalized by the frequency of vibration at the bottom of the well),

$$\varepsilon = \frac{kT}{\Delta U}$$

is dimensionless temperature (normalized by the potential barrier height), and $\dot{w}$ is standard Gaussian white noise. Kramers sought to calculate the reaction rate κ, which is the rate of escape of the Brownian particle from the potential well in which it is confined. In particular, he sought to determine the dependence of of κ on temperature T and on the viscosity (friction) γ, and to compare the values found with the results of transition state theory [87].

8.1 The overdamped Langevin equation

The Langevin equation (8.1) serves as a model for many activated processes [146], for which the escape rate determines their time evolution. It is one of the most extensively studied equation in statistical physics [100]. Its solution $x(t)$ is not a diffusion process, not even a Markov process, but the pair $(x(t), \dot{x}(t))$ is a two-dimensional diffusion process (see Exercise 2.16). It can be seen from Exercise 1.2, however, that for high damping their joint pdf breaks into a product of the stationary Maxwellian pdf of the velocity $v = \dot{x}$ and the time-dependent pdf of the displacement $x(t)$, which becomes nearly a diffusion process. This is the case not only for the linear Langevin equation, but holds in general.

Z. Schuss, *Theory and Applications of Stochastic Processes: An Analytical Approach*, Applied Mathematical Sciences 170, DOI 10.1007/978-1-4419-1605-1_8,

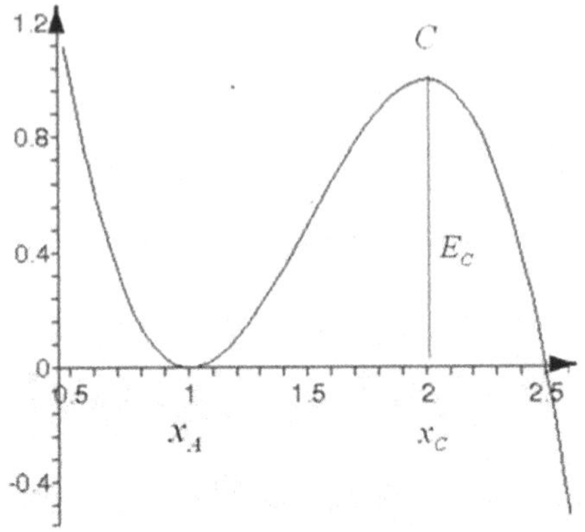

Figure 8.1. The potential $U(x) = 3(x-1)^2 - 2(x-1)^3$. The well is $x < x_C = 2$, the bottom is at $x = x_A = 1$, the top of the barrier is $C = (x_C, U(x_C))$, and the height of the barrier is $E_C = U(x_C) - U(x_A) = 1$.

Theorem 8.1.1 (The Smoluchowski limit). *As* $\gamma \to \infty$ *the trajectories* $x(t)$ *of the Langevin equation* (8.1) *converge in probability to these of the Smoluchowski equation*

$$\gamma \dot{x} + U'(x) = \sqrt{2\varepsilon\gamma}\,\dot{w}, \tag{8.2}$$

Proof. Writing the Langevin equation (8.1) as the phase space system

$$\dot{x} = v \tag{8.3}$$

$$\dot{v} = -\gamma v - U'(x) + \sqrt{2\varepsilon\gamma}\,\dot{w}, \tag{8.4}$$

we scale time by setting

$$t = \gamma s \tag{8.5}$$

and obtain the scaled Brownian motion $w(t) = \sqrt{\gamma} w^{\gamma}(s)$, where by the Brownian scaling (see Exercise 2.4) $w^{\gamma}(s)$ is a standard Brownian motion in time s. The scaled white noise is formally

$$\dot{w}(t) = \gamma^{-1} d\sqrt{\gamma} w^{\gamma}(s)/ds = \gamma^{-1/2} \overset{\circ}{w}{}^{\gamma}(s).$$

We denote $x^{\gamma}(s) = x(\gamma t)$, $v^{\gamma}(s) = v(\gamma t)$, and note that $\dot{x}(t) = \gamma^{-1} \overset{\circ}{x}{}^{\gamma}(s)$ and $\dot{v}(t) = \gamma^{-1} \overset{\circ}{v}{}^{\gamma}(s)$. It follows that eq. (8.4) can be written as

$$\overset{\circ}{v}{}^{\gamma}(s) + \gamma^2 v^{\gamma}(s) = -\gamma U'(x^{\gamma}(s)) + \gamma\sqrt{2\varepsilon}\,\overset{\circ}{w}{}^{\gamma}(s).$$

Solving for $v^\gamma(s)$, we obtain

$$v^\gamma(s) = v^\gamma(0)e^{-\gamma^2 s} + \gamma \int_0^s e^{-\gamma^2(s-u)} \left[-U'(x^\gamma(u))\,du + \sqrt{2\varepsilon}\,dw^\gamma(u)\right].$$

Equation (8.3) has the form $\dot{x}(t) = \gamma^{-1} \overset{\circ}{x^\gamma}(s) = v^\gamma(s)$, so that

$$x^\gamma(s) = x^\gamma(0) + \gamma \int_0^s v^\gamma(z)\,dz = x^\gamma(0) + v^\gamma(0)\frac{1-e^{-\gamma^2 s}}{\gamma}$$
$$+ \gamma^2 \int_0^s \int_0^z e^{-\gamma^2(z-u)} \left[-U'(x^\gamma(u))\,du + \sqrt{2\varepsilon}\,dw^\gamma(u)\right] dz.$$

Using the fact that $\lim_{\gamma\to\infty} \int_0^\infty \gamma^2 e^{-\gamma^2 s} f(s)\,ds = f(0)$ for any function f continuous at $s = 0$, such that for some $\delta > 0$ the function $e^{-\delta s} f(s)$ is integrable, we obtain

$$\lim_{\gamma\to\infty} x^\gamma(s) = x^\infty(0) + \int_0^s \left[-U'(x^\infty(u))\,du + \sqrt{2\varepsilon}\,dw^\infty(u)\right],$$

where $w^\infty(u)$ is Brownian motion. That is,

$$dx^\infty(s) = -U'(x^\infty(s))\,ds + \sqrt{2\varepsilon}\,dw^\infty(s). \tag{8.6}$$

Returning to the original time scale, eq. (8.6) becomes (8.2), which means that eq. (8.6) is the Langevin equation (8.1) without the inertia term $\ddot{x}$. This means that the limit exists on every trajectory. It remains to show convergence in probability (see Exercise 8.1 below). □

Exercise 8.1 (Smoluchowski convergence in probability). Prove convergence in probability in Theorem 8.1.1. □

Exercise 8.2 (The Smoluchowski limit in higher dimensions). Generalize the proof of Theorem 8.1.1 to the Langevin equation in higher dimensions. □

Example 8.1 (The Smoluchowski limit of a free Brownian particle). The motion of a free Brownian particle is described by the Langevin equation (8.1) with $U(x) = 0$. In the large friction limit the Smoluchowski approximation equation (8.2) becomes $dx = \sqrt{2k_BT/\gamma m}\,dw$, or $x(t) = \sqrt{2D}\,w(t)$, where the diffusion coefficient is given by the Einstein relation $D = k_BT/\gamma m$. □

8.1.1 The overdamped limit of the GLE

We consider below systems in thermal equilibrium; that is, systems in which all net fluxes vanish. For example, a system described by the Langevin equation (8.1) in the entire phase space with potential $U(x)$ such that $e^{-U(x)/\varepsilon}$ is integrable, reaches an equilibrium state, in which the pdf in phase space is the Maxwell–Boltzmann density $p(x,\dot{x}) = \mathcal{N}\exp\{-\dot{x}^2/2 - U(x)/\varepsilon\}$ ($\mathcal{N}$ is a normalization constant). The net flux in every direction vanishes everywhere. Similarly, a system described by the Smoluchowski limit (8.2) reaches equilibrium with the Boltzmann density $p(x) = \mathcal{N}\exp\{-U(x)/\varepsilon\}$, whose net flux vanishes everywhere. Systems in which the net flux across each hyperplane vanishes are said to be in *detailed balance*.

The Langevin equation (8.1) was derived under the assumption that the heavy Brownian particle moves in an equilibrium thermal bath of much lighter particles, and interacts with one light particle at a time at the instant of collision. Such a system is characterized by two time scales: the correlation time of the heavy particle and a much shorter time characterizing the bath. This model, however, is not valid in certain physical situations, when collisions between the particle and the thermal bath do not occur instantaneously, but rather take place over an interval of time characterizing the interaction between the particle and a group of bath particles. Another example is that of frequency-dependent conductivity, in which the Fourier transform of the dissipation term is given by $\hat{\gamma}(\omega)\hat{v}(\omega)$, where $v(t)$ is voltage and $\hat{\gamma}(\omega)$ is the frequency-dependent conductivity (see [61] and references therein). We denote $\gamma(t) = \gamma\varphi(t)$ with γ a constant. In each case a better description is given by the *generalized Langevin equation* (GLE)

$$\ddot{x} + \gamma\int_0^t \varphi(t-s)\dot{x}(s)\,ds + U'(x) = \xi(t), \tag{8.7}$$

where $\xi(t)$ is a zero mean stationary Gaussian process, white noise. The solution $x(t)$ is in general neither Markovian nor a component of a finite-dimensional Markov process (see Exercise 8.3 below). The GLE can be derived from statistical mechanics with the *generalized fluctuation-dissipation principle*

$$\mathbb{E}\xi(t_1)\xi(t_2) = \frac{\gamma k_B T}{m}\varphi(|t_1 - t_2|). \tag{8.8}$$

The generalized Langevin equation can be derived also from a model of a particle in a potential field, coupled linearly to a bath of harmonic oscillators (see [100] for references). Consider, for example, the Hamiltonian $\mathcal{H} = p^2/2M + U(x) + \mathcal{H}_{\text{bath}}(q_1,\ldots,q_N,p_1,\ldots,p_N)$, where

$$\mathcal{H}_{\text{bath}} = \frac{1}{2}\sum_{i=1}^{N} m_i\left[\dot{q}_i^2 + \omega_i^2\left(q_i + \frac{C_i}{m_i\omega_i^2}x\right)^2\right].$$

Although each oscillator may perturb the particle only weakly, the combined effect of all the bath modes on the particle motion may be significant. The coupling to the bath can cause strong dissipation and strong fluctuations of the particle's trajectory.

The equations of motion are given by

$$\dot{x} = \frac{\partial \mathcal{H}}{\partial p} = \frac{p}{M}, \quad \dot{p} = -\frac{\partial \mathcal{H}}{\partial x} = -U'(x) - \sum_{i=1}^{N} C_i \left(q_i + \frac{C_i}{m_i \omega_i^2} x \right)$$

$$m_i \dot{q}_i = \frac{\partial \mathcal{H}}{\partial p_i} = p_i, \quad \dot{p}_i = -\frac{\partial \mathcal{H}}{\partial q_i} = -m_i \omega_i^2 \left(q_i + \frac{C_i}{m_i \omega_i^2} x \right).$$

Solving the equations of motion for the forced harmonic oscillators, we obtain the generalized Langevin equation (8.7) with memory kernel

$$\varphi_N(t) = \sum_{i=1}^{N} \frac{C_i}{m_i \omega_i^2} \cos \omega_i t$$

and noise

$$\xi_N(t) = -\sum_{i=1}^{N} C_i \left[\left(q_i(0) + \frac{C_i}{m_i \omega_i^2} x(0) \right) \cos \omega_i t + \frac{\dot{q}_i(0)}{\omega_i^2} \sin \omega_i t \right].$$

If we assume that at time $t = 0$ the bath is in thermal equilibrium, such that the initial bath distribution in phase space is given as

$$\Pr\left\{ q_i(0) = q_i,\, \dot{q}_i(0) = \dot{q}_i \right\} = C \exp\left\{ -\frac{\mathcal{H}_{\text{bath}}}{k_B T} \right\},$$

where C is a normalization constant, we find that $\mathbb{E}\xi_N(t) = 0$ and

$$\mathbb{E}\xi_N(t_1)\xi_N(t_2) = \frac{k_B T}{M} \varphi_N(|t_1 - t_2|),$$

so that the generalized fluctuation-dissipation principle (8.8) is satisfied. The spectral density of the noise is given by

$$S_N(\omega) = \frac{\pi}{2} \sum_{i=1}^{N} \frac{C_i}{m_i \omega_i^2} \delta(\omega - \omega_i),$$

so the memory kernel can be represented as

$$\varphi_N(t) = \frac{2}{\pi} \int_{-\infty}^{\infty} \frac{S_N(\omega)}{\omega} \cos \omega t \, d\omega$$

with the Laplace transform

$$\hat{\varphi}_N(s) = \int_0^{\infty} e^{-st} \varphi_N(t)\, dt = \frac{2}{\pi} \int_{-\infty}^{\infty} \frac{S_N(\omega)}{\omega} \frac{s}{s^2 + \omega^2}\, d\omega.$$

If the coefficients of the random initial conditions are chosen in an appropriate way and N is increased to infinity, the noise $\xi_N(t)$ can be made to converge to any stationary Gaussian process with sufficiently "nice" power spectral density function $S(\omega)$. The Langevin equation (8.1) corresponds to $\varphi(t) = \delta(t)$.

Unlike the Langevin equation (8.1), in the phase space representation of eq. (8.7),

$$\dot{x} = v, \quad \dot{v} = -\gamma \int_0^t \varphi(t-s)v(s)\,ds - U'(x) + \xi(t), \tag{8.9}$$

the two-dimensional process (x, v) is not Markovian, unless $\varphi(t) = \delta(t)$. It can, however, be approximated by a component of a higher-dimensional Markov process. Indeed, if the Laplace transform $\hat{\varphi}(s)$ of the memory function is an analytic function, it can be represented as a continued fraction

$$\hat{\varphi}(s) = \cfrac{\Delta_1^2}{s + \gamma_1 + \cfrac{\Delta_2^2}{\ddots}}, \quad \gamma_i \geq 0, \quad \Delta_i^2 > 0. \tag{8.10}$$

If the continued fraction is approximated by a constant, then $\varphi(t) = \delta(t)$. In general, however, the expansion is infinite. For example, the conductivity of an ionic crystal is often assumed to vanish above a certain cutoff frequency ω, so that $\varphi(t) = \sin \omega t/\omega t$ and the expansion is infinite. This is the case in other applications as well [91], [209], [2].

A model based on eq. (8.7) can be understood as the limit of a sequence of models defined by eq. (8.7) with $\varphi(t)$ replaced by $\varphi_N(t)$. Given a function $\varphi(t)$, whose Laplace transform is given by eq. (8.10), an approximating sequence, $\varphi_N(t)$, can be constructed by truncating the continued fraction at level N. Thus the approximating sequence has memory kernels whose Laplace transforms are rational functions. In addition, the boundedness of $\varphi(t)$ implies that

$$\lim_{s\to\infty} \hat{\varphi}_N(s) = 0. \tag{8.11}$$

Note that a scalar stationary Gaussian process, whose autocorrelation function is $\gamma k_B T \varphi_N(t)/m$, can be constructed as the output of a linear system of stochastic differential equations. Indeed, assume

$$\frac{\gamma k_B T}{m} \hat{\varphi}_N(s) = \frac{a_1 s^{N-1} + a_2 s^{N-2} + \cdots + a_N}{s^N + b_1 s^{N-1} + \cdots + b_N},$$

where a_j, b_j are real constants. Define

$$dx_j = [-b_j x_j(t) + x_{j+1}(t)]\,dt + a_j\,dw_j, \quad j = 1, 2, \ldots, N, \tag{8.12}$$

where $x_{N+1}(t) = 0$ and $w_j(t)$ are independent Brownian motions. The process $x(t) = x_1(t)$ is the required scalar stationary Gaussian process, whose autocorrelation function is $(\gamma k_B T/m)\varphi_N(t)$ [213].

Exercise 8.3 (**$x_1(t)$ is a scalar stationary Gaussian process**)**.** Write the system (8.12) in matrix form and prove that $x(t) = x_1(t)$ is a scalar stationary Gaussian process, whose autocorrelation function is $(\gamma k_B T/m)\varphi_N(t)$. □

If the memory function is exponential, $\varphi(t) = \alpha e^{-\alpha t}$, the fluctuation-dissipation principle (8.8) implies that the noise $\xi(t)$ in the generalized Langevin equation (8.7) is colored noise (the Ornstein–Uhlenbeck process, see Section 1.4).

Theorem 8.1.2 (Smoluchowski limit of the generalized Langevin equation with colored noise). *In the limit $\gamma \to \infty$ the trajectories of the generalized Langevin equation* (8.7) *with colored noise converge to these of the Smoluchowski equation* (8.2).

Proof. Scaling $\xi(t) = \sqrt{\gamma}\alpha v(t)$, we find that $\dot{v}(t) = -\alpha v(t) + \sqrt{2k_B T/m}\,\dot{w}$. Defining

$$u = -\gamma \int_0^t \exp\{-\alpha(t-s)\}\, v(s)\, ds + v(t),$$

we transform (8.9) to the system

$$\dot{x} = v, \quad \dot{v} = \sqrt{\gamma}\alpha u - U'(x), \quad \dot{u} = -\sqrt{\gamma}v - \alpha u + \sqrt{2\varepsilon}\,\dot{w}. \tag{8.13}$$

Thus the system of integro-differential equations (8.9) for the non-Markovian process (x, v) has been transformed into the system of stochastic differential equations (8.13) for the Markov process (x, v, u).

To investigate (8.13) in the Smoluchowski limit of $\gamma \to \infty$, we change the time according to eq. (8.5). This transforms the system (8.13) to

$$\overset{\circ}{x}(s) = \gamma v(s) \tag{8.14}$$

$$\overset{\circ}{v}(s) = \gamma^{3/2}\alpha u(s) - \gamma U'(x(s)) \tag{8.15}$$

$$\overset{\circ}{u}(s) = -\gamma^{3/2}v(s) - \gamma\alpha u(s) + \sqrt{2\gamma\varepsilon}\,\overset{\circ}{w}(s), \tag{8.16}$$

where $w(s)$ is Brownian motion.

Equations (8.15), (8.16) can be solved for $v(s)$ and $u(s)$, as functions of $x(s)$ and s. Then $x(s)$ is determined from eq. (8.14). In order to find the leading term in the expansion of the process $x(t)$ for large γ, we take the limit of (8.14) as $\gamma \to \infty$. We first solve the system (8.15), (8.16) to find $v(s)$. The fundamental matrix of the system is given by

$$\boldsymbol{M} = \begin{bmatrix} M_{11}(s) & M_{12}(s) \\ M_{21}(s) & M_{22}(s) \end{bmatrix} = e^{-\alpha s/2} \begin{bmatrix} (a/b)\sin cs + \cos cs & (2\gamma^{3/2}a/b)\sin cs \\ -(2\gamma^{3/2}a/b)\sin cs & -(a/b)\sin cs - \cos cs \end{bmatrix},$$

where $a=\gamma\alpha$, $b=\sqrt{4\gamma^3\alpha-\gamma^2\alpha^2}$, and $c=b/2$. Therefore $v(s)$ is given by

$$v(s)=M_{11}(s)A_0+M_{12}(s)B_0-\gamma\int_0^s M_{11}(s-\tau)U'(x(\tau))\,d\tau$$
$$+\left(2\sqrt{2\varepsilon}\,\gamma^{5/2}\alpha/b\right)\int_0^s M_{12}(s-\tau)\overset{\circ}{w}(\tau)\,d\tau,$$

where A_0 and B_0 are constants determined by initial conditions. Because

$$\lim_{\gamma\to\infty}M_{11}(s)=\lim_{\gamma\to\infty}M_{12}(s)=0,\quad \lim_{\gamma\to\infty}\gamma^2M_{11}(s-\tau)=\delta(s-\tau)$$
$$\lim_{\gamma\to\infty}\gamma^{3/2}M_{12}(s-\tau)=\delta(s-\tau),$$

we obtain the limiting equation for the process from eq. (8.14), exactly as in the non-memory case, in the form (8.6); hence (8.2) [61]. □

Exercise 8.4 (Convert memory into auxiliary variables). Consider the case

$$\varphi(t)=\frac{1}{2}\left[\alpha_1e^{-\alpha_1t}+\alpha_2e^{-\alpha_2t}\right].$$

Show that the fluctuation-dissipation principle (8.8) implies that

$$\xi(t)=\sqrt{\frac{\gamma}{2}}\,\alpha_1v_1(t)+\sqrt{\frac{\gamma}{2}}\,\alpha_2v_2(t),$$

where $v_1(t)$ and $v_2(t)$ are the colored noises defined by

$$\dot v_i(t)=-\alpha_iv_i(t)+\sqrt{\frac{2k_BT}{m}}\,\dot w_i\ \text{ for } i=1,2$$

and $w_i(t)$ are independent Brownian motions. Define

$$u_i=-\sqrt{\frac{\gamma}{2}}\int_0^t\exp\{-\alpha_i(t-s)\}\,v(s)\,ds+v_i(t)$$

and transform the generalized Langevin equation into the system

$$\dot x=v,\quad \dot v=\sqrt{\frac{\gamma}{2}}\,\alpha_1u_1+\sqrt{\frac{\gamma}{2}}\,\alpha_2u_2-U'(x)$$
$$\dot u_i=-\sqrt{\frac{\gamma}{2}}\,v-\alpha_iu_i+\sqrt{\frac{2k_BT}{m}}\,\dot w_i\quad (i=1,2).$$

Find the Smoluchowski limit of $x(t)$. □

Exercise 8.5 (Approximate kernels). Use the construction in Exercise 8.3 to construct a sequence of generalized Langevin equations with memory kernels $\varphi_N(t)$, whose Laplace transforms are rational functions, and driving stationary Gaussian noises $\xi_N(t)$ that satisfy the fluctuation-dissipation principle eq. (8.8). Find the Smoluchowski limit for each N. Can the limits $N\to\infty$ and $\gamma\to\infty$ be interchanged? (See [61]). □

8.2 Smoluchowski expansion in the entire space

The high damping limit of the Langevin equation can be studied through the Fokker–Planck equation for the system (8.3), (8.4). The joint transition pdf of the pair $(x(t), v(t))$ is the solution of the FPE (7.102) (see also (5.31))

$$\begin{aligned}\frac{\partial p(x,v,t)}{\partial t} =& -\frac{\partial v p(x,v,t)}{\partial x} + \frac{\partial}{\partial v}\left[\gamma v + U'(x)\right] p(x,v,t)\\ &+ \frac{\gamma k_B T}{m} \frac{\partial^2 p(x,v,t)}{\partial v^2}, \end{aligned} \tag{8.17}$$

where $U(x)$ is potential per unit mass and k_B is Boltzmann's constant. We consider first approximations of $p(x,v,t)$ in the entire phase space in the more general case of state-dependent damping

$$\gamma(x) = \Gamma\gamma_0(x), \tag{8.18}$$

where $\Gamma = \max_x \gamma(x) \gg 1$. Such state dependence is expected, for example, in ionic channels, where the diffusion coefficient may vary along the pore [66].

Theorem 8.2.1 (Smoluchowski limit). *If the initial value problem for the Smoluchowski equation*

$$\frac{\partial P^0(x,t)}{\partial t} = \frac{\partial}{\partial x}\left\{\frac{1}{\gamma(x)}\left[\frac{k_B T}{m}\frac{\partial P^0(x,t)}{\partial x} + U'(x)P^0(x,t)\right]\right\} \tag{8.19}$$

$$P^0(x,0) = \varphi(x) \tag{8.20}$$

has a unique solution for $\varphi(x) \in L^1(\mathbb{R}) \cap L^\infty(\mathbb{R})$ *such that*

$$\frac{\partial P^0(x,t)}{\partial x} + \frac{mU'(x)}{k_B T}P^0(x,t)$$

is bounded for all $t > t_0$ *and* $x \in \mathbb{R}$, *then for all* (x,v,t) *such that*

$$\gamma(x) \gg \left| v\frac{\partial \log P^0(x,t)}{\partial x} + \frac{mU'(x)}{k_B T}\right|, \tag{8.21}$$

the pdf $p(x,v,t)$ *has the asymptotic representation*

$$\begin{aligned} &p(x,v,t) \\ &= \sqrt{\frac{m}{2\pi k_B T}} e^{-mv^2/2k_B T}\left\{P^0(x,t) - \frac{v}{\gamma(x)}\left[\frac{\partial P^0(x,t)}{\partial x} + \frac{mU'(x)}{k_B T}P^0(x,t)\right]\right.\\ &\left. + O\left(\frac{1}{\gamma^2(x)}\right)\right\}. \end{aligned} \tag{8.22}$$

Proof. With the scaling $t = \Gamma s$ the Fokker–Planck equation (8.17) takes the form

$$\frac{1}{\Gamma}\frac{\partial p}{\partial s} = -v\frac{\partial p}{\partial x} + \frac{\partial\left[\Gamma\gamma_0(x)v + U'(x)\right]p}{\partial v} + \frac{\Gamma\gamma_0(x)k_B T}{m}\frac{\partial^2 p}{\partial v^2}, \tag{8.23}$$

which can be written in operator form as

$$\Gamma L_0 p + L_1 p + \frac{1}{\Gamma} L_2 p = 0, \tag{8.24}$$

where

$$L_0 p = \gamma_0(x)\left[\frac{\partial}{\partial v} v p + \frac{k_B T}{m}\frac{\partial^2 p}{\partial v^2}\right], \quad L_1 p = -v\frac{\partial p}{\partial x} + \frac{\partial}{\partial v} U'(x) p, \quad L_2 p = -\frac{\partial p}{\partial s}.$$

Assuming the expansion $p = p^0 + \Gamma^{-1} p^1 + \Gamma^{-2} p^2 + \cdots$, we obtain from (8.24) the hierarchy of equations

$$L_0 p^0 = 0 \tag{8.25}$$

$$L_0 p^1 = -L_1 p^0 \tag{8.26}$$

$$L_0 p^2 = -L_1 p^1 - L_2 p^0 \tag{8.27}$$

$$\vdots$$

From eq. (8.25), we find that the only L^1 (integrable) solution is $p^0(x,v,s) = P^0(x,s)e^{-mv^2/2k_BT}$, where $P^0(x,s)$ is yet an undetermined function. Because the homogeneous equation (8.25) has a nontrivial solution $p^0(x,v,s)$, a necessary condition for the inhomogeneous equation (8.26) to have an integrable solution is that the right-hand side, $-L_1 p^0$, satisfies the solvability condition $\int L_1 p^0 \, dv = 0$, which is satisfied (see Section 7.3.1). The integrable solution of eq. (8.26) is given by

$$\begin{aligned} p^1(x,v,s) = &-\frac{v}{\gamma_0(x)} e^{-mv^2/2k_BT}\left[\frac{\partial P^0(x,s)}{\partial x} + \frac{mU'(x)}{k_BT} P^0(x,s)\right] \\ &+ P^1(x,s)e^{-mv^2/2k_BT}, \end{aligned}$$

where $P^1(x,s)$ is yet an undetermined function. The solvability condition for eq. (8.27) is that $\int \left[L_1 p^1 + L_2 p^0\right] dv = 0$; that is,

$$\begin{aligned} &\int \exp\left\{-\frac{mv^2}{2k_BT}\right\}\left\{-\frac{\partial P^0(x,s)}{\partial s}\right. \\ &\left. + v^2 \frac{\partial}{\partial x}\frac{1}{\gamma_0(x)}\left[\frac{\partial P^0(x,s)}{\partial x} + \frac{mU'(x)}{k_BT}P^0(x,s)\right]\right\} dv - \int U'(x)\frac{\partial p^1}{\partial v}\, dv = 0, \end{aligned}$$

or equivalently,

$$\frac{\partial P^0(x,s)}{\partial s} = \frac{\partial}{\partial x}\frac{1}{\gamma_0(x)}\left[\frac{k_BT}{m}\frac{\partial P^0(x,s)}{\partial x} + U'(x)P^0(x,s)\right]. \tag{8.28}$$

Scaling γ back into eq. (8.28), we obtain the Smoluchowski–Fokker–Planck equation (8.19). This is the Fokker–Planck equation corresponding to the Smoluchowski–Langevin equation (8.2). Proceeding as above, we find that $P^1(x,s) = 0$ in one dimension, hence it follows that the Smoluchowski approximation is given by (8.22). □

The unidirectional probability flux density in the x-direction in phase space is $J_x(x,v,t) = vp\,(x,v,t)$ (see Exercise 5.20).

Corollary 8.2.1 (Unidirectional Smoluchowski flux). *The probability flux densities in the x-direction are*

$$\begin{aligned} J_{LR,RL}(x,t) &= \int_0^{\pm\infty} vp\,(x,v,t)\,dv \qquad (8.29)\\ &= \int_0^{\pm\infty} \sqrt{\frac{m}{2\pi k_BT}}e^{-mv^2/2k_BT}\Bigg\{vP^0(x,t)\\ &\quad - \frac{v^2}{\gamma(x)}\left[\frac{\partial P^0(x,t)}{\partial x} + \frac{mU'(x)}{k_BT}P^0(x,t)\right]\\ &\quad + O\left(\frac{1}{\gamma^2(x)}\right)\Bigg\}\,dv\\ &= \frac{k_BT}{m}P^0(x,t) \mp \frac{1}{2\gamma(x)}\left[\frac{k_BT}{m}\frac{\partial P^0(x,t)}{\partial x} + U'(x)P^0(x,t)\right]\\ &\quad + O\left(\frac{1}{\gamma^2(x)}\right), \end{aligned}$$

and remain finite in the diffusion limit $\gamma(x) \to \infty$.

The net flux density in the x-direction is given by

$$\begin{aligned} J_x(x,t) &= J_{LR}(x,t) - J_{R,L}(x,t) = \int vp\,(x,v,t)\,dv\\ &= -\frac{1}{\gamma(x)}\left[\frac{k_BT}{m}\frac{\partial P^0(x,t)}{\partial x} + U'(x)P^0(x,t)\right] + O\left(\frac{1}{\gamma^2(x)}\right), \end{aligned} \qquad (8.30)$$

which agrees with the one-dimensional flux (8.32). Thus the Smoluchowski approximation (8.22) can be written as

$$\begin{aligned} &p\,(x,v,t)\\ &= \sqrt{\frac{m}{2\pi k_BT}}e^{-mv^2/2k_BT}\left\{P^0(x,t) + \frac{m}{k_BT}J_x(x,t)v + O\left(\frac{1}{\gamma^2(x)}\right)\right\}. \end{aligned} \qquad (8.31)$$

The Smoluchowski–Fokker–Planck equation (8.19) can be written in divergence form (5.32) as $P_t^0(x,t) = -\nabla \cdot J(x,t)$, where the flux density is given by

$$J(x,t) = -\frac{1}{\gamma(x)}\left[\frac{k_BT}{m}\frac{\partial P^0(x,t)}{\partial x} + U'(x)P^0(x,t)\right]. \qquad (8.32)$$

In higher dimensions, we obtain $P_t^0(\boldsymbol{x},t) = -\nabla \cdot \boldsymbol{J}(\boldsymbol{x},t)$, where

$$\boldsymbol{J}(\boldsymbol{x},t) = -\frac{1}{\gamma(\boldsymbol{x})}\left[\frac{k_BT}{m}\nabla P^0(\boldsymbol{x},t) + P^0(\boldsymbol{x},t)\nabla U(\boldsymbol{x})\right]. \qquad (8.33)$$

The expansion (8.31) shows that away from equilibrium the probability density function depends on flux, no matter what the friction is, so that the velocity distribution is not Maxwellian. When the flux is imposed, as in most experimental situations, then both terms, viz. $P^0(x,t) + (m/k_BT)J_x(x,t)v$, must be retained in the expansion (8.31). The presence of both terms insures that (8.31) is valid for all values of flux, thus for all barrier shapes.

8.3 Boundary conditions in the Smoluchowski limit

We consider a one-dimensional free Langevin particle diffusing in the interval $[0, d]$, starting at x_0 with velocity v_0, with a prescribed boundary behavior (see below). The Langevin equation for its trajectory is

$$\ddot{x} + \gamma\dot{x} = \sqrt{2\gamma\varepsilon}\,\dot{w}, \tag{8.34}$$

where γ is the friction parameter, ε is the noise strength, and w is independent white (Gaussian) noise. The Langevin equation (8.34) is equivalent to the phase space dynamics

$$\dot{x} = v, \quad \dot{v} = -\gamma v + \sqrt{2\gamma\varepsilon}\,\dot{w}, \tag{8.35}$$

with the initial conditions $x(0) = x_0$, $v(0) = v_0$. The transition probability density function of a free particle is the solution of the Fokker–Planck equation

$$\frac{\partial p}{\partial t} = \mathfrak{L}_{x,v}p = -v\frac{\partial p}{\partial x} + \frac{\partial\,(\gamma v p)}{\partial v} + \varepsilon\gamma\frac{\partial^2 p}{\partial v^2}, \tag{8.36}$$

with the initial condition $p\,(x,v,t=0) = \delta(x-x_0)\delta(v-v_0)$. In the case of reflecting boundaries at $x = 0$, d the probability density function satisfies the reflecting boundary conditions

$$p\,(0,v,t) = p\,(0,-v,t) \text{ for } v>0, \quad p\,(d,v,t) = p\,(d,-v,t) \text{ for } v>0, \tag{8.37}$$

whereas in the case of absorbing boundaries at $x = 0$, d the probability density function satisfies the absorbing boundary condition

$$p\,(0,v,t) = 0 \text{ for } v>0, \quad p\,(d,v,t) = 0 \text{ for } v<0. \tag{8.38}$$

The case of mixed boundaries, for example, an absorbing boundary at $x = 0$ and reflecting boundary at $x = d$ can be solved by the method of reflection; that is, by solving the problem in an interval of twice the length $[0, 2d]$ with absorbing boundary conditions at both ends. Here we find the long- and short-time behavior of the probability density function $p\,(x,v,t \mid x_0, v_0)$.

Our aim is to find boundary conditions for the Smoluchowski limit of the FPE. Boundary layers arise in the Smoluchowski expansion of the FPE and affect the calculation of such important quantities as the probability density function, the MFPT to the boundary, and the probability density function of the recurrence time of the (free) Brownian particle.

The solution of the initial boundary value problem (8.37) or (8.38) for the FPE (8.36) is constructed by separation of variables in the infinite series form

$$p(x,v,t)=\sum_{n=0}^{\infty}a_n e^{-\lambda_n t}p_n(x,v), \tag{8.39}$$

where $p_n(x,v)$ are the eigenfunctions of the Fokker–Planck operator

$$\mathfrak{L}_{x,v}p_n(x,v)+\lambda_n p_n(x,v)=0,$$

that also satisfy the boundary condition (8.37) or (8.38). We construct the eigenfunctions in the form $p_n(x,v) = e^{-v^2/2\varepsilon}q_n(x,v)$, and we rescale the velocity $\hat{v}=v/\sqrt{\varepsilon}$. Upon dropping the hat we obtain an eigenvalue equation for q_n,

$$q_{vv}-vq_v-\frac{\sqrt{\varepsilon}}{\gamma}vq_x+\frac{\lambda}{\gamma}q=0, \tag{8.40}$$

with suitable boundary conditions.

In the case of reflecting walls the smallest eigenvalue is $\lambda_0=0$, with the corresponding eigenfunction $q_0(x,v)=1$ or $p_0(x,v)=e^{-v^2/2\varepsilon}$; that is, a Maxwellian velocity distribution and a uniform displacement distribution, as expected in the steady-state equilibrium. Finding the other eigenvalues and eigenfunctions is not as straightforward a task.

It was shown in Section 8.2 that the Smoluchowski approximation of large damping to the free particle Fokker–Planck equation (8.36) reduces to

$$p_{\text{approx}}(x,v,t)=e^{-v^2/2\varepsilon}p(x,t),$$

where $p(x,t)$ is the solution of the diffusion equation $p_t=Dp_{xx}$, and $D=\varepsilon/\gamma$ is the diffusion coefficient. The reflecting boundary conditions $p_x(0,t)=p_x(d,t)=0$, or the absorbing boundary conditions $p(0,t)=p(d,t)=0$, are overdetermined, because the well-posed boundary conditions (8.38) are limited only to the half-lines $x=0, v>0$ and $x=d, v<0$ and not to the entire lines $x=0, -\infty<v<\infty$ and $x=d, -\infty<v<\infty$ that $p_{\text{approx}}(x,v,t)$ satisfies. This indicates the formation of a boundary layer at the endpoints $x=0$ and $x=d$.

It turns out that the determination of the boundary conditions for the Smoluchowski approximation and the analysis of the local behavior there is an extraordinarily hard analytical problem. To determine these conditions, we note first that separation of variables in the reduced Smoluchowski equation gives

$$p^R(x,t)=\frac{A_0}{2}+\sum_{n=1}^{\infty}A_n e^{-Dn^2\pi^2t/d^2}\cos\frac{n\pi x}{d}$$

$$p^A(x,t)=\sum_{n=1}^{\infty}B_n e^{-Dn^2\pi^2t/d^2}\sin\frac{n\pi x}{d},$$

where p^R, p^A are the solutions of the Smoluchowski equation with reflecting and absorbing boundaries, respectively. The long time behavior of solutions is determined by the smallest (nonvanishing) eigenvalue $D\pi^2/d^2 = O(\gamma^{-1})$. The Smoluchowski approximation $p^{R,A}_{\text{approx}}(x,v,t) = e^{-v^2/2\varepsilon} p^{R,A}(x,t)$, however, does not satisfy the boundary conditions (8.37) or (8.38), respectively.

Theorem 8.3.1 ([166], [96]). *The smallest eigenvalue of the absorbing boundary value problem* (8.38) *for the FPE* (8.36) *is given in the limit* $\gamma d/\sqrt{\varepsilon} \gg 1$ *by*

$$\lambda_1 = \frac{\pi^2 \varepsilon}{\gamma d^2}\left[1 + 2\zeta\left(\frac{1}{2}\right)\sqrt{\varepsilon}/\gamma d + O\left(\frac{\varepsilon}{\gamma^2 d^2}\right)\right],$$

where $\zeta(\cdot)$ *is Riemann's zeta function. The first eigenfunction* $p_1(x,v)$ *is given up to a transcendently small error by*

$$\begin{aligned} & p_1(x,v) \\ &= e^{-v^2/2\varepsilon}\left[\frac{e^{\sqrt{\lambda_1/\gamma}v/\sqrt{\varepsilon}}}{\alpha_0^- N(\alpha_0^-)} e^{-i\sqrt{\lambda_1}x/\sqrt{\varepsilon}} + \frac{e^{-\sqrt{\lambda_1/\gamma}v/\sqrt{\varepsilon}}}{\alpha_0^+ N(\alpha_0^+)} e^{i\sqrt{\lambda_1}x/\sqrt{\varepsilon}}\right] \\ &= A_- e^{-(v-\sqrt{\lambda_1\varepsilon/\gamma})^2/2\varepsilon - i\sqrt{\lambda_1/\varepsilon}\,x} + A_+ e^{-(v+\sqrt{\lambda_1\varepsilon/\gamma})^2/2\varepsilon + i\sqrt{\lambda_1/\varepsilon}\,x}, \end{aligned} \tag{8.41}$$

where $A_\pm$ *are known constants.*

Proof. We construct a boundary layer solution at the left boundary. We introduce a boundary layer variable $\tilde{x} = \gamma x/\sqrt{\varepsilon}$, so eq. (8.40) is rewritten as

$$q_{vv} - v q_v - v q_{\tilde{x}} + \frac{\lambda}{\gamma} q = 0. \tag{8.42}$$

Separation of variables in eq. (8.42); that is, $q(\tilde{x}, v) = A(\tilde{x})V(v)$, leads to an eigenvalue problem $V'' - vV' + (\lambda/\gamma)V = \alpha v V$, and $A' = \alpha A$ which gives $A(\tilde{x}) = Ce^{\alpha\tilde{x}}$. Substituting $V(v) = e^{v^2/4} f(v)$ in this equation results in the parabolic cylinder equation for f [1],

$$f'' - \left[\frac{1}{4}(v+2\alpha)^2 - \left(\frac{1}{2} + \alpha^2 + \frac{\lambda}{\gamma}\right)\right] f = 0. \tag{8.43}$$

The solution of (8.43) with the appropriate asymptotic behavior as $v \to +\infty$ is

$$f_1(v) = U\left(-\left(\frac{1}{2} + \alpha^2 + \frac{\lambda}{\gamma}\right), v + 2\alpha\right), \tag{8.44}$$

where $U(\cdot,\cdot)$ is the parabolic cylinder function with index $-(\frac{1}{2} + \alpha^2 + \lambda/\gamma)$. Similarly, the appropriate solution as $v \to -\infty$ is

$$f_2(v) = U\left(-\left(\frac{1}{2} + \alpha^2 + \frac{\lambda}{\gamma}\right), -(v + 2\alpha)\right). \tag{8.45}$$

We seek a solution that has the desired asymptotic behavior in both $v \to \pm\infty$, therefore we require the two solutions f_1 and f_2 to be linearly dependent, which holds whenever

$$\alpha^2 + \frac{\lambda}{\gamma} = n = 0, 1, 2, \ldots . \tag{8.46}$$

In this case the parabolic cylinder functions are closely related to the Hermite polynomials

$$U\left(-\left(n+\frac{1}{2}\right), v+2\alpha\right) = e^{-(v+2\alpha)^2/4} He_n(v+2\alpha), \tag{8.47}$$

so that

$$V(v,\alpha) = e^{-\alpha^2} e^{-\alpha v} He_n(v+2\alpha). \tag{8.48}$$

The long-time decay behavior is determined by the smallest nonvanishing eigenvalue $\lambda_1 \ll 1$, because it is of the order of γ^{-1}. Therefore,

$$\alpha^{\pm}(0,\lambda_1) = \pm i\sqrt{\frac{\lambda_1}{\gamma}}, \quad \alpha^{\pm}(n,\lambda_1) = \pm\sqrt{n - \frac{\lambda_1}{\gamma}}, \quad n = 1, 2, 3, \ldots, \tag{8.49}$$

and the corresponding eigenfunctions are given by

$$\begin{aligned} V_0^{\pm}(v) &= e^{\lambda_1/\gamma} e^{\mp\sqrt{\lambda_1/\gamma}\, v} \\ V_n^{\pm}(v) &= e^{-n+\lambda_1/\gamma} e^{\mp\sqrt{n-\lambda_1/\gamma}\, v} He_n(v \pm 2\sqrt{n-\lambda_1/\gamma}), \quad n = 1, 2, 3, \ldots . \end{aligned}$$

The boundary layer solution is given by the infinite superposition

$$\begin{aligned} q(\tilde{x}, v) =& C_0^+ V_0^+(v) e^{i\sqrt{\lambda_1/\gamma}\,\tilde{x}} + C_0^- V_0^-(v) e^{-i\sqrt{\lambda_1/\gamma}\,\tilde{x}} \\ &+ \sum_{n=1}^{\infty} C_n^+ V_n^+(v) e^{\sqrt{n-\lambda_1/\gamma}\,\tilde{x}} + \sum_{n=1}^{\infty} C_n^- V_n^-(v) e^{-\sqrt{n-\lambda_1/\gamma}\,\tilde{x}}. \end{aligned} \tag{8.50}$$

For the boundary layer solution to match the outer solution, we must have $C_n^+ = 0$ for $n = 1, 2, \ldots$, so that exponentially growing solutions in $\tilde{x}$ are ruled out

$$\begin{aligned} &q(\tilde{x}, v) \\ &= C_0^+ V_0^+(v) e^{i\sqrt{\lambda_1/\gamma}\,\tilde{x}} + C_0^- V_0^-(v) e^{-i\sqrt{\lambda_1/\gamma}\,\tilde{x}} + \sum_{n=1}^{\infty} C_n^- V_n^-(v) e^{-\sqrt{n-\lambda_1/\gamma}\,\tilde{x}}. \end{aligned} \tag{8.51}$$

The coefficients C_0^+, C_0^- and C_n^- are determined from the boundary condition $q(\tilde{x} = 0, v) = C_0^+ V_0^+(v) + C_0^- V_0^-(v) + \sum_{n=1}^{\infty} C_n^- V_n^-(v)$, so that in the case of a reflecting boundary we get for $v > 0$

$$\begin{aligned} &C_0^+ V_0^+(v) + C_0^- V_0^-(v) + \sum_{n=1}^{\infty} C_n^- V_n^-(v) \\ &= C_0^+ V_0^+(-v) + C_0^- V_0^-(-v) + \sum_{n=1}^{\infty} C_n^- V_n^-(-v), \end{aligned}$$

and in the case of an absorbing boundary, for $v > 0$,

$$C_0^+ V_0^+(v) + C_0^- V_0^-(v) + \sum_{n=1}^{\infty} C_n^- V_n^-(v) = 0.$$

This is a half-range expansion problem, as we are interested in an expansion of a function defined on the semi-infinite axis by only half the eigenfunctions. We apply the method of the half-range expansion [96] to solve the problem at hand.

Applying the Laplace transform with respect to the variable $\tilde{x}$ to eq. (8.51) gives

$$\hat{q}(s,v) = \frac{C_0^+ V_0^+(v)}{s - i\sqrt{\dfrac{\lambda_1}{\gamma}}} + \frac{C_0^- V_0^-(v)}{s + i\sqrt{\dfrac{\lambda_1}{\gamma}}} + \sum_{n=1}^{\infty} \frac{C_n^- V_n^-}{s + \sqrt{n - \dfrac{\lambda_1}{\gamma}}}, \tag{8.52}$$

which has simple poles at $\alpha_0^\pm = \pm i\sqrt{\lambda_1/\gamma}$ and $\alpha_n^- = -\sqrt{n - \lambda_1/\gamma}$, $n = 1, 2, \ldots$. Doing the same with respect to $\tilde{x}$ to the problem (8.42) and solving it, leads to a comparison of the solution with (8.52). We have

$$\hat{q}_{vv} - v\hat{q}_v - sv\hat{q} + \frac{\lambda_1}{\gamma}\hat{q} = 0, \quad \text{for } v > 0, \tag{8.53}$$

in the case of the absorbing boundary, and

$$\hat{q}_{vv} - v\hat{q}_v - sv\hat{q} + vq(\tilde{x} = 0, v) + \frac{\lambda_1}{\gamma}\hat{q} = 0,$$

in the case of the reflecting boundary.

We confine the analysis to the case of an absorbing boundary. The solution of eq. (8.53) is

$$\hat{q}(s,v) = e^{v^2/4} U(v + 2s) F(s), \tag{8.54}$$

where U denotes the parabolic cylinder function of index $-\left(\frac{1}{2} + s^2 + \lambda_1/\gamma\right)$, and the function $F(s)$ is yet undetermined. The two representations must have the same singularities in s, because for $v > 0$, eqs. (8.52) and (8.54) represent the same function. We also note that U is an entire function of s. Therefore

$$F(s) = \frac{E(s)}{\left(s^2 + \dfrac{\lambda_1}{\gamma}\right) N(s)}, \tag{8.55}$$

where $E(s)$ is an entire function, and

$$N(s) = \prod_{n=1}^{\infty} c_n (s - \alpha_n^-), \tag{8.56}$$

with coefficients c_n chosen so the product (8.56) converges, and it has suitable behavior for $s \gg 1$,

$$N(s) = \prod_{k=1}^{\infty} \left(\sqrt{1 - \frac{\lambda_1/\gamma}{k}} + \frac{s}{\sqrt{k}} \right) e^{-2s(\sqrt{k} - \sqrt{k-1})} \left(\frac{k+1}{k} \right)^{(s^2 + \lambda_1/\gamma)/2}. \tag{8.57}$$

Here the first factor gives $N(s)$ a simple zero at $s = -\sqrt{k - \lambda_1/\gamma} = \alpha_k^-$, and the last two terms are needed to make the product converge. (Note that the kth term goes as $1 + O\left(k^{-3/2}\right)$.) These particular convergence factors were chosen because they conveniently form a telescoping product. Clearly, $N(s)$ is an entire analytic function, because the product converges uniformly in every bounded region in the complex s plane and the only zeros of $N(s)$ are the simple zeros at $s = \alpha_n^-$, $n = 1, 2, \ldots$.

In order to determine $E(s)$ we analyze (8.54) for $|s| \gg 1$. It can be shown that the parabolic cylinder function and Hagan's special function $N(s)$ have the asymptotic behavior [166], [96], [127], [221]

$$U(v+2s) \sim \sqrt{2\pi}\, e^{-v^2/4} s^{1/3+\lambda_1/\gamma} \exp\left\{ s^2 \log s - \frac{1}{2} s^2 \right\} \mathrm{Ai}(s^{1/3} v)$$

$$N(s) \sim (2\pi)^{-1/4} s^{-(1/2-\lambda_1/\gamma)} \exp\left\{ s^2 \log s - \frac{1}{2} s^2 \right\}$$

$$\times \exp\left\{ \sum_{j=1}^{\infty} \frac{(-1)^{j-1}}{j s^j} \zeta\left(1 - \frac{\lambda_1}{\gamma}, -j/2\right) \right\},$$

where Ai denotes the Airy function. Therefore

$$\hat{q}(s,v) = e^{v^2/4} U(v+2s) F(s) \sim \frac{(2\pi)^{3/4}}{s^{7/6}} \mathrm{Ai}(s^{1/3} v) E(s). \tag{8.58}$$

On the other hand, from eq. (8.52) it follows that

$$\hat{q}(s,v) \sim \frac{1}{s} \left\{ C_0^+ V_0^+(v) + C_0^- V_0^-(v) + \sum_{n=1}^{\infty} C_n^- V_n^-(v) \right\} \quad \text{as } |s| \to \infty. \tag{8.59}$$

For (8.58) to be consistent with (8.59), especially at $v = 0$, we must have $|E(s)| \leq$ constant $\cdot\ |s|^{1/6}$ as $|s| \to \infty$. The Liouville theorem implies that it is a constant, $E(s) \equiv E$, because $E(s)$ is an entire analytic function. We have established, so far, that $\hat{q}(s,v) = e^{v^2/4} U(v+2s) E / (s^2 + \lambda_1/\gamma) N(s)$. Next, we invert the Laplace transform $\hat{q}(s,v)$ by integrating along the Bromwich contour B,

$$q(\tilde{x}, v) = \frac{1}{2\pi i} \int_B e^{s\tilde{x}} \hat{q}(s,v)\, ds \quad \text{for } v \geq 0.$$

Cauchy's theorem shows that for any $\tilde{x} \geq 0$ we can deform B into the Hankel contour H. This reduces the integral to the sum of residues at the poles. We find that

$$q(\tilde{x}, v) \tag{8.60}$$
$$= E \left\{ \frac{V_0^-(v)}{2\alpha_0^- N(\alpha_0^-)} e^{\alpha_0^- \tilde{x}} + \frac{V_0^+(v)}{2\alpha_0^+ N(\alpha_0^+)} e^{\alpha_0^+ \tilde{x}} + \sum_{n=1}^{\infty} \frac{V_n^-(v)}{n N'(\alpha_n^-)} e^{\alpha_n^- \tilde{x}} \right\},$$

is the boundary layer solution near the left boundary, because

$$e^{v^2/4}U(v+2s)\Big|_{s=\alpha_n^-} = V_n^-(v).$$

We construct the right boundary layer solution near $x=d$ in a similar way. We introduce a boundary layer variable $\bar{x} = \gamma(d-x)/\sqrt{\varepsilon}$, and reverse the velocity $\bar{v} = -v$, so eq. (8.40) is rewritten as $q_{\bar{v}\bar{v}} - \bar{v}q_{\bar{v}} - \bar{v}q_{\bar{x}} + (\lambda/\gamma)q = 0$, with the boundary condition $q(\bar{x}=0,\bar{v}) = 0$ for $\bar{v} > 0$. The previous analysis of this section shows that the solution is given by

$$q(\bar{x},v) = M\left\{\frac{V_0^-(-v)}{2\alpha_0^- N(\alpha_0^-)}e^{\alpha_0^-\bar{x}} + \frac{V_0^+(-v)}{2\alpha_0^+ N(\alpha_0^+)}e^{\alpha_0^+\bar{x}} + \sum_{n=1}^{\infty}\frac{V_n^-(-v)}{nN'(\alpha_n^-)}e^{\alpha_n^-\bar{x}}\right\}. \tag{8.61}$$

Next, we construct a uniform approximation to the first eigenfunction. The boundary layer solutions (8.60) and (8.61) have an oscillatory part in the form of $e^{\alpha_0^\pm \tilde{x}}$ and exponentially decaying part $e^{\alpha_n^- \tilde{x}} (n \geq 1)$. Outside the boundary layer, the exponentially decaying part may be neglected (transcendently small error), so only the oscillatory parts should be matched. Note that there is no need to construct an outer solution, as it is just the oscillatory part of the boundary layer solutions. The matching determines the first eigenvalue.

Matching the oscillatory parts of (8.60) and (8.61) gives the system of equations

$$\frac{E}{2\alpha_0^- N(\alpha_0^-)} = \frac{Me^{\alpha_0^+\gamma d/\sqrt{\varepsilon}}}{2\alpha_0^+ N(\alpha_0^+)}, \quad \frac{E}{2\alpha_0^+ N(\alpha_0^+)} = \frac{Me^{\alpha_0^-\gamma d/\sqrt{\varepsilon}}}{2\alpha_0^- N(\alpha_0^-)},$$

which has a nontrivial solution if and only if

$$e^{2\alpha_0^+\gamma d/\sqrt{\varepsilon}} = N^2(\alpha_0^+)/N^2(\alpha_0^-),$$

or equivalently, if and only if

$$2\alpha_0^+(\gamma d/\sqrt{\varepsilon}) = 2\left(\log N(\alpha_0^+) - \log N(\alpha_0^-)\right) + 2\pi i m$$

for $m = 0, 1, 2, \ldots$. Using the leading-order approximation (8.65) (see Appendix), we obtain

$$\sqrt{\frac{\lambda_1}{\gamma}}\left[\frac{\gamma d}{\sqrt{\varepsilon}} - 2\zeta\left(\frac{1}{2}\right)\right] = \pi m.$$

The smallest eigenvalue is obtained for $m=1$ as

$$\lambda_1 = \frac{\pi^2\gamma}{\left[\frac{\gamma d}{\sqrt{\varepsilon}} - 2\zeta\left(\frac{1}{2}\right)\right]^2},$$

which gives for $\gamma d/\sqrt{\varepsilon} \gg 1$

$$\lambda_1 = \frac{\pi^2\varepsilon}{\gamma d^2}\left[1 + 2\zeta\left(\frac{1}{2}\right)\frac{\sqrt{\varepsilon}}{\gamma d} + O\left(\frac{\varepsilon}{\gamma^2 d^2}\right)\right].$$

The leading-order term $\pi^2\varepsilon/\gamma d^2 = \pi^2 D/d^2$ is the first eigenvalue of Smoluchowski's approximating diffusion problem (8.19). Using only the oscillatory part, the first eigenfunction $p_1(x, v)$ is obtained from (8.60) as (8.41). □

Exercise 8.6 (The marginal density of the displacement [166], [96], [127]). Find the boundary values of $p(x,t) = \int_{-\infty}^{\infty} p(x,v,t)\,dv$ (the marginal density) and their Smoluchowski limit. □

Exercise 8.7 (The MFPT [166], [96], [127]). Find the leading-order approximation to the MFPT to the boundary of the interval. □

Equation (8.41) can be interpreted as follows. The eigenfunction is a superposition of two "traveling waves". The traveling waves have a shifted Maxwellian velocity distribution, so their mean velocity is $\pm\sqrt{\lambda_1\varepsilon/\gamma}$ which is to leading order $\pm\pi\varepsilon/\gamma d = \pm\pi D/d$, where the sign depends on the direction of wave propagation. In the Smoluchowski diffusion approximation the velocity is no longer a state variable and the spatial eigenfunctions are $\sin n\pi x/d$, which is a standing wave. It is obtained in the limit $\gamma \to \infty$, because the wave velocities tend to zero.

8.3.1 Appendix

This Appendix lists the results of [166], [250], [96], [94], [95], [128], [127], [221]. Hagan's special function $N(s)$ that appears in (8.57) in the proof of Theorem 8.3.1 is defined by $N(s) = N\left(s, i\sqrt{\lambda_1/\gamma}\right)$, where

$$N(\mu,\nu) = \prod_{k=1}^{\infty} e^{-2\mu(\sqrt{k}-\sqrt{k-1})}\left(\frac{k+1}{k}\right)^{(\mu^2-\nu^2)/2}\left(\sqrt{1+\frac{\nu^2}{k}} + \frac{\mu}{\sqrt{k}}\right). \tag{8.62}$$

The function N, which appears in [96], is $N(s, 0)$. Its asymptotic behavior is given in the following

Theorem 8.3.2 (Asymptotics of Hagan's function). *The asymptotic behavior of $N(\mu,\nu)$ for large $|\mu|$ is given by*

$$N(\mu,\nu) \sim (2\pi)^{-1/4}\mu^{-(1/2+\nu^2)}\exp\left\{\mu^2\log\mu - \frac{1}{2}\mu^2\right\}$$
$$\times \exp\left\{\sum_{j=1}^{\infty}\frac{(-1)^{j-1}}{j\mu^j}\zeta(1+\nu^2, -j/2)\right\}, \tag{8.63}$$

where $\zeta(x,s)$ is the Hurwitz zeta function. Consequently,

$$N(s) \sim (2\pi)^{-1/4} s^{-(1/2-\lambda_1/\gamma)} \times \exp\left\{ s^2 \log s - \frac{1}{2}s^2 \right\}$$
$$\times \exp\left\{ \sum_{j=1}^{\infty} \frac{(-1)^{j-1}}{js^j} \zeta(1-\lambda_1/\gamma, -j/2) \right\}. \tag{8.64}$$

For small $\mu = \pm\nu$

$$\log N(\pm\nu, \nu) = \pm\zeta\left(\frac{1}{2}\right)\nu + O(\nu^3). \tag{8.65}$$

8.4 Low-friction asymptotics of the FPE

The underdamped Langevin equation exhibits features much different from the moderately or overdamped equation. For example, it can have steady-states far from equilibrium, thus giving rise to fluctuations that cannot be described by the Boltzmann distribution. Actually, such systems do not necessarily have a well-defined energy and thus defy physical intuition in a large measure. One such system is the noisy physical pendulum forced by constant torque. Another is the point shunted Josephson junction. Both are discussed below.

An underdamped Brownian particle in a potential well is described by the (dimensionless) Langevin equation

$$\ddot{x} + \gamma\dot{x} + U'(x) = \sqrt{2\gamma\varepsilon}\,\dot{w}, \tag{8.66}$$

where $U(x)$ is a potential that forms a single well, for example, $U(x) = x^2/2 - x^3/3$ (this is the potential in Figure 8.1 moved one unit to the left and scaled by the factor $1/6$)). We assume that at the bottom of the well $U(x_W) = 0$ and at the top of the barrier the energy is $U(x_B) = E_B$. The noiseless dynamics in phase plane, given by

$$\dot{x} = y, \quad \dot{y} = -\gamma y - U'(x), \tag{8.67}$$

has a stable equilibrium point on the x-axis at the bottom of the well and its domain of attraction D is bounded by the separatrix Γ_B. Figure 8.2) shows the energy contours $y^2/2 + U(x) = E$ for the potential $U(x) = x^2/2 - x^3/3$. The critical energy contour $E = E_B = 0$ closes at the saddle point $B = (1,0)$. The domain of attraction D of the stable point $W = (0,0)$ is bounded by the separatrix $G = \Gamma_B$ (the thick black curve G). The contour $E = E_C > E_B$ can be chosen as any one that intersects the x-axis at a point $x < x_W$ and the line $x = x_0 > x_B$ outside D. The closed curve Γ_W can be chosen as any one of the contours about W inside D.

In the limit $\gamma \to 0$ the phase plane trajectories of the system inside the well are the closed contours

$$\frac{y^2}{2} + U(x) = E, \tag{8.68}$$

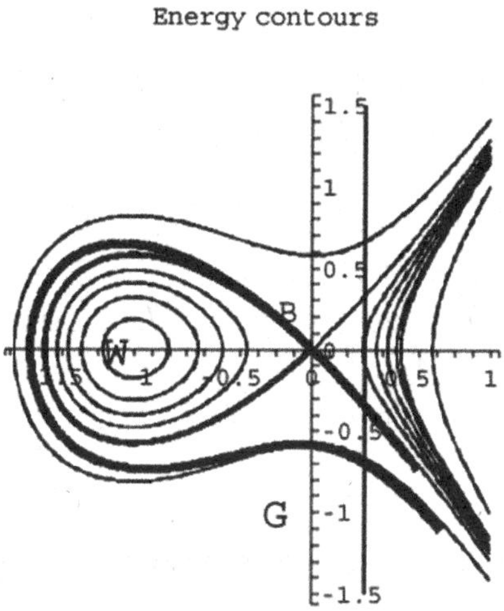

Figure 8.2. The energy contours $y^2/2 + U(x) = E$ for the potential $U(x) = x^2/2 - x^3/3$.

where the energy E varies in the interval $0 < E < E_B$. In the limit of small γ, the separatrix Γ_B shrinks to the critical energy contour $E = E_B$.

Obviously, the steady-state density vanishes on any compact subset of the phase plane, because the trajectories of (8.67) leave the domain of attraction of the attractor at the bottom of the well and drift to infinity. However, prior to their escape time, their pdf can be approximate by the solution of a much simplified FPE.

The FPE is given by

$$\gamma \frac{\partial p(x,y,s)}{\partial s} = \gamma\varepsilon \frac{\partial^2 p(x,y,s)}{\partial y^2} - y\frac{\partial p(x,y,s)}{\partial x} + \frac{\partial\left[\gamma y + U'(x)\right] p(x,y,s)}{\partial y}, \tag{8.69}$$

where the dimensionless time is $s = t\gamma$. The initial condition is

$$p(x,y,s) = \delta(x - x_0, y - y_0). \tag{8.70}$$

Theorem 8.4.1 (Low friction approximation). *The low friction approximation of the pdf $p(x,y,s)$ inside the critical energy contour $E = E_B$ is given by*

$$p(x,y,s) = p^0(x,y,s) + \gamma p^1(x,y,s) + \cdots, \quad \gamma \ll 1, \tag{8.71}$$

where the leading term is a function of energy, $p^0(x,y,s) = p^0(E,s)$, and satisfies *the reduced Fokker–Planck equation*

$$T(E)\frac{\partial p^0(E,s)}{\partial s} = \varepsilon\frac{\partial}{\partial E}I(E)\frac{\partial}{\partial E}p^0(E,s) + \frac{\partial}{\partial E}I(E)p^0(E,s) \text{ for } E < E_B, \tag{8.72}$$

where the period of the undamped motion $\ddot{x}+U'(x) = 0$ *on an energy contour* (8.68) *is* $T(E) = \oint_E dx/y$ *and* $I(E) = \int_0^{T(E)} y^2\,d\tau = \oint_E y\,dx$ *is the area enclosed by the contour. The initial and boundary conditions for* $p^0(E,s)$ *are*

$$\begin{aligned} p^0(E,0)\,dE &= \delta(E-E_0)\,dE, \text{ where } E_0 = \frac{y_0^2}{2} + U(x_0) < E_B \\ p^0(E_B,s) &= 0. \end{aligned} \tag{8.73}$$

Proof. We construct an asymptotic expansion of the solution for small γ, assuming that this is the smallest parameter in the problem. Thus, we use the regular expansion (8.71) in eq. (8.69) to obtain

$$L_0 p^0(x,y,s) = -y\frac{\partial p^0(x,y,s)}{\partial x} + U'(x)\frac{\partial p^0(x,y,s)}{\partial y} = 0.$$

This first-order linear partial differential equation can be written as the directional derivative

$$\begin{aligned} \frac{d}{d\tau}p^0(x(\tau),y(\tau),s) &= \dot{x}(\tau)\frac{\partial p^0(x,y,s)}{\partial x} + \dot{y}(\tau)\frac{\partial p^0(x,y,s)}{\partial y} \\ &= y\frac{\partial p^0(x,y,s)}{\partial x} - U'(x)\frac{\partial p^0(x,y,s)}{\partial y} \\ &= -L_0 p^0(x,y,s) = 0 \end{aligned}$$

on the trajectories of the noiseless dynamics (8.67). It follows that $p^0(x,y,s)$ is constant on the contours (8.68), so that $p^0(x,y,s) = p^0(E,s)$. The points where the contour intersects the x-axis are denoted $x_L(E)$ and $x_R(E)$.

The next order equation is

$$L_0 p^1(x,y,s) = \frac{\partial p^0(E,s)}{\partial s} - \frac{\partial y p^0(E,s)}{\partial y} - \varepsilon\frac{\partial^2 p^0(E,s)}{\partial y^2}. \tag{8.74}$$

The solvability condition for eq. (8.74) follows from the fact that the integral of the total derivative of the periodic function $p^1(x(\tau),y(\tau),s)$ on every constant energy contour vanishes; that is,

$$\int_0^{T(E)} \frac{dp^1(x(\tau),y(\tau),s)}{d\tau}\,d\tau = 0. \tag{8.75}$$

First, we note that

$$\frac{\partial y p^0(E,s)}{\partial y} = p^0(E,s) + y\frac{\partial p^0(E,s)}{\partial y} = p^0(E,s) + y\frac{\partial E}{\partial y}\frac{\partial p^0(E,s)}{\partial E}$$
$$= p^0(E,s) + y^2\frac{\partial p^0(E,s)}{\partial E}$$

and

$$\frac{\partial^2 p^0(E,s)}{\partial y^2} = \frac{\partial p^0(E,s)}{\partial E} + y^2\frac{\partial^2 p^0(E,s)}{\partial E^2}.$$

Now, applying the solvability condition (8.75) to each term on the right-hand side of eq. (8.74), we obtain

$$\int_0^{T(E)} \frac{\partial p^0(E,s)}{\partial s}\,d\tau = T(E)\frac{\partial p^0(E,s)}{\partial s}, \qquad \int_0^{T(E)} p^0(E,s)\,d\tau = T(E)p^0(E,s)$$
$$\int_0^{T(E)} y^2\frac{\partial p^0(E,s)}{\partial E}\,d\tau = I(E)\frac{\partial p^0(E,s)}{\partial E}.$$

It follows that

$$\int_0^{T(E)} \frac{\partial y p^0(E,s)}{\partial y}\,d\tau = T(E)p^0(E,s) + I(E)\frac{\partial p^0(E,s)}{\partial E}.$$

Similarly, we obtain

$$\int_0^{T(E)} \frac{\partial^2 p^0(E,s)}{\partial y^2}\,d\tau = T(E)\frac{\partial p^0(E,s)}{\partial E} + I(E)\frac{\partial^2 p^0(E,s)}{\partial E^2}.$$

Thus, the solvability condition (8.75) converts eq. (8.74) into

$$T(E)\frac{\partial p^0(E,s)}{\partial s} = \varepsilon\left[T(E)\frac{\partial p^0(E,s)}{\partial E} + I(E)\frac{\partial^2 p^0(E,s)}{\partial E^2}\right] + I(E)\frac{\partial p^0(E,s)}{\partial E} + T(E)p^0(E,s). \tag{8.76}$$

Note, however, that due to the fact that $E - U(x_R(E)) = y(x_R(E)) = 0$, we have

$$\frac{d}{dE}I(E) = T(E). \tag{8.77}$$

It follows that eq. (8.76) can be written as (8.72). □

Exercise 8.8 (Proof of (8.77)). Prove (8.77) by using the identities

$$\frac{d}{dE}I(E) = \frac{d}{dE}\oint_E \pm\sqrt{2\left[E - U(x)\right]}\,dx = 2\frac{d}{dE}\int_{x_L(E)}^{x_R(E)} \sqrt{2\left[E - U(x)\right]}\,dx$$

$$= \sqrt{2\left[E - U(x_R(E))\right]}\frac{dx_R(E)}{dE} - \sqrt{2\left[E - U(x_L(E))\right]}\frac{dx_L(E)}{dE}$$

$$+ 2\int_{x_L(E)}^{x_R(E)} \frac{dx}{\sqrt{2\left[E - U(x)\right]}} = \oint_E \frac{dx}{y}$$

□

Exercise 8.9 (The averaged FPE in the action variable). Write the FPE (8.72) in the action variable $I = I(E)$. Show that $dE/dI|_{E=E_B} = 0$. □

Exercise 8.10 (Initial and boundary conditions for $p^0(E,s)$). Prove that the initial and boundary conditions for $p^0(E,s)$ are given by (8.73). (HINT: Because $\gamma T(E_0) \ll 1$, the energy on the trajectory $(x(t), y(t))$ does not deviate much from E_0 in short (dimensionless) "time" s). □

Exercise 8.11 (Small noise asymptotics). Consider the damped dynamics driven by state-dependent colored noise $\ddot{x} + \beta x + U'(x) = \sqrt{\beta\alpha}\, g(x)v(t)$, where $\dot{v} = -\alpha v + \sqrt{2\varepsilon\alpha}\,\dot{w}$, or equivalently, the three-dimensional system

$$\dot{x} = y, \quad \dot{y} = -U'(x) - \beta y + \sqrt{\beta\alpha}\, g(x)v(t), \quad \dot{v} = -\alpha v + \sqrt{2\varepsilon\alpha}\,\dot{w}. \tag{8.78}$$

The process $\sqrt{\beta\alpha}v(t)$ is a Gaussian colored noise of strength $\varepsilon\beta\alpha$ and of correlation time α. Find the asymptotic expansion of the stationary joint pdf of x, y, v for small β and small α, assuming it exists. First expand for small α and then for small β. In (vii) expand in the reverse order and show that the limits are interchangeable.

(i) Begin with the case of the harmonic potential (7.112) and $g(x) = 1$ and use the explicit solution (7.114) to show that the stationary pdf is to leading order in α a function of the energy

$$E = U(x) + \frac{y^2}{2} \tag{8.79}$$

only.

(ii) Use the transformations $\beta \to \beta^2$ and $u = \sqrt{\beta\alpha}v(t)$ to rewrite the system (8.78) in the form

$$\dot{x} = y, \quad \dot{y} = -U'(x) - \beta^2 y + \sqrt{\beta}g(x)u(t), \quad \dot{u} = -\alpha u + \sqrt{2\varepsilon\alpha}\,\dot{w} \tag{8.80}$$

and expand the solution of the stationary FPE corresponding to (8.80) in the WKB form $p = \exp\{-\psi/\alpha^2\}$, where $\psi \sim \psi_0 + \alpha\psi_1 + \cdots$. Expand ψ_0 for small β as $\psi_0 \sim \phi_0 + \sqrt{\beta}\,\phi_1 + \beta\,\phi_2 + \cdots$, obtain for ϕ_0 the equation $y\phi_{0,x} - U'(x)\phi_{0,y} =$

0, and conclude that $\phi_0 = \phi_0(E,u)$ with ϕ_0 as yet undetermined. Obtain $\phi_1 = -uG(x)\phi_{0,E} + f_1(E,u)$, where $\phi_{0,E} = \partial\phi_0/\partial E$ and $G'(x) = g(x)$.

(iii) Assume that $U(x) \to \infty$ as $|x| \to \infty$. Show that the solvability condition for ϕ_2 is that the function $-g(x)uy\left[-uG(x)\phi_{0,EE} + f_{1,E}\right] - \varepsilon\phi_{0,u}^2$ is orthogonal to the function $p = 1/T(E)y$. Conclude that ϕ_0 is a function of E only. Conclude that $\phi_2 = \frac{1}{2}u^2G^2(x)\phi_0'' - uG(x)f_{1,E} + f_2(E,u)$.

(iv) Find ϕ_3 and show that the solvability condition for ϕ_4 is

$$\left(\frac{1}{T}\oint_E y(x,E)\,dx\right)\phi_0' - \varepsilon\left(\frac{1}{T}\oint_E \frac{G^2(x)}{y}\,dx\right)\phi_0'^2 - \varepsilon f_{1,u}^2 = 0.$$

Conclude that f_1 must be a function of E only in order for ϕ_0 to be a real function for all values of E. Hence obtain

$$\phi_0 = \int_0^E \frac{\oint_E y\,dx}{\varepsilon\oint_E \dfrac{G^2(x)}{y}\,dx}\,dE. \tag{8.81}$$

(v) Denote by $(x(t), y(t))$ the closed trajectory (8.79) and note that these are periodic functions of t with period $T(E)$. Expand the function $g(x(t))y(t)$ in the Fourier series $g(x(t)y(t)) = \sum_{n=1}^{\infty} a_n \sin\omega t$, where $\omega = \omega(E)$ is the frequency of the motion on the closed trajectory. Define the function

$$h(E) = \frac{1}{(T(E))}\oint_E y(x,E)\,dx$$

and rewrite (8.81) as

$$\phi_0 = \int_0^E \frac{2h(E)}{\sum_{n=1}^{\infty}\dfrac{a_n^2}{n^2\omega^2}}\,dE.$$

(vi) Show that the leading-order approximation to the pdf is $p \sim \exp\{-\Phi(E)/\varepsilon\}$, where $\Phi(E) = \int_0^E [h(\sigma)/f(\sigma)]\,d\sigma$, and

$$\begin{aligned} f(E) &= \lim_{N\to\infty}\frac{\alpha}{NT(E)}\int_0^{NT(E)} g(x(t))y(t)\int_0^t e^{-a(t-s)}g(x(s))y(s)\,ds\,dt \\ &= \frac{\alpha^2}{2}\sum_{n=1}^{\infty}\frac{a_n^2}{\alpha^2 + n^2\omega^2}. \end{aligned}$$

(vii) Expand the pdf first for small β and then for small α and show that the limits are interchangeable. □

Exercise 8.12 (Undamped nonlinear oscillator driven by small colored noise). Consider an undamped nonlinear oscillator driven by small colored noise,

$$\dot{x} = y, \quad \dot{y} = -U'(x) + \sqrt{\varepsilon\alpha}\, u(t), \quad \dot{u} = -\alpha u + \sqrt{2\alpha}\, \dot{w}, \tag{8.82}$$

and assume $U(x) \to \infty$ as $|x| \to \infty$. Find the leading-order approximation to the stationary joint pdf of x, y, u for small ε. □

Exercise 8.13 (Quasi-stationary density for $E < E_B$). Show that $p^0(x, y, s)$ in Theorem 8.4.1 satisfies for $E < E_B$,

$$\lim_{s\to\infty} \frac{p^0(x, y, s)}{\displaystyle\int_{E<E_B} p^0(x, y, s)\, dx\, dy} = \mathcal{N} e^{-E/\varepsilon}, \tag{8.83}$$

where $\mathcal{N}$ is a normalization constant. What is the probabilistic interpretation of $\int_{E<E_B} p^0(x, y, s)\, dx\, dy$? □

The exit problem at low friction is to determine the mean time to escape the domain of attraction of the stable equilibrium point of the noiseless dynamics at the bottom of the potential well and the density of exit points on the *stochastic separatrix*. A trajectory that originates inside the well is said to have escaped the well when it reaches a line $x = x_0$ for some $x_0 > x_B$ (we assume the well is to the left of x_B, as in Figure 8.2).

Definition 8.4.1 (Stochastic separatrix). *The* stochastic separatrix *is the locus of initial points of trajectories whose probability to escape the well before reaching a given small neighborhood of the well's bottom is* $\frac{1}{2}$.

Theorem 8.4.2 (The stochastic separatrix at low friction). *The stochastic separatrix $y_{1/2}(x)$ at low noise and low friction has the asymptotic representation*

$$y_{1/2}(x) = 2\left[U(x_B) - U(x) - \varepsilon \log 2 + O(\varepsilon^2) + O(\gamma)\right]. \tag{8.84}$$

Proof. First, we show that in the limit of small γ and for small ε the stochastic separatrix is inside the separatrix of the noiseless dynamics; that is, that trajectories starting inside the separatrix, but outside a boundary layer near the separatrix, reach any neighborhood of the point $(x_W, 0)$ before they reach any neighborhood of a point outside the separatrix with probability nearly 1. To do so, we consider a neighborhood of the stable equilibrium point, N_W, bounded by the contour $\Gamma_W = \{(x, y) \,|\, E = E_0\}$ for some positive E_0, and any constant energy contour outside the well, $\Gamma_C = \{(x, y) \,|\, E = E_C\}$, where $E_C > E_B$. It should be noted that there are constant energy contours outside Γ_W with energy lower than E_B. For a point (x_0, y_0) inside Γ_C, but outside Γ_W, we have $E_0 < E(x_0, y_0) < E_B$ (see Figure 8.2).

Now, we consider the exit problem in the domain bounded between Γ_W and Γ_C. The probability of reaching Γ_W before Γ_C is the total flux, $\oint_{\Gamma_W} \boldsymbol{J} \cdot \boldsymbol{\nu}\, (x, y_W(x))\, d\sigma$

on Γ_W of the density $p(x,y) = \gamma^{-1}\int_0^\infty p(x,y,s)\,ds$, where $p(x,y,s)$ is the solution of the FPE (8.69) with the initial condition (8.70) and the absorbing boundary conditions $p|_{\Gamma_W} = 0$ and $p|_{\Gamma_B} = 0$. Denoting by τ_W the time to reach Γ_W and by τ_B the time to reach Γ_B, this can be written as

$$\lim_{\gamma\to 0} \Pr\{\tau_W < \tau_B \,|\, (x_0,y_0)\} = -\oint_{\Gamma_W} \varepsilon \frac{\partial}{\partial y} p^0(E)\,dx,$$

where $p^0(E)$ is the solution of

$$\varepsilon \frac{\partial}{\partial E} I(E) \frac{\partial}{\partial E} p^0(E) + \frac{\partial}{\partial E} I(E) p^0(E) = -\delta\left(E - E(x_0,y_0)\right)$$

with the boundary conditions $p^0(E_W) = p^0(E_B) = 0$. The solution is given by

$$p^0(E) = e^{-E/\varepsilon}\left[C\int_{E_W}^{E} \frac{e^{z/\varepsilon}}{\varepsilon I(z)}\,dz - H\left(E - E(x_0,y_0)\right)\int_{E(x_0,y_0)}^{E} \frac{e^{z/\varepsilon}}{\varepsilon I(z)}\,dz\right],$$

where $H(z)$ is the Heaviside step function and

$$C = \int_{E_W}^{E_C} \frac{e^{z/\varepsilon}}{I(z)}\,dz \bigg/ \int_{E(x_0,y_0)}^{E_C} \frac{e^{z/\varepsilon}}{I(z)}\,dz.$$

Note that $C \sim 1$ for $\varepsilon \ll 1$. It follows that the flux density on Γ_W is

$$\boldsymbol{J}\cdot\boldsymbol{\nu}\left(x, y_W(x)\right)\,d\sigma \sim \frac{y_W(x)}{I(E_W)}\,dx, \tag{8.85}$$

where $y_W(x) = \pm\sqrt{2\left[E_W - U(x)\right]}$, $\boldsymbol{\nu}\left(x, y_W(x)\right)$ is the unit outer normal to Γ_W, and $d\sigma$ is arclength element on Γ_W. We find that for small ε

$$\lim_{\gamma\to 0} \Pr\{\tau_W < \tau_B \,|\, (x_0,y_0)\} \sim \oint_{E_W} \frac{y_W(x)}{I(E_W)}\,dx = 1.$$

Next, we show that trajectories that start on Γ_C escape the well with probability nearly 1, so that the stochastic separatrix is inside Γ_C. Recall that a trajectory has escaped the well when it reaches a line $x = x_0$ for some $x_0 > x_B$. We denote by τ_0 the time to reach this line. Note that all constant energy contours with $E > E_B$ intersect this line. To investigate trajectories that start on Γ_C, we recall that the function

$$P(x_0,y_0) = \Pr\{\tau_W < \tau_0 \,|\, (x_0,y_0)\}$$

is the solution of the equation

$$\gamma\varepsilon \frac{\partial^2 P(x,y)}{\partial y^2} + y\frac{\partial P(x,y)}{\partial x} - \left[\gamma y + U'(x)\right]\frac{\partial P(x,y)}{\partial y} = 0 \tag{8.86}$$

in the domain enclosed between the line $x = x_0$ and the contour Γ_W with the boundary conditions $P|_{\Gamma_W} = 1$ and $P(x_0, y) = 0$ for all y. We construct an approximate solution to this boundary value problem in the form of an asymptotic series, as in (8.71), and find that outside Γ_B we get $P_{\text{out}}(E) = 0$. Inside Γ_B we obtain the averaged equation

$$\varepsilon I(E)P_{\text{in}}''(E) + \varepsilon I'(E)P_{\text{in}}'(E) - I(E)P_{\text{in}}'(E) = 0 \tag{8.87}$$

with the boundary condition $P_{\text{in}}(E_W) = 1$ and the matching condition $P_{\text{in}}(E_B) = P_{\text{out}}(E_B) = 0$. The solution is given by

$$P_{\text{in}}(E) = \int_E^{E_B} \frac{e^{z/\varepsilon}}{I(z)}\,dz \Bigg/ \int_{E_W}^{E_B} \frac{e^{z/\varepsilon}}{I(z)}\,dz. \tag{8.88}$$

It follows from (8.88) that $P_{\text{in}}(E) = \frac{1}{2}$ for $E = E_B - \varepsilon \log 2 + O(\varepsilon^2)$, which is (8.84). □

Corollary 8.4.1. *The stochastic separatrix is located inside Γ_B, so trajectories that reach the separatrix Γ_B are practically sure to escape the well.*

Exercise 8.14 (The separatrix at low friction). Find the asymptotic expansion for small γ of the separatrix Γ_B of the noiseless dynamics (8.67). Show that the separatrix Γ_B is given by $y_{\Gamma_B}(x) = y_B(x) - \gamma \int_{x_B}^{x} y_B(z)\,dz + o(\gamma)$, where $\frac{1}{2}y_B^2(x) + U(x) = U(x_B)$. Is $o(\gamma) = O(\gamma^2)$? □

Exercise 8.15 (The exit energy). Show that the mean energy of trajectories on Γ_B is $\gamma I(E_B)/2$ above the barrier energy E_B. □

Theorem 8.4.3 (Kramers' underdamped rate formula [140]). *The escape rate of an underdamped Brownian particle from a potential well at low noise is given by*

$$\kappa_{\text{underdamped}} = \frac{1}{\bar{\tau}(x_W, y_W)} \sim \frac{\gamma I(E_B)\omega_W}{2\pi\varepsilon} e^{-E_B/\varepsilon}. \tag{8.89}$$

Proof. Keeping the notation of Theorem 8.4.2, we recall that the MFPT $\bar{\tau}(x,y)$ from a point (x,y) in D to Γ_B is the solution of the backward equation

$$\gamma\varepsilon \frac{\partial^2 \bar{\tau}(x,y)}{\partial y^2} + y\frac{\partial \bar{\tau}(x,y)}{\partial x} - [\gamma y + U'(x)]\frac{\partial \bar{\tau}(x,y)}{\partial y} = -1$$

inside Γ_B with the boundary condition $\bar{\tau}(x,y) = 0$ on Γ_B. The asymptotic expansion of $\bar{\tau}(x,y)$ is of the form $\bar{\tau}(x,y) = \gamma^{-1}\bar{\tau}_0(x,y) + \bar{\tau}_1(x,y) + \cdots$. The leading-order equation gives $\bar{\tau}_0(x,y) = \bar{\tau}_0(E)$ and the next order gives

$$L_0\bar{\tau}_1(x,y) = \frac{d}{d\tau}\bar{\tau}_1(x(t), y(t)) = -1 - \varepsilon\frac{\partial^2 \bar{\tau}_0(E)}{\partial y^2} + y\frac{\partial \bar{\tau}_0(E)}{\partial y}.$$

As above, the solvability condition is that the integral over a period vanishes; that is,

$$\varepsilon I(E)\bar{\tau}_0''(E) + \varepsilon I'(E)\tau_0'(E) - I(E)\bar{\tau}_0'(E) = -T(E)$$

with the boundary conditions that $\bar{\tau}_0(0) < \infty$ and $\tau_0(E_B) = 0$. The solution is given by

$$\bar{\tau}_0(E) = \int\limits_E^{E_B} \frac{e^{s/\varepsilon}}{\varepsilon I(s)} \int\limits_0^s T(z)e^{-z/\varepsilon}\,dz\,ds.$$

If ε is small, we obtain

$$\bar{\tau}_0(E) = \frac{\varepsilon T(0)}{I(E_B)} e^{E_B/\varepsilon}(1 + o(1))$$

for all $E < E_B$ such that $(E_B - E)/\varepsilon >> 1$. Recall that $E = E_W = 0$ at the bottom of the well.

The period of motion $T(0)$ is the period of free oscillations at the bottom of the well, given by $T(0) = 2\pi/\omega_W$, where $\omega_W^2 = U''(x_W)$. The leading-order approximation to the mean lifetime in the well is

$$\bar{\tau}_E = \frac{2\pi\varepsilon}{\gamma I(E_B)\omega_W} e^{E_B/\varepsilon}, \tag{8.90}$$

because trajectories that reach Γ_B are almost sure to escape. Therefore, in view of Corollary 8.4.1, the escape rate is given by (8.89). □

Exercise 8.16 (Continuation of Exercise 8.13). Find an explicit expression for $\int_{E<E_B} p^0(x, y, s)\,dx\,dy$. □

8.5 The noisy underdamped forced pendulum

The noisy underdamped forced physical pendulum represents several important physical systems, including the current-driven point Josephson junction, charged density waves, and more [16]. It is an archetypical example of a multistable physical system that has many steady-states, some of which are equilibria and some are not. In the context of the Josephson junction, there are stable thermodynamical equilibrium states, in which no current flows through the junction, and a nonequilibrium steady-state, in which the junction carries a steady current. Due to the noise in the system, the current switches at random times between its two steady values, giving rise to an average current proportional to the probability of the conducting state. The random switching of the current on and off is analogous to gating in ionic channels of biological membranes [102].

The main difficulty in analyzing the thermal fluctuations and activated transitions between the steady-states is the fact that thermodynamics and the Boltzmann distribution of energies apply to this system only near the stable equilibria, whereas they do not apply near the nonequilibrium steady-state of the system. Thus the

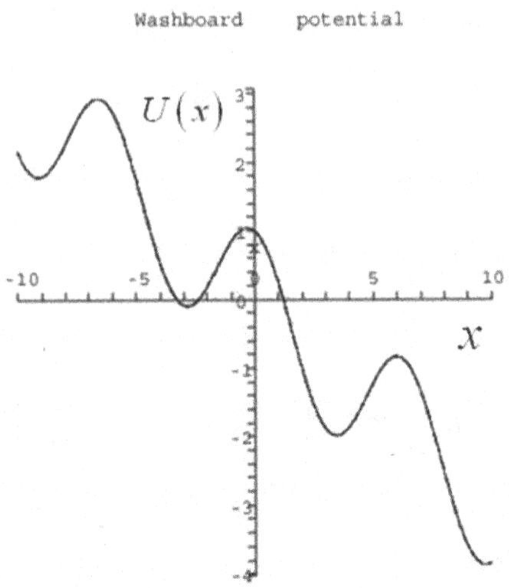

Figure 8.3. The washboard potential $U(x) = \cos x - Ix$ for $I = 0.3$.

Maxwell–Boltzmann distribution in phase space has to be replaced with a probability distribution in the space of the trajectories of the system. This section shows how stochastic differential equations and the Fokker–Planck equation can supplant the classical notions of energy and entropy in nonequilibrium statistical mechanics of certain physical systems. Some of the predictions of this analysis were actually discovered in laboratory experiments (see Section 8.5.4 below and [39]).

8.5.1 The noiseless underdamped forced pendulum

The stochastic dynamics of the noisy underdamped forced physical pendulum is described by the Langevin equation (8.66) with the potential

$$U(x) = -\cos x - Ix, \tag{8.91}$$

where x represents the deflection angle and I represents an applied constant torque (see Figure 8.3). To understand the behavior of the noiseless system, we examine first the phase space dynamics of the forced pendulum. The dynamics of the pendulum depend on the values of γ and I. There is a range of values of these parameters, $I_{\text{min}}(\gamma) < I < 1$ and $\gamma < \pi/4$, for which both stable equilibria and nonequilibrium stable steady-state solutions coexist. In the latter running state $\dot{x}$ is a periodic function of x and $x(t) \to \infty$ as $t \to \infty$.

To explore this range of parameters, we represent the phase space trajectories by

$$\frac{dy}{dx} = -\gamma + \frac{I - \sin x}{y}. \tag{8.92}$$

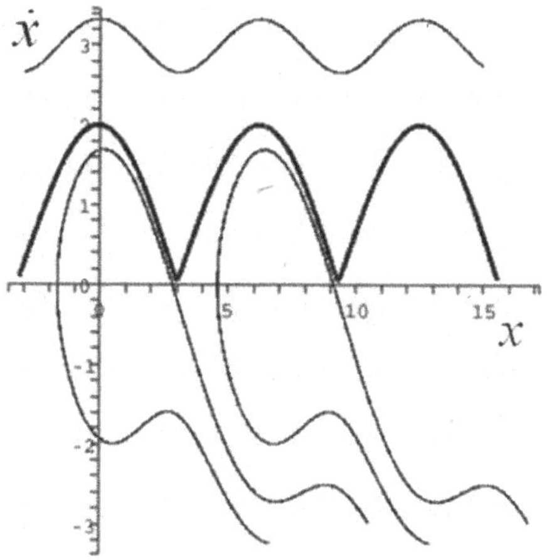

Figure 8.4. The stable running solution (top) and separatrices, 2π apart in the phase space (x, y), for $\gamma = 0.1$ and $I = 0.3$. The thick curve is the critical trajectory $S = S_C$ that touches the x-axis. The energy of the system on the running solution slides to $-\infty$ down the washboard potential in Figure 8.3.

Obviously, the system has stable equilibrium points on the x-axis, where $U'(x) = \sin x - I = 0$, at distances 2π apart, if $|I| < 1$. For $|I| > 1$ no such equilibria exist.

Exercise 8.17. (A stable periodic solution).

(i) Show that a stable periodic solution of eq. (8.92) can be constructed for small γ and $|I| < 1$ in the form

$$y \sim \frac{y_{-1}}{\gamma} + y_0 + \gamma y_1 + \cdots. \tag{8.93}$$

(ii) Use (8.93) into (8.92), compare coefficients of like powers of γ and use the periodicity condition to find the coefficients $y_{-1}, y_0, y_1, \ldots$. Obtain

$$y_S(x) = \frac{I}{\gamma} + \frac{\gamma}{I}\cos\left(x + \frac{\gamma^2}{I}\right) - \frac{1}{4}\left(\frac{\gamma}{I}\right)^3 \cos 2x + O\left(\frac{\gamma^5}{I^5}\right). \tag{8.94}$$

(iii) Show that the domains of attraction D_S of the two types of stable states of the system are separated by separatrices.

(iv) Draw the phase plane trajectories of the system (see Figure 8.4).

(v) Show that the running solution (8.94) disappears if γ is not sufficiently small.

(vi) Scale $I = \gamma I_0$ and expand $y \sim y_0 + \gamma y_1 + \cdots$ in (8.92) and then show that the first term, y_0, satisfies the undamped and unforced pendulum equation, so that

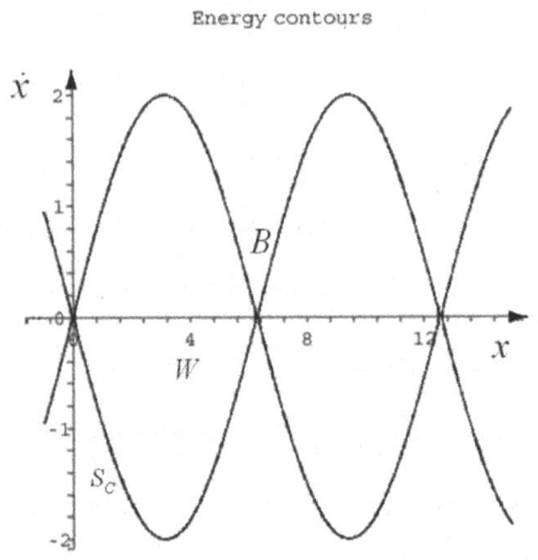

Figure 8.5. In the limit $\gamma \to \infty$ the stable running solution of Figure 8.4 disappears to infinity and the critical energy contour $S = S_C$ and the separatrices coalesce.

$y_0 = \sqrt{2 + 2\cos x} = 2\,|\cos x/2|$. Finally, obtain at the next order the periodicity condition $2\pi I_0 = \int_0^{2\pi} y_0\,dx = 8$. □

Exercise 8.18. (The critical trajectory).

(i) Show that $I_0 = 4/\pi$ on the critical trajectory $S = S_C$ that touches the x-axis. Conclude that the maximal value of γ for which the running solution (8.94) exists, for given $|I| < 1$, is $\gamma_M(I) = \pi I/4$ and the minimal value of I for which the running solution exists, for a given value of γ, is $I_{\min}(\gamma) \approx 4\gamma/\pi$.

(ii) Show that as $\gamma \uparrow \gamma_M$, the minimum of the periodic trajectory (8.94) approaches the x-axis. The critical trajectory S_C touches the separatrix at the unstable equilibrium point and has a cusp there. To this end set $\gamma = \gamma_M$ in eq. (8.92) and take the limit $x \to x_0$, where $I - \sin x_0 = 0$ (x_0 is the coordinate of the unstable equilibrium point). Obtain

$$y'(x_0) = -\gamma_M - \frac{\cos x_0}{y'(x_0)} = -\gamma_M - \frac{\sqrt{1 - I^2}}{y'(x_0)}.$$

(iii) Show that the critical trajectory has two different slopes at the point x_0 it touches the x-axis, by proving that

$$y'(x_0) = \frac{-\gamma_M \pm \sqrt{\gamma_M^2 + 4\sqrt{1 - I^2}}}{2}. \tag{8.95}$$

(iv) Show that in the limit $\gamma \to 0$ Figure 8.4 collapses to Figure 8.5 □

Exercise 8.19. (Loss of bistability).

(i) Show that if $|I| > 1$, then there is a stable running solution for all values of γ.

(ii) Show that for $\gamma >> 1$ and $|I| > 1$ an expansion in powers of $1/\gamma$ gives

$$y(x) = \frac{1 - \sin x}{\gamma} + \frac{I \cos x - \frac{1}{2} \sin 2x}{\gamma^3} + O\left(\frac{1}{\gamma^5}\right). \tag{8.96}$$

(iii) Obtain a uniform expansion of $y(x)$, valid for all γ and $|I| > 1$, by truncating the Fourier expansion of y, as

$$y(x) = \frac{1}{\gamma} + \frac{I\gamma \cos x}{I^2 + \gamma^4} - \frac{\gamma^3 \sin x}{I^2 + \gamma^4} + \cdots.$$

Show that this expansion reduces to (8.94) and (8.96) in the appropriate limits.

(iv) Show that on this trajectory $x(t)$ increases in time and $y(t)$ is a periodic function. Show that the time average of $y(t) = \dot{x}(t)$ is not zero. □

For $\gamma < \gamma_M$ and $I_{\text{min}}(\gamma) < I < 1$, stable equilibrium solutions and the stable nonequilibrium solution (8.94) can coexist and the system can exhibit hysteresis. In this range of parameters, phase space is divided into a basin of attraction D_S of the stable nonequilibrium steady-state S and a basin of attraction of each of the stable equilibrium states E, denoted generically by D_E. The basins are separated from each other by separatrices, which correspond to solutions of (8.92) that converge asymptotically to the unstable equilibrium points at $y = 0$ and the local maxima of $U(x)$.

For given values of $\gamma < \gamma_M$, as I is decreased toward $I_{\text{min}}(\gamma)$, the separatrices and the nonequilibrium steady-state S approach each other. When $I = I_{\text{min}}(\gamma)$, these curves coalesce, leading to the curve (8.75). Alternatively, for a given value of $|I| < 1$, when γ increases toward γ_M, the separatrices and S approach each other, as above. However, in this case, the unstable equilibrium points, which lie on the separatrices, do not move.

The phase space trajectory S can be characterized as the only periodic solution of the differential equation (8.92). For $I = I_{\text{min}}(\gamma)$, a first approximation to the critical stable periodic trajectory S_C; that is, the steady-state that has just coalesced with the separatrix, is given by

$$y(t) = \dot{x}(t) = 2\left|\cos \frac{x(t) - \Delta}{2}\right|,$$

where $\Delta = -\arcsin I_{\text{min}}(\gamma)$.

If the phase plane is wrapped on a cylinder; that is, the x-axis is reduced mod(2π), the stable nonequilibrium S becomes a stable limit cycle and the stable equilibria coalesce into a single stable equilibrium point. The domains of attraction of these stable states are separated by a separatrix. The noise-induced fluctuations about the limit cycle and rate of noise-induced transitions over the separatrix are the object of interest in the next sections.

8.5.2 Local fluctuations about a nonequilibrium steady state

The noisy forced pendulum is described by the Langevin equation (8.66) with the potential (8.91), so the stable states of the noiseless dynamics (8.67), now

$$\dot{x} = y, \quad \dot{y} = -\gamma y - \sin x + I, \tag{8.97}$$

as described above, become meta-stable due to noise-induced fluctuations and transitions between the domains of attraction of these states. Although the noisy system reaches steady-state, it is never in equilibrium, because it carries a steady current. Therefore, as mentioned above, its fluctuations and noise-induced transitions cannot be described by thermodynamics, which is an equilibrium theory. The absence of well-defined energy renders Kramers' formula (8.89) (see Exercise 6.4) inapplicable for transitions from the running state to equilibrium, although it applies in the reverse direction.

Specifically, the Fokker–Planck equation (8.69) with the potential (8.91) has the stationary solution

$$c = Ce^{-E/\varepsilon}. \tag{8.98}$$

This density actually represents the probability density function of fluctuations about an equilibrium state. It does not, however, represent the probability density function of fluctuations about nonequilibrium steady-states, because (8.98) is unbounded on S, and is not periodic in x. The phase space probability current density

$$\boldsymbol{J} = \begin{pmatrix} yp \\ -U'(x)p - \gamma yp - \gamma\varepsilon \dfrac{\partial p}{\partial y} \end{pmatrix}$$

vanishes for $p = p_B$, whereas in the nonequilibrium steady-state, we expect a nonzero probability current flowing in the direction of decreasing $U(x)$. Therefore, instead of (8.98), we seek a solution of the Fokker–Planck equation that is bounded, periodic in x with the same period as $U(x)$, and produces a nonzero current in the appropriate direction. Thus the Maxwell–Boltzmann distribution of steady-state fluctuations, in which velocity and displacement are statistically independent, has to be replaced with a different steady-state distribution in phase space.

For small ε we seek the steady-state distribution in phase space as a periodic density of local fluctuations about the nonequilibrium steady-state in the WKB form

$$p_S = p_0 e^{-W/\varepsilon}, \tag{8.99}$$

where the functions $W(x,y)$ and $p_0(x,y,\varepsilon)$ remain to be determined. The periodic density of $p_S(x,y)$ is normalized over $[0,2\pi]\times\mathbb{R}$. The eikonal function W replaces the notion of energy, which is not well defined for this damped not isolated system; W plays a role similar to that of energy in (8.98). Both W and p_0 must be periodic on S, and p_0 is assumed a regular function of ε at $\varepsilon = 0$. Substituting (8.99) into the stationary Fokker–Planck equation and expanding in powers of ε, we find that W satisfies the Hamilton–Jacobi–(eikonal-type) equation

$$\gamma\left(\frac{\partial W}{\partial y}\right)^2 + y\frac{\partial W}{\partial x} - [\gamma y + U'(x)]\frac{\partial W}{\partial y} = 0. \tag{8.100}$$

At the same time, to leading order in ε, p_0 is the 2π-periodic (in x) solution of the transport equation

$$\left[2\gamma\frac{\partial W}{\partial y} - \gamma y - U'(x)\right]\frac{\partial p_0}{\partial y} + y\frac{\partial p_0}{\partial x} + \gamma\left[\frac{\partial^2 W}{\partial y^2} - 1\right]p_0 = 0. \tag{8.101}$$

This approximation is valid throughout the basin of attraction D_S, as long as $\varepsilon \ll 1$; that is, as long as the usual Boltzmann thermal energy $\varepsilon = k_B T$ is much less than the potential barrier.

We first show that the contours of constant W in phase space correspond to the deterministic nonequilibrium steady-state trajectories of (8.67) for $0 \le \gamma \le \gamma_M(I)$. Using this property, we determine the function $W(x,y)$. To this end we consider the following equations in the phase space (x,y): the equations of motion (8.97), the parametric equations for the constant-W contours

$$\dot{x} = y, \quad \dot{y} = -\gamma y - U'(x) + \gamma W_y, \tag{8.102}$$

and the parametric equations for the characteristic curves of the eikonal equation

$$\begin{aligned} &\dot{x} = y, \quad \dot{y} = -\gamma y - U'(x) + \gamma W_y \\ &\dot{W}_y = -W_x + \gamma W_y, \quad \dot{W}_x = U''(x) W_y, \quad \dot{W} = \gamma W_y^2. \end{aligned} \tag{8.103}$$

Lemma 8.5.1. $W = const.$ *and* $\nabla W = 0$ *on* S.

Proof. Calculating the total derivative $\dot{W}$ on S from (8.97) and the eikonal equation (8.100), we find that $\dot{W} = -\gamma W_y^2 \le 0$. Hence, to keep W periodic on S the right-hand side must vanish identically, rendering $W = const.$ on S. Now it follows from (8.100) that $W_x = 0$ on S as well, hence $\nabla W = 0$ on S. □

Lemma 8.5.2 (W-contours). *The W-contours are the family of steady-state trajectories of*

$$\dot{x} = y, \quad \dot{y} = -\Gamma(W) y - U'(x), \tag{8.104}$$

where $\Gamma(W) = \gamma[1 - K(W)]$, *where* $K(W)$ *is given in* (8.106) *for* $0 \le \Gamma(W) \le \gamma_M(I)$, *and they are given by the approximate expression* (8.94) *with* γ *replaced by* Γ.

Proof. First, we express W_y on the W-contours in terms of W. Consider the function

$$H(x,y) = \frac{y^2}{2} + U(x) + \gamma \int_{x_0}^{x} (y - W_y)\, dx,$$

where the integral is a line integral along the W-contour that passes through the point (x,y). From eq. (8.102), it is easy to see that $H = const.$ on any W-contour. The rate of change of $H(W)$ on a characteristic curve (8.103) is $\dot{H} = \gamma y W_y + \gamma^2 \int_{x_0}^{x} W_y (1 - W_{yy})\, dx$. It follows that

$$H'(W) = \frac{dH}{dW} = \frac{\dot{H}}{\dot{W}} = \frac{y}{W_y} + \frac{\gamma}{W_y^2}\int_{x_0}^{x} W_y (1 - W_{yy})\, dx. \tag{8.105}$$

It is shown below that $W_{yy} = 1 + O(\gamma)$ for small γ (see (8.113)), so (8.105) gives

$$H'(W) = \frac{y}{W_y} + O(\gamma^2) = \frac{1}{K(W)} + O(\gamma^2). \tag{8.106}$$

Using this result, we can rewrite (8.102) for small γ as (8.104), which has the same form as (8.97), except that γ has been replaced by Γ. □

Note that (8.106) is actually valid only for values of K which are $O(1)$ for small γ, whereas for large values of K, for example, for K corresponding to the critical contour, the $O(\gamma^2)$ estimate of the integral in (8.105), which leads to (8.106), is no longer valid. However, the contribution of the integral to (8.104) is $O(1)$ only on short time intervals, because most of the time on the critical contour is spent near the stable equilibrium point. The influence of the integral on the solution of (8.104) is therefore negligible.

Theorem 8.5.1 (Nonequilibrium steady-state fluctuations). *The stationary probability density of fluctuations about S is given by*

$$p_S(x,y) \approx \exp\left\{-\frac{(\Delta A)^2}{2\varepsilon}\right\} \quad \text{for } \gamma \ll 1. \tag{8.107}$$

where the generalized action $A(\Gamma)$ of the steady-state trajectory through (x,y) (which corresponds to a different value of the friction constant Γ instead of γ) is given by

$$A(\Gamma) = \frac{1}{2\pi}\int_0^{2\pi} y\,dx \sim \frac{I}{\Gamma} \tag{8.108}$$

and $\Delta A = A(\Gamma) - A(\gamma)$.

Proof. In view of (8.99), we need to evaluate $W(x,y)$. We begin by determining its relation to Γ. First, we map the x,y plane to the W,x plane. Employing the relation $W_y = Ky + O(\gamma^2)$ and $\Gamma = \gamma(1-K)$, we obtain to leading order in γ,

$$W_\Gamma = \frac{\left(1 - \dfrac{\Gamma}{\gamma}\right) y}{\Gamma_y}. \tag{8.109}$$

Hence, to leading order in γ, we have

$$W(\Gamma) = \frac{1}{2}\int_0^\Gamma \left(1 - \frac{\Gamma}{\gamma}\right) \left(y^2\right)_\Gamma \, d\Gamma.$$

Integrating by parts, we obtain

$$W(\Gamma) = \frac{1}{2}\left[\left(1 - \frac{\Gamma}{\gamma}\right) y^2(\Gamma, x) + \frac{1}{\gamma}\int_0^\Gamma y^2(\Gamma, x)\, d\Gamma\right]. \tag{8.110}$$

The asymptotic expression (8.94) for $y(\gamma, x)$ gives in (8.110)

$$W(\Gamma) = \frac{1}{2}\left(\frac{I}{\Gamma} - \frac{I}{\gamma}\right)^2, \quad \gamma \leq \Gamma \leq \gamma_M, \tag{8.111}$$

because the right-hand side of (8.110) is independent of x on a W-contour. Thus eq. (8.111) can be written in the form

$$W(\Gamma) \sim \frac{1}{2}\left[A(\Gamma) - A_0\right]^2, \tag{8.112}$$

where $A_0 = A|_{\Gamma=\gamma}$.

Next, we show that for small γ the solution p_0 of (8.101) is to leading order a constant. Choosing p_0 to have the average value 1 on S and employing (8.103) in the transport equation (8.101), we see that $\dot{p}_0 = -\gamma(W_{yy}-1)p_0$ along the characteristic curves. Differentiating (8.111) and employing (8.94) differentiated with respect to y, we find that

$$W_{yy} = 1 + O(\gamma), \tag{8.113}$$

so that $p_0 = 1 + O(\gamma^2)$. □

Note that the quantity ΔA is the difference between the action associated with S and the action associated with the steady-state trajectory through (x, y). A finer resolution of the local fluctuations about S can be achieved by considering the probability density function p_S in local coordinates near S.

Theorem 8.5.2 (The pdf of local fluctuations). *The probability density function of the local fluctuations about S is given by*

$$p_S(x, y) \sim \exp\left\{-\frac{\alpha(x)\left[y - y_S(x)\right]^2}{2\varepsilon}\right\} \quad \text{for } \frac{\gamma}{I} \ll 1, \tag{8.114}$$

where

$$\alpha(x) = 1 + \left(\frac{\gamma}{I}\right)^2 (I\pi + \cos x) + O\left(\frac{\gamma^4}{I^4}\right). \tag{8.115}$$

The variance of the local fluctuations is

$$\sigma_y^2 = \varepsilon\left[1 - 2\pi\frac{\gamma^2}{I} + O\left(\frac{\gamma^3}{I^3}\right)\right]. \tag{8.116}$$

Proof. We introduce the variable $\delta = y_S(x) - y$, where $y_S(x)$ is given by (8.94). Recall that

$$W(x, y_S(x)) = W_x(x, y_S(x)) = W_y(x, y_S(x)) = 0. \tag{8.117}$$

It follow that Taylor's expansion of W near S, given in powers of δ, is

$$W(x,y) = \frac{1}{2}\alpha(x)\delta^2 + \frac{1}{6}\beta(x)\delta^3 + \cdots, \tag{8.118}$$

where $\alpha(x), \beta(x), \dots$, are yet undetermined functions. It follows that Taylor's expansions of $W_x(x,y)$ and $W_y(x,y)$ are given by

$$W_x(x,y) = -\frac{\nu_x(x,y)}{\nu_y(x,y)}\alpha(x)\delta + \frac{1}{2}\left[\alpha'(x) - \frac{\nu_x(x,y)}{\nu_y(x,y)}\beta(x)\right]\delta^2 + \cdots \tag{8.119}$$

and

$$W_y(x,y) = \alpha(x)\delta + \frac{1}{2}\beta(x)\delta^2 + \cdots. \tag{8.120}$$

The functions $\nu_x(x,y)$ and $\nu_y(x,y)$ are the components of the unit normal vector $\boldsymbol{\nu}(x,y)$ to S at $(x, y_S(x))^T$. The vector $\boldsymbol{\nu}(x,y)$ is orthogonal to the vector $(y_S(x), -\gamma y_S(x) - U'(x))$, which defines the flow (8.97). Hence,

$$\nu_x(x,y) = \frac{-[\gamma y_S(x) + U'(x)]}{\sqrt{y_S^2(x) + [\gamma y_S(x) + U'(x)]^2}}$$
$$\nu_y(x,y) = \frac{-\gamma y_S(x)}{\sqrt{y_S^2(x) + [\gamma y_S(x) + U'(x)]^2}}. \tag{8.121}$$

We substitute the expansions (8.119) and (8.120) of $W_x(x,y)$ and $W_y(x,y)$, respectively, into the eikonal equation (8.100) to get

$$\frac{\delta}{\nu_y(x,y)}\left[\alpha(x) + \frac{1}{2}\beta(x)\delta\right]\{\nu_y(x,y)[\gamma y_S(x) + U'(x)] - \nu_x(x,y)\gamma y_S(x)\}$$
$$+\,\delta^2\left\{\frac{1}{2}\alpha'(x)y_S(x) + \left[-\frac{\nu_x(x,y)}{\nu_y(x,y)} - \gamma\right]\alpha(x) + \gamma\alpha^2(x)\right\} = O\left(\delta^3\right).$$

We have $\{\nu_y(x,y)[\gamma y_S(x) + U'(x)] - \nu_x(x,y)\gamma y_S(x)\} = 0$, because this is the scalar product of the flow on S with its normal. Note that the function $\beta(x)$ is no longer needed for the calculation of $\alpha(x)$. It follows that $\alpha(x)$ satisfies the Bernoulli equation

$$\frac{1}{2}\alpha'(x)y_S(x) + \left[-\frac{\nu_x(x,y)}{\nu_y(x,y)} - \gamma\right]\alpha(x) + \gamma\alpha^2(x) = 0. \tag{8.122}$$

We first convert eq. (8.122) into a linear equation by the substitution $\alpha(x) = 1/\beta(x)$ to get

$$\beta'(x) - r(x)\beta(x) = t(x). \tag{8.123}$$

Here the coefficients are given by

$$r(x) = \frac{2}{y_S(x)}\left[\frac{\nu_x(x,y)}{\nu_y(x,y)} + \gamma\right] = \frac{2U''(x)}{y_S^2(x)}, \quad t(x) = \frac{2\gamma}{y_S(x)}. \tag{8.124}$$

The 2π-periodic solution of (8.123) is

$$\beta(x) = \frac{1}{R(x)}\left[\frac{R(0)}{1-R(0)}\int_0^{2\pi} t(z)R(z)\,dz + \int_0^x t(z)R(z)\,dz\right], \tag{8.125}$$

where $R(z) = \exp\left\{\int_z^{2\pi} r(u)\,du\right\}$.

The variance of the steady-state local fluctuations about S can be calculated from the above analysis. Noting that the function $\alpha(x)$ is the local frequency of the quasi-potential $W(x,y)$ in the y direction, we find that in the limit $\gamma/I << 1$, the variance of the local fluctuations of y, at fixed x, is given by

$$\sigma_y^2(x) \doteq \frac{\displaystyle\int \delta^2 \exp\{-\alpha(x)\delta^2/2\varepsilon\}\,d\delta}{\displaystyle\int \exp\{-\alpha(x)\delta^2/2/\varepsilon\}\,d\delta} = \frac{\varepsilon}{\alpha(x)}.$$

The average fluctuation of y is given by

$$\sigma_y^2 = \frac{\displaystyle\int dx \int \delta^2 \exp\{-\alpha(x)\delta^2/2\varepsilon\}\,d\delta}{\displaystyle\int dx \int \exp\{-\alpha(x)\delta^2/2\varepsilon\}\,d\delta} = \varepsilon\frac{\displaystyle\int \alpha^{-3/2}(x)\,dx}{\displaystyle\int \alpha^{-1/2}(x)\,dx}.$$

An expansion of $\alpha(x)$ in powers of γ/I yields (8.115), hence we obtain (8.116) for $\gamma/I << 1$ and the pdf of the local fluctuations about S is given by (8.114). □

8.5.3 The FPE and the MFPT far from equilibrium

We consider now the underdamped pendulum mod 2π in x; that is, we wrap the phase space on a cylinder of radius 1 about the y-axis. The domain D_S becomes one period of Figure 8.4 (and in the limit $\gamma \to 0-$ of Figure 8.5). Note that the tails of the domains of attraction of the stable equilibria form a long ribbon wrapped around the cylinder. We retain the notation D_S for the domain of attraction on the cylinder. The periodic solution of the FPE in the phase plane can be now normalized on the cylinder (on a single period). The nonequilibrium steady-state S becomes a stable limit cycle on the cylinder and the stable equilibria coalesce into a single one.

The *lifetime* of the nonequilibrium steady-state S is the first passage time from D_S to the separatrix. The mean lifetime is sufficiently long to establish a quasi-stationary probability density function in D_S. This pdf, p_S, is (8.99) mod 2π in x and as seen above, it is essentially constant on W-contours. Thus the process can be considered as a one-dimensional stationary diffusion process in the space of W-contours. The range of attraction D_S of S corresponds to Γ in the range $0 \leq \Gamma \leq \gamma_M$.

Theorem 8.5.3 (The averaged FPE far from equilibrium). *The leading term in the low-friction expansion of the stationary pdf in D_S is the solution of the one-dimensional FPE*

$$\frac{\partial p}{\partial s} = \frac{\partial}{\partial A}\left\{[A - A_0]\,p + \varepsilon\frac{\partial p}{\partial A}\right\} \quad \text{for } A > A(\gamma_M),\ \frac{\gamma}{I} \ll 1, \tag{8.126}$$

where $A = A(\Gamma)$ is defined in (8.108) *and $s = \gamma t$.*

Proof. The average of a function $F(x, y)$ over a W-contour is defined as

$$F(W) = \frac{1}{T_S}\int_0^{T_S} F(x(t), y(t))\,dt, \tag{8.127}$$

where t is the parameter of eq. (8.104), which represents time along the steady-state trajectory corresponding to $\Gamma(W)$, and $T_S = T_S(W)$ is the period of $y(t)$ on the trajectory.

We introduce W and x as coordinates in the FPE (8.69) and use the eikonal equation (8.100) to reduce it to

$$\gamma\frac{\partial p}{\partial s} = -y\frac{\partial p}{\partial x} + \gamma\varepsilon W_y^2\frac{\partial^2 p}{\partial W^2} + \gamma\left(W_y^2 + \varepsilon W_{yy}\right)\frac{\partial p}{\partial W} + \gamma p. \tag{8.128}$$

An expansion in powers of γ leads to averaging (8.128) on W-contours, as defined in 8.127), as a solvability condition, because $\left\langle W_y^2\right\rangle \sim 2W$ and $\langle W_{yy}\rangle \sim 1$. We obtain for an x-independent solution the averaged equation

$$\frac{\partial p}{\partial s} = 2\left\{\varepsilon W\frac{\partial^2 p}{\partial W^2} + \left(W + \frac{1}{2}\varepsilon\right)\frac{\partial p}{\partial W} + \frac{p}{2}\right\}. \tag{8.129}$$

A considerable simplification of (8.129) is achieved if we choose the action $A = A(\Gamma)$ as the independent variable in (8.129), rather than W. Using the relation $W \sim \frac{1}{2}\left[A(\Gamma) - A_0\right]^2$ (see (8.112)) in (8.129), we obtain (8.126). □

From Theorems 8.4.1, 8.5.3, and Exercise 8.13, we have the following

Corollary 8.5.1. *In the limit $\gamma \to 0$ the global stationary pdf in one period of the stationary underdamped dynamics is the stationary solution of the FPE*

$$\frac{\partial}{\partial A}\left\{\Phi'(A)p + \varepsilon\frac{\partial p}{\partial A}\right\} = 0 \quad \text{for } A > 0, \tag{8.130}$$

where

$$\Phi(A) = \begin{cases} \dfrac{(A - A_0)^2}{2} & \text{for } A > A(\gamma_M) \\ \mathcal{E}(A) & \text{for } 0 < A < A(\gamma_M), \end{cases} \tag{8.131}$$

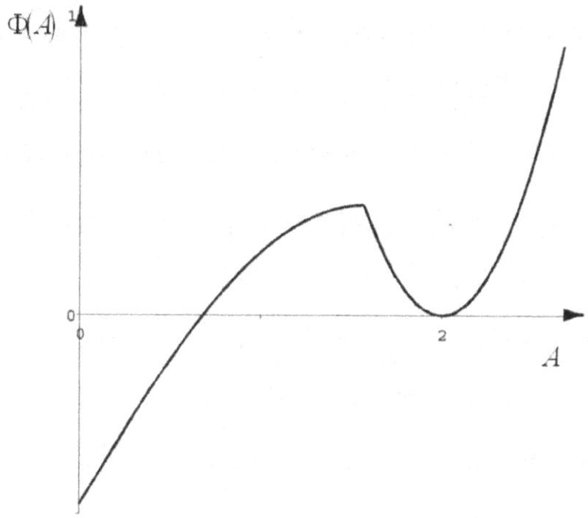

Figure 8.6. The effective global potential $\Phi(A)$ (8.131) in the steady-state FPE (8.130).

and where $A = I/4\pi$ *for* $0 < A < A(\gamma_M)$ *(in the domain* $E < E_B$*),* $\mathcal{E}(A) = E(I) - E_0$*,* $I = I(E)$ *is the action on the closed* E *contour, and*

$$E_0 = E_B - \frac{1}{2}\left(\frac{I(E_B)}{4\pi} - A_0\right)^2.$$

The connection of two local stationary WKB solutions of the FPE across the separatrix to form a global pdf is in general not a trivial matter, because the separatrix can be a caustic for the eikonal equation and the eikonal function may be discontinuous there. An obvious necessary connection criterion is that the net total probability flux across the separatrix vanishes [77], [90]. In the case at hand the fluxes of the WKB solutions on both sides of the separatrix vanish, so choosing A on either side so that its unilateral limits there are the same connects $\Phi(A)$ continuously, although not smoothly.

Exercise 8.20 (The critical W-contour). Show that

(i) The critical W-contour that first touches the separatrix does so at the unstable equilibrium point C of the noiseless dynamics (8.97). (HINT: C is a saddle point for W because $W_y(x_0, 0) = 0$ (see (8.109)), so $W_x \to 0$ as $(x, y) \to (x_0, 0)$. Thus W achieves its maximal value at (x_0, y) along any curve that reaches that point from the side of S_C that includes S, and W achieves its minimum value there along the separatrix.)

(ii) Show that the critical W-contour has a cusp at C and that $\nabla W = 0$ there (HINT:

Recall that this contour has two different slopes at this point and the slope q of the separatrix at this point is different from either slope of the W-contour, because q is given by (8.95) with the negative square root, and with γ_M replaced by γ.)

(iii) Show that W attains its minimum value along the separatrix at C, so that C is a saddle point of W. (HINT: Note that $\dot{W} = -\gamma W_y^2$ for motion on the separatrix, so the minimum of W on the separatrix is achieved at the unstable equilibrium point, toward which the separatrix converges.)

(iv) Show the characteristic curve through C is the most probable path out of D_S [77]. □

In addition to fluctuations about the nonequilibrium steady-state S, thermal noise can also cause transitions from the basin of attraction D_S of S, into one of the basins of attraction D_E of a stable equilibrium state.

Theorem 8.5.4 (The MFPT to S_C). *For $\gamma < \gamma_M$ and $I_{\min}(\gamma) < I < 1$ the MFPT from S to S_C is*

$$\bar{\tau}_S \approx \frac{\sqrt{\pi}}{\gamma}\sqrt{\frac{\varepsilon}{\Delta W}}\exp\left\{\frac{\Delta W}{\varepsilon}\right\} \quad \text{for } \frac{\varepsilon}{\Delta W} \ll 1, \tag{8.132}$$

where

$$\Delta W = \frac{1}{2}\left(\frac{I}{\gamma} - \frac{I}{\gamma_M(I)}\right)^2 \approx \frac{1}{2}\left(\frac{I}{\gamma} - \frac{4}{\pi}\right)^2 \quad \text{for } \frac{\gamma}{I} \ll 1. \tag{8.133}$$

Proof. We use (6.18) to evaluate the MFPT. The solution of the boundary value problem

$$\frac{\partial}{\partial A}\left\{\Phi'(A)p + \varepsilon\frac{\partial p}{\partial A}\right\} = -\delta(A - A_0) \text{ for } A > 0, \quad p\left(A(\gamma_M)\right) = 0, \tag{8.134}$$

is given by

$$p(A) = \frac{1}{\varepsilon}e^{-\Phi(A)/\varepsilon}\int\limits_{A(\gamma_M)}^{A} H(A_0 - z)e^{\Phi(z)/\varepsilon}\,dz.$$

The boundary flux is $F = -\varepsilon p'(A(\gamma_M)) = 1$ and the population is

$$N = \int\limits_{A(\gamma_M)}^{\infty} p(A)\,dA = \frac{\sqrt{2\pi\varepsilon}}{A(\gamma_M) - A_0}\exp\left\{\frac{(A(\gamma_M) - A_0)^2}{2\varepsilon}\right\}(1 + o(1)) \text{ as } \varepsilon \to 0,$$

hence, using (8.112) we obtain (8.132). □

Exercise 8.21 (Numerical solution of the eikonal equation). Integrate numerically the characteristic equations (8.103) to construct the solution of the eikonal equation (8.100), which is a 2π-periodic function of x.

(i) Explain why initial conditions have to be given at a finite distance from S. (HINT: The initial conditions for (8.103) are given by eq. (8.117) on S, however, S is a characteristic curve and a caustic; that is, an attractor of the characteristic curves.)

(ii) Use the expansion (8.118)–(8.120) of W, W_x, and W_y as initial conditions for (8.103) near S. For a given x_0, take the value of $y_S(x_0)$ as the approximation (8.94). Then use $y_S(x)$ in (8.94) and (8.121) to calculate $r(x)$ and $t(x)$ in (8.124) and calculate $\alpha(x)$ from (8.125) on a lattice $x_0 < x_1 < \cdots < x_n = 2\pi + x_0$. Note that although the lattice $(x_i, y_S(x_i) - \delta)$, for $i = 0, 1, \ldots, n$, which is parallel to the contour $W = 0$, can be used for assigning initial conditions, the vertical lattice $(x_0, y_S(x_0) - \delta(1 + i\Delta))$, for $i = 1, \ldots, k$ with small δ and Δ, is more efficient for covering D_S with characteristics uniformly. Determine the initial values of W, W_x, and W_y on this lattice from (8.118)–(8.120).

(iii) Evaluate the function W numerically to confirm the validity of the asymptotic approximations given above over a wide range of values of I and γ [142]. □

8.5.4 Application to the shunted Josephson junction

The shunted Josephson junction is often described, in dimensional units, by the phenomenological equation

$$C\frac{dV}{dt} + \frac{V}{R} + I_j \sin\theta = I_{\text{dc}}, \tag{8.135}$$

where C, R, and I_j are the capacitance, resistance, and the critical current of the junction, respectively. I_{dc} is the external current source, whereas θ and V are the phase difference and voltage across the junction, respectively. Employing the Josephson relation

$$\dot{\theta} = \frac{2e}{\hbar}V, \tag{8.136}$$

the circuit equation (8.135) reduces to (8.97) with

$$I = \frac{I_{\text{dc}}}{I_j}, \quad \gamma = \frac{1}{\omega_j RC}, \tag{8.137}$$

where the Josephson plasma frequency ω_j is given by

$$\omega_j^2 = \frac{2eI_j}{\hbar c} \tag{8.138}$$

and the time is measured in units of ω_j^{-1}. Here e is the electronic charge, c is the velocity of light propagation in vacuum, and $\hbar$ is Planck's constant divided by 2π.

The stable equilibrium state corresponds to $\dot{\theta} = 0$; that is, to $V = 0$; that is, there is no voltage across the junction while a current I is flowing through it. Thus the junction is in a superconducting state. In contrast, on the nonequilibrium steady-state S the mean voltage is not zero so the junction has normal resistance. In a typical measurement of the lifetime of each state the driving external current

I_{dc} is swept repeatedly through a certain range at relatively slow rate and the $I-V$ response is recorded.

When $\gamma > \gamma_M(I)$ and $|I| < 1$ there is no running state so the junction is superconducting and the $I-V$ characteristic lies on the I-axis. As I exceeds 1 the stable equilibrium disappears and the characteristic begins to grow and converges to the ohmic line. In the hysteretic regime, when $\gamma < \gamma_M(I)$ and $|I| < 1$, the average characteristic is detached from the I-axis and increases at a high slope and gradually converges to the ohmic line as I exceeds 1. This is interpreted as follows. During each sweep the response curve stays on the I-axis, in a superconducting state with zero voltage, for random times that are the lifetimes of the stable equilibrium state, as calculated above for the given potential (8.91), and in the conducting state, with finite voltage V_S, for times that are the lifetimes of the running state S. The superconducting and the conducting characteristics are averaged with weights $\bar{\tau}_E/(\bar{\tau}_E + \bar{\tau}_S)$ and $\bar{\tau}_S/(\bar{\tau}_E + \bar{\tau}_S)$, respectively, to give the experimentally observed response curve [15],

$$V = \frac{0\bar{\tau}_E + V_S\bar{\tau}_S}{\bar{\tau}_E + \bar{\tau}_S}. \tag{8.139}$$

In [39] the value of $\bar{\tau}_E$ was calculated from Kramers' underdamped rate formula (8.89) and $\bar{\tau}_S$ was calculated from the measured eq. (8.139) and compared with the theoretically predicted value (8.132) with (8.133) and gave reasonable agreement between the theoretical and experimental values of $\bar{\tau}_S$.

To understand the different approaches to the calculation of $\bar{\tau}_E$ and $\bar{\tau}_S$ it should be noted that the range of values of γ for which the system is underdamped in D_E is different from that in D_S. In D_S (8.132) is valid in practically the entire range of $|I| < 1$ and $\gamma < \gamma_M(I)$, for which the system is bistable.

The results of this section can be used to calculate the fluctuation in voltage across the Josephson junction for a given current I_{dc}. The voltage is related to the dimensionless velocity by

$$V = I_j R\gamma y \tag{8.140}$$

(see (8.136)), hence, using (8.116) and (8.140), and substituting (8.137) and (8.138),

$$\langle \Delta V^2 \rangle = \frac{k_B T}{C}\left[1 - 2\pi\frac{\gamma^2}{I} + O\left(\frac{\gamma^3}{I^3}\right)\right],$$

where k_B is Boltzmann's constant and T is absolute temperature in degrees Kelvin. This fluctuation is a measurable quantity. It corresponds to the width of error bars when measuring the voltage across the junction at a given current I_{dc}.

A measure of the relative stability of S and the stable equilibrium state E is the ratio of their expected lifetimes, $\bar{\tau}_S$ and $\bar{\tau}_E$, respectively. We have

$$\lim_{\varepsilon\to 0}\log\frac{\bar{\tau}_S}{\bar{\tau}_E} = W_{\max} - E_B = \frac{1}{2}\left[\frac{I}{\gamma_M(I)} - \frac{I}{\gamma}\right]^2 - 2\arcsin I - \pi I + 2\sqrt{1-I^2},$$

so that $\bar{\tau}_S/\bar{\tau}_E \propto \exp\{-[W_{\max} - E_B]/2\varepsilon\}$. The measurements in [39] were done with the parameters $C = 7.2nF, I_j = 4\mu A$, and $T = 4.2K$. For these parameters, if $W_{\max} - E_B = 1$, then the measure of relative stability is e^{26}, hence the region where the hysteresis can be observed experimentally is very narrow.

8.6 Annotations

Theorem 8.1.1 about pathwise convergence in the Smoluchowski limit was proved for finite time intervals in [213]. The GLE has been derived from statistical mechanics in [141], [177], [23], [170], [123], [91], [92], [93]. Theorem 8.1.2 about the Smoluchowski limit of the GLE was proved in [61] and more recently in [75].

Continuum theories of diffusive systems describe the concentration field by the Nernst–Planck equation with fixed boundary concentrations [102], [65], [64], [24], [191], [189], [190], [188], [219]. In these theories there is no time-dependence of the macroscopic boundary concentrations. The force field does not fluctuate and is usually calculated from a Poisson equation coupled to the Nernst–Planck equations. The huge voltage fluctuations in the salt solution and in the channel are averaged out in these theories.

The question of the boundary behavior of the Langevin trajectories, corresponding to fixed boundary concentrations, arises both in theory and in the practice of molecular simulations of diffusive motion [22], [29], [12], [181], [179], [180], [104]. The boundary behavior of diffusing particles in a finite domain has been studied in various cases, including absorbing, reflecting, sticky boundaries, and many other modes of boundary behavior [161], [117]. In [212] a sequence of Markovian jump processes is constructed such that their transition probability densities converge to the solution of the Nernst–Planck equation with given boundary conditions, including fixed concentrations and sticky boundaries.

The Smoluchowski boundary layer was studied in [166], [250], [96], [94], [95], [128], [127]. The study of the relevant special functions, and especially of the Hurwitz zeta function was done in [221].

The results of Section 8.4, about thermal activation of an underdamped Brownian particle over a potential barrier, go back to Kramers [140], [100]. The thermal activation of a steady-state system far from equilibrium, from a stable limit cycle across the separatrix, remained an open problem because Kramers' 1940 paper [140]. The need to find the expected time of the running state of the Josephson junction (the underdamped physical pendulum) arose with the advent of cryogenic devices based on the Josephson effect and related problems. The problem was solved in [15]–[19], [142], which are the basis for Section 8.5, and in the work of Risken and Vollmer [206], and [42].

Chapter 9

Large Deviations of Markovian Jump Processes

The diffusion approximations to Markovian jump processes, which are obtained by truncating the Kramers–Moyal expansion of the master equation after two terms, are often valid only on time scales that are significantly shorter than stochastic time scales of the jump process, such as the time to a rare event. These approximations may be useful as local approximations, but are not useful for estimating the probability of a large deviation or the mean first passage time to the boundary of a domain of attraction of a stable or meta-stable state. It is the purpose of this chapter to develop a methodology for constructing analytical approximations to the pdf of Markovian jump processes that are valid also in the tail regions and to mean first passage times to tail regions.

9.1 The WKB structure of the stationary pdf

We consider again the random walk (7.80),

$$x_{n+1} = x_n + \varepsilon\xi_n, \tag{9.1}$$

where

$$\Pr\{\xi_n = \xi \,|\, x_n = x,\, x_{n-1} = y,\, \ldots\} = w(\xi \,|\, x, \varepsilon)$$

ε is a small parameter, and x_0 is a random variable with a given pdf $p_0(x)$. As the scale of the jumps ε becomes smaller, a larger number of steps are required for the process x_n to reach a fixed point y, so visits to such a point become rare events and the probability of such a visit becomes a tail event. This is more the case if the averaged dynamics (9.1) has a stable attractor $\tilde{x}$. It becomes harder to get away from $\tilde{x}$ as ε decreases so that reaching a fixed point y becomes a large deviation.

The transition pdf $p_\varepsilon(y, n \,|\, x, m) = \Pr\{x_n = y \,|\, x_m = x\}$ satisfies the master

Z. Schuss, *Theory and Applications of Stochastic Processes: An Analytical Approach*,
Applied Mathematical Sciences 170, DOI 10.1007/978-1-4419-1605-1_9,

equation (forward Kolmogorov equation)

$$\begin{aligned}
&p_\varepsilon(y,n+1\,|\,x,m)-p_\varepsilon(y,n\,|\,x,m) &(9.2)\\
&=\int\limits_{\mathbb{R}}[p_\varepsilon(y-\varepsilon\xi,n\,|\,x,m)w(\xi\,|\,y-\varepsilon\xi,\varepsilon)-p_\varepsilon(y,n\,|\,x,m)w(\xi\,|\,y,\varepsilon)]\,d\xi\\
&=L_\varepsilon p_\varepsilon(y,n\,|\,x,m)
\end{aligned}$$

and the backward master equation

$$\begin{aligned}
&p_\varepsilon(y,n\,|\,x,m)-p_\varepsilon(y,n\,|\,x,m+1) &(9.3)\\
&=\int\limits_{\mathbb{R}}[p_\varepsilon(y,n\,|\,x+\varepsilon\xi,m+1)-p_\varepsilon(y,n\,|\,x,m)]w(\xi\,|\,x,\varepsilon)\,d\xi\\
&=L_\varepsilon^* p_\varepsilon(y,n\,|\,x,m),
\end{aligned}$$

with the initial or terminal condition

$$p_\varepsilon(y,m\,|\,x,m)=\delta(y-x). \tag{9.4}$$

To determine the probability of a large deviation, we denote the conditional moments of the jump process

$$m_n(x,\varepsilon)=\mathbb{E}\left[\xi_n\,|\,x_n=x\right]=\int\limits_{\mathbb{R}}\xi^n w(\xi\,|\,x,\varepsilon)\,d\xi$$

and instead of the diffusion scaling $n=t/\varepsilon^2$, used in Section 7.3, we scale $n=t/\varepsilon$ to obtain the drift equation $x(t+\varepsilon)-x(t)=\varepsilon m_1(x)+O\left(\varepsilon^2\right)$, which is the discretized ordinary differential equation

$$\dot{x}(t)=m_1(x). \tag{9.5}$$

We consider the case that there exists at least one stable equilibrium point $\tilde{x}$ such that $m_1(\tilde{x})=0$ and $m_1'(\tilde{x})<0$. Setting $\Delta t=\varepsilon$, we write the master equation (9.2) in the form

$$\begin{aligned}
&p_\varepsilon(y,t+\Delta t\,|\,x,m)-p_\varepsilon(y,t+\Delta t\,|\,x,m)\\
&=\int\limits_{\mathbb{R}}p_\varepsilon(x-\varepsilon\xi,t)w(\xi\,|\,x-\varepsilon\xi,\varepsilon)\,d\xi &(9.6)
\end{aligned}$$

and assume that it has a unique integrable solution satisfying the given initial condition (9.4). If there is a stationary solution,

$$p_\varepsilon(y)=\lim_{n\to\infty}p_\varepsilon(y,n\,|\,x,m),$$

independent of the initial variables (x, m), it satisfies the stationary forward Kolmogorov equation

$$L_\mathcal{E} p_\varepsilon(y) = \int\limits_{\mathbb{R}} [p_\varepsilon(y - \varepsilon\xi)w(\xi \,|\, y - \varepsilon\xi) - p_\varepsilon(y)w(\xi \,|\, y, \varepsilon)]\, d\xi = 0. \tag{9.7}$$

We assume, for simplicity, that $w(\xi \,|\, y, \varepsilon)$ is independent of ε and write

$$w(\xi \,|\, y, \varepsilon) = w(\xi \,|\, y). \tag{9.8}$$

For small ε, we seek an asymptotic solution to (9.7) in the WKB form

$$p_\varepsilon(y) \sim [K_0(y) + \varepsilon K_1(y) + \cdots]e^{-\psi(y)/\varepsilon}, \quad 0 < \varepsilon \ll 1. \tag{9.9}$$

The WKB approximation (9.9) is the *large deviations expansion* of the stationary pdf, because it approximates the pdf for all values of y.

To determine the components of the expansion (9.9), we substitute (9.9) in (9.7) and compare the coefficient of each power of ε to zero We obtain at the leading order the Hamilton–Jacobi type (eikonal) equation for $\psi(y)$, given by

$$\int\limits_{\mathbb{R}} \left[e^{\xi\psi'(y)} - 1\right] w(\xi \,|\, y, 0)\, d\xi = 0. \tag{9.10}$$

At the next order we find the $K_0(y)$ is the solution of the "transport" equation

$$\int\limits_{\mathbb{R}} \left\{\frac{\partial}{\partial y}[w(\xi \,|\, y)K_0(y)] + \frac{\xi w(\xi \,|\, y)}{2}\psi''(y)K_0(y)\right\} \xi e^{\xi\psi'(y)}\, d\xi = 0. \tag{9.11}$$

Lemma 9.1.1. *If*

$$(y - \tilde{x})m_1(y) < 0 \;\text{ for } y \neq \tilde{x}, \tag{9.12}$$

then the eikonal equation (9.10) has a unique solution $\psi = \psi(y)$ satisfying $\psi(\tilde{x}) = 0$, and $\psi(y) > 0$ for $y \neq \tilde{x}$.

Proof. The moment generating function $M_y(\theta)$ of the conditional jump density $w(\xi \,|\, y)$ is defined by

$$M_y(\theta) = \int\limits_{\mathbb{R}} e^{\xi\theta} w(\xi \,|\, y)\, d\xi.$$

equation *(9.10)* can be written in terms of $M_y(\theta)$ as

$$M_y(\psi'(y)) = 1. \tag{9.13}$$

The condition (9.12) implies that $(y - \tilde{x})M_y'(0) < 0$ and because $M_y''(\theta) > 0$ and $M_y(0) = 1$, the eq.(9.13) has a unique nonzero solution $\theta(y)$ for $y \neq \tilde{x}$, such that

$(y - \tilde{x})\theta(y) > 0$. Therefore $m_1(\tilde{x}) = 0$ implies that the solution $\theta(\tilde{x})$ is unique. It follows that the function

$$\psi(y) = \int_{\tilde{x}}^{y} \theta(s)\, ds \tag{9.14}$$

is well defined and has an absolute minimum at $\tilde{x}$. □

Given the solution $\psi(y)$ of the eikonal equation (9.10), the solution of the transport equation (9.11) is

$$K_0(y) \tag{9.15}$$
$$= \left[\int_{\mathbb{R}} \xi w(\xi \mid y) e^{\xi \psi'(y)}\, d\xi\right]^{-1/2} \exp\left\{-\frac{1}{2}\int_0^y \frac{\displaystyle\int_{\mathbb{R}} \xi \frac{\partial w(\xi \mid \eta)}{\partial \eta} e^{\xi\psi'(\eta)}\, d\xi}{\displaystyle\int_{\mathbb{R}} \xi w(\xi \mid \eta) e^{\xi\psi'(\eta)}\, d\xi}\, d\eta\right\}.$$

Higher order corrections $K_i(y)$ can be obtained in a straightforward manner.

Exercise 9.1 (Further expansion). Assume that

$$w(\xi \mid x, \varepsilon) \sim \sum_{i=0}^{\infty} \varepsilon^i w_i(\xi \mid x), \tag{9.16}$$

$$m_n(x, \varepsilon) \sim \sum_{i=0}^{\infty} \varepsilon^i m_{n,i}(x) \text{ for all } n > 0, \tag{9.17}$$

where $w_i(\xi \mid x)$, $m_{n,i}(x)$ are regular functions. Write down the hierarchy of transport equations for $K_i(y)$. □

A local analysis of the eikonal equation (9.10) near $\tilde{x}$ implies that $\psi'(\tilde{x}) = 0$, which implies that the WKB solution (9.9) is peaked about $\tilde{x}$ and locally has the form

$$p_{\text{local}}(y) = K_0(\tilde{x}) \exp\left\{-\frac{\psi''(\tilde{x})(y - \tilde{x})^2}{2\varepsilon}\right\}. \tag{9.18}$$

Exercise 9.2. (Local solution).
(i) Find the local behavior (9.18) by introducing the scaled variable $\zeta = (y - \tilde{x})/\sqrt{\varepsilon}$ into the stationary forward equation (9.7) and expanding the solution in powers of ε. Obtain the leading-order approximation as the solution to the Ornstein–Uhlenbeck type equation

$$\mathcal{L}_\zeta p_{\text{outer}}(\zeta) = -m_1'(\tilde{x}) \frac{\partial \zeta p_{\text{outer}}(\zeta)}{\partial \zeta} + \frac{1}{2} m_1(\tilde{x}) \frac{\partial^2 p_{\text{outer}}(\zeta)}{\partial \zeta^2} = 0. \tag{9.19}$$

Show that the solution of (9.19) is (9.18) with $\psi''(\tilde{x}) = -2m_1'(\tilde{x})/m_2(\tilde{x})$ and $K_0(\tilde{x})$ is a normalization constant.

(ii) Explain why the local solution (9.18) is a good approximation to the solution $p_\varepsilon(y)$ only for $y - \tilde{x} = O(\sqrt{\varepsilon})$.

(iii) Show that truncating the Kramers–Moyal expansion of the forward equation (9.7) after the second term gives the diffusion approximation

$$\mathcal{L}_{\text{diff}} p_{\text{KM}}(y) = -[m_1(y)p_{\text{KM}}(y)]' + \frac{\varepsilon}{2}[m_2(y)p_{\text{KM}}(y)]'' = 0, \tag{9.20}$$

whose WKB solution

$$p_{\text{KM}}(y) \sim H_0(y)e^{-\varphi(y)/\varepsilon}, \tag{9.21}$$

leads to the eikonal equation

$$m_1(y)\varphi'(y) + m_2(y)[\varphi'(y)]^2 = 0. \tag{9.22}$$

Show that the eikonal function $\varphi(y)$ is different from the eikonal function $\psi(y)$ of (9.10).

(iv) Conclude that the diffusion approximation is a good approximation to $p_\varepsilon(y)$ only for $y - \tilde{x} = O(\sqrt{\varepsilon})$ (HINT: Expand both eikonal functions in Taylor's series about $\tilde{x}$.) □

Exercise 9.3 (A drift-jump process with small rapid jumps). Instead of the diffusion scaling of the jump rate $\lambda(x)/\varepsilon^2$ in the continuous time Markov drift and jump process described in Example 7.10, scale

$$x_\varepsilon(t+\Delta t) = \begin{cases} x_\varepsilon(t) + A(x_\varepsilon(t))\Delta t + o(\Delta t) \\ \text{w.p. } 1 - \dfrac{\lambda(x_\varepsilon(t))}{\varepsilon}\Delta t + o(\Delta t) \\ \\ x_\varepsilon(t) + A(x_\varepsilon(t))\Delta t + \varepsilon\xi(t) + o(\Delta t) \\ \text{w.p. } \dfrac{\lambda(x_\varepsilon(t))}{\varepsilon}\Delta t + o(\Delta t), \end{cases} \tag{9.23}$$

where for $t > t_1 > \cdots$,

$$\Pr\{\xi(t) = \xi \,|\, x_\varepsilon(t) = x,\, x_\varepsilon(t_1) = x_1,\, \ldots\} = \tilde{w}(\xi \,|\, x, \varepsilon).$$

The process $x_\varepsilon(t)$ makes small jumps with the large rate $\lambda(x)/\varepsilon$.

(i) Show that the pdf $p_\varepsilon(y, t \,|\, x, s)$ of $x_\varepsilon(t)$ satisfies master equation

$$\begin{aligned} &\frac{\partial p_\varepsilon(y,t\,|\,x,s)}{\partial t} \\ &= -\frac{\partial A(y)p_\varepsilon(y,t\,|\,x,s)}{\partial y} \\ &\quad + \frac{1}{\varepsilon}\left[\int_{\mathbb{R}} \lambda(y-\varepsilon\xi)p_\varepsilon(y-\varepsilon\xi,t\,|\,x,s)\tilde{w}(\xi\,|\,y-\varepsilon\xi,\varepsilon)\,d\xi - \lambda(y)p_\varepsilon(y,t\,|\,x,s)\right] \end{aligned} \tag{9.24}$$

and the backward master equation

$$\frac{\partial p_\varepsilon(y,t\,|\,x,s)}{\partial s} = A(x)\frac{\partial p_\varepsilon(y,t\,|\,x,s)}{\partial x} + \frac{\lambda(x)}{\varepsilon}\int\limits_{\mathbb{R}}[p_\varepsilon(y,t\,|\,x+\varepsilon\xi,s) - p_\varepsilon(y,t\,|\,x,s)]\tilde{w}(\xi\,|\,x,\varepsilon)\,d\xi,$$

with the initial or terminal condition

$$p_\varepsilon(y,s\,|\,x,s) = \delta(y-x). \tag{9.25}$$

(ii) Show that the scaling of the rate $\lambda(x)/\varepsilon$ in (9.23) gives the drift equation

$$\dot{x}(t) = A(x) + \lambda(x)\tilde{m}_1(x), \tag{9.26}$$

where

$$\tilde{m}_k(x) = \int\limits_{\mathbb{R}} z^k \tilde{w}(x\,|\,x,\varepsilon)\,dz.$$

(iii) Construct a WKB type approximation (9.9) to the solution of the stationary forward equation (9.24). Derive the eikonal and transport equations

$$\begin{aligned} &A(y)\psi'(y) + \lambda(y)\int\limits_{\mathbb{R}}\left[e^{\xi\psi'(y)} - 1\right]\tilde{w}(\xi\,|\,y,0)\,d\xi = 0 \\ &\frac{\partial[A(y)K_0(y)]}{\partial y} + \int\limits_{\mathbb{R}}\left\{\frac{\partial[\lambda(y)\tilde{w}(\xi\,|\,y)K_0(y)]}{\partial y} + \frac{\xi w(\xi\,|\,y)\lambda(y)}{2}\psi''(y)K_0(y)\right\} \\ &\times \xi e^{\xi\psi'(y)}\,d\xi = 0. \end{aligned} \tag{9.27}$$

(iv) Prove a lemma analogous to Lemma 9.1.1 for $x_\varepsilon(t)$. □

In cases where the process is restricted to an interval D, such as the interval of attraction of the stable equilibrium of (9.5) or (9.26), the WKB solution (9.9) is a good approximation to $p_\varepsilon(y)$ only on a subset of D, up to a small neighborhood of ∂D. In this case the stationary WKB approximation does not necessarily represent the stationary solution of the forward equation, which may in fact not exist. The stationary WKB solution may approximate the n- or t-dependent solution of the forward equation on finite, though long time intervals (e.g., of order $O\left(\varepsilon^{-1}\right)$), which are, however, shorter than the mean exit time from D. This solution may be interpreted in this case as a quasi-stationary approximation.

9.2 The mean time to a large deviation

We now construct asymptotic approximations, for $\varepsilon \ll 1$, to the mean number of steps of the process x_n in (9.1) to exit the interval of attraction $D = (-A, B)$ of the equilibrium point $\tilde{x}$ of (9.5). Similar analysis applies to the asymptotic approximation of the mean exit time of the process $x_\varepsilon(t)$ in (9.23) from the interval of attraction of the stable equilibrium of (9.26). The random number of steps to exit D is

$$\tilde{n}_D = \min\{n \,:\, x_n \notin D\} \tag{9.28}$$

and the mean number is

$$n_D(x) = \mathbb{E}[\tilde{n}_D \,|\, x_0 = x]. \tag{9.29}$$

The mean number satisfies the equation

$$L_\varepsilon^* n_D(x) = \int_{\mathbb{R}} [n_D(x + \varepsilon\xi) - n_D(x)] w(\xi \,|\, x)\, d\xi - n_D(x) = -1 \tag{9.30}$$

with the condition

$$n_D(x) = 0 \ \text{ for } x \notin D. \tag{9.31}$$

Here L_ε^* is the backward Kolmogorov operator defined in (9.3). Equation (9.30) is derived as in the proof of Theorem 7.1.4, by noting that the pdf of the first exit time,

$$\Pr\{\tilde{n}_D = k \,|\, x_0 = x\} = \int_D [p_\varepsilon(y, k \,|\, x, 0) - p_\varepsilon(y, k+1 \,|\, x, 0)]\, dy, \tag{9.32}$$

satisfies the integrated backward master equation (9.3) with respect to y with $k = n - m$. Summing over n gives (9.30). The condition (9.31) reflects the fact that the mean time to exit D from a point x outside D is zero. We note that the boundary condition $n_D(-A) = n_D(B) = 0$ is incorrect, because $n_D(x)$ may suffer a discontinuity at ∂D, as shown in Theorem 7.4.1.

To construct an asymptotic expansion of $n_D(x)$ for $x \in D$ for $\varepsilon \ll 1$, we consider first the case $A = \infty$ and assume that $n_D(x) < \infty$ and that $\tilde{x}$ is a global attractor for (9.5) in D. We assume that $w(\xi \,|\, x)$ is a regular function of x. The following theorem generalizes Theorem 7.4.1.

Theorem 9.2.1 (The mean time to a large deviation). *Under the above assumptions, the expected number of steps $n_D(x)$ for the process x_n to leave D, has the asymptotic expansion*

$$n_D(x) \sim C_1(\varepsilon) U_0\left(\frac{B - x}{\varepsilon}\right) \quad \text{if } m_1(B) < 0, \tag{9.33}$$

where

$$C_1(\varepsilon) \sim \sqrt{\frac{2\pi}{\psi''(\tilde{x})} \frac{K_0(\tilde{x})}{K_0(B)}} \frac{e^{\psi(B)/\varepsilon}}{\int_{-\infty}^{0} e^{\eta\psi'(B)} \int_{-\infty}^{\eta} w(\xi \,|\, B) U_0(\eta - \xi)\, d\xi\, d\eta}, \tag{9.34}$$

$U_0(\eta)$ is the solution of the Wiener–Hopf equation

$$U_0(\eta) = \int_{0}^{\infty} w(\eta - \zeta \,|\, B) U_0(\zeta)\, d\zeta \ \text{ for } \eta > 0, \tag{9.35}$$

subject to

$$U_0(\eta) = 0 \ \text{ for } \eta \leq 0. \tag{9.36}$$

and

$$\lim_{\eta \to \infty} U_0(\eta) = 1. \tag{9.37}$$

If $m_1(B) = 0$, then

$$n_D(x) \tag{9.38}$$
$$\sim C_2(\varepsilon) \begin{cases} 2\sqrt{\dfrac{m_1'(B)}{\pi m_2(B)}} \displaystyle\int_0^{(B-x)/\sqrt{\varepsilon}} \exp\left\{ -\frac{m_1'(B) u^2}{m_2(B)} \right\} du \ \text{ for } \dfrac{B-x}{\varepsilon} \gg 1 \\ U_0\left(\dfrac{B-x}{\varepsilon} \right) \qquad \text{for } \dfrac{B-x}{\varepsilon} = O(1), \end{cases}$$

where

$$C_2(\varepsilon) \sim \frac{\pi}{\varepsilon} \frac{K_0(\tilde{x})}{K_0(B)} e^{\psi(B)/\varepsilon} \sqrt{\left| \frac{m_1'(\tilde{x}) m_1'(B)}{m_2(\tilde{x}) m_2(B)} \right|} \frac{1}{m_2(B)}. \tag{9.39}$$

The function $\psi(y)$ is the solution of the eikonal equation (9.10), *and $K_0(y)$ is given in* (9.15).

Proof. The proof is quite similar to that of Theorem 7.4.1. Using (9.31), we rewrite (9.30) as

$$\int_{-\infty}^{(B-x)/\varepsilon} [n_D(x + \varepsilon\xi)] w(\xi \,|\, x)\, d\xi - n_D(x) = -1. \tag{9.40}$$

For x away from the boundary, we extend the upper limit of integration to infinity and then use the Kramers–Moyal expansion of (9.40) to obtain the outer expansion of $n_D(x)$. That is, we expand $n_D(x + \varepsilon\xi)$ in powers of ε to get

$$\sum_{k=1}^{\infty} \frac{\varepsilon^k m_k(x)}{k!} \frac{d^k n_D(x)}{dx^k} = -1. \tag{9.41}$$

Clearly, as $\varepsilon \to 0$, we expect $n_D(x)$ to become infinite. Thus we scale

$$n_D(x) = C(\varepsilon)u(x), \tag{9.42}$$

where $C(\varepsilon) \to \infty$ as $\varepsilon \to 0$ and $\sup_{x<B} u(x) = 1$. We obtain from (9.41) and (9.42)

$$\sum_{k=1}^{\infty} \frac{\varepsilon^k m_k(x)}{k!} \frac{d^k u(x)}{dx^k} \sim 0 \ \text{ for } \varepsilon \to 0. \tag{9.43}$$

The regular expansion

$$u(x) \sim u_0(x) + \varepsilon u_1(x) + \cdots \tag{9.44}$$

gives the reduced equation

$$m_1(x)u_0'(x) = 0, \tag{9.45}$$

which implies $u_0(x) = 1$. The approximation

$$n_D(x) \sim C(\varepsilon)u_0(x) = C(\varepsilon) \tag{9.46}$$

satisfies (9.40) asymptotically for x bounded away from the boundary; that is, when the upper limit of integration can be extended to infinity, but fails to satisfy the condition (9.31). For x near B it is therefore necessary to construct a boundary layer correction to $u_0(x)$. We begin with the case $m_1(B) < 0$. We introduce the stretched variable

$$\eta = \frac{B-x}{\varepsilon} \tag{9.47}$$

and the scaled boundary layer function

$$U(\eta) = u(B - \varepsilon\eta) \tag{9.48}$$

into (9.40) to obtain

$$\int_{-\infty}^{\eta} U(\eta - \xi)w(\xi \,|\, B - \varepsilon\eta)\, d\xi - U(\eta) \sim 0 \tag{9.49}$$

Under the assumptions of the theorem, $w(\xi \,|\, x - \varepsilon\eta)$ can be expanded in powers of ε, and in particular, $w(\xi \,|\, B - \varepsilon\eta) = w(\xi \,|\, B) + o(1)$ as $\varepsilon \to 0$. The regular expansion

$$U(\eta) \sim U_0(\eta) + \varepsilon U_1(\eta) + \cdots \tag{9.50}$$

and the change of variables $\zeta = \eta - \xi$ in (9.49) gives the Wiener–Hopf equation

$$U_0(\eta) = \int_0^{\infty} w(\eta - \zeta \,|\, B)U_0(\zeta)\, d\zeta \ \text{ for } \eta > 0, \tag{9.51}$$

subject to

$$U_0(\eta) = 0 \quad \text{for } \eta \le 0. \tag{9.52}$$

In addition, $U_0(\eta)$ must match with the outer solution $u_0(x)$, so the matching condition is

$$\lim_{\eta\to\infty} U_0(\eta) = 1. \tag{9.53}$$

If $m_1(B) < 0$, the Fourier transform of $U_0(\eta)$ has a simple pole at the origin and therefore $\lim_{\eta\to\infty} U_0(\eta) = const.$, which satisfies the matching condition (9.53). In certain cases (see examples and exercises below) an explicit solution can be found. The uniform expansion of $n_D(x)$ now is

$$n_D(x) \sim C(\varepsilon)U_0\left(\frac{B-x}{\varepsilon}\right), \tag{9.54}$$

where $C(\varepsilon)$ is yet an unknown constant. To determine its value, we multiply (9.40) by the stationary solution $p_\varepsilon(x)$ of the forward equation (9.7) and integrate over D, to obtain

$$\begin{aligned}\int_{-\infty}^{B} p_\varepsilon(x)\,dx &= \int_{-\infty}^{B} n_D(x)p_\varepsilon(x)\,dx\\ &\quad - \int_{-\infty}^{B}\int_{-\infty}^{(B-x)/\varepsilon} n_D(x+\varepsilon\xi)w(\xi\,|\,x)\,d\xi p_\varepsilon(x)\,dx.\end{aligned} \tag{9.55}$$

Now we interchange the order of integration on the right-hand side of (9.55) and change variables, $x+\varepsilon\xi = z$, to obtain

$$\begin{aligned}\int_{-\infty}^{B} p_\varepsilon(x)\,dx &= \int_{-\infty}^{0}\int_{B}^{B-\varepsilon\xi} p_\varepsilon(z)w(\xi\,|\,z)n_D(z+\varepsilon\xi)\,dx\,d\xi\\ &\quad - \int_{-\infty}^{B} n_D(x)L_\varepsilon p_\varepsilon(x)\,dx,\end{aligned} \tag{9.56}$$

where L_ε is the forward operator defined in (9.7). Finally, using $L_\varepsilon p_\varepsilon(x) = 0$ and converting back to the original variables, (9.56) yields the identity

$$\int_{-\infty}^{B} p_\varepsilon(x)\,dx = \int_{B}^{\infty}\int_{-\infty}^{(B-x)/\varepsilon} n_D(x+\varepsilon\xi)w(\xi\,|\,x)\,d\xi\, p_\varepsilon(x)\,dx. \tag{9.57}$$

Now we replace $p_\varepsilon(x)$ in (9.57) by its WKB approximation (9.9)

$$p_\varepsilon(x) \sim \left[K_0(x) + \varepsilon K_1(x) + \cdots\right] e^{-\psi(x)/\varepsilon} \tag{9.58}$$

and $n_D(x)$ by its uniform approximation (9.54). Then, noting that the major contribution to the integral comes from the point $x = B$, where $p_\varepsilon(x)$ is maximal in $[b, \infty)$, we again introduce the stretched variable $\eta = (B - x)/\varepsilon$ and expand $p_\varepsilon(x)$ and $w(\xi \,|\, x)$ near $\eta = 0$. Finally, using the Laplace expansion of the integral on the left-hand side of (9.57) about the stable rest point at $x = \tilde{x}$ and solving for $C(\varepsilon)$, we find that

$$C(\varepsilon) \sim \sqrt{\frac{2\pi}{\psi''(\tilde{x})} \frac{K_0(\tilde{x})}{K_0(B)}} \frac{e^{\psi(B)/\varepsilon}}{\displaystyle\int_{-\infty}^{0} e^{\eta\psi'(B)} \int_{-\infty}^{\eta} w(\xi \,|\, B) U_0(\eta - \xi)\, d\xi\, d\eta}. \tag{9.59}$$

Hence the first part of (9.15) follows.

Next, we turn to the case that $m_1(B) = 0$. Again, a boundary layer analysis is required. As in the proof of Theorem 7.4.1, we introduce the scaled variable $\eta = (B - x)/\varepsilon$ and the boundary layer function $U(\eta) = u(B - \varepsilon\eta)$ into (9.51) to obtain the boundary layer equation (9.49), subject to the boundary condition (9.52) and to the matching condition (9.53). The solution, however, cannot satisfy the matching condition (9.53), because due to the assumption $m_1(B) = 0$, the Fourier transform of the leading term $U_0(\eta)$ in the regular expansion (9.50) of $U(\eta)$ has a double pole at the origin, hence $U_0(\eta)$ grows linearly for $\eta \gg 1$; that is,

$$U_0(\eta) \sim \hat{c}\eta \text{ as } \eta \to \infty. \tag{9.60}$$

Therefore an intermediate boundary layer correction is necessary to bridge between the original boundary layer $U_0(\eta)$ and the outer expansion $u_0(x) = 1$. We now introduce the new scaling and boundary layer function

$$\xi = \frac{B - x}{\sqrt{\varepsilon}}, \quad n_D(x) = C(\varepsilon)V(\xi), \tag{9.61}$$

where $C(\varepsilon) \to \infty$ as $\varepsilon \to 0$. Using (9.61) in (9.49), we obtain

$$V(\xi) = \int_{-\infty}^{\xi/\varepsilon} w(z \,|\, B - \sqrt{\varepsilon}\xi) V(\xi - \sqrt{\varepsilon} z)\, dz, \tag{9.62}$$

subject to the matching condition to the outer solution

$$\lim_{\xi \to \infty} V(\xi) = 1. \tag{9.63}$$

The second matching condition of the intermediate layer $V(\xi)$ is that for $\xi \ll 1$ it matches with $U_0(\eta)$ for $\eta \gg 1$ (see (9.60). For $\xi = O(1)$ we extend the upper limit of integration to infinity in (9.62) and expand all functions in powers of $\sqrt{\xi}$,

$$V(\xi) = V_0(\xi) + \sqrt{\varepsilon} V_1(\xi) + \cdots.$$

We obtain at the leading order

$$\frac{1}{2}m_2(B)V_0''(\xi)+m_1'(B)\xi V_0'(\xi)=0. \tag{9.64}$$

The solution of (9.64), (9.63) is given by

$$V_0(\xi)=\hat{\hat{c}}\sqrt{\frac{m_1'(B)}{\pi m_2(B)}}\int_0^{\xi}\exp\left\{-\frac{m_1'(B)u^2}{m_2(B)}\right\}du+1-\hat{\hat{c}}. \tag{9.65}$$

The approximation (9.65) is uniformly valid for $B-x\gg\varepsilon\gg 1$. To satisfy the second matching condition for $\xi\ll 1$, we note that

$$V_0(\xi)\sim 1-\hat{\hat{c}}+2\sqrt{\frac{m_1'(B)}{\pi m_2(B)}}\sqrt{\varepsilon}\eta\hat{\hat{c}}, \tag{9.66}$$

so to match (9.60) with (9.66), we have to choose

$$\hat{\hat{c}}=1,\quad \hat{c}=2\sqrt{\varepsilon}\sqrt{\frac{m_1'(B)}{\pi m_2(B)}}. \tag{9.67}$$

Thus a uniform asymptotic expansion of the solution of (9.40) is given by

$$n_D(x) \tag{9.68}$$
$$\sim C(\varepsilon)\begin{cases}2\sqrt{\dfrac{m_1'(B)}{\pi m_2(B)}}\displaystyle\int_0^{(B-x)/\sqrt{\varepsilon}}\exp\left\{-\frac{m_1'(B)u^2}{m_2(B)}\right\}du & \text{for } \dfrac{B-x}{\varepsilon}\gg 1\\ U_0\left(\dfrac{B-x}{\varepsilon}\right) & \text{for } \dfrac{B-x}{\varepsilon}=O(1),\end{cases}$$

where $U_0(\eta)$ is the solution of (9.51), (9.52). Finally, we evaluate $C(\varepsilon)$ by the procedure described above. Using the WKB expansion (9.9) and (9.68) in the identity (9.57), we obtain

$$C(\varepsilon)\sim\frac{\pi}{\varepsilon}\frac{K_0(\tilde{x})}{K_0(B)}e^{\psi(B)/\varepsilon}\sqrt{\left|\frac{m_1'(\tilde{x})m_1'(B)}{m_2(\tilde{x})m_2(B)}\right|}\frac{1}{m_2(B)}. \tag{9.69}$$

Equations (9.68) and (9.69) are (9.38) and (9.39), respectively.

If $D=(-A,B)$ is a finite interval and $\tilde{x}\in D$ is the only attractor of the dynamics (9.5) in the interval, then boundary layers have to be constructed at both ends to obtain a uniform asymptotic expansion of $n_D(x)$ in the interval. The constant $C(\varepsilon)$ is then determined by the procedure described above. If $\psi(B)<\psi(-A)$, the constant is given by (9.59) when $m_1(B)<0$, and by (9.69) when $m_1(B)=0$. If $\psi(B)>\psi(-A)$, then B has to be replaced by $-A$ in (9.59) when $m_1(-a)>0$ and in (9.69) when $m_1(-A)=0$. Also the boundary layer equations (9.51) and (9.62) have to be modified appropriately. The case $\psi(-A)=\psi(B)$ is left as an exercise. □

Remark 9.2.1. For processes x_n that must hit the boundary point $x = B$ to exit the interval $(-\infty, B)$ (i.e., processes that do not jump over the boundary), the Kramers–Moyal expansion (9.41) is valid up to the boundary and the solution $n_D(x)$ is continuous up to the boundary. Therefore the boundary condition (9.31) can be replaced by the simpler boundary condition $n_D(B) = 0$, as in the diffusion approximation.

Example 9.1 (The MFPT of a birth-death process). Consider the random walk (7.80), which jumps ε to the left and to the right at every point x with probabilities $l(x)$ and $r(x)$, respectively, where $l(x) + r(x) = 1$; that is,

$$w(\xi \,|\, x) = r(x)\delta(\xi - \varepsilon) + l(x)\delta(\xi + \varepsilon) + [1 - r(x) - l(x)]\delta(\xi). \tag{9.70}$$

The conditional moments $m_n(x)$ are given by

$$m_n(x) = r(x) + (-1)^n l(x)$$

so that

$$m_1(x) = r(x) - l(x) = -1 + 2r(x), \quad m_2(x) = r(x) + l(x) = 1.$$

This means that

$$x m_1(x) < 0;$$

that is, at each point the drift points toward the origin. The fluid approximation (9.5) indicates that the origin is an attractor for the averaged dynamics. In addition, we assume that the process is not trapped at the origin; that is, $r(0) = l(0) \neq 0$. Using (9.70) in the forward equation (9.2), we obtain the master equation

$$p_\varepsilon(x, n+1) = p_\varepsilon(x - \varepsilon, n) r(x - \varepsilon) + p_\varepsilon(x + \varepsilon, n) l(x + \varepsilon),$$

which can be written in the stationary regime as

$$\begin{aligned} L_\varepsilon p_\varepsilon(x) =& \, r(x - \varepsilon) p_\varepsilon(x - \varepsilon) + l(x + \varepsilon) p_\varepsilon(x + \varepsilon) - [r(x) + l(x)] p_\varepsilon(x) \\ =& \, 0. \end{aligned} \tag{9.71}$$

Assume that for $x > 0$, we have $r(x) < l(x)$ (or $r(x) < \frac{1}{2}$) and for $x < 0$, we have $l(x) < r(x)$ (or $r(x) > \frac{1}{2}$). The eikonal equation for the WKB solution (9.9) is

$$r(y)e^{\psi'(y)} + l(y)e^{-\psi'(y)} - [r(y) + l(y)] = 0 \tag{9.72}$$

with solution

$$\psi(y) = \int_0^y \log \frac{l(s)}{r(s)}\, ds. \tag{9.73}$$

The transport equation (9.11) becomes

$$\begin{aligned} &\left[r(y)e^{\psi'(y)} - l(y)e^{-\psi'(y)}\right] K_0'(y) \\ &+ \left\{r'(y)e^{\psi'(y)} - l(y)e^{-\psi'(y)} + \frac{\psi''(y)}{2}\left[r(y)e^{\psi'(y)} - l(y)e^{-\psi'(y)}\right]\right\} K_0(y) = 0, \end{aligned} \tag{9.74}$$

and the solution is given by

$$K_0(y) = \frac{C}{\sqrt{r(y)l(y)}}, \tag{9.75}$$

with C a normalization constant. The WKB solution,

$$p_\varepsilon(y) \sim \frac{C}{\sqrt{r(y)l(y)}} \exp\left\{-\frac{1}{\varepsilon}\int_0^y \log\frac{l(s)}{r(s)}\,ds\right\}, \tag{9.76}$$

is the leading term in the asymptotic expansion of the exact solution as $\varepsilon \to 0$, given in Exercise 7.11.

To calculate the MFPT of x_n from an interval $D = (-A, B)$, where $A, B > 0$, whose boundary consists of lattice points, we use (9.70) to write (9.30) as

$$\begin{aligned} L_\varepsilon^* n_D(x) &= r(x)n_D(x+\varepsilon) + lo(x)n_D(x-\varepsilon) - [l(x)+r(x)]n_D(x) \\ &= -1. \end{aligned} \tag{9.77}$$

Because x_n can jump one lattice point to the left or to the right, it has to hit the boundary to exit. It follows from Remark 9.31 that the boundary condition (9.31) can be replaced by the simpler condition

$$n_D(-A) = n_D(B) = 0. \tag{9.78}$$

In this case, the Kramers–Moyal expansion is valid up to the boundary and we write (9.77) as

$$\sum_{k=1}^{\infty} \frac{\varepsilon^k[r(x)+(-1)^k l(x)]}{k!} n_D^k(x) = -1. \tag{9.79}$$

Equation (9.46) is

$$n_D(x) \sim C(\varepsilon), \tag{9.80}$$

where $C(\varepsilon) \to \infty$ as $\varepsilon \to 0$. We construct the boundary layers at $-A$ and B for the two cases mentioned in Theorem 9.2.1, $l(B) > r(B)$, $l(-A) < r(-A)$ and $l(B) = r(B)$, $l(-A) = r(-A)$. In the former case, we seek a local solution

$$N(x) \sim C(\varepsilon)U(\eta), \tag{9.81}$$

where $\eta = (B-x)/\varepsilon$. Substituting (9.81) into the Kramers–Moyal equation (9.79), we obtain that the leading order boundary layer equation is

$$\sum_{k=1}^{\infty} \frac{[r(B)+(-1)^k l(B)]}{k!} U_0^{(k)}(\eta) = 0 \tag{9.82}$$

with the boundary condition (cf. (9.78))

$$U_0(0) = 0 \tag{9.83}$$

and the matching condition

$$\lim_{\eta\to\infty} U_0(\eta) = 1. \tag{9.84}$$

The solution of (9.82)–(9.84) is

$$U_0(\eta) = 1 - e^{-\beta\eta}, \tag{9.85}$$

where $\beta = \psi'(B) = \log[l(B)/r(B)]$. A similar expansion holds locally near $-A$. Thus the uniform expansion of the MFPT in D is

$$n_D(x) \sim C(\varepsilon)\left[1 - e^{-\alpha(x+A)/\varepsilon} - e^{-\beta(B-x)/\varepsilon}\right], \tag{9.86}$$

where $\alpha = \psi'(-A)$. The undetermined constant $C(\varepsilon)$ is determined by a procedure analogous to that used in the proof of Theorem 9.2.1, with L_ε and L_ε^* given in (9.71) and (9.77), respectively, and integration is replaced by summation over the lattice points. We then obtain

$$C(\varepsilon) \sim \frac{\sqrt{\dfrac{2\pi}{\varepsilon r(0)[l'(0) - r'(0)]}}}{\dfrac{|l(-A) - r(-A)|}{\sqrt{r(-A)l(-A)}} e^{-\psi(-A)/\varepsilon} + \dfrac{|l(B) - r(B)|}{\sqrt{r(B)l(B)}} e^{-\psi(B)/\varepsilon}}. \tag{9.87}$$

The MFPT is given by (9.86) with (9.87).

If $l(B) = r(B),\ l(-A) = r(-A)$, then the analysis is modified, as in the proof of Theorem 9.2.1, to obtain the uniform expansion

$$\begin{aligned} &n_D(x)\\ &\sim C(\varepsilon)\left[\operatorname{erf}\left(\frac{B-x}{\sqrt{\varepsilon}}\sqrt{\frac{r'(B) - l'(B)}{2r(B)}}\right) + \operatorname{erf}\left(\frac{A+x}{\sqrt{\varepsilon}}\sqrt{\frac{l'(-A) - r'(-A)}{2r(A+x)}}\right) - 1\right] \end{aligned}$$

and $C(\varepsilon)$ is found from (9.39) as

$$C(\varepsilon) \sim \frac{\dfrac{\pi}{\varepsilon\sqrt{r(0)[l'(0) - r'(0)]}}}{\sqrt{\dfrac{|l'(-A) - r'(-A)|}{r(-A)}} e^{-\psi(-A)/\varepsilon} + \sqrt{\dfrac{|l'(B) - r'(B)|}{r(B)}} e^{-\psi(B)/\varepsilon}}.$$

□

Exercise 9.4 (Calculation of the MFPT). Carry out the missing calculations in Example 9.1. □

Example 9.2 (A state-dependent M/M/1 queue). The unfinished work $U(t)$ in a queueing system was defined in Example 7.7 as the time it takes the queue to empty, if the arrival of new customers is stopped at time t. In an M/M/1 queue with *discouraged arrivals* customers arrive at exponential waiting times with rate $\lambda = \lambda(U(t))$, which decreases as the unfinished work in the queue increases, and demand work ξ, whose pdf is exponential with constant service rate μ. We assume that the queue is observed over a long time period of order $O\left(\varepsilon^{-1}\right)$, where ε is a small parameter. Thus, we scale time so that the arrival and the service rates appear to be fast and are given by $\lambda(U(t))/\varepsilon$ and μ/ε, respectively. The density of the interarrival times is exponential,

$$\Pr\{\text{arrival in } (t, t+\Delta t),\, U(t) = x\} = \frac{\lambda(x)}{\varepsilon}\Delta t + o(\Delta t) \text{ as } \Delta t \to 0, \tag{9.88}$$

and the density of the demanded service time s_n of the nth customer is given by

$$\frac{d}{ds}\Pr\{s_n \leq s\} = \frac{\mu}{\varepsilon}e^{-\mu s/\varepsilon}. \tag{9.89}$$

Because the work is measured in units of time, the server works at fixed rate 1. The dynamics (9.23) of the unfinished work is given by

$$U(t+\Delta t) = \begin{cases} U(t) - \Delta t & \text{w.p. } 1 - \dfrac{\lambda(U)}{\varepsilon}\Delta t + o(\Delta t) \\ U(t) + \varepsilon\xi & \text{w.p. } \dfrac{\lambda(U)}{\varepsilon}\Delta t + o(\Delta t), \end{cases}$$

where

$$\Pr\{\xi = y \,|\, U(t) = x,\, U(t_1) = x_1,\, \ldots,\, U(t_n) = x_n\} = \mu e^{-\mu y}H(y) = \tilde{w}(y \,|\, x)$$

and $H(y)$ is the Heaviside step function. Thus the drift is

$$A = \begin{cases} -1 & \text{if } x > 0 \\ 0 & \text{if } x = 0, \end{cases}$$

so that for $x > 0$ the drift equation (9.26) is

$$\dot{x} = -1 + \frac{\lambda(x)}{\mu}, \tag{9.90}$$

which we assume as a unique stable equilibrium point at $x = \tilde{x} > 0$. This condition also implies that the queue is stable in the sense that that there exists a stationary probability density for $U(t)$ (i.e., the traffic intensity, $r(x) = \lambda(x)/\mu < 1$ for $x > \tilde{x}$).

To calculate the stationary density of $U(t)$, we note that there is a positive probability that the queue is empty; that is, the density has a probability mass, A, at $y = 0$ so that

$$p_\varepsilon(y) = p(y) + A\delta(y). \tag{9.91}$$

The regular part of the pdf, $p(y)$, satisfies the stationary forward equation (9.24) for $y > 0$, which is the stationary Takács equation in Example 7.7

$$\begin{aligned} L_\varepsilon p(y) &= \frac{dp(y)}{dy} - \frac{\lambda(y)}{\varepsilon} p(y) + \frac{1}{\varepsilon} \int_0^{y/\varepsilon} p(y - \varepsilon\xi)\lambda(y - \varepsilon\xi)\mu e^{-\mu\xi}\, d\xi + \frac{Ap(0)}{\varepsilon^2}\mu e^{-\mu y/\varepsilon} \\ &= 0. \end{aligned} \tag{9.92}$$

The normalization of $p_\varepsilon(y)$ and boundary condition imply that

$$\int_0^\infty p(y)\, dy + A = 1 \tag{9.93}$$

and

$$\frac{\lambda(0)}{\varepsilon} A = p(0). \tag{9.94}$$

The nonhomogeneous term in the Takács equation (9.92) is due to customers arriving at an empty queue with y units of unfinished work. The boundary condition (9.94) represents the steady-state balance between the probability flux into the boundary (emptying the queue) and the probability flux of customers arriving at an empty queue.

The eikonal equation (9.27) for the stationary Takács equation (9.92) is

$$\psi'(y) + \lambda(y) = \frac{\lambda(y)\mu}{\mu - \psi'(y)}, \tag{9.95}$$

so that choosing $\psi(\tilde{x}) = 0$, we obtain the solution

$$\psi(y) = \int_{\tilde{x}}^{y} [\mu - \lambda(\xi)]\, d\xi. \tag{9.96}$$

The transport equation (9.28) for (9.92) gives

$$K_0(y) = const. = K_0, \tag{9.97}$$

which is determined by normalization. Evaluating the integral in the normalization condition (9.93) by the Laplace method and taking into consideration the fact that $A = o(1)$ as $\varepsilon \to 0$, we find that

$$p(y) \sim \sqrt{\frac{-\lambda'(\tilde{x})}{2\pi\varepsilon}} \exp\left\{ -\frac{1}{\varepsilon} \int_{\tilde{x}}^{y} [\mu - \lambda(\xi)]\, d\xi \right\}. \tag{9.98}$$

The boundary layer analysis of $p_\varepsilon(y)$, as given in the proof of Theorem 9.2.1 has to be modified, because (9.98) does not describe the behavior or $p_\varepsilon(y)$ near the boundary $y=0$. We introduce the local variable $\eta=y/\varepsilon$ into the stationary Takács equation (9.92) and define $P(\eta)=p\,(y/\varepsilon)$ to obtain the boundary layer equation

$$P(\eta)-\lambda(0)P(\eta)+\lambda(0)\int\limits_0^\eta P(\eta-\xi)\mu e^{-\mu\xi}\,d\xi+p\,(0)\mu e^{-\mu\eta}=0. \tag{9.99}$$

The boundary layer equation (9.99) is solved by applying the Laplace transform and then matching for large η with the outer solution expanded near $y=0$. The unknown constant $p\,(0)$ in the Laplace transform of $P(\eta)$ is determined by the matching procedure as

$$p\,(0)\sim\sqrt{\frac{-\lambda'(\tilde{x})}{2\pi\varepsilon}}\exp\left\{\frac{1}{\varepsilon}\int\limits_0^{\tilde{x}}[\mu-\lambda(\xi)]\,d\xi\right\}.$$

Using the boundary condition (9.94) we find that

$$A=\Pr\{U(t)=0\}=\sqrt{\frac{-\lambda'(\tilde{x})\varepsilon}{2\pi\lambda^2(0)}}\exp\left\{\frac{1}{\varepsilon}\int\limits_0^{\tilde{x}}[\mu-\lambda(\xi)]\,d\xi\right\}. \tag{9.100}$$

Thus the uniform expansion of the stationary pdf $p_\varepsilon(y)$ of $U(t)$ is

$$p_\varepsilon(y)\sim\sqrt{\frac{-\lambda'(\tilde{x})\varepsilon}{2\pi\lambda^2(0)}}\exp\left\{\frac{1}{\varepsilon}\int\limits_0^{\tilde{x}}[\mu-\lambda(\xi)]\,d\xi\right\}\delta(y)$$
$$+\sqrt{\frac{-\lambda'(\tilde{x})}{2\pi\varepsilon}}\exp\left\{-\frac{1}{\varepsilon}\int\limits_{\tilde{x}}^{y}[\mu-\lambda(\xi)]\,d\xi\right\}.$$

Note that the probability of finding an empty queue decays exponentially fast as $\varepsilon\to 0$, which means that the queue is "busy".

Next, we compute the mean time $\bar{\tau}_K(x)$ for $U(t)$ to reach a specified capacity $K>\tilde{x}$, given x units of unfinished work remaining in the queue. This can be interpreted as the mean time to lose a customer, if no more customers can be accepted into the queue once the workload exceeds the capacity K. The MFPT $\bar{\tau}_K(x)$ satisfies the backward master equation

$$L_\varepsilon^*\bar{\tau}_K(x)=-\bar{\tau}_K'(x)-\frac{\lambda(x)}{\varepsilon}\bar{\tau}_K(x)\int\limits_0^{(K-x)/\varepsilon}\bar{\tau}_K(x+\varepsilon\xi)\mu e^{-\mu\xi}\,d\xi=-1 \tag{9.101}$$

with the boundary condition at $x=0$

$$-\frac{\lambda(0)\bar{\tau}_K(0)}{\varepsilon}+\frac{\lambda(0)}{\varepsilon}\int\limits_0^{K/\varepsilon}\bar{\tau}_K(\varepsilon\xi)\mu e^{-\mu\xi}\,d\xi=-1 \tag{9.102}$$

and the condition

$$\bar{\tau}_K(x) = 0 \text{ for } x > K. \tag{9.103}$$

The latter was used to cut off the upper limit of integration in (9.101) and (9.102). As in the proof of Theorem 9.2.1, we find that away from the boundary $x = K$

$$\bar{\tau}_K(x) \sim C(\varepsilon) \to \infty \text{ as } \varepsilon \to 0. \tag{9.104}$$

We introduce into the backward master equation (9.101) the boundary layer variable $\eta = (K - x)/\varepsilon$ and the boundary layer function $T(\eta) = \bar{\tau}_K(K - \varepsilon\eta)$ and expand

$$T(\eta) \sim T_0(\eta) + \varepsilon T_1(\eta) + \cdots \tag{9.105}$$

to obtain the boundary layer equation

$$T_0'(\eta) - \lambda(K)T_0(\eta) \int_0^\eta T_0(\eta - \xi)\mu e^{-\mu\xi}\, d\xi = 0 \tag{9.106}$$

and the matching condition

$$\lim_{\eta\to\infty} T_0(\eta) = 1. \tag{9.107}$$

The solution of (9.106), (9.107) is

$$T_0(\eta) = 1 - \frac{\lambda(K)}{\mu} e^{[\lambda(K)-\mu]\eta}. \tag{9.108}$$

Note that $T_0(0) = 1 - \lambda(K)/\mu = 1 - r(K) \neq 0$, and (9.103) gives $\bar{\tau}_K(K+) = 0$. The procedure for determining the constant $C(\varepsilon)$ in the proof of Theorem 9.2.1 also has to be modified. We multiply the backward master equation (9.101) by the continuous part $p(y)$ of $p_\varepsilon(y)$, given in (9.98), and integrate over the interval $(0, K)$ to obtain

$$\int_0^K p(x)\, dx = \bar{\tau}_K(K)p(K) - \frac{\varepsilon}{\lambda(0)} p(0). \tag{9.109}$$

Because $\int_0^K p(x)\, dx \sim 1 - A$ and the boundary condition (9.94) gives $A = \varepsilon p(0)/\lambda(0)$, the identity (9.109) becomes

$$\bar{\tau}_K(K) \sim \frac{1}{p(K)}. \tag{9.110}$$

Locally, we have $\bar{\tau}_K(K) \sim C(\varepsilon)[1 - t(K)]$, so that

$$C(\varepsilon) \sim \frac{1}{p(K)} \frac{1}{1 - r(K)}. \tag{9.111}$$

Finally, the value of $p(K)$ is found from (9.98) and the MFPT is given by

$$\bar{\tau}_K(x) \sim \frac{\sqrt{-2\pi\varepsilon/\lambda'(\tilde{x})}}{1-r(K)} \exp\left\{\frac{1}{\varepsilon}\int_{\tilde{x}}^{K}[\mu-\lambda(\xi)]\,d\xi\right\}. \tag{9.112}$$

For a generalization to state-dependent M/G/1 queues see [134]. □

Exercise 9.5 (Exit from a finite interval). (i) Fill in the details of the proof for the case that $D = (-A, B)$ is a finite interval.

(ii) Find the asymptotic expansion of $n_D(x)$ if there is more than one attractor in D.
□

Exercise 9.6 (The mean exit time of a jump-drift process). Prove a theorem analogous to Theorem 9.2.1 for the jump-drift process (9.23). □

Exercise 9.7 (Unimolecular dissociation). Consider the following Markovian model of thermal excitation of large molecules, where the quantity of interest is the dissociation rate, which is the reciprocal of the mean number of collisions until dissociation [240], [35], [134]. In this model dissociation occurs when the molecule's energy E exceeds a certain threshold energy E_0. The molecule undergoes random collisions, which cause an increase or decrease in its energy. The Markov process is defined in terms of dimensionless energy $x = E/E_0$, whose jumps are scaled to $\xi = (E_1 - E_2)/kT$, where k is Boltzmann's constant and T is absolute temperature. The conditional jump density function of the jump process ξ_n is modeled by

$$w(\xi\,|\,x) = K\begin{cases} e^{a\xi} & \text{for } -\dfrac{x}{\varepsilon} < \xi < 0 \\ \left[\dfrac{1}{K} - a - b + ae^{-x/a\varepsilon}\right]\delta(\xi) & \text{for } \xi = 0 \\ e^{-\xi/b} & \text{for } \xi > 0. \end{cases}$$

The assumption of high activation energy E_0 is expressed by the assumption that the dimensionless parameter $\varepsilon = kT/E_0$ is small (relative to 1). The parameter K is the collision frequency and the parameters a and b are defined by $a = \alpha/kT$ and β/kT, where α and β are the average energy loss and gain per collision, respectively. Assume that the stationary distribution is Boltzmannian, $p_\varepsilon(x) \propto e^{-x/\varepsilon}$, which requires that $1/b - 1/a = 1$.

(i) Show that $m_1(x) \approx b - a$ for $x/\varepsilon \gg 1$.

(ii) Show that the mean number of collisions to dissociation, $n(x)$, satisfies the condition $n(x) = 0$ for $x \notin (0,1)$.

(iii) Derive the backward master equation

$$\int_{-x/\varepsilon}^{0} n(x+\varepsilon\xi)e^{\xi/a}\,d\xi + [-a-b+ae^{-x/a\varepsilon}]n(x) + \int_{0}^{(1-x)/\varepsilon} n(x+\varepsilon\xi)e^{-\xi/b}\,d\xi$$
$$= -\frac{1}{K}.$$

(iv) Solve the backward master equation exactly by converting it to the ordinary linear differential equation

$$[a+b-ae^{-x/a\varepsilon}]n''(x)+\frac{1}{\varepsilon}[-a-b+(a+2)e^{-x/a\varepsilon}]n'(x)=-\frac{1}{\varepsilon^2 abK}$$

with the boundary conditions

$$n'(0)=-\frac{1}{\varepsilon ab^2K},\quad [a+b-ae^{-1/a\varepsilon}]n'(1)+\frac{a+b}{\varepsilon a}n(1)=-\frac{1}{\varepsilon abK}.$$

(v) Prove that when $(1-x)/\varepsilon \gg 1$, the asymptotic expansion

$$n(x)\sim\frac{e^{1/\varepsilon}}{Kb^2(a+b)}$$

is a valid approximation to $n(x)$.

(vi) Use the proof of Theorem 9.2.1 to construct a uniform asymptotic expansion of $n(x)$ [137]. □

9.3 Asymptotic theory of large deviations

The estimation of the probability of rare events, what amounts to an asymptotic expansion of the tails of the pdf of a given random variable or stochastic process, is the subject of large deviation theory. For examples, the expansion of the pdf of a diffusion process with small noise and the exit problem are all problems of large deviations. Another class of large deviations is that of an ergodic[1] stationary Markov chain, $\{X_n\}$, with zero mean, and its sample averages

$$Y_n\equiv\frac{1}{n}\sum_{j=1}^{n}X_j\,. \tag{9.113}$$

The problem of large deviations here is to determine the uniform asymptotic behavior of the PDF of Y_n and the joint PDF of (X_n,Y_n) for large n. The case of independent i.i.d. $\{X_n\}$ falls into this category as a particular case. The central limit theorem, which applies in this case, gives a correct estimate of the tail probability of Y_n only for small deviations and misses tail events by orders of magnitude (see discussion in Sections 9.3.2 and 9.4). The purpose of this section is to develop a straightforward method for the explicit calculation of the full formal asymptotic expansion of the pdfs $p_n(y)$ and $p_n(x,y)$ for a discrete or continuous state stationary Markov chain.

The analysis of this problem is based on the observation that the pair $\{X_n,Y_n\}$ is a Markov chain so that its joint pdf (whenever it exists) satisfies the master equation (forward Kolmogorov) equation. The construction of an asymptotic solution to this

[1] That is, all its multidimensional densities can be determined from a single sample path of the process.

equation is based on the WKB method that was employed for the analysis of the Fokker–Planck equation in the case of diffusion with small noise.

We assume that $\{X_n\}$ is an ergodic stationary Markov process with zero mean, stationary pdf $\phi_0(x)$, and a finite moment generating function. The master equation for the transition probability density function of X_n is

$$p_{n+1}(y\,|\,x) = \int \rho(y,t)p_n(t\,|\,x)\,dt \equiv \boldsymbol{L}p_n(y\,|\,x), \tag{9.114}$$

where

$$\rho(y,t) \equiv \frac{\partial}{\partial y}\Pr\{X_{n+1} \le y\,|\,X_n = t\}. \tag{9.115}$$

We use the term pdf in the sense of distributions and the partial derivative in eqs. (9.178) and (9.115) is understood in this sense as well. Thus the pdf of a discrete variable is a sum of delta functions. The initial condition for (9.114) is

$$p_0(y\,|\,x) = \delta(x-y). \tag{9.116}$$

The first two stationary moments are denoted

$$m_1 \equiv \int t\phi_0(t)\,dt = 0, \quad m_2 \equiv \int t^2\phi_0(t)\,dt. \tag{9.117}$$

The operator $\boldsymbol{L}$ is bounded in $L^1(\mathbb{R})$ with $||\boldsymbol{L}||_1 \le 1$ and the stationary pdf, $\phi_0(x)$, is an eigenfunction of $\boldsymbol{L}$ corresponding to the eigenvalue $\mu_0 = 1$.

The transition density $\rho(x,t)$ defines a one parameter family of operators, $\boldsymbol{M}(\theta)$, acting on functions of x in $L^1(\mathbb{R})$, is defined for any fixed θ by

$$\boldsymbol{M}(\theta)f(x) \equiv e^{x\theta}\int \rho(x,t)f(t)\,dt \equiv e^{x\theta}\boldsymbol{L}f(x).$$

We assume that the integral operator has the greatest positive eigenvalue $\mu(\theta)$ with geometric multiplicity one, because the operator $\boldsymbol{M}(\theta)$ has a nonnegative kernel, and we also assume that the corresponding eigenfunction $q^0(x,\theta)$ is the only positive eigenfunction up to normalization ([40, p. 287], [72, vol. 2, p. 271], [105]),

$$q^0(x,\theta)\,\mu(\theta) = \int \rho(x,t)q^0(t,\theta)\,e^{x\theta}\,dt. \tag{9.118}$$

Setting in eq. (9.118) $\theta = \psi'(y)$, we can define a function $\psi(y)$ as the solution of the first-order differential equation

$$\mu(\psi'(y)) = e^{-\psi(y)+y\psi'(y)}, \tag{9.119}$$

or equivalently,

$$-\psi(y) + y\psi'(y) = \log\mu(\psi'(y)). \tag{9.120}$$

Equation (9.120) can be reduced to an implicit equation for $\psi'(y)$ by differentiating (9.120) with respect to y,

$$y = \frac{\mu'(\psi'(y))}{\mu(\psi'(y))}\,. \tag{9.121}$$

The case of independent random variables can be recovered from (9.121) as follows. If X_n are i.i.d. random variables, then $\rho(x,t)$ is independent of t. Therefore, integrating (9.128) with respect to x, it is found that $\mu(\psi'(y))$ is the moment generating function of $\rho(x)$ (which is the pdf of X_n); that is,

$$\mu(\psi'(y)) = \int e^{\psi'(y)x}\rho(x)\,dx\,.$$

Thus (9.120) reduces to the well-known result of large deviations theory [47].

Exercise 9.8 (Convexity of the $\psi(y)$). Prove that $\psi(y)$ is a convex function of y. □

Due to convexity, $\psi'(y)$, wherever it exists, is an increasing function and thus has an inverse in its range. It follows that $\mu(y)$ can be considered a function of $\psi'(y)$. We use interchangeably, with some abuse of notation, both $\mu(\psi'(y))$ and $\mu(y)$.

Theorem 9.3.1 (Asymptotic expansion of the joint and marginal pdf). *The joint pdf of $\{X_n, Y_n\}$ has the asymptotic representation*

$$p_n(x,y) \sim \left\{\sqrt{\frac{n\psi''(y)}{2\pi}}\tilde{q}(x,y)e^{-F(y)}\right\} e^{-n\psi(y)} \quad \text{for } n \gg 1 \tag{9.122}$$

and the marginal density of Y_n has the asymptotic representation

$$p_n(y) \sim \left\{\sqrt{\frac{n\psi''(y)}{2\pi}}e^{-F(y)}\right\} e^{-n\psi(y)} \quad \text{for } n \gg 1. \tag{9.123}$$

The functions $\tilde{q}(x,y)$ and $F(y)$ are defined in equations (9.134), (9.135), (9.144), *and* (9.145) *below.*

Proof. First, we note that Y_n satisfies the stochastic equation

$$Y_{n+1} = \frac{n}{n+1}Y_n + \frac{1}{n+1}X_{n+1}, \tag{9.124}$$

which implies that the pair (X_n, Y_n) is Markovian. It follows from (9.114) and (9.124) that the joint pdf of $\{X_n, Y_n\}$ satisfies the master equation (forward Kolmogorov equation)

$$p_{n+1}(x,y) = \frac{n+1}{n}\int \rho(x,\xi)p_n\left(\xi, y + \frac{y-x}{n}\right)\,d\xi. \tag{9.125}$$

An asymptotic solution of (9.125) is constructed in the WKB form

$$p_n(x,y) = K_n(x,y)e^{-n\psi(y)}, \tag{9.126}$$

where

$$K_n(x,y) = \sqrt{n}\left[q^0(x,y) + \frac{1}{n}q^1(x,y) + \frac{1}{n^2}q^2(x,y) + \cdots\right]. \tag{9.127}$$

Substituting (9.126) and (9.127) in (9.125) and expanding in negative powers of n, it is found at the leading order that

$$q^0(x,y)\,e^{-\psi(y)+y\psi'(y)} = \int \rho(x,t)q^0(t,y)\,e^{x\psi'(y)}\,dt, \tag{9.128}$$

which is (9.118). The eigenfunction $q^0(x,y)$ is determined up to a normalization factor that is a function of y. This function is usually found from the next order equation, obtained by comparing the coefficients of $n^{-1/2}$ in the expansion (9.126), (9.127) on both sides of the master equation (9.125). The resulting equation degenerates to $0 = 0$, as can be seen from eq. (9.141) below. To remove the degeneration, we define an approximate sample average

$$Y_{n+1}^{\varepsilon} = \frac{n+\varepsilon}{n+1}Y_n^{\varepsilon} + \frac{1}{n+1}X_{n+1}, \tag{9.129}$$

where ε is a small parameter, ultimately to be set equal to zero. The master equation (9.125) now takes the form

$$p_{n+1}^{\varepsilon}(x,y) = \frac{n+1}{n+\varepsilon}\int \rho(x,\xi)\,p_n^{\varepsilon}\left(\xi, y + \frac{(1-\varepsilon)y - x}{n+\varepsilon}\right)\,d\xi\,.$$

Consequently, the eikonal equation (9.128) takes the form

$$\{\mu(y) - \boldsymbol{M}(y)\}\,q_{\varepsilon}^0(x,y) = 0, \tag{9.130}$$

where

$$\mu(y) = e^{-\psi_\varepsilon(y)+(1-\varepsilon)y\psi_\varepsilon'(y)},$$

and eq. (9.121) becomes

$$(1-\varepsilon)y - \frac{\mu'(\psi_\varepsilon'(y))}{\mu(\psi_\varepsilon'(y))} = \frac{\varepsilon\,\psi_\varepsilon'(y)}{\psi_\varepsilon''(y)}.$$

To simplify notation, we drop the ε in the notation for functions of the perturbed process (9.129).

The next order equation can be written as

$$\begin{aligned}&\left\{\left(\frac{1}{2}-\varepsilon\right) + \varepsilon[(1-\varepsilon)y - x]\psi'(y) - \frac{1}{2}\psi''(y)[(1-\varepsilon)y - x]^2\right\}\mu(y)q^0(x,y)\\ &+ [(1-\varepsilon)y - x]\boldsymbol{M}(y)q_y^0(x,y) = [\mu(y) - \boldsymbol{M}(y)]q^1(x,y)\,.\end{aligned} \tag{9.131}$$

Equation (9.131) is handled by considering the adjoint operator to $\boldsymbol{M}(y)$, denoted $\boldsymbol{M}^*(y)$. The operator $\boldsymbol{M}^*(y)$ acts on functions in $L^\infty(\mathbb{R})$ and is defined by the pairing

$$\langle f, g\rangle = \int f(x)g(x)\,dx, \quad f \in L^\infty(\mathbb{R}),\ g \in L^1(\mathbb{R})$$

as

$$\boldsymbol{M}^*(y)f(t) \equiv \int \rho(x,t)\, e^{x\psi'(y)} f(x)\,dx.$$

The eigenvalue $\mu(y)$ is also the greatest positive eigenvalue of $\boldsymbol{M}^*(y)$ with a corresponding positive eigenfunction of multiplicity one, denoted $\tilde{p}(x,y)$. We conclude from eq. (9.128) that

$$\boldsymbol{M}^*(y)\tilde{p}(t,y) \equiv \int \rho(x,t)\, e^{x\psi'(y)}\tilde{p}(x,y)\,dx = e^{-\psi(y)+y\psi'(y)}\,\tilde{p}(t,y). \tag{9.132}$$

By orthogonality considerations

$$\int \tilde{p}(x,y)\,[\boldsymbol{M}(y) - \mu(y)]\,f(x)\,dx = 0 \tag{9.133}$$

for any function $f(x)$ in the domain of $\boldsymbol{M}(y)$. We write $q^0(x,y)$ as

$$q^0(x,y) = k_0(y)\tilde{q}(x,y), \tag{9.134}$$

where $\tilde{q}(x,y)$ is normalized by

$$\int \tilde{q}(x,y)\,dx = 1, \tag{9.135}$$

and $k_0(y)$ is determined from (9.131). We normalize $\tilde{p}(x,y)$ by

$$\int \tilde{q}(x,y)\tilde{p}(x,y)\,dx = 1. \tag{9.136}$$

Multiplying eq. (9.131) by $\tilde{p}(x,y)$ and integrating with respect to x, we obtain the differential equation

$$\begin{aligned}
&\left(\frac{1}{2} - \varepsilon\right)\mu(y)k_0(y) + \varepsilon\psi'(y)\int [(1-\varepsilon)y - x]\mu(y)q^0(x,y)\tilde{p}(x,y)\,dx \\
&- \frac{1}{2}\psi''(y)\int [(1-\varepsilon)y - x]^2\mu(y)q^0(x,y)\tilde{p}(x,y)\,dx \\
&+ \int [(1-\varepsilon)y - x]\boldsymbol{M}(y)q^0_y(x,y)\tilde{p}(x,y)dx = 0.
\end{aligned} \tag{9.137}$$

Equation (9.137) is used to determine the factor $k_0(y)$.

We need the following two identities. By differentiating eq. (9.130) with respect to y, multiplying the result by $\tilde{p}(x,y)$, and integrating with respect to x, we get

$$\int [(1-\varepsilon)y - x]\mu(y)q^0(x,y)\tilde{p}(x,y)dx = \varepsilon\frac{\psi'(y)}{\psi''(y)}\mu(y)k_0(y). \tag{9.138}$$

For $\varepsilon = 0$, this identity reduces to

$$\int (y-x)\mu(y)q^0(x,y)\tilde{p}(x,y)\,dx = 0\,. \tag{9.139}$$

Next, differentiating eq. (9.130) with respect to y twice, multiplying the result by $\tilde{p}(x,y)$, integrating with respect to x, and applying eq. (9.138), we get the identity

$$\begin{aligned}
&\int [(1-\varepsilon)y - x]\tilde{p}(x,y)\boldsymbol{M}(y)q_y^0(x,y)\,dx \\
&-\frac{1}{2}\psi''(y)\int [(1-\varepsilon)y-x]^2\mu(y)q^0(x,y)\tilde{p}(x,y)\,dx \\
&=\varepsilon\frac{\psi'(y)}{\psi''(y)}\int \tilde{p}(x,y)\boldsymbol{M}(y)q_y^0(x,y)dx + \frac{\varepsilon}{2}\int x\psi'(y)\mu(y)q^0(x,y)\tilde{p}(x,y)\,dx \\
&+\left\{\frac{\varepsilon(\varepsilon-1)}{2}y\psi'(y) - \frac{\varepsilon\psi'''(y)\psi'(y)}{2(\psi'')^2} - \frac{(1-2\varepsilon)}{2}\right\}\mu(y)k_0(y)\,.
\end{aligned} \tag{9.140}$$

Now, substituting eqs. (9.138) and (9.140) in eq. (9.137), we get

$$\begin{aligned}
&\int \frac{\varepsilon\psi'(y)}{\psi''(y)}\tilde{p}(x,y)\boldsymbol{M}(y)q_y^0(x,y)dx + \int \frac{\varepsilon}{2}\psi'(y)\,x\mu(y)q_y^0(x,y)\tilde{p}(x,y)\,dx \\
&=\left\{\frac{\varepsilon}{2}\,y\psi'(y) + \frac{\varepsilon}{2}\frac{\psi'''(y)\psi'(y)}{[\psi''(y)]^2} - \frac{\varepsilon^2}{2}\psi'(y)y - \frac{\varepsilon^2\psi'^2(y)}{\psi''(y)}\right\}\mu(y)k_0(y).
\end{aligned} \tag{9.141}$$

Dividing eq. (9.141) by ε, taking the limit $\varepsilon \to 0$, and applying (9.139), we obtain

$$\frac{1}{2}\frac{\psi'''(y)}{\psi''(y)}\mu(y)k_0(y) - \int \tilde{p}(x,y)\boldsymbol{M}(y)q_y^0(x,y)\,dx = 0\,, \tag{9.142}$$

for which the notation without ε is correct. We observe that by eq. (9.133)

$$\int \tilde{p}(x,y)\boldsymbol{M}(y)q_y^0(x,y)\,dx = \int \mu(y)q_y^0(x,y)\tilde{p}(x,y)\,dx\,. \tag{9.143}$$

Finally, using eq. (9.134) and eq. (9.143) in eq. (9.142), we obtain a simplified equation for $k_0(y)$,

$$\frac{k_0'(y)}{k_0(y)} = \frac{1}{2}\frac{\psi'''(y)}{\psi''(y)} - \int \tilde{q}_y(x,y)\tilde{p}(x,y)\,dx. \tag{9.144}$$

We denote

$$f(y) \equiv \int \tilde{q}_y(x,y)\tilde{p}(x,y)\,dx, \quad F(y) \equiv \int_0^y f(z)\,dz. \tag{9.145}$$

Then the solution of eq. (9.144) is

$$k_0(y) = C\sqrt{\psi''(y)}e^{-F(y)}\,,$$

where C is a constant. By the normalization requirement of $\sqrt{n}k_0(y)e^{-n\psi(y)}$ in the limit of large n, using Laplace's method for the asymptotic evaluation of integrals, one obtains that

$$k_0(y) = \sqrt{\frac{\psi''(y)}{2\pi}}e^{-F(y)}. \tag{9.146}$$

In the special case where X_n are i.i.d. random variables the expression (9.146) is simplified. By eq. (9.132) $\tilde{p}(x,y) = 1$, resulting in

$$f(y) = \int \tilde{q_y}(x,y)\tilde{p}(x,y)dx = 0,$$

in agreement with [133]. It is also found by (9.128) that for the i.i.d. case

$$\tilde{q}(x,y) = \frac{\rho(x)e^{\psi'(y)x}}{\mu(\psi'(y))}.$$

□

9.3.1 More general sums

Next, we consider the more general case of the sum

$$Y_n = \frac{1}{n}\sum_{i=1}^{n} f(X_i),$$

where $f(x)$ is a measurable bounded function. The schemes (9.124) and

$$Y_{n+1} = \frac{n}{n+1}Y_n + \frac{1}{n+1}f(X_{n+1}) \tag{9.147}$$

are particular cases of stochastic approximation [184], [144]. The master equation is given by

$$p_{n+1}(x,y) = \frac{n+1}{n}\int \rho(x,\xi)p_n\left(\xi, y + \frac{y - f(x)}{n}\right)d\xi. \tag{9.148}$$

We assume that $\{f(X_n)\}$ is an ergodic stationary process for which the first two moments exist. The asymptotic expansion for eq. (9.148) is constructed as above. The eikonal equation is

$$q^0(x,y)\,e^{-\psi(y)+y\psi'(y)} = \int \rho(x,t)q^0(t,y)\,e^{f(x)\psi'(y)}\,dt. \tag{9.149}$$

We find that $\boldsymbol{M}(y) \equiv e^{f(x)\psi'(y)}\boldsymbol{L}(y)$, its greatest eigenvalue $\mu(y)$, the normalized eigenfunctions $\tilde{q}(x,y)$ and $\tilde{p}(x,y)$, and the rate function $\psi(y)$ are all dependent on the function $f(x)$. We obtain

$$p_n(x,y) \sim \left\{\sqrt{\frac{n\psi''(y)}{2\pi}}\tilde{q}(x,y)e^{-F(y)}\right\}e^{-n\psi(y)}, \tag{9.150}$$

and

$$p_n(y) \sim \left\{ \sqrt{\frac{n\psi''(y)}{2\pi}} e^{-F(y)} \right\} e^{-n\psi(y)}, \tag{9.151}$$

where

$$F(y) \equiv \int_0^y \int_{-\infty}^{\infty} \tilde{q}_z(x,z)\tilde{p}(x,z)\,dx\,dz.$$

The explicit form of the pre-exponential factor in the expansion (9.151) can be used to improve the large deviations theory estimate of the sample size needed to achieve a given significance level in statistical tests or of the word length required to ensure a given bound on the decoding error in block coding.

The results obtained in the previous section can be generalized by considering $\{\boldsymbol{X}_n\}$ to be an N-dimensional ergodic stationary Markov process with stationary probability density function (pdf) $\phi_0(\boldsymbol{x}) \equiv \phi_0(x_1,\dots,x_N)$. The moments of the stationary pdf, up to second order, are denoted

$$\boldsymbol{m}_{k_1\cdots k_N} \equiv \int \cdots \int z_1^{k_1} \cdots z_N^{k_N} \phi_0(z_1 \dots z_N)\,dz_1 \cdots dz_N, \quad k_1 + \cdots + k_N \leq 2.$$

The master equation for the transition probability density function $p_n(\boldsymbol{y}\,|\,\boldsymbol{x})$ is given by

$$p_{n+1}(\boldsymbol{y}\,|\,\boldsymbol{x}) = \int \rho(\boldsymbol{y},\boldsymbol{z})p_n(\boldsymbol{z}\,|\,\boldsymbol{x})\,d\boldsymbol{z} \equiv \boldsymbol{L}p_n(\boldsymbol{z}\,|\,\boldsymbol{x}), \qquad p_0(\boldsymbol{y}\,|\,\boldsymbol{x}) = \delta(\boldsymbol{x}-\boldsymbol{y}),$$

where

$$\rho(\boldsymbol{y},\boldsymbol{z}) \equiv \frac{\partial^N \Pr\{X_{n+1} \leq \boldsymbol{y}\,|\,X_n = \boldsymbol{z}\}}{\partial y^1 \cdots \partial y^N}.$$

We consider the average

$$\boldsymbol{Y}_n \equiv \frac{1}{n}\sum_{j=1}^{n} \boldsymbol{f}(\boldsymbol{X}_j),$$

where $\boldsymbol{f}$ is a bounded measurable function on $\mathbb{R}^M$, for any natural M and we assume that $\{\boldsymbol{f}(\boldsymbol{X}_n)\}$ is an ergodic and stationary process for which the stationary moments up to second order exist. As in the scalar case, for a small parameter ε an average $\boldsymbol{Y}_n^{\varepsilon}$ is constructed as above and we obtain the master equation for the joint pdf of the pair $\{\boldsymbol{X}_n, \boldsymbol{Y}_n^{\varepsilon}\}$,

$$\begin{aligned} &p_{n+1}^{\varepsilon}(\boldsymbol{x},\boldsymbol{y}) \\ &= \left(\frac{n+1}{n+\varepsilon}\right)^N \int \cdots \int \rho(\boldsymbol{x},\boldsymbol{\xi})p_n^{\varepsilon}\left(\boldsymbol{\xi}, \boldsymbol{y} + \frac{(1-\varepsilon)\boldsymbol{y} - \boldsymbol{f}(\boldsymbol{x})}{n+\varepsilon}\right) d\xi_1 \cdots d\xi_N\,. \end{aligned}$$

The asymptotic solution now takes the form

$$p_n(\boldsymbol{x},\boldsymbol{y}) = K_n(\boldsymbol{x},\boldsymbol{y})\, e^{-n\psi(\boldsymbol{y})},$$

where

$$K_n(\boldsymbol{x},\boldsymbol{y}) = (\sqrt{n})^N \left[q^0(\boldsymbol{x},\boldsymbol{y}) + \frac{1}{n}q^1(\boldsymbol{x},\boldsymbol{y}) + \frac{1}{n^2}q^2(\boldsymbol{x},\boldsymbol{y}) + \cdots\right].$$

To simplify notation, we denote

$$\frac{\partial\psi(\boldsymbol{y})}{\partial y_j} \equiv \psi_j(\boldsymbol{y}) \equiv \psi_j$$

and summation over repeated indices is assumed. Approximation at leading order leads to the equation

$$\begin{aligned} &q^0(\boldsymbol{x},\boldsymbol{y})e^{-\psi(\boldsymbol{y})+(1-\varepsilon)\boldsymbol{y}\cdot\nabla\psi(\boldsymbol{y})} \\ &= \int\cdots\int \rho(\boldsymbol{x},\boldsymbol{\xi})q^0(\boldsymbol{\xi},\boldsymbol{y})e^{\boldsymbol{f}(\boldsymbol{x})\cdot\nabla\psi(\boldsymbol{y})}\, d\xi_1\cdots d\xi_N. \end{aligned} \tag{9.152}$$

For $\varepsilon = 0$ it reduces to

$$q^0(\boldsymbol{x},\boldsymbol{y})e^{-\psi(\boldsymbol{y})+\boldsymbol{y}\cdot\nabla\psi(\boldsymbol{y})} = \int\cdots\int \rho(\boldsymbol{x},\boldsymbol{\xi})q^0(\boldsymbol{\xi},\boldsymbol{y})e^{\boldsymbol{f}(\boldsymbol{x})\cdot\nabla\psi(\boldsymbol{y})}\, d\xi_1\cdots d\xi_N.$$

We set

$$\boldsymbol{\theta} \equiv \nabla\psi(\boldsymbol{y})$$

and consider a one parameter family of operators defined by

$$\boldsymbol{M}(\boldsymbol{\theta})\phi(\boldsymbol{x},\boldsymbol{y}) \equiv e^{\boldsymbol{f}(\boldsymbol{x})\cdot\boldsymbol{\theta}} \int\cdots\int \rho(\boldsymbol{x},t)\phi(t,\boldsymbol{y})\, dt_1\cdots dt_N \equiv e^{\boldsymbol{f}(\boldsymbol{x})\cdot\boldsymbol{\theta}}\boldsymbol{L}\phi(\boldsymbol{x},\boldsymbol{y}).$$

By the same considerations as in the scalar case, the operator $\boldsymbol{M}(\boldsymbol{\theta})$ possesses a positive largest eigenvalue, denoted $\mu(\boldsymbol{\theta})$. Thus $\psi(\boldsymbol{y})$ is a solution of the first-order partial differential equation

$$\mu(\boldsymbol{\theta}) = e^{-\psi(\boldsymbol{y})+(1-\varepsilon)\boldsymbol{y}\cdot\nabla\psi(\boldsymbol{y})}.$$

The function $q^0(\boldsymbol{x},\boldsymbol{y})$ is a positive eigenfunction of $\boldsymbol{M}(\theta)$ determined up to a factor $k_0(\boldsymbol{y})$, which is found from the next order equation

$$\begin{aligned} &[\mu(\boldsymbol{\theta}) - \boldsymbol{M}(\boldsymbol{\theta})]q^1(\boldsymbol{x},\boldsymbol{y}) \\ &= \left\{N\left(\frac{1}{2}-\varepsilon\right) + \varepsilon[(1-\varepsilon)y_i - f_i(\boldsymbol{x})]\psi_i(\boldsymbol{y}) \right. \\ &\quad \left. - \frac{1}{2}[(1-\varepsilon)y_i - f_i(\boldsymbol{x})]\,\psi_{ij}(\boldsymbol{y})\,[(1-\varepsilon)y_j - f_j(\boldsymbol{x})]\right\}\mu((\theta))q^0(\boldsymbol{x},\boldsymbol{y}) \\ &\quad + [(1-\varepsilon)y_i - f_i(\boldsymbol{x})]\boldsymbol{M}(\theta)q_i^0(\boldsymbol{x},\boldsymbol{y}) \end{aligned}$$

To determine $k_0(\boldsymbol{y})$, we follow the considerations used in the scalar case. The eigenfunction of the adjoint operator $\boldsymbol{M}^*(\boldsymbol{\theta})$ is denoted $p_0(\boldsymbol{x}, \boldsymbol{y})$ and is normalized together with the eigenfunction $q^0(\boldsymbol{x}, \boldsymbol{y})$ to $\tilde{q}(\boldsymbol{x}, \boldsymbol{y})$ and $\tilde{p}(\boldsymbol{x}, \boldsymbol{y})$ as in (9.135) and (9.136). The differentiation of eq. (9.152) yields two identities, as above. All these lead to the equation (summation over repeated indices)

$$(\psi_{lk})^{-1}\psi_l \left[\int \tilde{q}_k(\boldsymbol{x}, \boldsymbol{y})\tilde{p}(\boldsymbol{x}, \boldsymbol{y})\, d\boldsymbol{x} + \frac{\partial}{\partial y_k}\log k_0(\boldsymbol{y}) - \frac{1}{2}\psi_{ikj}(\psi_{ij})^{-1}\right] = 0. \quad (9.153)$$

The following property of the Hessian matrix can be easily verified

$$\left(\frac{\partial \psi_{ij}}{\partial y_k}\right)(\psi_{ij})^{-1} \equiv \frac{\partial\left[\log\left(\det\left[\psi_{ij}\right]\right)\right]}{\partial y_k}. \quad (9.154)$$

Using (9.154) in eq. (9.153) yields

$$\begin{aligned}(\psi_{lk})^{-1}\psi_l \frac{\partial}{\partial y_k}\left(\log\left[\frac{k_0(\boldsymbol{y})}{\sqrt{\det[\psi_{ij}]}}\right]\right) \\ = -(\psi_{lk})^{-1}\psi_l \int \frac{\partial}{\partial y_k}\tilde{q}(\boldsymbol{x}, \boldsymbol{y})\tilde{p}(\boldsymbol{x}, \boldsymbol{y})\, d\boldsymbol{x}. \end{aligned} \quad (9.155)$$

A normalized solution for (9.155) is

$$\begin{aligned} k_0(\boldsymbol{y}) = {} & \left(\frac{1}{2\pi}\right)^{N/2}\sqrt{\det[\psi_{ij}(\boldsymbol{y})]} \\ & \times \exp\left\{-\sum_{k=1}^{N}\int_0^{y_k}\left(\int\cdots\int \tilde{p}(\boldsymbol{x}, \boldsymbol{y})\frac{\partial}{\partial y_k}\tilde{q}(\boldsymbol{x}, \boldsymbol{y})\, d\boldsymbol{x}\right) dy_k\right\}. \end{aligned}$$

The large deviations result for the joint pdf of $\{\boldsymbol{X}_n, \boldsymbol{Y}_n\}$ is

$$p_n(\boldsymbol{x}, \boldsymbol{y}) \sim n^{N/2}k_0(\boldsymbol{y})\tilde{q}(\boldsymbol{x}, \boldsymbol{y})e^{-n\psi(\boldsymbol{y})},$$

where $\tilde{q}(\boldsymbol{x}, \boldsymbol{y})$ is the eigenfunction of the operator $\boldsymbol{M}(\boldsymbol{\theta})$, corresponding to the largest eigenvalue $\mu(\boldsymbol{\theta})$, normalized by (9.135). The asymptotic representation of the pdf of $\boldsymbol{Y}_n$ is given by

$$p_n(\boldsymbol{y}) \sim n^{N/2}k_0(\boldsymbol{y})e^{-n\psi(\boldsymbol{y})}. \quad (9.156)$$

Example 9.3 (A two-states Markov chain). We consider a Markov process X_n that stays at -1 with probability p_1 and jumps to 1 with probability $1 - p_1$, stays at 1 with probability p_2 and jumps to -1 with probability $1 - p_2$. The one step transition density is given by

$$\begin{aligned}\rho(x, t) = {} & \{p_1\delta(x+1) + (1-p_1)\delta(x-1)\}\,\delta(t+1) \\ & + \{(1-p_2)\delta(x+1) + p_2\delta(x-1)\}\,\delta(t-1).\end{aligned}$$

Hence $\boldsymbol{L}$ can be represented as the matrix

$$\begin{pmatrix} p_1 & 1-p_2 \\ 1-p_1 & p_2 \end{pmatrix},$$

operating on column vectors. The stationary pdf of X_n is given by

$$p\,(x) = \frac{1-p_2}{2-p_1-p_2}\,\delta(x+1) + \frac{1-p_1}{2-p_1-p_2}\,\delta(x-1).$$

The assumption (9.117) is $p_2 = p_1 \equiv p$, hence the stationary pdf of X_n is $p\,(x) = \frac{1}{2}\,\delta(x+1) + \frac{1}{2}\,\delta(x-1)$.

We denote $\theta = \psi'(y)$ and write the eikonal equation (9.128) in the form

$$\mu(\theta)\begin{bmatrix} q^0(-1,y) \\ q^0(1,y) \end{bmatrix} = \begin{bmatrix} e^{-\theta}p & e^{-\theta}(1-p) \\ e^{\theta}(1-p) & e^{\theta}p \end{bmatrix}\begin{bmatrix} q^0(-1,y) \\ q^0(1,y) \end{bmatrix},$$

and obtain that

$$\mu(\theta) = p\cosh(\theta) + \sqrt{(p-1)^2 + p^2\sinh^2\theta}\,. \tag{9.157}$$

The normalized eigenfunctions, with some abuse of notations, are given by

$$\begin{aligned}\tilde{q}(x,y) &\equiv \tilde{q}(-1,y)\,\delta(x+1) + \tilde{q}(1,y)\,\delta(x-1)\\ \tilde{p}(x,y) &\equiv \tilde{p}(-1,y)\,\delta(x+1) + \tilde{p}(1,y)\,\delta(x-1),\end{aligned}$$

where

$$\begin{bmatrix} \tilde{q}(-1,y) \\ \tilde{q}(1,y) \end{bmatrix} = \begin{bmatrix} e^{-\theta}(1-p) \\ \mu(\theta) - e^{-\theta}p \end{bmatrix}\frac{1}{e^{-\theta}(1-2p)+\mu(\theta)}$$

and

$$\begin{bmatrix} \tilde{p}(-1,y) \\ \tilde{p}(1,y) \end{bmatrix} = \begin{bmatrix} e^{\theta}(1-p) \\ \mu(\theta) - e^{-\theta}p \end{bmatrix}\frac{e^{-\theta}(1-2p)+\mu(\theta)}{(1-p)^2 + (\mu(\theta)-e^{-\theta}p)^2}\,.$$

Next, we determine $\psi'(y)$. We write eq. (9.121) in the form

$$\begin{aligned}&\sqrt{(p-1)^2 + (2p-1)\tanh^2\theta}\,(p\tanh\theta - yp)\\ &= y\left((p-1)^2 + (2p-1)\tanh^2\theta\right) - p^2\tanh\theta,\end{aligned}$$

and solve it with respect to the variable $\tanh\theta$ to obtain

$$\tanh\theta = \frac{y(1-p)}{\sqrt{y^2(1-2p)+p^2}},$$

so we get that

$$\psi'(y) \equiv \theta = \tanh^{-1}\left[\frac{y(1-p)}{\sqrt{y^2(1-2p)+p^2}}\right]. \tag{9.158}$$

The rate function $\psi(y)$ is

$$\psi(y) = y\psi'(y) - \frac{1}{2}\log[(1-y^2)p^2(1-2p)^2] + \frac{1}{2}\log\left|p^2 + y(1-2p) + (p-1)\sqrt{p^2+y^2(1-2p)}\right| + \frac{1}{2}\log\left|p^2 - y(1-2p) + (p-1)\sqrt{p^2+y^2(1-2p)}\right| .$$

To determine the factor $k_0(y)$ from eq. (9.146), we find the function $\sqrt{\psi''(y)/2\pi}$ by differentiating (9.158) with respect to y, which yields

$$\sqrt{\frac{\psi''(y)}{2\pi}} = \sqrt{\frac{1-p}{2\pi(1-y^2)}}\,\frac{1}{[y^2(1-2p)+p^2]^{1/4}}.$$

Differentiating $\tilde{q}(x,y)$ with respect to y, we get that

$$\begin{bmatrix} \tilde{q}_y(-1,y) \\ \tilde{q}_y(1,y) \end{bmatrix} = \begin{bmatrix} -1 \\ 1 \end{bmatrix} \frac{\psi''(y)e^{-\theta}\,(1-p)\,\mu(\theta)\,(1+y)}{[e^{-\theta}\,(1-2p)+\mu(\theta)]^2}$$

and thus

$$f(y) = \int_{-\infty}^{\infty} \tilde{q}_y(x,y)\tilde{p}(x,y)dx \equiv \langle \tilde{q}_y(x,y), \tilde{p}(x,y)\rangle = \frac{[e^{-\theta}(1-p)\psi''(y)\mu(\theta)(1+y)][\mu(\theta)-e^{\theta}(1-p)-e^{-\theta}p]}{[e^{-\theta}(1-2p)+\mu(\theta)][\theta(1-p)^2+(\mu(\theta)-e^{-\theta}p)^2]}.$$

The function $f(y)$ vanishes only for $p=\frac{1}{2}$ and then our approximation is the same as that for i.i.d. Bernoulli random variables. □

9.3.2 A central limit theorem for dependent variables

When y is sufficiently small, the large deviations results reduce to central limit theorem type (CLT) results. It should be borne in mind, however, that the standard form of the CLT requires X_n to be i.i.d. random variables. When they are correlated, as is the case at hand, a more refined version of the CLT is given in the following

Theorem 9.3.2 (CLT for a Markov chain). *If $\{X_n\}$ is an ergodic stationary Markov process with zero mean, with stationary pdf $\phi_0(x)$ and a finite moment generating function, then the asymptotic forms of the pdf of the average and of the joint pdf of the process and its average are given by*

$$p_n(y) \sim \sqrt{\frac{n}{2\pi S(0)}}\, e^{-ny^2/2S(0)}$$
$$p_n(x,y) \sim \sqrt{\frac{n}{2\pi S(0)}}\,\phi_0(x)\,e^{-ny^2/2S(0)}, \tag{9.159}$$

where $S(0)$ is the spectral density of X_n at $\omega=0$.

Proof. We can change y to a new variable $\theta = \psi'(y)$, which is zero when y is zero. Then eq. (9.128) can be solved for small y by expanding the eigenvalue $\mu(\theta)$, the corresponding eigenfunction $f(x,\theta)$, and the exponential function in powers of θ. We apply only quadratic-order Taylor expansion, so that small values of y are such that higher-order terms in Taylor's expansion are negligible relative to the terms retained. Thus, writing

$$\begin{aligned}\mu(\theta) &= \mu_0 + \theta\mu_1 + \theta^2\mu_2 + o\left(\theta^2\right) , \\ f(x,\theta) &= f_0(x) + \theta f_1(x) + \theta^2 f_2(x) + o\left(\theta^2\right) , \\ e^{x\theta} &= 1 + \theta x + \frac{\theta^2 x^2}{2} + o\left(\theta^2\right) ,\end{aligned}$$

and equating coefficients of like powers of θ in (9.128), we obtain

$$(\mu_0 \boldsymbol{I} - \boldsymbol{L}) f_0(x) = 0 , \tag{9.160}$$

$$(\mu_0 \boldsymbol{I} - \boldsymbol{L}) f_1(x) = -\left(\mu_1 \boldsymbol{I} - x\boldsymbol{L}\right) f_0(x) , \tag{9.161}$$

$$(\mu_0 \boldsymbol{I} - \boldsymbol{L}) f_2(x) = -\left(\mu_1 \boldsymbol{I} - x\boldsymbol{L}\right) f_1(x) - (\mu_2 \boldsymbol{I} - \frac{x^2}{2!}\boldsymbol{L}) f_0(x) . \tag{9.162}$$

From (9.160) it follows that $\mu_0 = 1$ and $f_0(x) = \phi_0(x)$. Integrating (9.161) with respect to x it is found from (9.117) that $\mu_1 = 0$ and that

$$(\boldsymbol{I} - \boldsymbol{L}) f_1(x) = x\phi_0(x).$$

Then, if $x\phi_0(x)$ is in the range of $\boldsymbol{I} - \boldsymbol{L}$, we can consider the operator $(\boldsymbol{I} - \boldsymbol{L})^{-1}$ on a subspace containing $x\phi_0(x)$ to obtain

$$f_1(x) = (\boldsymbol{I} - \boldsymbol{L})^{-1} x\phi_0(x) \equiv \sum_{i=0}^{\infty} \boldsymbol{L}^i x\phi_0(x) + c f_0(x), \tag{9.163}$$

where c is an arbitrary constant. Integrating (9.162) with respect to x, and using (9.163), it is found that

$$\mu_2 = \frac{m_2}{2} + \int x\boldsymbol{L} f_1(x)\, dx = \frac{m_2}{2} + \int \sum_{i=1}^{\infty} x\boldsymbol{L}^i x\phi_0(x)\, dx. \tag{9.164}$$

Next, the exponent in (9.119) is expanded in Taylor series in y,

$$e^{-\psi(y) + y\psi'(y)} = 1 + \frac{1}{2}\psi''(0)\, y^2 + o\left(y^2\right). \tag{9.165}$$

Setting $\theta = \psi'(y) = \psi''(0)\, y + o(y)$ in (9.165), we obtain $\mu(\theta) = 1 + \theta^2/2\psi''(0) + o\left(\theta^2\right)$ and thus $2\mu_2 = 1/\psi''(0)$ and by (9.164), we conclude that

$$\psi''(0) = \frac{1}{2\sum_{i=1}^{\infty} \int x\boldsymbol{L}^i x\phi_0(x)\, dx + m_2}. \tag{9.166}$$

It can be easily found that the denominator in (9.166) is the spectral density of the process $\{X_n\}$ at frequency 0, denoted $S(0)$. Indeed, the autocorrelation function of $\{X_n\}$ is given by

$$R(n) = \int\int xy\phi_0(x)p_n(y\,|\,x)\,dx\,dy.$$

From (9.114) and (9.116) it follows that $p_n(y\,|\,x) = \rho_n(y,x)$, for $n \geq 1$, where $\rho_1(y,x) = \rho(y,x)$ and $\rho_{n+1}(y,x) = \int \rho_n(y,t)\rho(t,x)\,dt$. With this notation the integrals in the denominator of (9.166) can be written as

$$\int x\boldsymbol{L}^i x\phi_0(x)\,dx = \int\int xy\phi_0(x)\rho_i(y,x)\,dx\,dy. \tag{9.167}$$

For $n \geq 1$ the autocorrelation function can be written as

$$R(n) = \int\int xy\phi_0(x)\rho_n(x,y)\,dx\,dy, \tag{9.168}$$

whereas for $n < 0$ we have $R(n) = R(-n)$ and for $n = 0$

$$R(0) = m_2. \tag{9.169}$$

Applying (9.167)–(9.169), we write (9.166) as

$$\psi''(0) = \frac{1}{\displaystyle\sum_{n=-\infty}^{\infty} R(n)} \equiv \frac{1}{S(0)},$$

where $S(0)$ is the spectral density at frequency 0. The asymptotic pdf of Y_n for small deviations is found by integrating (9.126) with respect to x and expanding for small y. Thus Y_n is asymptotically a zero mean normal variable with variance $S(0)/n$; that is, (9.159). □

Equations (9.159) generalize the standard CLT to the case of correlated random variables that form a stationary Markov process, as described above (see [25]).

Exercise 9.9. (Analysis of the next order terms).

(i) Find the full asymptotic expansion for $p_n(x,y)$ (9.126) in the form of an infinite series. Show that if the relations $q^{i+1}(x,y)/q^i(x,y)$ are bounded uniformly in x and y, the expansion (9.126) is indeed an asymptotic series and the leading-order approximation for $p_n(x,y)$ can be limited to only the first terms $\psi(y)$ and $q^0(x,y)$.

(ii) Show that in certain cases (e.g., when X_n have bounded support) the WKB expansion may fail at the boundary of the support and must be fixed there by a boundary layer (this problem was extensively studied in [48] for the case of i.i.d. random variables).

(iii) To evaluate the next term $q^1(x,y)$, show that the quotients $q^{i+1}(x,y)/q^i(x,y)$ are asymptotically similar to $q^1(x,y)/q^0(x,y)$. Substitute the approximation (9.126) in (9.125) and expand it in negative powers of n. Show that the zeroth-order equation is (9.128) and that the first-order equation is an equation for $q^1(x,y)$,

$$\begin{aligned}&[\mu(y)-\boldsymbol{M}(y)]q^1(x,y)\\&=\{1-\psi''(y)(y-x)^2\}\frac{1}{2}\mu(y)q^0(x,y)+(y-x)\boldsymbol{M}(y)q^0_y(x,y),\end{aligned}\tag{9.170}$$

which is eq. (9.131) for $\varepsilon=0$.

(iv) Solve $q^1(x,y)$ by differentiating eq. (9.128), to obtain the two identities

$$[\boldsymbol{M}(y)-\mu(y)]\,q^0_y(x,y)=(y-x)\psi''(y)\mu(y)q^0(x,y)\tag{9.171}$$

$$\begin{aligned}[\boldsymbol{M}(y)-\mu(y)]\,q^0_{yy}(x,y)=&\left\{\psi''(y)+(y-x)\psi'''(y)+(y-x)^2\psi''^2(y)\right\}\\&\times\mu(y)q^0(x,y)+2(y-x)\psi''(y)\mu(y)q^0_y(x,y).\end{aligned}$$

Hence obtain

$$q^1_p(x,y)=\frac{\psi'''(y)}{2\psi''^2(y)}q^0_y(x,y)-\frac{1}{2\psi''(y)}q^0_{yy}(x,y)$$

as a particular solution for eq. (9.170). Conclude that

$$q^1(x,y)=A(y)q^0(x,y)+q^1_p(x,y),\tag{9.172}$$

where $A(y)$ is determined from the next order equation (9.173) below.

(v) Show that the second-order equation is

$$\begin{aligned}&[\mu(y)-\boldsymbol{M}(y)]q^2(x,y)\\&=\mu(y)\left[\frac{1}{8}q^0(x,y)+\frac{1}{2}q^1(x,y)\right]\\&\quad+\left[\frac{1}{8}(y-x)^4\psi''^2(y)-\frac{1}{6}(y-x)^3\psi'''(y)-\frac{1}{2}(y-x)^2\psi''(y)\right]\mu(y)q^0(x,y)\\&\quad+\left[(y-x)-\frac{1}{2}(y-x)^3\psi''(y)\right]\boldsymbol{M}(y)q^0_y(x,y)+\frac{1}{2}(y-x)^2\boldsymbol{M}(y)q^0_{yy}(x,y)\\&\quad+(y-x)\boldsymbol{M}(y)q^1_y(x,y)+\left[1-\frac{1}{2}(y-x)^2\psi''(y)\right]\boldsymbol{M}(y)q^1(x,y).\end{aligned}\tag{9.173}$$

In order to determine $A(y)$, multiply eq. (9.173) by $\tilde{p}(x,y)$ and integrate with respect to x. Then simplify it by differentiating eq. (9.170) up to second order and eq. (9.128) up to fourth order. Substitute (9.172) for $q^1(x,y)$ (use *Mathematica* or *Maple*) to find that

$$A(y)=\frac{1}{8}\frac{\psi^{(iv)}(y)}{\psi''^2(y)}-\frac{5}{24}\frac{\psi'''^2(y)}{\psi''^3(y)},$$

and thus

$$q^1(x,y) = \left\{ \frac{1}{8}\frac{\psi^{(iv)}(y)}{\psi''^2(y)} - \frac{5}{24}\frac{\psi'''^2(y)}{\psi''^3(y)} \right\} q^0(x,y) + \frac{\psi'''(y)}{2\psi''^2(y)} q^0_y(x,y) - \frac{1}{2\psi''(y)} q^0_{yy}(x,y). \tag{9.174}$$

(vi) Substitute (9.134) in (9.174) and apply (9.135). Note that (9.135) also means that

$$\int \tilde{q}_y(x,y)\,dx = \int \tilde{q}_{yy}(x,y)\,dx = 0.$$

Obtain that

$$k_1(y) \equiv \int q^1(x,y)\,dx \tag{9.175}$$

$$= \frac{1}{8}k_0(y)\frac{\psi^{(iv)}(y)}{\psi''^2(y)} - \frac{5}{24}k_0(y)\frac{\psi'''^2(y)}{\psi''^3(y)} + k_0'(y)\frac{\psi'''(y)}{2\psi''^2(y)} - \frac{1}{2\psi''(y)}k_0''(y).$$

Finally, apply (9.146) in (9.175) and get that

$$\frac{k_1(y)}{k_0(y)} = \frac{1}{6}\frac{\psi'''^2(y)}{\psi''^3(y)} - \frac{1}{8}\frac{\psi^{(iv)}(y)}{\psi''^2(y)} + \frac{1}{2\psi''(y)}\int \tilde{q}_y(x,y)\tilde{p}_y(x,y)\,dx \tag{9.176}$$

$$+ \frac{1}{2\psi''(y)}\left[\int \tilde{q}_{yy}(x,y)\tilde{p}(x,y)\,dx - \left(\int \tilde{q}_y(x,y)\tilde{p}(x,y)\,dx\right)^2\right].$$

Show that the expansion for $p_n(y)$ remains asymptotic as long as this quotient is bounded. □

9.4 Annotations

The problem of large deviations arises in many problems in statistical decision theory, communications, information theory, and statistical physics [32], [54]. The large deviations principle [54], [68], [55], [234] characterizes the exponential decay of the tails of these PDFs by a rate function. The rate function controls the logarithmic asymptotics of the PDFs, but does not provide a full asymptotic expansion. In the case of i.i.d. $\{X_n\}$ a full asymptotic expansion was formulated by H. Cramér in [47] under the assumption that the moment generating function exists near zero. The expansion for the complementary PDF of Y_n was given in the form

$$1 - F_n(y) \sim \frac{1}{\sqrt{2\pi n}} e^{-n\psi(y)}\left[K_0(y) + \frac{1}{n}K_1(y) + \frac{1}{n^2}K_2(y) + \cdots\right]. \tag{9.177}$$

The exponential rate function $\psi(y)$ was determined as the solution of a differential equation. The explicit expressions for the higher-order functions K_j were later calculated in [8], [48], [9], [133] by different methods. In this case, the joint probability density function of $\{X_n, Y_n\}$, can be obtained from the pdfs $\rho(x)$ and $p_n(y)$ of X_n and Y_n, respectively, as

$$\begin{aligned} p_n(x,y) &= \frac{\partial \Pr(Y_n \leq y \,|\, X_n = x)}{\partial y} \rho(x) \\ &= \left(\frac{n}{n-1}\right) \rho(x) p_{n-1}\left(\frac{ny-x}{n-1}\right). \end{aligned} \tag{9.178}$$

For Markovian $\{X_n\}$ the joint pdf is not such a simple function of the marginals.

The Edgeworth expansion [72] is essentially a small deviations approximation in the sense that its asymptotic validity is limited to the region $y = o(n^{-1/3})$ [72] and thus can lead to a significant relative error in the tail region. For example, if X_n is supported in the interval $[-1, 1]$, the Edgeworth expansion results in an infinite relative error for $|y| > 1$.

The construction of a full asymptotic expansion in the case where X_n is a Markov process is a considerably harder problem than that for the i.i.d. case. The case of a finite state Markov chain was solved in [173] and [110] (see also [32]), and [110] also considers the general case. In [173] a restricted case of a continuous state space is considered and a full asymptotic expansion of the pdf of Y_n is given, but not that of the joint pdf of the pair $\{X_n, Y_n\}$. The expansions depend on the initial distribution of the chain.

The main results of this chapter are eq. (9.122) and eq. (9.123) for the leading terms in the asymptotic expansion of the pdfs. The calculations can be extended to higher-order terms in the expansion (see Appendix). These results reduce to those of [173], [109], [110] for a finite-state stationary Markov chain. The formulas eq. (9.122) and eq. (9.123) are convenient for calculations and give an explicit expression for the pre-exponential factor in the case of an asymmetric telegraph process. The expansion is formal and no proof of asymptotic convergence is presented.

The routine application of the WKB technique leads to a breakdown of the expansion scheme given in [133] so that a modification is introduced to remove the degeneration. The large deviations principle (9.126) with (9.119)–(9.121) is equivalent to that mentioned in [55, Ch. 4, Sect. 1] in the sense that $\psi(y)$ is the Legendre transform of the spectral radius of the operator $\boldsymbol{M}(y)$.

Chapter 10

Noise-Induced Escape From an Attractor

10.1 Asymptotic analysis of the exit problem

The analysis of the exit problem, as defined in Section 6.4 and analyzed in Chapter 6, was based on explicit integral representations of solutions to the one-dimensional Andronov–Vitt–Pontryagin boundary value problems (6.33) for the exit distribution and for the MFPT. Such representations are in general unavailable in higher dimensions so that there is no obvious generalization of these methods for the exit problem in higher dimensions. Another approach, that is based on constructing an asymptotic solution to the boundary value problems by the methods of singular perturbations, can be generalized to higher dimensions (see Section 10.2).

We begin with a primer of matched asymptotics. The approximations obtained in Section 6.4.2 can be derived directly from the equation by the method of singular perturbations [195], [121], [20]. The method is described in several texts on asymptotic methods and singular perturbation and only relevant topics of singular perturbation theory are reviewed here. We begin with a short primer on matched asymptotics through an example of an initial value problem for a first-order linear ordinary differential equation. Then, the second-order equations for the exit probability and for the FPT are analyzed.

Consider the problem

$$\varepsilon y_\varepsilon' + a(x)\, y_\varepsilon = f(x) \quad \text{for } x > 0 \tag{10.1}$$

$$y(0) = y_0. \tag{10.2}$$

It can, obviously, be solved explicitly and evaluated asymptotically for small ε by the Laplace method, as described in Section 6.4.2 (see Exercise 10.2 below). However, a direct and simpler method exists for the construction of a uniform asymptotic approximation for the solution, as described below. We assume that all functions are regular in x and that $a(x) > \delta > 0$, where δ is a constant. A naïve approximation

Z. Schuss, *Theory and Applications of Stochastic Processes: An Analytical Approach*, Applied Mathematical Sciences 170, DOI 10.1007/978-1-4419-1605-1_10,

is obtained by postulating that the solution has a regular asymptotic power series expansion of the form

$$y_{\text{outer}}(x) \sim y_0(x) + \varepsilon y_1(x) + \cdots, \tag{10.3}$$

where $y_j(x)$ are regular functions of x, independent of ε. Substituting the asymptotic series (10.3) in eq. (10.1) and comparing coefficients of like powers of ε on both sides of the equation, we obtain the hierarchy of equations

$$a(x)\, y_0(x) = f(x) \tag{10.4}$$

and

$$a(x)\, y_j(x) = -y'_{j-1}(x) \quad \text{for } j \geq 1. \tag{10.5}$$

Under the above regularity assumptions, the recurrence relation (10.5) defines all terms in the asymptotic series (10.3). The resulting series is not necessarily convergent, however, it can be shown (see Exercise 10.1 below) to be asymptotic to the solution $y_\varepsilon(x)$ for all $x >> \varepsilon$. In general, the asymptotic series $y_{\text{outer}}(x)$ cannot satisfy the initial condition (10.2). This is due to the reduced order of eq. (10.4) relative to that of eq. (10.1). This fact indicates that the solution $y_\varepsilon(x)$ undergoes a sharp change near x_0 that bridges the gap between the different values $y_0(0) = f(0)/a(0)$ and $y_\varepsilon(0) = y_0$. To bridge this gap both terms in eq. (10.1) have to be of the same order of magnitude when ε is small, unlike the case in the asymptotic series (10.3). To resolve the behavior of the solution in the region of rapid change the scaled variables $\xi = x\varepsilon^{-\kappa}$, $Y_\varepsilon(\xi) = y_\varepsilon(x)$ are introduced, with the positive constant κ chosen in such a way that all terms in eq. (10.1) become of comparable magnitude. With the variable ξ eq. (10.1) becomes

$$\varepsilon^{1-\kappa} Y'_\varepsilon(\xi) + a(\varepsilon^\kappa \xi) Y_\varepsilon(\xi) = f(\varepsilon^\kappa \xi), \quad Y_\varepsilon(0) = y_0. \tag{10.6}$$

The small region $x = O(\varepsilon^\kappa)$ is stretched into the region $\xi = O(1)$. Clearly, the three terms in eq. (10.6) become comparable if $\kappa = 1$. Expanding in asymptotic power series,

$$a(\varepsilon\xi) \sim a_0(\xi) + \varepsilon a_1(\xi) + \cdots \tag{10.7}$$

$$f(\varepsilon\xi) \sim f_0(\xi) + \varepsilon f_1(\xi) + \cdots \tag{10.8}$$

$$Y_\varepsilon(\xi) \sim Y_0(\xi) + \varepsilon Y_1(\xi) + \cdots, \tag{10.9}$$

we obtain the hierarchy of equations

$$\frac{dY_0(\xi)}{d\xi} + a_0(\xi) Y_0(\xi) = f_0(\xi), \quad Y_0(0) = x_0 \tag{10.10}$$

$$\frac{dYj(\xi)}{d\xi} + a_0(\xi) Y_j(\xi) = f_j(\xi) - \sum_{k=0}^{j-1} a_{j-k}(\xi) Y_k(\xi) \tag{10.11}$$

$$Y_j(0) = 0, \quad \text{for } j \geq 1.$$

Note that the series (10.7), (10.8) are regular Taylor expansions of regular functions and the sign $\sim$ (asymptotic to) can actually be replaced with = (equals). This is, however, not necessarily so for the series (10.9). The solution of (10.10) is

$$Y_0(\xi) = e^{-A(\xi)} y_0 + \int_0^\xi e^{-[A(\xi)-A(\sigma)]} f_0(\sigma)\, d\sigma,$$

where $A(\xi) = \int_0^\xi a_0(\eta)\, d\eta$. The equations (10.11) are readily solved in a similar manner. In terms of the original variable, we have $A(\xi) = a(0)\, x/\varepsilon$, so we obtain

$$\begin{aligned} Y_0\left(\frac{x}{\varepsilon}\right) &= e^{-a(0)x/\varepsilon} y_0 + \int_0^{x/\varepsilon} e^{-a(0)(x/\varepsilon - s)} f(0)\, ds \\ &= e^{-a(0)x/\varepsilon} y_0 + f(0) \frac{1 - e^{-a(0)x/\varepsilon}}{a(0)}. \end{aligned}$$

Note that although $Y_0(\xi)$ is a regular function of ξ, it has an essential singularity as a function of ε at the point $\varepsilon = 0$ for all $x > 0$.

It is apparent that $\lim_{\xi\to 0} Y_0(\xi) = y_0$ and $\lim_{\xi\to\infty} Y_0(\xi) = f(0)/a(0)$. We also observe that $\lim_{\xi\to\infty} Y_0(\xi) = f(0)/a(0) = y_0(0) = \lim_{\xi\to 0} y_{\text{outer}}(x)$. Thus the outer solution $y_{\text{outer}}(x)$ matches the boundary layer solution $Y_0(\xi)$ in the matching region between $x = O(\varepsilon)$ and $x = O(1)$. A uniform approximation is obtained by simply adding the two solutions and subtracting their common limit $f(0)/a(0)$; that is

$$y_{\text{unif}}(x) \sim e^{-a(0)x/\varepsilon} y_0 - \frac{f(0)}{a(0)} e^{-a(0)x/\varepsilon} + \frac{f(x)}{a(x)} + \cdots, \tag{10.12}$$

or $y_{\text{unif}}(x) \sim y_{\text{b.l.}}(x) + y_{\text{outer}}(x) - \ell$, where $y_{\text{b.l.}}(x) = Y_\varepsilon(x/\varepsilon)$ and ℓ is the common limit

$$\ell = \lim_{\xi\to\infty} y_{\text{b.l.}}(x) = \lim_{\xi\to 0} y_{\text{outer}}(x). \tag{10.13}$$

The first limit $\xi \to \infty$ in eq. (10.13) means that x is kept fixed and $\varepsilon \to 0$ whereas the limit $\xi \to 0$ means that ε is kept fixed and $x \to 0$. Equation (10.13) is called the *matching condition*. It means that the boundary layer approximation $y_{\text{b.l.}}(x)$ and the outer solution $y_{\text{outer}}(x)$ match as functions of the scaled variable ξ in the matching region between $x = O(\varepsilon)$ and $x = O(1)$.

Example 10.1 (uniform expansion). Consider the initial value problem

$$0.1y' + (x+1)y = 1 \text{ for } x > 0, \quad y(0) = 1. \tag{10.14}$$

The first three terms of the outer expansion are

$$y_{\text{outer}}(x) = \frac{1}{x+1} + \frac{0.1}{(x+1)^3} + \frac{3\times 0.1^2}{(x+1)^5} + \frac{15\times 0.1^3}{(x+1)^7} + \cdots$$

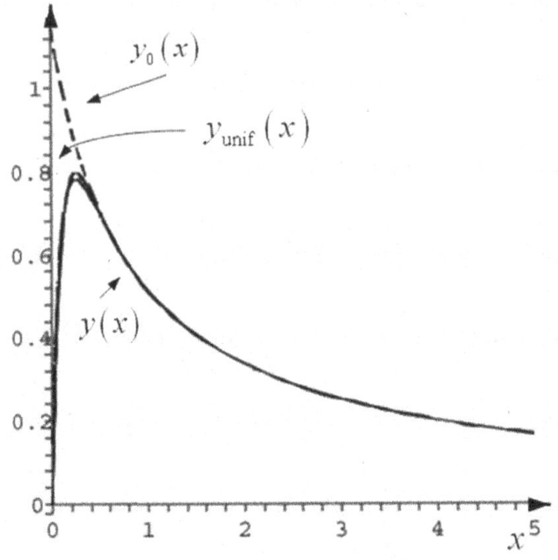

Figure 10.1. The three-term outer solution (dashed line), the uniform expansion (dotted line), and the exact solution (solid line).

(dashed line in Figure 10.1). The uniform expansion (10.12) that satisfies the matching condition (10.13) is given by

$$y_{\text{unif}}(x) = -1.145e^{-x/0.1} + \frac{1}{x+1} + \frac{0.1}{(x+1)^3} + \frac{3\times 0.1^2}{(x+1)^5} + \frac{15\times 0.1^3}{(x+1)^7}$$

(dotted line), includes the leading term of the boundary layer, and the exact solution is

$$y(x) = 10\int_0^x \exp\left\{-\frac{(x+1)^2+(s+1)^2}{0.2}\right\} ds$$

(solid line). □

Exercise 10.1 (Convergence of the asymptotic expansion). Construct the full asymptotic series (10.9) and determine the nature of its convergence. □

Exercise 10.2 (Expansion of the explicit solution). Construct an explicit solution of eq. (10.1), expand it by the Laplace method, and compare the obtained expansion with that obtained in Exercise 10.1. □

Next, we turn to the exit problem eqs. (6.44). We begin with the construction of the outer solution in the form of a regular asymptotic power series $u_{\text{outer}} \sim u_0(x) +$

$\varepsilon u_1(x) + \cdots$. The reduced problem is

$$L^{*0}u_0(x) = a(x)u_0'(x) = 0 \text{ for } \alpha < x < \beta \tag{10.15}$$

$$u_0(\alpha) = g_\alpha, \quad u_0(\beta) = g_b \tag{10.16}$$

and the higher-order terms satisfy the equations

$$L^{*0}u_j(x) = -\sigma(x)\,u_j''(x) \text{ for } j \geq 1, \quad u_j(\alpha) = u_j(\beta) = 0.$$

The solution of the reduced equation is $u_0(x) = u_0 = const.$ and the higher-order terms vanish. Both boundary conditions (10.16) cannot possibly be satisfied, unless $g_\alpha = g_\beta$. Thus boundary layers have to be constructed at both ends of the interval and the unknown constant u_0 has to be determined.

10.1.1 The exit problem for small diffusion with the flow

We consider first the case that the drift in the stochastic equation (6.35) carries the trajectories across the interval, from left to right, say. That is, we assume that $a(x) > \delta > 0$, $\sigma(x) > \delta > 0$. In this case the potential $\Phi(x)$ (see eq. (6.45)) is a monotone decreasing function. In this case the stochastic dynamics represents a particle sliding down an incline with slightly fluctuating velocity.

We construct a boundary layer approximation near $x = \alpha$. Introducing the scaled variables $\xi = (x - \alpha)/\varepsilon$ and $U_\varepsilon(\xi) = u_\varepsilon(x)$, the interval $[\alpha, \beta]$ is mapped onto the interval $[0, (\beta - \alpha)/\varepsilon]$, which is approximated by the entire positive axis. Equation (6.44) becomes

$$\sigma(\alpha + \varepsilon\xi)\,U_\varepsilon''(\xi) + a(\alpha + \varepsilon\xi)\,U_\varepsilon'(\xi) = 0.$$

Expanding the coefficients in Taylor's series and looking for a boundary layer approximation in the form $U_{\text{b.l.}}(\xi) \sim U_0(\xi) + \varepsilon U_1(\xi) + \cdots$, we find that $U_0(\xi)$ satisfies the equation

$$\sigma(\alpha)U_0''(\xi) + a(\alpha)U_0'(\xi) = 0$$

with the boundary condition

$$\lim_{\xi\to 0} U_0(\xi) = g_\alpha$$

and the matching condition

$$\lim_{\xi\to\infty} = u_0.$$

The general solution is

$$U_0(\xi) = A\exp\{-a(\alpha)\,\xi/\sigma(\alpha)\} + B,$$

where A and B are constants to be determined by the boundary and matching conditions. The matching condition gives $B = u_0$ and the boundary condition gives $A = g_\alpha - u_0$, because $a(\alpha)/\sigma(\alpha) > 0$. Thus the boundary layer function is to leading order

$$U_{\text{bl}}(\xi) \sim (g_\alpha - u_0)\exp\{-a(\alpha)\,\xi/\sigma(\alpha)\} + u_0.$$

Yet, the constant u_0 is still undetermined.

Similar analysis at the other boundary involves the variables $\eta = (\beta - x)/\varepsilon$ and $V_\varepsilon(\eta) = u_\varepsilon(x)$ and gives $V_0(\eta) = C\exp\{a(\beta)\eta/\sigma(\beta)\} + D$, where C and D are constants. The boundary condition is $\lim_{\eta\to 0} V_0(\eta) = g_\beta$ and the matching condition is $\lim_{\eta\to\infty} V_0(\eta) = u_0$. Because $a(\beta)/\sigma(\beta) > 0$, the matching condition can be satisfied only if $C = 0$ and $D = u_0$. The boundary condition then implies that $u_0 = g_\beta$. Thus $V_0(\eta) = g_\beta$ and there is no boundary layer at the right endpoint.

A uniform approximation to $u_\varepsilon(x)$ is obtained by adding the boundary layer to the outer solution and subtracting their common limit in the matching region. This gives the uniform approximation

$$u_\varepsilon(x) \sim u_{\text{unif}}(x) = g_\beta + (g_\alpha - g_\beta)\exp\left\{-\frac{a(\alpha)(\alpha - x)}{\varepsilon\sigma(\alpha)}\right\} + O(\varepsilon), \quad (10.17)$$

where $O(\varepsilon)$ is uniform in the interval $[\alpha, \beta]$.

Returning to the exit problem, Green's function is approximated by

$$G(x, y) \sim \delta(y - \alpha)\exp\left\{-\frac{a(\alpha)(\alpha - x)}{\varepsilon\sigma(\alpha)}\right\} + \delta(y - \beta)\left(1 - \exp\left\{-\frac{a(\alpha)(\alpha - x)}{\varepsilon\sigma(\alpha)}\right\}\right) + O(\varepsilon)$$

and the exit probability is

$$\Pr\{\tau_\beta < \tau_\alpha \mid x(0) = x\} = \left(1 - \exp\left\{-\frac{a(\alpha)(\alpha - x)}{\varepsilon\sigma(\alpha)}\right\}\right) + O(\varepsilon)$$

$$\Pr\{\tau_\alpha < \tau_\beta \mid x(0) = x\} = \exp\left\{-\frac{a(\alpha)(\alpha - x)}{\varepsilon\sigma(\alpha)}\right\} + O(\varepsilon).$$

This result agrees with eq. (6.60).

Exercise 10.3 (Full asymptotic expansion of the exit probability for positive drift). Obtain a full asymptotic expansion of the exit probability for the case $a(x) > \delta > 0$, $\sigma(x) > \delta > 0$. □

Exercise 10.4 (Asymptotics for drift that vanishes on the boundary). Consider the exit probability for the case $a(x) > 0$ for $x > \alpha$, but $a(\alpha) = 0$ and $\sigma(x) > \delta > 0$. □

A particular case of small diffusion with the flow is that of a flow directed toward the boundary with an unstable equilibrium point inside the interval, at a point ζ ($\alpha < \zeta < \beta$), say. This is represented by a drift such that $(x - \zeta)\, a(x) > 0$ for $x \neq \zeta$. In this case the potential $\Phi(x)$ has a maximum at ζ (like an inverted parabola) so that a particle placed anywhere in the interval slides down an incline toward the boundary, except at the point ζ, where it is at an unstable equilibrium. The outer solution may be discontinuous, because $a(x)$ changes sign at ζ and $a(\zeta) = 0$; that is, the solution to the reduced problem (10.15) may be one constant, u_α, say, for $\alpha \leq x < \zeta$, and another constant, u_β, say, for $\zeta < x \leq \beta$. In this case an *internal layer* at ζ has to be constructed to connect the two constants smoothly.

To do that, we assume that the local Taylor expansion of $a(x)$ near ζ is

$$a(x) = a'(\zeta)(x - \zeta) + O\left(|x - \zeta|^2\right) \tag{10.18}$$

with $a'(\zeta) > 0$ and introduce near ζ the scaled local variables $\xi = (x - \zeta)/\sqrt{\varepsilon}$ and $U_\varepsilon(\xi) = u_\varepsilon(x)$. Equation (6.44) becomes

$$\sigma\left(\zeta + \sqrt{\varepsilon}\xi\right) U_\varepsilon''(\xi) + \frac{1}{\sqrt{\varepsilon}} a\left(\zeta + \sqrt{\varepsilon}\xi\right) U_\varepsilon'(\xi) = 0.$$

Expanding $U_\varepsilon(\xi) \sim U_0(\xi) + \sqrt{\varepsilon} U_1(\xi) + \cdots$ and Taylor's expansion for the coefficients, we obtain the *internal layer equation*

$$\sigma(\zeta) U_0''(\xi) + \xi a'(\zeta) U_0'(\xi) = 0 \tag{10.19}$$

whose general solution is

$$U_0(\xi) = A \int_0^\xi \exp\left\{-\frac{s^2 a'(\zeta)}{2a(s)}\right\} ds + B,$$

where A and B are constants. The matching conditions for the internal layer function $U_\varepsilon(\xi)$ are $\lim_{\xi \to -\infty} U_\varepsilon(\xi) = u_\alpha$ and $\lim_{\xi \to \infty} U_\varepsilon(\xi) = u_\beta$. Because

$$\int_0^{\pm\infty} \exp\left\{-\frac{s^2 a'(\zeta)}{2\sigma(s)}\right\} ds = \pm\sqrt{\frac{\pi\sigma(\zeta)}{2a'(\zeta)}},$$

we find that

$$A = \frac{u_\beta - u_\alpha}{2}\sqrt{\frac{2a'(\zeta)}{\pi\sigma(\zeta)}}, \quad B = \frac{u_\beta + u_\alpha}{2}.$$

Under the given conditions, local boundary layer analysis indicates that there are no boundary layers so that the outer solution must satisfy the boundary conditions. That means that $u_\alpha = g_\alpha$, $u_\beta = g_\beta$ so that the leading term in the expansion of the internal layer function is

$$U_0(\xi) = \frac{g_\beta - g_\alpha}{2}\sqrt{\frac{2a'(\zeta)}{\pi\sigma(\zeta)}} \int_0^\xi \exp\left\{-\frac{s^2 a'(\zeta)}{2a(s)}\right\} ds + \frac{g_\beta + g_\alpha}{2}$$

and this is also the uniform approximation to the solution. Thus

$$u_\varepsilon(x) \sim \frac{g_\beta - g_\alpha}{2}\sqrt{\frac{2a'(\zeta)}{\pi\sigma(\zeta)}} \int_0^{(x-\zeta)/\sqrt{\varepsilon}} \exp\left\{-\frac{s^2 a'(\zeta)}{2a(s)}\right\} ds + \frac{g_\beta + g_\alpha}{2}.$$

It follows that in this case the exit probability is

$$\Pr\left\{\tau_\beta < \tau_\alpha | x(0) = x\right\} = \frac{1}{2} + \frac{1}{2}\sqrt{\frac{2a'(\zeta)}{\pi\sigma(\zeta)}} \int\limits_0^{(x-\zeta)/\sqrt{\varepsilon}} \exp\left\{-\frac{s^2 a'(\zeta)}{2a(s)}\right\} ds + O(\varepsilon) \tag{10.20}$$

$$\Pr\left\{\tau_\alpha < \tau_\beta | x(0) = x\right\} = \frac{1}{2} - \frac{1}{2}\sqrt{\frac{2a'(\zeta)}{\pi\sigma(\zeta)}} \int\limits_0^{(x-\zeta)/\sqrt{\varepsilon}} \exp\left\{-\frac{s^2 a'(\zeta)}{2a(s)}\right\} ds + O(\varepsilon).$$

We conclude that a trajectory that starts on top of the potential barrier at $x = \zeta$ has about equal chances to reach either boundary, but if it starts at a point on the left of ζ its probability to reach α is nearly 1.

The mean first passage time from a point x in the interval to the boundary, denoted $\bar{\tau}_\varepsilon(x)$, can be evaluated directly from the Andronov–Vitt–Pontryagin equation (4.70)

$$\varepsilon\sigma(x)\tau_\varepsilon''(x) + a(x)\tau_\varepsilon'(x) = -1, \quad \bar{\tau}_\varepsilon(\alpha) = \bar{\tau}_\varepsilon(\beta) = 0. \tag{10.21}$$

The equation can be integrated by quadratures and the integrals expanded for small ε (see Exercise 6.4). Alternatively, for $\alpha << x \leq \beta$ the solution can be expanded in a regular asymptotic power series

$$\bar{\tau}_\varepsilon(x) \sim T_0(x) + \varepsilon\bar{\tau}_1(x) + \cdots \tag{10.22}$$

where

$$T_0(x) = \int\limits_x^\beta \frac{ds}{a(s)}, \quad \bar{\tau}_1(x) = \int\limits_x^\beta \frac{\sigma(s)a'(s)\,ds}{a^3(s)},$$

and so on. A boundary layer is required near $x = \alpha$.

Exercise 10.5 (Construction of a boundary layer). Construct a boundary layer near $x = \alpha$ for the expansion (10.22). □

Exercise 10.6 (Asymptotics inside the boundary layer). Consider now the case that the maximum of the function $\Phi(x) = -\int_\alpha^x [a(s)/\sigma(s)]\,ds$ is achieved at an internal point ζ and $\Phi(x)$ is monotonically increasing in the interval $[\alpha, \zeta]$ and monotonically decreasing in the interval $[\zeta, \beta]$. For any point outside an ε-neighborhood of ζ the analysis of the previous case applies. However, for trajectories that start in an ε-neighborhood of ζ the situation is different. Find the asymptotic expansion of the MFPT for this case. □

Example 10.2 (The stochastic separatrix in one dimension and the transition state). The long time behavior of the deterministic dynamics

$$\dot{x} = -U'(x), \quad x(0) = x_0, \tag{10.23}$$

where $U(x)$ is a smooth potential in $\mathbb{R}$ such that for each $x_0 \in \mathbb{R}$ the solution of (10.23) exists for all $t > 0$, is quite simple [97]. A set $A \subset \mathbb{R}$ is an *invariant set* for (10.23), if $x(t) \in A$ for all $t > 0$ whenever $x_0 \in A$. The set A is a *maximal invariant set* if it is invariant, but $x(t) \notin A$ for all $t > 0$ whenever $x_0 \notin A$. For example, every critical point z of $U(x)$ (i.e., a point z such that $U'(z) = 0$) is an invariant set. A maximal invariant set that does not contain any proper maximal subsets is called a *proper maximal invariant set.* A proper maximal invariant set can be a point or an open (finite or infinite) interval. The entire line $\mathbb{R}$ is a (countable or uncountable) union of disjoint proper maximal invariant sets.

If $U(x)$ has a local minimum at $x = x_m$, the *domain of attraction* of x_m is the proper maximal invariant set that contains x_m; it is the potential well defined by x_m. The potential well consists of all points x_0 such that $\lim_{t\to\infty} x(t) = x_m$. If the domain of attraction has a boundary point x_b, then necessarily $U'(x_b) = 0$. If $x_m < x_b$, the domain of attraction may consist of half the line $x < x_b$, or of an interval $x_a < x < x_b$, where $U'(x_a) = 0$. The boundary of the domain of attraction is called the *separatrix* of the well. Thus the separatrix may consist of one or two points. If $U(x)$ has more than one local minimum, each one has its own domain of attraction.

According to Section 4.4.3, if the dynamics (10.23) is driven by small white noise,

$$\dot{x} = -U'(x) + \sqrt{2\varepsilon}\,\dot{w}, \quad x(0) = x_0, \tag{10.24}$$

almost all trajectories of (10.24), regardless of the initial value x_0, will reach every point in $\mathbb{R}$ in finite time. However, according to Section 10.1.1, the trajectories that start in the domain of attraction of the local minimum x_m will converge to x_m with overwhelming probability before they leave the domain of attraction. Consider, for example, the case that $U(x)$ has a single local minimum x_m and a single local maximum $x_M > x_m$. Then, for any $\delta \gg \sqrt{\varepsilon}$, trajectories with $x_0 < x_M - \delta$ will reach x_m before reaching $x_M + \delta$ with overwhelming probability. Setting $\bar{\tau}_{x_m}$ to be the FPT to x_m and $\bar{\tau}_{x_M+\delta}$ to $x_M + \delta$, we define the *stochastic separatrix* of the domain of attraction of x_m, for this case, as the point $x_0 = x_{SS}$ such that $\Pr\{\tau_{x_m} < \tau_{x_M+\delta} \,|\, x(0) = x_{SS}\} = \frac{1}{2}$. Thus the stochastic separatrix is the starting point for trajectories that are equally likely to reach the bottom of the well first as they are to reach a point outside the well first.

Note that x_M is not necessarily the boundary of the domain of attraction of x_m for the deterministic dynamics (10.23), because $U'(x)$ can have any number of zeros in the ray $x < x_M$, which are not local minima of $U(x)$. Although the deterministic dynamics (10.23) gets stuck at these zeros, the stochastic dynamics (10.24) simply slides past them, because of the noise.

We consider the exit problem from the interval $[x_m + \delta, x_M - \delta]$ in order to determine the stochastic separatrix. This is the problem of exit with the flow discussed in Section 10.1.1. Note that the location of the point δ does not affect the location of the stochastic separatrix to leading order.

The *transmission coefficient* is the probability that a trajectory that reaches the top of the barrier (the point x_M, say) will reach x_m before $x_M + \delta$; that is, that a

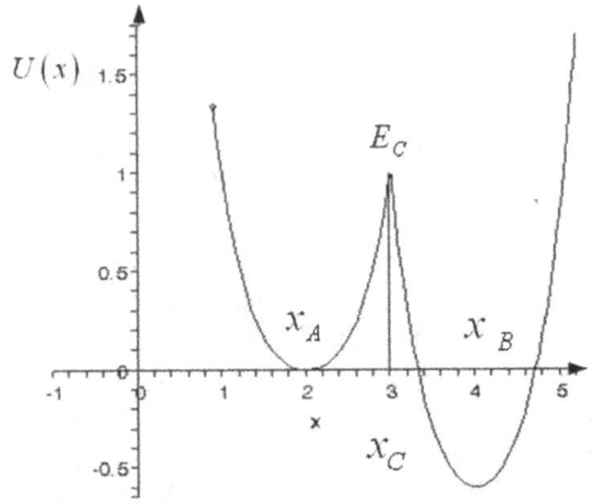

Figure 10.2. A potential $U(x)$ with two wells separated by a sharp barrier at x_C.

trajectory that reached the top actually exits the well. According to eq. (10.20), the transmission coefficient is to leading order frac12. The top of the barrier is often called the *transition state*. □

Exercise 10.7 (The stochastic separatrix). Expand the stochastic separatrix, denoted S, in the form $S = \alpha + \sqrt{\varepsilon} S_1 + \varepsilon S_2 + \cdots$; that is, calculate $S_1, S_2, \cdots$ from the equation $\Pr\{\tau_\beta < \tau_\alpha \,|\, x(0) = S\} = \frac{1}{2}$. □

Exercise 10.8 (The first correction). Find the first correction to the transmission coefficient. □

10.1.2 Small diffusion against the flow

First, we consider the case of sharp boundaries. That is, we assume that the drift points toward ζ at every point in the interval. That is, for all x in the domain of attraction

$$(x - \zeta)\, a\,(x) < 0 \quad \text{for } x \neq \zeta. \tag{10.25}$$

This is the case, for example, of a particle in a double well potential as in Figure 10.2. The attractor is $\zeta = x_A$ and the drift $a(x) = -U'(x)$ is positive for $x < x_A$ and negative for $x_A < x < x_C$ (see Figure 10.3).

The exit problem in this case is that of small diffusion against the flow. In this case the potential $U\,(x)$ has a minimum at ζ and a maximum at one of the boundaries. Thus $U\,(x)$ in eq. (6.45) forms a well in with sharp boundaries. To escape the interval a particle trapped in the well has to acquire from the noise sufficient energy to overcome the potential barrier. It is clear intuitively that it is more likely to escape the well at the side where the barrier is the lowest. If both barriers are of

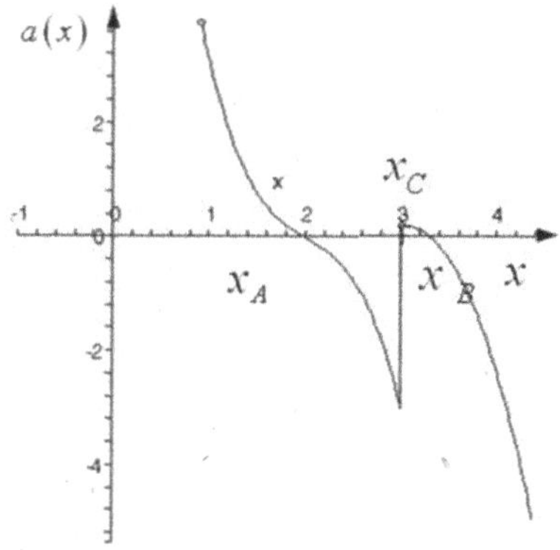

Figure 10.3. At the barrier of the left well of the potential in Figure 10.2 the (negative) drift $a(x_C-) = -U'(x_C-)$ points into the well (toward the attractor x_A).

equal height, a closer analysis is needed. To resolve this case, we assume the expansion (10.18) with $a'(\zeta) < 0$ and obtain an outer solution of the boundary value problem (6.44) as above. In this case, however, the internal layer function diverges to infinity on either side of ζ so that there is no internal layer and the outer solution is a constant throughout the interval. In this case there are boundary layers at both ends of the interval so that the uniform expansion has the form

$$u_{\text{unif}}(x) \sim u_0 + (g_a - u_0)\exp\left\{-\frac{a(\alpha)(x-\alpha)}{\varepsilon\sigma(\alpha)}\right\}$$
$$+ (g_\beta - u_0)\exp\left\{\frac{a(\beta)(\beta - x)}{\varepsilon\sigma(\beta)}\right\}. \qquad (10.26)$$

Note that $a(\alpha) > 0 > a(\beta)$. Note, further, that although the boundary and matching conditions are satisfied by $u_{\text{unif}}(x)$ the constant u_0 is still undetermined. To determine u_0 we use the following identity.

Lemma 10.1.1 (The Lagrange identity). *let $L^{*\varepsilon}$ be the operator defined in eq. (6.44), if v_ε is a solution of $L^\varepsilon v_\varepsilon = 0$ and u_ε is the solution of the adjoint equation $L^{*\varepsilon} u_\varepsilon = 0$, then*

$$\int_\alpha^\beta vL^{*\varepsilon}u\,dx = \varepsilon\,\sigma(x)\,v_\varepsilon(x)\,u_\varepsilon'(x)\Big|_\alpha^\beta - \varepsilon\left[\sigma(x)\,v_\varepsilon(x)\right]' u_\varepsilon(x)\Big|_\alpha^\beta = 0. \quad (10.27)$$

Proof. This identity is obtained by straightforward integration by parts. □

Choosing the solution $v_\varepsilon = e^{-U(x)/\varepsilon}$ and using the approximation (10.26) in

(10.27), we obtain

$$u_0 \sim \frac{e^{-U(\alpha)/\varepsilon}a(\alpha)\,g_a - e^{-U(\beta)/\varepsilon}a(\beta)\,g_\beta}{e^{-U(\alpha)/\varepsilon}a(\alpha) - e^{-U(\beta)/\varepsilon}a(\beta)},$$

having used the fact that

$$\exp\left\{-\frac{U(\beta)}{\varepsilon} + a(\alpha)\frac{-\beta+\alpha}{\varepsilon\sigma(\alpha)}\right\} << e^{-U(\beta)/\varepsilon}.$$

The uniform solution is now given by

$$\begin{aligned} u_{\text{unif}}(x) = & \frac{e^{-U(\alpha)/\varepsilon}a(\alpha)\,g_\alpha - e^{-U(\beta)/\varepsilon}a(\beta)\,g_\beta}{e^{-U(\alpha)/\varepsilon}a(\alpha) - e^{-U(\beta)/\varepsilon}a(\beta)} \\ & + \frac{a(\beta)e^{-U(\beta)/\varepsilon}}{e^{-U(\alpha)/\varepsilon}a(\alpha) - e^{-U(\beta)/\varepsilon}a(\beta)}\exp\left\{-\frac{a(\alpha)(x-\alpha)}{\varepsilon\sigma(\alpha)}\right\} \\ & + \frac{a(\alpha)e^{-U(\alpha)/\varepsilon}}{e^{-U(\alpha)/\varepsilon}a(\alpha) - e^{-U(\beta)/\varepsilon}a(\beta)}\exp\left\{\frac{a(\beta)(\beta-x)}{\varepsilon\sigma(\beta)}\right\} + O(\varepsilon). \end{aligned} \tag{10.28}$$

The probability of exit at β is the coefficient of g_β in (10.28), given by

$$\begin{aligned} & \Pr\{\tau_\beta < \tau_\alpha \,|\, x(0) = x\} \\ & = \frac{-e^{-U(\beta)/\varepsilon}a(\beta)}{e^{-U(\alpha)/\varepsilon}a(\alpha) - e^{-U(\beta)/\varepsilon}a(\beta)}\left(1 - \exp\left\{-\frac{a(\alpha)(x-\alpha)}{\varepsilon\sigma(\alpha)}\right\}\right) \\ & + \frac{a(\alpha)e^{-U(\alpha)/\varepsilon}}{e^{-U(\alpha)/\varepsilon}a(\alpha) - e^{-U(\beta)/\varepsilon}a(\beta)}\exp\left\{\frac{a(\beta)(\beta-x)}{\varepsilon\sigma(\beta)}\right\} + O(\varepsilon). \end{aligned}$$

If $U(\alpha) > U(\beta)$, the first term is exponentially close to 1, except for a boundary layer near α, and the second term is exponentially small throughout the interval. It follows that the exit probability at α is exponentially small in the interval, except for a boundary layer region near α, whereas the probability of exit at β is exponentially close to one, except for a boundary layer near α. This result agrees with (6.63) for the case $m = n = 0$.

If condition (10.25) is satisfied for $\alpha < x < \zeta < \beta$ and for $\alpha < \zeta < x < \beta$, but $a(\alpha) = 0$ or $a(\beta) = 0$, or both, the potential well is not cut off sharply at the edge but rather is cut off smoothly. This happens, for example, if the potential is defined in a larger interval containing $[\alpha, \beta]$ in its interior, it has a local minimum at ζ, and local maxima at α and β, say. This is the typical situation in modeling thermal activation (see Section 10.3). The analysis of this case proceeds in a similar manner to that of the case of sharp boundaries in the sense that the outer solution is still an unknown constant u_0, however, the boundary layer equations are different.

Exercise 10.9 (Flat barrier). Assume, as above, that $a(\alpha) = a'(\alpha) = \cdots = a^{(n-1)}(\alpha) = 0$, but $a^{(n)}(\alpha) < 0$, and $a(\beta) = a'(\beta) = \cdots = a^{(m-1)}(\beta) = 0$, but $a^{(m)}(\beta) > 0$, where m and n are positive integers. Construct the boundary layer near α by introducing the scaled variables $\xi = (x-\alpha)\varepsilon^{-\kappa}$, $U_\varepsilon(\xi) = u_\varepsilon(x)$ and write eq. (6.44) as

$$\varepsilon^{1-2\kappa}\sigma\left(\alpha+\varepsilon^{\kappa}\xi\right)\frac{d^2U_\varepsilon\left(\xi\right)}{d\xi^2}+\varepsilon^{-\kappa}a\left(\alpha+\varepsilon^{\kappa}\xi\right)\frac{dU_\varepsilon\left(\xi\right)}{d\xi}=0. \tag{10.29}$$

Expanding asymptotically all functions in powers of ε, obtain for $\varepsilon << 1$

$$U_\varepsilon\left(\xi\right)=U_0\left(\xi\right)+o\left(1\right),\quad a\left(\alpha+\varepsilon^{\kappa}\xi\right)=\frac{\left(\varepsilon^{\kappa}\xi\right)^n}{n!}a^{(n)}\left(\alpha\right)+O\left(\varepsilon^{\kappa+1}\right),$$
$$\sigma\left(\alpha+\varepsilon^{\kappa}\xi\right)=\sigma\left(\alpha\right)+o\left(1\right).$$

(i) Show that the leading-order term, $U_0\left(\xi\right)$, satisfies the equation

$$\varepsilon^{1-2\kappa}\sigma\left(\alpha\right)\frac{d^2U_0\left(\xi\right)}{d\xi^2}+\varepsilon^{(n-1)\kappa}\frac{\xi^n}{n!}a^{(n)}\left(\alpha\right)\frac{dU_0\left(\xi\right)}{d\xi}=0.$$

(ii) Show that the two terms in the equation are comparable only if $1-2\kappa = (n-1)\kappa$; that is, only if $\kappa = 1/\left(n+1\right)$.

(iii) Choosing this value of κ, obtain

$$U_0\left(\xi\right)=A+B\int_0^{\xi}\exp\left\{-\frac{a^{(n)}\left(\alpha\right)\eta^{n+1}}{(n+1)!\sigma\left(\alpha\right)}\right\}d\eta,$$

where A and B are constants.

(iv) Show that the boundary and matching conditions are $U_0\left(0\right) = g_\alpha$ and $U_0\left(\infty\right) = u_0$, respectively.

(v) Obtain a similar expression for a boundary layer function at β.

(vi) Construct the uniform approximation of $u_\varepsilon\left(x\right)$.

(vii) Note that the two terms in the Lagrange identity (10.27) are not of the same order of magnitude.

(viii) Use the asymptotic expansion in the Lagrange identity and find u_0.

(ix) Find $\Pr\left\{\tau_\beta < \tau_\alpha \,|\, x(0) = x\right\}$ for x in (α, β) such that $(x-\alpha)^{n+1} >> \varepsilon$ and $(\beta - x)^{m+1} >> \varepsilon$; that is, for x outside the boundary layers (see Section 6.4.2).

(x) Find the boundary layer behavior from the asymptotics of the incomplete Gamma function (see Section 6.4.2, eqs. (6.61) and (6.62)). □

Exercise 10.10 (Equal and unequal barriers). Consider the different possible values of $m \geq 0$ and $n \geq 0$ in the boundary layers for equal and unequal potential barriers. □

Exercise 10.11 (Matched asymptotics). Obtain the above results by constructing an asymptotic solution to the boundary value problem (6.48), (6.49) and then using (6.47). Use the method of matched asymptotics, as described above. □

Exercise 10.12 (The first eigenfunction). Prove the following results [168], [213]:
(i) Let $u_\varepsilon(x)$ be the solution of the boundary value problem

$$\varepsilon u'' + a(x) u' = 0 \text{ for } \alpha < x < \beta, \quad u(\alpha) = g_\alpha, \quad u(\beta) = g_\beta,$$

where $a(x) = -U'(x)$. Let $\alpha = y_0 < y_1 < y_2 < \cdots < y_n < y_{n+1} = \beta$ be the local maxima of $U(x)$ in the interval $[\alpha, \beta]$; denote by $s_1 < s_2 < \cdots < s_\ell$ those points y_j where $U(s_j) = \max_i U(y_i)$ $(j = 1, \ldots, \ell)$, and by $x_1 < x_2 < \cdots < x_m$ those points s_j where $a(x)$ vanishes to maximal order k (maximum with respect to j, $1 \le j \le m$). Assume that the local Taylor expansion of $a(x)$ about x_i is given by $a(x) = A_i (x - x_i)^k + \cdots$, where $A_i > 0$ and denote $B_i = (A_i)^{1/(k+1)}$. Then

$$u_\varepsilon(x) \sim \sum_{j=0}^{m-1} C_j \chi_{(x_j, x_{j+1})}(x) \text{ for } \varepsilon << 1 \text{ and fixed } x_j < x < x_{j+1},$$

where the indicator function is defined by

$$\chi_{(x_j, x_{j+1})}(x) = \begin{cases} 1 & \text{for } x_j < x < x_{j+1} \\ 0 & \text{otherwise,} \end{cases}$$

and

$$C_j = \frac{g_\alpha P_{m,j} + g_\beta Q_{m,j}}{P_m}, \quad P_{n,j} = \sum_{i=0}^{j} P_i, \quad Q_{n,j} = P_n - P_{n,j}$$

$$P_n = \sum_{j=0}^{n+1} \left[\left(\prod_{i=0}^{n+1} B_i \right) / B_{n-j+1} \right] = \sum_{j=0}^{n+1} P_j.$$

Construct boundary and internal layers to connect the constants C_j in adjacent intervals.
(ii) Show that

$$\Pr\{\tau_\alpha < \tau_\beta \,|\, x(0) = x\} \sim \sum_{i=0}^{m-1} \frac{P_{m,j}}{P_m} \chi_{(x_j, x_{j+1})}(x).$$

(see [218], [168], [213]). □

10.1.3 The MFPT of small diffusion against the flow

To obtain approximate expressions for the mean exit time and for its PDF, we return to the boundary value problem (6.48), (6.49) and to the expression (6.18) for the MFPT. To be specific, we assume that the potential $U(y)$ in eq. (6.45) forms a single well in the interval $[\alpha, \beta]$. First, we establish the asymptotic form of the solution to the boundary value problem (6.48), (6.49) by rewriting the explicit solution (6.50) in the form $p_\varepsilon(y\,|\,x) = \left[e^{-U(y)/\varepsilon}/\sigma\,(y)\right] q_\varepsilon(y\,|\,x)$. We obtain

$$q_\varepsilon(y\,|\,x) = \begin{cases} \dfrac{\displaystyle\int_x^\beta \exp\left\{\frac{U(s)}{\varepsilon}\right\} ds \int_\alpha^y \exp\left\{\frac{U(s)}{\varepsilon}\right\} ds}{\displaystyle\varepsilon \int_\alpha^\beta \exp\left\{\frac{U(s)}{\varepsilon}\right\} ds} & \text{for } \alpha < y < x < \beta \\[4ex] \dfrac{\displaystyle\int_\alpha^x \exp\left\{\frac{U(s)}{\varepsilon}\right\} ds \int_y^\beta \exp\left\{\frac{U(s)}{\varepsilon}\right\} ds}{\displaystyle\varepsilon \int_\alpha^\beta \exp\left\{\frac{U(s)}{\varepsilon}\right\} ds} & \text{for } \alpha < x < y < \beta. \end{cases}$$

10.1.4 Escape over a sharp barrier

If $U(y)$ forms a well and $U(\alpha) < U(\beta) \max_{[\alpha,\beta]} U(y)$, and the boundaries are sharp; that is, $U'(\alpha) < 0$, and $U'(\beta) > 0$, the function $q_\varepsilon(y\,|\,x)$ can be approximated asymptotically for small ε. For small ε

$$\frac{\displaystyle\int_x^\beta \exp\left\{\frac{U(s)}{\varepsilon}\right\} ds}{\displaystyle\int_\alpha^\beta \exp\left\{\frac{U(s)}{\varepsilon}\right\} ds} \sim 1,$$

so that for $\alpha < y < x < \beta$ and y close to α (so that $U\,(y) < U\,(\alpha)$)

$$q_\varepsilon(y\,|\,x) = \frac{\displaystyle\int_x^\beta \exp\left\{\frac{U(s)}{\varepsilon}\right\} ds \int_\alpha^y \exp\left\{\frac{U(s)}{\varepsilon}\right\} ds}{\displaystyle\varepsilon \int_\alpha^\beta \exp\left\{\frac{U(s)}{\varepsilon}\right\} ds}$$

$$\sim \frac{\exp\left\{\dfrac{U(\alpha)}{\varepsilon}\right\}}{-U'\,(\alpha)} \left(1 - e^{U'(\alpha)(y-\alpha)/\varepsilon}\right). \tag{10.30}$$

For $\alpha < x < y < \beta$ and y close to β

$$q_\varepsilon(y\,|\,x) = \frac{\displaystyle\int_\alpha^x \exp\left\{\frac{U(s)}{\varepsilon}\right\} ds \int_y^\beta \exp\left\{\frac{U(s)}{\varepsilon}\right\} ds}{\displaystyle\varepsilon \int_\alpha^\beta \exp\left\{\frac{U(s)}{\varepsilon}\right\} ds} \tag{10.31}$$

$$\sim \frac{\left(1 - e^{U'(\beta)(y-\beta)/\varepsilon}\right)}{\varepsilon} \int_\alpha^x \exp\left\{\frac{U(s)}{\varepsilon}\right\} ds$$

$$= \left(1 - e^{U'(\beta)(y-\beta)/\varepsilon}\right) \begin{cases} \dfrac{\exp\left\{\dfrac{U(x)}{\varepsilon}\right\}}{-U'(x)} & \text{if} \quad U(\alpha) < U(x) \\[2ex] \dfrac{\exp\left\{\dfrac{U(\alpha)}{\varepsilon}\right\}}{-U'(\alpha)} & \text{if} \quad U(x) < U(\alpha). \end{cases}$$

Thus, if $U(x) < U(\alpha)$, then for all $\alpha < y < \beta$

$$\lim_{\varepsilon\to 0} \exp\left\{-\frac{U(\alpha)}{\varepsilon}\right\} q_\varepsilon(y\,|\,x) = \frac{1}{-U'(\alpha)}.$$

That is, for y outside boundary layers and for x in the domain $U(x) < U(\alpha)$, this result means that

$$p_\varepsilon(y\,|\,x) \sim C_\varepsilon \frac{e^{-U(y)/\varepsilon}}{\sigma(y)} = p_{\text{outer}}(y), \tag{10.32}$$

where $C_\varepsilon = -e^{U(\alpha)/\varepsilon}/U'(\alpha)$. The result (10.32) means that for a potential $U(y)$ that forms a single well the outer solution to the boundary value problem (6.48), (6.49) is the solution to the asymptotic problem

$$\frac{\partial^2 \varepsilon\sigma(y)\, p_{\text{outer}}(y)}{\partial y^2} - \frac{\partial a(y)\, p_{\text{outer}}(y)}{\partial y} \sim 0 \text{ for } \alpha < x, y < \beta \tag{10.33}$$

given in eq. (10.32). According to eqs. (10.30) and (10.31), the boundary layers for $q_\varepsilon(y\,|\,x)$ have the form

$$q_\varepsilon(y\,|\,x) \sim \frac{\exp\left\{\dfrac{U(\alpha)}{\varepsilon}\right\}}{-U'(\alpha)} \left(1 - e^{U'(\alpha)(y-\alpha)/\varepsilon}\right) \text{ for } y \text{ near } \alpha \tag{10.34}$$

$$q_\varepsilon(y\,|\,x) \sim \frac{\exp\left\{\dfrac{U(\alpha)}{\varepsilon}\right\}}{-U'(\alpha)} \left(1 - e^{U'(\beta)(y-\beta)/\varepsilon}\right) \text{ for } y \text{ near } \beta; \tag{10.35}$$

that is, the uniform asymptotic expansion of $p_\varepsilon(y\,|\,x)$ in the domains $\alpha < y < x < \beta$, and $\alpha < x < y < \beta$, $\quad U(x) < U(\alpha)$, is given by

$$p_\varepsilon(y\,|\,x) \sim \frac{e^{-U(y)/\varepsilon}}{\sigma(y)} \frac{\exp\left\{\dfrac{U(\alpha)}{\varepsilon}\right\}}{-U'(\alpha)} \left(1 - e^{U'(\alpha)(y-\alpha)/\varepsilon}\right) \quad \text{for } \alpha < y < x < \beta$$

$$p_\varepsilon(y\,|\,x) \sim \frac{e^{-U(y)/\varepsilon}}{\sigma(y)} \frac{\exp\left\{\dfrac{U(\alpha)}{\varepsilon}\right\}}{-U'(\alpha)} \left(1 - e^{U'(\beta)(y-\beta)/\varepsilon}\right) \quad \text{for } \alpha < x < y < \beta.$$

It is argued below that this is the general asymptotic structure of Green's function of the Fokker–Planck equation in a domain D, with homogeneous conditions on its boundary ∂D, when the drift vector field has a single attractor in the domain. The construction of this expansion by direct asymptotic analysis of the differential equation is now straightforward. First, construct the outer solution, $p_{\text{outer}}(y)$, in the WKB form (10.32) and transform the FPE (10.32) to the backward Kolmogorov equation by the substitution $p_\varepsilon(y\,|\,x) = p_{\text{outer}}(y)\, q_\varepsilon(y)$ to obtain

$$\varepsilon\sigma(y)\, q_\varepsilon''(y) + a(y)\, q_\varepsilon'(y) \sim 0 \quad \text{for } y \in D,\ \varepsilon << 1 \tag{10.36}$$

with the matching and boundary conditions

$$q_\varepsilon(y) \sim C_\varepsilon \quad \text{for } y \in D,\ \varepsilon << 1, \quad q_\varepsilon|_{\partial D} = 0. \tag{10.37}$$

The boundary layers are constructed by introducing the stretched variable $\xi = \rho/\varepsilon$ and define $Q_\varepsilon(\xi) = q_\varepsilon(y)$, where $\rho = \text{dist.}(y, \partial D)$ and by expanding $Q_\varepsilon(\xi) = Q_0(\xi) + \varepsilon Q_1(\xi) + \cdots$. Thus, $\rho = y - \alpha$ near α and the backward Kolmogorov equation (10.36) takes the form

$$\sigma(\alpha)\, Q_0''(\xi) + a(\alpha)\, Q_0'(\xi) \sim 0, \quad Q_0(0) = 0, \quad \lim_{\xi\to\infty} Q_0(\xi) = C_\varepsilon.$$

The solution is given by $Q_0(\xi) = C_\varepsilon(1 - e^{U'(\alpha)\xi})$, or equivalently, (10.34). The expression (10.35) is obtained in an analogous manner. We rewrite these expressions in the unified form

$$\begin{aligned} p_\varepsilon(y\,|\,x) \sim p_{\text{unif}}(y\,|\,x) &= p_{\text{outer}}(y)\, q_\varepsilon(y) \\ &= C_\varepsilon \frac{e^{-U(y)/\varepsilon}}{\sigma(y)} \left(1 - e^{U_n \rho/\varepsilon}\right), \end{aligned} \tag{10.38}$$

where U_n is the outer normal derivative of $U(y)$ at the boundary point nearest y.

Finally, the MFPT is obtained from eq. (6.18) with the asymptotic values (10.38). The normal component of the total flux is given by

$$\begin{aligned} F_\varepsilon(x) &\sim -\varepsilon\sigma(\beta) \frac{\partial p_{\text{unif}}(\beta\,|\,x)}{\partial y} + \varepsilon\sigma(\alpha) \frac{\partial p_{\text{unif}}(\alpha\,|\,x)}{\partial y} \\ &= C_\varepsilon \left[U'(\beta)\, e^{-U(\beta)/\varepsilon} - U'(\alpha)\, e^{-U(\alpha)/\varepsilon}\right] \sim -C_\varepsilon U'(\alpha)\, e^{-U(\alpha)/\varepsilon}, \end{aligned} \tag{10.39}$$

because $e^{-U(\beta)/\varepsilon} << e^{-U(\alpha)/\varepsilon}$ if $U(\alpha) < U(\beta)$. The total population is evaluated by the Laplace method as

$$N_\varepsilon(x) \sim C_\varepsilon \int_\alpha^\beta \frac{e^{-U(y)/\varepsilon}}{\sigma(y)} \left(1 - e^{U_n \rho/\varepsilon}\right) dy \sim C_\varepsilon \sqrt{\frac{2\pi\varepsilon}{U''(\zeta)}} \frac{e^{-U(\zeta)/\varepsilon}}{\sigma(\zeta)}, \tag{10.40}$$

where ζ is the point of minimum of $U(y)$ in the domain. Note that neither the total flux $F(x)$ nor the total population $N_\varepsilon(x)$ depends on x to leading order, as long as x is not in a boundary layer. Now, eq. (6.18) gives

$$\tau_\varepsilon(x) = \frac{N_\varepsilon(x)}{F_\varepsilon(x)} \sim \sqrt{\frac{2\pi\varepsilon}{U''(\zeta)\sigma^2(\zeta)|U'(\alpha)|^2}} e^{[U(a)-U(\zeta)]/\varepsilon}. \tag{10.41}$$

The expression (10.41) is often written in the form

$$\tau_\varepsilon = \Omega^{-1} e^{\Delta E/\varepsilon}, \tag{10.42}$$

where $\Delta E = U(a) - U(\zeta)$ is the height of the lowest potential barrier, and

$$\Omega = \sqrt{\frac{U''(\zeta)\sigma^2(\zeta)|U'(\alpha)|^2}{2\pi\varepsilon}}$$

is the so-called *attempt frequency*. Recalling the definition (6.11) of the escape rate, we obtain $\kappa_\varepsilon = \tau_\varepsilon^{-1} = De^{-\Delta E/\varepsilon}$.

Note that the value of the constant C_ε does not enter the expression (10.42) for the MFPT to the boundary. The value $C_\varepsilon = -e^{U(\alpha)/\varepsilon}/U'(\alpha)$ can be easily deduced from eq. (10.39) by recalling that the total flux is 1,

$$1 = F(x) = C_\varepsilon \left[U'(\beta)e^{-U(\beta)/\varepsilon} - U'(\alpha)e^{-U(\alpha)/\varepsilon}\right] \sim -C_\varepsilon U'(\alpha)e^{-U(\alpha)/\varepsilon}.$$

10.1.5 The MFPT to a smooth boundary and the escape rate

At a smooth boundary $a(\alpha) = 0$, which usually happens when the potential $U(y)$ has a local maximum at the boundary of the well. In this case, we assume that $a'(\alpha) = a''(\alpha) = \cdots = a^{(n-1)}(\alpha) = 0$, but $a^{(n)}(\alpha) > 0$. We assume that $U(y)$ has a global maximum at $y = \alpha$. We proceed as in Section 10.1; that is, the local Taylor expansion of $a(y)$ about α is given by $a(y) = [a^{(n)}(\alpha)/n!](y-\alpha)^n + \cdots$ and the scaled variable $\xi = (y-\alpha)/\varepsilon^\kappa$ transforms the asymptotic boundary value problem (10.36)–(10.37) into

$$\varepsilon^{1-2\kappa}\sigma(\alpha)Q_0''(\xi) + \varepsilon^{(n-1)\kappa}a^{(n)}(\alpha)\xi^n Q_0'(\xi) \sim 0 \tag{10.43}$$

$$Q_0(0) = 0, \quad \lim_{\xi\to\infty} Q_0(\xi) = C_\varepsilon, \tag{10.44}$$

where $q_\varepsilon(y) = Q_\varepsilon(\xi) = Q_0(\xi) + \varepsilon^\kappa Q_1(\xi) + \cdots$. The two terms in eq. (10.43) are comparable if $1 - 2\kappa = (n-1)\kappa$; that is, if $\kappa = 1/(n+1)$. Then eqs. (10.43), (10.44) reduce to

$$\sigma(\alpha) Q_0''(\xi) + a^{(n)}(\alpha)\xi^n Q_0'(\xi) \sim 0, \quad Q_0(0) = 0, \quad \lim_{\xi\to\infty} Q_0(\xi) = C_\varepsilon,$$

whose solution is

$$Q_0(\xi) = \frac{C_\varepsilon A^{1/(n+1)}(n+1)}{\Gamma\left(\frac{1}{n+1}\right)} \int_0^\xi e^{-Ax^{n+1}}\,dx,$$

where $A = a^{(n)}(\alpha)/\sigma(\alpha)(n+1)$. Now,

$$\begin{aligned} Q_0'(0) &= \left(\frac{a^{(n)}(\alpha)}{\sigma(\alpha)}\right)^{1/(n+1)} \frac{(n+1)^{n/(n+1)}}{\Gamma\left(\frac{1}{n+1}\right)} \\ &= \left[U^{(n+1)}(\alpha)\right]^{1/(n+1)} \frac{(n+1)^{n/(n+1)}}{\Gamma\left(\frac{1}{n+1}\right)} \end{aligned}$$

and the total flux at the boundary in eq. (10.39), $F_\varepsilon|_{\partial D}$, is given by

$$\begin{aligned} F_\varepsilon|_{\partial D} &= -\varepsilon\sigma(y)\,p_{\text{outer}}(y) \left.\frac{\partial q_\varepsilon(y)}{\partial n_y}\right|_{\partial D} \\ &= C_\varepsilon \left[\varepsilon^{n/(n+1)} \exp\left\{-\frac{U(\alpha)}{\varepsilon}\right\} \left[U^{(n+1)}(\alpha)\right]^{1/(n+1)} \frac{(n+1)^{n/(n+1)}}{\Gamma\left(\frac{1}{n+1}\right)}\right. \\ &\quad \left. + \varepsilon^{m/(m+1)} \exp\left\{-\frac{U(\beta)}{\varepsilon}\right\} \left[U^{(m+1)}(\beta)\right]^{1/(m+1)} \frac{(m+1)^{m/(m+1)}}{\Gamma\left(\frac{1}{m+1}\right)}\right] \\ &\sim C_\varepsilon \varepsilon^{n/(n+1)} \exp\left\{-\frac{U(\alpha)}{\varepsilon}\right\} \left[U^{(n+1)}(\alpha)\right]^{1/(n+1)} \frac{(n+1)^{n/(n+1)}}{\Gamma\left(\frac{1}{n+1}\right)}. \end{aligned}$$

To calculate the total population, $N_\varepsilon(D)$, we assume that the local expansion of $U(y)$ near its point of minimum in the interval, ζ, is

$$U(y) = U(\zeta) + \frac{U^{(2\ell)}(\zeta)}{(2\ell)}(y-\zeta)^{2\ell} + \cdots,$$

where $U^{(2\ell)}(\zeta) > 0$ and ℓ is a positive integer. Then the integral in eq. (10.40) gives for sufficiently small ε

$$N_\varepsilon(D) \sim \frac{C_\varepsilon e^{-U(\zeta)/\varepsilon}}{\sigma(\zeta)} \int_\alpha^\beta \exp\left\{-\frac{U^{(2\ell)}(\zeta)}{(2\ell)!\varepsilon}(y-\zeta)^{2\ell}\right\} dy$$
$$\sim \frac{C_\varepsilon e^{-U(\zeta)/\varepsilon}}{\sigma(\zeta)} \left(\frac{(2\ell)!\varepsilon}{U^{(2\ell)}(\zeta)}\right)^{1/2\ell} \Gamma\left(\frac{1}{2\ell}\right).$$

The resulting MFPT is given by

$$\tau_\varepsilon(D) = \frac{N_\varepsilon(D)}{F_\varepsilon(D)}$$
$$\sim \frac{((2\ell)!)^{1/2\ell}\,\Gamma\left(\frac{1}{2\ell}\right)\Gamma\left(\frac{1}{n+1}\right)}{(n+1)^{\frac{n}{n+1}}\sigma(\zeta)} \frac{\varepsilon^{1/2\ell+1/(n+1)-1}}{\left[U^{(2\ell)}(\zeta)\right]^{1/2\ell}\left[U^{(n+1)}(\alpha)\right]^{1/(n+1)}}$$
$$\times \exp\left\{\frac{U(\alpha)-U(\zeta)}{\varepsilon}\right\}.$$

In the typical case of $n = \ell = 1$ this reduces to

$$\tau_\varepsilon(D) \sim \frac{\pi}{\omega_W \omega_B} e^{\Delta E/\varepsilon},$$

where the frequency at the bottom of the well and the imaginary frequency at the top of the potential barrier are given, respectively, by $\omega_W^2 = U''(\zeta)$ and $\omega_B^2 = -U''(\alpha)$. This gives the arrival rate from the domain D to the boundary, because

$$\kappa_\varepsilon(D) = \tau_\varepsilon^{-1}(D) = \frac{\omega_W \omega_B}{\pi} e^{-\Delta E/\varepsilon}.$$

Note that κ_e is not the *escape rate* from the potential well, but rather the rate at which trajectories reach the top of the barrier. The top of the barrier (the transition state) is to leading order the stochastic separatrix; that is, only 50% of the trajectories that reach the transition state return to the well before escaping into the next well or into the continuum outside. Thus the *mean escape time*, denoted $\bar{\tau}_{\text{escape}}$, is twice the MFPT $\tau_\varepsilon(D)$; that is, the mean escape time is

$$\tau_{\text{escape}} = 2\tau_\varepsilon(D) = \frac{2\pi}{\omega_W \omega_B} e^{\Delta E/\varepsilon}$$

and the escape rate, denoted κ_{escape}, is given by

$$\kappa_{\text{escape}} = \frac{1}{2}\kappa_\varepsilon(D). \tag{10.45}$$

Note that all variables in the above analysis are dimensionless. In physical units the Arrhenius chemical reaction rate in the overdamped regime is given by

$$\kappa_{\text{Arrhenius}} = \frac{1}{2}\kappa_\varepsilon(D) = \frac{\omega_W \omega_B}{2\pi\gamma} e^{-\Delta E/\varepsilon}, \tag{10.46}$$

where ω_W and ω_B have the dimension of frequency. The expression (10.46) was first derived by Kramers in 1940 [140] for a model of a chemical reaction (e.g., dissociation) as overdamped diffusion over a potential barrier (the so-called *thermal activation*).

10.1.6 The MFPT eigenvalues of the Fokker–Planck operator

If the potential $U(x)$ forms a single well in the interval $[\alpha, \beta]$, the exit time from the interval is exponentially large as a function of $1/\varepsilon$. It follows from Section 6.1 that the principal (the smallest) eigenvalue of the Fokker–Planck operator in the interval with absorbing boundary conditions is exponentially small. If the potential forms a sequence of wells in the interval, there may be several exponentially decaying eigenvalues for $\varepsilon << 1$. In particular, it can be shown [213] that the second eigenvalue has the form $\lambda_2 \sim \Omega_2 e^{-\Delta E_2/\varepsilon}$, where ΔE_2 is the height of the highest barrier a trajectory has to cross in order to reach the bottom of the deepest well in the interval. The coefficient Ω_2 can be expressed in terms of the second derivatives of U at its points of local minimum and local maximum.

Exercise 10.13 (The second eigenfunction). Use the method of Exercise 10.12 to construct an asymptotic expansion of the first and second eigenfunctions of the Fokker–Planck equation with absorbing boundary when the potential forms multiple wells (see [213]). □

10.2 The exit problem in higher dimensions

As in the one-dimensional case, we consider an autonomous dynamical system driven by small noise. We assume that the noiseless dynamics has a global attractor in a given domain. The exit problem is to determine the MFPT and the exit distribution of the random trajectories on the boundary of the domain. This problem leads to a singularly perturbed elliptic boundary value problem in the domain. The solution of the exit problem is based on the construction of a uniform asymptotic approximation to the solution of the stationary FPE with a source in the domain and absorption on its boundary. Specifically, we consider the autonomous multidimensional system

$$d\boldsymbol{x} = \boldsymbol{a}(\boldsymbol{x})\,dt + \sqrt{2\varepsilon}\,\boldsymbol{B}(\boldsymbol{x})\,d\boldsymbol{w}(t), \quad \boldsymbol{x}(0) = \boldsymbol{x}, \tag{10.47}$$

in a domain D in $\boldsymbol{R}^d$ for flows $\boldsymbol{a}(\boldsymbol{x})$ that cross the boundary ∂D of the domain. We assume that the noiseless dynamics

$$\dot{\boldsymbol{x}} = \boldsymbol{a}(\boldsymbol{x}) \tag{10.48}$$

has a unique critical point $\boldsymbol{x}_0$ in D and it is a global attractor. This means that $\boldsymbol{a}(\boldsymbol{x}_0) = \boldsymbol{0}$ and we assume that the eigenvalues of the matrix

$$\boldsymbol{A} = \left\{ \frac{\partial a^i(\boldsymbol{x}_0)}{\partial x^j} \right\}_{i,j=1}^{d} \tag{10.49}$$

of the linearized system $\dot{\boldsymbol{z}} = \boldsymbol{A}\boldsymbol{z}$, have negative real parts. Thus the trajectories of the system (10.48) that start in D cannot reach ∂D. The case of other attractors, such as limit cycles, is considered separately. We distinguish between the case that the flow on ∂D points into D, and the case that its normal component vanishes on ∂D. In the former case the boundary is called *noncharacteristic* and in the latter case it is *characteristic*. Denoting by $\boldsymbol{\nu}(\boldsymbol{x})$ the unit outer normal at the boundary, we distinguish between the two cases according to the inequalities

$$\boldsymbol{a}(\boldsymbol{x}) \cdot \boldsymbol{\nu}(\boldsymbol{x}) < 0 \text{ for } \boldsymbol{x} \in \partial D, \tag{10.50}$$

if ∂D is noncharacteristic, and

$$\boldsymbol{a}(\boldsymbol{x}) \cdot \boldsymbol{\nu}(\boldsymbol{x}) = 0 \text{ for } \boldsymbol{x} \in \partial D,. \tag{10.51}$$

if ∂D is characteristic.

The Fokker–Planck equation for the stationary pdf $p_\varepsilon(\boldsymbol{y} \,|\, \boldsymbol{x})$ of the solution $\boldsymbol{x}(t, \varepsilon)$ of eq. (10.47) with a source at $\boldsymbol{x}$ and absorption in ∂D, is

$$\begin{aligned}
-\sum_{i=1}^{d} \frac{\partial\left[a^i(\boldsymbol{y})\, p_\varepsilon(\boldsymbol{y} \,|\, \boldsymbol{x})\right]}{\partial y^i} &+ \sum_{i,j=1}^{d} \varepsilon \frac{\partial^2\left[\sigma^{i,j}(\boldsymbol{y})\, p_\varepsilon(\boldsymbol{y} \,|\, \boldsymbol{x})\right]}{\partial y^i \partial y^j} \\
= &- \delta(\boldsymbol{y} - \boldsymbol{x}),
\end{aligned} \tag{10.52}$$

where $\boldsymbol{\sigma}(\boldsymbol{y}) = \boldsymbol{B}(\boldsymbol{y})\, \boldsymbol{B}^T(\boldsymbol{y})$. It can also be written as the conservation law

$$\begin{aligned}
\nabla_{\boldsymbol{y}} \cdot \boldsymbol{J}(\boldsymbol{y} \,|\, \boldsymbol{x}) &= \delta(\boldsymbol{y} - \boldsymbol{x}) \\
J^i(\boldsymbol{y} \,|\, \boldsymbol{x}) &= a^i(\boldsymbol{y})\, p_\varepsilon(\boldsymbol{y} \,|\, \boldsymbol{x}) - \varepsilon \sum_{j=1}^{d} \frac{\partial\left[\sigma^{i,j}(\boldsymbol{y}) p_\varepsilon(\boldsymbol{y} \,|\, \boldsymbol{x})\right]}{\partial y^j}.
\end{aligned} \tag{10.53}$$

The function $p_\varepsilon(\boldsymbol{y} \,|\, \boldsymbol{x})$ satisfies the absorbing boundary condition

$$p_\varepsilon(\boldsymbol{y} \,|\, \boldsymbol{x}) = 0 \text{ for } \boldsymbol{y} \in \partial D,\ \boldsymbol{x} \in D. \tag{10.54}$$

It was shown in Chapter 6 that the exit density at a point $\boldsymbol{y}$ on the boundary of the trajectories of (10.47) that start at a point $\boldsymbol{x} \in D$ is given by

$$\Pr\left\{\boldsymbol{x}(\tau) \in \boldsymbol{y} + dS_{\boldsymbol{y}} \,|\, \boldsymbol{x}(0) = \boldsymbol{x}\right\} = \frac{\boldsymbol{J}(\boldsymbol{y} \,|\, \boldsymbol{x}) \cdot \boldsymbol{\nu}(\boldsymbol{y})\, dS_{\boldsymbol{y}}}{\displaystyle\oint_{\partial D} \boldsymbol{J}(\boldsymbol{y} \,|\, \boldsymbol{x}) \cdot \boldsymbol{\nu}(\boldsymbol{y})\, dS_{\boldsymbol{y}}}, \tag{10.55}$$

and the MFPT to the boundary is given by

$$\bar{\tau}_\varepsilon(\boldsymbol{x}) = \frac{\displaystyle\int_D p_\varepsilon(\boldsymbol{y} \,|\, \boldsymbol{x})\, d\boldsymbol{y}}{\displaystyle\oint_{\partial D} \boldsymbol{J}(\boldsymbol{y} \,|\, \boldsymbol{x}) \cdot \boldsymbol{\nu}(\boldsymbol{y})\, dS_{\boldsymbol{y}}}. \tag{10.56}$$

Thus a uniform approximation to $p_\varepsilon(\boldsymbol{y} \,|\, \boldsymbol{x})$ will provide a full solution to the exit problem through eqs. (10.55), (10.56).

10.2.1 The WKB structure of the pdf

The function $p_\varepsilon(\boldsymbol{y} \mid \boldsymbol{x})$ develops singularities in the domain and on its boundary as $\varepsilon \to 0$. We resolve these singularities by constructing an approximate solution that contains all the singularities of $p_\varepsilon(\boldsymbol{y} \mid \boldsymbol{x})$ in the limit of small ε. To this end, we first transform the Fokker–Planck equation (10.52) by seeking a solution in the WKB form [182]

$$p_\varepsilon(\boldsymbol{y} \mid \boldsymbol{x}) = K_\varepsilon(\boldsymbol{y} \mid \boldsymbol{x})\exp\left\{-\frac{\psi(\boldsymbol{y})}{\varepsilon}\right\} \tag{10.57}$$

with unknown functions $K_\varepsilon(\boldsymbol{y} \mid \boldsymbol{x})$ and $\psi(\boldsymbol{y})$. The essential singularity of $p_\varepsilon(\boldsymbol{y} \mid \boldsymbol{x})$ inside D is captured by the exponential term in (10.57) and that on ∂D by the pre-exponential factor $K_\varepsilon(\boldsymbol{y} \mid \boldsymbol{x})$. Substituting (10.57) in eq. (10.57) and collecting like powers of ε, we obtain at the leading order the first-order eikonal equation

$$\sum_{i,j=1}^{d} \sigma^{i,j}(\boldsymbol{y})\frac{\partial\psi(\boldsymbol{y})}{\partial y^i}\frac{\partial\psi(\boldsymbol{y})}{\partial y^j} + \sum_{i=1}^{d} a^i(\boldsymbol{y})\frac{\partial\psi(\boldsymbol{y})}{\partial y^i} = 0. \tag{10.58}$$

The eikonal equation has the form of a Hamilton–Jacobi equation and is solved by the method of characteristics [45], [229] or by optimizing an appropriate action functional, as done in large deviations theory [77], [76], [68], [55], [54].

The function $K_\varepsilon(\boldsymbol{y} \mid \boldsymbol{x})$ is a regular function of ε for $\boldsymbol{x}$ and $\boldsymbol{y}$ in the domain and develops singularities at ∂D. The boundary condition (10.54) implies the boundary condition

$$K_\varepsilon(\boldsymbol{y} \mid \boldsymbol{x}) = 0 \quad \text{for } \boldsymbol{y} \in \partial D,\ \boldsymbol{x} \in D. \tag{10.59}$$

To resolve these singularities, we decompose the function $K_\varepsilon(\boldsymbol{y} \mid \boldsymbol{x})$ further into the product

$$K_\varepsilon(\boldsymbol{y} \mid \boldsymbol{x}) = \left[K_0(\boldsymbol{y} \mid \boldsymbol{x}) + \varepsilon K_1(\boldsymbol{y} \mid \boldsymbol{x}) + \cdots\right] q_\varepsilon(\boldsymbol{y} \mid \boldsymbol{x}), \tag{10.60}$$

where $K_0(\boldsymbol{y} \mid \boldsymbol{x})$, $K_1(\boldsymbol{y} \mid \boldsymbol{x})$, ... are regular functions in D and on its boundary and are independent of ε, and $q_\varepsilon(\boldsymbol{y} \mid \boldsymbol{x})$ is a boundary layer function. The functions $K_j(\boldsymbol{y} \mid \boldsymbol{x})$ $(j = 0, 1, \dots)$, satisfy first-order partial differential equations and therefore cannot satisfy the boundary condition (10.59). The boundary layer function $q_\varepsilon(\boldsymbol{y} \mid \boldsymbol{x})$ satisfies the boundary condition

$$q_\varepsilon(\boldsymbol{y} \mid \boldsymbol{x}) = 0 \ \text{ for } \boldsymbol{y} \in \partial D,\ \boldsymbol{x} \in D, \tag{10.61}$$

the matching condition

$$\lim_{\varepsilon\to 0} q_\varepsilon(\boldsymbol{y} \mid \boldsymbol{x}) = 1 \ \text{ for all } \boldsymbol{x}, \boldsymbol{y} \in D,\ \boldsymbol{x} \neq \boldsymbol{y}, \tag{10.62}$$

and the smoothness condition

$$\lim_{\varepsilon\to 0} \frac{\partial^i q_\varepsilon(\boldsymbol{y} \mid \boldsymbol{x})}{\partial (y^j)^i} = 0, \ \text{ for all } \boldsymbol{x}, \boldsymbol{y} \in D,\ \boldsymbol{x} \neq \boldsymbol{y},\ i \geq 1,\ 1 \leq j \leq d. \tag{10.63}$$

The function $K_\varepsilon(\boldsymbol{y}\,|\,\boldsymbol{x})$ satisfies the equation

$$\begin{aligned}
&\varepsilon\sum_{i,j=1}^{d}\frac{\partial^2\sigma^{i,j}(\boldsymbol{y})K_\varepsilon(\boldsymbol{y}\,|\,\boldsymbol{x})}{\partial y^i\partial y^j}\\
&-\sum_{i=1}^{d}\left(2\sum_{j=1}^{d}\sigma^{i,j}(\boldsymbol{y})\frac{\partial\psi(\boldsymbol{y})}{\partial y^j}+a^i(\boldsymbol{y})\right)\frac{\partial K_\varepsilon(\boldsymbol{y}\,|\,\boldsymbol{x})}{\partial y^i}\\
&-\sum_{i=1}^{d}\left(\frac{\partial a^i(\boldsymbol{y})}{\partial y^i}+\sum_{j=1}^{d}\left(\sigma^{i,j}(\boldsymbol{y})\frac{\partial^2\psi(\boldsymbol{y})}{\partial y^i\partial y^j}+2\frac{\partial\sigma^{i,j}(\boldsymbol{y})}{\partial y^j}\frac{\partial\psi(\boldsymbol{y})}{\partial y^j}\right)\right)K_\varepsilon(\boldsymbol{y}\,|\,\boldsymbol{x})\\
=&-\delta(\boldsymbol{y}-\boldsymbol{x}).
\end{aligned}$$

The equation for $q_\varepsilon(\boldsymbol{y}\,|\,\boldsymbol{x})$ is derived and studied in Section 10.2.3 below.

10.2.2 The eikonal equation

The eikonal function can be constructed by solving the eikonal equation

$$\sum_{i,j=1}^{d}\sigma^{i,j}(\boldsymbol{y})\frac{\partial\psi(\boldsymbol{y})}{\partial y^i}\frac{\partial\psi(\boldsymbol{y})}{\partial y^j}+\sum_{i=1}^{d}a^i(\boldsymbol{y})\frac{\partial\psi(\boldsymbol{y})}{\partial y^i}=0 \tag{10.64}$$

by the method of characteristics [45], [229]. In this method a first-order partial differential equation of the form

$$F(\boldsymbol{x},\psi,\boldsymbol{p})=0, \tag{10.65}$$

with $\boldsymbol{p}=\nabla\psi(\boldsymbol{x})$, is converted into a system of ordinary differential equations as follows,

$$\begin{aligned}
\frac{d\boldsymbol{x}}{ds}=&\nabla_{\boldsymbol{p}}F\\
\frac{d\boldsymbol{p}}{ds}=&-\left(\frac{\partial F}{\partial\psi}\boldsymbol{p}+\nabla_{\boldsymbol{x}}F\right)\\
\frac{d\psi}{ds}=&\boldsymbol{p}\cdot\nabla_{\boldsymbol{p}}F.
\end{aligned} \tag{10.66}$$

The function $\psi(\boldsymbol{x})$ is defined by the third equation at each point $\boldsymbol{x}$ of the trajectory of the first equation. There is a neighborhood of the initial conditions (see below) that is covered by trajectories.

In the case at hand the function $F(\boldsymbol{x},\psi,\boldsymbol{p})$ in the eikonal equation (10.64) has the form

$$F(\boldsymbol{x},\psi,\boldsymbol{p})=\sum_{i,j=1}^{d}\sigma^{i,j}(\boldsymbol{x})p^ip^j+\sum_{i=1}^{d}a^i(\boldsymbol{x})p^i,$$

so that the characteristic equations (10.66) are

$$\frac{d\boldsymbol{x}}{ds} = 2\boldsymbol{\sigma}(\boldsymbol{x})\boldsymbol{p} + \boldsymbol{a}(\boldsymbol{x}) \tag{10.67}$$

$$\frac{d\boldsymbol{p}}{ds} = -\nabla_{\boldsymbol{x}}\boldsymbol{p}^T\boldsymbol{\sigma}(\boldsymbol{x})\boldsymbol{p} - \nabla_{\boldsymbol{x}}\boldsymbol{a}^T(\boldsymbol{x})\boldsymbol{p} \tag{10.68}$$

$$\frac{d\psi}{ds} = \boldsymbol{p}^T\boldsymbol{\sigma}(\boldsymbol{x})\boldsymbol{p}. \tag{10.69}$$

In deriving the characteristic equation (10.69) the eikonal equation (10.64) was used. First, we observe that the trajectories of the autonomous system (10.67), (10.68) that begin near the attractor $(\boldsymbol{x}_0, \boldsymbol{0})$ diverge. To see this, we linearize the system (10.67), (10.68) around this point and obtain $\boldsymbol{z}'(s) = 2\boldsymbol{\sigma}(\boldsymbol{x}_0)\boldsymbol{p}(s) + \boldsymbol{A}\boldsymbol{z}(s)$ and $\boldsymbol{\pi}'(s) = -\boldsymbol{A}\boldsymbol{\pi}(s)$, where $\boldsymbol{A}$ is defined in (10.49). It follows that $\boldsymbol{\pi}(s) = e^{-\boldsymbol{A}s}\boldsymbol{\pi}_0$, hence

$$\boldsymbol{z}(s) = e^{\boldsymbol{A}s}\boldsymbol{z}_0 + 2\int_0^s e^{\boldsymbol{A}(s-u)}\boldsymbol{\sigma}(\boldsymbol{x}_0)e^{-\boldsymbol{A}u}\boldsymbol{\pi}_0\,du.$$

Both $\boldsymbol{z}(s)$ and $\boldsymbol{\pi}(s)$ diverge as $s \to \infty$, because the eigenvalues of $-\boldsymbol{A}$ have positive real parts.

To integrate the characteristic equations (10.67), (10.68) initial conditions can be imposed near the attractor $(\boldsymbol{x}_0, \boldsymbol{0})$ by constructing $\psi(\boldsymbol{x})$ in the form of a power series. The truncation of the power series near the attractor provides an approximation to $\psi(\boldsymbol{x})$ and to $\boldsymbol{p} = \nabla\psi(\boldsymbol{x})$ whose error can be made arbitrarily small. Expanding $\psi(\boldsymbol{x})$, $\boldsymbol{a}(\boldsymbol{x})$, and $\boldsymbol{\sigma}(\boldsymbol{x})$ in powers of $\boldsymbol{z} = \boldsymbol{x} - \boldsymbol{x}_0$, we find from the eikonal equation (10.64) that $\nabla\psi(\boldsymbol{x}_0) = 0$ so that the power series expansion of $\psi(\boldsymbol{x})$ begins as a quadratic form

$$\psi(\boldsymbol{x}) = \frac{1}{2}\boldsymbol{x}^T\boldsymbol{Q}\boldsymbol{x} + o\left(|\boldsymbol{x}|^2\right), \tag{10.70}$$

and $\boldsymbol{Q}$ is the solution of the Riccati equation

$$2\boldsymbol{Q}\boldsymbol{\sigma}(\boldsymbol{x}_0)\boldsymbol{Q} + \boldsymbol{Q}\boldsymbol{A} + \boldsymbol{A}^T\boldsymbol{Q} = \boldsymbol{0}. \tag{10.71}$$

Obviously, the first term in the power series expansion of $\boldsymbol{p} = \nabla\psi(\boldsymbol{x})$ is given by

$$\boldsymbol{p} = \boldsymbol{Q}\boldsymbol{x} + O\left(|\boldsymbol{x}|^2\right). \tag{10.72}$$

In deriving eq. (10.71) use is made of the facts that $\boldsymbol{Q}$ and $\boldsymbol{\sigma}$ are symmetric matrices and that a quadratic form vanishes identically if and only if it is defined by an anti-symmetric matrix [213]. The solution of eq. (10.71) is a positive definite matrix [213], [81].

Exercise 10.14 (Square root of a positive definite symmetric matrix). Show that a positive definite symmetric matrix has a positive definite symmetric square root.
□

Exercise 10.15 (The Riccati equation). Reduce the Riccati equation (10.71) to

$$\boldsymbol{A}\boldsymbol{Y}+\boldsymbol{Y}^T\boldsymbol{A}^T=-\boldsymbol{I} \tag{10.73}$$

by the substitutions $\boldsymbol{X}=\boldsymbol{Q}\sqrt{\boldsymbol{\sigma}}$, where $\boldsymbol{X}$ is the solution of

$$2\boldsymbol{X}\boldsymbol{X}^T+\boldsymbol{X}\boldsymbol{A}+\boldsymbol{A}^T\boldsymbol{X}^T=\boldsymbol{0}$$

and $\boldsymbol{X}=-\frac{1}{2}\boldsymbol{Y}^{-1}$. Show that the solution of eq. (10.73) is a symmetric matrix given by $\boldsymbol{Y}=\int_0^\infty e^{\boldsymbol{A}t}e^{\boldsymbol{A}^Tt}\,dt$, and show that the integral converges. □

Taking the contour

$$\frac{1}{2}\boldsymbol{x}^T\boldsymbol{Q}\boldsymbol{x}=\delta, \tag{10.74}$$

for some small positive δ, as the initial surface for the system (10.67)–(10.69) and using the approximate initial values $\psi(\boldsymbol{x})=\delta$ and (10.72) at each point of the surface, we can integrate the system (10.67)–(10.69) analytically or numerically. Once the domain D is covered with characteristics, the approximate value of $\psi(\boldsymbol{x})$ can be determined at each point $\boldsymbol{x}\in D$ as the value of the solution $\psi(s)$ of eq. (10.69) at s such that the solution of eq. (10.67) satisfies

$$\boldsymbol{x}(s)=\boldsymbol{x}. \tag{10.75}$$

The initial condition on the surface (10.74) determines the unique trajectory of the system (10.67)–(10.69) that satisfies (10.75) for some s. It can be found numerically by the method of shooting.

10.2.3 The transport equation

As mentioned in Section 10.2.1, the function $K_\varepsilon(\boldsymbol{y}\,|\,\boldsymbol{x})$ satisfies the equation

$$\begin{aligned}
&\varepsilon\sum_{i,j=1}^d\frac{\partial^2\sigma^{i,j}(\boldsymbol{y})K_\varepsilon(\boldsymbol{y}\,|\,\boldsymbol{x})}{\partial y^i\partial y^j}-\sum_{i=1}^d\left(2\sum_{j=1}^d\sigma^{i,j}(\boldsymbol{y})\frac{\partial\psi(\boldsymbol{y})}{\partial y^j}+a^i(\boldsymbol{y})\right)\frac{\partial K_\varepsilon(\boldsymbol{y}\,|\,\boldsymbol{x})}{\partial y^i}\\
&-\sum_{i=1}^d\left(\frac{\partial a^i(\boldsymbol{y})}{\partial y^i}+\sum_{j=1}^d\left(\sigma^{i,j}(\boldsymbol{y})\frac{\partial^2\psi(\boldsymbol{y})}{\partial y^i\partial y^j}+2\frac{\partial\sigma^{i,j}(\boldsymbol{y})}{\partial y^j}\frac{\partial\psi(\boldsymbol{y})}{\partial y^j}\right)\right)K_\varepsilon(\boldsymbol{y}\,|\,\boldsymbol{x})\\
&=-\,\delta(\boldsymbol{y}-\boldsymbol{x}).
\end{aligned} \tag{10.76}$$

The function $K_\varepsilon(\boldsymbol{y}\,|\,\boldsymbol{x})$ cannot have an internal layer at the global attractor point $\boldsymbol{x}_0$ in D. This is due to the fact that stretching $\boldsymbol{y}-\boldsymbol{x}_0=\sqrt{\varepsilon}\boldsymbol{\xi}$ and taking the limit $\varepsilon\to 0$ (10.76) converts the transport equation to

$$\begin{aligned}
&\sum_{i,j=1}^d\frac{\partial^2\sigma^{i,j}(\boldsymbol{x}_0)K_0(\boldsymbol{\xi}\,|\,\boldsymbol{x})}{\partial\xi^i\partial\xi^j}-(2\boldsymbol{A}\boldsymbol{Q}+\boldsymbol{A})\boldsymbol{\xi}\cdot\nabla_{\boldsymbol{\xi}}K_0(\boldsymbol{\xi}\,|\,\boldsymbol{x})\\
&-\operatorname{tr}(\boldsymbol{A}+\boldsymbol{\sigma}(\boldsymbol{x}_0)\boldsymbol{Q})\,K_0(\boldsymbol{\xi}\,|\,\boldsymbol{x})=0,
\end{aligned}$$

whose bounded solution is $K_0(\boldsymbol{\xi}\,|\,\boldsymbol{x}) = const.$, because $\operatorname{tr}(\boldsymbol{A} + \boldsymbol{\sigma}(\boldsymbol{x}_0)\boldsymbol{Q}) = 0$. The last equality follows from the Riccati eq. (10.71) (left multiply by $\boldsymbol{Q}^{-1}$ and take the trace).

In view of eqs. (10.60)–(10.63), we obtain in the limit $\varepsilon \to 0$ the transport equation

$$\sum_{i=1}^{d}\left(2\sum_{j=1}^{d}\sigma^{i,j}(\boldsymbol{y})\frac{\partial\psi(\boldsymbol{y})}{\partial y^j} + a^i(\boldsymbol{y})\right)\frac{\partial K_0(\boldsymbol{y}\,|\,\boldsymbol{x})}{\partial y^i} \tag{10.77}$$
$$= -\sum_{i=1}^{d}\left(\frac{a^i(\boldsymbol{y})}{\partial y^i} + \sum_{j=1}^{d}\left(\sigma^{i,j}(\boldsymbol{y})\frac{\partial^2\psi(\boldsymbol{y})}{\partial y^i\partial y^j} + 2\frac{\partial\sigma^{i,j}(\boldsymbol{y})}{\partial y^j}\frac{\partial\psi(\boldsymbol{y})}{\partial y^j}\right)\right)K_0(\boldsymbol{y}\,|\,\boldsymbol{x}).$$

Because the characteristics diverge, the initial value (at $s = 0$) on each characteristic is given at $\boldsymbol{y} = \boldsymbol{x}_0$ as $K_0(\boldsymbol{x}_0\,|\,\boldsymbol{x}) = const.$ (e.g., $const. = 1$). With this choice of the constant the function $p_\varepsilon(\boldsymbol{y}\,|\,\boldsymbol{x})$ has to be renormalized.

Exercise 10.16 (The potential case). Show that if the diffusion matrix $\boldsymbol{\sigma}$ is constant and $\boldsymbol{a}(\boldsymbol{x}) = -\boldsymbol{\sigma}\nabla\phi(\boldsymbol{x})$ for some function $\phi(\boldsymbol{x})$, then $\psi(\boldsymbol{x}) = \phi(\boldsymbol{x})$ and the WKB solution of the homogeneous Fokker–Planck equation (10.52) is given by $p_\varepsilon(\boldsymbol{y}) = e^{-\psi(\boldsymbol{y})/\varepsilon}$; that is, the solution of the transport equation (10.76) is $K_0 = const.$ □

10.2.4 The characteristic equations

The transport equation has to be integrated numerically, together with the characteristic equations (10.67), (10.68). To evaluate the partial derivatives $\partial^2\psi(\boldsymbol{y})/\partial y^i\partial y^j$ along the characteristics, we use eqs. (10.70), (10.72), and $\partial^2\psi(\boldsymbol{y})/\partial y^i\partial y^j|_{\boldsymbol{y}=\boldsymbol{x}_0} = Q^{i,j}$ on the initial ellipsoid (10.74). The differential equations for $\partial^2\psi(\boldsymbol{y})/\partial y^i\partial y^j$ along the characteristics are derived by differentiating the characteristic equations (10.67), (10.68) with respect to the initial values $\boldsymbol{x}(0) = \boldsymbol{x}_0$. Writing

$$\boldsymbol{x}_j(s) = \frac{\partial\boldsymbol{x}(s)}{\partial x_0^j},\quad \boldsymbol{p}_j(s) = \frac{\partial\boldsymbol{p}(s)}{\partial x_0^j},\quad Q^{i,j}(s) = \frac{\partial^2\psi(\boldsymbol{x}(s))}{\partial y^i\partial y^j}, \tag{10.78}$$

we get the identity $\boldsymbol{p}_j(s) = \boldsymbol{Q}(s)\boldsymbol{x}_j(s)$. The initial conditions are

$$x_j^i(0) = \delta_{i,j} \tag{10.79}$$

$$p_j^i(0) = \left.\frac{\partial^2\psi(\boldsymbol{y})}{\partial y^i\partial y^j}\right|_{\boldsymbol{y}=\boldsymbol{x}_0} = Q^{i,j}(0) = Q^{i,j} \tag{10.80}$$

and the dynamics

$$\frac{d\boldsymbol{x}_j(s)}{ds} = \sum_{k=1}^{d}\left[2\frac{\partial}{\partial x^k}\boldsymbol{\sigma}(\boldsymbol{x}(s))\boldsymbol{p}(s) + 2\boldsymbol{\sigma}(\boldsymbol{x}(s))\boldsymbol{p}_k(s) + \frac{\partial}{\partial x^k}\boldsymbol{a}(\boldsymbol{x}(s))\right] x_k^j(s) \tag{10.81}$$

$$\begin{aligned}\frac{d\boldsymbol{p}_j(s)}{ds} = -\sum_{k=1}^{d}\Bigg[&\nabla_{\boldsymbol{x}}\boldsymbol{p}^T(s)\frac{\partial}{\partial x^k}\boldsymbol{\sigma}(\boldsymbol{x}(s))\boldsymbol{p}(s) + 2\nabla_{\boldsymbol{x}}\boldsymbol{p}_k^T(s)\boldsymbol{\sigma}(\boldsymbol{x}(s))\boldsymbol{p}(s)\\ &+ \nabla_{\boldsymbol{x}}\boldsymbol{a}^T(\boldsymbol{x}(s))\boldsymbol{p}_k(s) + \frac{\partial}{\partial x^k}\nabla_{\boldsymbol{x}}\boldsymbol{a}^T(\boldsymbol{x}(s))\boldsymbol{p}(s)\Bigg] x_k^j(s).\end{aligned} \tag{10.82}$$

The transport equation (10.77) can be written on characteristics as

$$\begin{aligned}&\frac{dK_0(\boldsymbol{x}(s)\,|\,\boldsymbol{x})}{ds} \qquad\qquad (10.83)\\ &= -\sum_{i=1}^{d}\left[\frac{a^i(\boldsymbol{x}(s))}{\partial y^i} + \sum_{j=1}^{d}\left(\sigma^{i,j}(\boldsymbol{x}(s))Q^{i,j}(s) + 2\frac{\partial\sigma^{i,j}(\boldsymbol{x}(s))}{\partial y^j}\boldsymbol{p}(s)\right)\right]\\ &\quad\times K_0(\boldsymbol{x}(s)\,|\,\boldsymbol{x}).\end{aligned}$$

In summary, the numerical integration of the eikonal and the transport equations consists in integrating numerically the differential equations (10.67)–(10.69), (10.81)–(10.83) with initial values of $\boldsymbol{x}_0 = \boldsymbol{x}(0)$ that cover the ellipsoid (10.74), with $\boldsymbol{p}(0)$ and $\psi(\boldsymbol{x}(0))$ given by $\boldsymbol{p}(0) = \boldsymbol{Q}\boldsymbol{x}(0)$ and $\psi(\boldsymbol{x}(0)) = \delta$, and the initial values (10.79), (10.80), and $K_0(\boldsymbol{x}(0)\,|\,\boldsymbol{x}) = 1$. Equations (10.78) have to be solved at each step of the integration to convert from $\boldsymbol{p}_j(s)$ to $\boldsymbol{Q}(s)$.

10.2.5 Boundary layers at noncharacteristic boundaries

Although the functions $K_j(\boldsymbol{x}(0)\,|\,\boldsymbol{x})$ in the expansion (10.60) are regular in the domain D and on its boundary, the boundary layer function $q_\varepsilon(\boldsymbol{y}\,|\,\boldsymbol{x})$ has an essential singularity at the boundary. Its normal derivatives at the boundary become infinite as $\varepsilon \to 0$ and its derivatives in the direction of the boundary vanish, because $q_\varepsilon(\boldsymbol{y}\,|\,\boldsymbol{x})$ vanishes there. Furthermore, the higher order normal derivatives of $q_\varepsilon(\boldsymbol{y}\,|\,\boldsymbol{x})$ at the boundary are larger than the lower order derivatives. More specifically, we postulate that for $\boldsymbol{y} \in \partial D,\ \boldsymbol{x} \in D$

$$\frac{\partial^k q_\varepsilon(\boldsymbol{y}\,|\,\boldsymbol{x})}{\partial \boldsymbol{\nu}_{\boldsymbol{y}}^k} = O\left(\varepsilon^{-k}\right) \quad \text{for } k = 0, 1, 2, \ldots. \tag{10.84}$$

Keeping this in mind, we keep in eq. (10.76) the second order partial derivatives of $q_\varepsilon(\boldsymbol{y}\,|\,\boldsymbol{x})$ and balance them with the first-order terms. The singularity of $q_\varepsilon(\boldsymbol{y}\,|\,\boldsymbol{x})$ is resolved by balancing terms of similar orders of magnitude in eq. (10.76) for $K_\varepsilon(\boldsymbol{y}\,|\,\boldsymbol{x})$ near the boundary.

To derive the boundary layer equation for $q_\varepsilon(\boldsymbol{y}\,|\,\boldsymbol{x})$, we introduced local coordinates near the boundary, $\rho(\boldsymbol{y}) = \operatorname{dist}(\boldsymbol{y}, \partial D)$ such that $\rho(\boldsymbol{y}) < 0$ for $\boldsymbol{y} \in D$ and

$d-1$ coordinates in the boundary $\boldsymbol{s}(\boldsymbol{y}) = (s_1, s_2, \ldots, s_{d-1})$. In the transformation $\boldsymbol{y} \to (\rho(\boldsymbol{y}), \boldsymbol{s}(\boldsymbol{y}))$, the point $\boldsymbol{y}' = (0, \boldsymbol{s}(\boldsymbol{y}))$ is the orthogonal projection of $\boldsymbol{y}$ on ∂D. The boundary is mapped into the hyper-plane $\rho = 0$. Then $\nabla \rho(\boldsymbol{y})\big|_{\rho=0} = \boldsymbol{\nu}(\boldsymbol{y})$ for $\boldsymbol{y} \in \partial D$, where $\boldsymbol{\nu}(\boldsymbol{y})$ is the unit outer normal to ∂D at $\boldsymbol{y}$.

Next, we introduce the stretched variable $\zeta = \rho/\varepsilon$ and define $q_\varepsilon(\boldsymbol{y}\,|\,\boldsymbol{x}) = Q(\zeta, \boldsymbol{s}, \varepsilon\,|\,\boldsymbol{x})$, and express the postulate (10.84) by assuming that the decomposition (10.60) becomes

$$K_\varepsilon(\boldsymbol{y}\,|\,\boldsymbol{x}) = \left(K_0(\rho, \boldsymbol{s}\,|\,\boldsymbol{x}) + \varepsilon K_1(\rho, \boldsymbol{s}\,|\,\boldsymbol{x}) + \cdots\right) Q(\zeta, \boldsymbol{s}, \varepsilon\,|\,\boldsymbol{x}).$$

Now, we expand all functions that appear in eq. (10.59) in an asymptotic series in powers of ε. Writing $Q(\zeta, \boldsymbol{s}, \varepsilon\,|\,\boldsymbol{x}) \sim Q^0(\zeta, \boldsymbol{s}\,|\,\boldsymbol{x}) + \varepsilon Q^1(\zeta, \boldsymbol{s}\,|\,\boldsymbol{x}) + \cdots$, we obtain for $Q^0(\zeta, \boldsymbol{s}\,|\,\boldsymbol{x})$ the boundary layer equation

$$\left(\sum_{i,j=1}^{d} \sigma^{i,j}(\boldsymbol{y}')\nu^i(\boldsymbol{y}')\nu^j(\boldsymbol{y}')\right) \frac{\partial^2 Q^0}{\partial \zeta^2} - \left[\sum_{i=1}^{d}\left(2\sum_{j=1}^{d} \sigma^{i,j}(\boldsymbol{y}')\frac{\partial \psi(\boldsymbol{y}')}{\partial y^j} + a^i(\boldsymbol{y}')\right)\nu^i(\boldsymbol{y}')\right]\frac{\partial Q^0}{\partial \zeta} = 0,$$

which we rewrite as $Q^0_{\zeta\zeta} - A(\boldsymbol{s})Q^0_\zeta = 0$, where

$$A(\boldsymbol{s}) = \frac{\displaystyle\sum_{i=1}^{d}\left(2\sum_{j=1}^{d} \sigma^{i,j}(\boldsymbol{y}')\frac{\partial \psi(\boldsymbol{y}')}{\partial y^j} + a^i(\boldsymbol{y}')\right)\nu^i(\boldsymbol{y}')}{\displaystyle\sum_{i,j=1}^{d} \sigma^{i,j}(\boldsymbol{y}')\nu^i(\boldsymbol{y}')\nu^j(\boldsymbol{y}')}. \tag{10.85}$$

The function $A(\boldsymbol{s})$ is positive on the boundary, because the denominator in eq. (10.85) is a positive definite quadratic form and the numerator is the normal component of the direction of the characteristics at the boundary. Because the characteristics exit D, their direction points away from D at ∂D. This means that

$$\sum_{i=1}^{d}\left(2\sum_{j=1}^{d} \sigma^{i,j}(\boldsymbol{y}')\frac{\partial \psi(\boldsymbol{y}')}{\partial y^j} + a^i(\boldsymbol{y}')\right)\nu^i(\boldsymbol{y}') > 0.$$

The boundary and matching conditions eqs. (10.61), (10.62) are expressed in the boundary layer function as $Q^0(0, \boldsymbol{s}\,|\,\boldsymbol{x}) = 0$ and $\lim_{\zeta\to-\infty} Q^0(\zeta, \boldsymbol{s}\,|\,\boldsymbol{x}) = 1$, so that the solution is

$$Q^0(\zeta, \boldsymbol{s}\,|\,\boldsymbol{x}) = 1 - e^{A(\boldsymbol{s})\zeta}. \tag{10.86}$$

The uniform expansion of the solution of the Fokker–Planck equation (10.52), valid up to the boundary ∂D, is given by

$$p_{\text{unif}}\,(\boldsymbol{y}\,|\boldsymbol{x}) = [K_0(\boldsymbol{y}\,|\,\boldsymbol{x}) + O(\varepsilon)]\exp\left\{-\frac{\psi(\boldsymbol{y})}{\varepsilon}\right\} \times \left[1 - \exp\left\{\frac{A(\boldsymbol{s})\rho(\boldsymbol{y})}{\varepsilon}\right\}\right]. \tag{10.87}$$

Equation (10.87) is a uniform approximation to $p\,(\boldsymbol{y}\,|\boldsymbol{x})$ for $\boldsymbol{x} \in D$ outside the boundary layer, all $\boldsymbol{y} \in D$, and $O(\varepsilon)$ is uniform for $\boldsymbol{x} \in D$ outside the boundary layer , all $\boldsymbol{y} \in D$.

To obtain a uniform approximation to $p\,(\boldsymbol{y}\,|\boldsymbol{x})$, valid for all $\boldsymbol{x}, \boldsymbol{y} \in D$, we have to solve for $\boldsymbol{x}, \boldsymbol{y} \in D$ the equation

$$\varepsilon \sum_{i=1}^{d}\sum_{j=1}^{d} \sigma^{ij}\,(\boldsymbol{x})\,\frac{\partial^2 p\,(\boldsymbol{y}\,|\,\boldsymbol{x})}{\partial x^i \partial x^j} + \sum_{i=1}^{d} a^i\,(\boldsymbol{x})\,\frac{\partial p\,(\boldsymbol{y}\,|\,\boldsymbol{x})}{\partial x^i} = -\,\delta(\boldsymbol{y} - \boldsymbol{x}) \tag{10.88}$$

with the boundary condition

$$p\,(\boldsymbol{y}\,|\,\boldsymbol{x}) = 0 \text{ for } \boldsymbol{x} \in \partial D, \boldsymbol{y} \in D.$$

The analysis of this case is straightforward. The outer solution is a constant and the equation for the leading term $p^0(\zeta', \boldsymbol{s}')$, in the boundary layer expansion in the variables $(\zeta', \boldsymbol{s}')$ defined above, is

$$\sigma(\boldsymbol{s}')\frac{\partial^2 p^0(\zeta', \boldsymbol{s}')}{\partial \zeta'^2} + a_n(\boldsymbol{s}')\frac{\partial p^0(\zeta', \boldsymbol{s}')}{\partial \zeta'} = 0 \text{ for } \zeta' < 0, \tag{10.89}$$

where

$$\sigma(\boldsymbol{s}') = \sum_{i,j=1}^{d} \sigma^{i,j}(0, \boldsymbol{s}')\nu^i(\boldsymbol{s}')\nu^j(\boldsymbol{s}'), \quad a_n(\boldsymbol{s}') = \sum_{i}^{d} a^i(0, \boldsymbol{s}')\nu^i(\boldsymbol{s}') < 0,$$

with the matching conditions

$$p^0(0, \boldsymbol{s}') = 0, \qquad \lim_{\zeta' \to -\infty} p^0(\zeta', \boldsymbol{s}') = 1. \tag{10.90}$$

The solution is

$$p^0(\zeta', \boldsymbol{s}') = 1 - e^{a_n(\boldsymbol{s}')\zeta'/\sigma(\boldsymbol{s}')}. \tag{10.91}$$

Now, the uniform approximation to $p\,(\boldsymbol{y}\,|\boldsymbol{x})$, valid for all $\boldsymbol{x}, \boldsymbol{y} \in D$, can be written as

$$p_{\text{unif}}\,(\boldsymbol{y}\,|\boldsymbol{x}) = [K_0(\boldsymbol{y}\,|\,\boldsymbol{x}) + O(\varepsilon)]\exp\left\{-\frac{\psi(\boldsymbol{y})}{\varepsilon}\right\}\left[1 - \exp\left\{\frac{A(\boldsymbol{s})\rho(\boldsymbol{y})}{\varepsilon}\right\}\right] \times \left[1 - \exp\left\{\frac{a_n(\boldsymbol{s}')\rho(\boldsymbol{x})}{\varepsilon\sigma(\boldsymbol{s}')}\right\}\right]. \tag{10.92}$$

10.2.6 Boundary layers at characteristic boundaries in the plane

We now assume that D is a bounded planar domain whose boundary ∂D consists of a finite number of piecewise smooth closed simple curves and write $x^1 = x,\ x^2 = y$. The boundary ∂D is characteristic if the drift vector $\boldsymbol{a}(x, y)$ is tangent to the boundary or vanishes there; that is, if eq. (10.51) holds. In either case the normal component of the drift vector $\boldsymbol{a}(\boldsymbol{x})$ vanishes at the boundary. At each point $(x, y) \in D$, near the boundary, we denote by (x', y') its orthogonal projection on the boundary. We denote by $\boldsymbol{\nu}(x, y)$ and $\boldsymbol{\tau}(x, y)$ the unit outer normal and unit tangent at the boundary point (x', y'), respectively. We define the signed distance to the boundary

$$\rho(x, y) = -\operatorname{dist}((x, y), \partial D) = -\sqrt{(x - x')^2 + (y - y')^2}, \quad \text{for } (x, y) \in D$$
$$\rho(x, y) = \operatorname{dist}((x, y), \partial D) = \sqrt{(x - x')^2 + (y - y')^2}, \quad \text{for } (x, y) \notin D.$$

The boundary corresponds to $\rho(x, y) = 0$. We denote by $s(x, y)$ arclength on a given component of the boundary, measured counterclockwise from a given boundary point to the point (x', y'). Thus the transformation $(x, y) \to (\rho, s)$, where $\rho = \rho(x, y),\ s = s(x, y)$ maps a strip near a connected component of the boundary onto the strip $|\rho| < \rho_0,\ 0 \le s \le S$, where $\rho_0 > 0$ and S is the arclength of the given component of the boundary. The transformation is given by $(x, y) = (x', y') + \rho\boldsymbol{\nu}(x, y)$, where (x', y') are functions of s. We write $\boldsymbol{\nu}(x, y) = \boldsymbol{\nu}(s)$.

We assume that in addition to eq. (10.51), we have in the strip $|\rho| < \rho_0$ the small ρ expansion

$$\boldsymbol{a}(x, y) = \left\{\rho^\alpha a^0(s)\boldsymbol{\nu}(s) + \rho^\beta B(s)\boldsymbol{\tau}(s)\right\}\{1 + o(1)\}, \tag{10.93}$$

for some $\alpha, \beta > 0$. We assume for the present analysis that $\alpha = 1,\ \beta = 0$ (other cases are considered in [167]).

If the tangential component $B(s)$ of the drift vector vanishes at a point s, we say that s is a *critical point* in ∂D. If there are no critical points on a given component of ∂D, $B(s)$ has a constant sign there, $B(s) > 0$, say, so that this component of ∂D is a limit cycle[1] for the noiseless dynamics

$$\frac{d}{dt}\begin{pmatrix} x \\ y \end{pmatrix} = \boldsymbol{a}(x, y). \tag{10.94}$$

We write eq. (10.93) in local coordinates as

$$\boldsymbol{a}(\rho, s) = a^0(s)\rho\nabla\rho + B(s)\nabla s + o(\rho). \tag{10.95}$$

The coefficient $B(s)$ is the speed of the deterministic motion on the boundary; we assume $B(s) > 0$ for all $0 \le s \le S,\ a^0(s) \ge 0$, and that the exit density has a limit as $\varepsilon \to 0$.

[1] In the case of a center all trajectories of the noiseless dynamics are closed

Before deriving the boundary layer equation, we turn to the analysis of the eikonal equation. First, we note that the solution of the eikonal equation is constant on the given component of the boundary. Indeed, with the obvious notation, the eikonal equation (10.58) can be written in local coordinates on ∂D as

$$\sum_{i,j=1}^{2} \sigma^{i,j}(0,s)\frac{\partial\psi(0,s)}{\partial x^i}\frac{\partial\psi(0,s)}{\partial x^j} + B(s)\frac{\partial\psi(0,s)}{\partial s} = 0, \tag{10.96}$$

where $x^1 = x$, $x^2 = y$. To be well defined on ∂D, $\psi(0,s)$ must be a periodic function of s with period S. However, eq. (10.96) implies that the derivative $\partial\psi(0,s)/\partial s$ does not change sign, because $B(s) > 0$ and the diffusion matrix $\sigma^{i,j}(0,s)$ is positive definite. Thus we must have

$$\psi(0,s) = \text{const.} = \hat{\psi}, \quad \nabla\psi(0,s) = 0 \text{ for all } 0 \le s \le S. \tag{10.97}$$

It follows that near ∂D

$$\psi(\rho,s) = \hat{\psi} + \frac{1}{2}\rho^2\frac{\partial^2\psi(0,s)}{\partial\rho^2} + o\left(\rho^2\right) \text{ as } \rho \to 0. \tag{10.98}$$

setting $\phi(s) = \partial^2\psi(0,s)/\partial\rho^2$, and using eqs. (10.95) and (10.98) in eq. (10.58), we see that $\phi(s)$ must be the S-periodic solution of the Bernoulli equation

$$\sigma(s)\phi^2(s) + a^0(s)\phi(s) + \frac{1}{2}B(s)\phi'(s) = 0, \tag{10.99}$$

where $\sigma(s) = \sum_{i,j=1}^{2} \sigma^{i,j}(0,s)\nu^i(s)\nu^j(s)$. Note that using eqs. (10.93) and eq. (10.58), the drift vector in eq. (10.76) can be written for our two-dimensional problem in local coordinates near the boundary, as

$$\begin{aligned} &2\sum_{j=1}^{2} \sigma^{i,j}(\boldsymbol{y})\frac{\partial\psi(\boldsymbol{y})}{\partial y^j} + a^i(\boldsymbol{y}) \\ &= 2\sum_{j=1}^{2} \sigma^{i,j}(0,s)\frac{\partial\psi(0,s)}{\partial x^j} + a^i(0,s) + o(\rho) \\ &= \rho\left(2\phi(s)\sum_{j=1}^{2} \sigma^{i,j}(0,s)\frac{\partial\rho}{\partial x^j} + a^0(s)\frac{\partial\rho}{\partial x^i}\right) + o(\rho). \end{aligned} \tag{10.100}$$

To derive the boundary layer equation, we introduce the stretched variable $\zeta = \rho/\sqrt{\varepsilon}$ and define $q_\varepsilon(x,y\,|\,x_0,y_0) = Q(\zeta,s,\varepsilon\,|\,x_0,y_0)$. Expanding

$$Q(\zeta,s,\varepsilon\,|\,x_0,y_0) \sim Q^0(\zeta,s) + \sqrt{\varepsilon}Q^1(\zeta,s) + \cdots, \tag{10.101}$$

and using eq. (10.100), we obtain the boundary layer equation

$$\sigma(s)\frac{\partial^2 Q^0(\zeta,s)}{\partial\zeta^2} - \zeta\left(a^0(s) + 2\sigma(s)\phi(s)\right)\frac{\partial Q^0(\zeta,s)}{\partial\zeta} - B(s)\frac{\partial Q^0(\zeta,s)}{\partial s} = 0. \tag{10.102}$$

As in the previous section, the boundary and matching conditions eqs. (10.61), (10.62) imply that

$$Q^0(0,\boldsymbol{s}\,|\,\boldsymbol{x})=0,\qquad \lim_{\zeta\to-\infty}Q^0(\zeta,\boldsymbol{s}\,|\,\boldsymbol{x})=1. \tag{10.103}$$

The solution to the boundary value problem eqs. (10.102), (10.103) is given by

$$Q^0(\zeta,s)=-\sqrt{\frac{2}{\pi}}\int\limits_0^{\xi(s)\zeta}e^{-z^2/2}\,dz, \tag{10.104}$$

where $\xi(s)$ is the S-periodic solution of the Bernoulli equation

$$\sigma(s)\xi^3(s)+\left(a^0(s)+2\sigma(s)\phi(s)\right)\xi(s)+B(s)\xi'(s)=0. \tag{10.105}$$

Setting $\xi_0(s)=\sqrt{-\phi(s)}$ in eq. (10.99), we see that $\xi_0(s)$ is the S-periodic solution of the Bernoulli equation

$$B(s)\xi_0'(s)+a^0(s)\xi_0(s)-\sigma(s)\xi_0^3(s)=0. \tag{10.106}$$

The solutions of the three Bernoulli equations (10.99), (10.105), (10.106) are related to each other as follows: $\xi_0(s)=\sqrt{-\phi(s)}=\xi(s)$.

The uniform expansion of the solution of the Fokker–Planck equation (10.52), valid up to the boundary ∂D, is given by

$$\begin{aligned} p_{\text{unif}}(x,y\,|\,x_0,y_0)=&\left[K_0(x,y\,|\,x_0,y_0)+O(\sqrt{\varepsilon})\right]\exp\left\{-\frac{\psi(x,y)}{\varepsilon}\right\}\\ &\times Q^0\left(\frac{\rho}{\sqrt{\varepsilon}},s\right), \end{aligned} \tag{10.107}$$

where $O(\sqrt{\varepsilon})$ is uniform in $(x,y)\in\bar{D}$ for all fixed $(x_0,y_0)\in D$.

10.2.7 Exit through noncharacteristic boundaries

The uniform expansion eqs. (10.87), (10.107) of $p_\varepsilon(\boldsymbol{y}\,|\,\boldsymbol{x})$ can be used for the asymptotic solution of the exit problem. First, we consider the exit density on ∂D. It is the normal component of the flux density vector (10.53) on the boundary,

$$\begin{aligned} &\boldsymbol{J}(\boldsymbol{y}\,|\,\boldsymbol{x})\cdot\boldsymbol{\nu}(\boldsymbol{y})\big|_{\boldsymbol{y}\in\partial D} \\ &\sim -K_0(\boldsymbol{y})\sum_{i,j=1}^d\varepsilon\sigma^{i,j}(\boldsymbol{y})\exp\left\{-\frac{\psi(\boldsymbol{y})}{\varepsilon}\right\}\frac{\partial q_\varepsilon(\boldsymbol{y}\,|\,\boldsymbol{x})}{\partial y^i}\nu^j(\boldsymbol{y}), \end{aligned} \tag{10.108}$$

where $q_\varepsilon(\boldsymbol{y}\,|\,\boldsymbol{x})$ is the boundary layer function and in the local coordinates $(\rho,\boldsymbol{s})$ near the boundary, $\boldsymbol{y}=(0,\boldsymbol{s})$ on ∂D.

If ∂D is a noncharacteristic boundary, we use the expansion eqs. (10.87) in eq. (10.108) and obtain

$$\begin{aligned} & \boldsymbol{J}(\boldsymbol{y}\,|\,\boldsymbol{x})\cdot\boldsymbol{n}(\boldsymbol{y})|_{\boldsymbol{y}\in\partial D} \\ & \sim K_0(\boldsymbol{y})\exp\left\{-\frac{\psi(\boldsymbol{y})}{\varepsilon}\right\}A(\boldsymbol{s})\sum_{i,j=1}^{d}\sigma^{i,j}(\boldsymbol{s})\nu^i(\boldsymbol{y})\nu^j(\boldsymbol{y}), \end{aligned} \tag{10.109}$$

where $A(\boldsymbol{s})$ is given in eq. (10.85). The latter simplifies eq. (10.109) into

$$\begin{aligned} & \boldsymbol{J}(\boldsymbol{y}\,|\,\boldsymbol{x})\cdot\boldsymbol{n}(\boldsymbol{y})|_{\boldsymbol{y}\in\partial D} \\ & \sim K_0(\boldsymbol{y})\exp\left\{-\frac{\psi(\boldsymbol{y})}{\varepsilon}\right\}\sum_{i=1}^{d}\left(2\sum_{j=1}^{d}\sigma^{i,j}(\boldsymbol{y})\frac{\partial\psi(\boldsymbol{y})}{\partial y^j}+a^i(\boldsymbol{y})\right)\nu^i(\boldsymbol{y}). \end{aligned}$$

Thus, the small ε asymptotic expansion of the exit density on the boundary, of trajectories that start at a fixed point $\boldsymbol{x}\in D$ (independent of ε), to exit at a point $\boldsymbol{y}\in\partial D$, is given by

$$\begin{aligned} & \Pr\left\{\boldsymbol{x}(\tau)\in\boldsymbol{y}+dS_{\boldsymbol{y}}\,|\,\boldsymbol{x}(0)=\boldsymbol{x}\right\} \\ & \sim \frac{K_0(\boldsymbol{y})\exp\left\{-\dfrac{\psi(\boldsymbol{y})}{\varepsilon}\right\}\displaystyle\sum_{i=1}^{d}\left(2\sum_{j=1}^{d}\sigma^{i,j}(\boldsymbol{y})\frac{\partial\psi(\boldsymbol{y})}{\partial y^j}+a^i(\boldsymbol{y})\right)\nu^i(\boldsymbol{y})\,dS_{\boldsymbol{y}}}{\displaystyle\oint_{\partial D}K_0(\boldsymbol{y})\exp\left\{-\frac{\psi(\boldsymbol{y})}{\varepsilon}\right\}\sum_{i=1}^{d}\left(2\sum_{j=1}^{d}\sigma^{i,j}(\boldsymbol{y})\frac{\partial\psi(\boldsymbol{y})}{\partial y^j}+a^i(\boldsymbol{y})\right)\nu^i(\boldsymbol{y})\,dS_{\boldsymbol{y}}}, \end{aligned} \tag{10.110}$$

where τ is the first passage time to the boundary.

The denominator in eq. (10.110) can be evaluated asymptotically for small ε by the Laplace method. Consider an isolated minimum point, $\boldsymbol{y}_k$, of $\psi(\boldsymbol{y})$ on ∂D. At this point the gradient, $\nabla\psi(\boldsymbol{y}_k)$, is parallel to the outer normal $\boldsymbol{\nu}(\boldsymbol{y}_k)$; that is

$$\boldsymbol{\nu}(\boldsymbol{y}_k)=\frac{\nabla\psi(\boldsymbol{y}_k)}{||\nabla\psi(\boldsymbol{y}_k)||}. \tag{10.111}$$

We can write the eikonal equation at $\boldsymbol{y}_k$ as

$$\sum_{i,j=1}^{d}\sigma^{i,j}(\boldsymbol{y}_k)\frac{\partial\psi(\boldsymbol{y}_k)}{\partial y^j}\nu^i(\boldsymbol{y}_k)+\sum_{i=1}^{d}a^i(\boldsymbol{y}_k)\nu^i(\boldsymbol{y}_k)=0. \tag{10.112}$$

This reduces the sums in both numerator and denominator of eq. (10.110) to

$$\sum_{i=1}^{d}\left(2\sum_{j=1}^{d}\sigma^{i,j}(\boldsymbol{y}_k)\frac{\partial\psi(\boldsymbol{y}_k)}{\partial y^j}+a^i(\boldsymbol{y}_k)\right)\nu^i(\boldsymbol{y}_k)=-\boldsymbol{a}(\boldsymbol{y}_k)\cdot\boldsymbol{\nu}(\boldsymbol{y}_k). \tag{10.113}$$

To evaluate the integral in the denominator of eq. (10.110) by the Laplace method, we denote $\tilde{\psi} = \psi(\boldsymbol{y}_k)$ and assume that there are K distinct points of absolute minimum of $\psi(\boldsymbol{y})$ on ∂D, denoted $\{\boldsymbol{y}_k\}_{k=1}^K$. Denoting by $H(\psi(\boldsymbol{y}_k))$ the $(d-1)$-dimensional Hessian of ψ in ∂D at the point $\boldsymbol{y}_k$, and using eq. (10.113), we obtain from the Laplace expansion that

$$\oint_{\partial D} K_0(\boldsymbol{y}) \exp\left\{-\frac{\psi(\boldsymbol{y})}{\varepsilon}\right\} \sum_{i=1}^d \left(2\sum_{j=1}^d \sigma^{i,j}(\boldsymbol{y})\frac{\partial \psi(\boldsymbol{y})}{\partial y^j} + a^i(\boldsymbol{y})\right) \nu^i(\boldsymbol{y})\, dS_{\boldsymbol{y}}$$
$$\sim -\,(2\pi\varepsilon)^{(d-1)/2} e^{-\tilde{\psi}/\varepsilon} \sum_{k=1}^K \boldsymbol{a}(\boldsymbol{y}_k)\cdot\boldsymbol{\nu}(\boldsymbol{y}_k) K_0(\boldsymbol{y}_k) H^{-1/2}\left(\psi(\boldsymbol{y}_k)\right). \quad (10.114)$$

Now, equation (10.110) gives

$$\lim_{\varepsilon\to 0} \Pr\left\{\boldsymbol{x}(\tau) \in \boldsymbol{y} + dS_{\boldsymbol{y}} \mid \boldsymbol{x}(0) = \boldsymbol{x}\right\}$$
$$= \frac{\sum_{k=1}^K \boldsymbol{a}(\boldsymbol{y}_k)\cdot\boldsymbol{\nu}(\boldsymbol{y}_k) K_0(\boldsymbol{y}_k) H^{-1/2}\left(\psi(\boldsymbol{y}_k)\right)\delta(\boldsymbol{y}-\boldsymbol{y}_k)\, dS_{\boldsymbol{y}}}{\sum_{k=1}^K \boldsymbol{a}(\boldsymbol{y}_k)\cdot\boldsymbol{\nu}(\boldsymbol{y}_k) K_0(\boldsymbol{y}_k) H^{-1/2}\left(\psi(\boldsymbol{y}_k)\right)}. \quad (10.115)$$

The MFPT is obtained from the equation (see Chapter 6)

$$\mathbb{E}\left[\tau \mid \boldsymbol{x}(0) = \boldsymbol{x}\right] = \frac{\displaystyle\int_D p_\varepsilon(\boldsymbol{y} \mid \boldsymbol{x})\, d\boldsymbol{y}}{\displaystyle\oint_{\partial D} \boldsymbol{J}(\boldsymbol{y} \mid \boldsymbol{x})\cdot\boldsymbol{\nu}(\boldsymbol{y})\, dS_{\boldsymbol{y}}} \quad (10.116)$$

by applying the Laplace expansion to both the volume integral in the numerator and the surface integral in the denominator. Denoting by $\mathcal{H}(\psi(\boldsymbol{x}_0))$ the d-dimensional Hessian of $\psi(\boldsymbol{x})$ at its absolute minimum in D, at the point $\boldsymbol{x}_0$, say, we obtain from eq. (10.114) the asymptotic approximation for small ε,

$$\mathbb{E}\left[\tau \mid \boldsymbol{x}(0) = \boldsymbol{x}\right] \sim \frac{\sqrt{2\pi\varepsilon}\,\mathcal{H}^{-1/2}\left(\psi(\boldsymbol{x}_0)\right)\exp\left\{\dfrac{\tilde{\psi}-\psi(\boldsymbol{x}_0)}{\varepsilon}\right\}}{\sum_{k=1}^K \boldsymbol{a}(\boldsymbol{y}_k)\cdot\boldsymbol{\nu}(\boldsymbol{y}_k) K_0(\boldsymbol{y}_k) H^{-1/2}\left(\psi(\boldsymbol{y}_k)\right)}$$
$$\times\left[1-\exp\left\{\frac{a_n(\boldsymbol{s}')\rho(\boldsymbol{x})}{\varepsilon\sigma(\boldsymbol{s}')}\right\}\right]. \quad (10.117)$$

Other cases of noncharacteristic boundaries are discussed in [213].

The WKB structure of the pdf and eqs. (10.116), (10.117), (10.108) give the large deviations theory result [54], that for $\boldsymbol{x} \in D$, outside the boundary layer,

$$\lim_{\varepsilon\to 0} \varepsilon \log \mathbb{E}\tau \, \lim_{\varepsilon\to 0} \varepsilon \left[\sup_{\boldsymbol{y}\in\partial D} \log \boldsymbol{J}(\boldsymbol{y} \mid \boldsymbol{x})\cdot\boldsymbol{\nu}(\boldsymbol{y}) - \sup_{\boldsymbol{y}\in D} \log p_\varepsilon(\boldsymbol{y} \mid \boldsymbol{x})\right].$$

This result is valid also for the case of a characteristic boundary.

Exercise 10.17 (Overdamped escape over a sharp potential barrier). The three-dimensional (or d-dimensional) overdamped motion of a Brownian particle diffusing in a field of force is described by the simplified Langevin–Smoluchowski equation (see [214])

$$\gamma \frac{d\boldsymbol{x}}{dt} + \nabla U(\boldsymbol{x}) = \sqrt{\frac{2\gamma k_B T}{m}} \frac{d\boldsymbol{W}}{dt}, \tag{10.118}$$

where γ is the dynamical viscosity (friction) coefficient, k_B is Boltzmann's constant, T is absolute temperature, m is the mass of the particle, and $\boldsymbol{W}(t)$ is three-dimensional (or d-dimensional) standard Brownian motion. Assume the potential $U(\boldsymbol{x})$ forms a well; that is, it has a single minimum at a point $\boldsymbol{x}_0$ in a simply connected domain $\mathcal{D}$ and $U(\boldsymbol{x})$ has no local maxima in $\mathcal{D}$. Assume the boundary $\partial\mathcal{D}$ has a continuous outer normal $\boldsymbol{\nu}(\boldsymbol{x})$ and $\partial U(\boldsymbol{x})/\partial\nu > 0$ for all $\boldsymbol{x} \in \partial\mathcal{D}$.

(i) What are the units of $\boldsymbol{W}(t)$?

(ii) Introduce dimensionless displacement and time to reduce eq. (10.118) to the form (10.47), where all parameters, variables, and functions are dimensionless. The domain $\mathcal{D}$ is mapped onto a domain D. This corresponds to the case that the diffusion matrix is $\boldsymbol{\sigma}(\boldsymbol{x}) = \boldsymbol{I}$ and the drift vector is $\boldsymbol{a}(\boldsymbol{x}) = -\nabla U(\boldsymbol{x})$. Choose ε and the unit of length such that $\Delta U = \max_{\partial D} U(\boldsymbol{x}) - \min_D U(\boldsymbol{x}) = 1$ and $D = \mathcal{H}^{1/2}(U(\boldsymbol{x}_0)) = 1$. Show that $\varepsilon = k_B T/m\Delta U$, where $\Delta U = \max_{\partial\mathcal{D}} U(\boldsymbol{x}) - \min_{\mathcal{D}} U(\boldsymbol{x})$. Small ε means high potential barrier or low temperature.

(iii) Show that $U(\boldsymbol{x})$ is the solution of the eikonal equation.

(iv) Find the small ε expansion of the exit density on ∂D and of the MFPT in terms of the potential $U(\boldsymbol{x})$ and its derivatives.

(v) Return to dimensional variables and express the Hessians in terms of vibration frequencies at the bottom of the potential well and at saddle points on the boundary.

(vi) Derive the relation $\lambda = 1/\mathbb{E}\tau$ between the escape rate and the MFPT for this case of sharp boundaries (see Section 10.1).

(vii) Express the pre-exponential term in the escape rate in terms of "attempt frequencies" at the bottom of the well and the frequencies of vibration in saddle points on the boundary.

(viii) Express the exponential part in terms of "activation energy".

(ix) Explain the effect of the normal derivative $\partial U(\boldsymbol{x})/\partial\boldsymbol{\nu}$ on the escape rate. □

Exercise 10.18 (Escape at critical energy [169]). Consider the random motion of a one-dimensional Brownian particle diffusing in a field of force. It is described by the Langevin equation (7.103),

$$\ddot{x} + \gamma\dot{x} + \tilde{U}'(x) = \sqrt{\frac{2\gamma k_B T}{m}}\,\dot{\tilde{w}}, \tag{10.119}$$

where $\tilde{U}(x)$ is the potential of the force. Assume that the potential forms a well and $\tilde{U}(x)$ has a single local minimum at a point x_0 and a single maximum at the origin.

Define the energy of a trajectory by $\mathcal{E}(t) = \dot{x}^2(t)/2 + \tilde{U}(x(t))$.

(i) Introduce non-dimensional variables so that the Langevin equation (10.119) can be written in the form

$$\frac{d^2\xi}{d\tau^2} + \beta\frac{d\xi}{d\tau} + U'(\xi) = \sqrt{2\beta\varepsilon}\,\frac{dw}{d\tau},$$

where ξ is dimensionless displacement, τ is dimensionless time, β is a dimensionless friction coefficient, $U(\xi)$ is dimensionless potential, and $w(\tau)$ is dimensionless Brownian motion. Assume that

$$\lim_{\xi\to\pm\infty} U(\xi) = \mp\infty, \quad U'(\xi_0) = 0, \quad U''(\xi_0) = \omega_0^2 \tag{10.120}$$
$$U(0) = U'(0) = 0, \quad U''(0) = -\omega_C^2,$$

and that the dimensionless variables are chosen so that $\Delta U = U(0) - U(\xi_0) = 1$, $\omega_0 = 1$. Define dimensionless energy $E(t) = \frac{1}{2}(d\xi/d\tau)2 + U(\xi)$.

(ii) Define $\eta = d\xi/d\tau$ and write the dimensionless Langevin equation as the phase plane system

$$\frac{d\xi}{d\tau} = \eta, \quad \frac{d\eta}{d\tau} = -\beta\eta - U'(\xi) + \sqrt{2\beta\varepsilon}\,\frac{dw}{d\tau}. \tag{10.121}$$

(iii) Linearize the noiseless dynamics

$$\frac{d\xi}{d\tau} = \eta, \quad \frac{d\eta}{d\tau} = -\beta\eta - U'(\xi) \tag{10.122}$$

around the critical points $(\xi_0, 0)$ and $(0, 0)$ and show that the former is an attractor and the latter is a saddle point. Find the eigenvalues and eigenvectors of the matrices $\boldsymbol{A}_0$ and $\boldsymbol{A}_C$ of the linearized system at both critical points, respectively.

(iv) Show that the boundary of the domain of attraction of the attractor $(\xi_0, 0)$ consists of the two unstable trajectories of (10.122) that emanate from the saddle point $(0, 0)$ in the direction of the eigenvector corresponding to the positive eigenvalue of $\boldsymbol{A}_C$. The domain of attraction is denoted D and its boundary, denoted Γ, is called the *separatrix* of the system (10.122). Draw Γ and the flow lines of the plane flow (10.122), inside and outside D and interpret the flow portrait in terms of the motion of a particle. Determine the slope and the outer unit normal to Γ at the saddle point $(0, 0)$.

(v) Define the energy $E(\xi, \eta) = \frac{1}{2}\eta^2 + U(\xi)$ and the critical energy contour $\Gamma_C = \{E(\xi, \eta) = E_C\}$, where $E_C = 0$. Determine the slope of Γ_C at the saddle point and draw the separatrix and the contour Γ_C. Show that Γ_C forms a noncharacteristic boundary of the domain $D_C = \{E(\xi, \eta) < E_C\}$, except for two critical points.

(vi) Construct a small ε asymptotic solution of the stationary FPE (10.52) in D with absorbing boundary conditions on E_C, valid away from the saddle point at $(0, 0)$. Can this expansion be valid up to the saddle point? Use the expansion to determine the distribution of exit points on Γ_C of the trajectories of (10.121) that start out in D_C. Calculate the MFPT to Γ_C. □

10.2.8 Exit through characteristic boundaries in the plane

In the case of a two-dimensional system with a characteristic boundary, we use the expansions of Section 10.2.5 in eqs. (10.108) and (10.116). Using eq. (10.104) in (10.108) gives

$$\boldsymbol{J}\cdot\boldsymbol{\nu}|_{\partial D}(s) \sim \sqrt{\frac{2\varepsilon}{\pi}} K_0(0,s)\xi(s)\sigma(s)e^{-\hat{\psi}/\varepsilon}, \tag{10.123}$$

hence

$$\Pr\left[(x(\tau),y(\tau))=(x,y)\,|\,(x_0,y_0)\right]ds \sim \frac{K_0(0,s)\xi(s)\sigma(s)\,ds}{\displaystyle\int_0^S K_0(0,s)\xi(s)\sigma(s)\,ds}. \tag{10.124}$$

The function $K_0(0,s)$ can be expressed in terms of the coefficients of the problem as follows. The function $K_0(0,s)$ is the solution of the transport equation (10.83). Using the assumption (10.95) and the eikonal equation (10.96), eq. (10.83) on the boundary becomes

$$\frac{d}{ds}K_0(0,s) = -\left[a^0(s)+B'(s)+\sigma(s)d\phi(s)\right]K_0(0,s). \tag{10.125}$$

Now we use the fact that $\phi(s) = -\xi^2(s)$ to rewrite eq. (10.125) in the separated form

$$\frac{dK_0}{K_0} = -\left(\frac{a^0}{B}+\frac{B'}{B}-\frac{\sigma\xi^2}{B}\right)ds. \tag{10.126}$$

Integrating eq. (10.126) and simplifying it with the aid of eq. (10.105), we obtain $K_0(0,s) = \hat{K}_0\sqrt{-\phi(s)}/B(s)$, where $\hat{K}_0 = const.$ Note that $\phi(s)$ cannot change sign, because if it vanishes at a point, it vanishes everywhere, as indicated by the Bernoulli equation (10.99). It has to be negative, because at a point s of local extremum, we have $\phi(s) = -a^0(s)/\sigma(s)$. The assumption that the boundary is an unstable limit cycle of the drift equation (10.48) implies that $a^0(s)/\sigma(s) \geq 0$.

Using these simplifications in eq. (10.124) gives the more explicit expression for the exit density

$$\Pr\left[(x(\tau),y(\tau))=(x,y)\,|\,(x_0,y_0)\right]ds \sim \frac{\left(\xi^2(s)\sigma(s)/B(s)\right)\,ds}{\displaystyle\int_0^S \left(\xi^2(s)\sigma(s)/B(s)\right)\,ds}. \tag{10.127}$$

The MFPT to the boundary from any fixed point $(x,y)\in D$ is calculated as above, but with the flux given by eq. (10.123). We obtain

$$\mathbb{E}\left[\tau\,|\,(x,y)\right] \sim \frac{\pi^{3/2}\sqrt{2\varepsilon}\mathcal{H}^{-1/2}\left(\psi(0,0)\right)}{\displaystyle\int_0^S K_0(0,s)\xi(s)\sigma(s)\,ds}\exp\left\{\frac{\hat{\psi}}{\varepsilon}\right\}, \tag{10.128}$$

where $\mathcal{H}(\psi(0,0))$ is the Hessian at the stable equilibrium point $(0,0)$.

Example 10.3 (Constant speed). If $a^0(s)/\sigma(s) = const. > 0$, then $\phi(s) = -a^0(s)/\sigma(s) = const.$ and $\xi(s) = \sqrt{-\phi(s)} = \sqrt{a^0(s)/\sigma(s)} = const.$, hence eq. (10.127) gives

$$\Pr\left[(x(\tau),y(\tau)) = (x,y) \mid (x_0,y_0)\right] ds \sim \frac{[\sigma(s)/B(s)]\,ds}{\int\limits_0^S [\sigma(s)/B(s)]\,ds};$$

that is, the density is inversely proportional to the local speed of motion of the drift equation (10.48) on the boundary.

If $B(s)$ changes sign on ∂D, then $\psi(0,s)$ is not constant in general. If $\psi(0,s)$ has absolute minima at a finite number of points, the total flux has to be evaluated by the Laplace method at these points. □

Exercise 10.19 (Characteristic boundary with critical points [169]). If the speed $B(s)$ vanishes at isolated points or changes sign on ∂D, the eikonal function on the boundary, $\psi(0,s)$, is not constant in general. Thus the following three classes of characteristic boundaries can be distinguished.
Type I: The speed doesn't vanish; $B(s) > 0$, as considered above.
Type II: The flow on the boundary is unidirectional with N unstable critical points;

$$B(s) \geq 0, \quad \frac{\partial^j B(s_i)}{\partial s^j} = 0, \quad 0 \leq j < k_i, \ \ i = 1,\ldots,N$$
$$\frac{\partial^{k_i} B(s_i)}{\partial s^{k_i}} \ \text{are all} > 0 \text{ or all} < 0, \quad k_i \ \text{even}.$$

Type III: The flow on the boundary has N stable and N unstable critical points; $B(s)$ changes sign. At stable points s_i

$$\frac{\partial^j B(s_i)}{\partial s^j} = 0, \quad 0 \leq j < k_i, \quad \frac{\partial^{k_i} B(s_i)}{\partial s^{k_i}} < 0, \ \ k_i \ \text{odd}, \ \ i = 1,\ldots,N.$$

At unstable points σ_i,

$$\frac{\partial^j B(\sigma_i)}{\partial s^j} = 0, \quad 0 \leq j < l_i, \quad \frac{\partial^{l_i} B(\sigma_i)}{\partial s^{l_i}} > 0, \ \ l_i \ \text{odd}, \ \ i = 1,\ldots,N.$$

(i) Retain terms of order $\sqrt{\varepsilon}$ and ε in the boundary layer equation and use the transformation $\eta = \xi(s)\zeta$, as in Section 10.2.5.

(ii) Introduce into the boundary layer equation the stretched variable $\phi = (s - s_i)\varepsilon^{-r}$, where r is chosen by balancing terms in the boundary layer equation, depending on the order of the zero of $B(s)$ at s_i. Choose the value of $\gamma(s_i)$ so that $\gamma'(s_i)$ remains bounded and S-periodic.

(iii) Retain the second derivative with respect to ϕ in the stretched boundary layer equation. The resulting boundary layer equation is separated with respect to η and

ϕ (or s). Solve it by separation of variables and obtain a singularly perturbed eigenvalue problem with periodic boundary conditions in the variable s.

(iv) Use the matching condition to determine the η-dependence of the boundary layer function.

(v) Find explicit expressions for the exit density and the MFPT for the exit problem. □

10.2.9 Kramers' exit problem

We consider the exit problem from a domain $D \subset \mathbb{R}^d$ whose boundary ∂D is characteristic and assume that $\psi(\boldsymbol{y})$ has a single absolute minimum on ∂D at a point $\boldsymbol{y}_k \in D$. As above, we introduce local orthogonal coordinates $(\rho, \boldsymbol{s})$ near $\boldsymbol{y}_k$, where ρ is the distance to ∂D and $\boldsymbol{s} = (s_1, \dots, s_{d-1})$ are coordinates in the tangent plane to ∂D, such that $\boldsymbol{y}_k$ is mapped to the origin. The decomposition (10.95) becomes

$$\boldsymbol{a}(\rho, \boldsymbol{s}) = a^0(\boldsymbol{s})\rho\nabla\rho + [\boldsymbol{B}_0 + O(\rho + |\boldsymbol{s}|)]\,\boldsymbol{s}, \tag{10.129}$$

where $a^0(\boldsymbol{s}) > 0$ and $\boldsymbol{B}_0$ is a $(d-1) \times (d-1)$ constant matrix whose eigenvalues are assumed negative. The boundary layer equation is again (10.104), but (10.106) takes the form

$$\boldsymbol{B}_0\boldsymbol{s} \cdot \nabla_{\boldsymbol{s}}\xi_0(\boldsymbol{s}) + a^0(\boldsymbol{s})\xi_0(\boldsymbol{s}) - \sigma(\boldsymbol{s})\xi_0^3(\boldsymbol{s}) = 0. \tag{10.130}$$

At the saddle point $\boldsymbol{s} = \boldsymbol{0}$ we have $\sigma(\boldsymbol{y}_k) = \boldsymbol{\nu}(\boldsymbol{y}_k) \cdot \boldsymbol{\sigma}(\boldsymbol{y}_k)\boldsymbol{\nu}(\boldsymbol{y}_k)$, and

$$\xi_0(\boldsymbol{0}) = \sqrt{\frac{a^0(\boldsymbol{y}_k)}{\sigma(\boldsymbol{y}_k)}},$$

so the flux density (10.108) near the saddle point $\boldsymbol{y}_k$ is

$$\begin{aligned}\boldsymbol{J} \cdot \boldsymbol{\nu}\big|_{\boldsymbol{y}\in\partial D}(s) &\sim -\, K_0(\boldsymbol{y}) \sum_{i,j=1}^{d} \varepsilon\sigma^{i,j}(\boldsymbol{y})\exp\left\{-\frac{\psi(\boldsymbol{y})}{\varepsilon}\right\} \frac{\partial q_\varepsilon(\boldsymbol{y}\,|\,\boldsymbol{x})}{\partial y^i}\nu^j(\boldsymbol{y})\\ &= \sqrt{\frac{2\varepsilon}{\pi}} K_0(\boldsymbol{y})\sqrt{a^0(\boldsymbol{y})\sigma(\boldsymbol{y})}\,e^{-\psi(\boldsymbol{y})/\varepsilon}\end{aligned} \tag{10.131}$$

and the boundary flux is

$$\begin{aligned}&\oint_{\partial D} \boldsymbol{J}(\boldsymbol{y}) \cdot \boldsymbol{\nu}(\boldsymbol{y})\, dS_{\boldsymbol{y}} \\ &\sim \frac{(2\pi\varepsilon)^{d/2} K_0(\boldsymbol{y}_k)\sqrt{a^0(\boldsymbol{y}_k)\sigma(\boldsymbol{y}_k)}\,\exp\left\{-\dfrac{\psi(\boldsymbol{y}_k)}{\varepsilon}\right\}}{\pi\, H^{1/2}(\boldsymbol{y}_k)}.\end{aligned} \tag{10.132}$$

Therefore, if $\psi(\boldsymbol{y})$ achieves its absolute minimum $\hat{\psi}$ on ∂D at points $\boldsymbol{y}_k$ ($k = 1, \ldots, K$), then for small ε the exit density is (see (10.115))

$$\Pr\{\boldsymbol{x}(\tau) = \boldsymbol{y} \,|\, \boldsymbol{x}\} \sim \frac{\sum_{k=1}^{K} K_0(\boldsymbol{y}_k) H^{-1/2}(\boldsymbol{y}_k) \sqrt{a^0(\boldsymbol{y}_k)\sigma(\boldsymbol{y}_k)}\, \delta(\boldsymbol{y} - \boldsymbol{y}_k)}{\sum_{k=1}^{K} K_0(\boldsymbol{y}_k) H^{-1/2}(\boldsymbol{y}_k) \sqrt{a^0(\boldsymbol{y}_k)\sigma(\boldsymbol{y}_k)}}.$$

Thus the exit density on the boundary is independent of the initial point $\boldsymbol{x} \in D$ for $\boldsymbol{x}$ outside the boundary layer.

To evaluate the MFPT, we use (10.132) and (10.57) in (10.56) and evaluate the integral in the numerator by the Laplace method to obtain

$$\begin{aligned} &\mathbb{E}\left[\tau \,|\, \boldsymbol{x}(0) = \boldsymbol{x}\right] \\ &\sim \frac{\pi \mathcal{H}^{-1/2}\left(\psi(\boldsymbol{x}_0)\right) \exp\left\{\tilde{\psi} - \psi(\boldsymbol{x}_0)\right\}}{\sum_{k=1}^{K} K_0(\boldsymbol{y}_k) \sqrt{a^0(\boldsymbol{y}_k)\sigma(\boldsymbol{y}_k)} H^{-1/2}(\boldsymbol{y}_k)}\, Q^0(\zeta', \boldsymbol{s}'), \end{aligned} \tag{10.133}$$

where

$$Q^0(\zeta', \boldsymbol{s}') = -\sqrt{\frac{2}{\pi}} \int_0^{\xi(\boldsymbol{s}')\zeta'} e^{-z^2/2}\, dz. \tag{10.134}$$

The multidimensional Kramers' problem of the escape of a Brownian particle over a potential barrier in the overdamped regime, as described in Exercise 10.17, is described by the Langevin-Smoluchowski-Kramers equation (10.118), which can be written in dimensionless variables as (see Exercise 10.17)

$$\dot{\boldsymbol{x}} + \nabla U(\boldsymbol{x}) = \sqrt{2\varepsilon}\, \dot{\boldsymbol{w}}, \tag{10.135}$$

where $U(\boldsymbol{x})$ is a potential that forms a well.

The well is the domain of attraction D of a local (or global) minimum point $\boldsymbol{x}_0$ of $U(\boldsymbol{x})$; that is, the well is the locus of all points $\boldsymbol{x}$ such that the trajectory of the dynamical system

$$\dot{\boldsymbol{x}} = -\nabla U(\boldsymbol{x}), \quad \boldsymbol{x}(0) = \boldsymbol{x} \tag{10.136}$$

converges to $\boldsymbol{x}_0$ as $t \to \infty$. Thus $\boldsymbol{x}(t) \notin D$ for all $t \geq 0$ if $\boldsymbol{x} \notin D$. We assume that the boundary ∂D of the domain of attraction contains a single saddle point $\boldsymbol{x}_S$, where $U(\boldsymbol{x}_S) =$
$min_{\boldsymbol{x} \in \partial D} U(\boldsymbol{x})$ (see figures 10.4–10.6).

The MFPT to ∂D for this case is given by (10.117), where $\psi(\boldsymbol{x}) = U(\boldsymbol{x})$ and $\boldsymbol{a}(\boldsymbol{x}) = -\nabla U(\boldsymbol{x})$ (see Exercise 10.16). The specialization of (10.133) to this case gives

$$\psi(\boldsymbol{y}) = U(\boldsymbol{y}), \quad K_0(\boldsymbol{y}) = 1, \quad a^0(\boldsymbol{y}_k) = -\left.\frac{\partial^2 U(\boldsymbol{y})}{\partial \rho^2}\right|_{\boldsymbol{y} = \boldsymbol{y}_k} = \omega_d^2(\boldsymbol{y}_k)$$

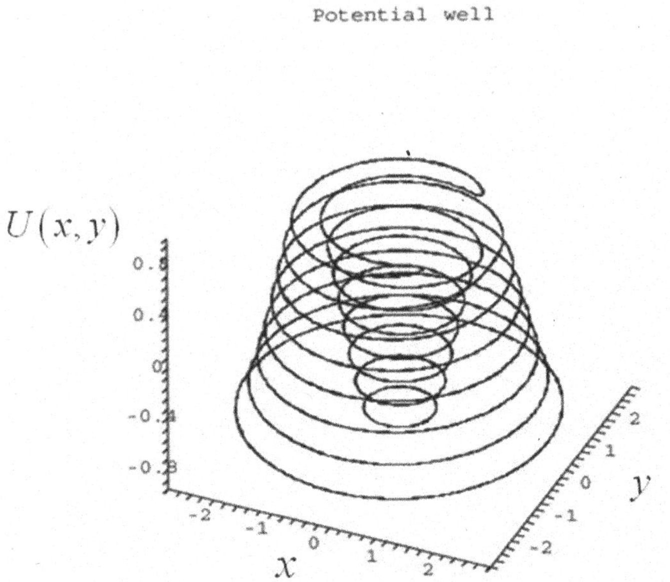

Figure 10.4. Potential well with a single saddle point on the boundary.

and

$$\mathcal{H}^{1/2}\left(\psi(\boldsymbol{x}_0)\right)=\prod_{i=1}^{d}\omega_i(0),\quad H^{1/2}\left(\psi(\boldsymbol{y}_k)\right)=\prod_{i=1}^{d-1}\omega_i(\boldsymbol{y}_k),$$

where $\omega_i^2(0)$ are the (positive) eigenvalues of the Hessian matrix of $U(\boldsymbol{y})$ at the bottom of the well $\boldsymbol{x}_0$ and $\omega_i^2(\boldsymbol{y}_k)$ are the $d-1$ positive eigenvalues of the $(d-1)\times(d-1)$ Hessian at the saddle point $\boldsymbol{y}_k$ and $-\omega_d^2(\boldsymbol{y}_k)$ is the negative eigenvalue there. For $d=1$ (see (6.66)) the only one eigenvalue of the Hessian $\mathcal{H}$ is ω_A^2 andthere is no saddle point, so $H=1$, and $\omega_d(y_k)=\omega_C$. Scaling γ into the MFPT, we see that (10.133) reduces to (6.66).

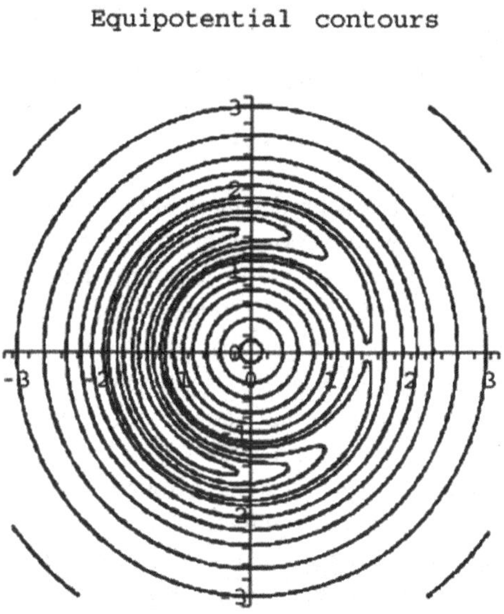

Figure 10.5. Equipotential contours of the potential in Figure 10.4.

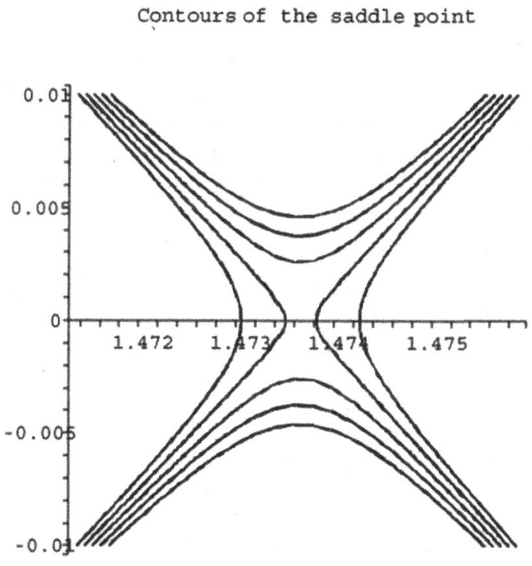

Figure 10.6. The saddle point region in Figure 10.5. The boundary is the trajectory that converges to the saddle point (on the x-axis). The unstable direction is the x-axis, which is normal to the boundary.

10.3 Activated escape in Langevin's equation

In Kramers' [140] model of the chemical dissociation reaction

$$AB \overset{\kappa}{\rightharpoonup} A + B \tag{10.137}$$

with rate κ, the molecule $\boldsymbol{AB}$ undergoes thermal collisions with the surrounding medium and therefore the relative motion of the two bound components can be represented as a Brownian motion in a field of force. In the one-dimensional model of the reaction the chemical bond is represented by a potential well of the (local) form of Figure 8.1, in which the chemically bound state $\boldsymbol{AB}$ is the equilibrium at the bottom of the well x_A. The dissociation reaction consists in thermal activation of the molecule over the potential barrier at x_C. As mentioned in Chapter 8, Kramers sought to determine the dependence of κ on temperature T and on the viscosity (friction) γ, and to compare the values found with the results of transition state theory, which recovers Arrhenius' law [87]. He solved the problem under the assumption that the activation energy E_C (per unit mass) is much higher than the thermal energy k_BT/m in the Smoluchowski limit of large γ (see Section 8.1) and in the underdamped limit (see Section 8.4).

Specifically, assuming

$$\varepsilon = \frac{k_BT}{mE_C} \ll 1,$$

the stochastic trajectories of eq. (8.1) in configuration space $x(t)$ stay inside the well for a long time, but ultimately escape. Kramers' exit problem is to (i) find an explicit asymptotic expression for the rate κ, as a function of ε and γ, valid for all $\gamma > 0$, in the limit of small ε, and (ii) find the density of exit points on the (stochastic) separatrix. Kramers solved (i) in the two limits $\gamma \to \infty$ and $\gamma \to 0$. The solution to Kramers' problems (i) and (ii) in the one-dimensional case of the Smoluchowski limit is given in Sections 6.4.1 and 6.4.2 and Kramers' result in the underdamped limit is (8.89). The activated escape from a nonequilibrium steady-state (e.g., a stable limit cycle) is given in Section 8.5.

The problem of a uniform approximation to κ for all values of γ remained unsolved for over 40 years (see [100]). This section presents the solution of (i) and (ii) in the phase plane and a uniform approximation to κ over the full range of the friction coefficient γ.

10.3.1 The separatrix in phase space

We consider the escape of a Brownian particle in a field of force of a potential $U(x)$, from the domain of attraction D of the stable equilibrium point A (see Figure 10.7). To describe the escape process the Langevin equation is converted to the phase plane system

$$\begin{aligned} \dot{x} &= y \\ \dot{y} &= -\gamma y - U'(x) + \sqrt{2\varepsilon\gamma}\,\dot{w}. \end{aligned} \tag{10.138}$$

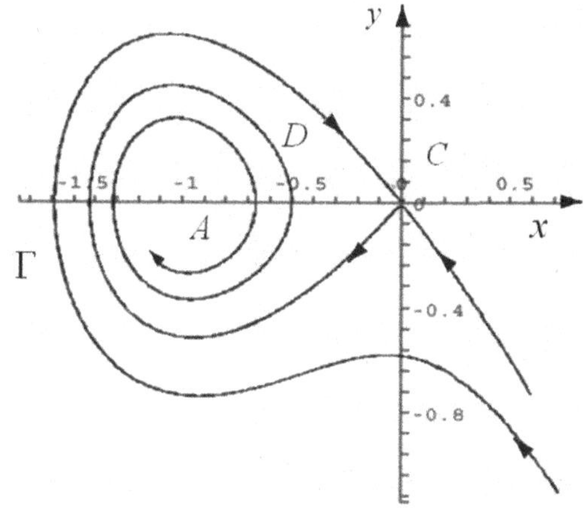

Figure 10.7. The domain of attraction D of the stable equilibrium point $A = (x_A, 0)$ in the phase plane (x, y) is bounded by the separatrix Γ. The saddle point is at $C = (x_C, 0)$. The potential is $U(x) = (x+1)^2/2 - (x+1)^3/3$.

The drift vector and the noise matrix of the stochastic system eq. (10.138), corresponding to the Langevin equation (8.1) in the phase plane, are denoted

$$\boldsymbol{b}(x,y) = \begin{pmatrix} b_1(x,y) \\ b_2(x,y) \end{pmatrix} = \begin{pmatrix} y \\ -\gamma y - U'(x) \end{pmatrix}$$

$$\boldsymbol{\sigma}(x,y) = \begin{pmatrix} 0 & 0 \\ 0 & \sqrt{2\varepsilon\gamma} \end{pmatrix}. \tag{10.139}$$

The point $A = (x_A, 0)$ is an attractor, while the point $C = (x_C, 0)$ is a saddle point. We assume for simplicity that $x_A < x_C = 0$, as in Figure 10.7. The domain of attraction of the attractor and its boundary are denoted D and Γ, respectively. The curve Γ is the separatrix of the noiseless dynamics and in the limit $\varepsilon \to 0$ it becomes the stochastic separatrix. We denote by $\omega_C = \sqrt{-U''(x_C)}$ the imaginary frequency at the top of the barrier. The energy at a phase plane point (x, y) is

$$E(x,y) = \frac{y^2}{2} + U(x)$$

and the depth of the potential well is $E_C = U(x_C) - U(x_A)$ (we assume, without loss of generality, that $U(x_A) = 0$). The critical energy contour in the phase plane is $E(x; y) = E_C$ and is given explicitly by

$$y = y_C(x) = \sqrt{2(E_C - U(x))}$$

(see Figure 8.2).

The separatrix Γ is given by $y = y_\Gamma(x)$, where $y_\Gamma(x)$ is the solution of the initial value problem [169]

$$y'_\Gamma(x) = -\gamma - \frac{U'(x)}{y_\Gamma(x)}, \qquad y_\Gamma(x_C) = 0. \tag{10.140}$$

Of the two solutions of eq. (10.140), $y_\Gamma(x)$ is the one with

$$y'_\Gamma(x_C) = -\frac{\gamma + \sqrt{\gamma^2 + 4\omega_C^2}}{2} = -\lambda. \tag{10.141}$$

The outer unit normal on Γ,

$$\boldsymbol{\nu}(x, y_\Gamma(x)) = \begin{pmatrix} \nu_1(x, y_\Gamma(x)) \\ \nu_2(x, y_\Gamma(x)) \end{pmatrix},$$

is given by

$$\boldsymbol{\nu}(x, y_\Gamma(x)) = \frac{1}{\sqrt{\left[\gamma y_\Gamma(x) + U'(x)\right]^2 + y_\Gamma^2(x)}} \begin{pmatrix} \gamma y_\Gamma(x) + U'(x) \\ y_\Gamma(x) \end{pmatrix}. \tag{10.142}$$

We denote by $\bar{\tau}_1(A)$ the MFPT from x_A to the critical energy contour E_C and by $\bar{\tau}_2(E_C)$ the MFPT from E_C to the separatrix Γ.

10.3.2 Kramers' exit problem at high and low friction

We assume that all variables are dimensionless and consider Kramers' exit problem in the domain $E(x, y) < E_C$ in the limit of small ε.

Theorem 10.3.1. *The probability density function of terminal points on $E = E_C$ is given by*

$$\Pr\{(x(\tau), y(\tau)) \in (x + dx, y_C(x)) \,|\, \xi, \eta\} \sim \frac{y_C(x)\, dx}{I_C}, \tag{10.143}$$

for (ξ, η) such that $E(\xi, \eta) < E_C$ outside the boundary layer, where τ is the FPT to E_C and

$$I_C = \oint_{E_C} y_C(x)\, dx$$

is the action of the critical energy contour, and the MFPT is given by

$$\bar{\tau}_1(A) \sim \frac{2\pi\varepsilon}{\gamma I_C \omega_A} e^{(E_C - E_A)/\varepsilon}. \tag{10.144}$$

Proof. To calculate $\bar{\tau}_1(A)$ and the density of exit points on E_C, we use the general "population over flux" expression $\bar{\tau} = N/F$, given explicitly as

$$\mathbb{E}\left[\tau \,|\, \boldsymbol{x}(0) = \boldsymbol{x}\right] = \frac{\displaystyle\int_D p_\varepsilon\left(\boldsymbol{y} \,|\, \boldsymbol{x}\right) d\boldsymbol{y}}{\displaystyle\oint_{\partial D} \boldsymbol{J}\left(\boldsymbol{y} \,|\, \boldsymbol{x}\right) \cdot \boldsymbol{\nu}(\boldsymbol{y})\, dS_{\boldsymbol{y}}}. \tag{10.145}$$

The probability density function $p\,(x, y,\,|\,\xi, \eta)$ is the solution of the FPE for subcritical energies $E(x, y), E(\xi, \eta) < E_C$

$$\varepsilon\gamma\frac{\partial^2 p}{\partial y^2} - y\frac{\partial p}{\partial x} + \frac{\partial[\gamma y + U'(x)]p}{\partial y} = -\,\delta(x-\xi)\delta(y-\eta) \tag{10.146}$$
$$p\,(x, y\,|\,\xi, \eta) = 0 \text{ for } E(\xi, \eta) < E_C,\ E(x, y) = E_C.$$

Writing

$$p\,(x, y\,|\,\xi, \eta) = e^{-E(x,y)/\varepsilon}q(x, y\,|\,\xi, \eta) \tag{10.147}$$

converts (10.146) into the boundary value problem

$$L^*_{x,y}q = \varepsilon\gamma\frac{\partial^2 q}{\partial y^2} + y\frac{\partial q}{\partial x} - [\gamma y + U'(x)]\frac{\partial q}{\partial y} \tag{10.148}$$
$$= -\,e^{E(\xi,\eta)/\varepsilon}\delta(x-\xi)\delta(y-\eta) \text{ for } \quad E(x, y), E(\xi, \eta) < E_C$$
$$q(x, y\,|\,\xi, \eta) = 0 \text{ for } E(\xi, \eta) < E_C,\ E(x, y) = E_C.$$

Note that the constant $e^{E(\xi,\eta)/\varepsilon}$ is arbitrary, because it cancels in eq. (10.145). For small ε the solution $q(x, y\,|\,\xi, \eta)$ is a boundary layer function that matches to 1 in the interior of E_C and vanishes on the (noncharacteristic) boundary E_C. Therefore, changing variables $(x, y) \to (x, \zeta)$, where

$$\zeta = \frac{E_C - E(x, y)}{\varepsilon},$$

writing $q(x, y\,|\,\xi, \eta) = Q(x, \zeta\,|\,\xi, \eta)$, and expanding $Q = Q^0 + \varepsilon Q^1 + \cdots$, we find that Q^0 is a function of z and obtain the leading-order boundary value problem

$$\frac{\partial^2 Q^0}{\partial \zeta^2} + \frac{\partial Q^0}{\partial \zeta} = 0 \text{ for } 0 < \zeta < \infty \tag{10.149}$$
$$Q^0(0) = 0, \quad \lim_{\zeta\to\infty} Q^0(\zeta) = 1, \tag{10.150}$$

whose solution is

$$Q^0(\zeta) = 1 - e^{-\zeta}. \tag{10.151}$$

Using the approximate solution

$$p\,(x, y,\,|\,\xi, \eta) \sim e^{-E(x,y)/\varepsilon}\left(1 - e^{(E(x,y)-E_C)/\varepsilon}\right)$$

in eq. (10.145), we obtain (10.143).

The MFPT is given by the asymptotic approximation

$$\bar{\tau}_1(A) \sim \frac{\displaystyle\iint_{E<E_C} e^{-E/\varepsilon}dx\,dy}{\displaystyle\varepsilon\oint_{E_C} e^{-E/\varepsilon}\gamma\frac{\partial Q^0}{\partial y}\,dx}. \tag{10.152}$$

Evaluating the integrals in (10.152) by the Laplace method, keeping in mind that $E = E_C$ in the denominator, we obtain (10.144). □

10.3.3 The MFPT to the separatrix Γ

Note that if γ is not small relative to $\sqrt{\varepsilon}$, the contour $E(x,y) = E_C$ is outside the boundary layer of q near Γ, except for points near the saddle point $(x_C, 0)$ (see Section 10.3.5 below). It follows that $q(x, y_C(x)\,|\,\xi,\eta) \sim 1$ for $E(\xi,\eta) < E_C$ and x away from x_C. Under this assumption we have the following for small ε.

Theorem 10.3.2 (Kramers' rate formula). *If* $\min(\gamma,1) \gg \varepsilon$, *then the escape rate is given by*

$$\kappa_{Kramers} = \frac{1}{2\bar{\tau}_\Gamma(x_A,0)}, \tag{10.153}$$

where

$$\bar{\tau}_\Gamma(x_A,0) \sim \frac{2\pi\omega_C e^{(E_C-E_A)/\varepsilon}}{\omega_A\left[\sqrt{\gamma^2+4\omega_C^2}-\gamma\right]}. \tag{10.154}$$

Proof. The MFPT $\bar{\tau}_\Gamma(x,y)$ is given by eq. (10.145) with $p\,(x,y,\,|\,\xi,\eta)$ the solution of the boundary value problem (10.146) inside the domain D, whose boundary is Γ; that is,

$$\begin{aligned}\varepsilon\gamma\frac{\partial^2 p}{\partial y^2} - y\frac{\partial p}{\partial x} + \frac{\partial[\gamma y + U'(x)]p}{\partial y} &= -\,\delta(x-\xi)\delta(y-\eta) \ \text{ for } (x,y),\ (\xi,\eta)\in D\\ p\,(x,y\,|\,\xi,\eta) &= 0 \ \text{ for } (\xi,\eta)\in D,\ (x,y)=\Gamma.\end{aligned} \tag{10.155}$$

As above, writing (10.147) converts (10.155) into the backward boundary value problem (10.148) in D. To construct the boundary layer , we change variables in eq. (10.148) to local variables near Γ; that is, $(x,y)\to(x,\rho)$, where $\rho(x,y) = \text{dist}((x,y),\Gamma)$, and then we stretch ρ by setting $\zeta = \rho/\sqrt{\varepsilon}$. Expanding

$$q(x,y\,|\,\xi,\eta) = Q^0(x,\zeta\,|\,\xi,\eta) + \sqrt{\varepsilon}Q^1(x,\zeta\,|\,\xi,\eta) + \cdots, \tag{10.156}$$

we obtain the leading-order boundary layer equation

$$\gamma\left(\frac{\partial\rho(x,0)}{\partial y}\right)^2\frac{\partial^2 Q^0}{\partial\zeta^2} + \zeta b_0(x)\frac{\partial Q^0}{\partial\zeta} + y_\Gamma(x)\frac{\partial Q^0}{\partial x} = 0 \tag{10.157}$$

for $\zeta > 0,\ x < x_C$, where the separatrix Γ is given by $y = y_\Gamma(x)$, which is defined in (10.140) and (10.141). Here we have used the representation of the flow vector (10.139) in local coordinates as

$$\boldsymbol{b}(x,y) = b_0(x)\left(\rho + O\left(\rho^2\right)\right)\boldsymbol{\nu} + (b_1(x)+o(1))\boldsymbol{t}$$

near $\rho = 0$, where $o(1)\to 0$ as $\rho\to 0$, and $\boldsymbol{\nu}$, $\boldsymbol{t}$ are the unit normal and tangent at Γ, respectively. As for (10.149), the boundary and matching conditions for (10.157) are (10.150).

To solve the boundary value problem (10.157), (10.150), we change variables by defining $\mu = \beta(x)\zeta$, where $\beta(x)$ is the solution of the Bernoulli equation

$$y_\Gamma(x)\beta'(x) + b_0(x)\beta(x) = \gamma\rho_y^2(x, y_\Gamma(x))\beta^3(x), \tag{10.158}$$

that satisfies the condition

$$\beta(x_C) = \left[\frac{b_0(x_C)}{\gamma\rho^2(x_C, 0)}\right]^{1/2}.$$

The functions $b_0(x)$ and $\rho_y^2(x, y_\Gamma(x))$ are, respectively, given by

$$b_0(x) = \frac{y_\Gamma(x)U'(x)[1 - U''(x)] - \gamma y_\Gamma(x)U''(x)}{y_\Gamma^2(x) + [\gamma y_\Gamma(x) + U'(x)]^2} \tag{10.159}$$

$$\rho_y^2(x, y_\Gamma(x)) = \frac{y_\Gamma^2(x)}{y_\Gamma^2(x) + [\gamma y_\Gamma(x) + U'(x)]^2}. \tag{10.160}$$

Then (10.157) becomes

$$\frac{\partial^2 Q^0}{\partial \mu^2} + \mu\frac{\partial Q^0}{\partial \mu} + \frac{y_\Gamma(x)}{\gamma\rho_y^2(x, 0)\beta^2(x)}\frac{\partial Q^0}{\partial x} = 0. \tag{10.161}$$

Equation (10.161) is a backward degenerate parabolic equation, whose unique solution is

$$Q^0(x, \mu) = \left(\frac{2}{\pi}\right)^{1/2}\int_0^\mu e^{-z^2/2}\,dz = \operatorname{erf}\left(\frac{\zeta(x, y)\beta(x)}{\sqrt{2}}\right). \tag{10.162}$$

Using (10.162) in (10.156), then in (10.147), and finally in (10.145), we obtain (10.154) from the Laplace expansions of the integrals. Equation (10.153) follows from (10.46). □

Exercise 10.20. (Kramers' overdamped limit).

(i) Express the assumption of Theorem 10.3.2 in dimensional variables.

(ii) Show that for large γ Kramers' (10.153) reduces to (6.68). □

10.3.4 Uniform approximation to Kramers' rate

Kramers' formula (10.154) in Theorem 10.3.2 was proved under the assumption the $\gamma \gg \sqrt{\varepsilon}$, whereas the proof of eq. (10.144) in Theorem 10.3.1 did not require this assumption. Actually, $y_\Gamma(x) \to y_C(x)$ as $\gamma \to 0$, so (10.144) is expected to become the leading term in the expansion of the MFPT in Kramers' exit problem in this limit. However, the small ε and small γ expansions of the MFPT are not always exchangeable. They are, however, in the large friction limit, where (10.153) with (10.154) is asymptotically the same as (10.46) for $\gamma \gg 2\omega_C$. This, however, is not the case in the limit of small γ, because (10.153) with (10.154) converges to the transition state theory result [87]

$$\kappa_{\text{TST}} = \frac{\omega_A}{2\pi} e^{-\Delta E/\varepsilon},$$

and the small γ rate (8.89) is

$$\kappa_{\text{underdamped}} = \frac{1}{\bar{\tau}_E} \sim \frac{\gamma I(E_B)\omega_W}{2\pi\varepsilon} e^{-(E_B - E_A)/\varepsilon}, \tag{10.163}$$

where (see (8.90))

$$\bar{\tau}_E = \frac{2\pi\varepsilon}{\gamma I(E_B)\omega_W} e^{(E_B - E_A)/\varepsilon}, \tag{10.164}$$

which is the same as the MFPT to the critical energy contour (10.144). Note that in (8.90) and (8.89) the energy at the bottom of the well is $E_A = 0$. Therefore a uniform approximation to the escape rate is needed, valid for all γ (see [100]).

The saddle point contour $E = E_B$ is the same as the critical energy contour $E = E_C$. According to (4.81), the density of points on $E = E_C$, calculated in eq. (10.143), is Green's function for the boundary value problem

$$\begin{aligned} L^*_{x,y} u(x,y) &= 0 \text{ for } E(x,y) < E_C \\ u(x, y_C(x)) &= f(x), \end{aligned}$$

denoted $G(x, y_C(x) \,|\, \xi, \eta)$ (for $E(\xi,\eta) < E_C$). Here $L^*_{x,y}$ is the backward Kolmogorov equation defined in (10.148). Writing $\bar{\tau}_\Gamma(x,y)$ for the MFPT from (x,y) to Γ and $\bar{\tau}_C(x,y)$ for the MFPT to $E(x,y) = E_C$, we have the following lemmas.

Lemma 10.3.1. *The MFPT from a point (x,y) inside or on E_C to Γ satisfies the renewal equation*

$$\bar{\tau}_\Gamma(x,y) = \bar{\tau}_C(x,y) + \oint\limits_{E=E_C} G(x,y \,|\, \xi,\eta)\bar{\tau}_\Gamma(\xi,\eta)\, ds_{\xi,\eta}. \tag{10.165}$$

Proof. The identity follows from the fact that both sides of (10.165) satisfy the equation $L^*_{x,y} u = -1$ for $E(x,y) < E_C$ and coincide for $E(x,y) = E_C$. □

Corollary 10.3.1 (Uniform approximation to Kramers' rate). *A uniform approximation to the thermal activation rate in Kramers' model, valid for all $\gamma > 0$, is given by*

$$\kappa_{\text{uniform}} \approx \frac{1}{\bar{\tau}_C(x_A, 0) + 2\bar{\tau}_\Gamma(E_C)}, \tag{10.166}$$

where

$$\bar{\tau}_C(x_A, 0) = \frac{2\pi\varepsilon}{\gamma I(E_C)\omega_A} e^{(E_C - E_A)/\varepsilon}, \tag{10.167}$$

and

$$\bar{\tau}_\Gamma(E_C) \sim \frac{2\pi\omega_C e^{(E_C - E_A)/\varepsilon}}{\omega_A\left[\sqrt{\gamma^2 + 4\omega_C^2} - \gamma\right]}. \tag{10.168}$$

Proof. We note that the second term in (10.165),

$$\bar{\tau}_\Gamma(E_C) = \oint_{E=E_C} G(x,y\,|\,\xi,\eta)\bar{\tau}_\Gamma(\xi,\eta)\,ds_{\xi,\eta}, \tag{10.169}$$

is the mean time to go from $E = E_C$ to Γ. Thus eq. (10.165) means that the time to go from $(x,y) \in E < E_C$ to Γ is the sum of the times to go from (x,y) to $E = E_C$ and the mean time to go from $E = E_C$ to Γ. Because Γ is the stochastic separatrix, trajectories that reach it return with probability $\frac{1}{2}$ to $E = E_C$ before exiting the domain. Approximating the time to reach Γ from $E = E_C$ by Kramers' expression (10.154)

$$\bar{\tau}_\Gamma(\xi,\eta) \sim \frac{2\pi\omega_C e^{(E_C-E_A)/\varepsilon}}{\omega_A\left[\sqrt{\gamma^2+4\omega_C^2}-\gamma\right]},$$

which is valid for $\min(\gamma,1) \gg \sqrt{\varepsilon}$ and $\varepsilon \ll 1$, we obtain a uniform approximation to the escape rate from the bottom of the well as (10.166), where $\bar{\tau}_C(x_A,0)$ is obtained from (10.164) as (10.167) and $\bar{\tau}_\Gamma(E_C)$ is approximated by (10.154) as (10.168). For $\gamma \ll \sqrt{\varepsilon}$ the first term $\bar{\tau}_C(x_A,0)$ in the denominator of (10.166) is large whereas the second remains bounded, so that $\kappa_{\text{uniform}} \approx \kappa_{\text{underdamped}}$ (see (10.163)), but as γ increases, $\bar{\tau}_C(x_A,0)$ decreases and the second term $\bar{\tau}_\Gamma(E_C)$ increases and dominates the sum. Thus for $\gamma \gg \sqrt{\varepsilon}$ the uniform rate κ_{uniform} reduces to Kramers' rate κ_{Kramers}, given in (10.153) (see Figure 10.8). □

10.3.5 The exit distribution on the separatrix

The specific exit problem for the phase plane system (10.138) is to determine the probability density function (pdf) of the points where escaping trajectories hit Γ.

Theorem 10.3.3 (The exit density). *The probability density function of exit points on the separatrix is approximately* Weibull$\left(2, \sqrt{2\varepsilon/(\omega_C^2+\lambda^2)}\right)$ *and the most likely exit point is* $(x_M, y_\Gamma(x_M))$*, where*

$$x_M = x_C - \sqrt{\frac{\varepsilon}{\omega_C^2+\lambda^2}}\left(1+O\left(\sqrt{\frac{\varepsilon}{\omega_C^2+\lambda^2}}\right)\right),$$

and λ *is given in* (10.141).

Proof. The probability density function of exit points on the separatrix Γ is determined from the solution of the stationary Fokker–Planck equation (FPE) in D corresponding to the system (10.138), with a unit source at a point $(x_0,y_0) \in D$ and absorbing boundary condition on Γ. The FPE

$$-y\frac{\partial p}{\partial x} + \frac{\partial}{\partial y}\left[(\gamma y + U'(x))\,p\right] + \varepsilon\gamma\frac{\partial^2 p}{\partial y^2} = -\delta(x-x_0)\delta(y-y_0) \tag{10.170}$$

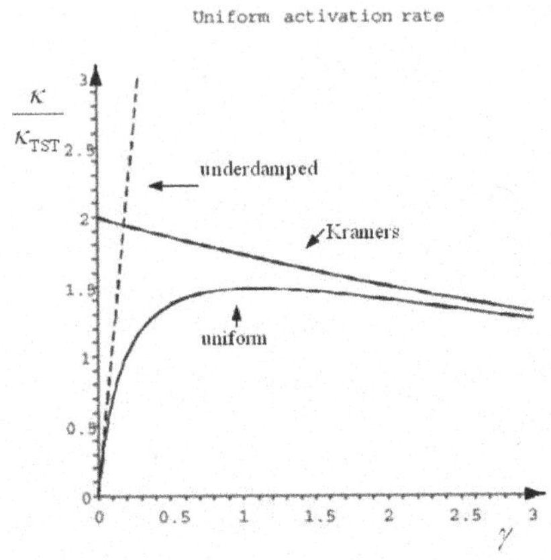

Figure 10.8. Comparison of formulas for reaction rate κ as a function of γ, for the potential $U(x) = 3x^2 - 2x^3$ with $\varepsilon = 0.2$. All rates are normalized by κ_{TST}. The underdamped (dashed) rate is (10.163), Kramers' rate (solid) is (10.153), and the uniform approximation (dot-dashed) is (10.166).

can be written as the conservation law

$$\nabla \cdot \boldsymbol{J}(x, y \,|\, x_0, y_0) = \delta(x - x_0)\delta(y - y_0), \tag{10.171}$$

where

$$\boldsymbol{J}(x, y \,|\, x_0, y_0) = \begin{pmatrix} y p_\varepsilon(x, y \,|\, x_0, y_0) \\ -\left[\gamma y + U'(x)\right] p_\varepsilon(x, y \,|\, x_0, y_0) - \varepsilon\gamma \dfrac{\partial p_\varepsilon(x, y \,|\, x_0, y_0)}{\partial y} \end{pmatrix}.$$

The solution of (10.170) in D that satisfies the boundary condition

$$p|_\Gamma = p_\varepsilon(x,\, y_\Gamma(x) |\, x_0, y_0) = 0 \tag{10.172}$$

has the flux density vector on Γ

$$\boldsymbol{J}(x, y_\Gamma(x) \,|\, x_0, y_0) = \begin{pmatrix} 0 \\ -\varepsilon\gamma \dfrac{\partial p_\varepsilon(x, y_\Gamma(x) \,|\, x_0, y_0)}{\partial y} \end{pmatrix}. \tag{10.173}$$

The probability density function per unit arc length of the exit points on Γ is given by

$$\tilde{p}_\Gamma(s \,|\, x_0, y_0)\, ds = \mathcal{N}^{-1} \boldsymbol{J}(x, y_\Gamma(x) \,|\, x_0, y_0) \cdot \boldsymbol{\nu}(x, y_\Gamma(x))\, ds \tag{10.174}$$

where the normalization constant is

$$\mathcal{N} = \oint_{\Gamma} \boldsymbol{J}(x, y \mid x_0, y_0) \cdot \boldsymbol{\nu}(x, y)\, ds. \tag{10.175}$$

This probability density function can be converted to a probability density function per unit x by the identities

$$\begin{aligned}\tilde{p}_\Gamma(s \mid x_0, y_0)\, ds &= -\mathcal{N}^{-1} \varepsilon\gamma \frac{\partial p_\varepsilon(x, y_\Gamma(x) \mid x_0, y_0)}{\partial y} \nu_2(x, y_\Gamma(x))\, ds \\ &= -\mathcal{N}^{-1} \varepsilon\gamma \frac{\partial p_\varepsilon(x, y_\Gamma(x) \mid x_0, y_0)}{\partial y}\, dx \\ &= p_\Gamma(x \mid x_0, y_0)\, dx,\end{aligned} \tag{10.176}$$

and the normalization constant can be written as

$$\mathcal{N} = -\oint_{\Gamma} \varepsilon\gamma \frac{\partial p_\varepsilon(x, y_\Gamma(x) \mid x_0, y_0)}{\partial y}\, dx.$$

Thus the exit density per unit x, denoted above $p_\Gamma(x \mid x_0, y_0)$, is the y component of the normalized flux density of the stationary probability density function $p_\varepsilon(x, y \mid x_0, y_0)$ on Γ.

Lemma 10.3.2. *The solution of the FPE* (10.170), *up to a normalization factor, has the asymptotic form*

$$p_\varepsilon(x, y \mid x_0, y_0) = e^{-E(x,y)/\varepsilon} q_\varepsilon(x, y \mid x_0, y_0), \tag{10.177}$$

where $E(x, y) = y^2/2 + U(x)$ *and for* $(x, y) \in D$, $(x_0, y_0) \in D$ *the function* $q_\varepsilon(x, y \mid x_0, y_0)$ *is the solution of the equation*

$$\begin{aligned}&-y\frac{\partial q_\varepsilon}{\partial x} + (-\gamma y + U'(x))\frac{\partial q_\varepsilon}{\partial y} + \varepsilon\gamma\frac{\partial^2 q_\varepsilon}{\partial y^2} \\ &= -e^{E(x,y)/\varepsilon}\delta(x - x_0)\delta(y - y_0)\end{aligned} \tag{10.178}$$

with the boundary and matching conditions

$$q_\varepsilon(x, y_\Gamma(x) \mid x_0, y_0) = 0, \quad \text{for } (x_0, y_0) \in D \tag{10.179}$$

$$\lim_{\varepsilon \to 0} q_0(x, y \mid x_0, y_0) = 1 \text{ for } (x, y) \in D,\ (x_0, y_0) \in D. \tag{10.180}$$

Proof. When a solution of the form (10.177) is substituted into the FPE (10.170), the resulting equation for $q_\varepsilon(x, y \mid x_0, y_0)$ is (10.178). To satisfy the absorbing boundary condition for the FPE the boundary condition for (10.178) has to be (10.179). To determine the matching condition, we note that because the right-hand side of eq. (10.178) can be written as $\delta(x - x_0)\delta(y - y_0)\exp\{E(x_0, y_0)/\varepsilon\}$, the constant $\exp\{E(x_0, y_0)/\varepsilon\}$ is only a scaling factor for $q_\varepsilon(x, y \mid x_0, y_0)$. Thus this constant can be replaced by an arbitrary nonzero constant that may depend on ε. To simplify the calculations, we choose a constant $C(\varepsilon)$ that is transcendentally small in ε. This does not affect the evaluation of the exit density, because this constant cancels out in the numerator and denominator of eq. (10.183) below. Thus, we rescale $q_\varepsilon(x, y \,|\, x_0, y_0)$ so that eq. (10.178) becomes

$$-y\frac{\partial q_\varepsilon}{\partial x} + (-\gamma y + U'(x))\frac{\partial q_\varepsilon}{\partial y} + \varepsilon\gamma\frac{\partial^2 q_\varepsilon}{\partial y^2} = -C(\varepsilon)\delta(x - x_0)\delta(y - y_0). \tag{10.181}$$

Keeping this in mind, we find that the outer expansion of $q_\varepsilon(x, y \,|\, x_0, y_0)$,

$$q_\varepsilon(x, y \,|\, x_0, y_0) = q_0(x, y \,|\, x_0, y_0) + \varepsilon q_1(x, y \,|\, x_0, y_0) + \cdots, \tag{10.182}$$

gives

$$-y\frac{\partial q_0}{\partial x} + (-\gamma y + U'(x))\frac{\partial q_0}{\partial y} = 0.$$

This is a first-order partial differential equation whose characteristics converge to the stable equilibrium point $A = (x_A, 0)$. Hence it follows that $q_0(x, y \,|\, x_0, y_0) = const.$ Scaling constant $C(\varepsilon)$ so that $q_0(x, y \,|\, x_0, y_0) = 1$, we obtain (10.180). □

Corollary 10.3.2. *The exit probability density function can be written in terms of the function $q_\varepsilon(x, y \,|\, x_0, y_0)$ as*

$$p_\Gamma(x \,|\, x_0, y_0)\, dx = \frac{\exp\left\{-\dfrac{E(x, y_\Gamma(x))}{\varepsilon}\right\} \dfrac{\partial q_\varepsilon(x, y_\Gamma(x) \,|\, x_0, y_0)}{\partial y}\, dx}{\displaystyle\oint_\Gamma \exp\left\{-\dfrac{E(x, y_\Gamma(x))}{\varepsilon}\right\} \dfrac{\partial q_\varepsilon(x, y_\Gamma(x) \,|\, x_0, y_0)}{\partial y}\, dx}. \tag{10.183}$$

Note that $p_\Gamma(x \,|\, x_0, y_0)$ in eq. (10.183) is independent of the scaling factor $C(\varepsilon)$.

Lemma 10.3.3. *Away from the saddle point C the noncharacteristic boundary layer is given in terms of the variables x and $\zeta = \rho(x, y)/\varepsilon$, where $\rho(x, y) = -dist\{(x, y), \Gamma\}$ for $(x, y) \in D$, is given by*

$$q_\varepsilon(x, y \,|\, x_0, y_0) = 1 - \exp\left\{\frac{y_\Gamma(x)\rho_x(x, y_\Gamma(x)) + [\gamma y_\Gamma(x) - U'(x)]\rho_y(x, y_\Gamma(x))}{\gamma \rho_y^2(x, y_\Gamma(x))}\zeta\right\}(1 + O(\varepsilon)). \tag{10.184}$$

Proof. The boundary Γ is noncharacteristic for the boundary value problem eqs. (10.178)–(10.180) in the sense that the drift vector

$$\tilde{\boldsymbol{b}}(x, y) = \begin{pmatrix} -y \\ -\gamma y + U'(x) \end{pmatrix}$$

points into D at Γ. Specifically, away from the saddle point $(0,0)$, the normal component of the drift vector is negative; that is,

$$\tilde{\boldsymbol{b}}(x,y_\Gamma(x))\cdot\boldsymbol{\nu}(x,y_\Gamma(x))=\frac{-2y_\Gamma^2(x)\,\gamma}{\sqrt{\left[-\gamma y_\Gamma(x)+U'(x)\right]^2+y_\Gamma^2(x)}}<0,$$

where $\boldsymbol{\nu}(x,y_\Gamma(x)) = \nabla\rho(x,y_\Gamma(x))$. We write the boundary layer function in terms of the noncharacteristic boundary layer variables x and ζ as $q_\varepsilon(x,y\,|\,x_0,y_0) = Q(x,\zeta)$, and, as remarked at the end of Section 10.3.2, we change the constant $e^{E(x_0,y_0)/\varepsilon}$ by a transcendentally small constant $C(\varepsilon)$. Then eq. (10.178) becomes

$$\begin{aligned}&\frac{\gamma\rho_y^2}{\varepsilon}Q_{\zeta\zeta}+\gamma\rho_{yy}Q_\zeta-yQ_x-\frac{y\rho_x}{\varepsilon}Q_\zeta+\left[-\gamma y+U'(x)\right]\frac{\rho_y}{\varepsilon}Q_\zeta\\&=C(\varepsilon)\delta(x-x_0)\delta(y-y_0).\end{aligned}\tag{10.185}$$

Expanding

$$Q(x,\zeta)=Q^0(x,\zeta)+\varepsilon Q^1(x,\zeta)+\cdots,\tag{10.186}$$

we obtain the leading-order boundary layer equation

$$\begin{aligned}&\gamma\rho_y^2(x,y_\Gamma(x))\,Q^0_{\zeta\zeta}(x,\zeta)-y_\Gamma(x)\,\rho_x(x,y_\Gamma(x))\,Q^0_\zeta(x,\zeta)\\&+\left[-\gamma y_\Gamma(x)+U'(x)\right]\rho_y(x,y_\Gamma(x))\,Q^0_\zeta(x,\zeta)=0.\end{aligned}\tag{10.187}$$

Note that $Q_x(x,\zeta)$ does not appear in eq. (10.187) so that x is a parameter in this equation. The boundary condition (10.179) and matching condition (10.180) give

$$Q(x,0)=0,\qquad\lim_{\zeta\to-\infty}Q(x,\zeta)=1.\tag{10.188}$$

The solution of eq. (10.187) that satisfies the boundary and matching conditions eq. (10.188) is given by

$$\begin{aligned}&Q^0(x,\zeta)\\&=1-\exp\left\{\frac{y_\Gamma(x)\rho_x(x,y_\Gamma(x))+\left[\gamma y_\Gamma(x)-U'(x)\right]\rho_y(x,y_\Gamma(x))}{\gamma\rho_y^2(x,y_\Gamma(x))}\zeta\right\}.\end{aligned}\tag{10.189}$$

□

We denote

$$r(x)=\frac{y_\Gamma(x)\rho_x(x,y_\Gamma(x))+\left[\gamma y_\Gamma(x)-U'(x)\right]\rho_y(x,y_\Gamma(x))}{\gamma\rho_y^2(x,y_\Gamma(x))}\tag{10.190}$$

and note that $r(x)>0$ for $x<0$ so that the exponent is negative for negative ζ. In particular, we use near the saddle point the linearized expressions

$$y_\Gamma(x)\sim-\lambda x,\quad\rho_x\sim\frac{\lambda}{\sqrt{1+\lambda^2}},\quad\rho_y\sim\frac{1}{\sqrt{1+\lambda^2}},\quad U'(x)\sim-\omega_C^2x.\tag{10.191}$$

Corollary 10.3.3. *At a distance $O(\sqrt{\varepsilon})$ away from the saddle point the boundary layer function is*

$$Q^0(x,\zeta) \sim 1 - \exp\left\{-2\lambda\sqrt{1+\lambda^2}x\zeta\right\}. \tag{10.192}$$

Lemma 10.3.4. *In a $\sqrt{\varepsilon}$-neighborhood of the saddle point the boundary layer function has the asymptotic form*

$$q_\varepsilon(x,y\,|\,x_0,y_0) = R_0(\xi,\eta) + \varepsilon R_1(\xi,\eta) + \cdots, \tag{10.193}$$

where $(\xi,\eta) = (x,\rho)/\sqrt{\varepsilon}$ are local variables,

$$R_0(\xi,\eta) = \sqrt{\frac{2}{\pi}}\int_0^{\chi(\xi,\eta)} e^{-z^2/2}\,dz, \tag{10.194}$$

and

$$\chi = \sum_{n=1}^{\infty}\frac{A_n(\xi)}{n!}\eta^n.$$

The coefficients $A_n(\xi)$ are determined recursively.

Proof. The noncharacteristic boundary layer expansion fails in the $\sqrt{\varepsilon}$ neighborhood of the saddle point, because $\lim_{x\to 0}\boldsymbol{b}(x,y_\Gamma(x))\cdot\boldsymbol{\nu}(x,y_\Gamma(x)) = 0$. Specifically, in this neighborhood the terms neglected in the powers series expansion eq. (10.186) of the solution of eq. (10.185) are of the same order of magnitude as the ones retained. Therefore a separate boundary layer analysis is required in this neighborhood. This expansion has to resolve the singularity of the boundary layer equation (10.185) at the saddle point. To do so, we introduce in eq. (10.178) the stretched variables (ξ,η) and expand the boundary layer function in the form (10.193). Then the leading term in eq. (10.178) satisfies

$$\frac{\partial^2 R_0}{\partial\eta^2} + (a\xi + b\eta)\frac{\partial R_0}{\partial\eta} + (c\xi + d\eta)\frac{\partial R_0}{\partial\xi} = 0, \tag{10.195}$$

where

$$a = 2\lambda^2\frac{\sqrt{(1+\lambda^2)}}{\gamma},\quad b = \frac{-\lambda+\gamma}{\gamma} < 0$$

$$c = \frac{\lambda}{\gamma}(1+\lambda^2) > 0,\quad d = -\frac{\sqrt{(1+\lambda^2)}}{\gamma} < 0$$

and

$$\lambda = \frac{\gamma + \sqrt{\gamma^2 + 4\omega_C^2}}{2} > 0.$$

The domain D is mapped onto the third quadrant in the (ξ, η)-plane and the separatrix Γ is mapped onto the line $\eta = 0,\ \xi < 0$. The boundary and matching conditions eq. (10.188) are now

$$R_0(\xi, 0) = 0, \qquad \lim_{\eta \to -\infty} R_0(\xi, \eta) = 1.$$

We seek a solution to eq. (10.195) in the form (10.194), where the function χ satisfies the equation

$$\frac{\partial^2 \chi}{\partial \eta^2} - \chi \left(\frac{\partial \chi}{\partial \eta} \right)^2 + (a\xi + b\eta) \frac{\partial \chi}{\partial \eta} + (c\xi + d\eta) \frac{\partial \chi}{\partial \xi} = 0 \tag{10.196}$$

and the boundary and matching conditions

$$\chi(\xi, 0) = 0, \qquad \lim_{\eta \to -\infty} \chi(\xi, \eta) = \infty \quad \text{for } \xi < 0.$$

A power series solution of (10.196),

$$\chi = \sum_{n=1}^{\infty} \frac{A_n(\xi)}{n!} \eta^n,$$

gives

$$\sum_{n=2}^{\infty} \eta^{n-2} \frac{A_n(\xi)}{(n-2)!} - \left(\sum_{n=1}^{\infty} \frac{A_n(\xi)}{n!} \eta^d \right) \left(\sum_{n=1}^{\infty} \frac{A_n(\xi)}{(n-1)!} \eta^{n-1} \right)^2$$

$$+ (a\xi + b\eta) \sum_{n=1}^{\infty} \frac{A_n(\xi)}{(n-1)!} \eta^{n-1} + (c\xi + d\eta) \sum_{n=1}^{\infty} \eta^d \frac{A_n'(\xi)}{n!} = 0,$$

which leads to the hierarchy of equations

$$A_2(\xi) + a\xi A_1(\xi) = 0 \tag{10.197}$$

$$A_3(\xi) - A_1^3(\xi) + bA_1(\xi) + a\xi A_2(\xi) + c\xi A_1'(\xi) = 0, \tag{10.198}$$

and so on. Substituting from eq. (8.92), we obtain eq. (10.198) in the form

$$A_3(\xi) - A_1^3(\xi) + bA_1(\xi) - a^2\xi^2 A_1(\xi) + c\xi A_1'(\xi) = 0. \tag{10.199}$$

The function $A_3(\xi)$ is determined by matching the solution eq. (10.194) with the boundary layer function $Q^0(x, \zeta)$ in the matching region. In the matching region we express the boundary layer function $Q^0(x, \zeta)$ in terms of the variables (ξ, η) as

$$Q^0(x, \zeta) = 1 - \exp\left\{ -2\lambda\sqrt{1 + \lambda^2}\xi\eta \right\} = 2\lambda\sqrt{1 + \lambda^2}\xi\eta + O((\xi\eta)^2) \quad \text{for small } \eta.$$

On the other hand, we have

$$R_0 = \sqrt{\frac{2}{\pi}} \{\chi + O(\chi^2)\} = \sqrt{\frac{2}{\pi}} \sum_{n=1}^{\infty} \frac{A_n(\xi)}{n!} \eta^n + O(\chi^2).$$

Matching $2\lambda\sqrt{1+\lambda^2}\xi\eta + O((\xi\eta)^2)$ and $\sqrt{2/\pi}\sum_{n=1}^{\infty} A_n(\xi)\eta^n/n! + O(\chi^2)$ as functions, for small $\xi\eta$, we find that $A_3(\xi)$ has to be chosen so that

$$\sqrt{\frac{2}{\pi}}A_1(\xi) = \frac{r(\sqrt{\varepsilon}\xi)}{\sqrt{\varepsilon}}. \tag{10.200}$$

Expanding

$$\sqrt{\frac{2}{\pi}}A_1(\xi) = \frac{r(\sqrt{\varepsilon}\xi)}{\sqrt{\varepsilon}} = -2\lambda\sqrt{1+\lambda^2}\xi + O(\xi^2), \tag{10.201}$$

we see that this can be accomplished by choosing

$$A_3(\xi) = A_1^3(\xi) + a^2\xi^2 A_1(\xi) - (b+c)A_1(\xi), \tag{10.202}$$

so that eq. (10.198) is reduced to

$$\xi A_1'(\xi) - A_1(\xi) = 0,$$

hence eq. (10.201) follows.

This matching gives a leading-order approximation to $\chi(\xi,\eta)$ as a power series in both ξ and η. In higher-order matching the coefficients in the boundary layer equation have to be expanded as power series in ξ and η and the matching condition in eq. (10.200) has to be satisfied at all orders. □

Finally, the proof of Theorem 10.3.3 follows from the lemmas. The exit density on Γ, as given in eq. (10.183), can be calculated from the boundary layer expansions eqs. (10.189) and (10.194). The former gives outside the $\sqrt{\varepsilon}$-neighborhood of the saddle point

$$\left.\frac{\partial p}{\partial y}\right|_{\Gamma} \sim -e^{-E/\varepsilon}\frac{r(x)p_y(x,y_\Gamma(x))}{\varepsilon},$$

and the latter gives inside this neighborhood

$$\left.\frac{\partial p}{\partial y}\right|_{\Gamma} \sim -e^{-E/\varepsilon}2\lambda\sqrt{1+\lambda^2}\,\frac{xp_y(x,y_\Gamma(x))}{\varepsilon}, \tag{10.203}$$

which match to leading order as functions in the matching region. It follows that the leading-order approximation to the exit probability density function on Γ is given by

$$p_\Gamma(x)\,dx \sim \frac{\exp\left\{-\dfrac{E(x,y_\Gamma(x))}{\varepsilon}\right\}r(x)\,dx}{\displaystyle\oint_\Gamma \exp\left\{-\dfrac{E(x,y_\Gamma(x))}{\varepsilon}\right\}r(x)\,dx}, \tag{10.204}$$

where $E(x,y_\Gamma(x)) = \frac{1}{2}y_\Gamma^2(x) + U(x)$ and $r(x)$ is defined in eq. (10.190).

Using (10.203) in (10.204) with the local linearization (10.191) gives the exit density for $x < 0$ near the saddle point $C = (0,0)$ as

$$p_\Gamma(x) \approx -\mathcal{N} x \exp\left\{-\left(\frac{\lambda^2 x^2}{2\varepsilon} + \frac{\omega_C^2 x^2}{2\varepsilon}\right)\right\}, \tag{10.205}$$

so the most likely exit point is

$$x_M = -\sqrt{\frac{\varepsilon}{\omega_C^2+\lambda^2}}\left(1 + O\left(\sqrt{\frac{\varepsilon}{\omega_C^2+\lambda^2}}\right)\right),$$

where λ is given in (10.141). The most likely exit point is located at a distance $O(\sqrt{\varepsilon})$ from the saddle point and the exit density on the separatrix is the distribution Weibull$\left(2, \sqrt{2\varepsilon/(\omega_C^2+\lambda^2)}\right)$, as asserted. □

10.4 Annotations

The exit problem in the theory of stochastic differential equations concerns the escape of the random trajectories of a dynamical system driven by noise from the domain of attraction of the underlying noiseless dynamics [140], [154], [162], [82], [27], [77], [172], [206], [100] (and references therein). Large deviations theory [77], [54] predicts that in the limit of vanishing noise escapes are concentrated at the absolute minima of an action functional on the separatrix (the boundary of the domain of attraction of an attractor of the noiseless dynamics). However, it has been observed in numerical simulations [213], [27], [118], [210] that for finite noise strength this is not the case and actually, escaping trajectories avoid the absolute minimum so that the escape distribution is spread on the separatrix away from the points predicted by large deviations theory. Some analytical results concerning this saddle point avoidance phenomenon were given in [213], [27], [52], [50], [51], [49], [156], [155], [158], [157], [159], [160], and more recently in [216] and [217], where the analytical results are also compared with results of simulations.

Kramers' model of activated escape [140] has become a cornerstone in statistical physics, with applications in many branches of science and mathematics. It has important applications in diverse areas such as communications theory [27], [244], stochastic stability of structures [118], [119], and even in the modern theory of finance [249].

A vast literature on exit problems has been accumulated [100] and the problem is still an active area of physical, chemical, biological, and mathematical research. The problem of distribution of the exit points is related to the distribution of energies of the escaping particles [172], [33], to the phenomenon of saddle point avoidance elaborated by Berezhkovskii et al. (see the review [186] and references therein), and to numerical simulations of escape problems (see, e.g., [26]). The problem of determining the distribution of exit points has been studied in different contexts, under various assumptions, and by a variety of methods, analytical, numerical, and experimental in [213], [156], [155], [158], [157], [159], [160]. The classical result

(10.144) was derived by Kramers [140] in the limit of small γ and in [169], for all γ. The Bernoulli equation (10.158) was derived in [162].

The unexpected phenomenon of saddle point avoidance was first observed in a class of noise-driven dynamical systems lacking detailed balance [27]. It was observed that the exit points on the boundary of the domain of attraction of the attractor of the noiseless dynamics is not necessarily peaked at the saddle point. This phenomenon, not being related to anisotropy in the noise or the dynamics [186], is counterintuitive and requires explanation. It was studied under a variety of assumptions in [52], [50], [51], [49], [156], [155], [158], [157], [159], [160], [186]. The significance of the problem in models of electronic signal tracking devices, such as RADAR, spread spectrum communications (as in cellular phones), and in various synchronization devices, is that the determination of the exit distribution on the boundary of the domain of attraction indicates where to tune the lock detector that determines if the signal is lost and has to be acquired afresh.

The realization that the exit point on the separatrix in the classical Kramers problem is not at the saddle point, even for large values of the damping coefficient, came as a surprise. This phenomenon in the Kramers problem was first observed in numerical simulations of the Langevin dynamics [118], [210] and was initially interpreted as a numerical instability of the simulation scheme. The problem of the asymptotic convergence of the boundary layer expansion of the exit problem is discussed in [112], [113]. The convergence of the expansion (10.92) is discussed in [114], [56].

Chapter 11

Stochastic Stability

The notion of stability in deterministic and stochastic systems is not the same. The solution $\boldsymbol{\xi}(t)$ of a deterministic system of differential equations

$$\dot{\boldsymbol{x}} = \boldsymbol{b}(\boldsymbol{x}, t) \tag{11.1}$$

is *stable* if for any positive number ε there exist two numbers, $\delta > 0$ and T, such that for any solution $\boldsymbol{x}(t)$ of (11.1)

$$|\boldsymbol{x}(t) - \boldsymbol{\xi}(t)| < \varepsilon \quad \text{for } t \geq T,$$

whenever

$$|\boldsymbol{x}(t_0) - \boldsymbol{\xi}(t_0)| < \delta \tag{11.2}$$

for some $t_0 \leq T$. The solution $\boldsymbol{\xi}(t)$ is said to be *asymptotically stable* if it is stable and, in addition,

$$\lim_{t\to\infty} |\boldsymbol{x}(t) - \boldsymbol{\xi}(t)| = 0 \tag{11.3}$$

for any solution $\boldsymbol{x}(t)$ satisfying (11.2). If eq. (11.3) holds for all solutions of eq. (11.1), then $\boldsymbol{\xi}(t)$ is said to be *globally stable.*

By setting

$$\boldsymbol{x}(t) = \boldsymbol{\xi}(t) + \boldsymbol{y}(t),$$

and assuming that $\boldsymbol{b}(\boldsymbol{x}, t)$ is smooth, we can reduce the problem of stability of $\boldsymbol{\xi}(t)$ to that of the problem of stability of the solution $\boldsymbol{y}(t) = \boldsymbol{0}$ for the system

$$\dot{\boldsymbol{y}} = \boldsymbol{B}(t)\boldsymbol{y} + \boldsymbol{C}(t, \boldsymbol{y}), \tag{11.4}$$

where

$$B^{ij}(t) = \left.\frac{\partial b^i(\boldsymbol{x}, t)}{\partial x^j}\right|_{\boldsymbol{x}=\boldsymbol{\xi}(t)}$$

and where $\boldsymbol{C}(t, \boldsymbol{y}) = o(|\boldsymbol{y}|)$ as $|\boldsymbol{y}| \to 0$. If the matrix $\boldsymbol{B}(t)$ is independent of t and the eigenvalues of $\boldsymbol{B}$ lie in the left half of the complex plane, the solution $\boldsymbol{y}(t) = \boldsymbol{0}$ is asymptotically stable; hence $\boldsymbol{\xi}(t)$ is asymptotically stable for eq. (11.1).

Z. Schuss, *Theory and Applications of Stochastic Processes: An Analytical Approach*, Applied Mathematical Sciences 170, DOI 10.1007/978-1-4419-1605-1_11,

Consider the autonomous case that $\boldsymbol{b}(\boldsymbol{x},t) = \boldsymbol{b}(\boldsymbol{x})$. Then a point $\boldsymbol{x}_0$ is called a *critical* or *equilibrium* point of the system (11.1) if $\boldsymbol{b}(\boldsymbol{x}_0) = \boldsymbol{0}$. In this case $\boldsymbol{\xi}(t) = \boldsymbol{x}_0$ is a solution of (11.1). If $\boldsymbol{x}_0$ is a critical point of (11.1), then $\boldsymbol{B}(t)$ in eq. (11.4) is constant.

Another method for examining the stability of the critical point $\boldsymbol{x}_0 = \boldsymbol{0}$ is due to Lyapunov.

Definition 11.0.1 (Lyapunov function). *A function $V(\boldsymbol{x})$ is called a* Lyapunov function *for (11.1) at $\boldsymbol{x}_0 = \boldsymbol{0}$ if (i) $V(\boldsymbol{x})$ is defined, continuous, and differentiable in a neighborhood of $\boldsymbol{0}$; (ii) $V(\boldsymbol{x}) > 0$ if $\boldsymbol{x} \neq \boldsymbol{0}$ and $V(\boldsymbol{0}) = 0$, and (iii) $\boldsymbol{b}(\boldsymbol{x}) \cdot \nabla V(\boldsymbol{x}) \leq 0$ in a neighborhood of $\boldsymbol{0}$.*

If the system (11.1) possesses a Lyapunov function, then $\boldsymbol{x} = 0$ is a stable solution, given that $\boldsymbol{b}(\boldsymbol{0}) = \boldsymbol{0}$. If

$$\boldsymbol{b}(\boldsymbol{x}) \cdot \nabla V(\boldsymbol{x}) < 0$$

for all $\boldsymbol{x} \neq \boldsymbol{0}$, then $\boldsymbol{x} = \boldsymbol{0}$ is asymptotically stable. The stability of a linear system

$$\dot{\boldsymbol{x}} = \boldsymbol{A}\boldsymbol{x}$$

is determined by a Lyapunov equation, which is a quadratic form

$$V(\boldsymbol{x}) = \boldsymbol{x}^T \boldsymbol{L}\boldsymbol{x}$$

with $\boldsymbol{L}$ a matrix that is a positive definite solution of the equation

$$\boldsymbol{A}^T \boldsymbol{L} + \boldsymbol{L}\boldsymbol{A} = -\boldsymbol{Q},$$

where $\boldsymbol{Q}$ is some positive definite matrix. Indeed,

$$\boldsymbol{A}\boldsymbol{x} \cdot \nabla V(\boldsymbol{x}) = \boldsymbol{x}^T \boldsymbol{L}\boldsymbol{A}\boldsymbol{x} + \boldsymbol{x}^T \boldsymbol{A}^T \boldsymbol{L}\boldsymbol{A}\boldsymbol{x} = -\boldsymbol{x}^T \boldsymbol{Q}\boldsymbol{x} < 0$$

for $\boldsymbol{x} \neq \boldsymbol{0}$.

Consider now a system of the form (11.1) with a globally asymptotically stable critical point at $\boldsymbol{x} = \boldsymbol{0}$. If the system is driven by an additive white noise,

$$d\boldsymbol{x} = \boldsymbol{b}(\boldsymbol{x})\,dt + \boldsymbol{\sigma}(\boldsymbol{x})\,d\boldsymbol{w}, \tag{11.5}$$

with positive definite $\boldsymbol{\sigma}(\boldsymbol{x})$, its mean first passage time out of any bounded domain is finite. It follows that the trajectories of the stochastic system (11.5) diverge from the origin to arbitrarily large distances with probability 1 in finite time, even if the noise is arbitrarily small. Thus the question of stability of the critical point $\boldsymbol{x} = \boldsymbol{0}$ for the stochastic system (11.5) is meaningful only if $\boldsymbol{\sigma}(\boldsymbol{0}) = \boldsymbol{0}$. There are several different kinds of stochastic stability of the critical point $\boldsymbol{x} = \boldsymbol{0}$.

Definition 11.0.2 (Stochastic stability). *The origin is* stochastically stable *for (11.5) if all trajectories that begin sufficiently close to the origin remain for all times in a neighborhood of the origin with the exception of a set of trajectories of arbitrarily small positive probability.*

This means that for any positive ε_1 and ε_2 there exists a positive number δ such that

$$\Pr\left\{\sup_{t>0}|\boldsymbol{x}(t)| < \varepsilon_1\right\} > 1 - \varepsilon_2 \tag{11.6}$$

if $|\boldsymbol{x}(0)| < \delta$, which means, in turn, that

$$\lim_{|\boldsymbol{x}(0)|\to 0} \Pr\left\{\sup_{t>0}|\boldsymbol{x}(t)| > \varepsilon\right\} = 0.$$

Definition 11.0.3 (Asymptotic stochastic stability). *If in addition to stochastic stability*

$$\lim_{|\boldsymbol{x}(0)|\to 0} \Pr\left\{\lim_{t\to\infty}|\boldsymbol{x}(t)| = \mathbf{0}\right\} = 1,$$

then the origin is said to be asymptotically stable.

Definition 11.0.4 (Global asymptotic stochastic stability). *If*

$$\Pr\left\{\lim_{t\to\infty}|\boldsymbol{x}(t)| = \mathbf{0}\right\} = 1$$

for all $\boldsymbol{x}(0)$, the origin is globally asymptotically stable.

Example 11.1 (Stability of stochastic linear systems). Consider the stochastic stability of the Itô system

$$d\boldsymbol{x} = \boldsymbol{A}\boldsymbol{x}\,dt + \sum_{i=1}^{n}\boldsymbol{B}_i\boldsymbol{x}\,dw_i, \tag{11.7}$$

where w_i are independent standard Brownian motions. If the equation

$$\boldsymbol{A}\boldsymbol{Z} + \boldsymbol{Z}\boldsymbol{A}^T + \sum_{i=1}^{n}\boldsymbol{B}_i\boldsymbol{Z}\boldsymbol{B}_i^T = -\boldsymbol{C} \tag{11.8}$$

has a symmetric positive definite solution $\boldsymbol{Z}$ for some symmetric positive definite matrix $\boldsymbol{C}$, then the function

$$V(\boldsymbol{x}) = \frac{1}{2}\langle\boldsymbol{Z}\boldsymbol{x},\boldsymbol{x}\rangle \tag{11.9}$$

is a Lyapunov function for the stochastic system (11.7) in the sense that it satisfies conditions (i) and (ii) of Section 1.2 and

$$L^*V(\boldsymbol{x}) \leq -kV(\boldsymbol{x}) \tag{11.10}$$

for some positive constant k, where L^* is the backward Kolmogorov operator,

$$L^*V(\boldsymbol{x}) = \sum_{i,j}A^{ij}x^j\frac{\partial V(\boldsymbol{x})}{\partial x^i} + \sum_{i,j,l,m,r}\frac{1}{2}B_l^{im}x^mB_j^{lr}x^r\frac{\partial^2V(\boldsymbol{x})}{\partial x^i\partial x^j}, \tag{11.11}$$

corresponding to (11.7). The condition (11.10) is obtained by substituting (11.9) in (11.11),

$$L^*V(\boldsymbol{x}) = -\langle \boldsymbol{C}\boldsymbol{x}, \boldsymbol{x}\rangle . \tag{11.12}$$

Because $\boldsymbol{Z}$ and $\boldsymbol{C}$ are positive definite matrices, there are positive constants, k_1 and k_2, such that

$$k_1 \langle \boldsymbol{Z}\boldsymbol{x}, \boldsymbol{x}\rangle \leq \langle \boldsymbol{C}\boldsymbol{x}, \boldsymbol{x}\rangle \leq k_2 \langle \boldsymbol{Z}\boldsymbol{x}, \boldsymbol{x}\rangle ,$$

therefore (11.10) follows from (11.12).

To show that (11.10) implies global asymptotic stability, we consider the probability of ever reaching the surface $V(\boldsymbol{x}) = V_C$, given that the initial condition is on the surface $V(\boldsymbol{x}) = V$, for arbitrary values of $0 < V < V_C$. Because the origin is a critical point for the stochastic system (11.7), this probability is the limit of the probability to reach the surface $V(\boldsymbol{x}) = V_C$ before reaching the surface $V(\boldsymbol{x}) = \delta$ as $\delta \to 0$. The latter probability is the solution of the equation

$$L^*P = 0 \tag{11.13}$$

for $\boldsymbol{x}$ such that $\delta < V(\boldsymbol{x}) < V_C$ with the boundary conditions

$$P = 1 \ \text{ for } V(\boldsymbol{x}) = V_C; \quad P = 0 \ \text{ for } \ V(\boldsymbol{x}) = \delta. \tag{11.14}$$

Changing variables $\boldsymbol{x} \mapsto (V, \boldsymbol{\theta})$, where $\boldsymbol{\theta}$ are variables on the surface $V = const.$, we find that eq. (11.13) has the form

$$\frac{1}{2} \sum_{i,j,l,m,r} B_l^{im} x^m B_j^{lr} x^r \frac{\partial^2 P}{\partial V^2} - \langle \boldsymbol{C}\boldsymbol{x}, \boldsymbol{x}\rangle \frac{\partial P}{\partial V} + t.d. = 0,$$

where $t.d.$ denotes terms that contain derivatives with respect to $\boldsymbol{\theta}$. A solution that depends only on V is found by averaging with respect to $\boldsymbol{\theta}$. The averaged equation is then

$$B(V)P''(V) - C(V)P'(V) = 0, \tag{11.15}$$

where $B(V) = O\left(V^2\right)$ and $C(V) = O(V)$ for $V \to 0$, and $B(V), C(V) > 0$. The solution of (11.15) with the boundary conditions (11.14) is given by

$$P = \frac{\displaystyle\int_\delta^V \exp\left\{ \int_{V_C}^U \frac{C(T)}{B(T)}\,dT \right\} dU}{\displaystyle\int_\delta^{V_C} \exp\left\{ \int_{V_C}^U \frac{C(T)}{B(T)}\,dT \right\} dU}.$$

Because the singularity of $C(T)/B(T)$ near $T = 0$ is of the order K/T, where K is a positive constant, we have

$$\exp\left\{ \int_{V_C}^U \frac{C(T)}{B(T)}\,dT \right\} = O\left(U^K\right) \ \text{ for } V \to 0.$$

It follows that the limit of P as $\delta \to 0$ exists and is given by

$$P(V) = \frac{\displaystyle\int_0^V \exp\left\{\int_{V_C}^U \frac{C(T)}{B(T)}\,dT\right\} dU}{\displaystyle\int_0^{V_C} \exp\left\{\int_{V_C}^U \frac{C(T)}{B(T)}\,dT\right\} dU}.$$

Thus $P(V) \to 0$ as $V \to 0$ so that the probability of ever reaching the surface $V(\boldsymbol{x}) = V_C$ is arbitrarily small if the initial state is sufficiently close to the origin; that is, condition (11.6) is satisfied and the origin is stable. □

Exercise 11.1. (Global asymptotic stability).

(i) Show that the origin is globally asymptotically stable for the system (11.7) under the given conditions.

(ii) Show that if the matrix $\boldsymbol{C}$ is negative definite, the origin is unstable for the system (11.7) in the sense that $V(\boldsymbol{x})$ becomes arbitrarily large in finite time and so does $\boldsymbol{x}(t)$. □

11.1 Stochastic stability of nonlinear oscillators

According to (11.5), a stochastically forced oscillator,

$$\ddot{x} + \gamma\dot{x}g(x,\dot{x}) + U'(x) = \xi(t), \tag{11.16}$$

where $\xi(t)$ is Gaussian white noise and $U(x)$ is a potential that forms a well, is unstable in the sense that its trajectories make arbitrarily large deviations in finite time with probability 1. Thus, the question of stability of a noisy oscillator is relevant only when the noise depends on the state of the oscillator in such a way that the noise vanishes at the equilibrium state of the oscillator, or more generally, on an attractor such as a limit cycle. This situation arises if the coefficients of the oscillator are noisy. For example, if a noisy vertical force is applied at the hinge of a mathematical pendulum, the equation of motion in the absence of friction is given by

$$\ddot{x} + k^2\left[1 + \xi(t)\right]x = 0, \tag{11.17}$$

where $\xi(t)$ is the applied force. Equation (11.17) also serves as a model of one-dimensional wave propagation in random media, Schrödinger's equation with a random potential, and other models [6], [151], [119]. The noise vanishes at the equilibrium point $x = \dot{x} = 0$ so that it cannot move the pendulum away from this state. The question of stability of the equilibrium point is whether the random force will drive the pendulum arbitrarily far away from equilibrium (in phase space), even if the initial displacement from equilibrium is arbitrarily small.

Equation (11.17) is a particular case of eq. (11.7),

$$\frac{d}{dt}\begin{pmatrix} x \\ y \end{pmatrix} = \begin{pmatrix} 0 & 1 \\ -k^2 & 0 \end{pmatrix}\begin{pmatrix} x \\ y \end{pmatrix} + \begin{pmatrix} 0 \\ k^2 \end{pmatrix}\dot{w}.$$

Lyapunov's function for this case is the energy

$$V(x,y) = \frac{k^2x^2}{2} + \frac{y^2}{2}$$

and condition (11.12) becomes

$$L^*V = y\frac{\partial V}{\partial x} - k^2x\frac{\partial V}{\partial y} + \frac{1}{2}k^4x^2\frac{\partial^2 V}{\partial y^2} = \frac{1}{2}k^4x^2 > 0.$$

In this case the matrix $\boldsymbol{C}$ is

$$\boldsymbol{C} = -\begin{pmatrix} k^2 & 0 \\ 0 & 0 \end{pmatrix},$$

and is semi-negative definite. This indicates that the equilibrium state of the parametrically excited pendulum may be unstable.

11.1.1 Underdamped pendulum with parametric noise

If a nonlinear oscillator contains parametric noise, its equation can still be written in the form (11.16), but now $\xi(t)$ depends on the state $(x, \dot{x})$. We assume that

$$\xi(t) = \sqrt{\gamma}\phi(x,\dot{x})v(t), \tag{11.18}$$

where $\phi(x,\dot{x})$ is a smooth function and $v(t)$ is a zero mean stationary Markov process defined by the forward Kolmogorov equation

$$\frac{\partial \tilde{p}}{\partial t} = L\tilde{p},$$

where $\tilde{p}$ denotes the transition pdf of $v(t)$ (see examples below). The stationary pdf $\Psi_0(v)$ of $v(t)$ is the eigenfunction corresponding to the eigenvalue 0 of the forward Kolmogorov operator

$$L\Psi_0(v) = 0, \tag{11.19}$$

normalized by

$$\int \Psi_0(v)\,dv = 1. \tag{11.20}$$

The zero mean assumption means that

$$\mathbb{E}v(t) = \int v\Psi_0(v)\,dv = 0, \tag{11.21}$$

and that the unique eigenfunction of the adjoint operator L^*, corresponding to the zero eigenvalue, is constant; that is,

$$L^* 1 = 0. \tag{11.22}$$

Assumption (11.22) is satisfied, for example, when $v(t)$ has a stationary distribution independent of the initial conditions. The process $(x, \dot{x}, v)$ is then Markovian and can be described by its joint transition probability density

$$\begin{aligned} &p\,(x, \dot{x}, v, x_0, \dot{x}_0, v_0, t) \\ &= \Pr\left\{x(t) = x,\ \dot{x}(t) = \dot{x},\ v(t) = v \,|\, x(t_0) = x_0,\ \dot{x}(t_0) = \dot{x}_0,\ v(t_0) = v_0\right\}. \end{aligned}$$

The function p is the solution of the master equation (or forward Kolmogorov equation)

$$\frac{\partial p}{\partial t} = -\dot{x}\frac{\partial p}{\partial x} + \frac{\partial}{\partial \dot{x}}\left[U'(x) + \gamma\dot{x}g(x, \dot{x}) - \sqrt{\gamma}\phi(x, \dot{x})v\right]p + Lp \tag{11.23}$$

for $t > t_0$ with the initial condition

$$p\,(x, \dot{x}, v, x_0, \dot{x}_0, v_0, t) \to \delta(x - x_0, \dot{x} - \dot{x}_0, v - v_0) \text{ as } t \to t_0.$$

The operator L acts on the variable v in general.

Example 11.2 (White and colored Gaussian noise). We consider two cases of Gaussian noise driving the pendulum.

(i) $v(t)$ = Gaussian white noise. In this case, the pair $(x, \dot{x})$ is Markovian, and therefore the term $-\sqrt{\gamma}\phi(x, \dot{x})v$ is dropped from (11.23). The operator L in (11.23) acts on the variable $\dot{x}$ in the (Stratonovich) form

$$Lp = \frac{1}{2}\gamma\frac{\partial}{\partial \dot{x}}\phi(x, \dot{x})\frac{\partial}{\partial \dot{x}}\phi(x, \dot{x})p.$$

(ii) $v(t)$ = Ornstein–Uhlenbeck process (Gaussian colored noise). In this case the operator L acts on v and has the form

$$Lp = -\alpha\frac{\partial}{\partial v}(vp) + \varepsilon\alpha^2\frac{\partial^2}{\partial v^2}p.$$

The process $v(t)$ is defined by the stochastic differential equation

$$\dot{v}(t) = -\alpha v(t) + \sqrt{2\varepsilon}\alpha\,\dot{w}(t), \tag{11.24}$$

where $w(t)$ denotes Brownian motion. Sometimes it is more convenient to define $v(t)$ by

$$\dot{v}(t) = -\alpha v(t) + \sqrt{2\varepsilon\alpha}\,\dot{w}(t). \tag{11.25}$$

Then (11.26) is replaced by

$$Lp = -\alpha\frac{\partial}{\partial v}(vp) + \varepsilon\alpha\frac{\partial^2}{\partial v^2}p. \tag{11.26}$$

and $\sqrt{\gamma}$ in (11.23) is replaced by $\sqrt{\alpha\gamma}$. In this case $\Psi_0(v) = \exp\{-v^2/2\varepsilon\}$. The parameter α represents the bandwidth of the power spectral density $S_v(\omega)$ of $v(t)$, and ε represents the "intensity", which is the spectral height of $S_v(\omega)$; that is, $S_v(0) = \varepsilon$. As $\alpha \to \infty$, $v(t) \to \sqrt{2\varepsilon}\dot{w}(t)$.

(iii) $v(t)$ = Markov jump process. In this case the operator L acts on v and has the form

$$Lp \qquad (11.27)$$
$$= -\frac{\partial}{\partial v}b(v)p + \int \lambda(v-u)w(u, v-u)p(x, \dot{x}, v-u, t)\,du - \lambda(v)p(x, \dot{x}, v).$$

The process $v(t)$ is defined by the random dynamics

$$v(t+\Delta t) = \begin{cases} v(t) + b(v(t))\Delta t + o(\Delta t) & \text{w.p. } 1 - \lambda(v)\Delta t + o(\Delta t) \\ v(t) + b(v(t))\Delta t + o(\Delta t) + \zeta(t) & \text{w.p. } \lambda(v)\Delta t + o(\Delta t) \end{cases}$$

as $\Delta t \to 0$, where $\zeta(t)$ is a random process, such that

$$\Pr\{\zeta(t) = u \,|\, v(t) = v, v(t_1) = v_1, \ldots, v(t_n) = v_n\} = w(u \,|\, v)$$

for all $t > t_1 > \cdots > t_n$. The function $\lambda(v)$ is the jump rate, the function $b(v)$ is the deterministic drift of $v(t)$, and $w(u \,|\, v)$ is the jump size conditional density.

For dichotomic noise, that takes two values, a_1 and a_2, say, with rates λ_1 and λ_2, respectively, we have

$$b(v) = 0, \quad \lambda(v) = \lambda_i \text{ if } v = a_i \quad (i = 1, 2) \qquad (11.28)$$

and

$$w(u \,|\, v) = \delta(u - a_1 - a_2 + 2v). \qquad (11.29)$$

Thus, we can write eq. (11.27) in the form

$$Lp = \lambda(a_1 + a_2 - v)p(a_1 + a_2 - v) - \lambda(v)p(v). \qquad (11.30)$$

If we scale $a_1 \to a_1/\varepsilon$ and $\lambda_i \to \lambda_i/\varepsilon^2$ (the so-called "white noise scaling"), then the effect of $v(t)$ on the oscillator in the limit $\varepsilon \to 0$ becomes that of white noise, as shown below.

The scaling (11.18) of the noise is assumed for convenience. It is further assumed that the functions $g(x, \dot{x})$ and $\phi(x, \dot{x})$ are independent of γ and that γ is small relative to the other parameters of the problem. The latter include the height of the potential barrier formed by $U(x)$, the frequency of the unperturbed oscillator

$$\ddot{x} + U(x) = 0$$

at any energy level, the correlation time, the jump moments of the noise, and so on. □

11.1.2 The steady-state distribution of the noisy oscillator

The steady-state probability density function of fluctuations in (11.16) is the stationary solution of the Fokker–Planck equation (11.23). Under the assumption $\gamma << 1$, we first derive an averaged form of (11.16) by the method of adiabatic elimination of fast variables. We scale time by

$$t = \frac{s}{\gamma}$$

and expand the solution of (11.16) in the asymptotic series

$$p \sim p^0 + \sqrt{\gamma} p^1 + \gamma p^2 + \cdots. \tag{11.31}$$

The leading-order equation is

$$L_0 p^0 = -\dot{x}\frac{\partial p^0}{\partial x} + U'(x)\frac{\partial p^0}{\partial \dot{x}} + L_0 p^0 = 0. \tag{11.32}$$

We consider the case when L acts on v alone. We then show that the integrable solution of (11.32) is given by

$$p^0(x, \dot{x}, v, s) = p^0(E, s)\Psi_0(v), \tag{11.33}$$

where the energy E is given by

$$E = \frac{\dot{x}^2}{2} + U(x) \tag{11.34}$$

and $\Psi_0(v)$ is the stationary pdf of $v(t)$. Indeed, because L acts on v alone, we can solve (11.32) by separation of variables. The first eigenvalue of the operator L is zero, the other eigenvalues have negative real parts, and the first eigenfunction is $\Psi_0(v)$ (see (11.19)). The eigenfunctions $p_i(x, \dot{x}, s)\Psi_i(v)$ of the problem thus satisfy

$$L\Psi_i = \mu_i \Psi_i, \quad \Re e\, \mu_i < 0 \quad (i > 1), \quad \mu_0 = 0$$

and consequently

$$L_1 p_i = -\mu_i p_i, \quad (i = 0, 1, 2, \ldots). \tag{11.35}$$

For $i = 0$ the solution of (11.35) is a function of E and s, whereas for $i > 1$

$$p_i = p_{i0} e^{-\mu_i t}$$

on the characteristics

$$\dot{x} = -y, \quad \dot{y} = U'(x),$$

of (11.35), where p_{i0} is constant on each characteristic. It follows that p_i is unbounded, unless $p_{i0} = 0$ for all $i \geq 1$.

In addition, because the normalization condition (11.20) is satisfied, we must have

$$\int \Psi_i(v)\, dv = 0 \text{ for } i > 1. \tag{11.36}$$

At the next order, we obtain

$$L_0 p^1 = v \frac{\partial \left[\phi(x,\dot{x})p^0\right]}{\partial \dot{x}}. \tag{11.37}$$

The solvability condition for (11.37) is

$$\frac{1}{T(E)} \oint_E \int p^* v \frac{\partial}{\partial \dot{x}} \left[\phi(x,\dot{x})p^0\right] \frac{dx}{\dot{x}} dv = 0, \tag{11.38}$$

where $T(E)$ is the period of motion on the constant energy contour (11.34) and p^* is any bounded eigenfunction of the adjoint equation

$$L_0^* p^* = \dot{x}\frac{\partial p^*}{\partial x} - U'(x)\frac{\partial p^*}{\partial \dot{x}} + L^* = 0. \tag{11.39}$$

Again, separating variables, using (11.22), and arguing as above, we find that

$$p^*(x,\dot{x},v,s) = p^*(E,s), \tag{11.40}$$

so that, in view of (11.33) and (11.21), eq. (11.38) is satisfied. The solution of (11.37) is constructed by expanding it in the eigenfunctions of L, as

$$p^1 = \sum_{n=0}^{\infty} b_n(x,\dot{x},s)\Psi_n(v) \tag{11.41}$$

where for $n = 0, 1, 2, \ldots$

$$L_1 b_n(x,\dot{x},s) + \mu_n b_n(x,\dot{x},s) = \frac{\partial}{\partial \dot{x}} \left[\phi(x,\dot{x})p^0(E,s)\right] m_n, \tag{11.42}$$

and

$$v\Psi_0(v) = \sum_{n=0}^{\infty} m_n \Psi_n(v). \tag{11.43}$$

Now (11.21) and (11.36) imply that $m_0 = 0$. Inasmuch as $\mu_0 = 0$

$$b_0(x,\dot{x},s) = b_0(E,s) \tag{11.44}$$

and for $n > 1$,

$$b_n(x,\dot{x},s) \tag{11.45}$$
$$= -m_n e^{\mu_n F(x,E)} \int_{x_1(E)}^{x} e^{-\mu_n F(x,E)} \frac{\partial}{\partial \dot{x}} \left[\phi(\bar{x},\dot{x}(\bar{x},E))p^0(E,s)\right] \frac{d\bar{x}}{\dot{x}(\bar{x},E)}$$

with

$$F(x,E) = \int_{x_1(E)}^{x} \frac{d\bar{x}}{\dot{x}(\bar{x},E)},$$

where $\dot{x}(x, E)$ is defined in (11.34) and $x_1(E)$ is a point such that

$$U(x_1(E)) = E. \tag{11.46}$$

At the next order in the expansion (11.31), we obtain

$$L_0 p^2 = \frac{\partial p^*}{\partial s} + v\frac{\partial}{\partial \dot{x}}\left[\phi(x,\dot{x})p^1(E,s)\right] - \frac{\partial}{\partial \dot{x}}\left[\dot{x}g(x,\dot{x})p^0\right]. \tag{11.47}$$

The solvability condition for (11.47) is given by

$$\frac{1}{T(E)}\oint_E \frac{dx}{\dot{x}}\int dv\left\{\frac{\partial p^0}{\partial s} + v\frac{\partial}{\partial \dot{x}}\left[\phi(x,\dot{x})p^1 - \frac{\partial}{\partial \dot{x}}\left[\dot{x}g(x,\dot{x})p^0\right]\right]\frac{dx}{\dot{x}}dv\right\} = 0.$$

Using (11.41), (11.43), and (11.45) in the above solvability condition, we obtain the averaged equation for $p^0(E, s)$ as

$$T(E)\frac{\partial p^0}{\partial s} = \frac{\partial}{\partial E}A(E)\frac{\partial p^0}{\partial E} + \frac{\partial}{\partial E}\left[B(E) + C(E)\right]p^0,$$

where

$$A(E) \tag{11.48}$$

$$= \lim_{N\to\infty}\frac{1}{N}\int_0^{NT(E)}\int_0^t \dot{x}(t)\phi(x(t),\dot{x}(t))\dot{x}(\bar{t})\phi(x(\bar{t}),\dot{x}(\bar{t}))\sum_{n=1}^{\infty} e^{\mu_n(t-\bar{t})}M_n d\bar{t}\,dt,$$

with $M_n = m_n \int v\Psi_n(v)\,dv$ and

$$B(E) = \lim_{N\to\infty}\frac{1}{N}\int_0^{NT(E)}\int_0^t \dot{x}(t)\phi(x(t),\dot{x}(t))\phi_x(x(\bar{t}),\dot{x}(\bar{t}))\sum_{n=1}^{\infty} e^{\mu_n(t-\bar{t})}M_n d\bar{t}\,dt$$

$$C(E) = \oint_E \dot{x}g(x,\dot{x})\,dx. \tag{11.49}$$

Here, $x(t)$ is the solution of (11.16), and the dependence of $x(t)$ and $\dot{x}(t)$ on E is given by (11.34). We have used the fact that on the trajectory (11.34), the function $F(x, E)$ represents time.

Two limits are of particular interest. The white noise limit in Example 11.2(iii) is obtained by scaling $v(t) \to v(t)/\varepsilon$ and $\lambda(v) \to \lambda(v)/\varepsilon^2$ and letting $\varepsilon \to 0$. Then

$$\mu_n \to \frac{\mu_n}{\varepsilon^2}, \quad M_n \to \frac{M_n}{\varepsilon^2},$$

and

$$A(E) \to D\oint_E \dot{x}^2\phi^2(x,\dot{x})\,dx \quad \text{as } \varepsilon \to 0 \tag{11.50}$$

$$B(E) \to \frac{1}{2}D\oint_E \dot{x}\phi_x\,dx \quad \text{as } \varepsilon \to 0,$$

with $D = -\sum_n M_n/\mu_n$ and $C(E)$ given in (11.49).

Example 11.3. (Random telegraph and colored noise).

(i) The dichotomic noise (or random telegraph process) defined in (11.28)–(11.30) leads to $\mu_n = 0, \quad \mu_1 = -(\lambda_1 + \lambda_2)$,

$$\Psi_0(v) = \frac{\lambda(a_1 + a_2 - v)}{\lambda_1 + \lambda_2}\left[\delta(v - a_1) + \delta(v - a_2)\right] \tag{11.51}$$

$$\Psi_1(v) = \delta(v - a_1) - \delta(v - a_2) \tag{11.52}$$

$$m_0 = \int v\Psi_0(v)\,dv = \lambda_1 a_1 + \lambda_2 a_2 = 0 \tag{11.53}$$

$$m_1 = \frac{a_1\lambda_2}{\lambda_1 + \lambda_2} \tag{11.54}$$

$$M_1 = m_1 \int v\Psi_1(v)\,dv = \frac{(a_1 - a_2)a_1\lambda_2}{\lambda_1 + \lambda_2} = -a_1 a_2, \tag{11.55}$$

by (11.32), and $D = -a_1 a_2/(\lambda_1 + \lambda_2)$ (see [126]).

The shot noise limit in the dichotomic noise example is obtained by scaling $a_1 \to a_1/\varepsilon$, $\lambda_1 \to \lambda_1/\varepsilon$. As $\varepsilon \to 0$ equations (11.49)–(11.50) hold with $D = -a_1 a_2/lambda a_1$.

(ii) The white noise limit in the colored noise (Ornstein–Uhlenbeck) example, defined in (11.25)–(11.26), is obtained by letting the bandwidth $\alpha \to \infty$. In this case $\mu_1 = -\alpha$, $M_1 = \varepsilon\alpha$, $M_n = 0$, $n > 1$ (the eigenfunctions are Hermite functions), and $D = \varepsilon$ (see [36], [124]). □

11.1.3 First passage times and stability

Assume that in the absence of noise the nonlinear oscillator

$$\ddot{x} + \gamma\dot{x}g(x,\dot{x}) + U'(x) = 0$$

has a stable steady-state solution whose domain of attraction in phase space is denoted D. If γ is small, D is approximately given by

$$D_E = \{(x,\dot{x}) \mid E < E_C\}$$

for some value E_C of the energy E. We denote by $\tilde{\tau}$ the first passage time of the energy E to the level E_C

$$\tilde{\tau} = \inf\left\{t \;\middle|\; \frac{\dot{x}^2(t)}{2} + U(x(t)) = E_C\right\}$$

and by $\tau(x,\dot{x})$ its conditional mean

$$\tau(x,\dot{x}) = E\left[\tilde{\tau} \mid x(0) = x,\ \dot{x}(0) = \dot{x}\right].$$

If $\tau(x,\dot{x})$ is finite, then $\tilde{\tau}$ is finite with probability one. That is, the energy of the oscillator reaches the critical value E_C in finite time with probability one.

Example 11.4 (Damped pendulum with random vertical force applied at the hinge). A linear damped oscillator with random vertical force applied at the hinge is described by the dynamics

$$\ddot{x} + \gamma\dot{x} + \omega^2 \left[1 + \sqrt{2\varepsilon\gamma}v(t)\right] x = 0, \tag{11.56}$$

where $v(t)$ is a stationary noise, as described in the previous section. If the energy of the pendulum exceeds a given critical level (above the initial energy), the deflection angle will take values that exceed a preset critical deflection (above the initial deflection angle). It follows that the pendulum absorbs energy from the noise and becomes unstable. If the pendulum is at equilibrium ($x = \dot{x} = 0$), no amount of noise will affect the deflection angle. Therefore the mean time to reach a critical energy level, starting with zero energy, is infinite, and the probability of ever reaching this level is zero. If the initial energy of the pendulum is not zero, it may absorb energy from the applied noise at a rate higher (lower) than the rate at which energy is dissipated and thus be unstable (stable). Thus $\tau < \infty$ implies instability of the oscillator. On the other hand, $\tau = \infty$ is insufficient, in general, to ascertain stability, because cases are known in which $\tau = \infty$, but $\Pr\{\tilde{\tau} < \infty\} = 1$. This is the case, for example, of the first passage time to the origin of a Brownian motion. The function $\tau(x, \dot{x})$ is the solution of the boundary value problem

$$L^*\tau = -1 \text{ for } E < E_C, \quad \tau = 0 \text{ for } E = E_C, \tag{11.57}$$

whenever τ is finite. If the noise does not vanish at the steady-state, then τ is finite for all $E < E_C$, and (11.57) has a unique bounded solution. If, however, the noise vanishes at the steady-state (e.g., equilibrium point or limit cycle), a trajectory that starts at the steady-state will never leave, so that τ is infinite at the steady-state, which implies that (11.57) may have a singularity at the steady-state. Such a singular solution may exist whether or not the steady-state is stable. We consider this case next. The (averaged) equation for the leading term τ^0 in the expansion

$$\tau \sim \frac{\tau^0}{\gamma} + \frac{\tau^1}{\sqrt{\tau}} + \tau^2 + \cdots$$

is given by (see (11.48)–(11.49))

$$\frac{d}{dE}A(E)\frac{d\tau^0}{dE} - [B(E) + C(E)]\frac{d\tau^0}{dE} = -T(E) \tag{11.58}$$

because τ^0 is a function of E only, as described above. One boundary condition for τ^0 is

$$\tau^0(E_C) = 0. \tag{11.59}$$

The boundary conditions for (11.58) that determine a unique solution, depend on whether the noise vanishes at the steady-state. If the noise does not vanish there, the condition (11.59) together with

$$\tau^0(0) < \infty$$

(we assume $E = 0$ is the steady-state) determine $\tau^0(E)$ uniquely from (11.58). In contrast, if the noise vanishes at the steady-state, then

$$\tau^0(0) = \infty. \tag{11.60}$$

Condition (11.60) cannot serve as a second boundary condition, because it does not determine $\tau^0(E)$ uniquely from (11.58) and (11.59). The question of boundary conditions for this exit problem is related to the nature of the boundary behavior of the diffusion process $E(t)$, according to Feller's classification [72], [117]. □

To discuss this situation, we consider the special case where $x = \dot{x} = 0$ is a stable equilibrium of (11.56) and $\phi(0,0) = 0$. Then, although (11.58)–(11.60) hold, they are insufficient for the determination of the finiteness of $\tau^0(E)$ for $0 < E < E_C$. Therefore, we consider the probability of ever reaching E_C, starting at $0 < E < E_C$. The function

$$p^0(E) = \Pr\{\tilde{\tau} < \infty \,|\, E(t=0) = E\}$$

is the solution of the equation

$$\frac{d}{dE} A(E) \frac{dp^0}{dE} - [B(E) + C(E)] \frac{dp^0}{dE} = 0 \tag{11.61}$$

with

$$p^0(E_C) = 1, \quad p^0(0) = 0, \tag{11.62}$$

(because $\Pr\{\tilde{\tau} = \infty \,|\, E(t=0) = 0\} = 1$). The solution of (11.61), (11.62) is given by

$$p^0(E) = \lim_{\delta \to 0} \left[1 - \frac{q(E)}{q(\delta)}\right],$$

where

$$q(E) = \int_E^{E_C} \exp\left\{ \int_{E_C}^{\tilde{E}} \frac{B(z) + C(z)}{A(z)} \, dz \right\} \frac{d\tilde{E}}{A(\tilde{E})}.$$

We now describe the stability criterion in terms of the Feller-type function $q(E)$. If $\lim_{\delta \to 0} q(\delta) < \infty$, then $p^0(E) < 1$ and the oscillator is stochastically stable, because $\lim_{E \to 0} p^0(E) = 0$ or, for every $\varepsilon_1, \varepsilon_2 > 0$ there exists $\delta > 0$ such that if the initial energy satisfies $0 < E < \delta$, then $\Pr\{\sup_{t>0} E(t) > \varepsilon_1 \,|\, E\} < \varepsilon_2$. If, however, $\lim_{\delta \to 0} q(\delta) = \infty$, then $p^0(E) = 1$, so the oscillator's energy reaches the value E_C (which may be chosen arbitrarily) in finite time with probability 1, starting at any positive energy. Then the oscillator is stochastically unstable. The mean time to reach E_C is then given by

$$\tau(E) \sim \frac{\tau^0(E)}{\gamma},$$

where $\tau^0(E)$ is the limit as $\delta \to 0$ of the solution of (11.58) in $\delta < E < E_C$ with $\tau^0(\delta) = \tau^0(E_C) = 0$, if the limit exists.

Example 11.5 (Continuation of Example 11.4). Consider the linear oscillator (11.56) in Example 11.4 with

$$U(x) = \frac{\omega^2 x^2}{2}, \quad \phi(x, \dot{x}) = \sqrt{\varepsilon} U'(x), \quad g(x, \dot{x}) = 1$$

and with $v(t)$ a Gaussian white noise. The noise vanishes at the equilibrium point $x = \dot{x} = 0$, because $U'(0) = 0$. The function $x(t)$ is given by

$$x(t) = \frac{\sqrt{2E}}{\omega} \cos \omega t, \quad T(E) = \frac{2\pi}{\omega}$$

and

$$\frac{A(E)}{T(E)} = \frac{\varepsilon E^2 \omega^2}{2} \sum_{n=1}^{\infty} \Re\mathrm{e} \frac{-M_n}{\mu_n + 2\omega i} = DE^2$$
$$\frac{B(E)}{T(E)} = 0, \quad \frac{C(E)}{T(E)} = 0.$$

The equilibrium point is stochastically stable if $D < 1$, because

$$q(\delta) = \frac{1}{E_C(1-D)} \left[1 - \left(\frac{\delta}{E_C} \right)^{(1-D)/D} \right].$$

The probability of ever reaching E_C, given an initial energy E_C, is given by

$$p^0(E) = \left(\frac{E}{E_C} \right)^{(1-D)/D}.$$

If $D > 1$, the equilibrium point is stochastically unstable, because $\lim_{\delta \to 0} q(\delta) = \infty$. In this case $\tilde{\tau}$ is finite for all $E > 0$, and the leading term in its expansion is given by

$$\tau^0(E) = \frac{1}{D-1} \log \frac{E_C}{E}. \tag{11.63}$$

Note that $\lim_{E \to 0} \tau^0(E) = \infty$. Indeed, there may be many solutions of (11.58)–(11.60). The correct solution (11.63) is chosen by considering the problem on $\delta < E < E_C$, and then letting $\delta \to 0$. Separate analysis shows that in the case $D = 1$ the pendulum is stochastically unstable; that is, $p^0(E) = 1$, although $\tau^0(E) = \infty$ for all E. □

Example 11.6 (Explicit calculation of the stability criterion). We now calculate the parameter D, which determines the stability criterion for two types of noise. If the noise is the Ornstein–Uhlenbeck process (colored noise) described by (11.25), (11.26), we obtain

$$D = \frac{\varepsilon \alpha^2 \omega^2}{2(\alpha^2 + 4\omega^2)}. \tag{11.64}$$

If the noise is dichotomic, as described by (11.28)–(11.30), we obtain (see (11.51)–(11.55))

$$D = \frac{-\varepsilon\omega^2\,(\lambda_1+\lambda_2)\,a_1a_2}{2\left[(\lambda_1+\lambda_2)^2+4\omega h2\right]}.$$

The white noise result is obtained by taking the limit $\alpha \to \infty$ in (11.64).

The stability criterion $D < 1$ can be understood as follows. The criterion $D < 1$ expresses the fact that the rate at which the oscillator dissipates energy is higher than the rate it absorbs energy from the noise, because D in (11.64) is proportional to the noise-to-friction ratio ε. It follows that the oscillator loses energy and remains near the equilibrium state [202], [6], [151], [119]. □

Example 11.7 (A nonlinear noisy oscillator). Consider the nonlinear noisy oscillator

$$\ddot{x} + \gamma\dot{x}|\dot{x}|^{m-1} + x|x|^{n-2}\left[1+\sqrt{2\varepsilon\gamma}v(t)\right] = 0, \tag{11.65}$$

where $m \geq 0$, $n > 0$, and $v(t)$ is Gaussian white noise. Note that $m = 0$ corresponds to Coulomb friction. The energy is given by

$$E = \frac{\dot{x}^2}{2} + \frac{|x|^n}{n}, \tag{11.66}$$

hence

$$x = C_0E^{1/n}\Psi(E^{1/2-1/n}t), \tag{11.67}$$

where C_0 is some constant and $\Psi(t)$ is a periodic function. It follows that

$$\frac{A(E)}{T(E)} = C_1E^{3-2/n}, \quad \frac{B(E)}{T(E)} = 0, \quad \frac{C(E)}{T(E)} = C_2E^{(m+1)/2},$$

where C_1 and C_2 are positive constants. Thus, $\lim_{\delta\to 0} q(\delta) = \infty\ (<\infty)$ if

$$m + \frac{4}{n} > 3\,(<3),$$

so the oscillator is stochastically unstable (stable). The case

$$m + \frac{4}{n} = 3,$$

which holds for linear oscillators, requires separate analysis. The stochastic stability of the oscillator then depends on the coefficients, as described above in the linear case. □

Example 11.8 (Growth rate in the unstable case). The growth rate of the random trajectories in the unstable case

$$m + \frac{4}{n} > 3$$

is related to the dependence of $\tilde{\tau}$ on E_C for large E_C. The solution of (11.58) with (11.59) and with the appropriate conditions is given by

$$\tau^0(E) = K \int_E^{E_C} x^{-5/2+1/n} e^{x^\zeta} \int_0^x s^{1/n-1/2} e^{-x^\zeta} \, ds \, dx,$$

where K is a constant and

$$\zeta = -\frac{5}{2} + \frac{m}{2} + \frac{2}{n}.$$

If $\zeta \leq 0$, then, for $0 < n < 2$,

$$\tau^0(E) \sim const. E_C^{-1+2/n} \quad \text{for } E_C >> 1;$$

if $n = 2$, then

$$\tau^0(E) \sim const. \log E_C; \tag{11.68}$$

if $n > 2$, then

$$\tau^0(E) \sim const. \left(E^{-1+2/n} - E_C^{-1+2/n}\right); \tag{11.69}$$

that is, the random trajectory blows up in finite time with probability one. Note that in (11.68) the Lyapunov exponent vanishes so that the stability criterion based on the Lyapunov exponent is inconclusive. In contrast, the exponent in (11.69) is positive. For $\zeta > 0$,

$$\tau^0(E) \sim c_1 e^{c_2 E_C^\zeta} E_C^{-6+m/2+3/n},$$

where c_1 and c_2 are constants. □

Exercise 11.2. (A nonlinearizable oscillator).

(i) Consider the nonlinearizable oscillator

$$\ddot{x} + \gamma \dot{x} |\dot{x}|^{m-1} + x|x|^{n-2} + \sqrt{\gamma} |x|^\alpha |\dot{x}|^\beta v(t) = 0, \tag{11.70}$$

with $m \geq 0$, $\alpha, \beta, n \geq 0$, and $v(t)$ a random process satisfying the above assumptions. Use the energy (11.66) and the deterministic trajectories (11.67) to obtain the non-dimensional equation for Ψ,

$$(\Psi')^2 + \frac{|\Psi|^n}{n} = 1,$$

whose solution is a periodic function of some period T, independent of E. Conclude that the period of $x(t)$ at energy E is

$$T(E) = TE^{-1/2+1/n}.$$

Show that

$$\frac{A(E)}{T(E)} = -2E^{2(\alpha+1)/n+\beta}\sum_{k=0}^{\infty} M_k \sum_{j=1}^{\infty} \frac{\mu_k|a_j|^2T^2}{j^2+T^2\mu_k E^{2/n-1}}$$

$$\frac{B(E)}{T(E)} = 2E^{(2\alpha+1)/n+\beta-1/2}\sum_{k=0}^{\infty} M_k \sum_{j=1}^{\infty} \frac{\Re\mathrm{e}\left[\left(\frac{ij}{T} - \mu_k E^{-1/2+1/n}\right)\bar{a}_j bj\right]}{\frac{j^2}{T^2} + \mu_k^2 E^{-1+2/n}}$$

$$\frac{C(E)}{T(E)} = const. \times E^{(m+1)/2},$$

where a_j and b_j are the Fourier coefficients of the functions

$$|\Psi(t)|^{\alpha}|\Psi'(t)|^{\beta+1}\mathrm{sgn}\,\Psi'(t)$$

and

$$|\Psi(t)|^{\alpha}|\Psi'(t)|^{\beta-1}\mathrm{sgn}\,\Psi'(t),$$

respectively, and the bar denotes complex conjugation. Obtain the following stability criteria:
(1) If $0 < n < 2$, then for

$$\frac{m+1}{2} - \frac{2(\alpha+1)}{n} - \beta + 1 < 0 \tag{11.71}$$

the oscillator is stable. If the inequality is reversed, the oscillator is stable if

$$\frac{-2(\alpha+1)}{n} - \beta + \frac{1}{2} - \frac{1}{n} + c_1 > -1,$$

and is unstable otherwise. Here c_1 is given by

$$c_1 = \sum_{j=1}^{\infty} \frac{a_j b_j}{j^2} \sum_{k=1}^{\infty} \frac{|a_k|^2}{k^2}$$

under the assumption that a_j and b_j are real.
(2) If $n > 2$, then for

$$\frac{m+1}{2} - \frac{2\delta}{n} - \beta < 0$$

the oscillator is stable if

$$\frac{2\alpha+1}{n} + \beta + \frac{1}{2} - c_2 < 1,$$

and unstable otherwise. Here c_2 is given by

$$c_2 = \sum_{j=1}^{\infty} a_j b_j \sum_{k=1}^{\infty} |a_k|^2.$$

(3) The case $n = 2$ corresponds to a harmonic potential. In this case, the above conditions hold with c_2 replaced by c_3, given by

$$c_3 = \sum_{j=1}^{\infty}\sum_{k=1}^{\infty}\frac{M_k\mu_k a_j b_j}{j^2+T^2\mu_k}\sum_{j=1}^{\infty}\sum_{k=1}^{\infty}\frac{M_k\mu_k a_j^2}{j^2+T^2\mu_k}$$

(ii) The nonlinear pendulum with noise applied at the hinge, described by (11.70), corresponds to $\beta = 0$, $\alpha = n-1$, so that $B(E) = 0$. Show that if $v(t)$ is Gaussian white noise with diffusion coefficient D, then the stability condition is

$$m + \frac{4}{n} < 3,$$

as above. The linear case corresponds to $m = 1$, $n = 2$, whereas the case of Coulomb friction corresponds to $m = 0$. Note that the stability condition for harmonic potentials depends on the noise, but the conditions in (i) and (ii) do not. The latter is the case of the undamped linear oscillator.

(iii) Show that for white, colored, shot, and finite state noise, all the series that appear in calculation are convergent [125]. □

11.2 Stabilization with oscillations and noise

It was shown in the previous section that a large class of noisy parametric perturbations destabilizes deterministically stable systems if there is insufficient damping. There is, however, a class of noisy parametric perturbations that may stabilize a deterministically unstable systems, such as the inverted pendulum.

11.2.1 Stabilization by high-frequency noise

We consider a class of dynamical systems of the form

$$\frac{d\boldsymbol{x}}{dt} = \boldsymbol{X}_0(\boldsymbol{x}) + \frac{1}{\sqrt{\varepsilon}}\boldsymbol{X}_1\left(\boldsymbol{x},\boldsymbol{\xi}_1\left(\frac{t}{\varepsilon}\right)\right) + \frac{1}{\varepsilon}\boldsymbol{X}_2\left(\boldsymbol{x},\boldsymbol{\xi}_2\left(\frac{t}{\varepsilon}\right)\right), \tag{11.72}$$

where

$$\boldsymbol{x} \in \mathbb{R}^n, \quad \boldsymbol{X}_0 : \mathbb{R}^n \to \mathbb{R}^n, \quad \boldsymbol{\xi}_i \in \mathbb{R}^{m_i}$$
$$\boldsymbol{X}_i : \mathbb{R}^n \times \mathbb{R}^{m_i} \to \mathbb{R}^n,\ i = 1, 2, \quad 0 < \varepsilon << 1.$$

Here $\boldsymbol{X}_0(\boldsymbol{x})$ represents the deterministic unperturbed dynamics and the processes $(1/\sqrt{\varepsilon})\,\boldsymbol{X}_1(x,\boldsymbol{\xi}_1(t/\varepsilon))$ and $(1/\varepsilon)\,\boldsymbol{X}_2(x,\boldsymbol{\xi}_2(t/\varepsilon))$ are intended to model wideband and high-frequency (state-dependent) perturbations, respectively. We assume that for each fixed $\boldsymbol{x} \in \mathbb{R}^n$ the vector $(1/\sqrt{\varepsilon})\,\boldsymbol{X}_1(x,\boldsymbol{\xi}_1(t/\varepsilon))$ is a stationary ergodic wideband random process and the vector process $(1/\varepsilon)\,\boldsymbol{X}_2(x,\boldsymbol{\xi}_2(t/\varepsilon))$ is either a stationary ergodic high-frequency random process; that is, a process whose power spectral density vanishes, at least quadratically, at the origin, or an almost periodic function. We assume that for each $\boldsymbol{x} \in \mathbb{R}^n$, we have

$$\mathbb{E}\boldsymbol{X}_i\left(\boldsymbol{x}, \boldsymbol{\xi}_i\left(\frac{t}{\varepsilon}\right)\right) = 0, \quad i = 1, 2. \tag{11.73}$$

Although (11.73) holds for each fixed $\boldsymbol{x} \in \mathbb{R}^n$, there may be nonzero correlation between the state $\boldsymbol{x}(t)$ and the processes $\boldsymbol{\xi}_i(\tau)$, $(i = 1, 2)$ in eq. (11.72). This would imply, in particular, that even if $\boldsymbol{X}_0(\boldsymbol{x}) = \boldsymbol{A}\boldsymbol{x}$,

$$\frac{d\mathbb{E}\boldsymbol{x}(t)}{dt} \neq \boldsymbol{X}_0\left(\mathbb{E}\boldsymbol{x}(t)\right)$$

and therefore, the averaging would not reveal the deterministic counterpart of the stochastic system.

In the limit $\varepsilon \to 0$ the system is asymptotically averaged and the correlations become apparent. The stability properties of the averaged system can be elucidated through a correlation-free form of the limiting diffusion.

11.2.2 The generating equation

First, we reduce the system (11.72) to a form suitable for asymptotic analysis by introducing the generating equation [13]

$$\frac{d\boldsymbol{x}}{d\tau} = \boldsymbol{X}_2\left(\boldsymbol{x}, \boldsymbol{\xi}_2(\tau)\right), \quad \tau = \frac{t}{\varepsilon}. \tag{11.74}$$

We assume that eq. (11.74) has a unique solution,

$$\boldsymbol{x}(\tau) = \boldsymbol{h}(\tau, \boldsymbol{x}_0), \tag{11.75}$$

defined for every initial condition $\boldsymbol{x}_0 \in \mathbb{R}^n$ for all $\tau \geq 0$. We assume furthermore that $\boldsymbol{X}_2\left(\boldsymbol{x}, \boldsymbol{\xi}_2(\tau)\right)$ is differentiable with respect to $\boldsymbol{x}$. Now, we use the substitution

$$\boldsymbol{x}(t) = \boldsymbol{h}(\tau, \boldsymbol{y}(t)) \tag{11.76}$$

in eq. (11.72) and obtain the standard form

$$\frac{d\boldsymbol{y}(t)}{dt} = \boldsymbol{Y}_0\left(\boldsymbol{y}, \frac{t}{\varepsilon}\right) + \frac{1}{\sqrt{\varepsilon}}\boldsymbol{Y}_1\left(\boldsymbol{y}, \frac{t}{\varepsilon}, \boldsymbol{\xi}_1\left(\frac{t}{\varepsilon}\right)\right), \tag{11.77}$$

where for $i = 0, 1$ and $\boldsymbol{\xi}_0 = \boldsymbol{0}$

$$\boldsymbol{Y}_i\left(\boldsymbol{y}, \frac{t}{\varepsilon}, \boldsymbol{\xi}_i\left(\frac{t}{\varepsilon}\right)\right) = \left[\frac{\partial \boldsymbol{h}}{\partial \boldsymbol{y}}\left(\frac{t}{\varepsilon}, \boldsymbol{y}\right)\right]^{-1} \boldsymbol{X}_i\left(\boldsymbol{h}\left(\frac{t}{\varepsilon}, \boldsymbol{y}\right), \boldsymbol{\xi}_i\left(\frac{t}{\varepsilon}\right)\right). \tag{11.78}$$

We make the following assumptions about the functions $\boldsymbol{X}_i$ and the noise processes $\boldsymbol{\xi}_i(\tau)$: the functions $\boldsymbol{X}_i$ are smooth with respect to $\boldsymbol{x}$ and continuous with respect to $\boldsymbol{\xi}_i$ and furthermore, the functions $\boldsymbol{X}_0, \boldsymbol{X}_1$, and $\boldsymbol{h}$ satisfy the following conditions. There exists a constant $C > 0$, independent of τ and $\boldsymbol{\xi}_i(\tau)$, such that for all $\boldsymbol{x} \in \mathbb{R}^n$

(i) $|\boldsymbol{X}_1(\boldsymbol{x}, \boldsymbol{\xi}_1)| + |\boldsymbol{h}(\tau, \boldsymbol{x})| \leq C(1 + |\boldsymbol{x}|)$.

(ii) $\left|\dfrac{\partial \boldsymbol{X}_0(\boldsymbol{x})}{\partial \boldsymbol{x}}\right| + \left|\dfrac{\partial \boldsymbol{X}_i(\boldsymbol{x}, \boldsymbol{\xi}_i)}{\partial \boldsymbol{x}}\right| + \left|\dfrac{\partial \boldsymbol{h}(\tau, \boldsymbol{x})}{\partial \boldsymbol{x}}\right| \leq C.$

(iii) Higher-order derivatives with respect to $\boldsymbol{x}$ of $\boldsymbol{h}$ and $\boldsymbol{X}_i$ $(i = 0, 1)$, are bounded by powers of $|\boldsymbol{x}|$ uniformly in τ and $\boldsymbol{\xi}_1$.

(iv) $\boldsymbol{\xi}_1(\tau)$ and $\boldsymbol{\xi}_2(\tau)$ are independent processes.

(v) $\boldsymbol{\xi}_1(\tau)$ is an ergodic stationary diffusion process, whose transition PDF, $P(t, \boldsymbol{\xi} \,|\, \boldsymbol{\eta})$ converges to a stationary PDF-$P(\boldsymbol{\xi})$ so that the recurrent potential kernel

$$Q(\boldsymbol{\xi} \,|\, \boldsymbol{\eta}) = \int_0^\infty [P(t, \boldsymbol{\xi} \,|\, \boldsymbol{\eta}) - P(\boldsymbol{\xi})]\, dt$$

exists and maps smooth bounded functions of $\boldsymbol{\xi}$ into themselves [23].

(vi) The function $\boldsymbol{h}(\tau, \boldsymbol{x})$ satisfies the ergodicity condition

$$\lim_{T\to\infty} \frac{1}{T} \int_0^T \boldsymbol{h}(t, \boldsymbol{x}_0)\, dt = \bar{\boldsymbol{h}}(\boldsymbol{x}_0), \quad \text{for all } \boldsymbol{x}_0 \in \mathbb{R}^n,$$

where $\bar{\boldsymbol{h}} : \mathbb{R}^n \to \mathbb{R}^n$ is a deterministic function.

11.2.3 The correlation-free equation

To determine the limiting behavior of the solution of eq. (11.72), $\boldsymbol{x}(t)$, we determine first the limiting behavior of the solution of eq. (11.77), $\boldsymbol{y}(t)$, and then we use eq. (11.76) to return to $\boldsymbol{x}(t)$.

For each fixed $\boldsymbol{x}_0$, the function $\boldsymbol{h}(\tau, \boldsymbol{x}_0)$, defined in eqs. (11.74), (11.75), is a stochastic process whose trajectories are determined by the trajectories of the noise $\boldsymbol{\xi}_2(\tau)$. If we condition eq. (11.77) on a given trajectory of $\boldsymbol{h}(\tau, \boldsymbol{x}_0)$, the conditional process $(\boldsymbol{y}(t), \boldsymbol{\xi}_1(\tau))$ is a diffusion process on $\mathbb{R}^n \times \mathbb{R}^{m_1}$, because $\boldsymbol{\xi}_1(\tau)$ and $\boldsymbol{\xi}_2(\tau)$ are independent. Its conditional transition pdf,

$$p(t, \boldsymbol{y}, \boldsymbol{\xi}_1) = p(t, \boldsymbol{y}, \boldsymbol{\xi}_1 |\, \boldsymbol{h}(s, \cdot),\, 0 \leq s < \infty),$$

satisfies the backward and forward Kolmogorov (Fokker–Planck) equations. We assume that $p(t, \boldsymbol{y}, \boldsymbol{\xi}_1)$ is a function of slow and fast times t and $\tau = t/\varepsilon$ and write the backward Kolmogorov equation in the form

$$
\begin{aligned}
Lp &= \frac{\partial p}{\partial t} + \frac{1}{\varepsilon}\frac{\partial p}{\partial \tau} + \sum_{i=1}^{n}\left(Y_0^i + \frac{1}{\sqrt{\varepsilon}}Y_1^i\right)\frac{\partial p}{\partial y_i} + \frac{1}{\varepsilon}Kp \\
&= \frac{\partial p}{\partial t} + \boldsymbol{Y}_0 \cdot \nabla_{\boldsymbol{y}} p + \frac{1}{\sqrt{\varepsilon}}\boldsymbol{Y}_1 \cdot \nabla_{\boldsymbol{y}} p + \frac{1}{\varepsilon}\left(\frac{\partial p}{\partial \tau} + Kp\right) = 0,
\end{aligned}
\tag{11.79}
$$

where K is the differential generator of the process $\boldsymbol{\xi}_1(\tau)$. Next, we derive an averaged equation for $\boldsymbol{y}$ by expanding $p(t, \boldsymbol{y}, \boldsymbol{\xi}_1)$ in powers of $\sqrt{\varepsilon}$. Rewriting eq. (11.79) in the obvious operator notation

$$
Lp = L_0 p + \frac{1}{\sqrt{\varepsilon}} L_1 p + \frac{1}{\varepsilon} L_2 p = 0
$$

and expanding

$$
p \sim p_0 + \sqrt{\varepsilon} p_1 + \varepsilon p_2 + \cdots, \tag{11.80}
$$

gives at order $1/\varepsilon$

$$
L_2 p_0 = \frac{\partial p_0}{\partial \tau} + K p_0 = 0. \tag{11.81}
$$

It follows from the ergodicity of $\boldsymbol{\xi}_1$ that the only bounded solution of eq. (11.81) is constant with respect to τ and $\boldsymbol{\xi}_1$. Thus

$$
p_0 = p_0(\boldsymbol{y}, t). \tag{11.82}
$$

At order $1/\sqrt{\varepsilon}$, we obtain

$$
L_2 p_1 = -L_1 p_0 = -\boldsymbol{Y}_1 \cdot \nabla_{\boldsymbol{y}} p_0. \tag{11.83}
$$

The solvability condition for eq. (11.83) is that

$$
\lim_{T \to \infty} \frac{1}{T} \int_0^T L_1 p_0(\boldsymbol{y}, t)\, dt = 0. \tag{11.84}
$$

Assuming that the limit

$$
\lim_{T \to \infty} \frac{1}{T} \int_0^T \mathbb{E}\left[\boldsymbol{Y}_1\left(\boldsymbol{y}, \tau, \boldsymbol{\xi}_1(\tau)\right)\right] d\tau = \boldsymbol{0}
$$

exists, uniformly in $\boldsymbol{y}$, we obtain (11.84). The solution of (11.83) is given by

$$
p_1 = -L_2^{-1} \boldsymbol{Y}_1 \cdot \nabla_{\boldsymbol{y}} p_0.
$$

The operator L_2^{-1} can be expressed in terms of the pdf $P(t, \boldsymbol{\xi} \mid \boldsymbol{\eta})$ [23]. At order 1, we obtain

$$
L_2 p_2 = -\frac{\partial p_0}{\partial t} - \boldsymbol{Y}_0(\boldsymbol{y}, \tau) \cdot \nabla_{\boldsymbol{y}} p_0 - \boldsymbol{Y}_1(\boldsymbol{y}, \tau, \boldsymbol{\xi}_1(\tau)) \cdot \nabla_{\boldsymbol{y}} L_2^{-1} \boldsymbol{Y}_1 \cdot \nabla_{\boldsymbol{y}} p_0. \tag{11.85}
$$

The solvability condition for eq. (11.85) gives the equation

$$\frac{\partial p_0(\boldsymbol{y},t)}{\partial t} + \bar{\boldsymbol{Y}}_0(\boldsymbol{y})\nabla_{\boldsymbol{y}} p_0(\boldsymbol{y},t) + \sum_{i,j=1}^{n} a^{ij}(\boldsymbol{y})\frac{\partial^2 p_0(\boldsymbol{y},t)}{\partial y_i \partial y_j}, \tag{11.86}$$

where

$$\begin{aligned}\bar{\boldsymbol{Y}}_0(\boldsymbol{y}) &= \lim_{T\to\infty}\frac{1}{T}\int_0^T \mathbb{E}\left[\boldsymbol{Y}_0(\boldsymbol{y},\tau)\right] d\tau \\ &+ \lim_{T\to\infty}\frac{1}{T}\int_0^T \mathbb{E}\left[\boldsymbol{Y}_1(\boldsymbol{y},\tau,\boldsymbol{\xi}_1(\tau))\cdot\nabla_{\boldsymbol{y}}\int_0^\tau \boldsymbol{Y}_1(\boldsymbol{y},\sigma,\boldsymbol{\xi}_1(\sigma))\,d\sigma\right] d\tau \\ a^{ij}(\boldsymbol{y}) &= \lim_{T\to\infty}\frac{1}{T}\int_0^T \mathbb{E}\left[\boldsymbol{Y}_1(\boldsymbol{y},\tau,\boldsymbol{\xi}_1(\tau))\cdot\int_0^\tau \boldsymbol{Y}_1(\boldsymbol{y},\sigma,\boldsymbol{\xi}_1(\sigma))\,d\sigma\right] d\tau. \end{aligned} \tag{11.87}$$

It follows that $\boldsymbol{y}(t)$ converges weakly to a diffusion process $\bar{\boldsymbol{y}}(t)$ whose evolution is governed by the backward Kolmogorov equation (11.86) with coefficients defined by (11.87). The process $\boldsymbol{x}(t)$ converges weakly to a process $\bar{\boldsymbol{x}}(t)$ defined by $\bar{\boldsymbol{x}}(t) = \boldsymbol{h}(\tau,\bar{\boldsymbol{y}}(t))$.

11.2.4 The stability of (11.72)

Assume that $\boldsymbol{x}=\boldsymbol{0}$ is a critical point for the system (11.72). Then $\boldsymbol{y}=\boldsymbol{0}$ is a critical point for the averaged dynamics

$$d\boldsymbol{y} = \bar{\boldsymbol{Y}}_0(\boldsymbol{y})\,dt + \bar{\boldsymbol{Y}}_1(\boldsymbol{y})\,d\boldsymbol{w}, \tag{11.88}$$

where $\bar{\boldsymbol{Y}}_1(\boldsymbol{y})$ is a matrix such that $\bar{\boldsymbol{Y}}_1(\boldsymbol{y})\bar{\boldsymbol{Y}}_1^T(\boldsymbol{y}) = \boldsymbol{A}(\boldsymbol{y})$ with $\boldsymbol{A}(\boldsymbol{y}) = \left\{a^{ij}(\boldsymbol{y})\right\}_{i,j}^n$ and $\boldsymbol{w}(t)$ is m-dimensional Brownian motion.

The stability of the critical point of the averaged system (11.88) is inherited by that of the original system (11.72). Indeed, by assumption $\boldsymbol{X}_2(\boldsymbol{x},\boldsymbol{\xi}_2)$ is a smooth function of $\boldsymbol{x}$ and, thus, so is $h(\tau,\boldsymbol{y})$. Furthermore, by assumption, $|\partial\boldsymbol{h}(\tau,\boldsymbol{x})/\partial\boldsymbol{x}| \leq C$, and higher-order derivatives of $\boldsymbol{h}(\tau,\boldsymbol{x})$ are bounded by powers of $|\boldsymbol{x}|$. Therefore,

$$|h(\tau,\boldsymbol{y}_1) - h(\tau,\boldsymbol{y}_2)| \leq C\,|\boldsymbol{y}_1 - \boldsymbol{y}_2| + o\left(|\boldsymbol{y}_1 - \boldsymbol{y}_2|\right)$$

as $|\boldsymbol{y}_1 - \boldsymbol{y}_2| \to 0$, uniformly in τ. This means that

$$|\boldsymbol{x}_1(t) - \boldsymbol{x}_2(t)| \leq C\,|\boldsymbol{y}_1(t) - \boldsymbol{y}_2(t)| + o\left(|\boldsymbol{y}_1(t) - \boldsymbol{y}_2(t)|\right).$$

Example 11.9 (Damped harmonic oscillator with high-frequency parametric noise). Consider the damped harmonic oscillator with random high-frequency force applied at the hinge,

$$\ddot{x} + \gamma\dot{x} + \left[\omega^2 + \xi(t)\right]x = 0, \tag{11.89}$$

where

$$\xi(t) = \frac{1}{\sqrt{\varepsilon}} F\left(\xi_1\left(\frac{t}{\varepsilon}\right)\right) + \frac{1}{\varepsilon}\xi_2\left(\frac{t}{\varepsilon}\right),$$

$\xi_1(\tau)$ is a one-dimensional Ornstein–Uhlenbeck process, $d\xi_1 = -\xi_1\, d\tau + dw$ and $F : \mathbb{R} \to (a, b)$ (a bounded interval) such that $\mathbb{E}\left[F(\xi_1)\right] = 0$; the process $\xi_2(\tau)$ is any locally integrable, stationary, ergodic, high-frequency process whose integral, $\eta(\tau) = \int_0^\tau \xi_2(\sigma)\, d\sigma$, is also a stationary, ergodic, and bounded process.

The generating equation (11.74) for the oscillator (11.89) is

$$\dot{x}_1 = 0, \quad \dot{x}_2 = \frac{1}{\varepsilon}\xi_2\left(\frac{t}{\varepsilon}\right) x_1$$

and the substitution (11.74) becomes

$$x_1 = y_1, \quad x_2 = \eta(\tau) y_1 + y_2.$$

The phase space equation becomes

$$\dot{y}_1 = y_2 + \eta\left(\frac{t}{\varepsilon}\right) y_1 \tag{11.90}$$

$$\dot{y}_2 = -\left[2\gamma + \eta\left(\frac{t}{\varepsilon}\right)\right] y_2 - \left[\omega^2 + \eta^2\left(\frac{t}{\varepsilon}\right) - 2\gamma\eta\left(\frac{t}{\varepsilon}\right) + \frac{1}{\sqrt{\varepsilon}} F\left(\xi_1\left(\frac{t}{\varepsilon}\right)\right)\right] y_1.$$

The averaged equation corresponding to (11.89) is

$$\frac{d}{dt}\bar{y}_1 = \bar{y}_2, \quad \frac{d}{dt}\bar{y}_2 = -2\gamma\bar{y}_2 - \left(\omega^2 + \alpha^2\right)\bar{y}_1 + \sigma\bar{y}_1\dot{w}, \tag{11.91}$$

where $\dot{w}$ is standard Gaussian white noise, α^2 is the variance of $\eta(\tau)$,

$$\alpha^2 = \lim_{T\to\infty} \frac{1}{T}\int_0^T \eta^2(\tau)\, d\tau,$$

and

$$\sigma^2 = \int_{-\infty}^{\infty} \mathbb{E}\left[F(\xi_2(s))F(\xi_2(0))\right] ds = \int_{-\infty}^{\infty} R(s)\, ds.$$

The differential generator of the averaged diffusion process (11.91) is

$$\bar{L} = \bar{y}_2\frac{\partial}{\partial\bar{y}_1} - \left[2\gamma\bar{y}_2 + \left(\omega^2 + \alpha^2\right)\bar{y}_1\right]\frac{\partial}{\partial\bar{y}_2} + \frac{\sigma^2}{2}\bar{y}_1^2\frac{\partial^2}{\partial\bar{y}_2^2}.$$

To analyze the stability properties of (11.91) and, therefore, of (11.90) and (11.89) for sufficiently small ε, we construct a Lyapunov function by solving the equation

$$\bar{L}V(\bar{\boldsymbol{y}}) = -(c_1\bar{y}_1^2 + c_2\bar{y}_2^2), \quad c_1, c_2 > 0$$

in the form of a quadratic function

$$V(\bar{\boldsymbol{y}}) = m_{11}\bar{y}_1^2 + 2m_{12}\bar{y}_1\bar{y}_2 + m_{22}\bar{y}_2^2. \tag{11.92}$$

A simple calculation shows that the quadratic form (11.92) is positive definite for all $c_i > 0$ $(i = 1, 2)$ if and only if $2\left(\omega^2 + \alpha^2\right)\gamma > sigma^2/2$. Writing this condition as

$$4\gamma > \frac{\sigma^2}{\omega^2 + \alpha^2}, \tag{11.93}$$

we observe that wideband noise (large σ^2) requires a large damping coefficient γ to stabilize the oscillator, whereas high-frequency noise (large α^2) reduces the bound (11.93) on γ. We conclude that the wideband noise destabilizes (11.89) and the high-frequency noise is stabilizing, because (11.93) is a necessary condition for stability of (11.89). □

Example 11.10 (Inverted pendulum). An interesting special case is obtained if ω^2 is replaced by $-\omega^2$ in (11.89), which corresponds to an inverted pendulum. The stability condition (11.93) becomes

$$4\gamma > \frac{\sigma^2}{-\omega^2 + \alpha^2}$$

and can be satisfied if α^2 is sufficiently large. This explains why a broomstick can be stabilized on a fingertip by juggling it up and down at high-frequency. Note that if the only driving force at the hinge is $(\alpha/\varepsilon)\sin(t/\varepsilon)$, and if no damping is present $(\gamma = 0)$, the averaged dynamics (11.91) becomes

$$\frac{d}{dt}\bar{y}_1 = \bar{y}_2, \quad \frac{d}{dt}\bar{y}_2 = -\left(\omega^2 + \alpha^2\right)\bar{y}_1 = -\Omega^2\bar{y}_1,$$

so that even if ω^2 is replaced by $-\omega^2$, the oscillator stays stable, if α^2 is chosen sufficiently large. In this case,

$$x \sim \bar{y}_1 = A\cos(\Omega t + \phi) + O(\sqrt{\varepsilon})$$
$$\dot{x} \sim \eta(\tau)\bar{y}_1 + \bar{y}_2 = -\alpha\cos\left(\frac{t}{\varepsilon}\right) A\cos(\Omega t + \phi) - A\Omega\sin(\Omega t + \phi);$$

that is, the inverted pendulum oscillates with frequency $\Omega = \sqrt{\alpha^2 - \omega^2}$ with small fast vibrations about this slow motion. □

Example 11.11 (Convergence to the Ornstein–Uhlenbeck process). Consider the linear system

$$\dot{x}_1 = -x_1 + \frac{1}{\sqrt{\varepsilon}}F\left(\xi_1\left(\frac{t}{\varepsilon}\right)\right), \quad \dot{x}_2 = -2x_2 + \frac{1}{\varepsilon}\sin\left(\frac{t}{\varepsilon}\right)x_1, \tag{11.94}$$

where $F\left(\xi_1\left(t/\varepsilon\right)\right)$ is a wideband process as in Example 11.9. The generating equation for (11.94) is

$$\dot{x}_1 = 0, \quad \dot{x}_2 = \frac{1}{\varepsilon}\sin\left(\frac{t}{\varepsilon}\right)x_1$$

and the substitution (11.76) is

$$x_1 = y_1, \quad x_2 = -(\cos\tau)y_1 + y_2.$$

The system in standard form is

$$\begin{aligned} \dot{y}_1 &= -y_1 + \frac{1}{\sqrt{\varepsilon}} F\left(\xi_1\left(\frac{t}{\varepsilon}\right)\right) \\ \dot{y}_2 &= -2y_2 + \cos\left(\frac{t}{\varepsilon}\right) y_1 + \frac{1}{\sqrt{\varepsilon}} F\left(\xi_1\left(\frac{t}{\varepsilon}\right)\right) \cos\left(\frac{t}{\varepsilon}\right) \end{aligned} \tag{11.95}$$

and the averaged system is

$$\frac{d}{dt}\bar{y}_1 = -\bar{y}_1 + \sigma\dot{w}_1, \quad \frac{d}{dt}\bar{y}_2 = -2\bar{y}_2 + \frac{\sigma}{\sqrt{2}}\dot{w}_2, \tag{11.96}$$

where $\dot{w}_1$ and $\dot{w}_2$ are independent white noise processes. The noise process $\dot{w}_2$ is obtained as the limit of the modulated wideband process $(1/\sqrt{\varepsilon})\, F\left(\xi_1\left(t/\varepsilon\right)\right)$ as $\varepsilon \to 0$. Obviously, $(\sigma/\sqrt{2})\dot{w}_2$ is the generalized diffusion correction.

The limiting system (11.96) is a pair of ergodic, independent, Ornstein–Uhlenbeck processes. Hence, it has a unique stationary distribution and $\boldsymbol{y}(t)$ and $\bar{\boldsymbol{y}}(t)$ are close in distribution for ε sufficiently small and all $t \geq 0$. Furthermore, it follows that the solution of (11.94) is close in distribution to

$$x_1 = \bar{y}_1, \quad x_2 = -\cos\left(\frac{t}{\varepsilon}\right)\bar{y}_1 + \bar{y}_2$$

for ε sufficiently small and all $t \geq 0$. □

Example 11.12 (Convergence to the Rayleigh oscillator). Consider the Rayleigh equation with parametric oscillations and forcing

$$\ddot{x} + \mu\left(\frac{\dot{x}^3}{3} - \dot{x}\right) + \left[1 + \frac{1}{\sqrt{\varepsilon}} F\left(\xi_1\left(\frac{t}{\varepsilon}\right)\right)\right] x = \frac{1}{\varepsilon}\xi_2\left(\frac{t}{\varepsilon}\right), \tag{11.97}$$

or equivalently, the system

$$\begin{aligned} \dot{x}_1 &= x_2 \\ \dot{x}_2 &= -\mu\left(\frac{x_2^3}{3} - x_2\right) - \left[1 + \frac{1}{\sqrt{\varepsilon}} F\left(\xi_1\left(\frac{t}{\varepsilon}\right)\right)\right] x_1 + \frac{1}{\varepsilon}\xi_2\left(\frac{t}{\varepsilon}\right). \end{aligned} \tag{11.98}$$

We assume that $F(\xi_1)$ and ξ_2 satisfy the assumptions of Example 11.9. The change of coordinates for (11.98) is

$$x_1 = y_1, \quad x_2 = y_2 + \eta(\tau)$$

and the equation in standard form is

$$\begin{aligned} \dot{y}_1 &= y_2 + \eta(\tau) \\ \dot{y}_2 &= -\mu\frac{\left(y_2 + \eta\left(\frac{t}{\varepsilon}\right)\right)^3}{3} - \left(y_2 + \eta\left(\frac{t}{\varepsilon}\right)\right) - \left[1 + \frac{1}{\sqrt{\varepsilon}} F\left(\xi_1\left(\frac{t}{\varepsilon}\right)\right)\right] y_1. \end{aligned} \tag{11.99}$$

The averaged equation corresponding to (11.99) is

$$\frac{d}{dt}\bar{y}_1 = \bar{y}_2, \quad \frac{d}{dt}\bar{y}_2 = -\mu\left(\frac{\bar{y}_2^3}{3} + (\alpha^2 - 1)\bar{y}_2\right) - \bar{y}_1 + \sigma\bar{y}_1\dot{w}. \tag{11.100}$$

The systems (11.99) and (11.100) have an equilibrium point at $(0,0)$. To investigate the stability properties of (11.99) in a neighborhood of zero, we linearize (11.100) around $(0,0)$. The resulting linear system is

$$\frac{d}{dt}z_1 = z_2, \quad \frac{d}{dt}z_2 = -\mu(\alpha^2 - 1)z_2 - z_1 + \sigma z_1\dot{w}. \tag{11.101}$$

A simple calculation shows that (11.101) is uniformly, stochastically exponentially stable if

$$\alpha^2 > 1 + \frac{\sigma^2}{2\mu}. \tag{11.102}$$

Therefore it is asymptotically stable in a neighborhood of the origin if ε is sufficiently small and if (11.102) holds. Finally, because $x_1 = y_1$ and $x_2 = y_2 + \eta(t/\varepsilon)$, we conclude that if (11.102) is satisfied, then (11.98) has an asymptotically stable ergodic solution. Thus, the high-frequency oscillations in (11.97) have resulted in a transition of the unstable equilibrium point $(0,0)$ of the system

$$\ddot{x} + \mu\left(\frac{\dot{x}^3}{3} - \dot{x}\right) + \left[1 + \frac{1}{\sqrt{\varepsilon}}F\left(\xi_1\left(\frac{t}{\varepsilon}\right)\right)\right]x = 0$$

into an asymptotically stable ergodic solution. □

11.3 Stability of columns with noisy loads

A thin elastic structure, forced at one end by a noisy load, can represent a tall building or a drilling tower in an earthquake, a thin long rocket pushed by a jet engine, a high-pressure ruptured water pipe in a nuclear reactor or power station, and many other situations. The structure tends to absorb energy from the random load and become unstable, whereas internal or external dissipation mechanisms tend to check its oscillations. Stability criteria are needed for safe design of such structures.

Elasticity theory describes structures with any degree of rigidity by partial differential equations with fourth order spatial derivatives. The addition of noisy loads leads to partial differential equations, which are stochastic in time and fourth order in displacement. Rather than developing a theory of stochastic partial differential equations, we convert the problem into an N-dimensional system of Itô stochastic differential equations by approximating the structure with a series of multiple pendulums, whose energy distribution converges to that of the elastic structure. Stability criteria are established for the Itô equations by the mathematical methods developed in the previous chapters and are evaluated in the limit $N \to \infty$.

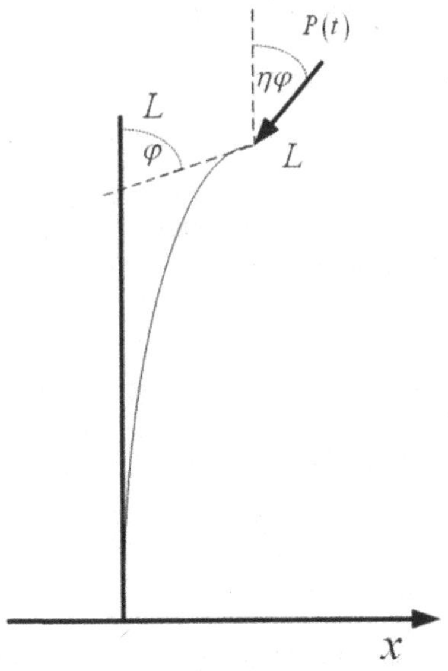

Figure 11.1. The angle of the load $P(t)$ is $\eta\varphi$ relative to vertical; that of the tangent line at the end of the deflected column is φ.

11.3.1 A thin column with a noisy load

Consider an elastic prismatic column built in at one end and loaded at the other. Let l be the length of the column, φ be the angle between the tangent at the end of the column and the vertical direction, $P(t)$ be the magnitude of the load, and $\eta\varphi$ be the angle between its direction and the vertical direction.

Assume the magnitude of the load is constant; that is, $P(t) = P_0$. Then, for $\eta = 0$ (vertical load) the system is an Euler column. It can be described by a potential, because it is conservative. In this case, the Euler buckling load is given by [252]

$$P_E = \frac{\pi^2 \alpha}{4l^2}, \tag{11.103}$$

where α is the flexural rigidity of the column. For $\eta = 1$ (follower load), the system is no longer conservative and the buckling load is given by

$$P_B = 2.031 \frac{\pi^2 \alpha}{l^2}.$$

Assume that the loading force $P(t)$ contains small random fluctuations of impulsive Poisson type (shot noise) or of white noise type. Thus, the random load is given by

$$P(t) = P_0 \left[1 + \varepsilon v(t)\right], \tag{11.104}$$

where P_0 is the mean load. Assume further that the process $v(t)$ is either a standard Gaussian white noise whose autocorrelation function is

$$\langle v(t+s)v(t)\rangle = b\delta(s) \tag{11.105}$$

and ε is a small parameter, which measures the noise intensity, or $v(t)$ is shot noise, which takes values ± 1 with probability $\frac{1}{2}$ at exponential waiting times.

In the case of the vertically loaded (Euler) column (i.e., $\eta = 0$) and in the absence of damping, linear elasticity theory describes the system by a parametrically excited fourth-order partial differential equation

$$\alpha\frac{\partial^4 W}{\partial x^4} + P(t)\frac{\partial^2 W}{\partial x^2} + \mu\frac{\partial^2 W}{\partial t^2} = 0, \tag{11.106}$$

where x is the coordinate along the column, $W(x,t)$ is the transversal deflection of the column from its equilibrium position, μ is the density, and α is the flexural rigidity.

In the presence of damping, the system is described by

$$\alpha\frac{\partial^4 W}{\partial x^4} + P(t)\frac{\partial^2 W}{\partial x^2} + \mu\frac{\partial^2 W}{\partial t^2} = F(W), \tag{11.107}$$

where $F(W)$ is a linear damping force. In the case of Newtonian damping, (e.g., external or internal friction) $F(W) = -\gamma W_t$; that is, the damping force is proportional to the local velocity of the column. Another example of internal damping is the Kelvin–Voigt model, $F(W) = -\gamma W_{xxt}$; that is, the damping force is proportional to the linearized rate of change in local curvature.

We assume the column is clamped at $x = 0$, so that the boundary condition at the built-in end is given by

$$W(0,t) = W_x(0,t) = 0 \tag{11.108}$$

and at the loaded end, $x = l$, is given by

$$W_{xx}(l,t) = \alpha W_{xxx}(l,t) + P(t)W_x(l,t) = 0. \tag{11.109}$$

A column with a follower load ($\eta = 1$) is described by eqs. (11.106)–(11.108) with the boundary condition (11.109) replaced by

$$W_{xx}(l,t) = W_{xxx}(l,t) = 0. \tag{11.110}$$

In the absence of damping and fluctuating forces ($\varepsilon = 0$ in eq. (11.106)), the vertically loaded column conserves energy. However, if $\varepsilon > 0$ the column absorbs energy from the random load, so the energy of the column may reach any level in finite time with high probability, as indicated by the example of the parametrically excited pendulum.

The stochastic stability and reliability of a structure containing the column as a component is measured by the time required by the energy of the column to reach a given critical value E_C. This level is determined by the performance specification of the structure, and can be sufficiently low to make the linear elasticity model (11.106)–(11.108) a valid description of the column. The *domain of reliability* of a vertically loaded column is defined as

$$D = \{W(x,t) \mid E(t) < E_C\}, \tag{11.111}$$

where $E(t)$ is the energy of the column with deflection $W(x,t)$ and E_C is a critical energy. The first time the energy of the system reaches the level E_C is defined by

$$\tau = \inf\{t \mid E(t) = E_C\}. \tag{11.112}$$

We define the *index of reliability* as the conditional expectation of τ, given an initial energy E_0 of the system. Thus

$$\bar{\tau} = \mathbb{E}[\tau \mid E(0) = E_C]. \tag{11.113}$$

For nonconservative systems the energy is replaced by an energy-like functional $\tilde{E}$, which is positive definite and is constant on trajectories of the deterministic system. In this case D, τ, and $\bar{\tau}$ are defined as in eqs. (11.111)–(11.113), but with $\tilde{E}$ replacing E.

Example 11.13 (Pendulum with vertical random force applied at the hinge). The simplest analogue of a vertically loaded column is a pendulum with vertical random force applied at the hinge, as described in Example 11.5. The example shows that the undamped pendulum is stochastically unstable for a large class of noises and that the underdamped pendulum is stable if the noise-to-friction ratio ε^* is below a threshold value and that the pendulum is unstable otherwise. More specifically, writing the noise intensity parameter in the form $\sqrt{\gamma\varepsilon^*}$, the stability criterion for white noise (see (11.105)) is $\varepsilon^* < 4/b$.

In this case, the probability that the energy exceeds E_C by time t, given an initial energy E_0, is given by [139]

$$p(E_C, t \mid E_0) \sim \left\{1 - \exp\left[-\frac{(a_1 - a_2)t}{4\gamma a_1}\right]\right\}\left(\frac{E_0}{E_C}\right)^{(a_1-a_2)/2a_1},$$

where

$$a_1 = \frac{\varepsilon^* b}{16}, \quad a_2 = -\frac{1}{2} + \frac{3\varepsilon^* b}{16}.$$

Otherwise, $\bar{\tau}$ is finite and is given by

$$\bar{\tau} = \frac{1}{2\gamma(a_2 - a_1)} \log\left(\frac{E_C}{E_0}\right),$$

so the pendulum is unstable. □

11.3.2 The double pendulum

The next approximation to an elastic column is a double pendulum that consists of two rigid bars of length $l/2$, connected by two elastic hinges [252], [254], [225]. Figure 11.2 shows the straight and deflected pendulum. The mass of the bottom bar is concentrated at distance $\gamma_1 L$ and that of the top bar at $\gamma_2 L$ from the hinge. The angle between the bottom bar and the vertical direction is θ_1, that of the top bar is θ_2, and that of the load is $\theta_3 = \eta\theta_2$. The spring constants at the hinges are C_1 and C_2. The masses of the bars, m_i, $(i = 1, 2)$, are concentrated at distances $\gamma_i l \;\; (i = 1, 2)$ from the hinges, respectively, the spring constants at the hinges are $C_i \; (i = 1, 2)$, θ_i are the angles between the bars and the vertical direction, P_0 is the magnitude of the load, and $\eta\theta_1$ is the angle between its direction and the vertical direction.

Consider the vertically loaded double pendulum ($\eta = 0$). For a constant load $P = P_0$ the system is conservative and its potential energy is given by

$$V = \frac{1}{2}\left[C_1\theta_1^2 + C_2(\theta_2 - \theta_1)^2\right] - \frac{P_0 l(\theta_1^2 + \theta_2^2)}{4} \tag{11.114}$$

and the kinetic energy is given by

$$T = \frac{l^2}{2}\left[m_1\gamma_1^2\dot{\theta}_1^2 + m_2\left(\frac{1}{2}\dot{\theta}_1 + \gamma_2\dot{\theta}_2\right)^2\right]. \tag{11.115}$$

The system is stable as long as (11.114) is positive definite; that is, as long as

$$\frac{P_0 l}{C_2} < 2 + \xi - \sqrt{4 + \xi}, \tag{11.116}$$

where $\xi = C_1/C_2$. This condition is independent of the ratio of the masses and of their locations on the bars. The only stable equilibrium point of the system is $\theta_i = \dot{\theta}_i = 0$, $(i = 1, 2)$. When P_0 does not satisfy (11.116) the system has positive eigenvalues and it buckles.

Can noisy subcritical loads destabilize (buckle) the double pendulum To answer this question, we assume that P is given by (11.104), where P_0 satisfies (11.116). In the absence of damping the equations of motion of the system are given by

$$\boldsymbol{M}\begin{pmatrix}\ddot{\theta}_1\\ \ddot{\theta}_2\end{pmatrix} + (\boldsymbol{C} + \boldsymbol{D})\begin{pmatrix}\theta_1\\ \theta_2\end{pmatrix} + \sqrt{\varepsilon}v(t)\boldsymbol{D}\begin{pmatrix}\theta_1\\ \theta_2\end{pmatrix} = \boldsymbol{0}, \tag{11.117}$$

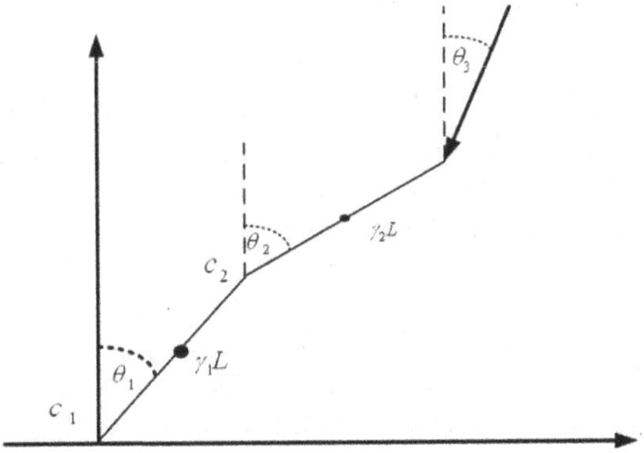

Figure 11.2. The straight and deflected pendulum.

where

$$\boldsymbol{M} = \begin{pmatrix} \left(m_1\gamma_1^2 + m_2/4\right) l^2 & m_2\gamma_2 l^2/2 \\ m_2\gamma_2 l^2/2 & m_2\gamma_2 l^2 \end{pmatrix} \tag{11.118}$$

$$\boldsymbol{C} = \begin{pmatrix} C_1 + C_2 & -C_2 \\ -C_2 & C_2 \end{pmatrix} \tag{11.119}$$

$$\boldsymbol{D} = \frac{P_0 l}{2}\begin{pmatrix} -1 & 0 \\ 0 & -1 \end{pmatrix}. \tag{11.120}$$

The damping forces in a linear model are represented by a term

$$-\gamma \boldsymbol{R}\begin{pmatrix} \dot{\theta}_1 \\ \dot{\theta}_2 \end{pmatrix},$$

where $\boldsymbol{R}$ is a positive definite matrix, and γ is a damping coefficient. In case of external damping $\boldsymbol{R}$ can be chosen as

$$\boldsymbol{R} = \frac{\boldsymbol{M}}{\mu},$$

where

$$\mu = \frac{m_1 + m_2}{l}.$$

If the external damping is proportional to the rate of change of local curvature, $\boldsymbol{R}$ can be chosen as

$$\boldsymbol{R} = \begin{pmatrix} 2 & -1 \\ -1 & 1 \end{pmatrix}.$$

Thus, the noisy, damped double pendulum is described by the system

$$M\begin{pmatrix}\ddot{\theta}_1\\ \ddot{\theta}_2\end{pmatrix}+\gamma R\begin{pmatrix}\dot{\theta}_1\\ \dot{\theta}_2\end{pmatrix}+(C+D)\begin{pmatrix}\theta_1\\ \theta_2\end{pmatrix}+\varepsilon v(t)D\begin{pmatrix}\theta_1\\ \theta_2\end{pmatrix}=\mathbf{0}. \quad (11.121)$$

First, we consider the undamped pendulum. Setting

$$\boldsymbol{X}=(\theta_1,\theta_2,\dot{\theta}_1,\dot{\theta}_2)^T,\quad \boldsymbol{w}(t)=(w_1(t),w_2(t),w_3(t),w_3(t))^T, \qquad (11.122)$$

where $\boldsymbol{w}(t)$ is a vector of independent Brownian motions, we get

$$d\boldsymbol{X}=\boldsymbol{A}\boldsymbol{X}\,dt+\sqrt{\varepsilon}\boldsymbol{B}\,d\boldsymbol{w}, \qquad (11.123)$$

with

$$\boldsymbol{A}=\begin{pmatrix}\mathbf{0} & \boldsymbol{I}\\ -\boldsymbol{K}(\boldsymbol{C}+\boldsymbol{D}) & \mathbf{0}\end{pmatrix},\quad \boldsymbol{B}=\begin{pmatrix}0&0&0&0\\0&0&0&0\\B_{31}&0&0&0\\B_{41}&0&0&0\end{pmatrix},\quad \boldsymbol{K}=\boldsymbol{M}^{-1}, \qquad (11.124)$$

where

$$B_{31}=\frac{P_0 l}{2}\left(K_{11}X_1+K_{12}X_2\right),\quad B_{41}=\frac{P_0 l}{2}\left(K_{21}X_1+K_{22}X_2\right). \qquad (11.125)$$

The reduced system of (11.123) ($\varepsilon=0$) is given by

$$d\boldsymbol{X}=\boldsymbol{A}\boldsymbol{X}\,dt \qquad (11.126)$$

and conserves energy. However, the system (11.123) absorbs energy from the noisy load and reaches any energy level in finite time with probability 1. The energy of the system (11.126) is the energy of the averaged system (11.123) and is given by

$$E=\frac{1}{2}\Big[m_1 l^2\gamma_1^2X_3^2+m_2l^2(X_3+X_4)^2+C_1X_1^2+C_2(X_2-X_1)^2 \\ -\frac{1}{2}P_0 l(X_1^2+X_2^2)\Big]. \qquad (11.127)$$

The surfaces containing the trajectories of (11.126) are four-dimensional ellipsoids in phase space. Setting

$$D=\{\boldsymbol{X}\,|\,E(\boldsymbol{X})<E_C\} \qquad (11.128)$$

to be the *domain of reliability* of (11.123) and setting

$$\tau=\inf\{t\,|\,E(\boldsymbol{X}(t))=E_C\},\quad \bar{\tau}(\boldsymbol{X})=\mathbb{E}[\tau\,|\,\boldsymbol{X}(0)=\boldsymbol{X}], \qquad (11.129)$$

$\bar{\tau}(\boldsymbol{X})$ is the *index of reliability*, given that the system (11.123) starts at $\boldsymbol{X}$. It is the solution of the boundary value problem

$$\frac{1}{2}\varepsilon^2L_1\bar{\tau}(\boldsymbol{X})+\boldsymbol{A}\boldsymbol{X}\cdot\nabla\bar{\tau}(\boldsymbol{X})=-1\ \text{ in } D,\quad \bar{\tau}(\boldsymbol{X})=0\ \text{ on } \partial D, \qquad (11.130)$$

where

$$L_1 = \left(\frac{P_0 l}{2}\right)^2 \left[(K_{11}X_1 + K_{12}X_2)^2 \frac{\partial^2}{\partial X_3^2} + (K_{11}X_1 + K_{12}X_2) \right. \tag{11.131}$$
$$\left. \times\ (K_{21}X_1 + K_{22}X_2) \frac{\partial^2}{\partial X_3 \partial X_4} + (K_{21}X_1 + K_{22}X_2)^2 \frac{\partial^2}{\partial X_4^2}\right].$$

To solve (11.130), we set

$$\bar{\tau}(\boldsymbol{X}) = \frac{\bar{\tau}_0(\boldsymbol{X})}{\varepsilon^2} + \bar{\tau}_1(\boldsymbol{X}) + \varepsilon^2 \bar{\tau}_2(\boldsymbol{X}) + \cdots \tag{11.132}$$

in (11.130) and compare the coefficients of $1/\varepsilon^2$ to zero, obtaining

$$\boldsymbol{AX} \cdot \nabla \bar{\tau}_0(\boldsymbol{X}) = 0. \tag{11.133}$$

Therefore $\bar{\tau}_0(\boldsymbol{X})$ is constant on all the trajectories of (11.126); that is, $\bar{\tau}_0(\boldsymbol{X})$ is constant on surfaces of constant energy and hence it is a function of E,

$$\bar{\tau}_0(\boldsymbol{X}) = \bar{\tau}_0(E). \tag{11.134}$$

At the next order, we get

$$\frac{1}{2} L_1 \bar{\tau}_0(\boldsymbol{X}) + \boldsymbol{AX} \cdot \nabla \bar{\tau}_1(\boldsymbol{X}) = -1. \tag{11.135}$$

The integrability condition for (11.134) is

$$\lim_{T\to\infty} \frac{1}{T} \int_0^T L_1 \bar{\tau}_0(E)\, dt = -1 \quad \text{in} \quad D. \tag{11.136}$$

Obviously,

$$\bar{\tau}_0(E_C) = 0 \tag{11.137}$$

and

$$\bar{\tau}_0(0) = \infty. \tag{11.138}$$

The condition (11.138) is due to the fact that a system that starts at its equilibrium state stays there forever.

To solve (11.136)–(11.138), we change to spherical coordinates (r, φ, η, Ψ), where

$$r = \sqrt{\frac{E}{E_C}} < 1. \tag{11.139}$$

From (11.134) we have that

$$\bar{\tau}_0(E) = T_0(r), \tag{11.140}$$

so that T_0 is the solution of the boundary value problem

$$\left\langle \sum_{i,j=3}^{4} a_{ij} \frac{\partial r}{\partial X_i} \frac{\partial r}{\partial X_j} \right\rangle T_0''(r) + \left\langle \sum_{i,j=3}^{4} a_{ij} \frac{\partial^2 r}{\partial X_i \partial X_j} \right\rangle T_0'(r) = -2 \tag{11.141}$$

for $0 < r < 1$,

$$T_0(1) = 0, \quad T_0(0) = \infty, \tag{11.142}$$

where

$$\begin{aligned}
a_{33} &= \left(\frac{P_0 l}{2}\right)^2 (K_{11}X_1 + K_{12}X_2)^2 \\
a_{34} = a_{43} &= \left(\frac{P_0 l}{2}\right)^2 (K_{11}X_1 + K_{12}X_2)(K_{21}X_1 + K_{22}X_2) \\
a_{44} &= \left(\frac{P_0 l}{2}\right)^2 (K_{21}X_1 + K_{22}X_2)^2 .
\end{aligned} \tag{11.143}$$

The averaging of the coefficients in (11.141) is done over all trajectories of constant energy. Thus, the coefficients are averaged over the surface $E = const$. We obtain

$$a_1(r)T_0''(r) + a_2(r)T_0'(r) = -1 \text{ for } 0 < r < 1, \tag{11.144}$$

where

$$a_1(r) = \frac{1}{2\pi^2} \int_0^{\pi}\int_0^{\pi}\int_0^{2\pi} \sum_{i,j=3}^{4} a_{ij} \frac{\partial r}{\partial X_i}\frac{\partial r}{\partial X_j} \sin\theta \sin^2\Psi \, d\varphi \, d\theta \, d\Psi \tag{11.145}$$

$$a_2(r) = \frac{1}{2\pi^2} \int_0^{\pi}\int_0^{\pi}\int_0^{2\pi} \sum_{i,j=3}^{4} a_{ij} \frac{\partial^2 r}{\partial X_i \partial X_j} \sin\theta \sin^2\Psi \, d\varphi \, d\theta \, d\Psi .$$

Evaluation of (11.145) yields

$$a_1(r) = a_1 r^2, \quad a_2(r) = a_1 r, \tag{11.146}$$

where, for $i = 1, 2$,

$$\begin{aligned}
a_i = \left(\frac{P_0 l}{2}\right)^2 b_i \Bigg[&\frac{m_1 l^2 \gamma_1^2}{N_1} (K_{11}\cos\delta + K_{12}\sin\delta)^2 \\
&+ \frac{m_2 l^2}{N_1} \left(\frac{1}{2}K_{11}\cos\delta + \frac{1}{2}K_{12}\sin\delta + \gamma_2 K_{21}\cos\delta + \gamma_2 K_{22}\sin\delta\right)^2 \\
&+ \frac{m_1 l^2 \gamma_1^2}{N_2} (-K_{11}\sin\delta + K_{12}\cos\delta)^2 \\
&+ \frac{m_2 l^2}{N_2} (-K_{11}\sin\delta + K_{12}\cos\delta - \gamma_2 K_{21}\sin\delta + \gamma_2 K_{22}\sin\delta)^2 \Bigg],
\end{aligned} \tag{11.147}$$

with

$$b_1 = \frac{1}{24}, \quad b_2 = \frac{5}{24}$$

$$N_1 = C_2 - \frac{P_0 l}{2} + \frac{C_1}{2} + \frac{h}{2}, \quad N_2 = C_2 - \frac{P_0 l}{2} + \frac{C_1}{2} - \frac{h}{2}$$

and

$$\sin\delta = \sqrt{\frac{h - C_1}{2h}}, \quad \cos\delta = \sqrt{\frac{h + C_1}{2h}}, \quad h = \sqrt{C_2^2 + 4C_2^2}.$$

It is obvious from (11.147) that $a_2/a_1 = 5$. Using (11.146) and (11.147) in (11.144), we get

$$a_1 r^2 T_0''(r) + a_2 r T_0'(r) = -2 \quad \text{for} \quad 0 < r < 1. \tag{11.148}$$

The boundary conditions (11.142) do not define a unique solution, because

$$T_0(r) = \frac{2}{a_1 - a_2}\log r + k(1 - r^{1-a_2/a_1}) \tag{11.149}$$

is a solution of (11.148), satisfying the boundary conditions (11.142) for any constant k. It can be shown, as above, that k must be zero. We conclude that the MFPT from E to E_C is finite and given by

$$\bar{\tau} \sim \frac{1}{\varepsilon^2\,(a_2 - a_1)}\log\sqrt{\frac{E_C}{E}}, \tag{11.150}$$

where a_1 and a_2 are defined in (11.147). Thus, the undamped vertically loaded double pendulum is unstable under fluctuating random loads of impulsive or white type.

11.3.3 The damped vertically loaded double pendulum

For small values of the damping coefficient γ we scale the noise intensity by $\varepsilon^2 = \gamma\varepsilon^*$, as in the Langevin equation. The equations of motion (11.121) are given by

$$\boldsymbol{M}\begin{pmatrix}\ddot{\theta}_1\\ \ddot{\theta}_2\end{pmatrix} + \gamma\boldsymbol{R}\begin{pmatrix}\dot{\theta}_1\\ \dot{\theta}_2\end{pmatrix} + (\boldsymbol{C}+\boldsymbol{D})\begin{pmatrix}\theta_1\\ \theta_2\end{pmatrix}$$
$$+\sqrt{\gamma\varepsilon^*}v(t)\boldsymbol{D}\begin{pmatrix}\theta_1\\ \theta_2\end{pmatrix} = \boldsymbol{0}, \tag{11.151}$$

where $\boldsymbol{M}, \boldsymbol{C}$, and $\boldsymbol{D}$ satisfy (11.118)–(11.120) and $\boldsymbol{R}$ is a positive definite matrix, as described in the previous section. Keeping the notation for X and $\boldsymbol{w}(t)$, we get

$$d\boldsymbol{X} = \boldsymbol{A}\boldsymbol{X}\,dt - \gamma\boldsymbol{G}\boldsymbol{X}\,dt + \sqrt{\gamma\varepsilon^*}\boldsymbol{B}\,d\boldsymbol{w}(t), \tag{11.152}$$

where $\boldsymbol{A}$ and $\boldsymbol{B}$ satisfy (11.124), (11.125), and

$$\boldsymbol{G} = \begin{pmatrix}\boldsymbol{0} & \boldsymbol{0}\\ \boldsymbol{0} & \boldsymbol{K}\boldsymbol{R}\end{pmatrix}. \tag{11.153}$$

Following the procedure of the previous section, we see that $\bar{\tau}$ is the solution of the boundary value problem

$$\frac{1}{2}\varepsilon^2 L_1\bar{\tau}(\boldsymbol{X}) - \gamma L_2\bar{\tau}(\boldsymbol{X}) + \boldsymbol{A}\boldsymbol{X}\cdot\nabla\bar{\tau}(\boldsymbol{X}) = -1 \text{ in } D \tag{11.154}$$

$$\bar{\tau}(\boldsymbol{X}) = 0 \quad \text{on } \partial D, \tag{11.155}$$

where L_1 satisfies (11.131) and

$$L_2 = (G_{33}X_3 + G_{34}X_4)\frac{\partial}{\partial X_3} + (G_{43}X_3 + G_{44}X_4)\frac{\partial}{\partial X_4}. \tag{11.156}$$

We assume the expansion

$$\bar{\tau}(\boldsymbol{X}) = \frac{\bar{\tau}_0(\boldsymbol{X})}{\gamma} + \bar{\tau}_1(\boldsymbol{X}) + \gamma\bar{\tau}_2(\boldsymbol{X}) + \cdots \tag{11.157}$$

and conclude, as in the previous section, that $\bar{\tau}_0(\boldsymbol{X}) = T_0(E)$. Using the method of averaging again, we obtain

$$\lim_{T\to\infty} \frac{1}{T}\int_0^T \left[\frac{1}{2}\varepsilon^* L_1 T_0(E) - L_2 T_0(E)\right] dt = -1 \text{ for } 0 < E < E_C$$

$$T_0(E_C) = 0, \quad T_0(0) = \infty.$$

Now, changing to spherical coordinates as above, we get the effective equation

$$\frac{\varepsilon^* a_1 r^2}{2} T_0''(r) + \frac{\varepsilon^* a_2 r}{2} T_0'(r) - a_3 r T_0'(r) = -1 \text{ for } 0 < r < 1 \tag{11.158}$$

$$T_0(1) = 0, \quad T_0(0) = \infty, \tag{11.159}$$

where a_1 and a_2 satisfy (11.147). The constant a_3 is defined by

$$a_3 = \frac{1}{2\pi^2}\int_0^\pi\int_0^\pi\int_0^{2\pi} \sum_{i,j=3}^4 G_{ij}X_j\frac{\partial r}{\partial X_i}\sin\theta\sin^2\Psi\, d\varphi\, d\theta\, d\Psi. \tag{11.160}$$

Evaluating (11.160) in this case, we obtain $a_3 = \frac{1}{4}(G_{33} + G_{44})$. The solution of (11.158) satisfying the first condition (11.159) is given by

$$T_0(r) = \frac{\log r}{a_3 - \varepsilon^*(a_2 - a_1)/2} + k\left[1 - r^{[\varepsilon^*(a_1-a_2)/2-a_3]/(\varepsilon^* a_1/2)}\right], \tag{11.161}$$

where k is any constant. As above, it can be shown that k must be zero. The second condition (11.159) can be satisfied only if

$$\frac{\varepsilon^*}{2} > \frac{a_3}{a_2 - a_1}. \tag{11.162}$$

In this case,

$$\bar{\tau} \sim \frac{1}{\varepsilon^2(a_2 - a_1)/2 - 2\gamma a_3}\log\left(\frac{E_C}{E}\right). \tag{11.163}$$

The boundary value problem(11.158)–(11.159) has no solution for

$$\frac{\varepsilon^*}{2} < \frac{a_3}{a_2 - a_1}. \tag{11.164}$$

As in the previous section, it can be shown that for ε^* satisfying (11.164) the origin is stochastically stable.

For a high dissipation rate a direct stability result can be obtained by using the Smoluchowski approximation. In this approximation the inertia term

$$M \begin{pmatrix} \ddot{\theta}_1 \\ \ddot{\theta}_2 \end{pmatrix}$$

is dropped from eq. (11.151). In this case the problem becomes two-dimensional rather than four-dimensional. Therefore, applying the Lyapunov function stability criterion, it can be easily shown that the origin is stochastically stable.

11.3.4 A tangentially loaded double pendulum (follower load)

A double pendulum with a follower load corresponds to $\eta = 1$ and is not conservative. Its deterministic kinetic stability criterion is given by [254]

$$\frac{P_0 l}{C_2} < \frac{1 + 4\bar{\mu}\gamma_1^2 + 4\gamma_2^2(1+\xi) + 4\gamma_2 - 8\gamma_1\gamma_2\sqrt{\bar{\mu}\xi}}{2\gamma_2^2 + \gamma_2}, \tag{11.165}$$

where $\bar{\mu} = m_1/m_2$ and $\xi = C_1/C_2$. Note that in contrast to the conservative case, this criterion depends on the mass ratios and on mass distribution.

We assume again that (11.104) holds with P_0 satisfying (11.165). The equations of motion of the system are now given by (11.117)–(11.119), and (11.120) is replaced by

$$D = \frac{P_0 l}{2} \begin{pmatrix} -1 & 1 \\ 0 & 0 \end{pmatrix}. \tag{11.166}$$

The dynamics of the system are described by (11.122)–(11.124), and (11.125) is replaced by

$$B_{31} = \frac{P_0 l}{2} K_{11}(X_1 - X_2), \quad B_{41} = \frac{P_0 l}{2} K_{21}(X_1 - X_2). \tag{11.167}$$

The reduced system (11.126) ($\varepsilon = 0$) is no longer conservative, but there exists a Lyapunov functional $\tilde{E}$ that is constant on the trajectories of (11.126). The functional $\tilde{E}$ is positive definite as long as P_0 satisfies (11.165) [245]. Thus, (11.126) is conservative with respect to the Lyapunov functional $\tilde{E}$ rather than with respect to the energy E.

The *index of reliability* of the system is now defined to be the expected time it takes the Lyapunov functional $\tilde{E}$ to reach a given critical value $\tilde{E}_C$. We seek a Lyapunov functional in the quadratic form

$$\tilde{E} = T_{11}X_1^2 + 2T_{12}X_1X_2 + T_{22}X_2^2 + T_{33}X_3^2 + 2T_{34}X_3X_4 + T_{44}X_4^2, \tag{11.168}$$

where $\boldsymbol{AX} \cdot \nabla\tilde{E} = 0$ for all points $\boldsymbol{X}$. Thus, the coefficients T_{ij} are the solutions of the linear system

$$\begin{aligned} T_{11} + A_{31}T_{33} + A_{41}T_{34} &= 0 \\ T_{12} + A_{32}T_{33} + A_{42}T_{34} &= 0 \\ T_{12} + A_{31}T_{34} + A_{41}T_{44} &= 0 \\ T_{22} + A_{32}T_{32} + A_{42}T_{44} &= 0. \end{aligned} \tag{11.169}$$

The Lyapunov function $\tilde{E}$ is positive definite only for P_0 satisfying (11.165). Note that there is more than one Lyapunov function for (11.126) in both the vertically and tangentially loaded double pendulum. The choice of a Lyapunov functional does not affect the reliability criterion, although the exit point on the surface $\tilde{E} = \tilde{E}_C$ may change.

The domain D and the times τ and $\bar{\tau}$ are defined again by (11.128)–(11.129) with the energy E replaced by the functional $\tilde{E}$. Therefore $\bar{\tau}$ is the solution of the boundary value problem in D,

$$\frac{\varepsilon^2}{2}\left(\frac{P_0 l}{2}\right)^2 (X_1 - X_2)^2 \left[K_{11}^2 \frac{\partial^2 \bar{\tau}}{\partial X_3^2} + 2K_{11}K_{12}\frac{\partial^2 \bar{\tau}}{\partial X_3 \partial X_4} + K_{12}^2 \frac{\partial^2 \bar{\tau}}{\partial X_4^2}\right] + \boldsymbol{AX}\cdot\nabla\bar{\tau} = -1 \tag{11.170}$$

with the boundary condition

$$\bar{\tau}(\boldsymbol{X}) = 0 \ \text{ on } \partial D. \tag{11.171}$$

We follow the same procedure as in the previous section to solve (11.170)–(11.171) with $\tilde{E}$ replacing E. Defining

$$r = \sqrt{\frac{\tilde{E}}{\tilde{E}_C}} < 1, \tag{11.172}$$

we obtain the equations (11.140)–(11.145) where $a_1(r)$ and $a_2(r)$ are given by (11.145) with a_{ij} defined now by

$$\begin{aligned} a_{22} &= \left(\frac{P_0 l}{2}\right)^2 K_{11}^2 (X_1 - X_2)^2 \\ a_{34} = a_{43} &= \left(\frac{P_0 l}{2}\right)^2 K_{11}K_{12}(X_1 - X_2)^2 \\ a_{44} &= \left(\frac{P_0 l}{2}\right)^2 K_{12}^2 (X_1 - X_2)^2. \end{aligned} \tag{11.173}$$

Evaluating $a_1(r)$ and $a_2(r)$, we find that

$$a_1(r) = a_1 r^2, \quad a_2(r) = a_2 r, \tag{11.174}$$

where a_1 and a_2 are constants (see below). We obtain

$$\bar{\tau} \sim \frac{1}{\varepsilon^2(a_2 - a_1)} \log\left(\frac{\tilde{E}_C}{\tilde{E}}\right).$$

The introduction of friction into the system leads to the stability criterion (11.164), where a_1 and a_2 are defined in (11.174) and a_3 is defined in (11.160). Note that r in this case is given by (11.172).

Exercise 11.3 (Stability criteria). Choosing $m_1 = m_2 = m$, $C_1 = C_2 = C$, and $\gamma_1 = \gamma_2 = 1/2$ corresponds to a system consisting of equal masses concentrated at the upper end of the bars and of identical springs at the hinges.

(i) Write down the equations of motion.

(ii) Show that the stability criterion is $P_0 l/C < 4$.

(iii) Construct a Lyapunov functional in the form of a quadratic function and calculate the MFPT to $\tilde{E}_C$. □

11.3.5 The N-fold pendulum and the continuous column

Finally, we approximate the Euler column by an N-fold pendulum consisting of N elements of length $\{\Delta x_i\}_{i=1}^N$ such that $\sum_{i=1}^N \Delta x_i = l$. Let $\{C_i\}_{i=1}^N$ be the spring constants at the hinges, and $\{\theta_i\}_{i=1}^N$ be the angles between the ith element and the vertical direction. Suppose that the mass of each element is concentrated at its upper end; then the kinetic energy of the pendulum is given by

$$T = \frac{\mu}{2}\sum_{i=1}^N \Delta x_i \left(\sum_{j=1}^i \Delta x_j \dot{\theta}_j\right)^2 \tag{11.175}$$

and the potential energy is given by

$$V = \frac{1}{2}\sum_{i=1}^N C_i(\theta_i - \theta_{i-1})^2 - \frac{P}{2}\sum_{i=1}^N \Delta x_i \theta_i^2. \tag{11.176}$$

If W_i is the deflection of the upper end of the ith element, then

$$W_i = \sum_{j=1}^i \Delta x_j \theta_j, \quad W_{i,x} \approx \theta_i \tag{11.177}$$

$$W_{i,xx} \approx \frac{\theta_i - \theta_{i-1}}{\Delta x_i}, \quad W_{i,t} = \sum_{j=1}^i \Delta x_j \dot{\theta}_j.$$

Using (11.177) in (11.175) and (11.176), we get

$$T \approx \frac{\mu}{2}\sum_{i=1}^N W_{i,t}\Delta x_i \tag{11.178}$$

and

$$V = \frac{1}{2}\sum_{i=1}^{N} C_i \Delta W_{i,xx}^2 \Delta x_i - \frac{P}{2}\sum_{i=1}^{N} W_{i,x}^2 \Delta x_i. \tag{11.179}$$

Now, letting $N \to \infty$ and $\Delta x_i \to 0$, we require that the energies of the pendulum converge to those of the column. Denoting by $W(x,t)$ the deflection of the continuous column, the kinetic and potential energies are given by

$$T = \frac{\mu}{2}\int_0^l W_t^2\, dx \tag{11.180}$$

and

$$V = \frac{\alpha}{2}\int_0^l W_{xx}^2\, dx - \frac{P}{2}\int_0^l W_x^2\, dx, \tag{11.181}$$

respectively. Note that the last integral in (11.181) is the work done by the external load P. We include it in the potential energy to describe the system, including external force. In the limit, as $N \to \infty$, we obtain

$$\alpha = \lim_{\Delta x_i \to 0} C_i \Delta x_i,$$

therefore, in order to model the column by an N-fold pendulum, we choose

$$C_i = \frac{\alpha}{\Delta x_i} \quad \text{for} \quad i = 1, 2, \ldots, N. \tag{11.182}$$

Suppose that the N-fold pendulum consists of equal elements of length l/N; then the spring constants are equal to $N\alpha/l$. The equations of motion are now given by

$$\boldsymbol{M}\dot{\boldsymbol{\theta}} + \boldsymbol{C}\boldsymbol{\theta} + \boldsymbol{D}\boldsymbol{\theta} = \boldsymbol{0}, \tag{11.183}$$

where

$$\boldsymbol{\theta} = (\theta_1, \theta_2, \ldots, \theta_N)^T, \quad M_{ij} = \frac{\mu}{N^3}\left[N + 1 - \max(i,j)\right]$$

$$D_{ij} = \frac{Pl}{N}\delta_{ij}.$$

$$\frac{l}{N\alpha}C_{ij} = \begin{cases} 2 & \text{if} \quad i = j \neq N \\ 1 & \text{if} \quad i = j = N \\ -1 & \text{if} \quad |i - j| = 1 \\ 0 & \text{if} \quad |i - j| > 1. \end{cases}$$

In the case of a constant load P_0, the system (11.183) is stable as long as the matrix $\boldsymbol{C} + \boldsymbol{D}$ is positive definite. This is the case if P_0 satisfies

$$\frac{P_0 l^2}{\alpha} < 4N_2 \sin^2\left(\frac{\pi}{4N+2}\right). \tag{11.184}$$

We observe that as $N \to \infty$, the right-hand side of (11.184) converges to $\pi^2/4$, which agrees with (11.103).

Now, we assume that P is given by (11.104) with P_0 satisfying (11.184). The energy of the deterministic system (11.183) ($\varepsilon = 0$) is given by

$$E = \frac{1}{2}\sum_{i=1}^{N}\sum_{j=1}^{N}\left(M_{ij}\dot{\theta}_i\dot{\theta}_j + C_{ij}\theta_i\theta_j - \frac{P_0 l}{N}\delta_{ij}\theta_i\theta_j\right). \tag{11.185}$$

Setting

$$D = \left\{(\boldsymbol{\theta},\dot{\boldsymbol{\theta}}) \,|\, E(\boldsymbol{\theta},\dot{\boldsymbol{\theta}}) < E_C\right\} \subset \mathbb{R}^{2n}, \quad \tau = \inf\left\{t \,|\, E(\boldsymbol{\theta}(t),\dot{\boldsymbol{\theta}}(t)) = E_C\right\},$$

the index of reliability is the conditional expectation of the MFPT

$$\bar{\tau}(\boldsymbol{\theta},\dot{\boldsymbol{\theta}}) = \mathbb{E}[\tau \,|\, (\boldsymbol{\theta}(0),\dot{\boldsymbol{\theta}}(0)) = (\boldsymbol{\theta},\dot{\boldsymbol{\theta}})]. \tag{11.186}$$

As above, $\bar{\tau}(\boldsymbol{\theta},\dot{\boldsymbol{\theta}})$ is the solution of the boundary value problem

$$\frac{\varepsilon^2}{2}\left(\frac{P_0 l}{N}\right)^2 \sum_{i,j,k,l} K_{ij}\theta_j K_{kl}\theta_l \frac{\partial^2\bar{\tau}}{\partial\dot{\theta}_i\partial\dot{\theta}_k} - \sum_{i,j,k,l} K_{ij}(C_{jk}+D_{jk})\theta_k \frac{\partial\bar{\tau}}{\partial\dot{\theta}_i}$$
$$+\sum_i \dot{\theta}_i \frac{\partial\bar{\tau}}{\partial\theta_i} = -1 \tag{11.187}$$

in D and $\bar{\tau}(\boldsymbol{\theta},\dot{\boldsymbol{\theta}}) = 0$ on the boundary of D, where $\boldsymbol{K} = \boldsymbol{M}^{-1}$. Proceeding as above, we expand $\bar{\tau}$ in powers of ε^2

$$\bar{\tau} = \frac{\bar{\tau}_0}{\varepsilon^2} + \bar{\tau}_1 + \varepsilon^2\bar{\tau}_2 + \cdots, \tag{11.188}$$

and find that $\bar{\tau}_0$ is a function of the energy and the solvability condition for $\bar{\tau}_1$ leads to the averaged equation

$$\left(\frac{P_0 l}{N}\right)^2 \lim_{T\to\infty}\frac{1}{T}\int_0^T \sum_{i,j,k,l} K_{ij}\theta_j K_{kl}\theta_l \frac{\partial^2 T_0(E)}{\partial\dot{\theta}_i\partial\dot{\theta}_k}\,dt = -2 \text{ for } 0 < E < E_C$$
$$T_0(E_C) = 0, \quad T_0(0) = \infty.$$

Next, we change coordinates to E and $N-1$ angles on the surfaces $E = const.$ so that the above averaged equation takes the form

$$\left(\frac{P_0 l}{N}\right)^2 \lim_{T\to\infty}\frac{1}{T}\int_0^T \left[\sum_{i,j}\theta_i\theta_j\dot{\theta}_i\dot{\theta}_j\, T_0''(E) + \sum_{i,j} K_{ij}\theta_i\theta_j\, dt\, T_0'(E)\right] dt$$
$$= -2 \text{ for } 0 < E < E_C. \tag{11.189}$$

Rather than proceeding with the calculation for N, we consider the case $N \to \infty$ to recover the continuous column. Thus, we note that

$$\lim_{N\to\infty} \left(\frac{l}{N}\right)^2 \sum_{i,j} \theta_i \theta_j \dot{\theta}_i \dot{\theta}_j = \frac{1}{4} \left[\int_0^l \left(\frac{\partial}{\partial t} W_x^2 \right) dx \right]^2 \tag{11.190}$$

and

$$\lim_{N\to\infty} \left(\frac{l}{N}\right)^2 \sum_{i,j,k,l} K_{ij} \theta_i \theta_j K_{kl} \theta_k \theta_l = \frac{1}{\mu} \int_0^l W_{xx}^2 \, dx. \tag{11.191}$$

Setting

$$\tilde{a}_1(E) = \lim_{T\to\infty} \frac{1}{T} \int_0^T \frac{P_0^2}{4} \left[\int_0^l \left(\frac{\partial}{\partial t} W_x^2 \right) dx \right]^2 dt \tag{11.192}$$

and

$$\tilde{a}_2(E) = \lim_{T\to\infty} \frac{1}{T} \int_0^T \frac{P_0^2}{\mu} \int_0^l W_{xx}^2 \, dx \, dt. \tag{11.193}$$

we rewrite (11.189) as

$$\tilde{a}_1(E) T_0''(E) + \tilde{a}_2(E) T_0'(E) = -2 \quad \text{for} \quad 0 < E < E_C \tag{11.194}$$

with the obvious definition of $T_0(E)$. We define

$$a_i(E) = \langle \tilde{a}_1(E) \rangle_{\text{i.c.}} \quad (i = 1, 2) \tag{11.195}$$

as the mean over initial conditions.

The continuous column theory is essentially a long wave theory. The initial data must therefore be restricted to the first few modes. Therefore, when averaging over initial conditions, we assign these modes much higher probability than to the higher ones. The analysis of the N-fold pendulum shows that

$$a_1(E) = a_1 E^2, \quad a_2(E) = a_2 E, \tag{11.196}$$

where a_1 and a_2 are determined by the weights assigned to the modes. Thus, the two-dimensional model essentially represents the qualitative behavior of the continuous column. It follows that (11.150) is also valid in the continuous case, and so is the stability criterion (11.164) with a_3 properly defined. We conclude that if the noise intensity is large relative to damping, the column loses its stability, and if the noise is relatively small, the column is stochastically stable.

Exercise 11.4 (Stability of a column with noisy tangential load). Find the limit $N \to \infty$ of the Lyapunov functional for a column with noisy tangential load. Argue that the behavior of the double or triple pendulum represents qualitatively that of the noisily loaded column in the nonconservative case as well. □

Bibliography

[1] Abramowitz, M. and I. A. Stegun. *Handbook of Mathematical Functions.* Dover, New York, 1972.

[2] Adelman, S.A. and J.D. Doll. "Generalized Langevin equation approach for atom/solid surface scattering: Collinear atom/harmonic chain model." *J. Chem. Phys.*, **61**, 4242–4245 (1974).

[3] Agmon, N. "Diffusion with back reaction." *J. Chem. Phys.*, **81**, 2811–2817 (1984).

[4] Allen, M.P. and D.J. Tildesley. *Computer Simulation of Liquids.* Oxford University Press, Oxford, 1991.

[5] Arnold, L. *Random Dynamical Systems.* Springer-Verlag, New York, 1998.

[6] Arnold, L., G. Papanicolaou, and V. Wihstutz. "Asymptotic analysis of the Lyapunov exponent and rotation number of the random oscillator and applications." *SIAM J. Appl. Math.*, **46**, 427–450 (1986).

[7] Asmussen, S., P. Glynn, and J. Pitman. "Discretization error in simulation of one-dimensional reflecting Brownian motion." *Ann. Appl. Prob.*, **5** (4), 875–896 (1995).

[8] Bahadur, R.P. and R. Ranga Rao. "On deviations of the sample mean." *Ann. Math. Stat.*, **31**, 1015–1027 (1960).

[9] Barndorf-Nielsen, O. and D.R. Cox. "Edgeworth and saddle point approximations with statistical applications." *J. Royal Statist. Soc. B*, **41** (3), 279–312 (1979).

[10] Batsilas, L., A.M. Berezhkovskii, and S.Y. Shvartsman. "Stochastic model of autocrine and paracrine signals in cell culture assays." *Biophys. J.*, **85**, 3659Ű-3665 (2003).

[11] Beccaria, M., G. Curci, and A. Vicere. "Numerical solutions of first-exit-time problems." *Phys. Rev E.*, **48** (2), 1539–1546 (1993).

[12] Belch, A.C. and M. Berkowitz. "Molecular dynamics simulations of tips2 water restricted by a spherical hydrophobic boundary." *Chem. Phys. Lett.*, **113**, 278–282 (1985).

[13] Bellman, R., J. Bentsman, and S.M. Meerkov. "Nonlinear systems with fast parametric oscillations." *J. Math. Anal. Appl.*, **97**, 572–589 (1983).

[14] Belopolskaya, Y.I. and Y.L. Dalecky. *Stochastic Equations and Differential Geometry*. Kluwer, Dordrecht, 1990.

[15] Ben-Jacob, E., D. Bergman, and Z. Schuss. "Thermal fluctuations and the lifetime of the nonequilibrium steady-state in a hysteretic Josephson junction." *Phys. Rev. B, Rapid Comm.*, **25**, 519–522 (1982).

[16] Ben-Jacob, E., D. Bergman, B.J. Matkowsky, and Z. Schuss. "Lifetime of oscillatory steady-states." *Phys. Rev. A*, **26**, 2805–2816 (1982).

[17] Ben-Jacob, E., D.J. Bergman, B.J. Matkowsky, and Z. Schuss. "Noise induced transitions in multi-stable systems," in *Fluctuations and Sensitivity in Nonequilibrium Systems*, Proceedings of an International Conference, University of Texas, Austin, Texas, March 12–16. Springer Proceedings in Physics 1, 1984.

[18] Ben-Jacob, E., D.J. Bergman, B.J. Matkowsky, and Z. Schuss. "Master-equation approach to shot noise in Josephson junctions." *Phys. Rev. B*, **34** (3), 1572–1581 (1986).

[19] Ben-Jacob, E., D.J. Bergman, Y. Imry, B.J. Matkowsky, and Z. Schuss. "Thermal activation from the fluxoid and the voltage states of dc SQUIDs. *J. Appl. Phys.*, **54**, 6533–6542 (1983).

[20] Bender, C.M. and S.A. Orszag. *Advanced Mathematical Methods for Scientists and Engineers*. McGraw-Hill, New York, 1978.

[21] Berezhkovskii, A.M., Y.A. Makhnovskii, M.I. Monine, V.Yu. Zitserman, and S.Y. Shvartsman. "Boundary homogenization for trapping by patchy surfaces." *J. Chem. Phys.*, **121** (22), 11390–11394 (2004).

[22] Berkowitz, M. and J.A. McCammon. "Molecular dynamics with stochastic boundary conditions." *Chem. Phys. Lett.*, **90**, 215–217 (1982).

[23] Berne, B.J. and R. Pecora. *Dynamic Light Scattering with Applications to Chemistry, Biology, and Physics*. Wiley-Interscience New York, 1976.

[24] Berry, R.S.,S. Rice, and J. Ross. *Physical Chemistry*. 2nd edition. Oxford University Press, Oxford. 2000.

[25] Billingsley, P. *Convergence of Probability Measures*. J. Wiley and Sons, New York, 1968.

[26] Bobrovsky, B.Z. and O. Zeitouni. "Some results on the problem of exit from a domain." *Stoch. Process. Appl.*, **41** (2), 241–256 (1992).

[27] Bobrovsky, B.Z. and Z. Schuss. "A singular perturbation method for the computation of the mean first passage time in a nonlinear filter." *SIAM J. Appl. Math.*, **42** (1), 174–187 (1982).

[28] Bossy, M., E. Gobet, and D. Talay. "A symmetrized Euler scheme for an efficient approximation of reflected diffusions." *J. Appl. Prob.*, **41**, 877–889 (2004).

[29] Brooks, C.L., III and M. Karplus. "Deformable stochastic boundaries in molecular dynamics." *J. Chem. Phys.*, **79**, 6312 (1983).

[30] Brush, S.G. *The Kind of Motion We Call Heat, I, II*. North Holland, Amsterdam, 1986.

[31] Bryan, G.H. "Diffusion with back reaction." *Proc. Camb. Phil. Soc.*, **7**, 246, (1891).

[32] Bucklew, J.A. *Large Deviation Techniques in Decision, Simulation and Estimation*. Wiley, New York, 1990.

[33] Büttiker, M., E.P. Harris, and R. Landauer. "Thermal activation in extremely underdamped Josephson-junction circuits." *Phys. Rev. B*, **28**, 1268 (1979).

[34] Cameron, R.H. and W.T. Martin. "The transformation of Wiener integrals by nonlinear transformations." *Trans. Amer. Math. Soc.*, **69**, 253Ű–283 (1949).

[35] Carmeli, B. and A. Nitzan. "First passage times and the kinetics of unimolecular dissociation." *J. Chem. Phys.*, **76**, 5321–5333 (1982).

[36] Carmeli, B. and A. Nitzan. "Non-Markovian theory of activated rate processes V, external periodic forces in the low friction limit." *Phys. Rev. A*, **32**, 2439–2454 (1985).

[37] Carslaw, H.S. and J. C. Jaeger. *Conduction of Heat in Solids*. 2nd edition. Oxford University Press, Oxford. 1959.

[38] Chandrasekhar, S. "Stochastic problems in physics and astronomy." *Rev. Mod. Phys.*, **15**, 2–89 (1943).

[39] Christiano, R. and P. Silvestrini. "Decay of the running state in Josephson junctions: Preliminary experimental results." *Phys. Lett. A*, **133** (2), 347 (1988).

[40] Cochran, J.A. *The Analysis of Linear Integral Equations*. McGraw-Hill, New York, 1972.

[41] Coddington, E. and N. Levinson. *Ordinary Differential Equations*. McGraw-Hill, New York, 1955.

[42] Coffey, W.T., Yu.P. Kalmykov, and J.T. Waldron. *The Langevin Equation: With Applications to Stochastic Problems in Physics, Chemistry and Electrical Engineering*. 2nd edition. Series in Contemporary Chemical Physics Vol. 14. World Scientific, New York, 2004.

[43] Collins, F.C. and G.E. Kimball. "Random diffusion-controlled reaction rates." *J. Colloid Sci.*, **4**, 425–437 (1949).

[44] Costantini, C., B. Pacchiarotti, and F. Sartoretto. "Numerical approximation for functionals of reflecting diffusion processes." *SIAM J. Appl. Math.*, **58** (1), 73–102 (1998).

[45] Courant, R. and D. Hilbert. *Methods of Mathematical Physics*. Wiley-Interscience, New York, 1989.

[46] Cox, D.R. and H.D. Miller. *The Theory of Stochastic Processes*. Chapman and Hall, London, 1984.

[47] Cramér, H. and M.R. Leadbetter. *Stationary and Related Stochastic Processes*. Wiley, New York, 1967.

[48] Daniels, H.E. "Saddle point approximations in statistics." *Ann. Math. Stat.*, **25**, 631–650 (1954).

[49] Day, M.V. "Conditional exits for small noise diffusions." *Ann. Prob.*, **20**, 1385–1419 (1992).

[50] Day, M.V. "Boundary local time and small parameter exit problems with characteristic boundaries." *SIAM J. Math. Anal.*, **20**, 222–248 (1989).

[51] Day, M.V. Large deviations results for the exit problem with characteristic boundary. *J. Math. Anal. Appl.*, **147**, 134–153 (1990).

[52] Day, M.V. "Some phenomena of the characteristic boundary exit problem," in *Diffusion Processes and Related Problems in Analysis*. Birkhäuser, Basel 1990.

[53] De Finetti, B. *Theory of Probability*. Wiley, New York, 3rd edition. 1993.

[54] Dembo, A. and O. Zeitouni. *Large Deviations Techniques and Applications*. Jones and Bartlett, Boston. 1993.

[55] Deuschel, J.D. and D.W. Stroock. *Large Deviations*. Academic Press, 1989.

[56] Devinatz, A. and A. Friedman. "The asymptotic behavior of the solution of a singularly perturbed Dirichlet problem." *Indiana Univ. Math. J.*, **27** (3), 527–537 (1978).

[57] Doob, J.L. *Stochastic Processes*. Wiley, New York, 1953.

[58] Doob, J.L. *Measure Theory*. Springer Verlag, New York, 1994.

[59] Doucet, A., N.D. Freitas, and N. Gordon. *Sequential Monte-Carlo Methods in Practice*. Springer, Berlin, 2001.

[60] Dvoretzky, A., P. Erdös, and S. Kakutani. "Nonincreasing everywhere of the Brownian motion process," in *Proceedings of the Fourth Berkeley Symposium on Mathematical Statistics and Probability II*, University of California Press, Bekeley. 1961.

[61] Dygas, M.M., B.J. Matkowsky, and Z. Schuss. "A singular perturbation approach to non-Markovian escape rate problems." *SIAM J. Appl. Math.*, **46** (2), 265–298 (1986).

[62] Dynkin, E.B., T. Kovary, and D. E. Brown. *Theory of Markov Processes*. Dover, New York, 2006.

[63] Einstein, A. *Investigations on the Theory of the Brownian Movement*. translated and reprinted by Dover, New York 1956.

[64] Eisenberg, R.S. "Ionic channels in biological membranes. Electrostatic analysis of a natural nanotube." *Contemp. Phys.*, **39** (6), 447–466 (1998).

[65] Eisenberg, R.S. "From structure to function in open ionic channels." *J. Membrane Biol.*, **171**, 1–24 (1999).

[66] Eisenberg, R.S., M.M. Kłosek, and Z. Schuss. "Diffusion as a chemical reaction: Stochastic trajectories between fixed concentrations." *J. Chem. Phys.*, **102** (4), 1767–1780 (1995).

[67] Eliezer, B. and Z. Schuss. "Stochastic (white noise) analysis of resonant absorption in laser generated plasma." *Phys. Lett. A*, **70** (4), 307–310 (1979).

[68] Ellis, R.S. *Entropy, Large Deviations and Statistical Mechanics*. Springer-Verlag, New York, 1985.

[69] Erban, R. and J. Chapman. "Reactive boundary conditions for stochastic simulations of reactionŰdiffusion processes." *Phys. Biol.*, **4**, 16Ű-28 (2007).

[70] Ethier, S.N. and T.G. Kurtz. *Markov Processes: Characterization and Convergence*. 2nd edition. Wiley Series in Probability and Statistics, New York, 2005.

[71] Feller, W. "Diffusion processes in one dimension." *Trans. AMS*, **77** (1), 1–31 (1954).

[72] Feller, W. *An Introduction to Probability Theory and Its Applications*, volumes I, II. 3rd edition. John Wiley & Sons, New York, 1968.

[73] Fichera, G. "On a unified theory of boundary value problems for-elliptic-parabolic equations of second order." in *Boundary Value Problems in Differential Equations*, University of Wisconsin Press, Madison. 1960.

[74] Freidlin, M. *Markov Processes and Differential Equations*. Birkhäuser, Boston, 2002.

[75] Freidlin, M. "Some remarks on the Smoluchowski–Kramers approximation." *J. Stat. Phys.*, **117** (3/4), 617–654 (2004).

[76] Freidlin, M.A. *Functional Integration and Partial Differential Equations*. Princeton University Press, Princeton, NJ, 1985.

[77] Freidlin, M.A. and A.D. Wentzell. *Random Perturbations of Dynamical Systems*. Springer-Verlag, New York, 1984.

[78] Friedman, A. *Partial Differential Equations of Parabolic Type*. Prentice-Hall, Englewood Cliffs, NJ, 1964.

[79] Friedman, A. *Foundations of Modern Analysis*. Dover, New York, 1983.

[80] Friedman, A. *Stochastic Differential Equations and Applications*. Dover, New York, 2007.

[81] Gantmacher, F.R. *The Theory of Matrices*. American Mathematical Society, Providenc, RI, 1998.

[82] Gardiner, C.W. *Handbook of Stochastic Methods*. 2nd edition. Springer Verlag, New York, 1985.

[83] Ghoniem, A.F. and F.S. Sherman. "Grid free simulation of diffusion using random walk methods." *J. Comp. Phys.*, **61**, 1–37 (1985).

[84] Gihman, I.I. and A.V. Skorohod. *Stochastic Differential Equations*. Springer-Verlag, Berlin, 1972.

[85] Gihman, I.I. and A.V. Skorohod. *The Theory of Stochastic Processes*, Vols. I, II, III. Springer-Verlag, Berlin, 1975.

[86] Girsanov, I.V. "On transforming a certain class of stochastic processes by absolutely continuous substitution of measures." *Theory Probab. Appl.*, **5**, 285–301 (1960).

[87] Glasstone, S., K.J. Laidler, and H. Eyring. *The Theory of Rate Processes*. McGraw-Hill, New York, 1941.

[88] Gobet, E. "Euler schemes and half-space approximation for the simulation of diffusion in a domain." *ESAIM Probab. Statist.*, **5**, 261–297 (2001).

[89] Goodrich, F.C. "Random walk with semiadsorbing barrier." *J. Chem. Phys.*, **22**, 588–594 (1954).

[90] Graham, R. and T. Tel. "Existence of a potential for dissipative dynamical systems." *Phys. Rev. Lett.*, **52**, 9–12 (1984).

[91] Grote, F.R. and J.T. Hynes. "The stable states picture of chemical reactions II: Rate constants for condensed and gas phase reaction models." *J. Chem Phys.*, **73**, 2715–2732 (1980).

[92] Grote, F.R. and J.T. Hynes. "Reactive modes in condensed phase reactions." *J. Chem Phys.*, **74**, 4465–4475 (1981).

[93] Grote, F.R. and J.T. Hynes. "Saddle point model for atom transfer reactions in solution." *J. Chem Phys.*, **75**, 2191–2198 (1981).

[94] Hagan, P.S., C.R. Doering, and C.D. Levermore. "Bistability driven by weakly colored Gaussian noise: The Fokker–Planck boundary layer and mean first-passage times." *Phys. Rev. Lett.*, **59**, 2129Ű–2132 (1987).

[95] Hagan, P.S., C.R. Doering, and C.D. Levermore. "The distribution of exit times for weakly colored noise." *J. Stat. Phys.*, **54** (5/6), 1321–1352 (1989).

[96] Hagan, P.S., C.R. Doering, and C.D. Levermore. "Mean exit times for particles driven by weakly colored noise." *SIAM J. Appl. Math.*, **49** (5), 1480–1513 (1989).

[97] Hale, J.K. *Ordinary Differential Equations*. Wiley, New York, 1969.

[98] Hänggi, P. *Activated Barrier Crossing: Applications in Physics, Chemistry and Biology*. World Scientific, Singapore, 1993.

[99] Hänggi, P. and P. Talkner. *New Trends in Kramers' Reaction Rate Theory*. Kluwer, Dordrecht 1995.

[100] Hänggi, P., P. Talkner, and M. Borkovec. "50 years after Kramers." *Rev. Mod. Phys.*, **62**, 251–341 (1990).

[101] Hida, T. *Brownian Motion*. Springer, New York, 1980.

[102] Hille, B. *Ionic Channels of Excitable Membranes*. 3rd edition. Sinauer, Sunderland, 2001.

[103] Honerkamp, J. *Stochastic Dynamical Systems: Concepts, Numerical Methods, Data Analysis*. VCH, New York, 1994.

[104] Im, B., S. Seefeld, and B. Roux. "A grand canonical Monte-Carlo-Brownian dynamics algorithm for simulating ion channels." *Biophys. J.*, **79**, 788–801 (2000).

[105] Iscoe, I., P. Ney, and E. Nummelin. "Large deviations of uniformly recurrent Markov additive processes." *Adv. Appl. Math.*, **6**, 373–412 (1985).

[106] Itô, K. and H.P. McKean, Jr. *Diffusion Processes and Their Sample Paths*. Classics in Mathematics. Springer-Verlag, New York, 1996.

[107] Jackson, J.D. *Classical Electrodymnics*. 2nd edition. Wiley, New York, 1975.

[108] Jazwinski, A.H. *Stochastic Processes and Filtering Theory*. Academic Press, New York, 1970.

[109] Jensen, J.L. "Saddle point expansions for sums of Markov dependent variables on a continuous state space." *Probab. Theor. Related Fields*, **89**, 181–199 (1991).

[110] Jensen, J.L. *Saddlepoint Approximations*. Oxford Statistical Science Series, Oxford, 1995.

[111] Johnson, J.B. "Thermal agitation of electricity in conductors." *Phys. Rev.*, **32**, 97–109 (1928).

[112] Kamienomostskaya, S. "On equations of elliptic type and parabolic type with a small parameter in the highest derivative." *Mat. Sbornik*, **31** (73), 703–708 (1952).

[113] Kamienomostskaya, S. "The first boundary value problem for elliptic equations containing a small parameter." *Izv. Akad. Nauk U.S.S.R.*, **27** (6), 935–360 (1955).

[114] Kamin, S. "On elliptic perturbation of a first order operator with a singular point of attracting type." *Indiana Univ. Math. J.*, **27** (6), 935–951 (1978).

[115] Karatzas, I. and S.E. Shreve. *Brownian Motions and Stochastic Calculus*. 2nd edition. Springer Graduate Texts in Mathematics 113, New York, 1991.

[116] Karlin, S. and H.M. Taylor. *A First Course in Stochastic Processes*, Vol. I, 2nd edition. Academic Press, New York, 1975.

[117] Karlin, S. and H.M. Taylor. *A Second Course in Stochastic Processes*, Vol. II, 2nd edition. Academic Press, New York, 1981.

[118] Katz, A. *Reliability of Elastic Structures Driven by Random Loads*. Ph.D. dissertation, Department of Mathematics, Tel-Aviv University, Israel, 1985.

[119] Katz, A. and Z. Schuss. "Reliability of elastic structures driven by random loads." *SIAM J. Appl. Math.*, **45** (3), 383–402 (1985).

[120] Keller, J.B. and D.W. McLaughlin. "The Feynman integral." *Amer. Math. Month.*, **82** (5), 451–576 (1975).

[121] Kevorkian, J. and J.D. Cole. *Perturbation Methods in Applied Mathematics*. Applied Mathematical Sciences. Springer-Verlag, Berlin and Heidelberg, 1985.

[122] Khasminskii, R.Z. and A.M. Il'in. "On equations of Brownian motion." *Theor. Probab. Appl.*, **9**, 421–444 (1964).

[123] Kim, S. and I. Oppenheim. "Molecular theory of Brownian motion in external fields." *Physica*, **57**, 469–482 (1972).

[124] Kłosek-Dygas, M.M., B.J. Matkowsky, and Z. Schuss. "Colored noise in dynamical systems." *SIAM J. Appl. Math.*, **48** (2), 425–441 (1988).

[125] Kłosek-Dygas, M.M., B.J. Matkowsky, and Z. Schuss. "A first passage time approach to stochastic stability of nonlinear oscillators." *Phys. Lett. A*, **130** (1), 11–18 (1988).

[126] Kłosek-Dygas, M.M., B.J. Matkowsky, and Z. Schuss. "Stochastic stability of nonlinear oscillators." *SIAM J. Appl. Math.*, **48** (5), 1115–1127 (1988).

[127] Kłosek, M.M. "Half-range expansions for analysis for Langevin dynamics in the high friction limit with a singular boundary condition: Non characteristic case." *J. Stat. Phys.*, **79** (1/2), 313–345 (1995).

[128] Kłosek, M.M., and P.S. Hagan. "Colored noise and a characteristic level crossing problem." *J. Math. Phys.*, **39**, 931–953 (1998).

[129] Kleinert, H. *Path Integrals in Quantum Mechanics, Statistics, and Polymer Physics*. World Scientific, New York, 1994.

[130] Kleinrock, L. *Queueing Systems*, Vols. I, II. Wiley, New York,, 1975, 1976.

[131] Kloeden, P.E. "The systematic derivation of higher order numerical schemes for stochastic differential equations." *Milan J. Math.*, **70**, 187Ű-207 (2002).

[132] Kloeden, P.E. and E. Platen. *Numerical Solution of Stochastic Differential Equations*. Springer-Verlag, New York, 1992.

[133] Knessl, C., B.J. Matkowsky, Z. Schuss, and C. Tier. "An asymptotic theory of large deviations for Markov jump processes." *SIAM J. Appl. Math.*, **45**, 1006–1028 (1985).

[134] Knessl, C., B.J. Matkowsky, Z. Schuss, and C. Tier. "Asymptotic analysis of a state dependent M/G/1 queueing system." *SIAM J. Appl. Math.*, **46** (3), 483–505 (1986).

[135] Knessl, C., B.J. Matkowsky, Z. Schuss, and C. Tier. "Boundary behavior of diffusion approximations to Markov jump processes." *J. Stat. Phys.*, **45**, 245–266 (1986).

[136] Knessl, C., B.J. Matkowsky, Z. Schuss, and C. Tier. "On the performance of state dependent single server queues." *SIAM J. Appl. Math.*, **46** (4), 657–697 (1986).

[137] Knessl, C., B.J. Matkowsky, Z. Schuss, and C. Tier. "A singular perturbations approach to first passage times for Markov jump processes." *J. Stat. Phys.*, **42**, 169–184 (1986).

[138] Knessl, C., B.J. Matkowsky, Z. Schuss, and C. Tier. "System crash in finite capacity M/G/1 queue." *Commun. Stat. - Stochastic Models*, **2** (2), 171–201 (1986).

[139] Korkotian, E. and M. Segal. "Spike-associated fast contraction of dendritic spines in cultured hippocampal neurons." *Neuron*, **30** (3), 751–758 (2001).

[140] Kramers, H.A. "Brownian motion in field of force and diffusion model of chemical reaction." *Physica*, **7**, 284–304 (1940).

[141] Kubo, R. "Statistical-mechanical theory of irreversible processes, I." *J. Phys. Soc. Japan*, **12**, 570–586 (1957).

[142] Kupferman, R., M. Kaiser, Z. Schuss, and E. Ben-Jacob. "A WKB study of fluctuations and activation in nonequilibrium dissipative steady-states." *Phys. Rev. A*, **45** (2), 745–756 (1992).

[143] Kushner, H.J. "A cautionary note on the use of singular perturbation methods for small-noise models." *Stochastics*, **6**, 116-Ű120 (1982).

[144] Kushner, H.J. and G. Yin. *Stochastic Approximation Algorithms and Applications*, revised 2nd edition. Springer-Verlag, Berlin and New York, 2003.

[145] Lamm, G. and K. Schulten. "Extended Brownian dynamics. II: reactive, nonlinear diffusion." *J. Chem. Phys.*, **78** (5), 2713–2734 (1983).

[146] Landauer, R. in *Self-Organizing Systems: The Emergence of Order*. Edited by F.E. Yates, D.O. walter, and G.B. Yates. Plenum Press, New York, 1983.

[147] Langevin, P. "Sur la théorie du mouvement Brownien." *C.R. Paris*, **146**, 530–533 (1908).

[148] Lee, R.C.K. *Optimal Estimation Identification and Control*. MIT Press, Cambridge, MA, 1964.

[149] Lépingle, D. "Euler scheme for reflected stochastic differential equations." *Math. Comput. Simul.*, **38**, 119–126 (1995).

[150] Lévy, P. *Le Mouvement Brownien*. Mémorial des Sciences Mathématiques, Fasc. 76, Paris, 1954.

[151] Li, G.W. and G.L. Blankenship. "Almost sure stability of linear stochastic systems with poisson process coefficients." *SIAM J. Appl. Math.*, **46**, 875–911 (1986).

[152] Lighthill, M.J. *Introduction to Fourier Analysis and Generalized Functions*. Cambridge University Press, New York, 1958.

[153] Liptser, R.S. and A.N. Shiryayev. *Statistics of Random Processes*, Vols. I, II. Springer-Verlag, New York, 1977.

[154] Ludwig, D. "Persistence of dynamical systems under random perturbations." *SIAM Rev.*, **17** (4), 605–640 (1975).

[155] Maier, R. and D.L. Stein. "Effect of focusing and caustics on exit phenomena in systems lacking detailed balance." *Phys. Rev. Lett.*, **71**, 1783–1786 (1993).

[156] Maier, R. and D.L. Stein. "Escape problem for irreversible systems." *Phys. Rev. E*, **48**, 931–938 (1993).

[157] Maier, R. and D.L. Stein. "Oscillatory behavior of the rate of escape through an unstable limit cycle." *Phys. Rev. Lett.*, **77**, 4860–4863 (1996).

[158] Maier, R. and D.L. Stein. "A scaling theory of bifurcation in the symmetric weak-noise escape problem." *J. Stat. Phys.*, **83**, 291–357 (1996).

[159] Maier, R. and D.L. Stein. "Limiting exit location distributions in the stochastic exit problem." *SIAM J. Appl. Math.*, **57**, 752–790 (1997).

[160] Maier, R., D.G. Luchinsky, R. Mannella, P.V.E. McClintock, and D.L. Stein. "Experiments on critical phenomena in a noisy exit problem." *Phys. Rev. Lett.*, **79**, 3109–3112 (1997).

[161] Mandl, P. *Analytical Treatment of One-Dimensional Markov Processes*. Springer Verlag, New York, 1968.

[162] Mangel, M. and D. Ludwig. "Probability of extinction in a stochastic competition." *SIAM J.Appl. Math.*, **33** (2), 256 (1977).

[163] Mannella, R. "Absorbing boundaries and optimal stopping in a stochastic differential equation." *Phys. Lett. A.*, **254** (5), 257–262 (1999).

[164] Mannella, R. "Integration of stochastic differential equations on a computer." *Int. J. Mod. Phys. C*, **13** (9), 1177–1194 (2002).

[165] Marchewka, A. and Z. Schuss. "Path integral approach to the Schrödinger current." *Phys. Rev. A*, **61**, 052107 (2000).

[166] Marshall, T.W. and E.J. Watson. "A drop of ink falls from my pen ... it comes to earth, I know not when." *J. Phys. A*, **18**, 3531–3559 (1985).

[167] Matkowsky, B.J. and Z. Schuss. "The exit problem for randomly perturbed dynamical systems." *SIAM J. Appl. Math.*, **33**, 365–382 (1977).

[168] Matkowsky, B.J. and Z. Schuss. "Eigenvalues of the Fokker–Planck operator and the approach to equilibrium in potential fields." *SIAM J. Appl. Math.*, **40**, 242–252 (1981).

[169] Matkowsky, B.J., Z. Schuss, and C. Tier. "Diffusion across characteristic boundaries with critical points." *SIAM J. Appl. Math.*, **43**, 673–695 (1983).

[170] Mazur, P. and I. Oppenheim. "Molecular theory of Brownian motion." *Physica*, **50**, 241–258 (1970).

[171] McKean, H.P., Jr. *Stochastic Integrals*. Academic Press, New York, 1969.

[172] Mel'nikov, V.I. and S.V. Meshkov. "Theory of activated rate processes: Exact solution of the Kramers problem." *J. Chem. Phys.*, **85**, 1018–1027 (1986).

[173] Miller, W.H. "Quantum-mechanical transition state theory and a new semi-classical model for reaction rate constants." *J. Chem. Phys.*, **61**, 1823–1834 (1974).

[174] Milshtein, G.N. "A method of second-order accuracy integration of stochastic differential equations." *Theor. Probab. Appl.*, **23**, 396–401 (1976).

[175] Milshtein, G.N. *Numerical Integration of Stochastic Differential Equations*. Mathematics and Its Applications. Kluwer, Dordrecht, 1995.

[176] Monine, M.I. and J.M. Haugh. "Reactions on cell membranes: Comparison of continuum theory and Brownian dynamics simulations." *J. Chem. Phys.*, **123**, 074908 (2005).

[177] Mori, H. "Transport, collective motion, and Brownian motion." *Prog. Theor. Phys.*, **33**, 423–454 (1965).

[178] Musiela, M. and M. Rutkowski. *Martingale Methods in Financial Modeling*. 2nd edition. Applications of Mathematics 36, Springer Verlag, New York, 1997.

[179] Nadler, B., T. Naeh, and Z. Schuss. "The stationary arrival process of diffusing particles from a continuum to an absorbing boundary is Poissonian." *SIAM J. Appl. Math.*, **62** (2), 433–447 (2002).

[180] Nadler, B., T. Naeh, and Z. Schuss. "Connecting a discrete ionic simulation to a continuum." *SIAM J. Appl. Math.*, **63** (3), 850–873 (2003).

[181] Naeh, T. *Simulation of Ionic Solutions*. Ph.D. dissertation, Department of Mathematics. Tel-Aviv University, Israel, 2001.

[182] Naeh, T., M.M. Kłosek, B. J. Matkowsky, and Z. Schuss. "Direct approach to the exit problem." *SIAM J. Appl. Math.*, **50**, 595–627 (1990).

[183] Natanson, I.P. *Theory of Functions of a Real Variable*. Ungar, New York, 1961.

[184] Nevel'son, M.B. and R.Z. Has'minski, *Stochastic Approximation and Recursive Estimation*, translated by Israel Program for Scientific Translations and B. Silver, AMS, Providence, RI, 1976.

[185] Nitzan, A. *Chemical Dynamics in Condensed Phases Relaxation, Transfer and Reactions in Condensed Molecular Systems*. Graduate Texts edition. Oxford University Press, New York, 2006.

[186] Nitzan, A. and Z. Schuss. "Multidimensional barrier crossing." in *Activated Barrier Crossings*, G.R. Fleming and P. Hänggi, Editors. World Scientific, Singapore, 1993.

[187] Noble, B. *Methods Based on the Wiener–Hopf Technique for the Solution of Partial Differential Equations*. AMS/Chelsea, Providence, RI, 1988.

[188] Nonner, W., D. Gillespie, D. Henderson, and R.S. Eisenberg. "Ion accumulation in a biological calcium channel: Effects of solvent and confining pressure." *J. Ph. Chem. B*, **105**, 6427–6436 (2001).

[189] Nonner, W., D.P. Chen, and R.S. Eisenberg. "Anomalous mole fraction effect, electrostatics, and binding in ionic channels." *Biophys. J.*, **74**, 2327–2334 (1998).

[190] Nonner, W., L. Catacuzzeno, and R.S. Eisenberg. "Binding and selectivity in l-type ca channels: A mean spherical approximation." *Biophys. J.*, **79**, 1976–1992 (2000).

[191] Nonner, W. and R.S. Eisenberg. "Ion permeation and glutamate residues linked by Poisson–Nernst–Planck theory in L-type calcium channels." *Biophys. J.*, **75**, 1287–1305 (1998).

[192] Nyquist, H. "Thermal agitation of electric charge in conductors." *Phys. Rev.*, **32**, 110Ű113 (1928).

[193] Øksendal, B. *Stochastic Differential Equations*. 5th edition. Springer, Berlin Heidelberg, 1998.

[194] Olver, F.W.J. *Asymptotics and Special Functions*. Academic Press, New York, and London, 1974.

[195] O'Malley, R.E., Jr. *Singular Perturbation Methods for Ordinary Differential Equations*. Springer-Verlag, New York, 1991.

[196] Oppenheim, I., K.E. Shuler, and G.H. Weiss. *Stochastic Processes in Chemical Physics: The Master Equation*. MIT Press, Cambridge MA, 1977.

[197] Paley, R.N. and N. Wiener. *Fourier Transforms in the Complex Plane*. AMS Colloquium Publicaions, Vol 19, New York, 1934.

[198] Paley, R.N., N. Wiener, and A. Zygmund. "Note on random functions." *Math. Z.*, **37**, 647–668 (1933).

[199] Papoulis, A. *Probability, Random Variables, and Stochastic Processes*. 3rd edition. McGraw Hill, New York, 1991.

[200] Perrin, J. "L'agitation moléculaire et le mouvement Brownien." *C.R. Paris*, **146**, 967–970 (1908).

[201] Peters, E.A.J.F. and Th.M.A.O.M. Barenbrug. "Efficient Brownian dynamics simulation of particles near walls. I: Reflecting and absorbing walls." *Phys. Rev. E*, **66**, 056701 (2002).

[202] Pinsky, R.G. "Asymptotics of the principal eigenvalue and expected hitting time for positive recurrent elliptic operators in a domain with a small puncture." *J. Funct. Anal.*, **200** (1), 177–197 (2003).

[203] Protter, M.H. and H.F. Weinberger. *Maximum Principles in Differential Equations*. Prentice-Hall, Englewood Cliffs, NJ, 1967.

[204] Protter, P. *Stochastic Integration and Differential Equations*. Springer-Verlag, New York, 1992.

[205] Ramo, S. "Currents induced by electron motion." *Proc. IRE.*, **9**, 584–585 (1939).

[206] Risken, H. *The Fokker–Planck Equation: Methods of Solutions and Applications*. 2nd edition. Springer-Verlag, New York, 1996.

[207] Robert, C. and G. Casella. *Monte-Carlo Statistical Methods*. Springer, Berlin, 1999.

[208] Rogers, L.C.G. and D. Williams. *Diffusions, Markov Processes, and Martingales*, Vols. I, II. Cambridge University Press, New York, 2000.

[209] Rubin, R.J. "Statistical dynamics of simple cubic lattices. model for the study of Brownian motion." *J. Math. Phys.*, **1**, 309–318 (1960).

[210] Ryter, D. and H. Meyr. "Theory of phase tracking systems of arbitrary order." *IEEE Trans. Inf. Theor.*, **IT-24**, 1–7 (1978).

[211] Schulman, L.S. *Techniques and Applications of Path Integrals*. Wiley, New York, 1981.

[212] Schumaker, M. "Boundary conditions and trajectories of diffusion processes." *J. Chem. Phys.*, **117**, 2469–2473 (2002).

[213] Schuss, Z. *Theory and Applications of Stochastic Differential Equations*. Wiley, New York, 1980.

[214] Schuss, Z. *Brownian Dynamics and Simulations*. Preprint, 2009.

[215] Schuss, Z. *Optimal Phase Tracking*. Preprint, 2009.

[216] Schuss, Z. and A. Spivak. "Where is the exit point?" *Chem. Phys.*, **235**, 227–242 (1998).

[217] Schuss, Z. and A. Spivak. "The exit distribution on the stochastic separatrix in Kramers' exit problem." *SIAM J. Appl. Math.*, **62** (5), 1698–1711 (2002).

[218] Schuss, Z. and B.J. Matkowsky. "The exit problem: A new approach to diffusion across potential barriers." *SIAM J. Appl. Math.*, **35** (3), 604–623 (1979).

[219] Schuss, Z., B. Nadler, and R.S. Eisenberg. "Derivation of PNP equations in bath and channel from a molecular model." *Phys. Rev. E*, **64**, (2-3), 036116–1–036116–14, 2001.

[220] Shockley, W. "Currents to conductors induced by moving point charge." *J. Appl. Phys.*, **9**, 635–636 (1938).

[221] Singer, A. *Diffusion Theory of Ion Permeation Through Protein Channels of Biological Membranes*. Ph.D. dissertation, Department of Mathematics, Tel-Aviv University, 2006.

[222] Singer, A. and Z. Schuss. "Brownian simulations and unidirectional flux in diffusion." *Phys. Rev. E*, **71**, 026115 (2005).

[223] Singer, A., Z. Schuss, A. Osipov, and D. Holcman. "Partially absorbed diffusion." *SIAM J. Appl. Math.*, **68** (3), 844–868 (2008).

[224] Skorokhod, A.V. "Stochastic equations for diffusion processes in a bounded region." *Theor. Probab. Appl.*, **6** (3), 264–274 (1961).

[225] Slemrod, M. "Stabilization of bilinear control systems with application in nonconservative problems in elasticity." *SIAM J. Control Opt.*, **16**, 131–141 (1978).

[226] Smoluchowski, M. von. "Studien über Molekularstatistik von Emulsionen und deren Zusammenhang mit der Brown'schen Bewegung." *Wien. Ber.*, **123**, 2381–2405 (1914).

[227] Smoluchowski, M. von. Drei Vorträge über Diffusion, Brownsche Bewegung und Koagulation von Kolloidteilchen." *Phys. Zeits.*, **17**, 557 (1916).

[228] Smoluchowski, M.R. von Smolan. "Zarys kinetycznej teorji ruchów Browna." *Rozprawy Wydziału Przyrodniczego Akademii Umiejętności (Kraków)*, **A46**, 257–282 (1906).

[229] Sneddon, I. *Elements of Partial Differential Equations*. McGraw-Hill International (Mathematics Series), 1985.

[230] Song, Y., Y. Zhang, T. Shen, C.L. Bajaj, J.A. McCammon, and N.A. Baker. "Finite element solution of the steady-state Smoluchowski equation for rate constant calculations." *Biophys. J.*, **86**, 2017–2029 (2004).

[231] Spitzer, F. "Some theorems concerning 2-dimensional Brownian motion." *Trans. AMS* **87**, 187–197 (1958).

[232] Stratonovich, R. *Topics in the Theory of Random Noise*. Mathematics and Its Applications, vols I, II. Gordon and Breach, New York, 1963.

[233] Stroock, D.W. *An Introduction to Markov Processes*. Springer Graduate Texts in Mathematics, New York, 2005.

[234] Stroock, D.W. and S.R.S. Varadhan. *Multidimensional Diffusion Processes*. Springer-Verlag, New York, Grundlehren der Mathematischen Wissenschaften 233, 1979.

[235] Sutherland, W. "A dynamical theory of diffusion for non-electrolytes and the molecular mass of albumin." *Philos. Mag.*, **9** (5), 781–785 (1905).

[236] Svedberg, T. "Über die Eigenbewegung der Teilchen in kolloidalen Lösungen I." *Z. Elektrochem. Angewandte Phys. Chem.*, **12**, 853–860 (1906).

[237] Svedberg, T. Nachweis der von der kinetischen Theorie Geforderten Bewegung gelöster Moleküle." *Zeits. f. physik Chemie*, **74**, 738 (1910).

[238] Szymczak, P. and A.J.C. Ladd. "Boundary conditions for stochastic solutions of the convection-diffusion equation." *Phys. Rev. E*, **68**, 036704 (2003).

[239] Tai, K., S.D. Bond, H.R. MacMillan, N.A. Baker, M.J. Holst, and J.A. McCammon. "Finite element simulations of acetylcholine diffusion in neuromuscular junctions." *Biophys. J.*, **84**, 2234Ű–2241 (2003).

[240] Troe, J. "Theory of thermal unimolecular reactions at low pressures: I. Solutions of the master equation." *J. Chem. Phys.*, **66**, 4745–4757 (1977).

[241] Üstünel, A.S. and M. Zakai. *Transformation of Measure on Wiener Space*. Springer, Berlin, 2000.

[242] van Kampen, N.G. *Stochastic Processes in Physics and Chemistry*. 3rd edition. North-Holland Personal Library, JAI Press, Greenwich, CT, 1992.

[243] Van Trees, H.L. *Detection, Estimation and Modulation Theory, Vols. I, II, III*. John Wiley, New York, 1970.

[244] Viterbi, A.J. *Principles of Coherent Communication*. McGraw-Hill, New York, 1966.

[245] Walker, J.A. "On the application of Lyapunov's direct method to linear lumped-parameter elastic systems." *J. Appl. Mech.*, **41**, 278–284 (1974).

[246] Westgren, A. "Die Veranderungsgeschwindigkeit der lokalen Teilchenkonzentration in Kolloiden Systemen, I." *Arkiv for Matematik, Astronomi, och Fysik*, **11** (14), 1–24 (1916).

[247] Westgren, A. "Die Veranderungsgeschwindigkeit der lokalen Teilchenkonzentration in Kolloiden Systemen, II." *Arkiv for Matematik, Astronomi, och Fysik*, **13** (14), 1–18 (1918).

[248] Wiener, N. "Differential space." *J. Math. Phys.*, **2**, 131–174 (1923).

[249] Wilmott, P. J. Dewynne, and S. Howison. *Option Pricing: Mathematical Models and Computation*. Oxford Financial Press, Oxford, 1994.

[250] Yu, A.W., G.P. Agrawal, and R. Roy. "Power spectra and spatial pattern dynamics of a ring laser,." *J. Stat. Phys.*, **54**, 1223 (1989).

[251] Zauderer, E. *Partial Differential Equations of Applied Mathematics*. Pure and Applied Mathematics: A Wiley-Interscience Series of Texts, Monographs and Tracts, 3rd edition. Wiley, New York, 2006.

[252] Ziegler, H. *Principles of Structural Stability*. Blaisdell, New York, 1968.

[253] Zwanzig, R. "Nonlinear generalized Langevin equations." *J. Stat. Phys.*, **9**, 215–220 (1973).

[254] Życzkowski, M. and A. Gajewski. "Optimal structural design in non conservative problems of elastic stability." *Instability of Continuous Systems*, H. Leipholz, Ed. IUTAM Symposium Herrnalb, Springer-Verlag, New York, 295–301 (1971).

Index

A

Abramowitz, M. 270
Absorption 59, 118, 154, 155, 175–184, 193, 202–206, 359, 360
  rate 182, 183
Adapted process 153, 161, 229, 232, 278–284, 308–328, 335–346, 351, 355, 359, 372
Adelman, S.A. 262
Agmon, N. 175
Agrawal, G.P. 275, 301
Allen, M.P., v
Andronov–Vitt–Pontryagin, vi, 104–110, 186–192, 250, 256, 339, 346
Arc-sine law 119
Arnold, L., v, 405, 416
Asmussen, S. 175
Asymptotic expansion 115, 120, 121, 171, 174, 207–212, 231–247, 252–261, 265–270, 280
  construction 322, 338
  boundary layer, vi, 152–162, 174, 175, 188, 248–256, 268–282, 310–320, 335, 341–361, 366–379, 385–398
  internal layer 345, 349, 352, 364
  outer expansion 25, 153, 162, 249, 250–253, 271, 309–312, 319, 341–345, 349, 354, 368, 392
  uniform expansion 161, 289, 311–322, 339–342, 350, 355, 360, 368, 372
Autocorrelation function 14, 15, 24, 43, 262, 263, 335
  Brownian bridge 60
  Brownian motion 43, 47
  continued fraction expansion of Laplace transform 262
  current fluctuations 20, 227, 229
  Gaussian process 14–16, 22, 41, 53, 81, 210–212
  linear system 24
  Markov process 335
  stochastic process 15
  white noise 16, 427
Autocovariance matrix 13, 22, 23, 45
  of Brownian motion 45
  of OU process 14

B

Backward
  degenerate parabolic equation 387
  differential 83, 84
  integral 73, 74, 82, 92, 123
  Kolmogorov
    equation 109, 120, 124, 141, 166, 167, 175, 177, 181, 219, 244, 355, 388, 420, 421
    operator 85, 105–112, 167, 179–181, 189, 221, 231, 308, 401
  master equation 303–321
  parabolic boundary value problem 110
  parabolic equation 105–109
  parabolic terminal boundary value problem 114
  parabolic terminal value problem 106
  stochastic equation 96
  variables141, 166, 167, 220, 221
Bahadur, R.P. 337

Bajaj, C.L. 174
Baker, N.A. 174
Barenbrug, Th.M.A.O.M. 174
Batsilas, L. 174
Bayes' rule 198, 201
Beccaria, M. 174
Belch, A.C. 301
Bellman, R. 418
Belopolskaya, Y.I., v, 168
Ben-Jacob, E. 216, 285, 299–301
Bender, C.M., vii, 146, 197, 339
Bentsman, J. 418
Berezhkovskii, A.M. 174
Bergman, D. 216, 285, 300, 301
Berkowitz, M. 301
Berne, B.J. 301, 419, 420
Berry, R.S. 301
Billingsley, P., v, 335
Blankenship G.L. 403, 414
Bobrovsky, B.Z. 397, 398
Bond, S.D. 174
Borel sets 207
Bossy, M. 175
Boundary
  conditions 17, 38, 57–60, 92, 111, 113, 133, 151, 166–178, 188–205, 222, 251, 255, 268, 269, 270, 278–285, 301, 322, 343, 345, 356, 359, 376, 378, 404, 413, 414, 436
  layer, iii, 114–121, 174, 175, 141, 185–210, 232–240, 252–291, 296–300
  outer expansion 25, 153, 162, 249, 250–253, 271, 309–312, 319, 341–345, 349, 354, 368, 392
Brooks, C.L. III 301
Brown, D.E. 158, 163
Brownian
  bridge 60, 198, 202
  dynamics,, vi, vii, 174
  filtration 30–34, 66
  motion 1–15, 22–97, 100–123, 175, 170, 175, 193, 201, 202, 241, 258–264, 374, 375, 382, 401, 405, 411, 421, 431
    absorbed 58, 59
    Mathematical 35–94, 110–178
  Free Brownian particle 6–10, 14, 63, 93, 259
    discrete time 29, 41
    reflected 57, 62
    rotation 45, 89
  scaling 46, 97, 258, 406
    nondifferentiable 53, 63, 93
  sampled trajectory 29, 33–41, 51, 134
Brush, S.G. 2
Bryan, G.H 175
Bucklew, J.A., v, 337, 338
Büttiker, M. 397

C

Cameron, R.H. 175
Campbell's theorem 229
Carmeli, B. 321, 410
Carslaw, H.S. 175
Casella G., v
Catacuzzeño, L. 301
Cauchy process 114, 208–212
Chandrasekhar, S., v, 7, 12, 13
Chapman. J. 152
Chapman–Kolmogorov
  backward differential equation 221
  equation 55, 56, 101, 122–125, 154, 159, 169, 207, 208, 219–221
Chen, D.P. 301
Christiano, R. 286, 300
Cochran, J.A. 323
Coddington, E. 191
Coffey, W.T., v, 301
Cole, J.D., vii, 339
Collins, F.C. 173–175
Conservation law 144–150, 168, 360, 390
Continuous time 58, 222
  process 249
    Markov process 238, 306

Convergence 22, 47–51, 64–71, 76–81, 133–154, 158– 175, 211, 214, 221, 238, 245–249, 259, 273, 301, 338, 342, 398, 423, 424
Costantini, C. 175
Covariance matrix 8–13, 22, 24
Courant, R. 214, 361, 362
Cox, D.R., v, 208
Cramér, H., v, 324, 337
Curci, G. 174
Cylinder set 28–44, 169–173, 201, 202

D
Dalecky, Y.L., v, 168
Daniels, H.E 335, 337
Day, M.V., v, 397, 398
De Finetti, B., v, 30, 36
Dembo, A., v, 337, 361, 373, 397
Devinatz, A. 398
Diffusion
  process 102–104, 151, 168, 173, 198–203, 218, 219, 230, 240, 257, 295, 322, 412, 419–422
    boundary behavior133, 150, 151, 159, 175, 247, 268, 301, 412
    (infinitesimal) coefficient 2–11, 20, 21, 102, 123, 159, 170–174, 185, 189, 208, 235, 245, 259, 265, 269, 417
    matrix 85, 103, 104, 187, 198–200, 219, 220, 365, 370, 374
    trajectory v–xii, 25–102, 104–200, 202–301, 343–348, 358–403, 409–415, 419, 428–436
    with killing
    measure 203, 204
    rate 108, 119, 203
Displacement process 10, 56, 93
Doering, C.R. 270–275, 301
Doll, J.D. 262
Doob, J.L., v, 30, 36, 67
Doucet, A., v
Drift 102, 122, 123, 159, 170–175, 193, 200, 214, 215, 232–240, 256, 303–317, 343–349, 406
  equation 376, 377
  (infinitesimal) coefficient 89, 138, 186, 218, 265, 301, 351
  vector 103, 104, 198–200, 219, 220, 355, 369–374, 383, 392, 393
  -jump process 238, 256, 306
Dvoretzky, A. 53
Dygas, M.M. 260, 264, 301
Dynamical system 92, 185, 187, 213, 230, 232, 359, 379, 397, 398, 417
Dynkin, E.B. 211, 219

E
Eigenvalue 22, 23, 178, 179, 191, 221, 230–235, 269–275, 323–334, 359, 363, 375–380, 399, 404–407, 429
Eikonal
  equation 246, 290–298, 304–309, 314, 318, 325–332, 361–366, 370–376
  function 290, 297, 306, 362, 377, 422
Einstein, A. 1–12, 25–30, 38, 259
Eisenberg, R.S. 265, 301
Eliezer, B. 13
Ellis, R.S, v, 337, 361
Equilibrium, v, 1–7, 241, 259–261, 268, 269, 276, 285–296, 301, 308, 382, 400–404, 411–414, 425, 427, 432
  nonequilibrium, v, 285–299, 382
  stable 186, 256, 276, 282–308, 317, 377–383, 392, 412, 429
  unstable 288, 289, 297, 298, 344, 425
Erban, R. 152
Erdös, P. 53
Ethier, S.N. 211
Exit

density 184, 360, 369–396
probability 185, 188, 339, 344, 346, 350, 392, 396
problem 185, 187, 282, 339–348, 359, 360, 378–397, 412
Eyring, H. 257, 382, 388

F
Feller, W., v, 11, 58, 151, 174, 323, 338, 412
Feynman–Kac formula, vi, 104–108, 119, 141
Fichera, G. 198
First exit time 34, 94, 95, 129, 249, 255, 308, 359
Fluctuations 290, 295, 300, 301
–dissipation principle 260–264
Fokker–Planck–Kolmogorov
equation 114–133, 139, 143–169, 173–193, 203–206, 219, 238, 239, 244–247, 255, 265–269, 277–296, 323, 355, 359–367, 371, 375, 38–391, 407
operator 120, 178
–Stratonovich equation 122, 123, 235
Freidlin, M.A., v, 170, 175, 297, 301, 361, 397
Freitas, N.D., v
Friedman, A., v, vii, 58, 76, 109, 118, 121, 139, 231, 398
Functional, vi, vii, 15, 22, 67, 104, 106, 241, 397, 436, 437

G
Gajewski, A. 429, 436
Gambler's Ruin "Paradox" 51
Gantmacher, F.R. 363
Gardiner, C.W., v, 148, 219, 220, 221, 397
Gaussian
process 13–16, 23, 47, 60, 98, 260–263
variable 8–12, , 42, 46, 53, 134
Ghoniem, A.F. 174
Gihman, I.I., v, 94, 109, 110, 132, 170, 208, 211, 219
Gillespie, D. 301
Girsanov, I.V. 170, 171, 175
Glasstone, S. 257, 382, 388
Glynn, P. 175
Gobet, E. 175
Gordon, N., v
Goodrich, F.C. 174
Graham, R. 297
Grote, F.R. 262, 301
Green's function, vi, 111–118, 183, 184, 189, 213, 344, 355, 388

H
Hagan, P.S. 270–275, 301
Hagan's function 275
Hale, J.K. 347
Hamilton–Jacobi equation 290, 304, 361
Hamiltonian 260
Hänggi, P., v, 257, 260, 301, 382, 388, 397
Harmonic
oscillator 13, 245, 260, 261, 421
potential 245, 280, 417
Harris, E.P. 397
Haugh, J.M. 174
Henderson, D. 301
Hermite
function 410
polynomial 271
Hida, T., v, 51, 52, 54, 56, 67
Hilbert, D. 214, 361, 362
Hille, B. 285, 301
Holcman, D. 173
Holst, M.J. 174
Honerkamp, J., v, 174
Hurwitz zeta function 276
Hynes, J.T. 262, 301

I
Il'in, A.M. 241, 244
Im, B. 301
Imry, Y. 301

Inequality
  Cauchy–Schwarz 48, 73, 87, 125, 126
  Chebyshev's 65, 127, 137
  Gronwall's 174, 137
  submartingale 68, 113
Initial value problem 25, 39, 94, 107, 120–123, 138–143, 154, 174, 213, 232, 237, 248, 339, 384
Instantaneous
  absorption rate 177
  re-injection 182
  reflection 158, 174
  termination 150, 154
  truncation 221
  velocity 17
Invariant set 347
Iscoe, I. 323
Itô, K., v, 38, 39, 46, 52–56
  calculus 63, 81, 83
  equation 95–100, 158, 166, 173, 201
  formula 85–91, 97–102, 107, 110, 114, 115, 120, 128, 137, 173
  integral 63–83, 92, 93, 102, 103, 126–129, 134, 137, 170
    construction 66, 68, 74, 83

J
Jackson, J.D. 17
Jaeger, J.C. 175
Jazwinski, A.H., v
Jensen, J.L., v, 338
Johnson, J.B. 16
Joint pdf 12, 13, 24, 42, 46, 55, 204, 218, 224, 232, 235, 242–244, 257, 280, 282, 322–331, 338
Josephson junction 216, 276, 285, 299
Jump
  –diffusion process 219, 221
  process 27, 152, 157–160, 166, 173, 210–214, 230–238, 249, 254, 301–306, 321, 406
  rate 208–213, 232, 306, 406

K
Kaiser, M. 299, 301
Kakutani, S. 53
Kalmykov, Yu.P., v, 301
Kamienomostskaya (Kamin), S. 398
Karatzas, I., v, 60–67, 76, 81, 104
Karlin, S., v, 27, 60, 119, 151, 198, 200–202, 301, 412
Karplus, M. 301
Katz, A. 397–403, 414
Keller, J.B. 174
Kevorkian, J., vii, 339
Khasminskii, R.Z. 241, 244, 328
Kim, S. 301
Kimball, G.E. 173–175
Kleinert, H. 175
Kleinrock, L. 215
Kloeden, P.E., v, 106, 139, 175
Kłosek, M.M. 206, 265, 273, 275, 301, 361
  –Dygas, M.M. 245, 410, 417
Knessl, C. 152, 216, 249, 256, 321, 322, 328, 337, 338
Kolmogorov
  forward equation, vi, 120, 221, 303, 304, 322, 324, 404, 405, 419
  formula 113
  forward operator 120, 178
  master equation 244–266, 324
Korkotian, E. 428
Kovary, T. 211, 219
Kramers, H.A. 7, 197, 257, 284, 301, 359, 382, 397, 398
  formula 197, 387
  –Moyal expansion 236–243, 248–255, 302, 306, 309, 314, 315
  problem 378–382
  rate 197, 386–390
    underdamped 284
    uniform 341
Kubo, R. 301
Kupferman, R. 299, 301
Kurtz, T.G. 211
Kushner, H.J. 238, 249, 328

L
Ladd, A.J.C. 174
Laidler, K.J. 257, 382, 388
Lamm, G. 174, 175
Landauer, R. 257, 399
Langevin, P. 6
Langevin's equation 5–13, 25, 30, 63, 92, 143, 156, 243, 247, 256–260, 266, 268, 276, 286, 301–375, 379–383, 398
  construction 25
  generalized 259–264, 301
  overdamped 20, 53, 196, 245, 257–359, 374, 379, 387
  underdamped 276, 285, 286, 300, 382, 390, 428
Large deviations 185, 302, 303, 308, 322, 403
  theory 226, 241, 304, 322, 324, 329–338, 361, 373, 397
Last passage time 32, 34, 200
Law of large numbers 175, 226
Leadbetter, M.R., v, 324, 337
Lee, R.C.K., v
Li, G.W. 403, 414
Lépingle, D. 175
Levermore, C.D. 270–275, 301
Levinson, N. 191
Lévy, P., v, 27, 46, 49
Li, G.W. 356, 365
Lifetime 285, 295–300, 443
Lighthill, M.J. 16, 25, 36, 121
Lindeberg's condition 208–210
Liptser, R.S., v
Luchinsky, D.G. 397, 398
Ludwig, D. 397, 398
Lyapunov function 400, 401, 422, 436–438, 441
  equation 23, 24

M
MacMillan, H.R. 174
Maier, R. 397, 398
Makhnovskii, Y.A. 174
Mandl, P., v, 151, 301
Mangel, M. 397, 398
Mannella, R. 174, 397, 398
Marchewka, A. 173, 174, 175
Markov
  chain 151, 165, 207, 222, 322, 331, 338
  process 34, 55–60, 101, 102–104, 169, 207, 208, 219, 221, 230, 232, 238, 249, 257, 260, 263, 321, 323, 329, 331, 333, 338, 404
  time 32, 33, 34, 56, 94, 110
    discrete time 207, 214, 219
  (stopping) time 32
Marshall, T.W. 270–275, 301
Martin, W.T. 175
Martingale 60–89, 94, 128, 170
  measure 170
Master equation
  backward 303–308, 319–321
  forward 303, 322–329, 407
Mathematical Brownian motion (MBM) 39, 41–46, 51–61, 74, 93, 95, 111, 114, 133, 142, 173, 193, 201, 208, 209, 219
  construction 26, 27, 39, 46
    Lèvy's 49–52
    Paley–Wiener 46
Matkowsky, B.J. 152, 206, 216, 245, 249, 256, 260, 264, 285, 301, 321, 322, 328, 337, 338, 352, 361, 369, 374, 377, 384, 398 410, 417
Mazur, P. 301
McCammon, J.A. 301
McClintock, P.V.E. 397, 398
McKean, H.P., Jr, v, 38, 39, 46, 52–56, 67, 74, 168
McLaughlin, D.W. 174
Mean
  exit time, vi, 111, 222, 307, 308, 321, 352, 359
  first passage time, vi, 110, 156, 174–200, 221, 249–256, 268, 284, 314–321, 339, 346, 352–

360, 373–388, 400, 434, 438, 440
over potential barrier 196, 197, 257, 291, 346–359, 374, 379, 382, 406
Measure, v, 26, 30–44, 52, 169–173, 202, 254, 300
of a cylinder set 28, 145, 127, 175, 229
Wiener's, v, 38, 41, 133, 169, 170
Meerkov, S.M. 418
Mel'nikov, V.I. 397
Meshkov, S.V. 397
Meyr, H. 397, 398
Miller, H.D., v, 208
Miller, W.H. 338
Milshtein, G.N. 175
Monine, M.I. 174
Mori, H. 301
Musiela, M. 60, 170

N
Nadler, B. 222, 301
Naeh, T. 206, 222, 301, 361
Natanson, I.P. 30, 49
Net flux 133, 145–149, 174, 267
Nevel'son, M.B. 328
Ney, P. 323
Nitzan, A., v, 321, 397, 398, 410
Noble, B. 153
Noise
classical theory 13
colored 14–23, 123, 245, 246, 263, 264, 280, 282, 405–413
in linear systems 21
in Josephson junction 176, 223, 230, 241
in nonlinear oscillator 414–417
parametric 404
stochastic stability 354
in nonlinear pendulum 305
Johnson 16
current fluctuations 20
matrix 65, 288
voltage fluctuations 16
shot 227, 410, 426, 427
white 15–23, 52, 53, 63, 66, 89, 92, 93, 186, 230, 244, 247, 257–260 , 347, 400–430
wideband 14, 244, 423
Non-Markovian process 196
Nonner, W. 301
Nummelin, E. 323
Nyquist, H. 17, 21

O
Olver, F. W. J., vii
O'Malley, R.E., Jr, vii, 339
Oppenheim, I., v, 301
Ornstein–Uhlenbeck process 14, 20, 23, 100, 111, 119, 123, 193, 263, 305, 405, 413, 422–424
Orszag, S.A., vii, 146, 197, 339
Osipov, A. 173
Øksendal, B. , v, 168

P
Pacchiarotti, B. 175
Paley, R.N. 15, 25, 46, 53
Papanicolaou, G. 405, 416
Papoulis, A., v, 27, 57, 59
Parabolic cylinder function 270–272
Partially absorbing boundary condition 157, 158
Path integral 38, 92, 94, 133, 151, 152, 169, 171–175
Pecora, R. 301, 419, 420
Perrin, J. 2, 11
Peters, E.A.J.F. 174
Pinsky, R.G. 414
Pitman, J. 175
Platen, E., v, 139, 175
Poisson process 210, 211, 223, 224, 225
rate 156, 166, 179
Power spectral density of stationary process 14, 15, 24, 262, 418
current fluctuations 21, 227, 229

height 15
in generalized Langevin equation 261
Lorentzian 15
Markov process 335
Ornstein–Uhlenbeck process 406
total output 15, 16
voltage fluctuations 16
white 16
width 15

Probability
discrete 323
flux density, vi, 145–150, 156, 174, 184, 267
discrete 119
measure 26, 30, 36–38, 151, 170, 197
space 26–37, 44, 46, 47, 173

Protein channel 145, 197, 198, 265, 301

Protter, M.H. 139

Protter, P., v, 67

Q

Queueing theory (examples) 215
M/M/1 queue 317
Takács equation 186, 275
unfinished (virtual) work 215, 317–319

R

Radon–Nikodym theorem 36, 37
derivative 35, 36, 173

Ramo, S. 17

Random
collisions 6, 321
force 6, 7, 403, 428
function 32, 37
increments 8, 11, 16, 25, 26, 39, 41–47, 53, 61–68, 101, 134
trajectory, v, vi, 133, 185, 359, 397, 414
variable 8, 12, 25, 30–39, 46–49, 62, 65, 82, 93, 134, 158, 170, 171, 207, 222–225, 236, 302, 322–335
i.i.d. (independent identically distributed) 8, 46, 49, 222–227, 322–328, 333–338
walk 111, 174, 175, 236, 240, 247, 254, 302, 314

Ranga Rao, R. 337

Rate 178, 180, 189, 203

Recurrent process 58, 116–222, 269, 340, 419

Relative stability 300

Renewal
equation 224, 388
process 222–227

Representation formula, vi, 104–110, 176

Rice, S. 301

Riemann
integral 67, 71,
sum 8, 19, 70
zeta function 270

Risken, H., v, 221, 301, 397

Robert, C., v

Rogers, L.C.G., v, 38, 52, 54, 60, 211

Ross, J. 301

Roux, B. 212

Roy, R. 275, 301

Rubin, R.J. 262

Rutkowski M. 60, 170

Ryter, D. 397, 398

S

Sample path 25

Sample space 27, 201

Sartoretto, F. 175

Scaling 159, 232–244, 255, 263–266, 303–307, 312, 380, 391, 392, 406–410

Shen, T. Shen

Shockley, W. 13

Schottky's formula 229

Schulman, L.S. 175

Schulten, K. 174, 175

Schumaker, M. 175, 301

Schuss, Z., v–viii, 13, 30, 53, 145, 152, 157, 165, 166, 173–

175, 206, 216, 222, 245, 249, 256, 260–265, 285, 299, 301, 321, 322, 328, 337, 338, 352, 359–363, 369, 373–377, 384, 397–403, 410–417
Seefeld, S. 301
Segal, M. 428
Semi-Markov process 222
Sherman, F.S. 174
Shiryayev, A.N., v
Shockley, W. 17
Shreve, S.E., v, 60–67, 76, 81, 104
Shuler, K.E., v, 301
Shvartsman, S.Y. 174
Sigma-algebra 29, 30, 44
Silvestrini, P. 286, 300
Simulation, vi, vii, 21–30, 38, 46, 49, 92, 111, 114, 133, 139, 145, 151, 160, 167, 174, 175, 197, 200, 397, 398
connecting to continuum, vi, 133, 145
Euler's method 25, 26, 93, 106, 133–143, 151–159, 165, 167, 173–175, 195, 196, 213, 426, 438
discrete trajectories 159
Singer, A. 145, 157, 173–175, 275, 301
Skorohod, A.V. v, 94, 109, 110, 132, 134, 151, 170, 175, 208, 211, 219
Slemrod, M. 429
Smoluchowski, M. von 2, 3, 175
approximation 436
equation 263, 265
expansion 265, 268
limit 258–266, 275, 382
Sneddon, I. 361, 362
Song, Y. 174
Spitzer, F. 114
Spivak, A. 397
Stability 91, 230, 239, 244, 397, 398
Stegun, I. A. 201
Stein, D.L. 397, 398
Stochastic
approximation 328
differential equation, v, 6, 74, 88–139, 151, 167–169, 176, 186, 200–204, 230–239, 244, 262, 263, 286, 397, 405
dynamics, vi, 27, 92, 104–108, 116–123, 142, 170, 174, 187–189, 198, 218, 286, 343, 347
discrete 133
Euler scheme construction 93
boundary behavior 119, 154, 215, 265, 362
process 27, 33–47, 55–67, 82, 94, 137, 170, 197, 207, 222, 231, 241, 322, 419
adapted 31, 59-82, 113-120, 146
ergodic 322–329, 333, 417–422
Gaussian 14, 54, 86, 222
stationary 13–24, 42, 118, 221, 231, 240–246, 260–264, 292, 295, 322–333, 338, 404, 411, 417–422
stability 399–403, 414, 427
stabilization with noise 417, 421, 423
inverted pendulum 423
Steady state 20, 178–184, 269, 276, 285–301, 318, 382, 407–412
Stratonovich, R. 229
construction 68
stochastic differential equation 122, 245
integral 69, 72, 74, 82, 89
Stroock, D.W., v, 168, 211, 337
Submartingale 61, 62
inequality 76, 77, 128
Survival probability, vi, 119, 155, 156, 177, 203, 206, 221
Sutherland, W. 4
Svedberg, T. 2–6
Szymczak, P. 174

T

Talay, D. 175
Tai, K. 174
Takács equation 215, 318, 319
Taylor, H.M., v, 27, 60, 119, 151, 198, 200–202, 301, 412
Tel, T. 297
Terminal value problem 105–109
Tier, C. 152, 216, 249, 256, 321, 322, 328, 337, 338, 374, 377, 384, 398
Tildesley, D.J., v
Time spent at a point 156
Total population 155, 177–182, 356, 357
Transition probability density 5, 55–57, 98–124, 167, 169, 177, 254, 268, 323, 329, 405
Transport equation 246, 291, 293, 305, 307, 314, 318, 364–366, 376
Troe, J. 321

U
Unidirectional probability flux density 133, 145–155, 174, 267
    discrete 159
    instantaneous 156
Üstünel, A.S., v, 175

V
van Kampen, N.G, v, 240
Van Trees, H.L. 22
Varadhan, S.R.S., v, 168, 337
Velocity process 5–14, 25, 31, 39, 53, 56, 93
Vicere, A. 174
Viterbi, A.J. 397

W
Waldron, J.T., v, 301
Walker, J.A. 436
Watson, E.J. 270–275, 301
Weak solution 121
Weinberger, H.F. 139
Weiss, G.H., v, 301
Wentzell, A.D., v, 297, 361, 397
Westgren, A. 3
Wiener, N. 15, 25, 38, 46, 53
Wiener–Hopf equation 153, 161, 248, 252, 253, 309, 310
    method, vi, 164
Wihstutz, V. 405, 416
Williams, D., v, 38, 52, 54, 60, 211
Wilmott, P. 397
WKB method, vi, 280, 290, 302–315, 323, 325, 335, 338, 355, 361, 365, 373
Wong–Zakai correction 69–73, 83, 90–96, 246, 247

Y
Yin, G. 238, 328
Yu, A.W. 275, 301

Z
Zakai, M., v, 69–73, 83, 90–96, 175, 246, 247
Zauderer, E. , vii
Zeitouni, O., v, 337, 361, 373, 397
Zeta function
    Hurwitz 237-242
    Riemann 232
Zhang, Y. Zhang
Ziegler, H. 426, 429
Zitserman, V.Yu. 174
Zygmund, A. 15, 25, 53
Zwanzig, R. 174
Życzkowski, M. 429, 436